3rd EPI International Conference on Science and Engineering (EICSE2019)

IOP Conference Series: Materials Science and Engineering
Volume 875

Makassar, Indonesia
24 - 25 September 2019

Editors:

Faisal Mahmuddin
Rosmalina Hanafi
Sufriadin
Muhammad Ramli
Achmad Yasir Baeda
Amil Ahmad Ilham
Indar C. Gunadin
Lukmanul Hakim Arma
Intan Sari Areni
Muhammad Zubair M. Alie
Muh. Anshar
Yashinta Kumala S. Dewi
Asniawaty Kusno
Irwan Setiawan

ISBN: 978-1-7138-4741-0
ISSN: 1757-8981

Printed from e-media with permission by:

Curran Associates, Inc.
57 Morehouse Lane
Red Hook, NY 12571

proceedings.com

Some format issues inherent in the e-media version may also appear in this print version.

Institute of Physics
Dirac House, Temple Back
Bristol BS1 6BE UK

Phone: 44 1 17 929 7481
Fax: 44 1 17 920 0979

techtracking@iop.org

Additional copies of this publication are available from:

Curran Associates, Inc.
57 Morehouse Lane
Red Hook, NY 12571 USA
Phone: 845-758-0400
Fax: 845-758-2633
Email: curran@proceedings.com
Web: www.proceedings.com

TABLE OF CONTENTS

Preface

Conference Committee

Peer review statement

Development of the Architectural Styles of the Reformation Era in Makassar, South Sulawesi, Indonesia 1
A Harisah, R Wikantari, M Mochsen Sir, A M Radja, S Syam, A K Deapati, Badri

Aerial Cable Car in the City Centre of Makassar: The Potential Routes, Technology and Station Locations 8
Y K D Sutopo, M Y Jinca, M F Azmy, M V H Maramis, M D Amin

The Study of Public Open Space effectiveness in Makassar Waterfront City using Good Public Space Index (GPSI) 18
S A Ekawati, M Ali, S Trisutomo, R C A Ghani

Adaptive Thermal Comfort of Elementary School Student (A Case study of the West Coastal Area of South Sulawesi) 29
R Mulyadi, B Hamzah, M T Ishak, Y R F Taufik

Local-Wisdom and Its Influence to Disaster Mitigation on the Spatial Configuration of Lakkang Waterfront Settlement 35
E Syarif, Asniawaty, N Nadjmi, A Syadzwina

Comparison of Building Envelope Models to Illuminance Level in Phinisi Tower Building 44
N Jamala, R Rahim, R Mulyadi

Coconut Leaf Midribs as an Acoustical Panel – Feasibility Study through Impedance Tube Method 56
A Kusno, M T Ishak, R Rahim, B Hamzah, R Mulyadi, N Jamala, R Anggraeni

The Structure System of Bugis Traditional House in Disasters Perspective 65
Hartawan, V Sampebulu, N Junus, Imriyanti, P Musharl, D Kuddu, I Alim

Post Occupancy Evaluation in the Senate Room, Faculty of Engineering, Hasanuddin University, Gowa 78
R A Ishak, T Martosenjoyo, S Beddu, S Latif

Characteristic of Student Informal Gathering Space (SIGS) Faculty of Engineering Hasanuddin University at Gowa Campus 91
A M Radja, R Wikantari, A Harisah

Post-Occupancy Evaluation on Infrastructure Service Performance of Mariso Low Cost Apartment (Rusunawa), Makassar 99
N Amri, Idawarni, E Syarif, S Amin, Y Siradjuddin, R Hiromi

Factors Affecting Thermal Comfort of Elementary Schools' Students in Makassar 111
B Hamzah, R Mulyadi, M T Ishak

Relationship between Regional Development Level and Accessibility Index in Makassar Coastal Area 119
A Akil, A Yudono, S Wahyuni, R Ramadhani, I Kamaruddin

Value of Concrete Compressive Strength on 28 days with Variation of Candlenut Shell Applicated to Plate on a Non-destructive Test Using Hammer Test 130
Nasruddin, V Sampebulu, P Mushar

Identification of Child-Friendly Integrated Public Spaces in Makassar City (Case Study: Losari Beach) 137
M Ali, S E Aliah, M Fachrul R, Musdalifah

Travel Behavior of the Engineering Students of UNHAS and UIN Alauddin in Meeting the Needs of Shopping Facilities 146
W W Osman, S Wunas, M Arifin, G Lakatupa, Misra

The Effectiveness of Incinerator at the Integrated Waste Treatment Plant in the Campus of Engineering Faculty - Hasanuddin University 154
A T Lando, I Djamaludin, A N Arifin, N Oktorina, R Danah, M F Sulistyowati

Soil-Compost Mixture as Alternative Material for Soil Cover Landfill 165
K Sari, I Ridwan, N A Putry M

Mechanical Properties of Concrete Using Plastic Waste 172
R Irmawaty, H Parung, R Djamaluddin, A A Amiruddin, M P Faturrahman

Investigation of Air Pollution Impact on Kinta River Water Quality at a Tropical Region 178
N A Mokhtar, W A Madhoun, M Y D Alazaiza, A L Balogun, N K B E M Yahaya, F C Ros

Appraisal of Indoor Microbial Pollutants at University Research Labs and Offices under Mediterranean Climate 183
E Y Abed, W A Madhoun, A A Elmanama, H Kim, X Xu, F C Ros

Impact Assessment of Traffic Emission on the Respiratory System of Non-Smoking Traffic Policemen in Palestine 189
W A Madhoun, E Salem, A Eljedi, H A Isiyaka, F C Ros

A Study of Assessment and Mapping of Carbon Footprints to Campus Activities in Hasanuddin University Faculty of Engineering 194
R Zakaria, S H Aly, M Hustim, A D M Oja

Model Microsimulation Roundabout Utilities in Makassar 203
M Pasra, S Hamid, A Faisal, H Yatmar

Phytoremediation of Fe and Mn Metal in Acid Mine Drainage Using Typha Angustifolia 211
R Ibrahim, A Zubair, M Selintung, A R Dwirachma

Experimental Study of Wave Reflection in Breakwater Overtopping Catcher Model 218
M P Hatta, A I D Puspita, M A Thaha, R Karamma, S Pongmanda, A S Mustari, M Ibrahim

An Estimation of Origin-Destination Matrices for a Public Transport Network in Makassar using Macrosimulation Visum 224
M I Ramli, D Runtulalo, H Yatmar, A Mangessi

The Airport Development due to the Demand of Passengers (Case: Mutiara Palu, Indonesia) 231
S A Adisasmitha, H Yatmar, S Rauf, M Hustim

Experimental Study of Natural Materials Utilization as a Stabilization Agent on Soft Soil 238
T Harianto, N Marfu'ah, I Rauf, T B C Leatemia

Type and Quantity of the Residual Construction Material of Low-rise Building Construction in Makassar 244
M A Asad, U L Rusdi, A Rosmariani, H Suharman, A Evi, O Yulio

Analysis of Sediment Distribution at the Intake Structure 250
B Bakri, S Pallu, R Lopa, F Maricar, A Sumakin, M F Maricar, Ridwan

Preliminary Study on Compressive Strength of Porous Asphalt Containing Modified Buton Asphalt, Waste Plastic and Limestone Powder 259
M W Tjaronge, Fakhruddin, M A Caronge, M H Zaifullah, M F Rahmat

Prototype of Electrical Generator Device from Piezoelectric with Serial Configuration - Based on Mechanical Pressure 265
Yusran, A A Madia, Z Zainuddin

Potential of Heart Disease Detection Based on Iridology 270
C Yohannes, I Nurtanio, K C Halim

Allocation of Wind Power Plant Construction Using Adaptive Neuro-Fuzzy Inference System 278
H Jasman, S M Said, I Nurtanio

Design and Implementation Monitoring and Booking Systems for Smart Parking at Engineering Faculty Campus 288
Muh Anshar, R S Sadjad, Dewiani, M Hanan, R Prayudha, M Abry

Performance Evaluation of Mini Single Board Computer in Hadoop Big Data Cluster 297
Adnan, Z Tahir, C Yohannes, Ariel

Partial Discharge Analysis on Crossed-linked Polyethylene (XLPE) Cable 150 KV 303
M Salama, K Ikhlas, S M Wahyu

Implementation of Clustering and Similarity Analysis for Detecting Content Similarity in Student Final Projects 310
A A Ilham, A Bustamin, I Aswad, F Armin

Interactive Learning Media Based on Android for Computer Programming Course 321
A D Achmad, A Achmad, Z M Putra

Effects of 500 kV EHV Transmission Lines on Location Determination and Capacity of Shunt Reactor in South Sulawesi Systems 327
I Kitta, A Suyuti, Z Muslimin, S Manjang, W Santoso, M T Hidayat

Smart City Application: Community Survey System of Urban Comfort Level 332
E Warni, Z Zainuddin, M F Amperiento

Comparison of Voltage Stability Index Before and After Wind Turbine Penetrated to Sulseltrabar Interconnection Power System Using Modal Analysis Method 341
I C Gunadin, Z Muslimin, A M Ilyas, A Siswanto

Control Design of Automatic Voltage Regulator to Improve the Voltage Stability at Sengkang Power Plant, South Sulawesi Indonesia 352
I C Gunadin, Z Muslimin, A M Ilyas, A Siswanto

Analysis of Transmission Line Stability for Sulselbar Interconnection System with the Penetration of Renewable Energy to Prevent Voltage Collapse 362
I C Gunadin, A Suyuti, A M Ilyas, A Siswanto

Performance of Routing Protocol OLSR and BATMAN in Multi-hop and Mesh Ad Hoc Network on Raspberry Pi 369
Wardi, Dewiani, M Baharuddin, S Panggalo, M F B Gufran

Design and Implementation of REST API for Academic Information System 378
A A Prayogi, M Niswar, Indrabayu, M Rijal

Stability Analysis and Fault Changes on Wind Turbine Effect in Multi Machine Power System 384
B A Ashad, A Siswanto, I C Gunadin, Yusran

Determination of Geochemical Correlation of Cu, Pb, Zn in Stream Sediment Samples in Barru Area, South Sulawesi 390
A Tonggiroh, Kaharuddin, U R Irfan, A Akbar

Sensitivity Analysis of Net Present Value due to Optimal Pit Limit in PT Ceria Nugraha Indotama, Kolaka Regency, Southeast Sulawesi Province 397
A V Anas, R Amalia, N F Qaidahiyani, Djamaluddin, S R D Herin

Flood Risk Mapping of Jakarta Using Genetic Algorithm Rule-Set Production (GARP) and Quick Unbiased Efficient Statistical Tree (QUEST) Methods 408
R R Yusya, M R Septyandy, T L Indra

Analysis of Soil Erosion on Mine Area 419
M Ramli, Purwanto, M Thamrin, Maemuna, M Asrafil

Extraction of Nickel and Cobalt from Sulawesi Limonite Ore in Nitric Acid Solution at Atmospheric Pressure 427
Sufriadin, S Widodo, I Nur, A Ilyas, M Y Ashari

Sediment Profiles and Distribution of Losari beach Ujungpandang District Makassar City 433
H Sirajuddin, Kaharuddin, Sultan

Ultrahigh-Pressure Metamorphic Rocks in the Bantimala Mélange Complex and Its Implication to Cretaceous Tectonics in the South Arm of Sulawesi 444
M Mawaleda, H Umar, J R Husain, A Maulana, A Jaya, U R Irvan, D D Permana

Identification of Landslide Disaster Potential based on Weathering Grade of Rock in Parepare City South Sulawesi, Indonesia 457
Busthan, Safruddim, H Pachri

The Roughness Characteristic of AA6061-F, AA6061-O and AA6061-T6 after Machining Process 464
H Arsyad, L H Arma, Januari

Cognitive Work Analysis to Comprehend Operations in Self Check-in Counter 470
Lukmandono, H C Suroso, A H Nasution

Characteristic of Variable Compression Ratio Diesel Engine Operating on Diesel and Biodiesel B20 478
N Amaliyah, B Mire, A E E Putra, M Syam, H Zakariah

Energy Absorption Analysis of aluminum Filled Foam Tube Under Axial Load using Finite Element Method with Cross Section Variations. 484
I Renreng, F Djamaluddin, F Furqani

Potential of Work Related Musculoskeletal Disorder in Traditional Salt Farmers 496
I Bakri, R D Mudiatuti, Mulyadi, A Darmawan, Akmal

Hardness Improvement of Aluminum Reinforced with Carbon Nanotubes through Forging and Friction Surfacing Process 504
L H Arma, H Arsyad, M Syahid, E Iskandar, Nurhudaedah, Asruddin

Inventories Analysis of Animal Feed Raw Materials by Using the Silver Meal Method and Wagner within Algorithm (Case Study of PT. XYZ Makassar) 510
S Asmal, I Setiawan, N Ikasari, Y Adriani

Fatigue Strength Improvement of Low Carbon Steel through Carburizing Process with Coconut Shell Charcoal 520
M Syahid, A Hayat, S Arief, Rudi

Six Sigma Implementation in Quality Evaluation of Raw Material: A Case Study 527
A Darmawan, S Bahri, A T B Putra

Distribution Route Optimization of a Capacitated Vehicle Routing Problem by Sweep Algorithm 534
R Hanafi, M Rusman, F Mardin, S M Parenreng, A Azzazli

Improvement of Production Scheduling Performance using Simulated Annealing Algorithm 540
S Mangngenre, N Syamsul, S Asmal, I P Marco, N I Syamsul, A B R Indah

The Performance of Clay Furnace with Variation in the Diameters of the Briquette Burning Chamber 548
Z Djafar, N Amaliyah, S Suluh, M Isra, W H Piarah

The Effect of Tool Geometry of Hole Towards Tool Life in Hard Material Holes 554
A Y Aminy, S Arief, Sulaiman

Structural Modeling Using 2D Shell Element to Predict the Initial Plastic Condition of Tubular Frame Structures 561
Asykur, Y S Hadiwidodo, Handayanu

Plate Deflection Analysis with Different Mesh 571
M Z M Alie, Juswan, D Paroka, M R S Rakib

NC3 CPP Jacket Launching Analysis by Using Numerical Simulation (Case Study of NC3 Gas Field, SK316 Block Bintulu Sarawak Malaysia) 578
M H M Saeful, Juswan, D Paroka

Analysis of the Ship and Cargo Service Facilities at the Local Port of Sorong 587
S Dewa, M Idrus, A S Chairunnisa, W Djafar, A H Djalante, A Karyadi

Strength Behavior Analysis of Insert Plate for Placing Corner Fitting due to Container Load 596
A M Nugraha, Hamzah, G Sitepu, A Ardianti, Ifhanuddin, S Baso

Study of Longshore Sediment Transport on Erosion and Sedimentation Beaches in South Sulawesi 610
H Umar, A Y Baeda, F Husain, Taufiqurrahman

Investigating the Performance Characteristics of a Semi-Planing Ship Hull at High Speed 616
S Baso, L Bochary, Rosmani, M Hasbullah, A D E Anggriani, A Ardianti

An Improvement for the Design of the Small Fiberglass Seaweed Farmers Boat to Minimize the Sinkable Risk 624
W Mustafa, S Asri, M R Firmansyah, F Fachruddin, G Apelaby

Risk Assessment for Fishing Boats Operating in the Makassar Strait 632
M R Alwi, A H Muhammad, Rahimuddin, H Hasan

Effect of Different Stern Flap Position and Size as an Energy Saving Device (ESD) for Increasing Fishing Boat Speed ... 638
S Klara, F Mahmuddin, S Hariyanto, G Sitepu, A A Aman

Cost Comparison for the Installation of Steel Frames vs Wooden Frames on a Traditional Wooden Fishing Boat in South Sulawesi ... 649
L Bochary, M R Firmansyah, S Asri, G Sitepu, W Djafar, Zulkifli, A H Djalante, M Arfah

Performance of a New Floating Breakwater ... 659
S Rahman, A Y Baeda, A Achmad, R F Jamal

The Analytical Solution of Waves Transmission Coefficient on Semi Fixed Floating Breakwater ... 664
C Paotonan, T Rachman, Ashury, T. P. Putra

Modeling Origin Destination Matrix in the Inter Island Cluster of North Liukang Using Gravity Model ... 670
A S Chairunnisa, S Asri, L Bochary, M R Firmansyah, Zulkifli

Effect of Wave Power Device Arrangement in Double-Water-Chamber Type Seawall towards Wave Power Extraction Rate and Wave Dissipation ... 680
F Husain

The Performance of Single and Double Intake Designs of Propeller Flow Cooling System (PFCS) ... 688
F Mahmuddin, A Suhardi, S Hariyanto, Syarifuddin, M Tasrief

EPIC (Education Priority Compact) Strategy as A Solution for Education of Post-Disaster Refugee Children in Palu City, Central Sulawesi ... 696
R Mahful, A Algifari, J Jacobus, A Haqiq

Fuzzy Logic Method as Determining of Internet Network VSAT performance in Black Spot Region at District of Enrekang ... 702
S Buwarda, K H Rasyid, Muhlis, M Hidayat

A Comparative Study using SAW, TOPSIS, SAW-AHP, and TOPSIS-AHP for Tuition Fee (UKT) ... 709
W Firgiawan, N Zulkarnaim, S Cokrowibowo

Web Page Extraction and Classification Using JSOUP and Naïve Bayes ... 714
S Cokrowibowo, N Nur, Irmayanti

Implementation of Connected Component Labelling for Calculation Amount of Eggs on the Laying Pullet Cage ... 720
A A A Cirua, S Cokrowibowo, M F Rustan

Effects of Mineral Admixtures on Pore Structure and Compressive Strength of Mortar Contaminated Chloride ... 725
D Patah, H Hamada, A Dasar

Online News Analysis of Majene Public Figure Electability with NLP (Natural Language Processing) ... 731
Sundari, S Cokrowibowo, N Zulkarnaim

Wireless Sensor Network Implementation for IoT-Based Environmental Security Monitoring ... 737
A Hendra, E Palantei, Syafaruddin, M S Hadis, N Zulkarnaim, M F Mansyur

Parking Slot Detection Using GLCM and Similarity Measure ... 742
A P Irfan, N Nur, F Wajidi

Author Index

IOP Conf. Series: Materials Science and Engineering **875** (2020) 011001 doi:10.1088/1757-899X/875/1/011001

- **Preface**

We are pleased that the proceeding of the 3rd EPI International Conference on Science and Engineering 2019 (EICSE2019) has been published. The conference is an annual international conference held by the Center of Technology (COT), Faculty of Engineering, Hasanuddin University. The 2019 conference was held on September 24-25, 2019 at Engineering Faculty Campus, Hasanuddin University, Gowa, Indonesia. Besides by Hasanuddin University, the conference was also supported by the Japan International Cooperation Agency (JICA) through C-BEST (Capacity Building in Engineering, Science, and Technology) project. The conference was also co-hosted by Engineering Faculty, Universitas Sulawesi Barat.

The purpose of the conference was stressed on the collaboration of 3 (three) parties which are University, Industry, and Government (UIG) known as "triple helix" to contribute to the development of a science technopark especially in the eastern part of Indonesia. Therefore, the adopted theme was "Strengthening UIG-Based Research and Innovations in Developing an Advanced Science Technopark". With this theme, we hope the research results presented and published in the conference could give a significant positive contribution to the acceleration of the development of this science technopark.

This time in the conference, we were very honored to have Prof. Yoshihiro Narita from Hokkaido University who was also the academic adviser of JICA C-BEST Project, Prof. Faizal Arya Samman from Hasanuddin University, Dr. Wesam Al Madhoun from University Teknologi Petronas, Prof. Kazuyuki Maeda from National Fisheries University, and Dr. Muh. Zulkifli Mochtar from PROSPEC AZ Inc. as the keynote speakers. We highly appreciate the outstanding speeches by the five keynote speakers delivered at this conference.

Finally, we would like to express our gratitude to all authors for their manuscript's contribution, proceeding editors and reviewers, and also to participants of the conference. We also would like to thank all other parties including our university and faculty leaders and staff for their support on conference organization and other matters.

On behalf of the EICSE2019 Committee,
Conference Chair

Dr. Faisal Mahmuddin

IOP Conf. Series: Materials Science and Engineering **875** (2020) 011001 doi:10.1088/1757-899X/875/1/011001

- **Proceeding Editors and Reviewers**
  - Dr. Faisal Mahmuddin
  - Dr. Rosmalina Hanafi
  - Dr. Sufriadin
  - Dr. Muhammad Ramli
  - Dr. Achmad Yasir Baeda
  - Dr. Amil Ahmad Ilham
  - Dr. Indar C. Gunadin
  - Dr. Lukmanul Hakim Arma
  - Dr. Intan Sari Areni
  - Dr. Muhammad Zubair M. Alie
  - Dr. Muh. Anshar
  - Dr. Yashinta Kumala Dewi S.
  - Dr. Asniawaty Kusno
  - Dr. Irwan Setiawan

- **Documentations**

IOP Conf. Series: Materials Science and Engineering **875** (2020) 011002 doi:10.1088/1757-899X/875/1/011002

Conference Committee

➢ International Scientific Committee (ISC)

- Prof. Yoshihiro Narito, Hokkaido University (Japan)
- Prof. Shinya Honda, Hokkaido University (Japan)
- Prof. Katsuhiko Sasaki, Hokkaido University (Japan)
- Prof. Jun Ando, Kyushu University (Japan)
- Prof. Shigeru Kashihara, Nara Institute of Science and Technology (Japan)
- Prof. Akio Miyara, Saga University (Japan)
- Prof. Tsubasa Otake, Hokkaido University (Japan)
- Prof. Hideaki Yasuhara, Ehime University (Japan)
- Prof. Masao Furusho, Kobe University (Japan)
- Prof. Satrio Soemantri Brodjonegoro, Institut Teknologi Bandung (Indonesia)
- Prof. Herman Parung, Hasanuddin University (Indonesia)
- Prof. Dato Nor Aieni binti Haji Mokhtar, University Malaysia Terengganu (Malaysia)
- Prof. Wan Mohd Norsani Bin Wan Nik, University Malaysia Terengganu (Malaysia)
- Prof. Ahmad Fitriadhy, University Malaysia Terengganu (Malaysia)
- Prof. Nasaruddin Salam, Hasanuddin University (Indonesia)
- Prof. Imran Umar, Hasanuddin University (Indonesia)
- Prof. Ramli Rahim, Hasanuddin University (Indonesia)
- Prof. Slamet Trisutomo, Hasanuddin University (Indonesia)
- Dr. Mahmoud M Masoud, Queensland University of Technology (Australia)

➢ Organizing Committee (OC)

- Steering board :
  – Dr. Ir. Muhammad Arsyad Thaha, MT.
  – Prof. Baharuddin Hamzah, ST., M.Arch., Ph.D.
  – Dr. Eng. Muhammad Rusman, ST., MT.
  – Mukti Ali, ST., MT., Ph.D.
- Coordinator : Dr. Eng. Nasruddin ST., MT.
- Chair : Dr.Eng. Faisal Mahmuddin, ST., M.Inf.Tech., M.Eng
- Secretary : Muhammad Zubair Muis Alie ST., MT., Ph.D
- Treasurer : Nurfitriani Seilah, ST.
- Members :
  – Dr. Eng. Ir. Muhammad Ramli, ST., MT.
  – Ir. Syerly Klara, MT.
  – Dr. Eng. Intan Sari Areni, ST., MT
  – Dr. Indar Chaerah Gunadin, ST., MT.
  – Dr. Eng. Purwanto, ST., MT.
  – Dr. Eng. Muralia Hustim, ST., MT.
  – Siti Hijraini Nur, ST., MT.
  – Dr. Eng. Abd. Mufti Radja, ST., MT.
  – Hamzah, ST., MT.
  – Ir. Zaenab, MT.
  – Dr. Eng. Ilham Alimuddin, ST., M.ArcGIS
  – Dr. Eng. Amil Ahmad Ilham, ST., MT
  – Dr. Ir. Zuryati Djafar, MT.

IOP Conf. Series: Materials Science and Engineering **875** (2020) 011002 doi:10.1088/1757-899X/875/1/011002

– Dr. Eng. Muh. Anshar, ST., MT.
– Dr. Eng. Ulva Ria Irvan, MT.
– Dr. Ir. Rosmalina Hanafi, M.Eng.
– Sabaruddin Rahman, ST., MT, P.h.D
– Dr. Eng. Achmad Yasir Baeda, ST., MT.
– Amos Datu Bala, S.Kom.
– Jumadil, ST., M.M.
– Herawati, S.Sos
– Susanti Alwi, S.Sos
– Jayadi, S.Sos
– Syamsir Alam, SKM
– Hasrawati, SE.
– Supriani, ST.
– Baharuddin
– Kaharuddin
– Samsuddin
– Dirman Abdillah, S.Pd, BASc
– Juliana Palambean, A.Md

IOP Conf. Series: Materials Science and Engineering **875** (2020) 011003 doi:10.1088/1757-899X/875/1/011003

Peer review statement

All papers published in this volume of *IOP Conference Series: Materials Science and Engineering* have been peer reviewed through processes administered by the proceedings Editors. Reviews were conducted by expert referees to the professional and scientific standards expected of a proceedings journal published by IOP Publishing.

IOP Conf. Series: Materials Science and Engineering **875** (2020) 012001 doi:10.1088/1757-899X/875/1/012001

Development of the Architectural Styles of the Reformation Era in Makassar, South Sulawesi, Indonesia

A Harisah[1]*, R Wikantari[1], M Mochsen Sir[1], A M Radja[1], S Syam[1], A K Deapati[1], and Badri[1]

[1]Architectural History and Theory Laboratory, Department of Architecture, Engineering Faculty, Hasanuddin University, Makassar, Indonesia

*Email: ifahhussein@yahoo.co.id

Abstract. The beginning of the reformation era was the post-New Order era marked by the resignation of President Soeharto in 1998, and the process is still ongoing. The purpose of this study was to identify trends in the development of architectural styles in Makassar, outlined in the form of a timeline for the development of architectural styles in the era. The used research method was the historical-diachronic analysis with data collection techniques through exploration on the internet using google maps, interviews, observations, and documents. The research population was phenomenal buildings in Makassar. Phenomenal means the amount was large/the shapes were prominent/it had a certain style/it had a political background in its emergence. The sampling technique used criterion-purposive-snowball sampling technique. The collected data was analyzed and interpreted descriptively. The assessment of validity and reliability of the data were done through the triangulation of qualitative data collection techniques. This research found that there were diverse architectural styles that developed in the reform era in Makassar, namely: No Style, My Style, Classic, Modern, Minimalist, Perforated Skin, Double Skin, Corporation, Iconic, Container Box, Colonial, Postmodern, and Tropical. The background of the emergence of these styles was influenced by various external factors: globalization, local strongmen, local/national/ international corporations, local iconic interpretation, decentralization, industry, and the availability of materials and experts, local and national government, political, economic, social and cultural dynamics. Furthermore, there were also internal factors that influence the styles, namely the taste of the architects, and the taste as well as economic capabilities of the users/owners.

1. Introduction

In general, the birth of architectural style is usually preceded by large works that become role models, or sources of inspiration which are then followed, interpreted, and/or made into architectural vocabulary by people afterward. A more detailed opinion is described by Melvin. According to Melvin [1], architectural style can be understood through its relation to ism (the term 'ism' according to the author is more correct when it is said 'trigger factor'). There are at least five trigger factors that influence the emergence of a style: 1) that arises because of widespread cultural tendencies, 2) movements determined by the artist, 3) labeling applied retrospectively, 4) representation of ideology, and 5) regional trends and national. If you look at the current conditions, Melvin's opinion has a point, even though there are 'trigger factors' besides these, including political factors in the period of government that affect the presence of architectural work.

Furthermore, according to Hasan [2], the period of Indonesian architectural development is based on the period of government. This shows that there is a strong relationship between architecture and politics. This is in line with the notion of Wiryomartono [3], and Ardiyanto [4], where the periodization of architecture is determined by the change of government, similarly, Hasan [2], Heryanto [5], and Kurniawan [6]. According to Heryanto, the face of the city has a strong relationship with the shape of the building and is a manifestation of the policies of the city authorities.

In the case of Makassar City, in a previous study by Harisah, there were quite phenomenal building in Makassar, for example: the emergence of container architecture in 2014 became a trend until now [7]. This is in line with the opinion of Jaharuddin et. al saying that there is an influence of the Mesjid Raya facade toward the architecture of small mosques in Makassar [8], it means that there is a trend style. Furthermore, also at the empirical level in the reformation era, there is an indication of diversity of architecture style in Makassar, and this becomes interesting to explore. In addition, historical research about architectural style of the reform era is very rarely studied. Therefore, two objects are studied: the timeline of the development of architectural styles in the era and the influential factors.

To answer the research objectives, it is used the rationalistic paradigm, where the approach is holistic, deductive-inductive, and uses theory as a direction to explore the field [9]. The research method used the historical-diachronic method where historical events were reviewed through layer by time or from year to year [10]. The research population was phenomenal buildings in Makassar. Phenomenal means many numbers/prominent shapes/indicated to have a certain style/have a political background in its emergence. The sampling technique used criterion-purposive-snowball sampling. Data collection techniques used observation, exploration with Google maps, interviews, and documents. The validity and reliability of the data used the triangulation of data collection techniques.

2. Discussion

How an architectural style is commonly identified? It can be identified by understanding of the definition of the style itself. A style is a type of architecture that is distinguished according to its structural characteristics and ornaments [11]. It is a way, model, or medium to express spiritual values and civilization and humanity, also a language that can be developed and adapt with the times [12]. It can be essentially identified visually and does not need to be related to the function of a building [13]. A more detailed explanation was offered by Wiryomartono [14], that a typical style or style is born from the architect's ideas, and/or the emotional involvement of design decision-makers. The essence of the style is to provide an identity that is based on perfecting the convention system through the techniques of composition, proportions, sequence, rhythm, detailed formulation, motives and so on. Thus, style can be identified visually based on structure, ornamentation, shape, building material, regional character, composition, proportion, and rhythm.

The results of the study showed (in Table 1) that there are at least 13 architectural styles that have emerged and are dominant since 1998 until now. The examination showed more diverse styles emerged. This is inseparable from the influence of globalization, where access to information from various parts of the world can be obtained via the internet, which use has been very widespread in Indonesia. Furthermore, six styles that appear almost every time, namely No Style, My Style, Classic, Modern, Corporation, and Tropical (Figure 2). These six architectural styles also dominate the atmosphere of Makassar City. It needs to Be explained here, for No Style is very dominant at the level of residences and kiosks that are scattered from the edge of the main road till the alleys of Makassar City. This No Style lacks specific and repetitive visual characteristics in other buildings.

The second category, My Style (see Figure 2) is also commonly found in residential houses and shophouses, especially those belong to medium to high income classes are already well established. The creation of My Style is usually influenced by the tastes of the owners and the architects. However, houses with My Style that were not designed by an architect, are mostly influenced by the tastes of the owner and the trend of other styles.

Table 1. Timeline of Development of Architectural Styles of Reformation Era in Makassar

Architectural Styles	1998-2002	2003-2007	2008-2012	2013-2019
No Style				
My Style				
Classic				
Modern				
Minimalist				
Perforated Skin				
Double Skin				
Corporation				
Iconic				
Container Box				
Colonial				
Postmodern				
Tropical				

Another dominant style is the Classical style, especially in the houses of people of high economic class. The Classical style in Makassar is identical to the use of classic columns, arch, the ornamentation of wrought iron on balconies and fences as found in the Art-Nouveau style, which is free-flowing wright iron, as well as a combination of open gable and a hip roof.

See Figure 1, Modern style refers more to the international style because it is only a game of boxes, the material is commonly used in various places, without ornaments and certain symbols, signs or icons, nothing stands out in a way that is specifics so the shape is very general. This modern style is especially visible in houses, hotels, shopping centers, and offices.

The Minimalist style can be found in many mass housings in the city. The style has a minimal line, shape, and color of the facade. It is also a trend in shophouses, which are mostly built and scattered on along the main roads of the city.

The Perforated Skin style (Figures 1 & 2) is predominantly seen in many mosques in the city, influenced by the architectural style of Masjid Raya in Andalas Street. The style not only has textured and ornate skin but also has ventilation holes that covers and cools the walls and rooms. The perforated ornamentation of the style is dominated by variation of geometric shapes that refers to Islamic ornaments. Besides, there are also the Double Skin style. It is mainly found in private offices, predominantly using composite aluminum or metal materials. The use of double skin or perforated skin is becoming a trend nowadays.

Figure 1. Left to right: Modern, Iconic and Double skin, and Corporation style

This Corporation style is mostly found in buildings that function as banks, supermarkets, minimarkets, private offices, and hotels. Generally, it uses certain colors, symbols, and signs as

IOP Conf. Series: Materials Science and Engineering **875** (2020) 012001 doi:10.1088/1757-899X/875/1/012001

identity. Buildings with this style have similar shape and can be easily recognized by a simple look as it represents a typical corporation/company building's design, for examples: Alfa Marts, Alfa Midis, Indomarets, KFCs, and Mc. Donalds.

The Iconic style (see Figure 1) refers to certain icons, such as pens (see Gedung Graha Pena) especially local icons, such as Phinisi ships, Toraja roofs and so on. This icon takes the silhouette of a local icon. There are many high-rise buildings in the form of offices, and hotels using the icon. The processed part is the upper floor, and or the building body. This can be seen in the UNM Tower, Kalla Group, and Karebosi Condotel Hotel.

Container Box style is especially dominant in buildings that function as cafes, restaurants, and food stalls. The forms are a modification of various container boxes that create an interesting look. The style is often finished with colors play. This can be seen in the Liberica Cafe, Baruga Culinary Park, UMI Canteen Campus Cabin, Urbanist Box, OTW Food Street, and several chicken-seller container boxes in front of minimarkets, and/or shophouses.

Colonial style refers to the combination of Classical-Dutch Architectural style with Tropical style, to create buildings that are adaptive to tropical climate. The Colonial Style is not much in number, but quite striking because of the differences with the buildings around it. This Colonial style is found in cafes, restaurants, and residences, for example, Bon Cafe which is inspired by the beauty of Fort Rotterdam Makassar.

Figure 2. Left to right: Perforated Skin, Postmodern, and My own style

The Postmodern style (Figure 2) is a blend of modern architecture with classic icons, so it is not a local icon. The classic icons used are classic columns but have undergone various modifications, but still leaving a classic silhouette, so the postmodern impression is thick. This style is found in offices and hotels. An example can also be seen in the office of Kejaksaan Tinggi Makassar.

The Tropical style had been around a long time before the reformation and was phenomenal in the late 20th century. Tropical style is very easy to find in residential houses, its characteristic is the sloping roof, many openings and use natural lighting and air during the day. Until now, the tropical style remains in demand, but the phenomena did not emerge in the reform era as in the minimalist style. Furthermore, there is a blend of tropical and classical styles, but the amount is not so many.

It should also be explained here, that among the 13 styles, some apply a specific combination such as the UNM (Universitas Negeri Makassar) building, in addition to the first skin layer of a hollow building, the second layer, the overall shape refers to the phinisi ship icon. This building is one of the most unique buildings in Makassar because it is very different from the surrounding buildings.

In Figure 3, it can be explained the influential factors toward the styles. Basically, it is accumulation of various factors, and each factor does not stand alone. These factors can be grouped into two groups: internal and external. The internal consist of three factors: architect's taste, taste and economic ability of the owner, while external are sub grouped into five factors: 1) Globalization and World Corporation, 2) National Scale (central government, national corporation, and the existence of centralization policy), 3) Dynamic Situation (political, economic, social, and cultural), 4) Local Scale (locality: local architectural elements, influential local strongmen, local government, local corporation, decentralization, local materials, and local personnel expert), and 5) Others (foundation, non-local materials, non-local personal expert, and so forth).

The influence of globalization and world corporation toward architectural styles can be seen in the dominance of modern styles, in this case, the international style; where the shape of buildings cannot be detected by the existence of local values, symbols, signs, and contents. The condition can be seen in modern buildings owned by world corporations such as Pizza Hut, KFC, Mc. Donald and so on in Makassar, so it is similar to buildings in other countries. This is in line with the opinion of Hidayatun et al. [15] saying that globalization results in the loss of specific things and changes in the order of life that removes geographical boundaries.

The intervention of the central government can be detected by its centralization policy through the construction, addition, and rehabilitation of educational buildings such as SD (Sekolah Dasar) Inpres, SDN (Sekolah Dasar Negeri), SMPN (Sekolah Menengah Pertama Negeri), and SMAN (Sekolah Menengah Atas Negeri), which are floor and form plans seem similar including the standard of dimensions and materials used so as to create architectures that are the same or similar form. As for, the influence of national and international corporations can be seen through the existence of various buildings both owned and rented by private parties, such as Mc. Donald, KFC (see Figure 1), Alfa Mart, Carrefour, Pizza Hut, Amaris Hotel, Ciputra Housing, and others.

Political and economic dynamics influence the import policy of building materials from outside Indonesia, the policy of the domestic building materials industry, and the direction of technological development. Furthermore, Makassar as a trading city provides building materials from outside and inside, especially for Sulawesi Island, so that the trend of building material and technology in Makassar can be seen in outside too, and on the contrary.

The socio-cultural dynamics influence the trends of architectural styles. In general, Makassar people tends to imitate beautiful buildings that suits their tastes visually both consciously and unconsciously, so that they become a part of the common visual culture. This is in line with Melvin's opinion (see the previous discussion about ism arising from widespread cultural tendencies), and Sachari [16] saying that one of the most obvious aesthetic values is the visual style that forms a major part in the design work that is relating to the exploration of comfort and beauty through the process of community intellectualization, or it can also be an unconscious process, for example modernism in the west countries was adopted for decades after Indonesian independence.

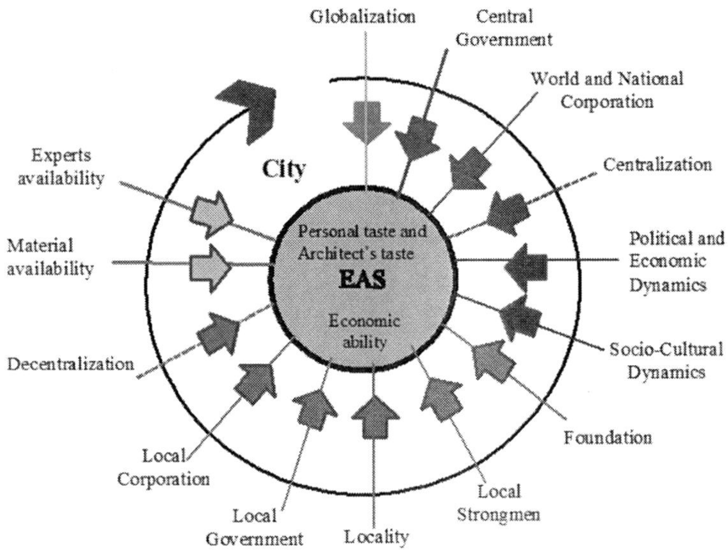

Figure 3. EAS (Emerging Architectural Styles) in Makassar and the Influential Factors

The many educational foundations in Makassar also influence the developing architectural styles, especially those managed by certain foundations that are identical with certain colors and shapes of buildings, for example, educational foundations under Wahda Islamiyah and UMI (Indonesian Muslim University) are identical with green and white, including buildings that are under the institution label.

Decentralization that occurred at the beginning of the reforms indirectly affected architectural styles in Makassar. According to Agustino et.al, decentralization means delegating a portion of power to regional/local government, this then creates local corporations, local governments, local strongmen. The presence of these local strongmen as an economic or political leader can also carry out social control [17]. This opinion is in line with some cases in the city. This can be seen by the presence of phenomenal buildings such as Graha Pena is belongs to the local corporation: Fajar group. Clarion hotel, Dalton hotel, Almadera hotel, and the Rinra and Phinisi Point are belonging to Willianto Tanta. Meanwhile, the governor of South Sulawesi: Syahrul Yasin Limpo realizes the phenomenal building in Losari beach: Mosque having 99 Domes, and Wisma Negara. The Mayor of Makassar: Ramdhan Pomanto succeeded to build Government Offices such as the District Offices, Damkar stations and others that are identical with the orange and green color. The pieces of evidence are also in line with the opinion of Hasan [2] and Kurniawan [6] about the presence of political identity through various symbols and signs in the works of architecture in Indonesia. Besides that, the reformation also presents the phenomenal buildings that represented by political parties through certain colors, especially and temporary approaching the general election. This event is a part of the face of architecture in Indonesia.

The availability of building materials and experts also affects the architectural styles, for example: modern architectural styles are strongly colored by glass materials, aluminum composites, conventional bricks, and lightweight bricks. According to Hidayatun et.al [18], this material factor is a part of the Indonesian Architecture identity. Another feature is that the roof of the building tends to be flat, or if it is sloping, it usually use tile roofs, zinc, or spandex. Besides, many building structures use conventional structures such as reinforced columns and beams of iron, beam-column structures with core, steel structure, and combination of steel and concrete structure. The condition is caused by the capability of building structure experts. Nowadays, many high-rise building structures can be found in Makassar.

3. Conclusion

The development of architectural styles of the reform era has 13 phenomenal styles: No Style, My Style, Classic, Modern, Minimalist, Perforated Skin, Double Skin, Corporation, Iconic, Container Box, Colonial, Postmodern, and Tropical. Among the 13 styles, the most phenomenal styles are No Style, My Style, Tropical style, Corporation style, and Modern style. In the timeline in Table 1, from year of 1999 to 2007, three new style emerged, and then from 2011 to 2019, four new style emerged. The styles that emerged in the reformation era were indicated more diverse than in the previous era. Whether we realize it or not, this diversity is the face of Indonesian architecture today. Moreover, it can be underlined that the emergence of a particular style is influenced by many factors, both internal and external factors. The three internal factors: architect's taste, taste and economic ability of the owner. The five external factors: 1) Globalization and World Corporation, 2) National Scale (central government, national corporation, and centralization policy), 3) Dynamic Situation (political, economic, social, and cultural), 4) Local Scale (locality: local architectural elements, local materials, and local personnel expert, influential local strongmen, local government, local corporation, decentralization,), and 5) Others (foundation, non-local materials, non-local experts, and so forth).

Acknowledgment
We would like to appreciate and state that this research has been funded by the Labo-based Education Research Funding Program 2019 of Engineering Faculty of Hasanuddin University.

References

[1] Melvin, J. 2018, ISMS, Understanding Architectural Styles, Universe Publishing, New York, pp. 7.
[2] Hasan, I. 2009, Architecture and Politics of Identity in Indonesia, Disertation, Adelaide University, Australia.
[3] Wiryomartono, B. 2013, Soejoedi and Architecture in Modern Indonesia, A Critical Post.
[4] Ardiyanto, A., Djunaedi, A. Ikaputra, Suryabrata, D. Adi, 2014, The Concept of Modern Dutch Colonial Architecture to The Development of Javanese Architecture, Dimensi: Journal of Architecture and Built Environment Vol 41, July, p. 37-42.
[5] Heryanto, B. 2013, Roh dan Citra Kota, Peran Perancangan Kota sebagai Kebijakan Publik, Brillian Internasional, Surabaya.
[6] Kurniawan, K. R., 2017, Dinamika Arsitektur Indonesia dan Representasi 'Politik Identitas' Pasca Reformasi, NALARs Jurnal Arsitektur Volume 17 Nomor 1 Januari 2017: pp. 65-78
[7] Harisah, H., 2018, Munculnya Arsitektur Kontainer di Makassar dan Sekitarnya, "Manifesto" Arsitektur Hijau? Seminar Arsitektur Nusantara IPLBI 2018, Universitas Hasanuddin. Makassar.
[8] Jaharuddin, W., Wikantari R.R, Harisah, A., 2017, Pengaruh Fasad Masjid Raya Terhadap Arsitektur Masjid-Masjid Kecil di Makassar, National Academic Journal of Architecture, pp. 39-49.
[9] Muhadjir, N., 2002, Metodologi Penelitian Kualitatif, Edisi IV, Penerbit Rake Sarasin.
[10] Kuntowidjoyo, 2003, Metodologi Sejarah, Edisi Kedua, Tiara Wacana, Yogyakarta.
[11] Oxford Dictionary dalam John C. Poppeliers S., Allen Chambers, JR. Nancy B. Schwartz, 1983. What Style Is It, A Guide to American Architecture. The Preservation Press, pp. 11.
[12] Kazim, J.A. in Mustafa, M.M., 2010, The Characteristics of Architectural Style of the Traditional Houses in The Mosul City, Analytical Study. American Journal of Engineering and Applied Sciences 3 (2) pp. 380-389.
[13] Poppeliers S., J.C., Chambers, JR. A., Schwartz, NB., 1983. What Style Is it, A Guide to American Architecture. The Preservation Press, pp. 11.
[14] Wiryomartono, B, 2016, Komposisi Arsitektur, Apresiasi dan Analisis Kasus di Indonesia, Penerbit Erlangga, Jakarta.
[15] Hidayatun, M. Prijotomo, J. dan Rachmawati, M., Arsitektur di Indonesia Dalam Perkembangan Jaman, Sebuah Gagasan untuk Jati Diri Arsitektur di Indonesia, Institut Teknologi Sepuluh November, Surabaya.
[16] Sachari, A., 2007, Budaya Visual Indonesia, Penerbit Erlangga, Jakarta.
[17] Agustino, L. dan Yusoff, MA., 2010, Politik Lokal di Indonesia: Dari Otokritik ke Reformasi Politik, Jurnal Ilmu Politik, Edisi 21, pp.12-13.
[18] Hidayatun, M. Prijotomo, J. and Rachmawati, M., 2013, Materials as Part of the Identity of Indonesian Architecture.

IOP Conf. Series: Materials Science and Engineering **875** (2020) 012002 doi:10.1088/1757-899X/875/1/012002

Aerial Cable Car in the City Centre of Makassar: The Potential Routes, Technology and Station Locations

Y K D Sutopo[1]*, M Y Jinca[1], M F Azmy[1], M V H Maramis[1], and M D Amin[1]

[1]Department of Urban and Regional Planning, Engineering Faculty, Hasanuddin University, Makassar, Indonesia

*E-mail: yashintasutopo@yahoo.com

Abstract. Aerial cable car is one of unique and efficient transportation modes that the world has today. This research aimed to explore the potential application of this mode in the urban areas of Makassar City, specifically in its city centre. It is designed to answer preliminary planning questions with regard to the best potential routes, technology, and station locations for the implementation of this mode in the near future. The research was conducted in 5 months (from April to August 2019). The data was collected through literature study and field survey including a traffic counting. The spatial analysis utilized satellite image map of Google Pro Earth, mapping, illustration and descriptive explanation. The research identified at least three trip purposes that the aerial cable car could potentially serve the city: for daily transportation, tourism, and shopping/business centres trips. It was concluded that trips to Lae-lae and Kayangan Resort Islands, the Losari City Tour, and trips from shopping mall to mall, are among potential routes that deserve for further comprehensive researches and feasibility studies. Monocable Detachable Gondola (MDG) is considered the best choice as a starting point in introducing this type of mode to the city and its people. The locational analysis based on defined criteria resulted five to nine potential station locations.

1. Introduction

Makassar is the capital city of South East Sulawesi Province and among the biggest cities in Indonesia. Its city center holds important role not only for the city but also regionally. It is where the city government, central business district, and national and international ports are located; the place of historical sites, traditional buildings and villages; a prominent coastal zone that connects the scatter small islands the city has to the mainland. It becomes the biggest attraction internally and externally that creates problems of heavy traffic during days and nights. Having a very compact and dense urban structure, it gives little flexibility for the government and planners for road widening project. Innovative ideas therefore are vitally needed not only for traffic management but also for a new efficient transportation mode to be proposed that consumes less land and environmentally friendlier.

In Europe, mass rapid transit, such as bus and train system, has a very long history as an effective solution for public transportation in urban areas. Creatively though, in the city center, the size and width of the bus may be modified in order to fit its very narrow streets. Building rails and roads above and under the ground have also proved efficient in expanding and improving the transport services without compromising the wellbeing of the existing on the ground traffic. Implementing aerial transportation such as cable car that requires very less amount of space for the towers and stations building without the need to drastically change the urban landscape is also a favorite choice in many European cities.

This research aimed to elaborate one of the above concepts, i.e., the aerial cable car, as one of potential solutions in Makassar City Center for better traffic management and to improve public transportation service quality. The research was designed to answer preliminary planning questions with regard to the best potential routes, technology, and station locations for the implementation of this mode in the near future. The result would hopefully be beneficial for the local urban and transport planners as well as the city government and to encourage further studies related to its application in other Indonesian cities.

2. Methodology

This research was conducted in 5 months (April to August 2019). The selected case study is the City Center of Makassar. The data was collected through literature study and field surveys including a traffic counting on daily passengers travelling by existing sea transportation mode to Lae-lae Resort Island. The traffic counting was conducted in August within 7 days comprises all working and weekend days (Wednesday/14th to Thursday/15th, continued on Saturday/17th to Tuesday/20th, and ended on Friday/23rd) effectively during 6 hours per day (from 8 am to 4 pm). The spatial analysis was accomplished with the help of satellite image map obtained from Google Earth Pro and using the method of mapping, drawing and descriptive explanation.

3. Result and Discussion

The aerial cable car is a transportation system that relies on cables to pull a motorless vehicle (cabin) on a steady speed along cable that spans meters away between tower to tower connecting station to station where an off-board motor is located. Its operation requires routes, vehicles (cabins), towers, simple stations, supporting cables, motors, information and ticketing system, and a reliable schedule, safety and evacuation system. It runs exclusively, neither undisturbed by- nor creates disturbance to- existing road traffic, thus it could stay consistent and dependable in its service and time schedule regardless the severity of road traffic congestion. These traits make aerial cable a popular choice in big cities with traffic problems and especially with a strong tourism sector.

Table 1. Comparison between technologies of aerial lifts (ropeways) [1]

	Tramways		Gondolas		
	Single-Haul Aerial Tramways	Dual-Haul Aerial Tramways	MDG	BDG	TDG
Illustration Photo					
Cabin capacity	Large 20-200 passengers	Large Up to 100 passengers	Small-medium 4-15 passengers	Small-medium 4-15 passengers	Small-medium Up to 35 passengers
Line capacity	500-2800 passengers per hour per direction	Up to 2000 passengers per hour per direction	Up to 300 passengers per hour per direction	Up to 3600 passengers per hour per direction	Up to 6000 passengers per hour per direction
Grips	Fixed grips	Fixed grips	Detachable grips	Detachable grips	Detachable grips
Operating speed	Up to 43.2 (km/h)	Up to 27 (km/h)	Up to 21.6 (km/h)	Up to 21.6 (km/h)	Up to 30.6 (km/h)

The aerial cable car system consists of two types of technology: tramway and gondola lifts. The technology of aerial tramway consists of a single- and a dual-haul. The technology of gondola lifts consists of Monocable Detachable Gondola (MDG), Bicable Detachable Gondola (BDG) and Tricable Detachable Gondolas (3S). The main differences between tramway and gondola, especially related to cabin and line capacity, operating speed and distance spans can be seen in Table 1.

3.1. Consideration in Selecting the Best Route and Defining The Planning Purposes

Selecting the best route for public transportation, especially aerial cable car, takes into consideration the following factors: 1) the planning purposes; 2) object of interest; 3) final destination; 4) potential traffic demand; 5) the shortest distance/time; 6) location of transit points; 7) intermodality; 8) complexity of urban structure: the land availability and existing structure flexibility to accommodate the building of infrastructure, i.e., towers and stations; 9) especially for the tourism purpose, it is important to consider routes that passing through as many touristic objects and scenic landscape as possible.

Aerial cable car can be built for a single as well as multiple trip purposes, for example:

- mainly for daily public transportation, such as the Complexo do Alemao Teleférico - Rio de Janeiro in Brazil [2], the Medellin Metrocable in Kolombia, and the Caracas Metrocable in Venezuela [3];
- mainly for tourism transportation, such as the Taman Mini Indonesia Indah (TMII) cable car in Jakarta, the Kumala Island Garden Resort Cable Car in East Kalimantan, and the Gondola Timang Beach in Yogjakarta, which was initially built to transport the lobster fisherman from the coast to lobster nests in Timang Island [4] [5];
- (dual purposes), primarily for public transportation and secondary for tourism, such as the Roosevelt Island Tramway in USA [3];
- (dual purposes), primarily for tourism and secondary for public transportation such as the Sentosa Island Gondola in Singapore [2];
- for a temporary construction to support a high-profile event, which later on been made permanently, such the Emirate Air Line in London for the Olympic Game event in 2012 [3] and the Rhine Ropeway Koblenz in Germany for the Hortikultura BUGA event in 2011 [6];
- to serve transportation demand in specific zone, such as the Polinka Cableway in Polandia that connects different parts of the University of Wroclaw separated by the Oder River [7]; and
- for multi-purposes transportation, such as the Mountain Village Gondola – Telluride in Colorado, which does not only serve the public and the tourists, but also the ski players and shopping trips, as well as providing access to reach parking areas [3].

3.2. Planning the Aerial Cable in the City Centre of Makassar

The demand for passenger transportation in the City Centre of Makassar comprises of, at least, three trip purposes: daily transportation, tourism, and shopping business, whereas the aerial cable car, as described above, has the capacity to serve for all these trip purposes.

With regard to the tourism purpose, the aerial cable car is an object of attraction in itself and could potentially induce more attraction to the existing ones. Its planning would encourage a better intermodality of all public transportation that would ease the tourists to reach various destinations from the nearest harbors and ports, from the far away airport, and from the surrounding regions. As Makassar has resort islands, it is a much cheaper solution to connect them to the mainland with aerial cables rather than building concrete bridges or roads. With regard to the business purpose, aerial cable car would enable businessmen and shoppers moving directly and easily from one office building or shopping center to the other. As for daily transportation, its implementation would encourage more utilization of environmental friendlier means of transport such as pedestrian, bicycles, and electrical vehicles, of which would potentially lower the use of private motorized vehicles. This is beneficial especially for a very dense urban structure such as Makassar City Centre. Its user friendliness would help diffable people by giving them more flexibilities and mobilities in travelling to fulfil their daily needs.

The tourism purpose in Makassar City Centre includes trips to resort island, city tours, and shopping centres. Among 12 small islands in Makassar City (table 2), Samalona, Lae-lae and Kayangan are well known as resort destinations that attract local tourist and foreigners due to their nature and recreational facilities. Among them, Lae-lae and Kayangan are the shortest in distance to the mainland and that may lead to higher possibility for the application of aerial cable for less technical challenges that they may cause. Short distance also would make short trips: half day or even couple hours trips, possible.

Table 2. Islands that administratively belong to Makassar city

No	Makassar Small Island	Administration	Characteristic
1	Lae-lae		Tourism
2	Kayangan		Tourism
3	Samalona	District of Ujung Pandang	Tourism
4	Gusung Tallang		Tourism
5	Kodingareng Keke		Tourism
6	Barrang Lompo	Subdistric of Barrang Lompo, District of Sangkarrang Archipelago	Settlement
7	Barrang Caddi		Settlement
8	Lumu-lumu	Subdistrict of Barrang Caddi,	Settlement
9	Bone Tambung	District of Sangkarrang	Settlement
10	Langkai	Archipelago	Settlement
11	Lanjukang		Settlement
12	Kodingareng Lompo	Subdistric of Kodingareng, District of Sangkarrang Archipelago	Settlement

3.3. Trips to the Lae-lae Resort Island

Lae-lae Island is located less than 500 meters above sea level, in the District of Ujung Pandang. It has total area of 0,22 km^2 which is divided into 10 RW (community units) and 3 RT (neighborhood units) and a home for 1.784 inhabitants (about 8.109 inhabitants/km^2 density) [8].

Currently, the inhabitants and tourists take journey to this island using the only means of public and private transport available, i.e., traditional fisherman ships and local speed boats. The trips started from Kayu Bangkoa, the local harbor in the mainland, crossing the distance of about 1.08 km. With the speed rate of about 10 knots (±18.52 km/hour) theoretically, the travel would only take less than 5 minutes, however in reality due to frequent detour, wind, and technical problems, it may take 10 minutes or even longer to arrive.

The location of Lae-lae Island and three possible alternative routes to travel to this island from the mainland by means of aerial cable car are shown in Figure 1. The first alternative (about 1.08 km) is a journey started from the Kayu Bangkoa Harbour, exactly from where the current sea transportation services are also started. The intermodality between the land, sea and aerial transportation in this point would be an interesting subject to work on. The second alternative (about 1.40 km) is started from the public open space called the Anjungan Toraja. The third alternative (0.48 km) is the shortest distance that started from the Centre Point of Indonesia, which is a newly built settlement resulted from the still ongoing reclamation project.

IOP Conf. Series: Materials Science and Engineering **875** (2020) 012002 doi:10.1088/1757-899X/875/1/012002

Figure 1. Aerial cable car alternative routes from Makassar city mainland to Lae-lae resort island [9]

The selection of the best technology (between aerial tramways or gondola lifts) takes into consideration the following factors: 1) the amount of travel demand, which is highly influenced by the public interests to travel by means of cable car system, especially true when it is a totally new system applied in the city/region; 2) whether it is built for a permanent or temporary use, such as for a project exhibition aimed to assess the public acceptance before being made permanently; and 3) the minimum free span and other technical challenges and requirements that have to be met in the planning field.

Whether to implement aerial tramway or gondola lifts, it depends on, among other things, potential demand and public acceptance. High cabin and line capacity such as tramway seem more suitable for high traffic demand. In new application, it could be an optimistic option to be applied in communities that are used to mass transit and cable kind of transportation as it would be relatively easier to gain public acceptance which may lead to an adequate traffic demand to cover the investment as well as daily operational and maintenance costs.

As for Makassar City where bus transportation does not survive well while rail and cable transportation do not exist, the tramway cabin could be too generous. Gondola, in the other hand, seems a rather perfect mode for a trial implementation by which the community's interests and rate of utilization could later on be assessed to define the continuity of its services in longer term. Monocable Detachable Gondola (MDG) system that is capable to serve up 15 passengers cabin capacity, 300 passengers per hour per direction line capacity and 21.6 km/h operating speed [1] would be the best choice as a starting point in introducing this type of transportation to the city and its people.

Table 3. The travel time to Lae-lae Island by existing transportation mode and aerial cable Car

Alternative route for cable car	Distance from mainland to Lae-lae Resort Island (km)	Travel time (minutes)		
		by existing public/rented boat (±10 knot)		by aerial cable car (MDG 21.6 km/h)
		Calculated	Real time	Calculated
1	±0.48	±2.00	-	1.33
2	±1.40	±6.00	-	3.89
3	±1.08	±4.60	±10.00	3.00

IOP Conf. Series: Materials Science and Engineering **875** (2020) 012002 doi:10.1088/1757-899X/875/1/012002

Table 4. Capacity comparison between existing public/rented boats with MDG technology

Survey Days	Capacity by existing public/rented boat			Capacity by aerial cable car with MDG (up to 15 psg/cabin)	
	Number of boats	Number of Passengers per day			
		Per 6 hours	Per hour	Cabin	Line Capacity
Monday	21	166	28	2	up to 300 pphpd
Tuesday	25	203	34	3	
Thursday	28	325	55	4	
Wednesday	24	254	43	3	
Friday	15	120	20	2	
Saturday	20	127	22	2	
Sunday	19	71	12	1	
Average	22	181	31	3	

Table 5. Aerial cable car built above water body, crossing ships and active harbor [7]

System	Application/ Purpose	Water Body/ Harbour	Length (Km)	City	Technology	Speed (m/s)	Cabins
Singapore Cable Car	Tourist/Urban	Keppel Harbour	1.7	Singapore	MDG	5	67
Wuhan Cable Car	Tourist/Urban	Hanjiang River	0.9	Wuhan	MDG	5	18
Vinpearl Cable Car	Tourist/Semi-Urban	Nha Trang Bay	3.3	Nha Trang	MDG	6	65
Nizhny Novgorod Cable Car	Transit/Urban	Bor River	3.6	Nizhny Novgord	MDG	6	28 (up to 56)
Emirate Air Line	Tourist/Urban	Thames River	1.0	London	MDG	6	34
Yeosu Cable Car	Tourist/Urban	Namhae Sea	1.5	Yeosu	MDG	5	50
Ngong Ping 360	Tourist/Urban	Tung Chung Bay	5.7	Hongkong	BDG	7	112
Kolner Seilbahn	Tourist/ Urban	Rhein River	0.9	Cologne	BDG	28	44
Roosevelt Island Tram	Tourist/Urban	East River	1.0	New York	Dual Haul	7.5	2
Koblenz Seilbahn	Tourist/ Urban	Rhein River	0.9	Koblenz	3S	7	18
Spokane Falls Skyride	Tourist/ Urban	Spokane River	0.4	Spokane Falls	Pulsed Gondola		6
Yangtze River Cable Car	Transit/Urban	Yangtze River	1.2	Chongqing	Aerial Tram	10	2
Polinka Cableway	Transit University)/ Urban	Oder River	0.3	Wroclaw	Aerial Tram	5	2
Port Vell Aerial Tramway	Tourist/ Urban	Port Vell	1.3	Barcelona	Aerial Tram	3	2

The technical planning challenges in implementing MDG, among other things, is to build towers that span between 100-300m, in Europe spans >400m are possible under a certain condition [7]. It is however much less than Tricable Detachable Gondolas (3S) which have larger spans between 1,000-3,000m with the consequence of larger towers [7]. Other technical planning challenges especially in implementing cable cars that crosses water body, ships and active harbors are: 1) clearance distance; 2) navigation marks and signs; 3) flight paths (in case of interference with airplane and helicopter); 4) maximum length of free span; 5) towers built in water; 6) regulating ships movement around cable car; 7) cable car installation, testing and maintenance; 8) fire, safety precautions and rescue system [7]. The examples of aerial cable car that were built above water body, crossing ships and active harbor can be seen in Table 5.

3.4. Trips to the Kayangan Resort Island

Kayangan Island is part of Bulogading Subdistrict in Ujung Pandang District. It has a total area of about 1 ha, which is much smaller than the size of Lae-lae Island. Considering its limitation in land size to accommodate large visitor and the building of more recreational facilities, applying aerial cable car to serve transportation to this island would be an extravagant plan. However, its natural beauty and aerial views could make the journey a favorable and exclusive one. This is the reason why this research includes it as one of the potential routes for aerial cable car.

Traditional fisherman ships and local speed boats are also the main transportation to reach this island. The trips could be started from the Kayu Bangkoa Harbor or Benteng Pannyua Harbor or Traditional Paotere Port. From Kayu Bangkoa Harbor, the journey crosses the distance of about 2.77 km, about less than 10 minutes theoretically but it could reach to even 15 minutes in reality. Travel route from the Paotere Port is more favorable to people as the distance is shorter and thus faster in travel time (about 2.25km, about 7 minutes theoretically but 10 minutes or even more in reality).

The following figure shows the location of Kayangan Island in satellite google earth and two alternative routes to travel to this island from the mainland by means of aerial cable car. The first alternative, similar to the previous concept for Lae-lae, is to initiate the aerial journey from the Kayu Bangkoa Harbour (about 2.77 km), exactly from where the current sea transportation services are also started. The intermodality between the land, sea and aerial transportation in this point become an advantage. The second alternative is the shortest distance, about 0.86 km, where the station may be integrated within the PELNI private harbour, or, as other possible concept is, fused to one of the commercial buildings nearby.

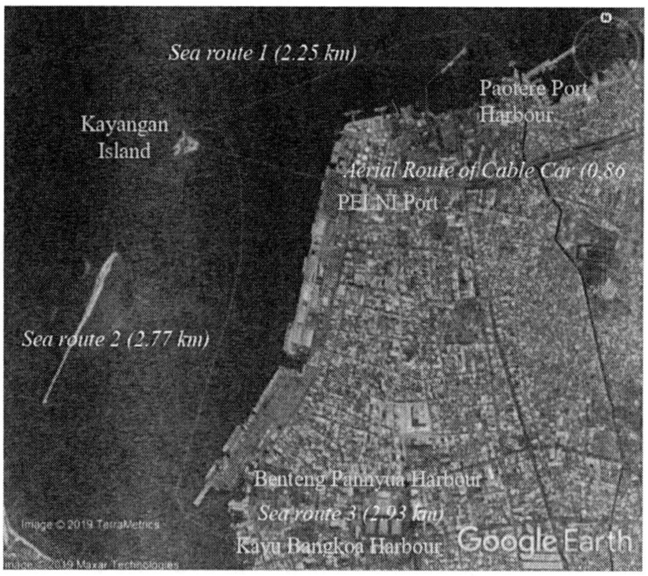

Figure 2. Aerial cable car potential route from Makassar city mainland to Kayangan resort island [9]

Table 6. The travel distance and time to Kayangan island

Routes	Distance from the mainland to Kayangan Resort Island (km)	Travel time (minutes)		
		by existing public/rented boat (±10 knot)		by aerial cable car (MDG 21.6 km/h)
		Calculated	Real time	Calculated
1	±2.25	±7.29	10	-
2	±2.77	±8.97	15	-
3	±2.93	±9.49	15	-
By Aerial	±0.86	±2.79	-	±2.39

3.5. The Losari City Tour

This type of trip is designed as a recreational and vocational trip aimed to present the historical, traditional mixed with the modern culture, and attraction, as well as the dynamic lives that embedded in the City Centre of Makassar. Currently it is a very common trip offered by many travel agents in Makassar for the incoming visitors and tourists. Currently such short trips are conducted by means of public/private/rented buses/minibuses/vehicles. The use of aerial cable car would offer even better and spectacular view of the city landscape, the urban structure and the coastal panoramic.

Losari is the name of a beach, along the coastal area exactly where Makassar City Center is radially located (from the Penghibur Street to Nusantara Street). This particular area is packed with enormous public facilities (shown in Figure 3), including the mayor and governmental offices, traditional harbors, national and international ports, jetty facilities, museums, business and banking services, radio and television centers, stadium and sport centers, hotels, malls and restaurants. It is indeed the prime zone of Makassar City and Regionally.

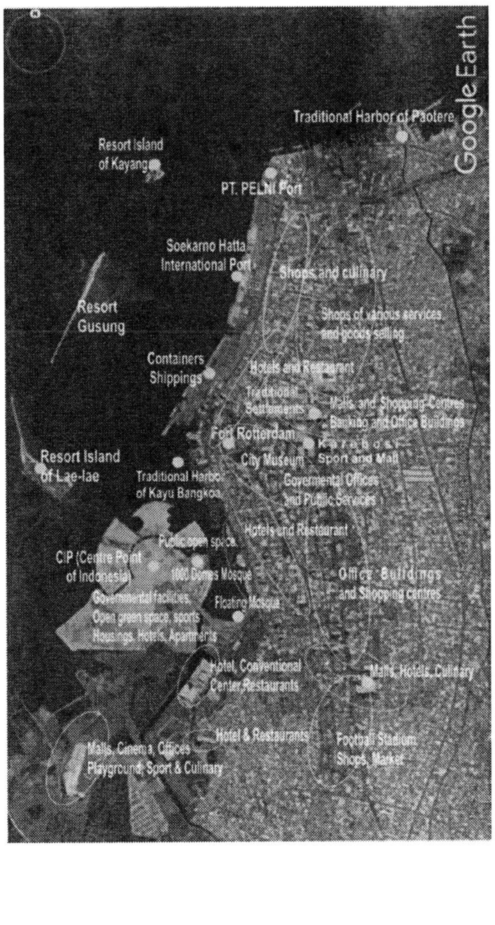

Figure 3. Objects of attraction along the Losari beach and Makassar city centre [9]

3.6. Shopping Trip from Mall to Mall

It is nowadays a common and favourite habit especially among the young people and women in Makassar City and also in other Indonesian cities. Mall is a type of shopping centre and today it is intentionally designed to be a one stop shopping place as it provides almost all kinds of goods and services of daily needs. As the number of malls iss increasing rapidly, people become not satisfied to only visit one mall at a time, but instead in making decision on what best to buy, they feel the need to go and see other options in other malls. This is one of the reasons why City of Brussel offers a free shopping bus line that transport its people from one shopping centre to another. At least 7 shopping centres served by this line from Saturday to Sunday. Besides as part of traffic demand management program and to ease the journey of the people, it is also a way to support the growing businesses and commercials in a very smart and positive way.

Aerial cable car can even be a better option than bus to serve this kind of trips, especially when the shopping centers are located a line (see figure 4). The MTC Mall in Makassar City, the underground Karebosi Link Mall and the Karebosi Kondotel (condominium/shopping/business center) are all clustered together in a very close distance and even have direct pedestrian way, horizontally and vertically, to one another. Right in direct line 2.3 km to the south, there is Ratu Indah Mall which is also clustered together with the surrounding shops, hotels, restaurants and business center. Connected these two very busy clusters with aerial cable car can reduce the use of motorized transport in between and thus ease off the on the ground traffic congestion.

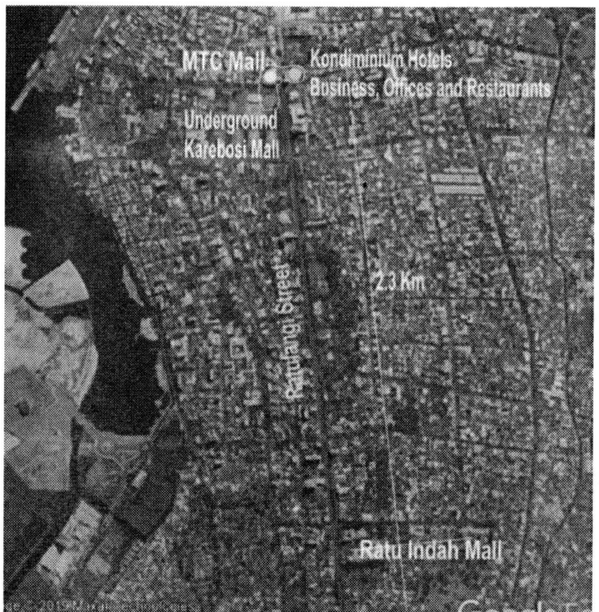

Figure 4. Route of mall to mall shopping trips [9]

The selection of the best location for aerial cable car station takes into consideration the following factors: 1) Along the selected routes; 2) Points where the potential demand is high or even the highest; 3) Points where the intermodality is high or even the highest; 4) Within standard of maximum distance from one station to the other; 5) The availability of land for its construction or existence of tall building to be integrated.

4. Summary and Conclusion

The research identified at least three purposes that this mode would potentially serve the city: for daily transportation, tourism and shopping/business centres trips. The analysis concluded that trips to Lae-lae and Kayangan Resort Island, tours along the Losari Beach, and journeys between shopping malls are among routes that deserve for further feasibility studied. Monocable Detachable Gondola (MDG) that is capable to serve up 15 passengers cabin capacity, 300 passengers per hour per direction line capacity and 21.6 km/h operating speed is considered the best choice as a starting point in introducing this type of transportation to the city and its people. The locational analysis based on defined criteria resulted in five to nine best potential locations for future stations.

Acknowledgment

The research was supported by Engineering Faculty of Hasanuddin University through Labo-based Education Research Funding Program 2019.

References

[1] Elyaris, Ghada Mohammad. 2017. "*Aerial Ropeways as Catalysts for Sustainable Public Transit in Egypt*". International Journal of Innovative Research in Science, Engineering and Technology. Vol. 6. ISSN (Online) 2319-8753. ISSN (Print) 2347-6710.

[2] Alshalalfah, B., dkk. 2015. "Aerial Ropeways Transit- Exploring its Potential for Makkah". Center of Reasearch Excellence in Hajj and Omrah. Link: http://civmin.utoronto.ca/wp-content/uploads/2015/08/Makkah-Project-Report-Part-1.pdf (last accessed on 27th June 2019)

[3] MPO (Miami-Dade Metropolitan Planning Organization). 2016. Final Report on *Aerial Cable Transit Feasibility Study*.
Webpage: https://k-learn.adb.org/system/files/materials/2017/03/201703-urban-ropeways-solution-congested-cities.pdf (last accessed on 27th June 2019)
Olivers, Pedro dan Roberto Ameneiro. 2013. "Los Metrocable de Caracas - Aerial Cable Cars as an innovatives solutions for urban transport". Web page: https://sustainabledevelopment.un.org/content/documents/3738olivares.pdf (last accessed on 27th June 2019).

[4] Gondola Projects. 2018. Timang Beach Gondola – World's most Exciting and Expensive 30-second Ropeway Ride? Website:http://gondolaproject.com/2018/08/24/timang-beach-gondola-worlds-most-exciting-and-expensive-30-second-ropeway-ride/ (last accessed on 30th July 2019).

[5] Tempo. 2017. Uji Nyali Naik Gondola Tradisional di Pantai Timang Gunungkidul, Website: https://travel.tempo.co/read/877028/uji-nyali-naik-gondola-tradisional-di-pantai-timang-gunungkidul (last accessed 30th July 2019).

[6] EURIST. 2015. "*Urban Ropeways in Europe*". Hamburg: Germany. Website: https://www.researchgate.net/publication/283266715_Urban_Ropeways_in_Europe_creating_opportunities_for_urban_development (last accessed on 27th June 2019).

[7] Transport Department, State of Gotherburg. 2016. Gothenburg – Cable Car Best Practice Report. Web page: https://www2.trafikkontoret.goteborg.se/resourcelibrary/underlagsrapport_best-practices.pdf (last accessed on 23rd July 2019).

[8] Central Bureau of Statistic of Makassar City. 2018. Kecamatan Ujung Pandang dalam Angka. Web page: https://makassarkota.bps.go.id/publication/2018/09/26/37b385523 1a20c5e5c44474d/kecamatan-ujung-pandang-dalam-angka-2018.html (last accessed on 23rd July 2019).

IOP Conf. Series: Materials Science and Engineering **875** (2020) 012003 doi:10.1088/1757-899X/875/1/012003

The Study of Public Open Space effectiveness in Makassar Waterfront City using Good Public Space Index (GPSI)

S A Ekawati[1]*, M Ali[1], S Trisutomo[1], and R C A Ghani[1]

[1]Laboratory of Waterfront City Development and Planning, Department of Urban and Regional Planning, Engineering Faculty, Hasanuddin University, Makassar, Indonesia

*Email: aliah.sriekawati@gmail.com

Abstract. Anjungan Pantai Losari is one of the famous public spaces in Makassar; it is a primary landmark in the city. Nevertheless, this public space is considered not optimal because it is only crowded with people's activities in the afternoon and night. On the other hand, some theories argued that a good public open space has the ability to accommodate various people all day long. This study was conducted to evaluate and to study the effectiveness of public open space using the Good Public Space Index (GPSI). There are six variables used in GPSI: intensity use, the intensity of social use, people's duration of stay, temporal diversity use, variety of use, and diversity of users. This study found that the average of GPSI value shows the high score of Anjungan Pantai Losari effectiveness. However, two variables, intensity of use and people duration of social use were still at a low level. It was expected that this research could be a reference to develop spatial planning, especially the improvement of public open space quality.

1. Introduction

Public open space is a universal urban facility. Every citizen has the right to access, utilize and enjoying activities in it. In this case, a good public open space is a place that can be accessed by various characteristics of people including ages, gender, education; it could be utilized all day long [1-4]. Therefore, it is necessary for a city to provide adequate public open spaces to accommodate social activities for its citizens [5].

The effectiveness of public open spaces is not only indicated by the existing plazas or parks in the urban area. More than that, the quality of public spaces could be defined by the advantages that can be felt by people. A good public space can be seen from the number of people who use the place [6]. The more amount and characterized the visitors, the better the quality of public open space. Not only the subject or the visitor but also the diversity of activities become the indicators of good public open space [4,7]. Lack of interaction in public spaces indicates the failure of the space in facilitating people's needs [3].

Makassar, as a coastal area, had been providing public open spaces to accommodate people's activities as well as enjoying the sea view and the urban atmosphere. Anjungan Pantai Losari is the landmark of Makassar City which is used as a primary place for various events. Nevertheless, based on initial observation, Anjungan Pantai Losari looks quiet in the morning and in the afternoon. This condition is contrary to the theory about ideal public open space which requires the accessibility for all of the people every time.

A study about performance evaluation in Anjungan Pantai Losari as public space is required. This evaluation is focused on the assessment of function and characteristics of public open space. This study used Good Public Open Space Inde with its six variables: intensity use, intensity of social use, people's duration of stay, temporal diversity use, variety of use, diversity of users. It is expected that the results of this study could be references to develop spatial planning particularly in improving the public open space effectiveness.

2. Literature Review

2.1. An Overview of Waterfront Public Open Space

Public open space is understood as sharing a place in an urban context. Human is a social creature who needs interaction both to the other people and also to the environment. The people dependency on public open space becomes the reason why a city should provide a good public space as one of green infrastructure. Indirectly, public open spaces influence people's behavior, health and prosperity [7].

Public open spaces which are located in the coastal area have their own uniqueness. The water element, such as sea, rivers, lakes, and canals, become points of interest for people. In spatial planning, public open space should collaborate with the water element and hardscape in the land. The people must have access to touch and play with water. In addition, all of buildings that obstruct sea views must be removed. The utilization of natural elements, the availability of facilities and the hygiene of the environment especially the water, are the requirements for the high quality of public spaces in the coastal areas.

Public open spaces can be categorized into two groups: green space and non-green space [8]. Green open space includes city parks, city forests, cemetery, green lanes in the middle of roads and border areas. Meanwhile, the non-green spaces could be plazas, streets, pedestrian ways, and bicycle lanes. Open spaces are also grouped in natural and artificial spaces [9]. Natural open spaces are dominated by natural elements like grass, vegetations, sands, and water. Artificial open spaces consist of concrete and bordered by buildings.

2.2. Characteristic of Public Open Space

In order to reach the effectiveness of public open spaces, at least there are three values that must be owned by the place: responsive, democratic and meaningful [10].

2.2.1. Responsive. Responsive value refers to the ability of public open spaces to accommodate people's needs. Public spaces play a role as gathering points and places for various events [11]. There are three variables to indicate a responsive public open space: activities, time, and user intensity. A good public space becomes a place for citizen's activities [2, 12], both active and passive activities [1]. Passive activities such as sitting and relaxing, walking while enjoying the atmosphere of public space, and observing the situation. Meanwhile, active activities refer to interaction with other people or communities, to communicate, or to discuss.

Activities could be divided into three groups [4]:

- Process activities: the movements from one place to another (cycling, driving, walking);
- Physical contacts: the interaction between two persons or more (talking, trading, playing together, doing sports together, baby-sitting);
- Traditional activities: the individual activities (sitting, standing, playing alone, talking at phone, watching/observing, reading, doing individual sports).

The duration of public space utilization also becomes an indicator of responsive public space. A good public space can certainly be used by the community in the morning, afternoon, evening and night. The expansive use of public open space encourages the emergence of creative activities. Besides activities and time, intensity of users is also used as a benchmark to identify a responsive public space [13]. A good public space could accommodate people in a large amounts. It is related to the total area of public space.

2.2.2. Democratic. A democratic public open space means the ability of the space to protect citizen's rights. Public spaces are places for discussion for all people [3]. Therefore, public spaces should be utilized freely without discrimination [2, 6]. The public space effectiveness depends on its ability to accommodate people from various backgrounds: age, gender, education, and works. Good accessibility in public space gives an opportunity for people to enjoy urban public space [1, 13]. A democratic public space facilitates activities for disabled people. It is related to special lanes and amenities provision. Accessibility is also related to permeability; it is an ability for public space to provide various access [12]. Therefore, public spaces could be reached by driving, cycling or walking.

2.2.3. Meaningful. Meaningful refers to an opportunity for people in public spaces to interact with themselves, to other people, and to the environment. Individual activities provide a chance to explore the sense of the place. The various meanings can be obtained from the form and function of public spaces elements [12]. Interaction with others and to the environment efforts meaningful creative activities.

3. Research Method

3.1. Location of Study and Data Collection

This research was conducted in Anjungan Pantai Losari, Makassar City (Figure 1). Located in the coastal area, Anjungan Pantai Losari is a primary plaza utilized by people to various local, national and international events. Besides the commercial street, Jalan Penghibur, Anjungan Pantai Losari is also boarded by shops, restaurants, and hotels to support activities in public open space.

Primary and secondary data were needed to answer the issue of public space. There are several data that were used in this study such as the number of visitors, the number of groups and communities, duration of activities, the variation of activities, and visitors' characteristics. Observation and questionnaires were conducted to collect data. The data collection was held seven days, started from morning until night.

Figure 1. Anjungan Pantai Losari where the study was conducted

3.2. Analysis Method

Good Public Space Index (GPSI) method was used to analyze the data. The GPSI method as a tool to measure the ability of public open space to work as its function. The effectiveness of public space was indicated by score from 0 to 1. There are five level of the effectiveness of public space, such as: very low (0-0.20); low (0.21-0.40); sufficient (0.41-0.60); high (0.61-0.80); and very high (0.81-1). Before defining the level of public open space effectiveness, it was needed to calculate the index of each variable of GPSI.

a. Intensity of use (IU)
IU refers to the amount of people who do activities in public open space that is observed. The more people amount, the better public open space performance.

$$IU = \frac{average\ amount\ of\ visitors}{the\ highest\ amount\ of\ visitors} \quad (1)$$

b. Intensity of social use (ISU)
ISU refers to the number of people involved in groups or communities. In this study, one group consists of two or more people. The more groups and communities, the better the performance of the public open space.

$$ISU = \frac{number\ of\ people\ who\ involve\ in\ groups}{the\ highest\ amount\ of\ visitors} \quad (2)$$

c. People's duration of stay (PDS)
PDS refers to how long people use public open spaces. The more duration of public space utilization, the better the performance of the place.

$$PDS = \frac{average\ duration\ of\ utilization}{the\ longest\ duration} \quad (3)$$

d. Temporal diversity of use(TDU)
TDU refers to the separation of activities that happen in one period of observation. A good public opens space avoid domination of time and space utilization. That means the amount of by that occur in morning, afternoon, evening and night are equal. This variable is measured by Simpson's Diversity Index.

$$\text{Simpson's Diversity Index} = (1 - D)$$

$$D = \frac{N(N-1)}{total\ n(n-1)} \quad (4)$$

where:
n = the amount of particular activities
N = the total amount of activities in all categories

e. Variety of use (VoU)
In this study, a variety of use was divided into four groups: process activities, physical contact, transitional activities, and optional activities. The more variation in utilization, the better the performance of public open space. Same to TDU, this variable is also calculated by Simpson's Diversity Index.

f. Diversity of users (DoU)
DoU refers to various people's characteristics. In this study, the diversity of users was grouped based on gender and ages. This variable is also calculated by Simpson's Diversity Index.

3.3. Research Variable
Research variables in this study were determined based on the GPSI theory. Table 1 shows the variables of GPSI as well as their sub-variables that are used as tools in data collecting and analyzing.

Table 1. Research variables

Variabel	Sub Variabel
Intensity use	-
Intensity of social use	-
People's duration of stay	-
Temporal diversity use	Morning, afternoon, evening, night.
Variety of use	Process activities (cycling, walking) Physical contact (talking, trading, playing with children, baby-sitting, doing sports with friends) Transitional activities (sitting, standing, playing alone, talking at phone, cleaning up, watching/observing, reading, doing individual sports)
Diversity of users	Women (age: 0-5; 6-15; 16-20;21-35;36-55; >55) Men (age: 0-5; 6-15; 16-20;21-35;36-55; >55)

4. Findings and Discussion

4.1. Intensity of Use

Table 2 shows the number of visitors of Anjungan Pantai Losari in one week during morning, afternoon, evening and night, as well as the results of intensity of use analysis using GPSI. The data shows that night was the favorite time for people to visit Anjungan Pantai Losari. The number of people who came to Anjungan Pantai Losari at night was 351 and reach the highest number on Saturday night with the amount of 569 people. In contrast, only 31 people visited Anjungan in the morning and afternoon. At the end of the table, the intensity of the use index shows 0.311 scores.

The intensity of use index score indicated the low level of effectiveness of Anjungan Pantai Losari to accommodate people with a great amount in the morning and afternoon. The gap between utilization in morning-afternoon and evening-night was caused by the lack of shelter to protect people from sun heat. The hot weather in the morning and afternoon rise the discomfort and unpleasant for people. The next reason was working time. Generally, people in Makassar works from 8 a.m. until 5 p.m. from Monday to Friday.

Table 2. Intensity of use

Time	Amount of Visitors (people)							Total
	Mo	Tue	Wed	Thur	Fri	Sat	Sun	
Morning	40	32	37	177	103	48	114	551
Afternoon	17	26	18	23	20	15	31	150
Evening	31	95	178	211	203	242	314	1274
Night	351	442	427	375	398	569	417	2979
							Average	176,929
							Highest amount	569
							IU	0.311

4.2. Intensity of Social Use

Table 3 shows the data and analysis result of the number of people involve in groups. Based on data, the number of individuals in groups was 4.544 people from 4.954 people who visited Anjungan Pantai Losari. It is clearly read that these groups do their activities in the evening and night. The index of ISU analysis resulted in a score 0.917.

The score resulted from the GPSI analysis means that the effectiveness of Anjungan Pantai Losari was at a very high level in accommodating groups and community. Generally, people come with their family and friends. They will be several small groups based on people's communities. Physically, Anjungan Pantai Losari area, the comfortable and romantic ambiance in evening and night become reasons why people want to spend their time in the place. The appearance of communities indicated that Anjungan Pantai Losari, as public open space, is successfully giving the meaning and the sense of place for people.

Table 3. Intensity of social use

Time	Amount of Visitors (people)							Total
	Mo	Tue	Wed	Thur	Fri	Sat	Sun	
Morning	37	25	32	153	91	42	97	477
Afternoon	15	20	14	20	15	14	27	125
Evening	29	90	151	193	188	229	295	1175
Night	325	421	384	349	370	532	386	2767
				Amount of people involved in groups				176,929
						Total amount of visitors		569
							ISU	0.311

4.3. People's Duration of Stay

Table 4 shows the average of visitors' duration time in Anjungan Pantai Losari. Generally, people spend time about only fifteen minutes in the morning and afternoon. Exception on Sunday morning, citizens spends about three hours because of car-free day event. In the evening and night, people spend their time one until three hours. On Saturday night, people could stay longer for five hours. At the end of the table, PDS index shows 0.275 scores; and it was categorized at a low level.

Based on PDS index calculation, it could be said that Anjungan Pantai Losari still not effective in accommodating temporal diversity use. The visitors were still not comfortable to linger for activities, especially in the morning and afternoon. In line with the result of the intensity of use analysis description, the quietness of Anjungan Pantai Losari in the morning and afternoon was caused by the lack of shelters and people's work-time. Generally, the visitors who come in the morning and afternoon were the local tourists who want to take pictures. Special on Sunday morning, the car-free day event becomes a trigger for people to spend their time longer. This event started at 6 A.M. until 10 A. M., provides various activities and a tent for shelter.

Table 4. People's duration of stay

Time	Amount of Visitors (people)							Total
	Mo	Tue	Wed	Thur	Fri	Sat	Sun	
Morning	0.25	0.25	0.25	0.25	0.25	1	3	5.25
Afternoon	0.25	0.25	0.25	0.25	0.25	0.5	0.5	2.25
Evening	1	2	1	1	2	3	3	13
Night	2	2	2	2	2	5	3	18
							Average	176,929
						Longest duration		569
							PDS	0.311

4.4. Temporal Diversity Use

Table 5 shows the data related to the number of temporal diversity activities in Anjungan Pantai Losari. Compared to the activities in the morning and afternoon, people's activities in the evening and night were more variable. The Simpson's Index Diversity analysis for TDU resulted in 0.739 scores, and it was categorized at a high level. It means Anjungan Pantai Losari was effective in providing space for various activities. Based on observation, people tend to use Anjungan for sports in the evening. At night they prefer playing and relaxing.

Table 5. Temporal diversity use

Time	Amount of Temporal Diversity Use (n)							Total n	n-1	n(n-1)
	Mon	Tue	Wed	Thur	Fri	Sat	Sun			
Morning	6	7	7	14	9	7	12	62	61	3782
Afternoon	6	6	7	6	7	6	7	45	44	1980
Evening	5	12	11	12	12	11	14	77	76	5852
Night	12	13	12	11	13	14	12	87	86	7482
							Total (N)	271		
							N-1	270		
							N(N-1)	73170		
									Total n(n-1)	19096
									D= Total n(n-1)/N(N-1)	0.261
									Simpson's Index of Diversity (1-D)	0.739

4.5. Variety of Use

From table 6, some activities that occur in Anjungan Pantai Losari could be identified. In this research, activities were divided into four categories: process activities, physical contact, transitional activities, and other activities. Based on observation, five most activities in Anjungan Pantai Losari were talking (3.005), sitting and relaxing (2.958), taking a picture (972), standing and enjoying the panorama (666), and walking (300). Other activities that also occur were trading, babysitting and playing. At the bottom of the table, Simpson's Index Diversity of VoU result shows 0.760 scores.

The value of VoU index means that Anjungan Pantai Losari has high effectiveness as public open spaces for activities. The total area of this place could accommodate various activities of citizens. In addition, the high accessibility of Anjungan become a reason why people choose Anjungan Pantai Losari for relaxing and recreating. However, generally, people prefer to do passive activities such as sitting, talking and enjoying sea views. The more active activities like playing, doing sports or cycling were rarely seen. It is because of the lack of facilities for active activities.

4.6. Diversity of Users

Table 7 shows the diversity of users in Anjungan Pantai Losari based on gender and age. Data was collected from observation in a week. Based on data, it was identified that there are more female visitors than males. Furthermore, both from female and male groups, the average ages of people were 21 until 35 years old. There were about 1002 females and 819 males who visited Anjungan Pantai Losari during the observation. The significant differences were showed in the number of toddlers (0 – 5 years old) and elderly people (above 65 years old). It can be seen in the table that the number of toddlers was 150 people; meanwhile, the number of elderly people was only 119 people. At the end of the table, the analysis resulted in the Simpson's Index Diversity of DoU; it was 0.885.

The analysis result of DoU (0.885) identified that the effectiveness of Anjungan Pantai Losari to accommodate the diversity of users is categorized at a very high level. Besides the beauty of the sea view, some interesting activities could attract people. Supported by various activities such as music and culinary festival, Anjungan Pantai Losari become one of the favorite destinations for people of every age.

Although Anjungan Pantai Losari had been successfully reaching a very high level, the data also showed that this public space was still not children-friendly and not elderly-friendly. Based on short interviews and observation, it was identified that facilities for these two groups of age are still negative. The elderly people need shelters to protect themselves from the sun heat; as well as the facilities for them who use a wheelchair. The broken pedestrian ways and the stairs in Anjungan Pantai Losari make these wheelchairs users difficult to move from one place to another.

Table 6. Variety of use

Variety of Use	Total n	n(n-1)
Proces activities		
Cycling	21	420
Walking	300	89700
Physical contact		
Talking	3005	9027020
Trading	157	24492
Baby-sitting	115	13110
Playing with friends	168	28056
Doing sports with friends	6	30
Transitional activities		
Sitting	2958	8746806
Standing	666	442890
Playing alone	174	30102
Talking at phone	138	18906
Cleaning up	6	30
Watching/observing	267	71022
Reading	0	0
Taking care animals	8	56
Doing sports alone	12	132
Other activities		
Fising	1	0
Busking	11	110
Taking pictures	972	943812
Sleeping	8	56
Total (N)	8993	
N-1	8992	
N(N-1)	80865056	
Total n(n-1)		19436750
D= Total n(n-1)/N(N-1)		0.240
Simpson's Index Diversity = 1-D		0.760

Table 7. Diversity of users table

Subject (years old)	Total n	n(n-1)
Female		
0-5	71	4970
6-15	353	124256
16-20	536	286760
21-35	1002	1003002
36-55	400	159600
56-65	195	37830
> 65	65	4160
Male		
0-5	79	6162
6-15	370	136530
16-20	453	204756
21-35	819	669942
36-55	383	146306
56-65	175	30450
> 65	54	2862
Total (N)	4955	
N-1	4954	
N(N-1)	24547070	
Total n(n-1)		2817586
D= Total n(n-1)/N(N-1)		0.115
Simpson's Index Diversity = 1-D		0.885

The average of GPSI value of Anjungan Pantai Losari shows the high level. Table 8 consists of the resume of each variable index. Intensity of social use and diversity of social use index indicate a very high level. Furthermore, temporal diversity use and variety of use are at a high level. However, intensity of use and people's duration of social use is still categorized at a low level.

Table 8. The Average of good public space index Anjungan Pantai Losari

No	Variable	Index	Level
1	Intensity of Use	0.311	Low
2	Intensity of Social Use	0.917	Very high
3	People Duration of Social Use	0.275	Low
4	Temporal Diversity Use	0.739	High
5	Variety of Use	0.792	High
6	Diversity of Users	0.885	Very high
	Total	3.920	
	Average Index	0.653	High

In addition, the analysis of the relation of GPSI variables and the public open space characteristic was conducted (Table 9). Based on potential and problems description, it was identified that Anjungan Pantai Losari had been success to be a democratic public open space. However, this public space was still not responsive to its visitors; as well as meaningless to them. An improvement planning is required to raise the effectiveness of Anjungan Pantai Losari as public open space.

Table 9. The matrix of good public space index variables and public open space characteristic

No	Variable	Information	Characteristic		
			Responsive	Democratic	Meaningful
1	Intensity of Use	It is visited only in the evening and night	X		
		Lack of shelters			X
2	The of Social Use	There are some groups and communities			√
		Adequate area	√		
		Comfortable in evening and night			√
3	People Duration of Social Use	Lack of utilization in the morning and afternoon	X		
		Car free day on Sunday morning as a positive trigger.		√	√
4	Temporal Diversity Use	Diversity of activities only occur at afternoon and nigh	X		
		Different activities at morning-afternoon and evening-night	√		
5	Variety of Use	Dominated by passive activities	X		
		Ease access (high permeability)		√	
		Various activities		√	
6	Diversity of Users	Unfriendly for children and elderly people			X
		Dominated by the youth (21-35 years old.	√		
		Interesting seaview			√
		Visited by various age groups of people		√	

5. Conclusion

The result of this study that the effectiveness of Anjungan Pantai Losari as a public open space was categorized at a high level. It can be seen from its democratic value. The ease of accessibility, the diversity of activities and the ability to accommodate people from various backgrounds are indicators of democratic public open space. Nevertheless, there are some gaps needed to follow up. Anjungan Pantai Losari was considered to be lacking in responding to citizens' activities. Furthermore, people still cannot touch the meaning of the place while they do their activities in Anjungan Pantai Losari. Generally, the identified problems relating to responsive value are the lack of activities in the morning and the afternoon. It is caused by the lack of shelter, such as trees, to protect people from the sun's heat. The other problem is the domination of passive activities, such as sitting, chatting, playing on gadgets, and taking a picture. The lack of supporting facilities for children to play and to do sport becomes a reason why people simply stay in public open space. It is not parallel to the public open spaces purpose which requires an active and rousing atmosphere.

References

[1] Carmona, M., Tiesdell, S., Heath, T., Oc, T. (2003).*Public Places, Urban Space: The Dimensions of Urban Design.* London: Architectural Press.

[2] Darmawan, E. (2009). *Ruang Publik dalam Arsitektur Kota.* Semarang: Badan Penerbit Universitas Diponegoro.

[3] Siahaan, J. (2010). *Ruang Publik: Antara Harapan dan Kenyataan.* Buletin Tata Ruang, Edisi IV (Juni-Juli 2010), page. 11-16.

[4] Zhang, W., dan Lawson, G. M. (2009). *Meeting and greeting: activities in public outdoor spaces outside high-density urban residential communities.* URBAN DESIGN International,

Vol. 14, No. 4, page 207-214.

[5] Siregar, J. P. (2014). Assessment of Public Space Quality Using Good Public Space Index (Case Study of Merjosari Sub District, Municipality of Malang, Indonesia). Procedia Social and Behavioral Sciences 135 (2014), page 10-17.

[6] Gumano, H. N., citizens, T., dan Nur, H. (2016). *Kajian Tingkat Efektifitas Ruang Publik yang Tersedia Pada Pusat Kota-Kota di Provinsi Sumatera Barat Berdasarkan Metode "Good Public Space Index (GPSI)"*. Journal of Faculty of Civil and Planning Engineering, Bung Hatta University Vol. 1 No. 3.

[7] Aljosha, E. Z., Eriawan, T., dan Nur, H. (2016). *Menilai Tingkat Keragaman Ruang Publik Pada Taman Imam Bonjol di Kota Padang sebagai Masukan dalam Perbaikan Kualitas Ruang*. Jurnal Fakultas Teknik Sipil da Perencanaan, Universitas Bung Hatta, Vol. 1 No. 3

[8] Undang-Undang Nomor 26 tahun 2007 tentang Penataan Ruang.

[9] Carmona, M., Magalhaes, S., Hammond, L. (2008). *Public Space, The Management Dimension.* Routledge, London and New York: Taylor and Francis Group.

[10] Carr, S., Frances, M., Rivlin, L. G., and Stone, A. M. (1992). *Public Space.* Cambridge University Press.

[11] Mandipour, A. (1996). Design of Urban Space: Inquiry into a Social-Spatial Process. Chichester: John Wiley& sons Ltd.

[12] Bentley, I., Alcock, A. Murrain, P., McGlynn, S., and Smith, G. (1985). *Responsive Environment. A Manual for Designers.* Burlington: Architectural Press.

[13] Parkinson, J. R. (2012). Democracy and Public Space. The Physical Sites of Democratic Performance. Oxford: University Press.

IOP Conf. Series: Materials Science and Engineering **875** (2020) 012004 doi:10.1088/1757-899X/875/1/012004

Adaptive Thermal Comfort of Elementary School Student (A Case study of the West Coastal Area of South Sulawesi)

R Mulyadi[1]*, B Hamzah[1], M T Ishak[1], and Y R F Taufik[1]

[1]Laboratory of Building Science & Technology, Department of Architecture, Faculty of Engineering, Universitas Hasanuddin. Jalan Poros Malino km 6. Bontomarannu, 92172 – Gowa, Indonesia

*E-mail: rosady@unhas.ac.id

Abstract. This research aims to study the adaptive thermal comfort in elementary school buildings in the west coastal area of South Sulawesi. It was conducted by field survey method to obtain personal data (clothing and metabolic rate) and environmental parameters such as air temperature and humidity, mean radiant temperature (MRT) and the airflow. During the measurements of environmental parameters, the students filled out questionnaires to find out their perceived comfort level by the Thermal Sensation Vote (TSV) and Thermal Comfort Vote (TCV). In addition, respondents are also asked to give an opinion on their desired thermal preference. The results of the environmental parameters measurements were used as references to calculate the Predicted Mean Vote (PMV), Operative Temperature (T_{op}) and Predicted Percentage of Dissatisfied (PPD). The results show that the condition of the elementary school classrooms along the west coast of South Sulawesi are generally in high temperatures. However, most respondents still feel comfortable and able to adapt to these conditions.

1. Introduction

It is currently developing two approaches related to thermal comfort. The first approach is called the static approach which refers to the results of the study of the thermal perception of respondents. It is based on the results of research PO Fanger in the 1970s [1] and later on adopted by ASHRAE (American Society of Heating, Refrigerating, and Air-Conditioning Engineers) as a standard of prediction of thermal comfort (ASHRAE Standard 55) [2] and ISO 7730 [3] in the form of an index of thermal comfort PMV-PPD.

The second approach is called adaptive approaches. This approach using respondents from building users, which has been adapted to the climatic conditions. Basically, the study of adaptive thermal comfort is an attempt to determine the thermal neutrality, acceptability of thermal conditions (thermal acceptability), and the preference of thermal conditions (thermal preference) of respondents. The prediction of adaptive thermal comfort models assumes that the occupant or user of the building (respondents) rather than as a passive recipient of the thermal environment, such as in the case of experimental subjects climate chamber, but on the contrary, have an important role in creating conditions comfort condition by three types of adaptation, namely: behavioral setting, physiological, and psychological [4].

Thermal comfort standards such as ASHRAE Standard 55 [2] and ISO 7730 [3] has been widely used as a standard thermal comfort in various countries. However, this standard is more widely used for a room with artificial conditioning (AC/Air-Conditioning). For buildings with natural conditioning, which is no less relevant standards for use [17]. ASHRAE Standard 55 using a seven-point scale to measure the thermal sensation perceived (Thermal Sensation Vote/TSV). The seven-point scale according to ASHRAE are rated +3 (hot), +2 (warm), +1 (slightly warm), 0 (neutral), -1 (slightly cool), -2 (cool), and -3 (cold). It also conducted a survey of TSV (Thermal Sensation Vote) using Bedford scale of thermal comfort. Bedford scales better known by the name of the response of comfort (comfort response) [13]. Thermal Comfort Vote (TCV) using a scale of Bedford which also consists of seven-points value is +3 (much too warm), +2 (too warm), +1 (comfortably warm), 0 (comfortable), -1 (comfortably cool), -2 (too cool), and -3 (much too cool).

Indoor thermal comfort will increase work productivity. A long study says that there is an influence of the thermal quality of a classroom with student achievement or student [5]. There is a correlative relationship between the quality of the room includes thermal comfort and student achievement. Research on thermal comfort in classrooms have been carried out in various places in the world as conducted by Buratti and Ricciardi [6], Corgnati, Ansaldi, and Filippi [7], Corgnati, Filippi, and Viazzo [8], Hwang, Lin, and Kuo [9], Kwok and Chun [10], Mors et al [11], Teli, Jentsch, and James [12] included in the tropics, namely in Singapore by Wong and Khoo [13].

Several recent studies also showed a positive relationship between the quality of the room (including the thermal conditions) with student achievement by Mendell and Heath [14], and Sensharma, Woods and Goodwin [15], and Baharuddin et al [16]. Therefore, thermal comfort becomes very important to the study, in order to achieve a conducive learning process and give satisfactory results for students.

School buildings with comfortable thermal conditions can help students achieve optimal learning outcomes. But on the contrary, the achievement of optimal learning outcomes will be difficult to produce in uncomfortable school buildings. This research is intended to analyze the thermal comfort of students in elementary school buildings on the west coastal area of South Sulawesi by using the adaptive approach method as described above.

2. Method

The survey was conducted to obtain data on the condition of elementary students which include: personal data (clothing and metabolic rate) and the measurement of environmental parameters: air temperature, air humidity, mean radiant temperature (MRT) and the airflow. At the same time, students are asked to fill out questionnaires asking the perceived comfort level of occupants using the Thermal Sensation Vote (TSV) and Thermal Comfort Vote (TCV). TSV is measured using ASHRAE scale, while TCV measured using Bedford scale. In addition, respondents are also asked to give an opinion on the desired thermal conditions (thermal preference), if the thermal environment at the time it was acceptable or not. The results of measurements of environmental parameters will be used as a reference to calculate the Predicted Mean Vote (PMV), Operative Temperature (T_{op}) and Predicted Percentage of Dissatisfied (PPD). The result of this calculation will be compared with the comfort perceived by the user of the room (the students).

Measurements were taken at a height of 1 meter above the floor following the measurement techniques of Feriadi and Wong [17]. Measurements were performed for 5 hours, ranging from 08: 00 – 12: 00 and 13: 00 – 14: 00 pm, with intervals of 1 hour for each respondent. It also will be measured against a dry bulb temperature, humidity, air velocity, and MRT.

The respondents were given a questionnaire to assessing the thermal conditions, acceptability of thermal conditions, and preferences of thermal conditions. The collection of questionnaires was conducted in conjunction with the collection of physical measurement data of the environment.

Analysis of the thermal conditions was made by simple linear regression analysis using SPSS software. Values of thermal neutrality are outlined in the form of graphs and tables. Furthermore, analysis acceptability of thermal conditions is done by plotting the average percentage of respondents in the comfortable temperature range. The results were elaborated in the form of graphs and tables. Furthermore, analysis of thermal preferences was done by analyzing the results of the respondent's responses to the questionnaire.

3. Result and discussion

Data collected from 798 respondents, where 47.4% is boys and 52.6% is girls. The ages varied from 8-13 years old. Table 1 describes the characteristics of respondents.

Table 1. Characteristics of respondents

Variables		Sum (n)	Percentage (%)
Age (Years)	8	13	1.6
	9	190	23.8
	10	256	32.1
	11	243	30.5
	12	82	10.3
	13	14	1.8
Sex	Male	378	47.4
	Female	420	52.6
Clothes	Sport uniforms	68	8.5
	Sports uniforms with hijab	25	3.1
	Boy scout uniforms	293	36.7
	Boy scout uniforms with jacket	1	0.1
	Boy scout uniforms with hijab	95	11.9
	Long-sleeved boy scout uniforms	10	1.3
	National (red-white) uniforms	160	20.1
	National (red-white) uniforms with hijab	109	13.7
	National (red-white) uniforms with vest	30	3.8
	Long-sleeved National (red-white) uniform	1	0.1
	National (red-white) uniforms with jacket and hijab	1	0.1
	National (red-white) uniforms with hijab and vest	5	0.6

Thermal sensation votes based on scale hot (3), warm (2), slightly warm (1), neutral (0), slightly cool (-1), cool (-2), and cold (-3) can be seen at figure 1. The votes indicate that about 35.1% of respondents feel slightly warm, 34.2% feel comfort, and about 17.5% feel slightly cool.

Thermal comfort vote determined by much too warm (3), too warm (2), comfortably warm (1), comfortable (0), comfortably cool (-1), too cool (-2), and much too cool (-3). Most of the respondents (51.1%) feel comfort. About 27.9% feel comfortably warm, and 10.8% of respondents feel comfortably cools as can be seen in figure 2.

Thermal preference measured by the scale of more-hot (1), no change (0), and more-cool (-1). As can be seen in figure 3. Most of the respondents (58.9%) want more-cool. About 39.8% of respondents want no change and feel comfortable with the current condition. Only 1.3% of respondents want more-hot of the current condition.

Thermal acceptance indicates that the current thermal condition can be accepted (1) by respondents or not (0). Figure 4 shows the acceptance of 67.4% of respondents, while 32.6% of respondents have not accepted the condition.

The air velocity vote measured the air velocity condition feels by respondents. It used a scale of no velocity (-2), less velocity (-1), enough velocity (0), fast velocity (1), and very fast velocity (2). Figure 5 shows that 44.5% of respondents feel less air velocity (-1), 42% feel enough air velocity (0) and 11,7% of respondents feel no air velocity.

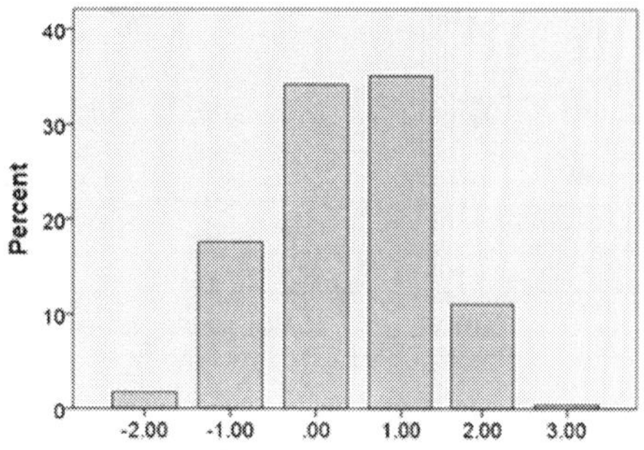

Figure 1. Percentage of thermal sensation vote

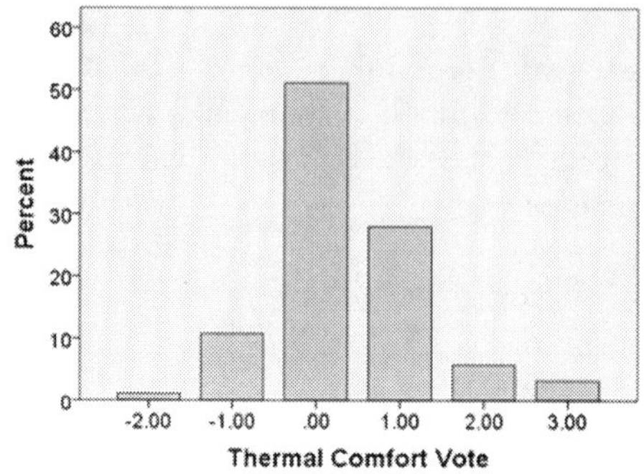

Figure 2. Percentage of thermal comfort vote

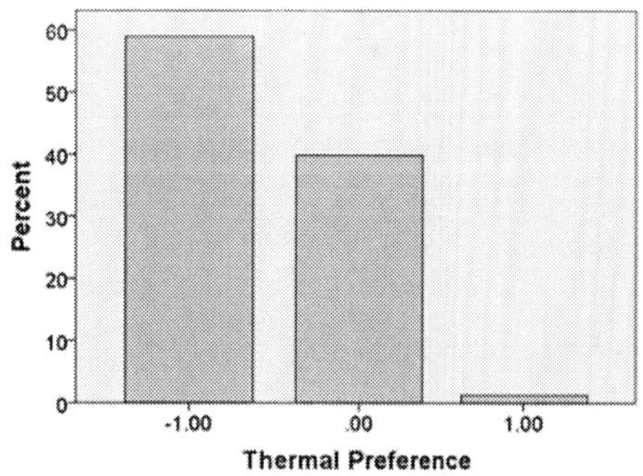

Figure 3. Percentage of thermal preference

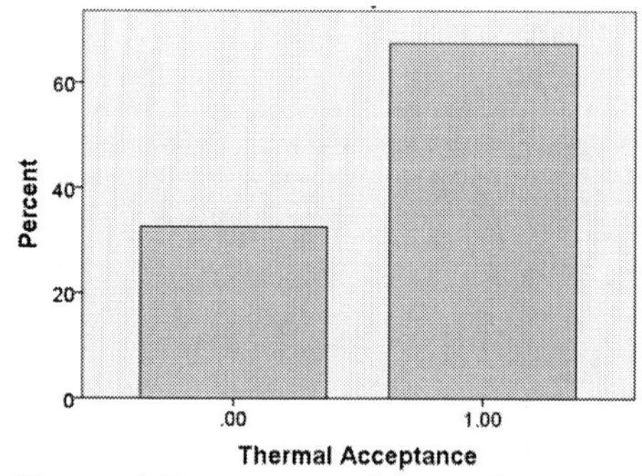

Figure 4. Percentage of thermal acceptance

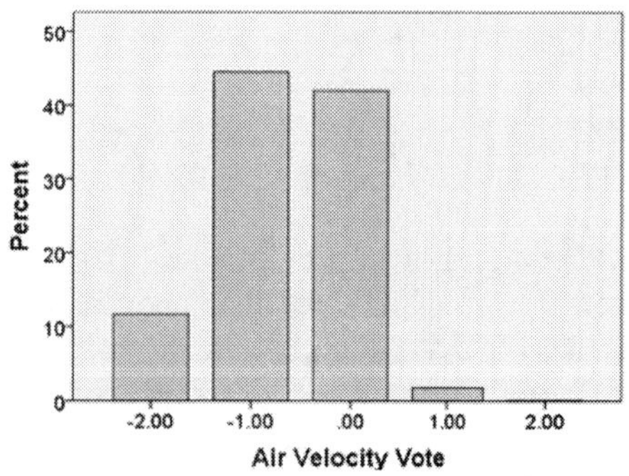

Figure 5. Percentage of air velocity vote

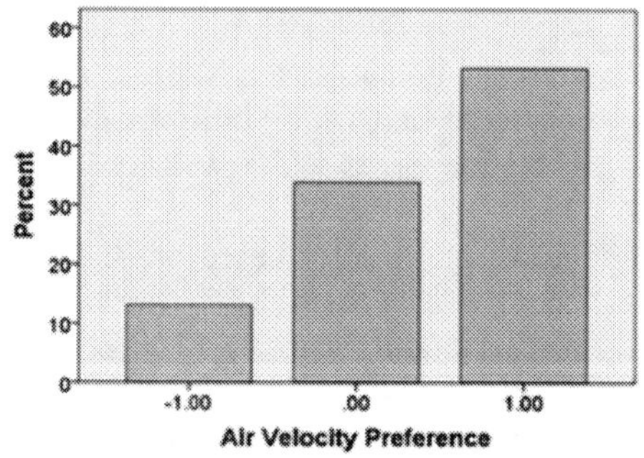

Figure 6. Percentage of Air Velocity Vote

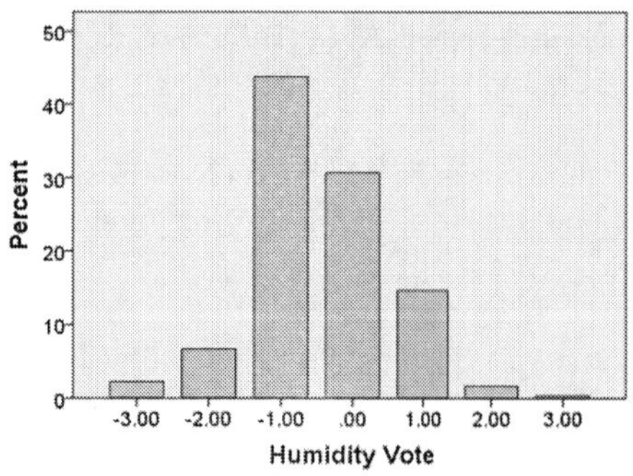

Figure 7. Percentage of Humidity Vote

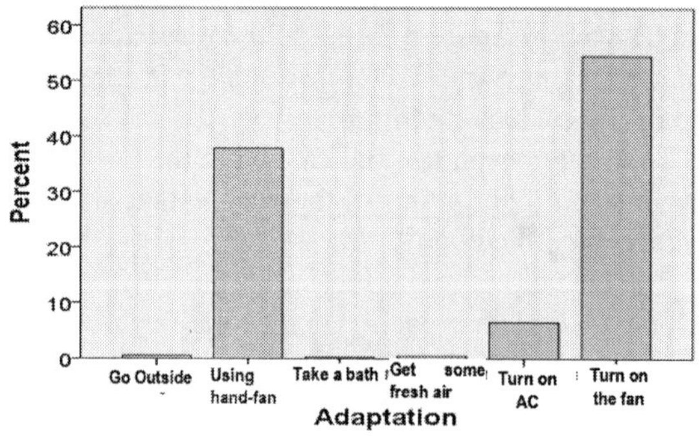

Figure 8. Percentage of thermal adaptation vote

Air velocity preference measured the preference of respondents to the air velocity. Most of the respondents (53.1%) want more air velocity (1), 33.8% respondents feel air velocity is enough (1), and 13% respondents want less air velocity as can be seen in figure 6. There are seven option on questionnaires to assess respondents to the relative air humidity (humidity vote) is extremely too humid (score -3), very humid (value -2), slightly moist (value -1), appropriate (value of 0), slightly dry (value of 0), somewhat dry (value 1), very dry (value of 2), it is too dry (value 3). Figure 7 shows that the majority of 43.7% of respondents felt the room was a bit humid (value -1), there are 30.7% of respondent feel it is appropriate (value of 0), and there are 14.7% of respondents felt a bit dry (value of 1).

There are three options on the questionnaire as an adaptation if the correspondent experiencing high-temperature conditions, namely the fans, turn on the fan, and defeated air-condition (AC). Figure 8 shows that the majority of respondents 54.5% chose turned on the fan if the room air temperature is high, there are 37.8% of respondents felt the fans choose, and there are 6.5% of the respondents chose the air-conditioning. There is another option which is 0.5% opted out of the room, 0.4% would like to get some fresh air, and 0.3% wanted a shower.

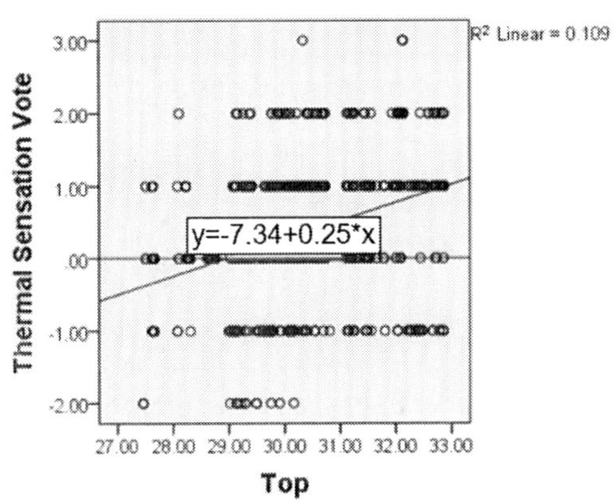

Figure 9. Regression between TSV and T_{op}

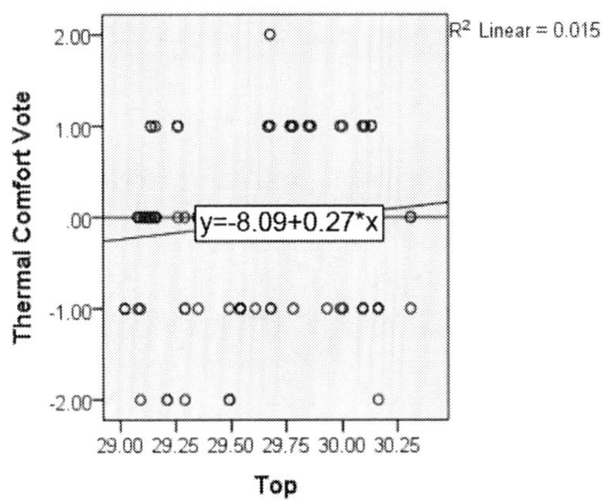

Figure 10. Regression between TCV and T_{op}

Figure 9 shows the relationship between the TSV with operative temperature (T_{op}). The graph shows that the T_{op} ranged between 27.46 – 32.86 °C. The neutral temperature ranges 29.36 °C and considered as high. A high value is caused by the physicality of these children - that are already familiar with the existing thermal conditions so they are able to accept the thermal conditions. Figure 10 shows the relationship between the value of TCV with operative temperature (T_{op}). The graph shows that the T_{op} ranged between 27.46 – 32.86 °C. The comfortable temperature range to 29.25 °C. This temperature is lower than the neutral temperature TSV models.

4. Conclusion

The general condition of classrooms in schools along the west coast of South Sulawesi showed high-temperature conditions mainly before noon has reached 35 °C. Although the measurement results indicate a high temperature, the survey results showed that most respondents still feel comfortable and able to adapt to these conditions. TSV model results in this study show the airflow is higher than the TCP model, it is advisable to conduct additional studies with more data. Therefore, the results of the study, indicating that despite the high temperatures, most respondents still feel comfortable.

Acknowledgment
The authors would like to thank the Laboratory-Based Education (LBE) Research Grant provided by the Faculty of Engineering Universitas Hasanuddin which supporting the financing of this research. Also, many thanks to the students of the Laboratory of Building Science & Technology Department of Architecture for their valuable contribution to this research.

References
[1] P. O. Fanger, *Thermal Comfort: Analysis and Application in Environmental Engineering*. New York: McGraw-Hill, 1973.
[2] ASHRAE, *Thermal Environmental Condition for Human Occupancy (ASHRAE Standard 55)*. Atlanta: American Society of Heating, Refrigerating, and Air-Conditioning Engineers, 2004.
[3] ISO 7730:2005, *Moderate Thermal Environments—Determination of the PMV and PPD Indices and Specifications for Thermal Comfort*, 2nd ed. Geneva: International Organization for Standardization, 2005.
[4] R. De Dear *et al.*, "Developing an adaptive model of thermal comfort and preference," *ASHRAE Trans.*, vol. 104, no. Part 1, pp. 1–18, 1998.
[5] R. D. Pepler and R. E. Warner, "Temperature and Learning: an experimental study," *ASHRAE Trans.*, vol. 74(21), pp. 211–224, 1968.
[6] C. Buratti and P. Ricciardi, "Adaptive analysis of thermal comfort in university classrooms: Correlation between experimental data and mathematical models," *Build. Environ.*, vol. 44, no. 4, pp. 674–684, 2009.
[7] S. P. Corgnati, R. Ansaldi, and M. Filippi, "Thermal comfort in Italian classrooms under free running conditions during mid seasons: Assessment through objective and subjective approaches," *Build. Environ.*, vol. 44, no. 4, pp. 785–792, 2009.
[8] S. P. Corgnati, M. Filippi, and S. Viazzo, "Perception of the thermal environment in high school and university classrooms: Subjective preferences and thermal comfort," *Build. Environ.*, vol. 42, no. 2, pp. 951–959, 2007.
[9] R. L. Hwang, T. P. Lin, and N. J. Kuo, "Field experiments on thermal comfort in campus classrooms in Taiwan," *Energy Build.*, vol. 38, no. 1, pp. 53–62, 2006.
[10] A. G. Kwok and C. Chun, "Thermal comfort in Japanese schools," *Sol. Energy*, vol. 74, no. 3, pp. 245–252, 2003.
[11] S. ter Mors, J. L. M. Hensen, M. G. L. C. Loomans, and A. C. Boerstra, "Adaptive thermal comfort in primary school classrooms: Creating and validating PMV-based comfort charts," *Build. Environ.*, vol. 46, no. 12, pp. 2454–2461, 2011.
[12] D. Teli, M. F. Jentsch, and P. A. B. James, "Naturally ventilated classrooms: An assessment of existing comfort models for predicting the thermal sensation and preference of primary school children," *Energy Build.*, vol. 53, pp. 166–182, 2012.
[13] N. H. Wong and S. S. Khoo, "Thermal comfort in classrooms in the tropics," *Energy Build.*, vol. 35, no. 4, pp. 337–351, 2003.
[14] M. J. Mendell and G. A. Heath, "Do indoor pollutants and thermal conditions in schools influence student performance? A critical review of the literature," *Indoor Air*, vol. 15, no. 1, pp. 27–52, 2005.
[15] N. P. Sensharma, J. E. Woods, and A. K. Goodwin, "4164 (RP-700) -- Relationships Between the Indoor Environment and Productivity: A Literature Review," *ASHRAE Trans.*, vol. 104, no. pt. 1A, 1998.
[16] B. Hamzah, Z. Gou, R. Mulyadi, and S. Amin, "Thermal Comfort Analyses of Secondary School Students in the Tropics," *Buildings*, vol. 8, no. 4, 2018.
[17] H. Feriadi and N. H. Wong, "Thermal comfort for naturally ventilated houses in Indonesia," *Energy Build.*, vol. 36, no. 7, pp. 614–626, 2004.

IOP Conf. Series: Materials Science and Engineering **875** (2020) 012005 doi:10.1088/1757-899X/875/1/012005

Local-Wisdom and Its Influence to Disaster Mitigation on the Spatial Configuration of Lakkang Waterfront Settlement

E Syarif[1]*, Asniawaty[1], N Nadjmi[1], and A Syadzwina[1]

[1]Architecture Department, Engineering Faculty, Hasanuddin University, Gowa, Indonesia

*Email: edosyarif@yahoo.com

Abstract. This paper discussed the concept of dwelling and its influence to the disaster-responsive mitigation on the spatial configuration of Lakkang waterfront settlement, as settlement is formed by the local cultural. This research was conducted by using space syntax analysis techniques and supported by the disaster mitigation concept. The result of this study shows that dwelling concepts of Lakkang communities is developed from natural dwelling to collective dwelling. The spatial configuration of Lakkang settlement is developed by collective dwelling. Collective dwelling will form a more integrated space of the space which is formed by natural dwelling. Dwelling culture of the Lakkang people has formed an open space configuration so that the spaces are integrated with each other. The Lakkang settlement is adapted to the natural environment so that disaster can be minimized. The spatial configuration of Lakkang settlement is in accordance with the disaster mitigation concept. The layout of settlements and ecological carrying capacity are the determining factors for disaster response settlements. This paper could be a concept for the development of the waterfront settlements based on local wisdom.

1. Introduction

Lakkang village is a settlement which is located on the coast. This settlement grew up spontaneously along the river area and it was forming an unplanned settlement. Initially the settlement was formed by the local community, in this case the fishing communities. Lakkang settlement developed due to the concept of dwelling by societies which is related to the river. The dependence on the river causes the houses to form and the spatial order settlement described the social community conditions. Man dwells when he can orientate himself within and identify himself with an environment [1].

Lakkang settlements is inhabited by the majority of Makassar ethnic community, so that the activity of the society is still influenced by Makassar culture. Similarly, the concept of dwelling community is influenced by Makassar culture as well. In Makassar community, the concept of dwelling is influenced by values of solidarity which are known as *sipakatau* culture. *Sipakatau* culture is mutual respect and sense of togetherness. The sense of solidarity the Makassar community is considered influencing the formation of Lakkang settlement.

Although Lakkang settlement is surrounded by rivers, this settlement has never flooded. This is due to the settlement layout which is arranged in accordance with the dwelling culture of local communities. Therefore, this paper aims to describe the concept of dwelling Lakkang communities, and its influences on spatial configuration and disaster mitigation. This paper could be a concept for the development of the waterfront settlements based on local wisdom.

2. Literature Review

2.1. Cultural Living

Shelter and settlements as a product of human culture can not be separated from those influences. Occupancy not just as a mere means of life, but rather a living process, namely the presence of humans as occupants in creating living spaces in homes and surrounding environment [2]. Shelter and settlement are the mediator between humans and the environment, the transformation from one to the other or form a relationship between them [3]. Culture of dwelling is a residence that has more meaning than just a form of buildings, but more than that is a very close relationship between occupancy and human, which includes all aspects of life for all residents: ideas, place and a physical expression of the activities carried out [4].

Dwelling by giving a three-part definition [1], first, dwelling means to meet others for exchange of products, ideas and feelings, that is, to experience life as a multitude of possibilities. Second, it means to come to an agreement with others, that is, to accept a set of common values. Finally, it means to be oneself, in the sense of having a small chosen world of our own. Dwelling principles consist of natural, collective, public, and private dwelling [1].

- Natural dwelling
  This method begins with the arrival of humans on a place to stay and settle. This method is the beginning of the human form a place for living. In other words, the first place was formed by humans as called settlement, it is only establish a relationship with the surrounding environment. In this natural dwelling the occupants orient themselves in their environment and identify the environment as a place for living. The settlement as a naturally form.
- Collective dwelling
  This method begins with the arrival of humans together to forming someplace. Besides as a place for living, this place serves as a place to meet and gather. It is a place where human beings come together in diversity. This method of community oriented and identify the places by creating an identity. Space and buildings formation reflects the shared values that exist in society. In other words, collective dwelling describe the influence of social in the process of living and the space form.
- Public dwelling
  This method shows the harmony patterns that illustrate various of togetherness structures, and then implications for the values and rules in society. This way influenced by their common belief in society that formed the communal spaces. Space and buildings formed oriented values, rules and beliefs so that forming the identity of a place. Public dwelling describe the influence of culture in the process of living and the space form.
- Private dweeling
  This method is shown by the activities of the inhabitants are separated from each other. These activities form the residence and show personal identity.

Based on the above explain, that concept of dwelling starts from the formation of a natural dwelling for orientation to the environment. This natural dwelling evolved forming the collective dwelling due to togetherness values in the community. Collective dwelling evolved into public dwelling because of the influence of the values and beliefs in society to form a communal space. Then, there are activities of the inhabitants of which requires it to separate in communities will be forming the private dwelling.

2.2. Spatial Pattern of Waterfront Settlement

The process of formation the waterfront area affected by aspects of environmental conditions, socio-cultural aspects, economic aspects, aspects of population and aspects of government policy. Aspects of most influence in the formation of the waterfront settlements are socio-cultural of society, so that culture of waters will be reflected on the settlement form. On the other side the dependence of community on the beach is an aspect that affects the coastal settlement patterns [5]. It describes that dependence of community on the beach influenced to the waterfront settlement patterns.

Spatial pattern of waterfront settlement is influenced by the physical aspects of the environmental and socio-cultural community. There are five types of the settlement patterns of coastal villages which are; inland water village, outward water village, parallel water village, water village and river mouth water village. [6]. The existence of coastal villages is very unique to typify their traditional urban settlements.

The spatial pattern of the coastal settlements has different forms according to the social cultural and ecological characteristics. Former research show that the spatial pattern of the coastal settlements generally formed elongated pattern, clustered pattern and scattered pattern [7, 8]. Figure 1 shows a sketch of spatial patterns of coastal settlement.

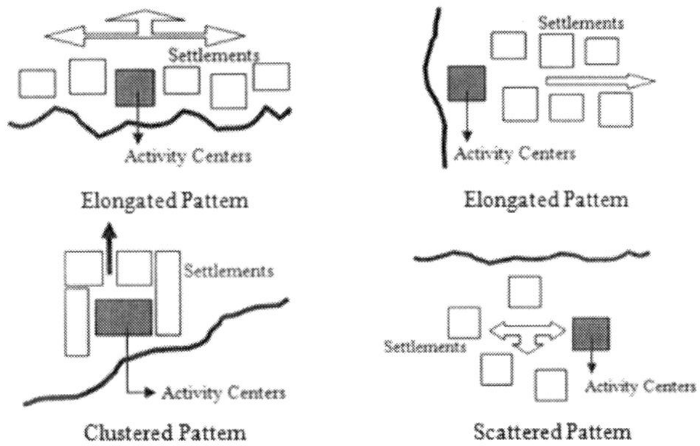

Figure 1. The Spatial pattern of coastal settlement [7, 8]

Figure 1 explains that there are four essential elements that exist in coastal settlements, namely: buildings, activity centers, open space and water as elements forming environment. These four elements should be integrated in the spatial arrangement of environmental settlements that do not conflict with the social cultural community and coastal ecological.

2.3. Disaster Mitigation Concept

Disaster mitigation is defined as the efforts made to prevent or reduce the impact of disasters. Disaster mitigation is actually related to the disaster management cycle in the form of efforts to handle it before a disaster occurs [9]. On the other hand, mitigation is defined as the efforts and activities carried out to reduce and minimize the consequences caused by disasters which include preparedness and alertness [10].

The planning of land use based on disaster response includes:

- Designation of Dangerous Areas as a Function of Open Space
  Allocation of hazardous natural disaster areas to be used as agricultural areas, parks and recreation areas as a land use planning strategy [11]. This strategy is designed so that development in hazardous areas can be as minimal as possible.
- Population and Building Density
  Strict regulation of building types and land use permitted in hazard areas. Regional planning and allocation must use population density rules or very large land allocations (for example a minimum of 4 hectares) so that only low density residential functions may be in hazardous areas [12]. Only low density settlements are allowed in dangerous areas.
  Figure 2 shows the density of buildings in hazardous areas

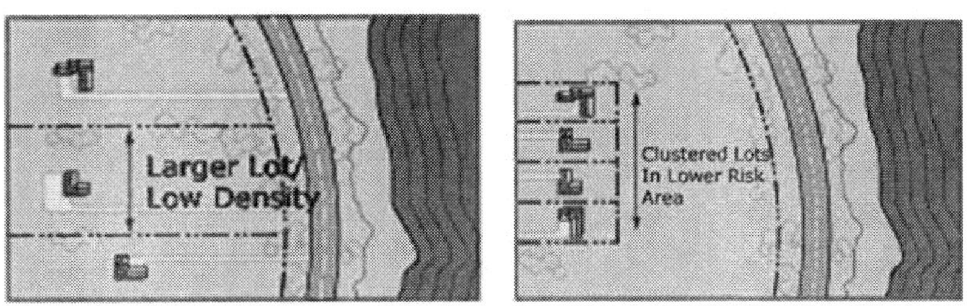

Figure 2. Density and risk level in settlement areas [12]

IOP Conf. Series: Materials Science and Engineering **875** (2020) 012005 doi:10.1088/1757-899X/875/1/012005

- Safe Zone Planning
  The safe zone can be in the form of open space that functions as green land, such as fields, forests or public facilities that function as refugee shelters. To deal with extreme waves and tsunamis, it is necessary to build Escape Hill, an altitude area that can accommodate people in areas that can serve, accommodate when extreme waves occur. The location must be easily traversed and equipped with road facilities, and the altitude exceeds the extreme wave heights that have occurred. While the safe zone area must be able to accommodate the surrounding community. Besides that, it must consider the location of the distance, for example 10 to 20 minutes, the community has been able to reach the zone, so that the community can save themselves from disasters [13].
- Utilization of Green Belts in Coastal Areas
  There is a boundary between the water area and the housing area which can reduce the level of incoming sea water during extreme waves. The barrier can be in the form of hard plants or green belts that are built parallel to the waters with suitable vegetation types as barriers [11]. Recommended types of vegetation are those that have wide canopies that are planted on the city belt, district belt and neighborhood belt.

The study of local wisdom and disaster mitigation in traditional communities in Indonesia is seen in relation to natural resources and human resources. In traditional (local) societies, humans and nature are a unity because they are both the almighty creations. Nature and humans are believed to have the same spirit. Nature can be friendly if humans treat wisely and vice versa will be angry if we damage it. If nature becomes angry so that natural disasters occur in the form of floods, landslides, volcanic eruptions and so on, then traditional communities generally also have local knowledge and ecological wisdom in predicting and mitigating natural disasters in their areas [14].

3. Method

This study described the influence of dwelling culture to the disaster-responsive mitigation on the spatial configuration of Lakkang waterfront settlement. This study used the diachronic reading technique analysis and was supported by the space syntax method [15]. To explain the formation of the settlement spatial configuration, a diachronic reading technique analysis was used based on map, observations and interviews from several sources. The space syntax method was used to analyze and explain the influence of dwelling culture to the characteristics of the settlement spatial configuration.

The results are then used to interpret the spatial configuration settlement formed. Then, the spatial configuration settlement formed is analyzed based on the disaster mitigation concept. The results explain the influenced of spatial configuration settlement to the disaster mitigation aspects in the Lakkang waterfront settlement.

This study was conducted on Lakkang settlements in Makassar, South Sulawesi Province as growing settlements on the waterfront described in Figure 3.

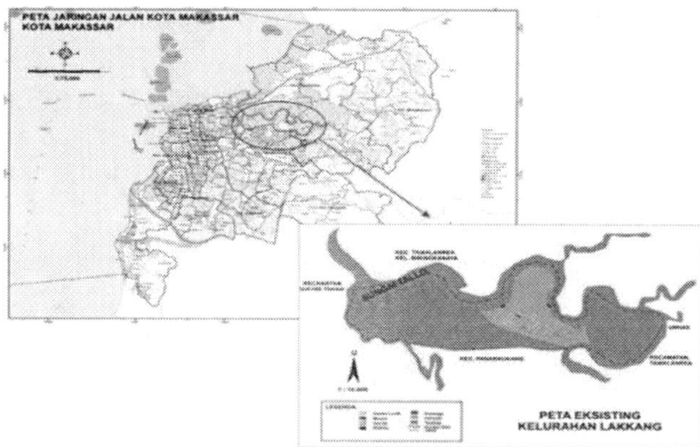

Figure 3. Research location [16]

4. Result and Discussion

4.1. Dwelling Concept to Housing Layout

Lakkang settlement is beginning to be formed by community related on the river, in this case by the fishing communities. So that the layout of Lakkang settlement describes the dwelling culture conducted by the society. The formation of individual houses on the water describes the dwelling culture of society is related on the beach. It is influenced by the natural dwelling conducted by the society as part of the adaptation on the river. This natural dwelling caused the origin of Lakkang settlement is formed. In other words, the first place formed by Lakkang community because of human relationship with the around environment. In this natural dwelling Lakkang community is oriented and identified itself on its environment as a place for living. It can be said that initially the Lakkang settlement is formed naturally. The natural dwelling causes the formation of houses spreaded on the water, directly reaching the river or the work place.

The influence of natural dwelling to the Lakkang settlement is shown in Figure 4.

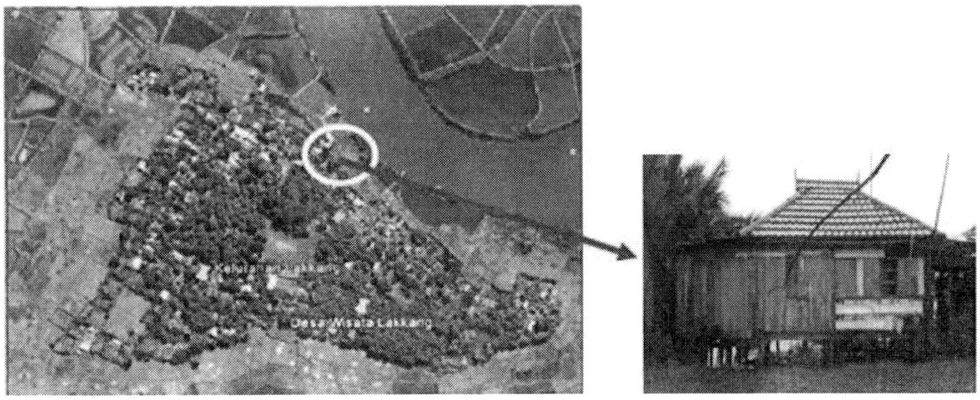

Figure 4. The Natural dwelling on the Lakkang Settlement

The influence of natural dwelling to house formation Lakkang settlement described in Table 1.

Table 1. Natural dwelling in house formation of Mariso Settlement

House Form	Access Graph	Relative Asymmetry Calculation
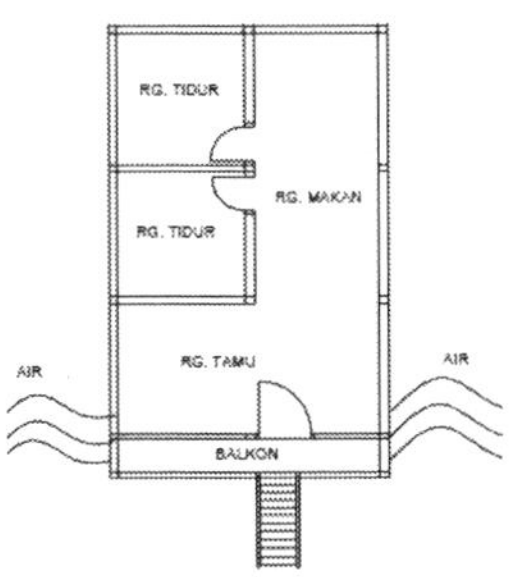 Layout of house	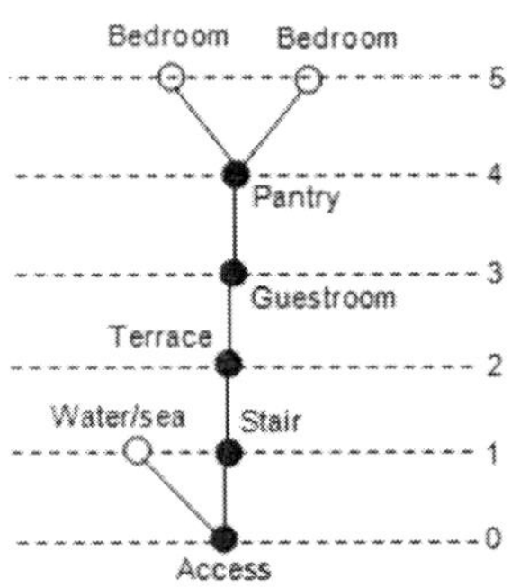	Main depth value : MD = 21/(8-1) = 3.00 Integrasi value : RA = 2 (3.0–1) /(8-2) = 0.667
	The spatial configuration is non-distributed formed, because there is no route choice for moving from one space to another space. It is the closed settlement formed.	The average settlement depth value of 3.00, it indicates that the settlement is relatively not easy to reach from outside the settlement. Spatial integration value of 0.667, it indicates residential spaces that are not integrated with each other.

In its development, the influence of kinship system of Makassar community is forming the collective dwelling. This occurs after a natural dwelling formed, then forming a collective dwelling in settlements. The togetherness tradition of Lakkang community for always wanted to stay close has been influencing natural dwelling formation. This method oriented on sense of similarly, so that it is forming the places identity. In other words, the collective dwelling for people Lakkang not only means a more meaningful place to stay but as a place collective that describes the influence of culture in the formation of space. Collective dwelling of Lakkang community have been forming the groups of homes in the settlements.

Lakkang settlement developments are influenced by collective dwelling of communities. It is occurred due to kinship of solidarity. Kinship solidarity has caused growing the family houses in one plot, so form the cluster-houses in the Lakkang settlement. The influence of collective dwelling to the Lakkang settlement forming is shown in Figure 5.

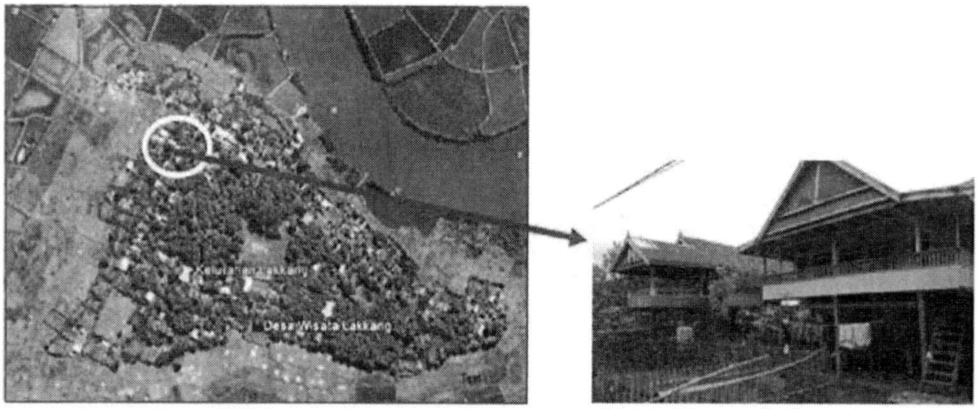

Figure 5. The Collective dwelling on the Lakkang Settlement

The influence of collective dwelling to house formation Lakkang settlement is described in Table 2.

Table 2. Collective dwelling in house formation of Lakkang Settlement

House Form	Access Graph	Relative Asymmetry Calculation
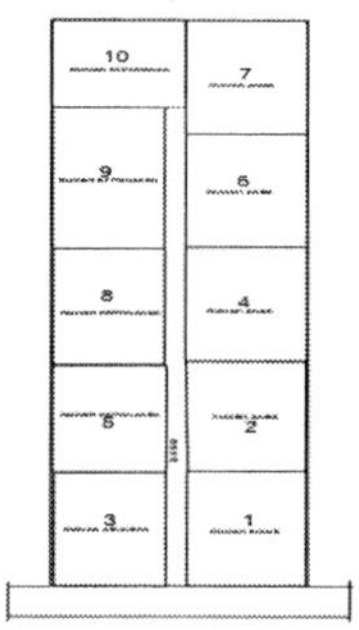	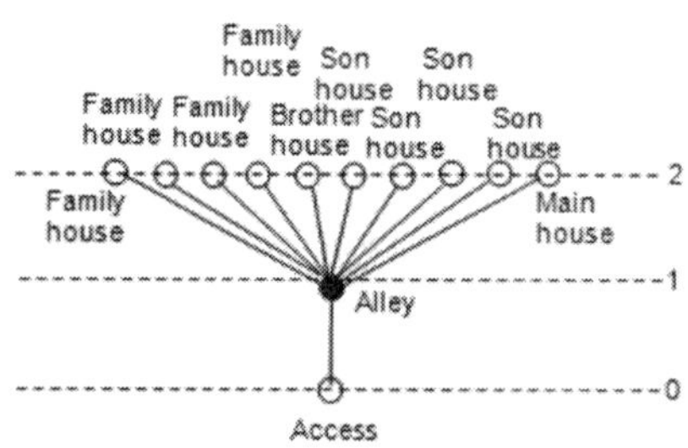	Main depth value : MD = 21/(12-1) = 1.91 Integrasi value : RA = 2 (1.91–1) /(12-2) = 0.182
Layout of house	The spatial configuration is non-distributed formed, because there is no route choice for moving from one space to another space. It is the closed settlement formed.	The average settlement depth value of 1.91, it indicates that the settlement is relatively easy to reach from outside the settlement. Spatial integration value of 0.182, it indicates residential spaces that are integrated with each other.

Based on the analysis above, the house formed by the natural dwelling has the integration value of 0.667, while the house formed by the collective dwelling has the integration value of 0.182. This is to indicate that the space formed by the collective dwelling will form more integrated spaces than the space formed by the natural dwelling. On the other side, elongated patterns are more integrated than scattered patterns

4.2. Spatial Configuration of Lakkang Settlement

Layout of Lakkang settlement is shown in Figure 6.

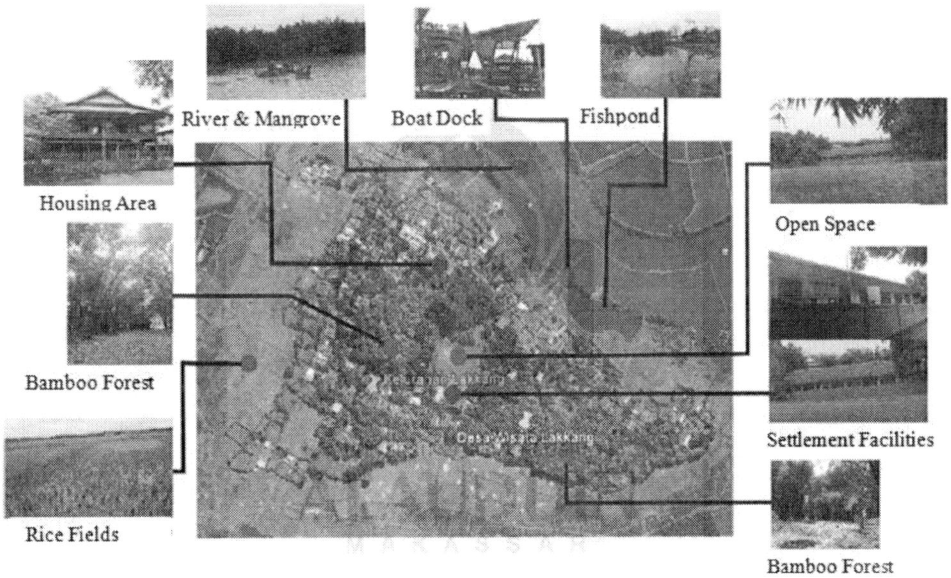

Figure 6. Lakkang settlement layout

Table 3. Spatial configuration of Lakkang Settlement

Layout Settlement Map	Access Graph	Relative Asymmetry Calculation
	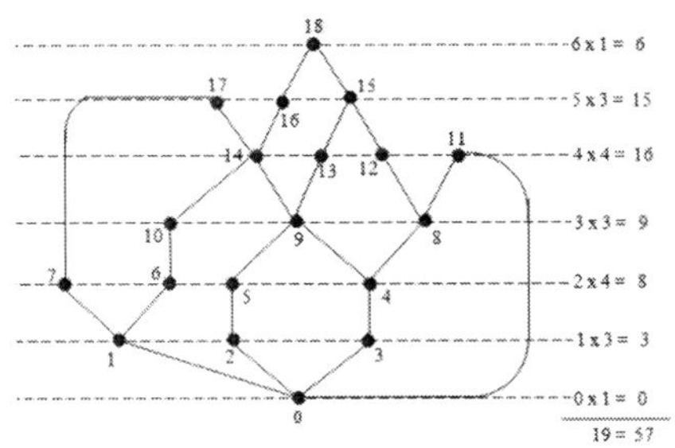	Main depth value : MD = 57/(19-1) = 3.167 Integrasi value : RA = 2 (3.167–1) /(19-2) = 0.255
	The settlement patterns are open because of the access around the settlements. The spatial configuration is distributed, so there is a route choice for moving from one place to another place.	The average settlement depth value of 3.167, it indicates that the settlement is relatively easy to reach from outside the settlement. Spatial integration value of 0.255, it indicates residential spaces that are integrated with each other.

Figure 6 explains that the Lakkang settlement is surrounded by waters and mangrove forests. Outdoor spatial configuration consists of fishponds and rice fields. Indoor spatial configuration consists of housing, open space and bamboo forests. This explains that the houses are located far from the water and are very protected by the fishpond and rice fields. The influence of cultural dwelling to the spatial configuration of Lakkang settlement described in Table 3.

Table 3 shows that the dwelling culture of the Lakkang people has formed an open space configuration so that the spaces are integrated with each other. This causes the Lakkang settlement to adapt to the natural environment so that disaster can be minimized.

4.3. The Influenced of Spatial Configuration on Disaster Mitigation

- Designation of Dangerous Areas as a Function of Open Space
  In hazardous areas used as ponds and rice fields. This indicates that the layout of the Lakkang settlement has minimized the area affected by the disaster.
- Population and Building Density
  The area of Lakkang village is around 1.65 km2. The total population of 2018 is 925 consisting of 504 men and 448 women. Number of houses 261 units. The population of 925 and the number of buildings 261 in an area of 1.65 km2 makes Lakkang village a low density (SNI Standards). On the other hand, houses that are formed in dangerous zones are only buildings that are related to livelihoods. This indicates that the Lakkang settlement density level is in accordance with the concept of disaster mitigation.
- Safe Zone Planning
  Placement of houses and facilities are placed in the middle of settlements that are far from the edge of the water and far from dangerous areas. Besides that, houses and settlement facilities are protected by bamboo forests. On the other hand, evacuation routes and safe gathering points are fields located in the center of settlements and in the highlands. This indicates that the layout of buildings and Lakkang settlement facilities have entered the safe zone area.
- Utilization of Green Belts in Coastal Areas
  Green belts in the Lakkang settlement are mangrove forests, rice fields and fish ponds, which function to reduce the level of tidal currents that enter during extreme waves. This indicates that the Lakkang settlement has a green belt that serves as disaster mitigation.

Based on the explanation above, the spatial configuration of Lakkang settlement is in accordance with the disaster mitigation concept.

5. Conclusion

The culture of Makassar community has changed the Lakkang settlement. Dwelling concepts of Lakkang communities is developed from natural dwelling to collective dwelling. The spatial configuration of Lakkang settlement is developed by collective dwelling. Collective dwelling will form a more integrated space of the space which is formed by natural dwelling. Therefore, kinship culture will form a more integrated space. On the other side, elongated patterns are more integrated than scattered patterns. Dwelling culture of the Lakkang people has formed an open space configuration so that the spaces are integrated with each other. This causes the Lakkang settlement adapts to the natural environment so that disaster can be minimized. This indicates that the spatial configuration of Lakkang settlement is in accordance with the disaster mitigation concept. Therefore, it is suggested that the arrangement of the waterfront settlements should pay more attention to local cultures. So it will emerge particularities or specificities of tradition and then disaster can be minimized.

Acknowledgements

This paper is part of a joint research with students of architecture department Hasanuddin University on the coastal and island settlements study in Lakkang village. The author thanks to students for their participation and assistance in surveying and collecting data. The author would like to express special thanks to the Lakkang District Head and Lakkang people are for the permission to survey in the location and the provision of data related to the Lakkang village, so that this research can be carried out.

References

[1] Schulz, Christian Norberg, 1985, The *Concept of Dwelling*, Rizolli, New York.
[2] Schulz, Christian Norberg, 2000, *Architecture: Presence, Language and Place*, Skira Architecture Library, Milan.
[3[Crowe, Norman, 1995, *Nature and the Idea of a Man-Made World*, MIT Press, New York.
[4] Oliver, Paul, 1987, *Dwelling, The House Across the World*, Phaidon Press, Limited, Oxford.
[5] Sairinen, Rauno and Kumpulainen, Satu, 2006, *Assessing Social Impacts In Urban Waterfront Regeneration*, Journal Elsevier, Environmental Impact Assessment Review 26.
[6] Hassan, Ahmad Sanusi, 2010, *Reviews On Old City Landscape With Reference to Traditional Fishing Village Settlements in Western Coastal Region, Peninsular Malaysia*, Journal of Human Settlements, Vol. 2 July 2010, Bandung.
[7] Kostof, Spiro, 1991, *The City Shaped*, MIT Press, New York.
[8] Darjosanjoto, Endang TS., 2007, *Permeability Maps of Residential Settlemets Within The Coastal Area of Surabaya, Indonesia*, Proceedings, 6th International Space Syntax Symposium, İstanbul.
[9] Subiyantoro, I., 2010, *Selayang Pandang tentang Bencana*, Jurnal Dialog Penanggulangan Bencana, 1, 43-46.
[10] Departemen Dalam Negeri, 2003, *Keputusan Menteri Dalam Negeri RI No. 131 tahun 2003*, Jakarta.
[11] Sriyana, 2005, *Pendekatan Model Pengendalian untuk Mengurangi Risiko Akibat Gelombang Gelombang ekstrim,* Jurnal Media Komunikasi Teknik Sipil, Volume 13, No. 2, Universitas Diponegoro, Semarang.
[12] NOAA et al., 2005, *Menghadapi Gelombang ekstrim: Tujuh Prinsip Perencanaan dan Perancangan. Masyarakat Lingkungan Binaan Untuk Komisi Darurat Kemanusiaan untuk Nanggroe Aceh Darussalam dan Sumatera Utara.*
[13] Samsirina, HeruW. Poerbo, dan Syahyudesrina, 2008, *Perencanaan Permukiman Kota Pasca Bencana dalam Mewujudkan Kota Tropis, Seminar Nasional Peran Arsitektur Perkotaan dalam Mewujudkan Kota Tropis*, 6 Agustus 2008, UNDIP Semarang.
[14] Raden Cecep Eka Permana, Isman Pratama Nasution, dan Jajang Gunawijaya, 2011, *Kearifan Lokal Tentang Mitigasi Bencana Pada Masyarakat Baduy*, Jurnal Makara, Sosial Humaniora, Vol. 15, No. 1, Juli 2011
[15] Hillier, Bill. Hanson, Julienne, 1984, *The Social Logic of Space*, Cambridge University Press, London.
[16] Pusat Statistik, Badan, 2018, *Kota Makassar Dalam Angka*, BPS, Makassar, Indonesia.

Comparison of Building Envelope Models to Illuminance Level in Phinisi Tower Building

N Jamala[1]*, R Rahim[1], and R Mulyadi[1]

[1]Department of Architecture, Engineering Faculty, Hasanuddin University, Makassar, Indonesia

*Email: nuruljamala@yahoo.co.id

Abstract. The hyperbolic parabolic facade is the appearance of a Phinisi Tower Building. The building facade characterizes this building and is also the aesthetic value of this building. Building facade models are increasingly developing, but need to pay attention to their association with visual for space users. Visual comfort is related to the standardization of the level of illuminance, which has been recommended by Indonesia National Standards. Quantitative research methods that used describe data measurement results in the workspace, which are directly related to the opening of the building envelope. The level of illuminance measurements are carried out in the workspace and in open areas in medium sky conditions. The results of the study are that the level of illuminance in the workspace is very different, because of the different facade models. vertical and horizontal strip facade models have a higher level of illuminance than the hyperbolic parabolic facade, whereas in spaces that do not use a facade but shaded by sunscreen are the highest. The results of the study concluded the building facade model influence of the level of illuminance in the workspace. The contribution of this research can be a guide in designing the facade shape of buildings, by paying attention to the distribution of lighting into the space, so that it can reduce the use of energy in terms of lighting and can create energy efficient building designs.

1. Introduction

The building facade needs to be considered in planning a building, because this is the aesthetic value of the building. The facade model is the appearance of a building, but it should be noted its relationship with the distribution of natural lighting entering into the building. Building facade models need to be adapted to the climatic conditions in which the building will be planned.

The city of Jakarta has issued governor regulation No. 38 of 2012 concerning green buildings [1], where the current development has implemented a reduction in energy use to achieve sustainable development goals in the future. Building facade model as building envelope design has an important role to protect the building in terms of visual and thermal convenience. Solar thermal (thermal), glare and brightness can be avoided by using building facades.

This research analyzed the Phinisi building that functions as the administration of Makassar State University. This research is related to the level of illuminance in the building workspace. The front view of the building uses a form of hyperbolic paraboloid facade which is a futuristic expression and application of sophisticated science and technology. In addition, this building also uses a horizontal and vertical strip facade. Shape of façade building have effect to level illuminance in the room with direct area to building envelope were develop [2]. According to Soegijanto [3] sky conditions based on the number and type of clouds can be grouped into: (a) Clear sky is the sky without clouds; (b) Overcast sky is a sky completely covered in white or gray white clouds or partial or complete partial clouds; (c) intermediate sky is a partially clouded sky with a variety of darker clouds and numbers: and (d) Uniform sky is the sky with the same luminance in all positions not dependent on the geographic latitude and the height of the sun. The study concluded the decline percentage in the value of the illuminance after the building using the building facade is 49% -74% and a mean value of 60.3%, so it can be concluded that the building facade effects on the natural lighting [4].

This study analyzed the level of illuminance in the space located in the area of the building envelope opening using the building facade. This is related to the design of energy-efficient buildings that is by maximizing the use of natural lighting, while still considering visual comfort and thermal comfort. This study only focused on visual comfort in accordance with recommendation on natural and artificial light namely SNI 03-6575-2001 [5] and SNI 03-2396-2001 [6]. The United Nations Environment Program [7] states that the level illuminance standard recommendation for the office room, based on values recommended by IES (Illuminating Engineers Society) [8] and CIE (Commision International De I'Eclaire) [9]. The Illuminance level of daylight depends on the sky condition and it have affect the light distribution into the building. The increased light distribution at the height of the floor. As well as of window opening on the building envelope (top lighting and side lighting) have effect on the level of illuminance in the building [10]. The height and depth of the space affect the distribution of natural light come into the room [11]. The building orientation have the effect of iluminance level come into the room. The Level of illuminance in the oriented space to the north are higher than south orientation. the other than, level illuminance has an influence the distance of the measurement point from the window opening on building envelope, the further away from the building envelope the light distribution decreases [12]. The distribution light will decrease, if the point measurement farther away from the building envelope and natural light come into the building affects the sun's radiation in the morning, afternoon and evening [13]. Rahim stated that CIE (International Illumination Commission) reference sky conditions namely clear sky and overcast sky [14]. In addition these, Nakamura proposed the sky condition third called i.e. overcast, clear and intermediate sky. Trying to define the luminance distribution of the intermediate sky [15].

2. Methodology

This study uses quantitative methods that describe the measurement data and analyze data in the form of tables, bar charts and graphs. The level of illuminance measurements carried out are outside and inside the building, at the planned placement of measurement points. The number of measuring points in space is adjusted to the area of the selected object. Measurements were made on 17-18 July 2019 at 10:00 to 12:00 o'clock with intermediate sky condition. The level of illuminance outside the building at the time of measurement ±100,000 lux. Building orientation to the west-south-west (WSW) and east-north-east (ENE)

3. Result and Discussion

The Phinisi Tower Building consists of 19 floors, but in this study only analyzed the workspace located on the second floor. On the 2nd floor there is a workspace with different spatial orientations. The workspace is directly related to openings in the building envelope, with different facade forms, namely hyperbolic paraboloid facades, horizontal and vertical strips.

The work space on the 2nd floor has different functions, namely (1) zone-1: administration room, the interests and reasoning section room, student and welfare room and Public Relations Room; (2) Zone-2: Education room; and (3) Zone-3: Asset management room and internship room. These three zones were selected as objects in this research, with consideration of using different forms of façade (Figure 1).

IOP Conf. Series: Materials Science and Engineering **875** (2020) 012006 doi:10.1088/1757-899X/875/1/012006

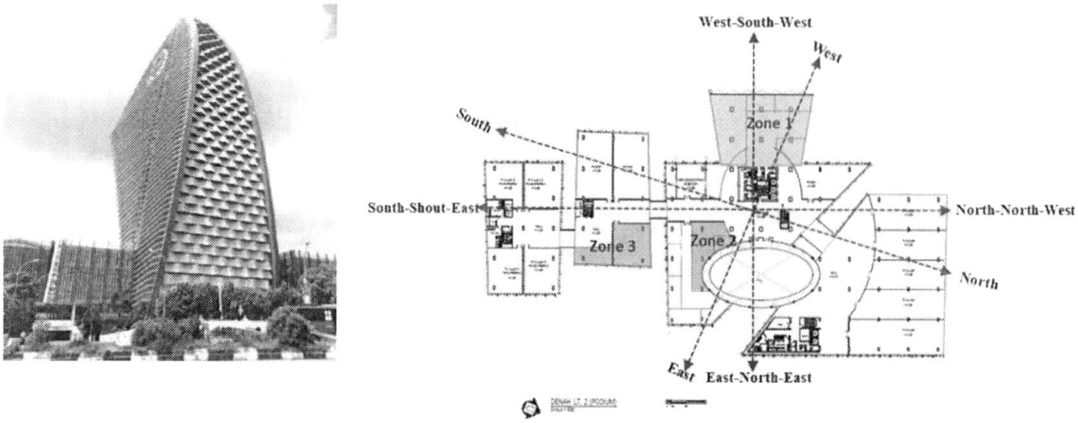

Figure 1. Orientation of the Phinisi Tower building

The shapes of façade in room on zone-1, zone-2 and zone-3 have building façade shape is the different. The three zone, have facade diagonal, horizontal strip and fertikal as show in Figure 2 below.

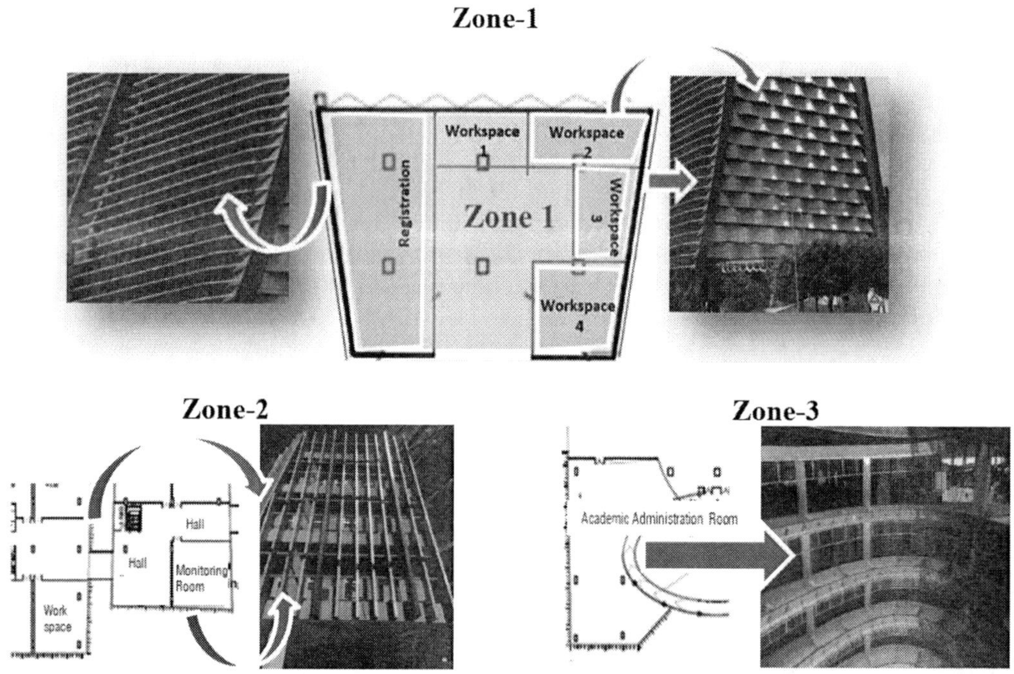

Figure 2. Shape of facade on zone-1, zone-2 and zone-3

In zone-1 the form of the building envelope uses a diagonal and horizontal facade, zone-2 forms the vertical facade and zone-3 of the building envelope using glass windows that are directly related to the open void. These third facades are able to distribute natural lighting that is different, so it is necessary to analyze this.

IOP Conf. Series: Materials Science and Engineering **875** (2020) 012006 doi:10.1088/1757-899X/875/1/012006

3.1. Analysis of Zone-1 (The façade Model of Strip Horizontal and Diagonal)

Figure 3. Administration room in second floor

In zone-1 it consists of several workspaces that are directly related to the openings of the diagonal shaped building and horizontal strips. This study analyzes the level of illuminance in the office room, as shown in Figure 4 below.

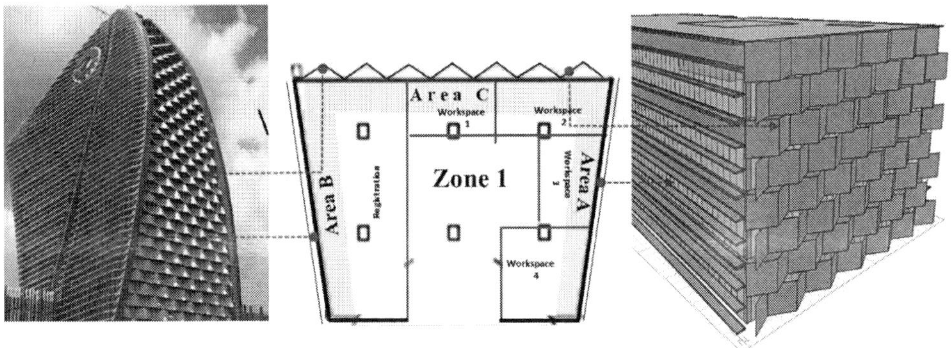

Figure 4. The shape of façade on the envelope building

In zone-1 it is divided into 3 areas, namely (1) Area-A: workspace oriented towards north-north-west; (2) Area-B: orientation towards south-south-east; and (3) Area-C: work space towards west-south-west. Area A and B have horizontal strip facade while Area C is hyperbolic paraboloid. Area-A and area-B have the same facade shape but different building orientations (Figure 4). The above area measured and analyzed by the measurement point in that space.

The second floor workspace has openings in the building envelope in the form of glass windows (Figure 5). Window openings are not entirely the envelope in the building, but there is a distance between the windows as shown in Figure 5 below. Measurements were made with intermediate sky weather conditions and artificial lighting (lamps) were turned on in each measured space.

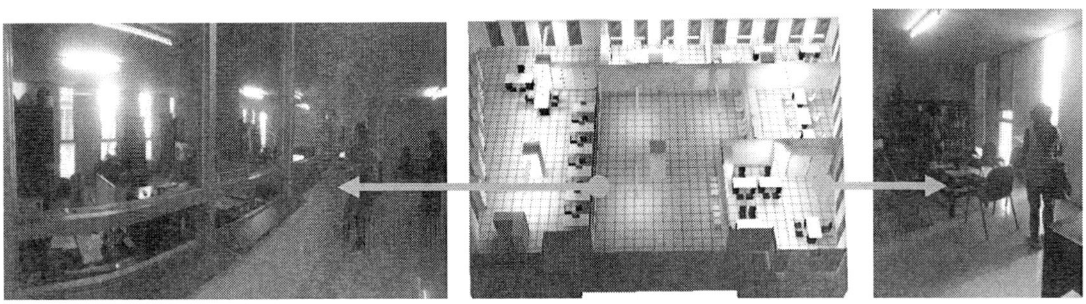

Figure 5. Layout design of windows on building envelope (Zone-1/area-A, area-B and area-C)

In Zone-1 there are several workspaces and when the was measured the condition of lights. Placement of the measuring points as shown in Figure 6 below and the measuring point notation is A-I and 1-9, as shown in Figure 6. Analysis carried out in zone-1 is divided into 3 areas, namely area-1 (horizontal facade model oriented towards the north-north-west), area-B (horizontal facade model oriented towards the south-south-east) and area-C (paraboloid hyperbolic facade model oriented towards the west-south-west). Measurement of the mean value is based on the placement of measuring points in the area of the building envelope, namely in area-1 (measuring points I1-I9), area-2 (measuring points A1-A9) and area-3 (measuring points A1-I1), as shown in Figure 6.

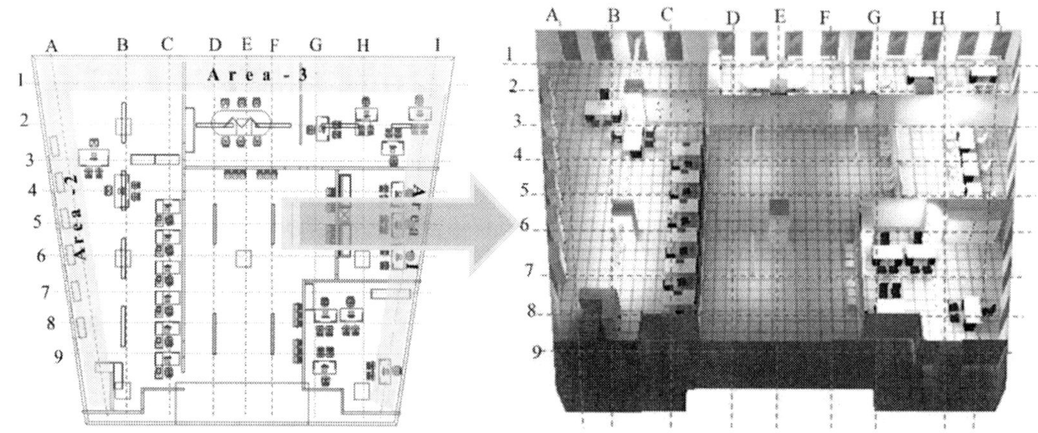

Figure 6. Placement of the measurement point in Zone 1 (A-I and 1-9)

Before analyzing the mean values in the area of the building envelope, namely area-1, area-2 and area-3, the distribution of measurement results data is carried out across all spaces in zone-1, as shown in Figure 7 below. The light source comes from natural and artificial lighting, but the level of illuminance is only high in the area near the openings of the building envelope, even though it has turned on the lights. The mean values in zone-1 area differ in each workspace, because the distribution of natural light is not the same as the percentage that enters the space.

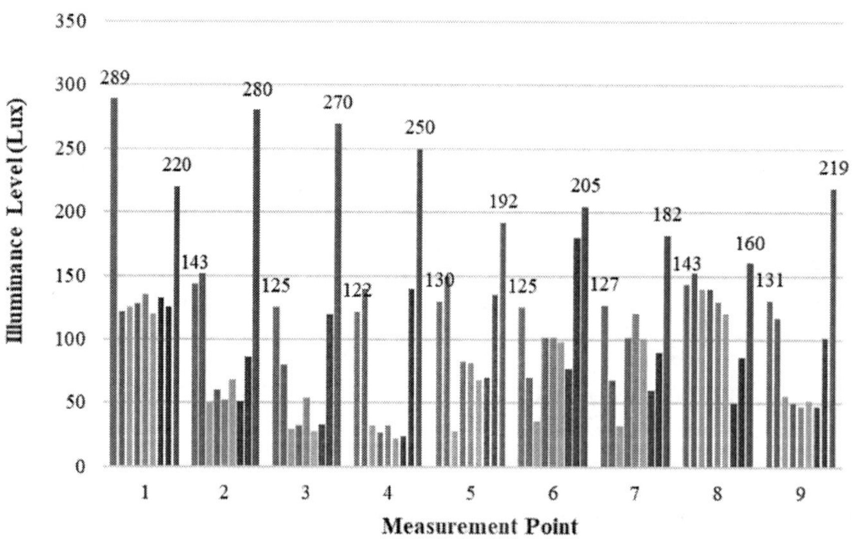

Figure 7. Level of illuminance on Zone-1

The level of illuminance at the measuring points I-1 to I-9 (area-1) is a room with a horizontal grid-shaped building façade 280-160 lux. The illumination value in area-1 has the highest illumination value compared to area-2 (A-1 to A-9) amount 289-70 lux and area-3 (B-2 to H-1) amount 135-120 lux, as shows in Figure 8.

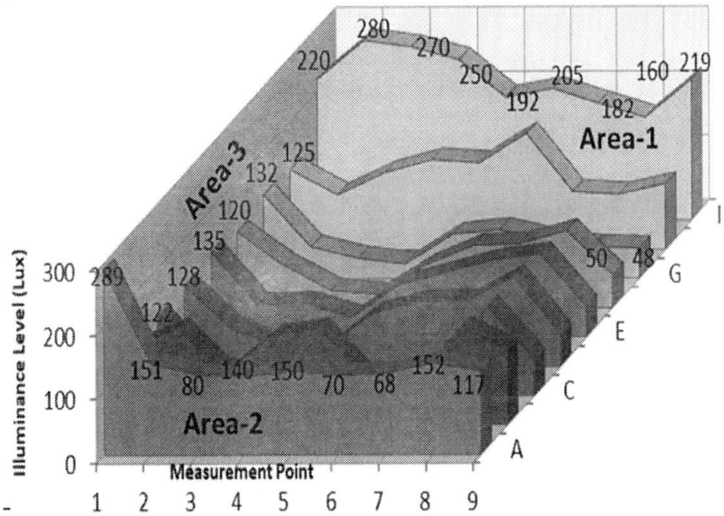

Figure 8. Illuminance level graph in zone-1 area

Figure 8 shows a graph of illuminance level in zone 1 area. This graph shows that the level of illuminance in the area near the building envelope openings is a maximum of 280 lux and a minimum of 70 lux. The level of illuminance in this area varies greatly, so it is necessary to analyze the level of illuminance at the measuring point located along the building envelope.

3.2. Analysis of the Illuminance Level in the Horizontal Facade Model

In the area-1 there are several spaces that are directly related to the openings of the building envelope with a horizontal facade model as shown in Figure 9 below. Workspace orientation in areas-1 and 2 with different orientations but the same facade model is horizontal facade as shown in the following.

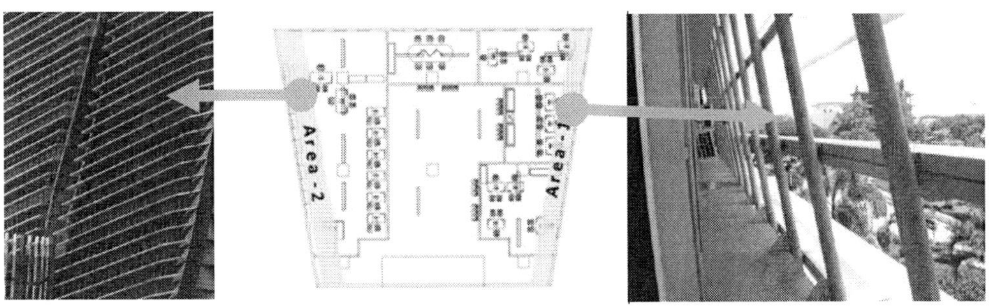

Figure 9. Administration room with horizontal facade on the 2nd floor

In the area-2 there are several spaces that are directly related to the openings of the building envelope with a horizontal facade model as shown in Figure 10 below. Workspace orientation in areas 1 and 2 with different orientations but the same facade model is horizontal facade as shown in the Figure 10.

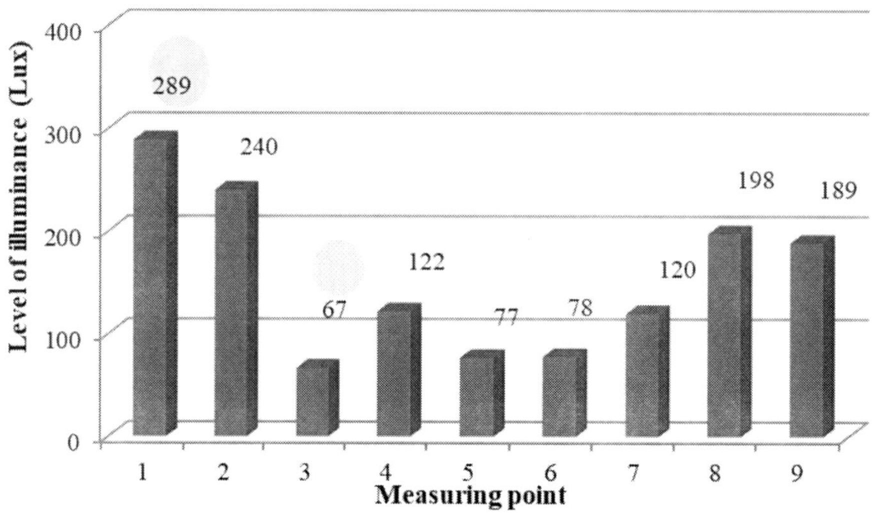

Figure 10. Level Illuminance chart on envelope building area of horizontal model (Area-1)

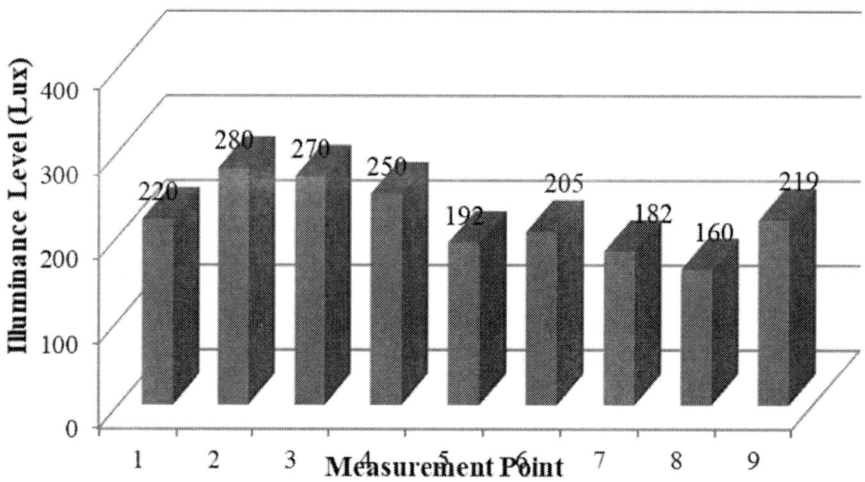

Figure 11. Level Illuminance chart on envelope building area of horizontal model (Area-2)

The level of illuminance in this workspace, is not uniformity because there are cabinets in the area of the building envelope so that the light distribution decreases. The average value in this area is 219 lux. Whereas in area 2 it has an average value of 222 lux. These two areas have the same facade shape but different orientations. The average level of illuminance in these two areas is 224 lux. Based on this analysis, it can be seen that the percentage of light distribution in the Phinisi building in the form of a horizontal facade that is in the intermediate sky conditions (18.000 lux) is 1.25 percent.

3.3. Analysis of Illuminance Level in the Diagonal Façade Model

Area-3 has a diagonal oriented facade west-south-west. The room is directly related to the building envelope as shown in the following Figure 12.

IOP Conf. Series: Materials Science and Engineering **875** (2020) 012006 doi:10.1088/1757-899X/875/1/012006

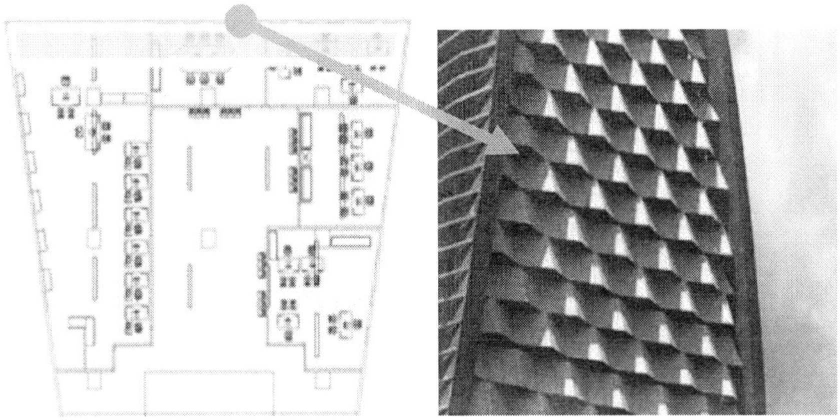

Figure 12. Administration room with hyperbolic paraboloid facade

Illuminance level on Area-3 has a diagonal façade, maximal 289 lux and minimal 101 lux. Average value of level illuminance amount 157 lux in Area-3. Measurement point A (289 lux) and point I (268 lux) have the highthes of illuminance level compare to others, because it is located in the corner of the room which has windows on two sides. While the other (point B-H) has a low illuminance value, because it has a smaller window area, namely between 101 lux to135 lux (Fig. 13).

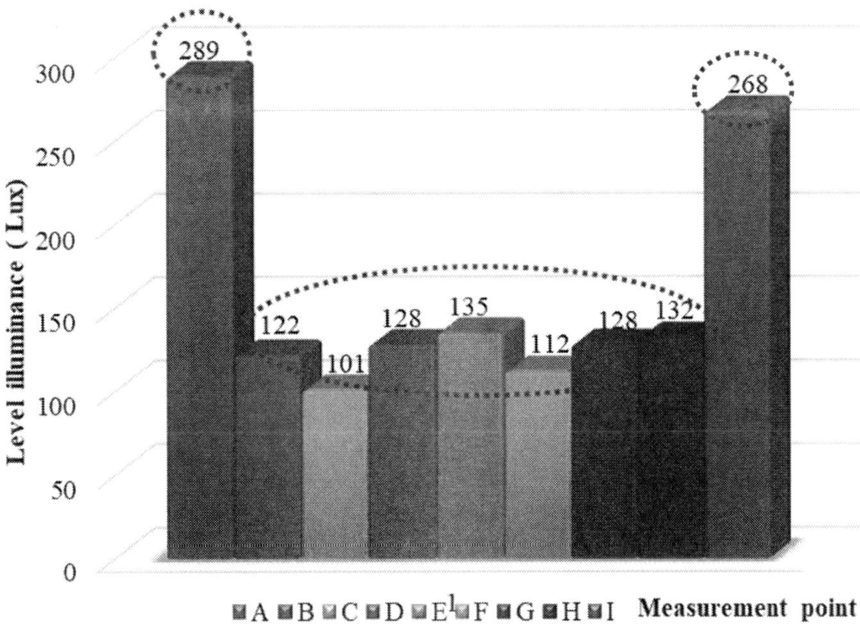

Figure 13. Level illuminance in Area-3 has a hyperbolic paraboloid facade

3.4. Analysis of Area-2 (Without Building Facade)

Area-2 functions as a student and educational room. In area-2, the administration room is directly related to the natural light source. the building envelope are glass and bordered by voids outside the building as shown in Figure 14 below. The building envelope does not use a facade, but the light source is protected by a sunscreen that functions as a corridor in the outdoor building.

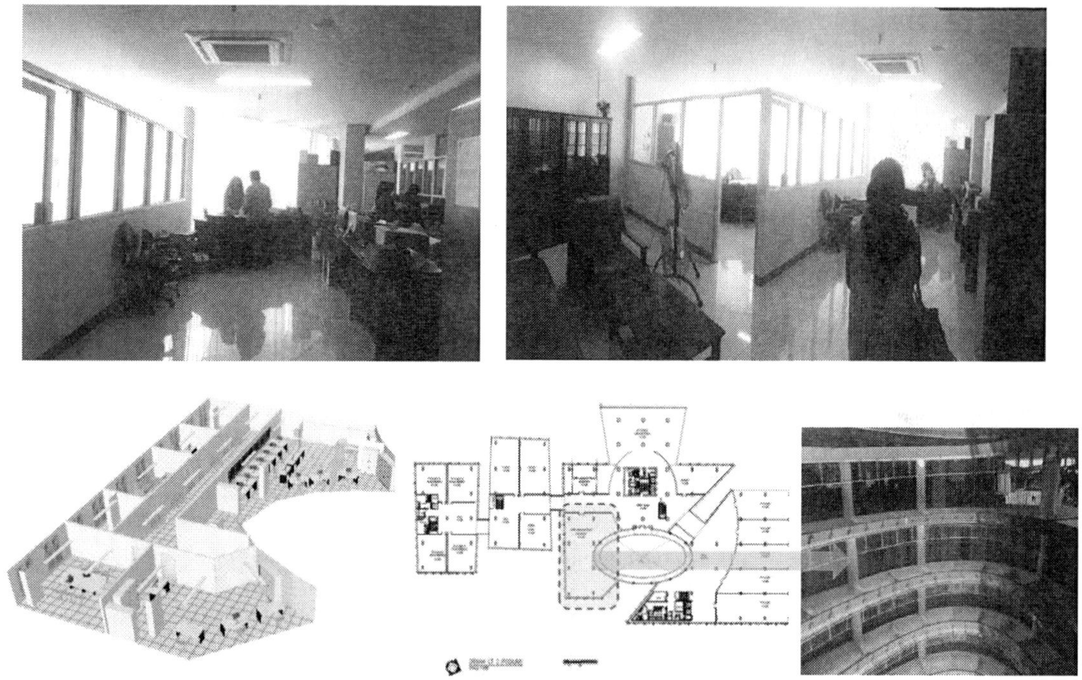

Figure 14. Zone-2 on administration and education room

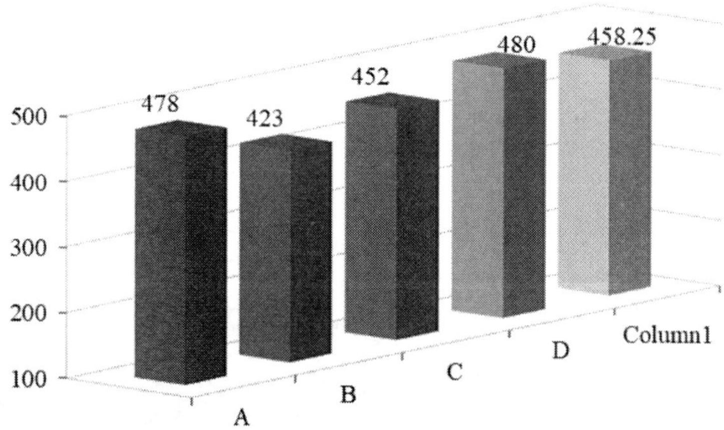

Figure 15. Illuminance level chart on envelope building without façade

The average value in the area of the building envelope is very high, because the building envelope is glass material that is directly related to the void, so that the light distribution can be maximized. Based on the measurement results, the average value of the five measurement points along the building envelope is equal to 458 lux.

3.5. Zone-3 Analysis (Strip Vertical Façade)

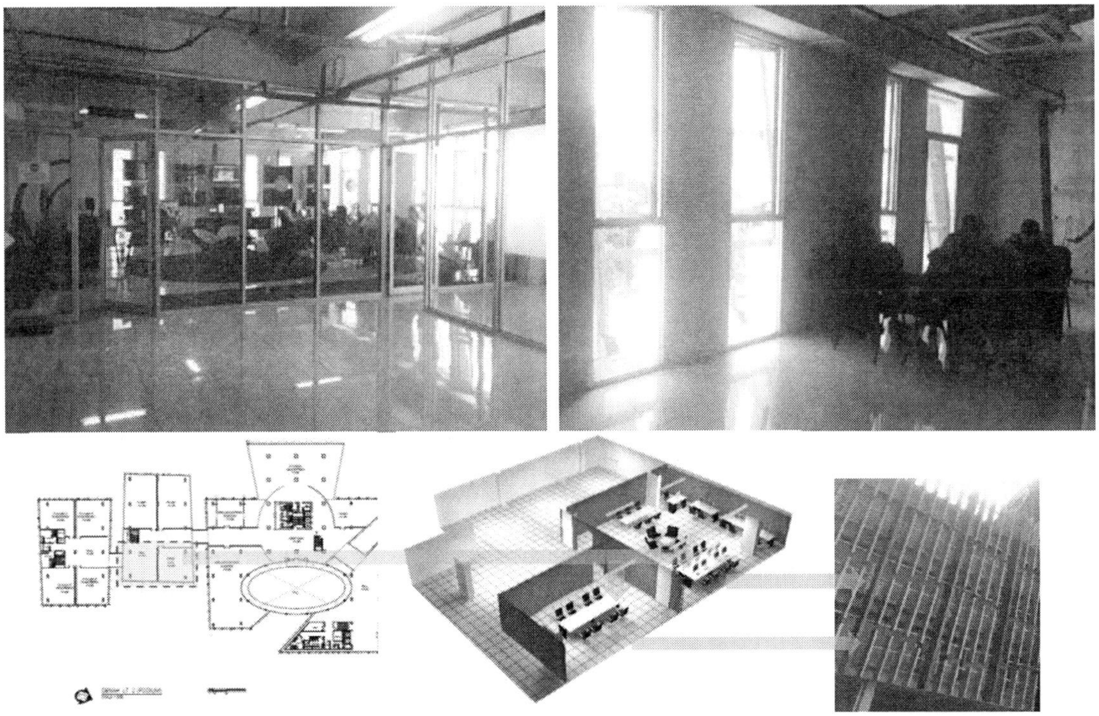

Figure 16. The shape of the vertical grid façade (zone 3th)

On the second floor there is an asset management room and an internship room. The space is directly related to the building envelope by using the vertical facade (figure 16). The average value of a number of points which have a distance of 50 cm from the building envelope is 723 lux while 150 meters from the building envelope is 230 lux. This space uses massive glass throughout the building envelope, but also uses a vertical façade.

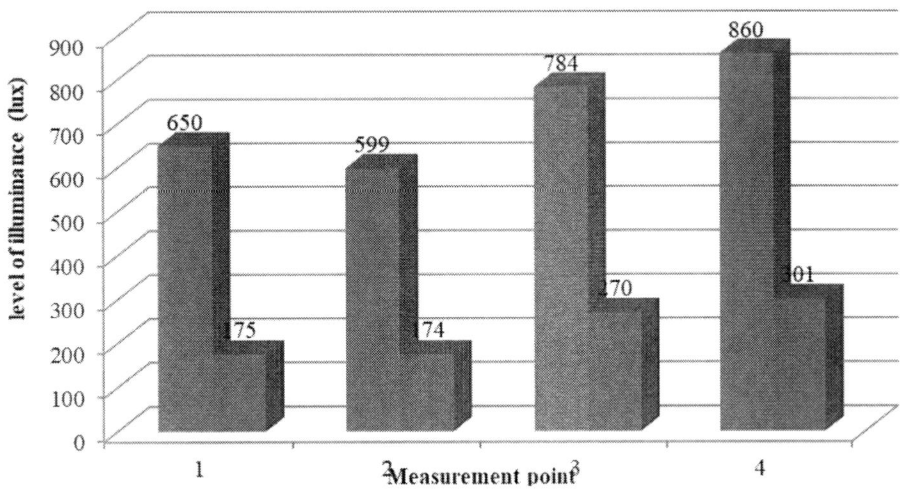

Figure 17. Illuminance level chart on workspace zone 3th

3.6. Analysis of Light Distribution in Horizontal, Vertical and Hyperbolic Paraboloid Facade Spaces

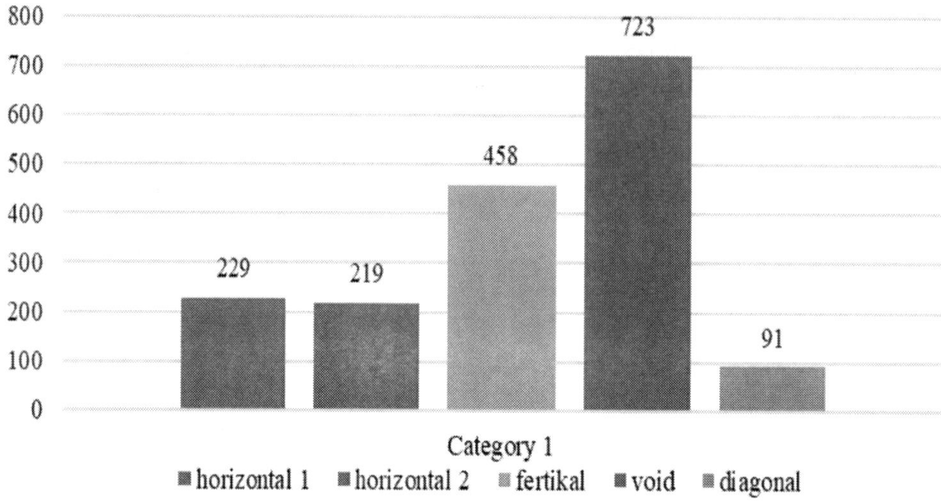

Figure 18. Average value of Illuminance level in space with various form of facade

The picture above shows a graph of mean values in spaces using horizontal, vertical, hyperbolic paraboloid facades and no façade. The level illuminance in spaces that do not use the façade has the highest value compare to the others. The level illuminance in spaces using horizontal 1 and 2 is the same.

4. Conclusion

This study analyzes the effect of building facade models on the level of illuminance in buildings that is directly related to openings in the building envelope. In the phinisi tower building there are three building facade models, namely horizontal, fertile and hyperbolic paraboloid strips, as well as comparing the level of illuminance in a room that does not use a facade but in the area of the building envelope covered by sunscreen. The results showed that the average level of illuminance rate in the space that uses a horizontal facade that is oriented towards the north is 299 lux and the space that leads to the south is 219 lux. Whereas the space using diagonal facade is 91 lux and the fertile facade model is 723 lux and the room without facade uses sunscreen of 458 lux.

Based on this description, it can be concluded that the building facade model influences the level of illuminance in space. The level of illuminance in space with horizontal stripes facades. lower than the vertical strip and hyperbolic paraboloid forms lower than both models. The level of illuminance in a room without a facade, but there are sunscreens on all sides of the building envelope, this space has a higher level of illuminance than the three forms mentioned above, this is because the distance between the sunscreen is greater than the distance between the vertical and horizontal shaped facades. And has wider openings between the three models. So on the second floor, this workspace has the highest level of illuminance. However, it is necessary to consider the negative effects of glare, thermal and brightness. This study analyzes based on the facade model on the Phinisi Tower building. Furthermore, this research can analyze the shape of building facade models by using echotech software to develop knowledge about the distribution of natural light to the facade models.

References

[1] Jakarta Regional Regulation Number 38 of 2012 *Concerning Green Building* article 1 p. 3.
[2] Jamala N 2017 *The Effect building Façade on Natural lighting (Case Study; Building Of Phinisi Tower UNM)* 7th ICMME Proceeding by AIP Publishing 978-0-7354-1499-0 (American Institute of Physics) Vol.183-No.02006.
[3] Sugijanto 1998 *Building in Indonesia with a Tropical Climate Humidity from Aspect Building Physics* Jakarta.
[4] Jamala N 2017 *The Effect of Building Façade model on light distribution (case study: Menara Phinisi Building)* Journal of Architecture and Built Environtment Vol. 44 No.2 ISSN 2338-7858, DOI: 10.9744/dimensi 44.2.149-154.
[5] Indonesian National Standard 03-6575-2001 *The Design of The System of Artificial Light in The Building*. [in Bahasa]
[6] Indonesian National Standard 03-2396-2001 *The Design of The System of Natural Light in The Building*. [in Bahasa]
[7] United Nations Environment Programme (UNEP), 2006, *Energy Efficiency Guide for Industry in Asia* (India: *www.energyefficiensiasia.org*), p. 1-43.
[8] Illuminating Engineering Society (IES) 1977 *Code for Interioir Lighting* (London: IES Lighting Handbook).
[9] Commision International De I'Eclaire (CIE) 1973 *Standardization of Luminance Distribution on Clear Skies* CIE Publication No. 22, TC- 4.2.
[10] Jamala N 2019 *Analysis of Natural Light in the Building* IOP Conf. Series: Material Science and Engineering Vol. 619 No.012024.
[11] Jamala. N, Annajma. N and Kusno. A, 2019, *Analysis of Daylight Distribution on Building Height and Space Depth* IOP Conf. Series: Material Science and Engineering vol.619 No. 012044.
[12] Jamala N. Rahim R and Mulyadi R 2019 *Analysis of Illuminance Level on Phinisi Tower Building*, Journal of Design + Built Vol. 12 No.1.
[13] Jamala N. Rahim R. Latief, S and Ramli H 2019 *Light Distribution Analysis on Buildings Located on the Coastal* IOP Conf. Series: Materials Science and Engineering Vol. 676 No. 012036.
[14] Rahim R et. al 2004 *Classification of Daylight and Radiation Data into Three Sky Condition by Cloud Ratio and Sunshine Duration* Journal Energy and Building Elsevier Vol. 36 issue 7 pages 660-666.
[15] Nakamura et. al 1985 *A Study on the Estimation of the Relative Frequency of Occurences of the Clear Sky the Intermediate Sky and the Overcast Sky in Japan* Journal of Light and Vision Environment 9 (2) 22-31.

IOP Conf. Series: Materials Science and Engineering **875** (2020) 012007 doi:10.1088/1757-899X/875/1/012007

Coconut Leaf Midribs as an Acoustical Panel – Feasibility Study through Impedance Tube Method

A Kusno[1]*, M T Ishak[1], R Rahim[1], B Hamzah[1], R Mulyadi[1], N Jamala[1], and R Anggraeni[1]

[1]Laboratory of Building Science and Technology, Department of Architecture, Faculty of Engineering, Hasanuddin University, Gowa, Indonesia

*Email: asniawaty@unhas.ac.id

Abstract. This research aims to determine sound absorption coefficients of absorber materials making from coconut leaf midribs. Absorber material has been made with different arrangement and thickness. Nine samples have been made which consists of 3 configurations and each configuration consists of 3 thicknesses. Absorption coefficients were measured by using impedance tube method. The experiment results show that coconut leaf midrib is up to standard for absorber material according to ISO 11654 with α over 0.15. The absorption coefficient obtained by the whole sample is in the range 0.02 to 0.65. Fluctuating values are affected by differences forms and thicknesses. Absorber materials that have been made are also compatible with marketed products.

1. Introduction

Research on sound absorbers made from acoustic materials is generally distinguished in natural materials and recycled materials. Some of them are like rice straw [1], sugar cane [2], sago midribs [3]. Fibrous sound absorbers are characterized by the presence of pores, where sound absorption is expressed in the coefficient of sound absorption as one of the characteristics in the design of acoustic materials. Porous sound absorbers can be produced by utilizing natural fiber waste into composite boards [4], where the sound absorption of the material depends on variable thickness, density, and fiber orientation.

In this study, the natural material chosen was midribs of coconut leaf, because there are still limited research references that choose the material as the object of research, especially acoustic material. Its hard porous outer texture allows this material to be used as an acoustic board material by measuring the absorption coefficient of sound so that it can be as a soundproofing material. The configurations and thickness of the material are used as research variables.

2. Research Method

The study uses an experimental methodology approach (experimental research) which will be divided into four stages of implementation. These stages are the process of getting raw materials, sample making, testing of sample characteristics and analysis of test results to the conclusion of the study.

Material Making:

1) Coconut leaves are separated using a knife
2) The material is then dried in direct sunlight

Figure 1. Separation of leaves and midrib using a knife

Figure 2. Coconut leaf midribs are dried for approximately four days

The sample making procedure conducted as follow:

1) Making an impedance tube pattern with a diameter of 10 cm to be used as a template for specimen raw material.
2) Raw materials are glued using PVAc adhesives according to templates with thicknesses varying from 25-50 to 75 mm.

Testing the characteristics of the next sample through the impedance tube method to get the absorption coefficient (α) of the sample with a variation of the configuration and thickness that have been determined.

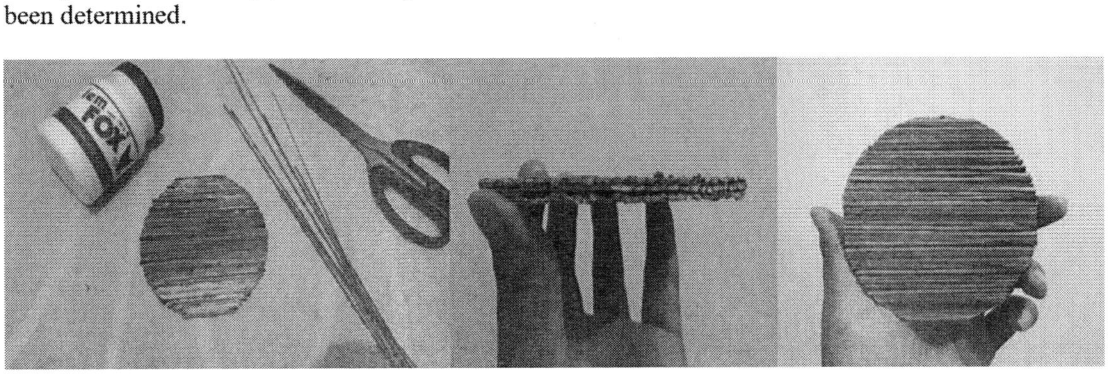

Figure 3. Coconut leaf midribs are glued together one by one using PVAc adhesives

The last method is a standard method commonly carried out by researchers or acousticians to get the absorption coefficient including the tube's sensor [5]-[9] This tube is intended to measure acoustic parameters by using a small sample placed on one end of the tube. A sound source emitted by a computer speaker is amplified and then forwarded into the impedance tube which is then captured by the two microphones. Results are recorded and processed using the PULSE labshop software version 16.1. Data of recorded sound waves in impedance tubes will produce absorption coefficient at frequencies from 100 Hz to 1600 Hz. The records then processed by *KaleidaGraph* and displayed in curves.

3. Measurement Setup

Figure 4. Samples of coconut leaf midribs

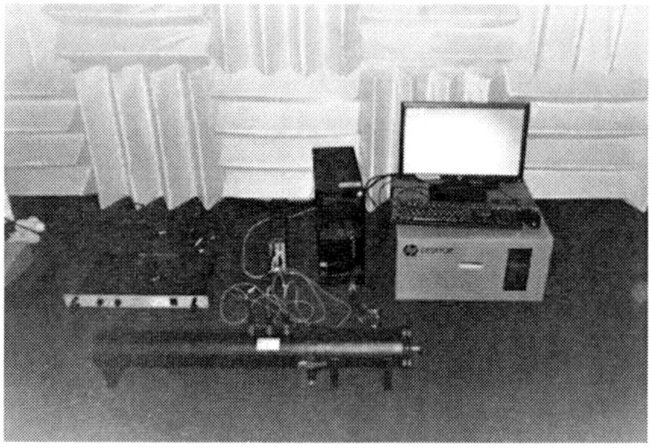

Figure 5. Measurement set up

Figure 6. Sample of coconut leaf midribs and acoustical gypsum with variety of thickness and configurations

Three types of coconut leaf midribs in the form of composite board thickness of 25-50-75 mm were measured. Composite board materials were cut in a circular shape with a diameter of 10 cm as the sample for measurement of absorption coefficient, see Figure 4. Similar measurements of acoustical gypsums with similar thickness are also performed for comparison purposes. Sample of coconut leaf midribs and acoustical gypsum with all variety of thickness and configurations that used in this study can be seen in Figure 6. The absorption coefficient of the coconut leaf midrib composite board is measured by the two microphone methods on impedance tube as shown in Figure 5.

4. Results and Discussions

The following figures of absorption coefficients according to configuration and thickness:

1) Parallel (A)
2) Cross (B)
3) Star (C)

As mentioned in the previous section, the normal sound absorption coefficients of the prepared Coconut leaf midribs (CM) for 25 mm 50 mm and 75 mm thick samples were measured in the impedance tube. Figures 7, 8 and 9 respectively show the measured results of CM with various configurations i.e. parallel, cross, and star, as well as 25 mm, 50 mm and 75 mm thicknesses. The curves in Figure 7 indicate the values of all CM samples of the same configurations (indicated as CM A is parallel) with different thickness (indicated as two numbers following the CM A such as 25, 50 and 75).

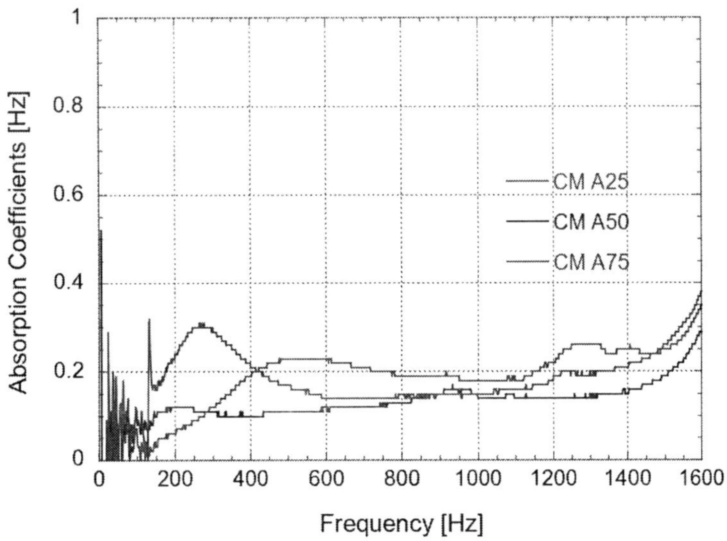

Figure 7. Absorption coefficients by variety of thicknesses of coconut midribs (CM) with parallel configurations

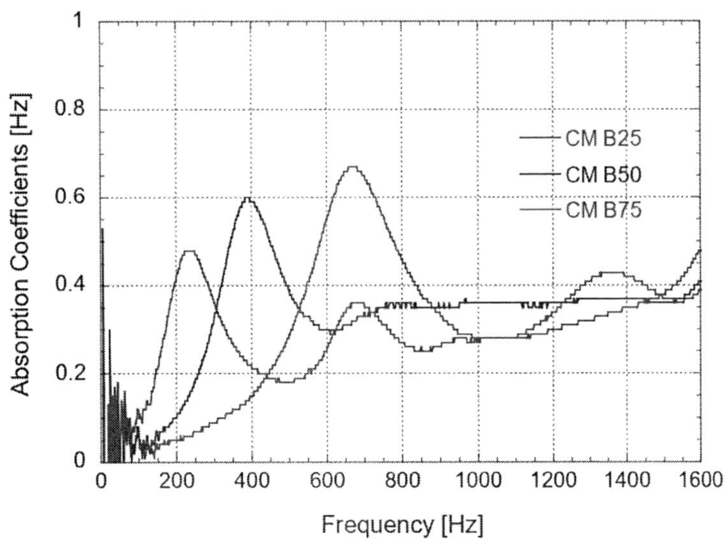

Figure 8. Absorption coefficients by variety of thicknesses of coconut midribs (CM) with cross configurations

In the Figure appears that all α of CM A25, 50 and 75 generally fluctuates below 0.4 with maximum value at frequency 1600 Hz. The curves quite stable at a middle frequency from about 500 Hz to 1000 Hz. Figure 8 shows all absorption coefficients of CM with cross configuration and thicknesses of 25 mm, 50 mm and 75 mm. All CM samples with cross configurations show the high fluctuation of absorption coefficient from low frequencies to frequencies of 1000 Hz, while absorption coefficients at high frequencies that are above 1000 Hz to 1600 are more stable.
Similarities in tendency in Figure 8 also occur in CM with the star configuration shown in Figure 9, where absorption coefficient fluctuations occur at frequencies below 1000 Hz and are more stable at frequencies above 1000 Hz.

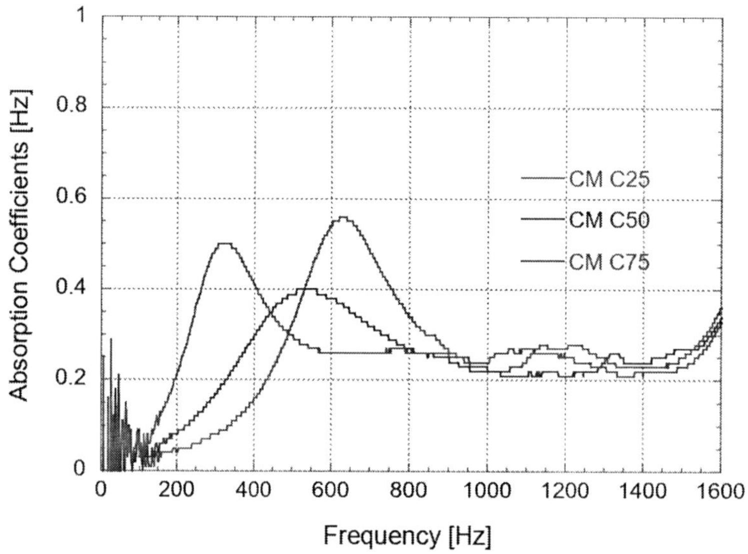

Figure 9. Absorption coefficients by variety of thicknesses of coconut midribs (CM) with cross configurations

From the overall graph presented it can be seen that the thicker the sample, the maximum sound absorption coefficient (α = 0.67) shifts to a lower frequency. This is because porous material can provide greater sound absorption when it is in a position where the sound wave particle velocity will reach a maximum value at a distance of λ/4, 3⁄4 λ and so on. The absorption coefficient has a slight decrease in the distance λ/2, λ and so on. The thicker the sample, the absorbed waves are waves that have a greater wavelength. Based on the theory in the following equation:

$$\alpha = \frac{ln_{I_0} - ln_I}{x}$$

The relationship between the thickness and the coefficient of sound absorption is that the greater the thickness of the absorbent medium, the value of the sound absorption coefficient gets smaller, and vice versa [1]. This theory is only fulfilled in samples with the form of configuration B which is evident in the comparison Graphs on Fig 7, where the sample with a thickness of 25 mm absorbs the most optimal sound followed by a sample with a thickness of 50 mm and the last is a sample with a thickness of 75 mm.
The same thing also happened in the sample configuration C wherein Fig. 8, the most optimal sound absorption sample sequence from the high to low-frequency range, which is 25 mm then 50 mm and the last is 75 mm. For the form of configuration A this theory is not fulfilled in the graph of Fig. 6

explaining sample A. Samples with a thickness of 25 mm of the most optimal sound absorption are in the highest frequency range but are followed by samples with a thickness of 75 mm and finally on a sample thickness of 50 mm.

For the form of configuration A, this theory is not fulfilled because it sees that thickness of 75 mm which absorbs optimally. The thickness of 75 mm is the greatest thickness in this study. Followed by a thickness of 25 mm and 50 mm. The form of configuration C also does not meet the existing theory. The results shown in the graphs show fluctuating values for the three graphs presented. On the whole graph that has been described, it can be seen that in configuration B at a thickness of 25 mm the optimal sound absorption coefficient is at the middle frequency as well as being the most optimal form of absorption coefficient compared to the other arrangement. As for the thickness of 50 mm and 75 mm the optimal noise absorption coefficient is at low frequency. So it can be confirmed that all forms of configuration (A, B and C) are suitable treatments because they meet the sound absorption standards that are good for use depending on the absorption coefficient desired.

The minimum absorption value of sound dampening material according to ISO (International Organization for Standardization) 11654 is 0.150. From the value of sound absorption coefficient based on thickness variation, it is obtained that the value of sound absorption coefficient is the largest and meets the standard that is at 25 mm thickness, whereas for variations in the composition of the material it is found that the crossing configuration with the highest sound absorption coefficient value.

Comparison with Gypsum Acoustic Materials

The following graph of Figures 10, 11 and 12 are the coconut leaf midribs samples compared with commercial gypsum acoustic boards according to midribs configuration.

Figure 10 shows the absorption coefficient of CM with variations in parallel configurations (A), cross (B) and star (C) and acoustic gypsum with 25 mm thickness. From the obtained graphic data of acoustic gypsum experimentations and the tested material with similar thicknesses, there were far differences of the resulted sound absorption values. Observation on both graphs showed that the sample of CM A25, CM B25 and CM C25 have increased in the middle frequencies while GA25 increased in high frequency ranges. The highest coefficient value achieved by the coconut midribs is 0.67 at a frequency of 650 Hz. For GA samples, the highest sound absorption coefficient achieved is 0.36 at frequency 1200 Hz. The difference between the highest absorption coefficient values of the two samples were 0.31.

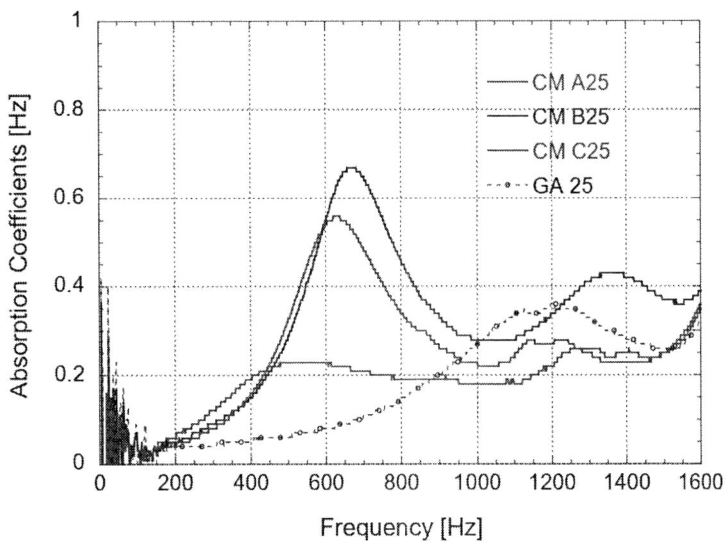

Figure 10. Comparison of absorption coefficients between coconut leaves midribs (CM) with variety of configurations and acoustica gypsum (GA) 25 mm thickness

The measurement results as seen in Figure 11 showed that the acoustic samples of gypsum material (GA50) and the tested material (CM A50, CM B50 and CM C50) with same thicknesses had difference of sound absorption coefficient values. Observation on both form of graphs, showed that the coconut midribs with configuration of A, B and C experience a fluctuated absorption coeficient values along the frequencies, while for GA50 tend to be stable and have an increasing tendency during high frequency ranges. The highest coefficient value achieved of the coconut midribs with all variety is 0.60 at a frequency of 40 Hz. While for GA50 samples the highest sound absorption coefficient achieved was 0.18 at 920 Hz frequency. The difference between the highest absorption values of the two samples was 0.42.

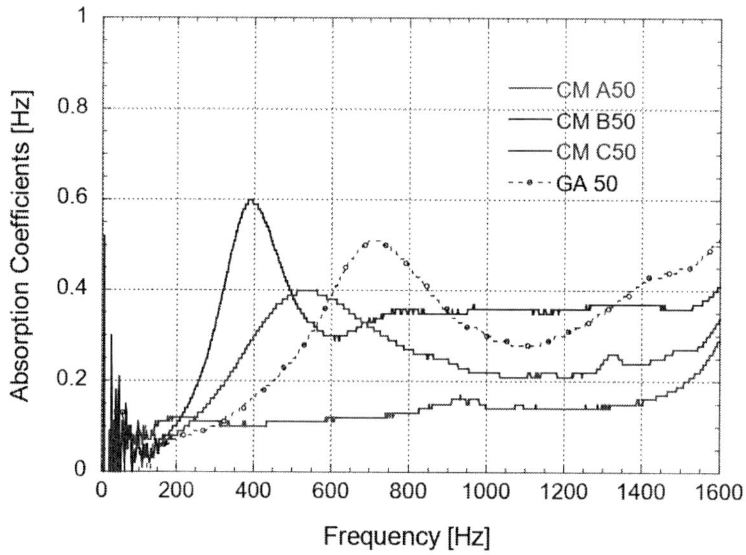

Figure 11. Comparison of absorption coefficients between coconut leaves midribs (CM) with variety of configurations and acoustical gypsum (GA) 50 mm thickness

Figure 12 shows the absorption coefficient of CM with variations in parallel configurations (A), cross (B) and star (C) and acoustic gypsum with 75 mm thickness. The test results showed that the acoustic samples of gypsum material and the tested material with same thicknesses had significance differences of sound absorption values. Observation on both form of graphs, showed that the coconut midribs samples of CM A75, CM B75 and CM C75 experience a fluctuated absorption coeficient values along the frequencies, while for GA75 tend to be stable and have an increasing tendency during high frequency ranges. The highest coefficient value achieved of the coconut midribs material is 0.49 at frequency 300 Hz. While for GA samples the highest sound absorption coefficient achieved was 0.29 at frequency 400 Hz. The difference between the highest absorption coefficient values of the two samples were 0.20.

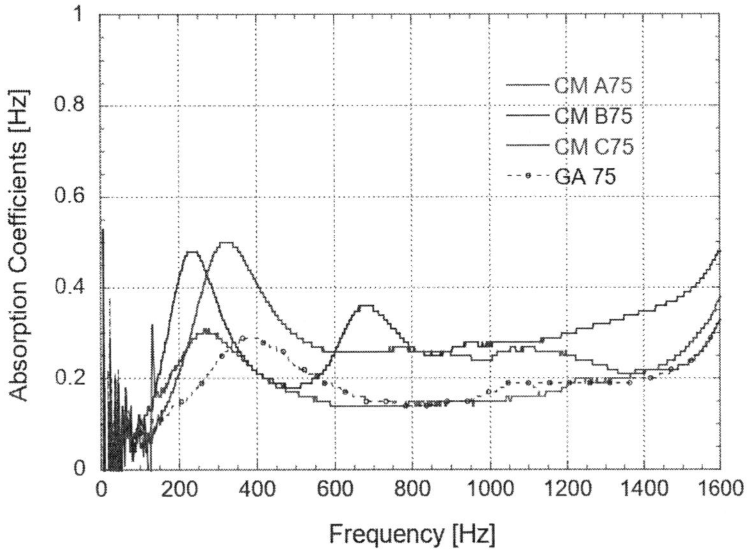

Figure 12. Comparison of absorption coefficients between coconut leaves midribs (CM) with variety of configurations and acoustical gypsum (GA) 75 mm thickness

5. Conclusions

The acoustic material from the coconut leaf midrib used in this study can be categorized as a good sound absorber because it has an absorption coefficient value that is better than the ISO 11654 standard, which is a minimum of 0.15. Based on the optimum absorption coefficient at each frequency (0 - 1600 Hz) with a frequency interval of 100 Hz at low-frequency intervals (0 - 500 Hz) the best results come from samples of crossing configurations of 50 mm and 75 mm thick. For intermediate frequency intervals (500 - 1000 Hz) the best results are obtained by crossing configurations of 25 mm thick. Whereas at high-frequency intervals (1000 - 1600 Hz) the best results from the crossing configurations sample are 50 mm. From the whole frequency interval, the cross-sample is the best sample compared to the other two forms. Sound absorption coefficient value based on variations in the thickness of acoustic material is obtained that the value of the greatest sound absorption coefficient and meets the standard that is at a thickness of 25 mm. The sound absorption coefficient value obtained by the whole sample is in the range of 0.02 to 0.65. This fluctuating value is obtained because of the influence of the midrib configuration and thickness of the test sample. Based on the value of the sound absorption coefficient obtained, the sample acoustic material from midrib coconut leaf material with PVAc adhesives is feasible to be an alternative acoustic material.

Acknowledgment
This research is supported by Laboratory Based Education (LBE) of Engineering Faculty-Hasanuddin University, 2019.

References

[1] C. E. Mediastika, "Acoustical quality of wall panels made of rice straw," *Dimens. (Jurnal Tek. Arsitektur)*, vol. 36, no. 1, pp. 20–27, 2009. [In Bahasa]
[2] A. Putra, Y. Abdullah, H. Efendy, W. M. Farid, M. R. Ayob, and M. S. Py, "Utilizing sugarcane wasted fibers as a sustainable acoustic absorber," *Procedia Eng.*, vol. 53, pp. 632–638, 2013.
[3] A. Kusno, S. Asriani, K. Kamil, and N. Jamala, "Preliminary study on Acoustical Characteristics

IOP Conf. Series: Materials Science and Engineering **875** (2020) 012007 doi:10.1088/1757-899X/875/1/012007

of Sago Midrib Composite," *IOP Conf. Ser. Mater. Sci. Eng.*, vol. 619, no. 1, 2019.

[4] M. Connelly and M. Hodgson, "Experimental investigation of the sound absorption characteristics of vegetated roofs," *Build. Environ.*, vol. 92, pp. 335–346, 2015.

[5] A. Kusno, K. Sakagami, T. Okuzono, M. Toyoda, T. Otsuru, R. Mulyadi, and K. Kamil, "A pilot study on the sound absorption characteristics of chicken feathers as an alternative sustainable acoustical material," *Sustain.*, vol. 11, no. 5, pp. 1–11, 2019.

[6] P. S. Liu, H. B. Qing, and H. L. Hou, "Primary investigation on sound absorption performance of highly porous titanium foams," *Mater. Des.*, vol. 85, pp. 275–281, 2015.

[7] J. F. Allard, R. Bourdier, and A. M. Bruneau, "The measurement of acoustic impedance at oblique incidence with two microphones," *J. Sound Vib.*, vol. 101, no. 1, pp. 130–132, Jul. 1985.

[8] A. Kusno, T. Otsuru, R. Tomiku, N. Okamoto, and N. Bin Che Din, "The effect of humidity on the stability of pressure - Velocity sensors: A measurement method of absorption characteristics of building materials by using ensemble averaging," *AIJ J. Technol. Des.*, vol. 19, no. 41, 2013.

[9] R. Bartolini, D. Borelli, and C. Schenone, "Flow Resistance and Sound Absorption of Polyester Fibre Material With Different Coatings," pp. 2–3.

IOP Conf. Series: Materials Science and Engineering **875** (2020) 012008 doi:10.1088/1757-899X/875/1/012008

The Structure System of Bugis Traditional House in Disasters Perspective

Hartawan[1*], V Sampebulu[1], N Junus[1], Imriyanti[1], P Musharl[1], D Kuddu[1], and I Alim[1]
[1]Architecture Department, Engineering Faculty, Hasanuddin University, Makassar.

*Email: hartawan@unhas.ac.id

Abstract. The existence of traditional house until now could be believed as proof of a good human protector. Bugis traditional house is one of them. The research problem is how Bugis behavior and building structure performance to struggle in flood. The goal is to reveal the performance of house structure and Bugis ancestor behavior to survive after disasters. The methods are logical argumentation to explain the architectural knowledge, and modeling simulation to explain building structural performance. The results are Bugis ancestor has been leave inheritance a good behavior in survive after disasters through they architectural space arrangement. They prepare space in a safe area at their house for food storage, so when disasters happen they still have food stock without waiting for aids. The structure system as the stilt house its find in safe for the occupant because of space for human living lifted more than two meters above the ground surface. The structural performance indicates the displacement in early dangerous occur when flood loads higher than human living space. The best knowledge can be adopted in this research is the availability of food stock at own house. The better construction method to apply in flood area is the stilt house.

1. Introduction

Disaster always followed by damage and victims. There are related to human and build environment. Building as the main places for human activities is the properties which expected as they protector. Therefore, human safety should be a guarantee when its harms happen in its.

Disaster is still a mystery in human life, coming uninvited and leaving uncommand. It's come in embody of earthquake, landslide, tornado, tsunami, tidal wave, flood, etc. They go anywhere by leave damages, sadness, harms, victims, and the death.

The great world threat disaster has found. This threat is noted at a map disaster and has been established called the ring of fire. The republic of Indonesia in the geographic location includes in these maps. The country which is listed in these maps has a serious threat of disasters, especially earthquake. Disasters like an earthquake, tsunami, tornado, flood, tide wave, etc., is the part of the life cycle of the world. It's an unsolved mystery even from sophisticated technology. The world is running across time parallel with disasters event. The latest even the largest impact of disasters. The earlier existence of the world seems to be the earliest disasters occur so that disasters are natural. The world and disasters are twin brothers. This event sometimes aggravated by human being lost control to manage development. The absences of human awareness in the activity can cause disasters which bring sadness and misery to them self. Its follows by disadvantages like properties damage, injures and some lost life. Disasters occur by two major ways, natural and human errors.

Indonesian National Board for Disaster Management record 3741 disasters in 2018/2019. This disaster includes floods, lands slide, tornado, tide wave, burning forest, tsunami, eruption, and the earthquake. In this period, South Sulawesi Province record 147 disasters caused 10 dead, 87 injures, and 699 building damages.

Built a house or building generally have done a proper way in the planning and design process. The exact calculation should be done to get a good performance and stability from any load include load caused by natural disasters. The basic rule of building design and planning is a part of the Indonesian constitution no 28 in 2002. The main consideration in this regulation is the priority of human safety.

Bugis traditional house has been proved their existence in a long period, and serve their function in a cross-generation. This house recorded in history as a cultural invention and its noted in *lontaraq* (Bugis script). This traditional house has been existed since *lontaraq* time, estimated by an expert in the same age with Chines ceramics period in 7 centuries. This means that the house estimated has been serving as the living space since as long as 1200 years ago. In this long time aged could be determined that it has been expert experiencing in some disasters. The ability of the house to survive in a long timeline believed that it has a specific construction method. This specific construction method known as local wisdom and this way should be revealed to develop as the new approach for a better life in the modern era.

The ability of Bugis traditional house survive in long period can be stated in simple ways that this building type trustworthy in facing disasters. The ingenuity of traditional building to withstand in the disaster has been proved by the expert. Traditional building in Kampung Panjalin West Java sowed they best ability to withstand from earthquake compared to the modern building. The construction method of this building has a similarity to Bugis traditional house as a stilt house.

South Sulawesi Province inhabit by people who used the stilt house as their daily life. In this prone, to disasters area, this house type spread until in remote region. This stilt house has been experiencing diversity disasters generally wind storm and flood. In this year, 2019 at the end of January the flood has been filled 6 districts. Flood disasters caused 9 people lost their life and 7 was unknown their existence. The same disasters come again at the end of June 2019. The east coastal region Wajo, Sidrap, and Soppeng submerge by rainwater. The quality and quantity of disasters seem to be bigger than before. So that, the hard effort should be forced to eliminate the negative effect of this tragedy.

This research is very important for setting preparing to handle an unpredictable disaster. The results will reveal the Bugis local wisdom which has been proved them self to survived in disasters for a long time. Bugis traditional house is the real evidence as a good protector to their occupant. It is consisting of two main entity in emboded, structure and architecture. Structurally has been show the ability to bring optimal protection and architecturally has been set in good room coordination for survive. What the secret thing hiding in that two factors will be reveal in this research.

2. Methods

This research will reveal the Bugis traditional house in facing disasters. There are two goals will be discovered, namely architectural local wisdom and house structural performance. Both of goals are discussing in disasters perspective. This research goal will bring to us the lesson from the past about the strategy to survive in disasters.

Logical argumentation method will describe the Bugis traditional house in facing disasters. Room function and composition will be analysed in the role of room arrangement in disasters perspective. The explanation composes in the logic description.

Structure performance in defended disasters especially flood will be explained by modeling and simulation methods. This method will be supported by computer software SAP 2000 version 14. Displacement node in structure caused by the flood described in the graphic. The role of this method in solving research problems needs additional setting to mirroring the real structure of Bugis traditional house. This effort caused by some of the joint in structure is not available in software default. Therefore, modification should be done to create similarities between model and reality. Structure models show at Figure 1 point (a).

Bugis traditional house has a specific structure system. Generally, this structure made from wood composed in two structure categories, primary structure, and secondary structure. The primary structure consists of beam and column to support all building as a frame. The primary structure covered by the secondary structure like floor, wall, and roof structure. Primary structure assembled similar to (H) shape in alphabet letter. Secondary structure installed to the primary structure by specific joints. The primary structure uses a fixed joint. Primary structure and secondary structure connect as pinned joints. Structure system of Bugis traditional house is displayed in detail in Figure 5.

Wood is the main materials for built Bugis traditional house. There for wood mechanic properties should be use for modeling and simulation. Wood classify as the orthotropic material class has specific mechanic properties. Wood species used for primary structure is *kayu keruing* (Dipterocarpus spp.) and wood species used for secondary structure is *kayu kapur* (Dryobalanops spp.). the first wood has 11.800 Mpa in elasticity modulus and the second wood has 11.000 Mpa in elasticity modulus. Both of modulus reached in green condition [1]. The quantity of load obtained from Indonesia regulation in load for building. There are 200 kg/m^2 for life load in housing function, wood weight is 1000 kg/m^3 [2].

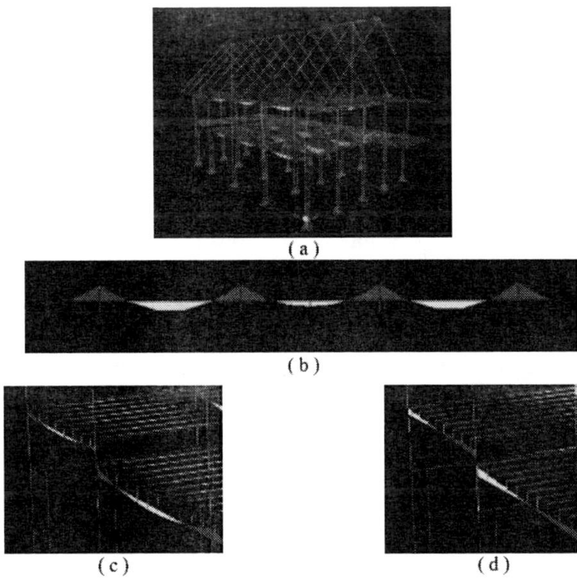

Figure 1. Modeling result of structure system of Bugis traditional house

The joint system which used in structure system of Bugis traditional house consists of 3 types; fixed joint, pinned joint, and freestanding joint. the fixed joint and the pinned joint are commonly known in modern mechanic science but the freestanding joint is rare. The last joint system has been used by Bugis ancient technician to build their house. The principle of a joint is delivered dead load only, moment and torsion are free effects to other bars. This joint principle applied at the bottom of the column and floor structure.

The column bottom sits on the wood pad or the stone pad. End of the column is not connected to earth directly. This end column allowed to move or shift position freely. Load distribution in this principle is delivered dead load only, moment and torsion not transferred in full ways.

The similarity of model and reality should be done and need to be modified is the floor structure. Representation of a model in the software setting has done by creating imitation connector between floor girder to the main beam. The connection this both of beam is free-standing joint or dead load joint.

The column of Bugis traditional structure can be divided into two-part, the upper part and bottom part. Bottom part column experienced load water pressure by the flood. The column in the general calculation principle is the specific task for bearing axial load only but not in this case. The column experience load pressure from the flood load. Load pressure replication for column needs modification. This modification setting by changed column to be a plane strain. This column set like a thin wall and force by 100 kN/m.

3. Results and Discussions

Research results and discussions describes in order of flood in general view. The explanation covered by architectural perspective, structural system perspective, and node displacement in structure caused by flood loads. The end of the discussion is closed by suggestion application for policy development in preventing harm in disasters. Complete discussion display below.

3.1. General View in Research Disaster

Research in disasters area is wide on the topic. Based on the trigger of disasters there are two main sources, disasters caused by the human errors by failed in control of their environment, and disasters caused by natural directly. The first disasters caused by a man easier solved than the second one. It is related to the victim's quantity and the spread impact area.

The research result becomes a new guide to the future generation in how to survive when facing disasters and How to prepare to survive after disasters. This effort will be explained by analyzed the building system structure performance when the flood hit the structure. The other way is exploring the Bugis ancestor traditional behavior to arrange the room function in their house.

Disaster events always bring a negative impact on people's lives. The effects of disasters will be felt individually and collectively by the community. Efforts to improve conditions and reduce negative effects provide considerable opportunities in the field of research. Research in terms of the disaster becomes an open-source, includes elements of management and technical elements. Many elements of management research have been carried out as in the following description, but the building structural behavior is less, especially exploration from the past in case of disasters.

Disaster management and efforts to improve community service affected by disasters have been carried out in Cuba [3]. Handling and disaster management are arranged in such a way about the duties and responsibilities of each interested party. The involvement of individuals, community groups, local, regional and state-level governments has a certain authority. These authorities include handling protection, emergency, rehabilitation, reconstruction, and long-term development. The environmental disaster recovery approach made in Cuba includes 5 main things, namely: development of the construction industry, housing development, systematic urban and regional planning, early warning and evacuation plans.

Disaster control management that generally occurs after a disaster event was developed. Research on the restoration of library services at Mzuzu University after experiencing fires in 2015. Data collected from 18 library officials and 64 students praised the recovery of post-disaster library services in terms of physical repair and the procurement of the latest library resources. The problems faced are limited reading space, limited internet access, and lack campus information services [4].

Construction activities towards community protection from disasters are interesting topics. This happens because the people in coastal areas will be faced with multiple problems [5]. Do you choose the beauty of the beach environment in this case environmental sustainability or priority safety? This difficult choice attracts attention to reveal a way out of 2 contradictory development choices. Facing the potential for disasters on the coast of Japan such as rising tides, tidal waves and tsunamis Japan has planned structures and coastal constructions such as breakwaters and retaining walls along the affected path. As a result of development activities such as this, there is often a conflict of interest between the community and coastal safety. The results of the study show that people who often visit the coast prefer ecosystem conservation compared to the construction of water retaining walls. People who have experienced disasters are more interested in the construction of water retaining construction. Finally, this research reveals that the community is more confident in the information about coastal management research findings. In this case, efforts to reduce the negative effects of water retaining construction on coastal ecosystems and consider other options such as disaster management based ecosystems.

Research on the effect of column end collisions on the floor surface on a flat plate structure. Structural damage at the top of the column surface which is connected directly to the floor plate can be avoided by several methods including [6]; a). plate thickness around the column needs to be carefully calculated. b) addition of reinforcement plates around the area around the column. c). additional elements in the form of a frame structure to distribute wall loads. d). calculation of slab resistance needs to be done in such a way that the placement of reinforcement assistance is obtained optimally to reduce load carrying capacity, plate span, or the addition of plate thickness, e). load that occurs during implementation, especially horizontal load and centralized load when the implementation process that uses fast time must always be calculated, f). the time of opening the mold must be determined, g). the scaffolding that is right in the area of the top of the column must be following the shape of the column and plate planned, h). must use a flat scaffold and a temporary diagonal at the time of execution.

Disaster prevention and recovery that occurs in the implementation of construction development requires efforts to avoid casualties and material losses [7]. Special agencies or institutions to deal with disasters in the implementation of construction are needed to plan disaster prevention which includes estimates of risks and hazards that may occur, crisis management, including management of companies that coordinate with government agencies. Estimates of disasters that may occur in the implementation of construction must be prevented.

A new method for simultaneously evaluating the advantages of temporary housing configuration. This method consists of 4 design models for analysis and optimizing decisions on housing after a disaster. The models include socio-economic, financing model, socioeconomic profit level, model optimization [8].

Disaster maps in Russia were developed for parties concerned to take appropriate decisions in disaster management and guidance efforts. Disaster management has been formulated statistically in terms of the magnitude of the disaster and consequently according to the risk element in an integrated manner [9].

Examine the community's temporary housing after experiencing a disaster. The location of the study was conducted in Chile and Peru [10]. This research provides an overview of the behavior of settler families in meeting their needs without the support of external elements. The results of the study indicate the importance of completing a temporary settlement design with instructions on how to safely change such temporary settlements. Likewise, instructions on the compatibility between their additional building structures are temporary and permanent structures.

The frequency of disasters tends to increase and seems to be a warning of increased vulnerability to danger. Disaster events are directly proportional to the increase in trauma for those who escaped the disaster [11]. The increase in people's inability to increase income after the earthquake occurred in Nepal after the earthquake. Continuity of the inability of the community needs to be resolved as a strategy to deal with natural disasters in the future. Post-disaster recovery is needed by medical staff and physical therapists.

Disaster management after the 2012 tornado disaster in southern Indiana. The selection of infrastructure after the occurrence of the disaster could only be implemented about 1 year after the event, namely in 2013. Factors that affected the rapid pace of post-disaster treatment were found in this study. The delay in handling disasters is caused by the quality of disaster, the greater the disaster, the slower the recovery. Delay in the information received by insurance companies, the number of recovery services. The speed of recovery services is determined by the government's concern in terms of the proximity of the government to affected communities, the strength of networks between communities, the presence of assistance from adjacent areas, the presence of previous experiences in the same disaster [12]. Disaster events that are related to various aspects have led to various types of research. The research is basically to find the best solution so that the negative consequences of disasters can be avoided or at least minimized.

Research develop in disasters is varied as described above. The scope includes management and optimal solution to minimize the negative effect of construction erection. That recent research differs to this one in case of the theme. This research present is accomplishment from the early research. The uniqueness of this research is the focus in learning from Buginese society in handle disaster. The research state of the art is the ingenuity of Bugis traditional house especially in the structural system and the architectural space coordination in disasters perspective. Learning from the past to develop a new better life is the specific discussion in this explanation

3.2. General Disasters in South Sulawesi

South Sulawesi Province has got mercy by its location in relatively safe from great nature disasters threat. Generally, this region hit by flood and tornado just low in frequency and impact. This disaster occurs periodically in a year. It's covered a wide area. Figure 2 below sawed the recent spreading flood in Sengkang City Wajo Regency The other great natural disasters like earthquake, eruption, landslide, mostly rare occurred.

Although this region relatively safe from great disaster, the preparation and planning should be equipped by a good strategy to minimize victims. Rarely occur does not mean free from disasters, there is no area in this world free from disaster. Flood disaster caused by heavy rainfalls occur two times in a year. In December to January occur in at the west area of the peninsula and in Mei June occur at the east area of the peninsula.

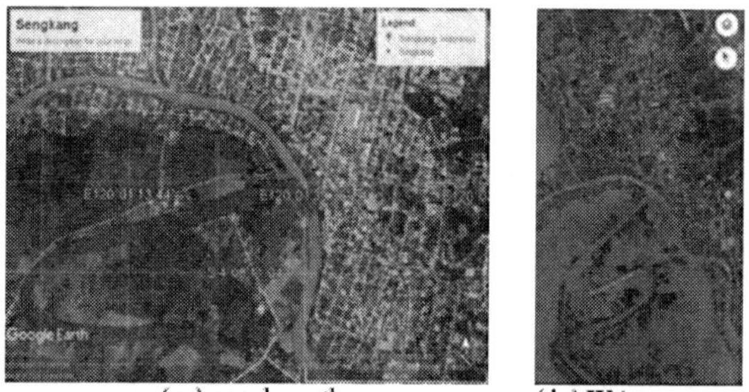

(a) google earth (b) WA group

Figure 2. Sengkang city aerial view in custom (a) and flooding situation (b)

The flood impact on the settlement caused the house inundated in the undefined certain time range. This condition forced a new adaptation for their society to continue their daily activity. Accessibility and distribution of their basic needs automatically disrupted.

Figure 3. Bugis traditional house settlement at the river bank(a and b) and flooding (c and d)

Figure 3 point (a and b) showed the Bugis settlement who lives in the river bank or waterfront, point (c and d) showed Bugis traditional house in flood situation. That houses in rainy seasons will be covered by the water as shown in the figures. The Bugis people in this situation still feel safe and doing their daily activity in the normal way. They can protect their properties and prepared their food without hesitant. This condition at a glance view, look like in a dangerous situation but not by Bugis society. Environment covered by water can be predicted in simple ways if the landed house built in it. It is

IOP Conf. Series: Materials Science and Engineering **875** (2020) 012008 doi:10.1088/1757-899X/875/1/012008

ensured will bring a great disaster. Bugis ancestor has brought us the wise lesson to live harmony even in a dangerous situation.

3.3. Architectural Space Arrangement in Disasters Perspective

Bugis traditional house is shown in the picture below. The figure is the cross and longitudinal section of space coordination and function in a vertical manner. Figure 4 point (a) display the vertical room function composition and point (b) display the rear part of the house in room function, also in a vertical manner.

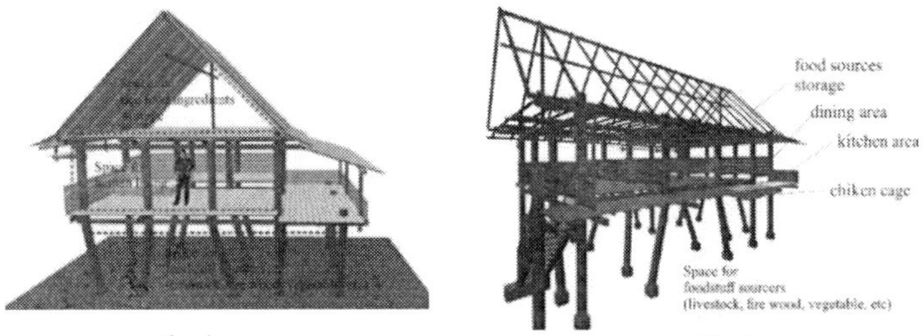

(a) (b)

Figure 4. Space and room function arrangement of Bugis traditional house

Architecturally, Bugis traditional house commonly known as show as the above picture in room function and it's the arrangement. Vertically divided in three-part; lower part floor, middle part floor, and upper part floor. The lower part floor use as life stock storage, food supplies materials, like banana, coconut, firewood, etc. the middle part floor use as daily living space. The upper part floor use as storage of rice. Rice in Bugis believes as a representation of God to maintain human being life.

The sustainable living concept expressed from this good arrangement. Main source food at the top mixed with the food source from below at the middle space. Kitchen and dining area placed at the rear part of the house. Under the dining room placed the chicken cage. The symbiosis mutualism among creature well organizes in this house. Leftover food from dining falls directly to the cage area, so no waste food left.

Bugis society to sustain their life done by preparing a stock of rice as the main food sources. These materials put at the best space. Its save at the highest space at the house in the attic. They have two places for store the food, at the attic called *rakkeang* and on the main floor called *paberesseng*. *Rakkeang* is for raw material of rice and *pabaresseng* is for rice ready for the cook.

A side dish of food and firewood prepare at the lower part floor. This behavior is the most important requirement to prevent life so that Bugis maintain the sustain this stock. The stock should be available in certain range time, commonly adjusted to their periodic livelihood. Farmer keep store the stock for the next year.

The architectural room arrangement described above brings to us a good lesson in handling disasters. Its contained wisdom value. The availability of enough food source is a valuable thing after disasters. This condition guarantees the people survive by them self without waiting for help or first aids. The common phenomenon after the disaster recently is the lack of food stock sources. Lack of food stock sources after disasters can be caused by famine. Famine is the most horrifying than the disasters itself. The lack of stock food reserve in the house which in disasters experiencing will be caused the community depends on the availability of helping aid. The person who saves from disasters need the basic thing in life, they need to eat first.

Description above bring to our generation the best learning in architecture principle, we called architecture good preparation after disasters. A sequence arrangement room of Bugis traditional house become a mirror from the past in surviving to struggle against disasters. Application this architecture concept automatically sett the people to be self-help to survive after disasters.

3.4. The Structural system in disasters perspective

The specific identity of Bugis traditional house in structure system is as a stilt house. This house was constructed by lifted floor elevation at a certain high level above the ground for human living space. It is high approximately three-meter high. This situation allows occupant got some benefit, such as; safe from reptile attack; free from the spreading of contamination germs. Release from the negative effect of earth temperature circulation; and safe from floodwater in a certain condition.

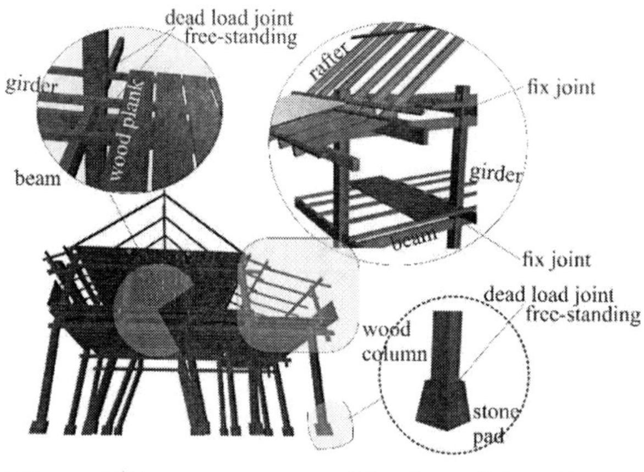

Figure 5. Structure system of Bugis traditional house

Learning from Bugis structure system in disaster perspective bring to us some good knowledge such as; the easy way for repaired part of the house after the disaster; the common damage part is the secondary structure, and settlement of ground is not affected foundation. Explanation of the above value is described below.

Damage part of the house structure caused by disaster allowed repaired easily. The common structure part damage caused by the disaster is the secondary structure. Secondary structure made from materials which low in quality and small in dimension compere to the primary structure. Roof and wall as a secondary structure are vulnerably destroyed by the wind. This part is easy for repair because of the joint and material dimension. The joint between the primary structure and the secondary structure are setting by pin joint and materials made from small dimension.

The end of the column is not tight to the ground in a fixed way. This condition allows the column hanging without destroying the structure. The ground settlement below the end of the column is possible to repair by adding existing elevation with the fill of earth. This principle is not only good for land settlement but also for an earthquake. The column of the house is unattached to the earth so that it's free for vertical and horizontal movement. The number of outside force that could give structure displacement when the force is more than the total weight of the house.

Figure 6 point (a) showed the column is hanging caused by land settlement at the column base point of the house site. This situation has not affected a disturbance in the structure. Figure in point (b) showed the people who replaced the roof sheet after hitting by the tornado. Replacing roof sheet at the Bugis house is an easy way because space below the roof which used as rice storage is the floor in strong enough to bearing large load. This floor is using as the place to stand for sheet roof assemble. It is no need for the special strut to prepare.

(a) (b)

Figure 6. Column hanging and roof sheet repaired

3.5. Structural Displacement in Disasters Perspective

Node displacement of the Bugis traditional house structure, analyze by modeling and simulation methods support by SAP 2000 version 14 software. The testing model has done by applied flood load in the wide direction of the house. This side used as load direction because of it wider than the front side. The Wide side of the house is more vulnerable than the front side because of the size. The side area is broad in size than the front area. Figure below shows the model of Bugis traditional structure in reality and software model. The upper left figure point (a) is the real Bugis house performance. The upper right figure point (b) is the Bugis primary structure. The lower figure point (c) is the structure model by software by load line, and the right figure point (d) is the structure section performance displacement in load disaster reach the half high upper column.

The real model of Bugis traditional performance and its structure system are used as the base of model in software. This model treated by load force and dead load to explain the structure performance in handling force. Load focus in this research is dead and flood. This load combine to make sure the structure behaviour same to the natural situation when experiencing real disasters.

Software output simplicity an easy way to explain Bugis traditional structure displacement caused by flood loads. Node measurement focused at a certain point of structure vertically. Point A is the end of the column bottom, point B is the lower connection structure of the column and the first main beam from below. Point C is the second beam and column connection from below. Pont D is beam and connection over the main floor level. Point E is the upper joint between the column and beam. Point F is the fringe of the roof. Point G is the top roof.

Node displacement performance when the structure hit by flood load is display in the graphic. Structure performance elucidates in three models of flood load. The first flood load type is flood load applied at a half lower column. The second flood load type applied in full length of the lower column. The third type load applied at a half-length of the upper column. The graphic line I show structure displacement caused by the first load type. Graphic line II shows structure displacement caused by the second load type. Graphic line III show structure displacement caused by the third load type.

Structure displacement according to the I line and the II line explains the structure node moves in save way. Its displace in range of 0.19 cm at the top of the roof. This situation is not enough power to bring damage situation to the house.

The early dangerous situation detected when the flood load is reached the half-high of upper column. Node displacement in this situation according to the III line is. It is explained the movement of structure node will be happening in reverse direction. Movement from point C in positive (+) direction change to the negative (-) direction at point D. Point C and D move in a different direction as far as 2.42 cm. The contrary direction movement effect is caused by flood load crash to the wide wall area on the upper floor.

IOP Conf. Series: Materials Science and Engineering **875** (2020) 012008 doi:10.1088/1757-899X/875/1/012008

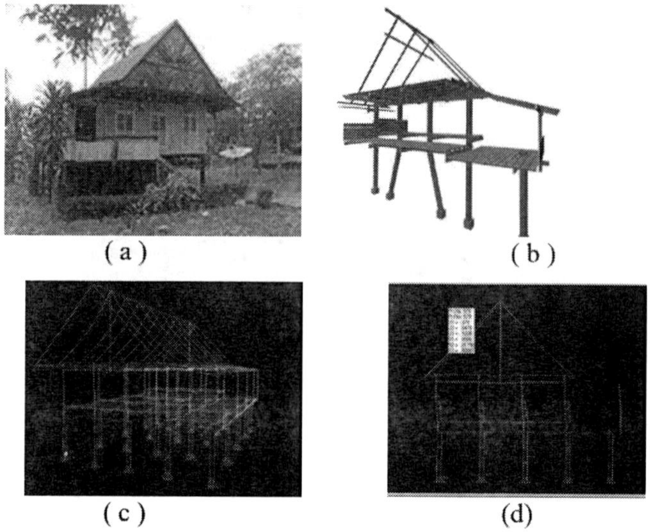

Figure 7. Software model of Bugis traditional house

House structure in dangerous condition occurred when the water high level reaches over the floor level at a half high of upper column. In this situation, space is inconveniently for living. The man who stays at the house should leave the house to find the new save places, caused by the floor has been immerse by the water. Therefore, the human safety in Bugis traditional house has been done in a good manner.

Learning from Bugis traditional structure system safely manner for a better life could be explained in three argumentations. Such as the joint principle, the structure performance in a disaster situation and the structure elevation.

Focused on connection method, Bugis ancestor simply used joint principle. Their use joint system in a proper way. This joint called freestanding joint or dead load joint. This joint has a good implication to structure system because force distribution is limited. Moment and torsion are not transferred in this joint, the only load to transfer is the dead load. The floor structure as the primary structure constructs in this manner. The primary structure movement experience not affected to the floor structure in full power. The floor is kept quiet, that means comfortable to stay still exist although primary structure moved

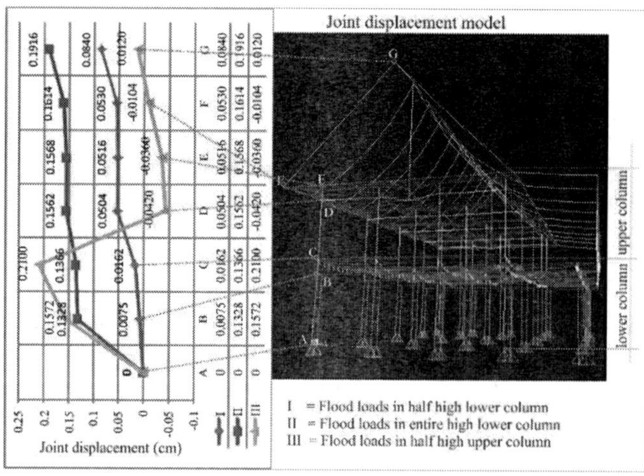

Figure 8. Joint displacement of Bugis traditional house caused by food load

The lesson from Bugis house structure system performance indicated that the dangerous situation of the structure occurs when flood level high reach at above the floor level. When the floodwater level high reaches this level, it is simply understood that space is not possible and unsatisfied for stay. This phenomenon seems to be an early warning system to the occupant to leave the house. It is like an announcement that the house in a dangerous situation. Time and opportunity to leave the house for a safe self is good enough.

The lesson from an elevated floor of Bugis structure system as a stilt house that using two-floor level structure can be explained. The first floor for living area and upper floor at the under roof (attic room) for rice storage function space for temporary evacuation. Floor for daily activity level house located at $\pm$3 meters high above the ground. The floor under roof or attic located $\pm$ 5.5 meters above the ground. The elevation at the high position like this simple known exist in save from flood threat. Figure 3 point (c&d) is the evidence of Bugis traditional house as a stilt house safer than the landed house from flood.

3.6. Policy Development Suggestion

Based on the explanation in the discussion of Bugis traditional house, the policy suggestion for helping people to save from disaster can be arranged. Suggestion can be a divide into two categories, land use policy, and community awareness. Land use policy suggestion to create proper regulation related to the best building structure in water immerse region. Structure system classification suggested applied proper to the hierarchy of water immerse in the region. the immerse region can divide into three zones; such as the high land in a free immerse zone; the middle-high in periodically immerse zone; and the low land prone immerse zone. Free immerse zone suitable for the landed house structure. Periodic immerse zone suitable for semi stilt house structure, and prone immerse zone suitable for the stilt house structure. Implementation structure system in proper to the topography as like this suggestion allow settlement will be free from load disaster or at least minimize damage. Figure 2 point (b) is the evidence of spreading flood water following the land topography surrounding the river.

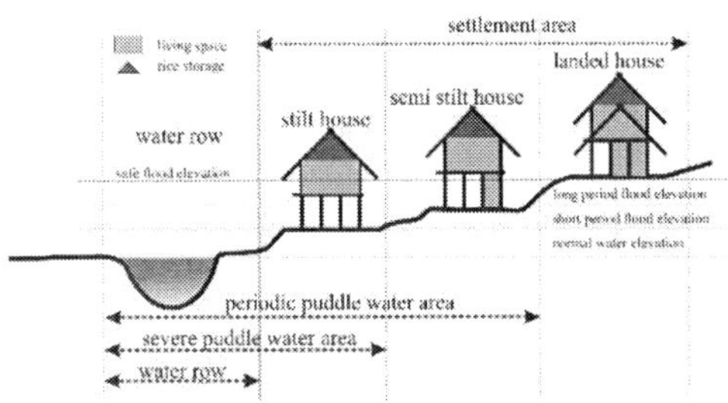

Figure 9. Policy suggestion for building structure and land use.

Community awareness related to living style. Daily habit to manage a good life is better to follow Bugis tradition to sustain their life. Each home should have a food stock reserve and the component of making food. This stock should be available in a certain range time according to livelihood income. The main food source should be guaranteed the existence is rice, ingredient, a side dish of food materials and firewood. The availability of this stock will be valuable when unpredictable disaster happened. The post-disaster phenomenon which experience recently is how to survive after disaster become the horrifying thing. The people who escape from disaster will require food or eat to sustain their life. The imaginable condition can be depicted, what happens if disasters occur in a remote area without food stock?. Infrastructure facility impossible allowed helping teams to delivered assistance quickly. The assistance late can be caused by the death toll by famine, not by the disasters itself. The death toll caused by famine is more horrible than the death cause of disasters.

4. Conclusions

Ancient society has been exceeding their generation by filling it with culture and technology invention according to the ability level at the era. This masterpiece is the base of the development of knowledge invention in the modern era. The technology invention in the past is full of wise meaning. Structure system of Bugis traditional house has been proof they ingenuity in high quality to serve human live inconvenient way. It is filled with pure value to maintain comfort and safety event from disasters. This value is applied in architecture arrangement and structure system. Architecturally, this house is setting the room as learning to survive after disasters by equipped food storage in the safest place at the highest elevation. Structure system function as the early warning system to safe live through their performance in displacement caused by flood load. Structure in an emergency when the floodwater rises at a half high column at the upper floor level. Frame structure system provides living space at a high elevation. This space constructed about 3 meters high above the ground surface. This condition brings some benefit to the occupant especially in free from flood disasters. The lower part of the house constructed from a slender column which allows water flow in freeway without obstruction. Free space at the bottom house is not only for water flow but also for flood protectors. Specifically, Bugis ancestor has been teaching to the next generation in how to prepare and how to survive in disasters. Room coordination in architecture arrangement of Bugis traditional house which prepares a special place for food stock is the fundamental principle in surviving from disasters. The application of free-standing joint allows floor structure not affected by the main structure displacement. Structure system design in consideration of easy maintenance is good learning.

Acknowledgments

This research was carried out with funding support from the policy of the head of Engineering Faculty Hasanuddin University (Unhas) through a program called Laboratory Based Education (LBE). For this reason, the authors would like to thank and hope to continue this program in a sustainable manner as Unhas Engineering Faculty identity.

References

[1] Usda 1999. Wood Handbook, Wood As An Engineering Material.

[2] ...., (1987). Indonesia Load Guide Line for Building and Housing. Public Work Department. Yayasan Badan Penerbit P.U.

[3] G. Lizarralde., 2015, A systems approach to resilience in the built environment: the case of Cuba', *Disasters*, 39, pp. s76–s95.

[4] W.D. Chawinga., and F. Majawa., 2018, An Assessment of Mzuzu University Library after a Fire Disaster, *African Journal of Library, Archives & Information Science*, 28(2), pp. 183–194. Available at: http://e-resources.perpusnas.go.id:2124/ login.aspx? direct=true&db= lih&AN=134012354&site=eds-live (Accessed: 5 April 2019).

[5] K. Imamura., K. Takano, K, N. Mori., T. Nakashizuka., and S. Managi., 2016, Attitudes toward disaster-prevention risk in Japanese coastal areas: analysis of civil preference, *Natural Hazards*, vol. 82, no. 1, pp. 209–226, viewed 5 April 2019, <http://e-resources.perpusnas.go.id:2124/login.aspx? direct=true&db=rkh&AN=114308726& site= ehost-live>.

[6] S. Kobielak., E. Hutnik., and Z. Zamiar.. 2016, Construction Disasters Caused by Punching in the Support Zone of Reinforced Concrete Slab-And-Column Ceilings as Well as Means of Their Repair and Reinforcement, *Journal of Science of the Gen. Tadeusz Kosiuszko Military Academy of Land Forces*, 179(1), pp. 151–167.

[7] W.W. Dukes., S.G. Sudler., and G.T. Culbreath., 2010, Management of Catastrophic Industrial or Construction Disasters: From Prevention to Preservation, *FDCC Quarterly*, 61(1), pp. 38–67. Available at: http://e-resources.perpusnas.go.id:2124 /login.aspx?direct =true&db= tsh&AN= 60588517&site=eds-live (Accessed: 5 April 2019).

[8] O. El-Anwar., 2013, Maximising the net social benefit of the construction of post-disaster alternative housing projects, *Disasters*, 37(3), pp. 489–515.
[9] V. Osipov.. *et al.* 2017, Methodology of natural risk assessment in Russia, *Natural Hazards*, 88, pp. 17–41.
[10] E. Wagemann., 2017, Need for adaptation: transformation of temporary houses, *Disasters*, 41(4), pp. 828–851.
[11] M.D. Landry., 2016, The 2015 Nepal Earthquake(s): Lessons Learned From the Disability and Rehabilitation Sector's Preparation for, and Response to, Natural Disasters, *Physical Therapy*, 96(11), p. 1714.
[12] A.M. Sadri., 2018, The role of social capital, personal networks, and emergency responders in post-disaster recovery and resilience: a study of rural communities in Indiana, *Natural Hazards*, 90(3), pp. 1377–1406.

IOP Conf. Series: Materials Science and Engineering **875** (2020) 012009 doi:10.1088/1757-899X/875/1/012009

Post Occupancy Evaluation in the Senate Room, Faculty of Engineering, Hasanuddin University, Gowa

R A Ishak[1]*, T Martosenjoyo[1], S Beddu[1], and S Latif[1]

[1]Architecture Design Laboratory, Engineering Faculty, Hasanuddin University, Makassar, Indonesia

*Email: rahmi1403@gmail.com

Abstract. In higher education institutions, an effective and efficient communication is an important competitive advantage. Most of the communication is carried out in various meeting room environments. In terms of creating the right space for meetings, flexibility, easy use, and supporting university productivity, meeting rooms must be specially designed according to the architectural standard of space requirements. The Senate Room in the Faculty of Engineering, of Hasanuddin University, is one of the many rooms that are designed especially for meeting activities within the faculty. It is located on the 2nd floor of the Center of Technology (CoT) Building, with a capacity of 64 seats, and an area of 180 m^2. This room is arranged in a two-way theater or proscenium type, with lined tables and flat floors. This study aims to: 1) Identify the characteristics of the elements that shape and fill the room; 2) Analyze the effect of technical space planning on the functional aspects of users, and 3) Evaluate spatial design performance in terms of functional aspects. The research method used is the Post Occupancy Evaluation with qualitative data analysis. The population in this study is the users of the room as the subjects 0f space as the subject of research and the Senate Room of Engineering Faculty as the object of research. The results showed a discrepancy between the standards and needs of users both functional and technical aspects. The results of this study are expected to contribute practical insights to architectural design, specifically related to the creation of meeting room interiors.

1. Introduction

The meeting room is a place that generally functions as a gathering place to discuss a matter. It is a requirement for every institution to have a meeting room facility, as a place to gather and express opinions in making decisions for the common good. Basic implementation of restructuring for the meeting room is the Minister of Health regulation, Number 288 / Menkes / SK / III / 2003 on Guidelines for Health Services and General Building [1], as follows: meeting room as office facilities otherwise qualified environmental health if it meets the needs of physiological and psychological, and also user comfort that does not cause injury if used for long periods.

Engineering Faculty Senate Room as the meeting place for representatives of the lecturer who serves as one of the organs of the Faculty. The duties and authority of the senate is to give consideration and supervision related to academic norms, the academic community code of ethics, and other academic provisions. Discussion and meeting activities often require a long time to sit and discuss, so it takes space and furniture that can support the physical and psychological needs of users.

Post-Occupancy Evaluation (POE) is the process of understanding how well buildings meet the needs of clients and occupants. POE can also overcome complex cultural problems such as identity, atmosphere, and ownership [2]. In the field of architecture, POE has been widely used in design studies, building systems, and environmental behavior. This Post Occupation Evaluation is an evaluation of the effectiveness of the design after occupied. POE is intended to look for facts, not to find fault with the results of design. Research findings are used as input for higher quality design results, both in the planning, programming, design, construction, and occupancy building stages [3].

According to Hay R [2], the building has a good performance by determining what it looks like, in a series of quality targets. Ideally, the POE principle is SMART: specific, measurable, achievable, realistic and time-bound. Three factors are inseparable from POE, namely; 1) The effectiveness of the procurement process itself; 2) Performance of fabric detail and the built construction; 3) The effectiveness of building operations, in this case, the 'user' activity needs to be defined.

Post-Occupancy Evaluation in the Senate Room Faculty of Engineering is intended to systematically evaluate the performance of space, based on technical and functional aspects after space occupied this research discusses the technical aspects of space; including spatial elements (i.e, floor, ceilings, walls), space conditioning elements (lighting, air conditioning, acoustics, circulation, and spatial planning); and functional aspects include elements of furniture, as well as user comfort related to circulation and space planning. The purpose of this study is to provide information, understanding, and knowledge related to the design and application of meeting room that can support the activities in the Senate Room, Faculty of Engineering, Hasanuddin University This study is also intended to provide an overview of the room performance related to technical, and functional aspects.

2. Literature Review

2.1. Meeting Room Type

The meeting room layout in the Senate Room is a theater type. This type is used for large capacity meetings, similar to a classroom type, but in the middle is provided an open circulation that separates participants who sit on the left and right side. Usually a large room applied to accommodate each part. This room type provides desk facilities, so it can be used for material with many notes. Because this type can only be applied in large places, it is therefore necessary to have a sound system, lighting and additional screens to make it easy for participants who sit on the top or back seat to keep up with the material of the speaker. Theater design is usually limited and even minimal interaction of participants, therefore only focused on the speaker and the participants only become passive listeners.

Table 1. Type of meeting room [4, 5]

Type	Features	Picture
U-Shape 	• Chair pattern shaped like the U letter • Usually used for medium meetings with a capacity of up to 20 people that makes it easy for the speaker to interact and also communicate with each participant. • Seating around three sides of the room – good for presentations from front. Square layout conducive to discussion. Presentation space is the middle of the room.	
Boardroom 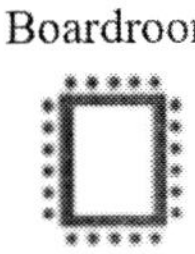	• Usually uses a long table in the middle of the participants sitting face to face. • Can reach participants slightly to medium, allows participants to concentrate more fully to pay attention to the speaker. • Centrally located table. Classic layout ideal for debate and discussion. Popular for smaller meetings	

Type	Features	Picture
Classroom 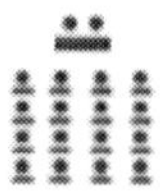	• Arranged with tables and chairs lined back with one main focus facing the speaker. • Accommodate more and maximum participants, but cannot cover every participant present. • Used to present to small to medium groups. Delegates (in ones or twos) have own workspace. Ideal for testing and individual training.	
Theater 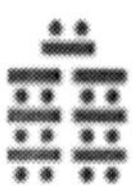	• Usually applied in a large room to accommodate each part (left and right). • Used for large capacity meetings, but usually limited and even minimal interaction of participants. • Used for product launches, presentations, displays. Used to present to large numbers of delegates. Allows for optimal room occupancy.	
Banquet 	• Divide each division into one round table, divide the discussion team or separate by section or position. • Functioned as a meeting and lunch, generally requires sufficient space and more space consuming. • All delegates are facing front centre on round tables. Large space in the middle of the room. Ideal for small-group work	

2.2. *Furniture Layout*

An effective furniture layout must meet functional and visual criteria. Functional criteria evaluate how well the layout supports human activities that occur in space, such as conversation, rest, or movement. Visual criteria concern the perception of layout as a visual composition [6]. The study of statistical distribution of human physical characteristics, such as body size and shape, is known as anthropometrics. This study sets guidelines for the distance needed around the object and for the exact distance and angle between the objects [7].

Other functional criteria come from the study of environmental psychology, which examines how room layout influences human activities [8]. Activities that are affected by furniture layout are conversation and circulation. Conversation is greatly influenced by the placement of chairs, whose settings must support comfortable eye contact and normal volume. Circulation requires sufficient space to walk comfortably to all parts of the room. Each piece of furniture requires a range of space around it to be accessible and functional. An effective furniture layout must support circulation through the room and access to all furniture. Human physiology influences how objects should be positioned in pairs, such as chairs and tables of the same type.

Visual criteria concern the perception of the layout of furniture as a visual composition [9]. The main visual rules used by interior designers are visual balance, alignment, and dominant points of emphasis. According to Arnheim '[9]' the best known principle of visual composition is visual balance. The principle is to place the average visual weight distribution at the center of the composition. The visual weight of an element is its perceptual salience. The general assumption is that larger objects carry more visual weight. Because of the arrangement of three-dimensional furniture, visual balance refers to the appearance of the arrangement from various points of view. In arranging furniture, the alignment mainly concerns the orientation of furniture items relative to each other and to the walls of the room.

Another principle of visual presentation that plays a role in interior design is emphasis / accent [10]. Generally the accent is desirable to have a dominant focal point in the interior, so that the eye can rest without competition for visual attention. In the interior of a meeting room, the focal point is often a protruding object such as a monitor screen.

2.3. Post-Occupancy Evaluation (POE)

According to Preiser [11], post-occupancy evaluation is the process of evaluating buildings in a systematic and meticulous manner, after they have been built and have been used for some time. The focus of the POE is on usage and usage requirements, thus providing in-depth knowledge of the consequences of decisions from the past and from the results of building performance. This knowledge becomes a good basis for creating better buildings in the future. Furthermore Preiser [11] divides three elements of building performance that can be identified and applied to POE, including:

- *Technical elements* consist of aspects of health, safety and security of buildings. These elements include; fire hazard, building structure, sanitation and ventilation, electricity, building walls, roofs, interior finishes, lighting, and acoustics.
- *Functional elements*, are the ability of residents to operate buildings effectively and efficiently. This element is related to human factors that affect physical dimensions, and configuration of space and furniture, as well as communication factors and user activity flow, ease of doing activities, and building specialization factors.
- *Behavioral Elements*, are social and psychological aspects of building occupant satisfaction levels. These aspects include the privacy and interaction of residents, environmental perceptions, sense of ownership, understanding and design of buildings, as well as the cognition and orientation of the residents' environment.

This research focus on functional elements (physical dimensions, room configurations, and furniture), as well as other functional standards specific to meeting rooms, and technical elements (lighting, audiovisual, acoustics, and space conditioning).

Based on the level of implementation, POE can be divided into three types [11], namely:

- *POE Indicative*, implemented in a short time (2-3 hours or 1-2 days). Aims to find indications of failure and the main success of a building's performance, so we need researchers who are familiar with the building to be evaluated. Data collection methods, through; evaluation of secondary data (archives and documents); performance issues (open questionnaire); walk-through evaluation (direct observation); interview people related to the object.
- *POE Investigative*, requires a longer time (160-240 hours), usually a continuation of indicative POE. Using the same data collection method as indicative POE, but with good and detailed data collection and analysis techniques.
- *POE Diagnostics*, is a thorough and in-depth investigation. The method used is the same as the POE indicative (questionnaire, survey, observation, and physical measurement), but then carried out a comparative evaluation with similar buildings using many variables. POE diagnostic seeks to find relationships between physical, environmental, and behavioral performance measures, so as to produce long-term recommendations for future improvements to the same type of buildings.

This study uses the POE Indicative type to find indications of failure and major successes in the performance of the Senate Room, Faculty of Engineering, Gowa Campus, based on the technical and functional aspects of space.

3. Research Methods

This paper draws on a qualitative study conducted in the Senate Room of Engineering Faculty, located on the 2nd floor of the CoT Building, Gowa Campus. The method used is Post-Occupancy Evaluation (POE) indicative with evaluation room-use by observing how occupants use the room to meet their needs, which can help the room to be run more efficiently. Collecting data based on measurement of room size, capacity, and dimensions of furniture, as well as anthropometric standards. In this study data analysis uses descriptive qualitative techniques that are used to analyze data by describing the data that has been collected, and data analysis using architectural graphics. Qualitative data include all theories or information and standard facilities, activities, and the needs of the function of the Senate Room.

IOP Conf. Series: Materials Science and Engineering **875** (2020) 012009 doi:10.1088/1757-899X/875/1/012009

The data collection is carried out in various ways, namely:

- *Literature study* is carried out by reading and recording information that contains theories and meeting room standards to obtain data that supports problem solving in research.
- *Data collection* is carried out with observation techniques that are based on direct experience that allows seeing and observing themselves, then notes according to what the user sees and feels.
- *Interviews*, observations and photo documentation are also techniques for accurate data collection, include an interview with room users.
- *Observation and photo documentation* conducted at the location of study.

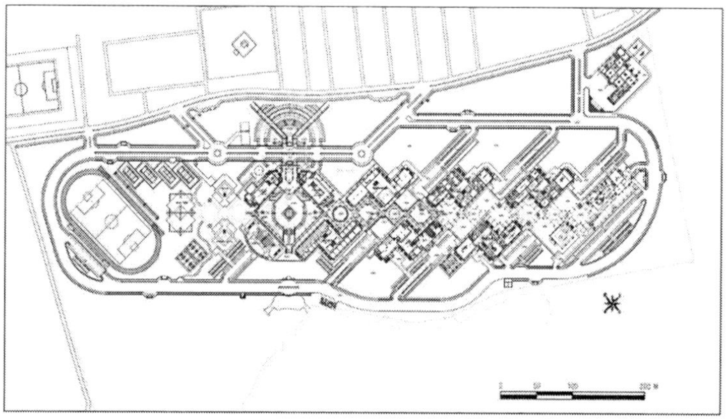

Figure 1. Site of the Faculty of Engineering Unhas, Gowa campus

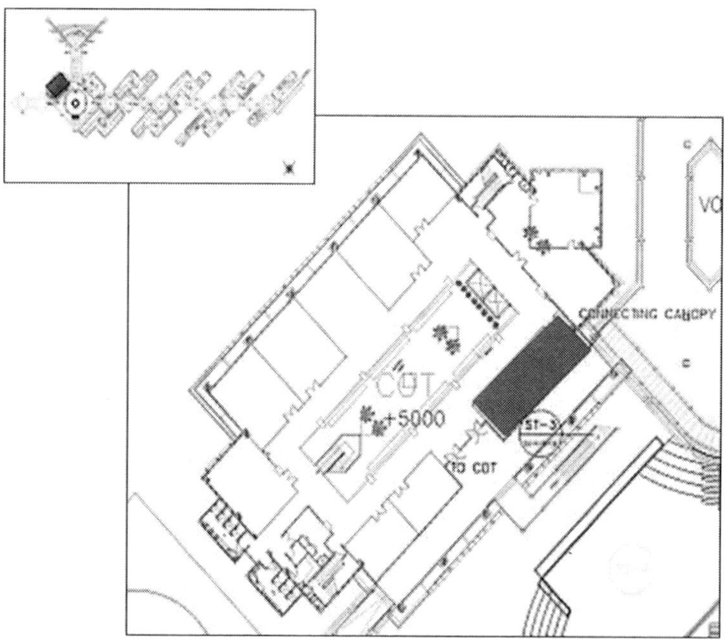

Figure 2. Location of the senate room on the 2nd floor of the CoT building of Faculty of Engineering of Unhas in Gowa

4. Results and Discussion

4.1. Characteristics of the Senate Room's Interior Elements

The first stage of post-occupancy evaluation in the Senate Room is to identify the characteristics of the interior elements (floor, wall, ceiling), the technical elements of the room (lighting, air conditioning, acoustics), and the functional elements of the space including the space filling elements (furniture design) and space planning (furniture layout). Identification is carried out through observation with documentation, measurements, drawing, and interviews of space users.

4.1.1. Technical Aspects. Identification of the interior elements on the technical aspects, including the space-shaping components (floors, ceilings, walls), while the technical element includes a thermal, audial and visual component. Tables 2 and 3 show the characteristics of each element of the Senate Room.

Table 2. Characteristics of interior elements

Components	Attribute	Pictures
Room shaping	Floor Covered with a 6 mm thick carpet, milano style, red and blue. Has good sound absorption.	
	Wall • Multiplex coated walls as high as 80 cm, wood motif and texture • Cream wall color • The windows and vents 12 units (6 fixed glass windows, 6 awnings window) • Vertical Blind, cream color	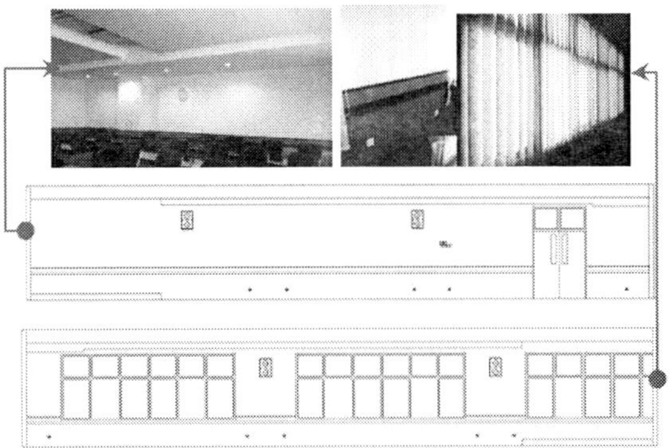
	Ceiling • Gypsum board material, drop ceiling type, white color, with edge trim, 291 cm high.	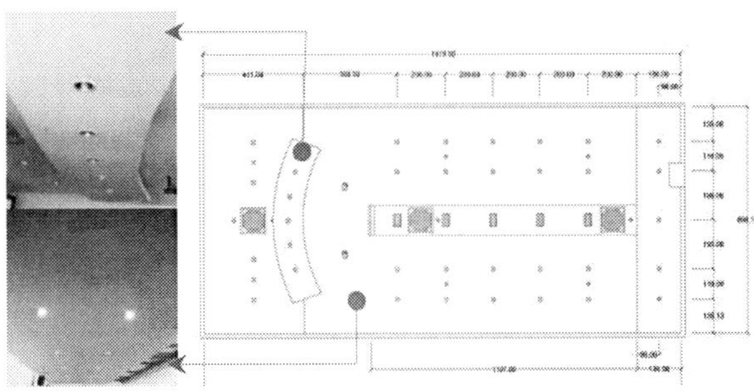

Table 3. Characteristics of technical elements

Components	Attribute	Pictures
Thermal	• Cassette type Air Conditioning, 2 PK (3 units)	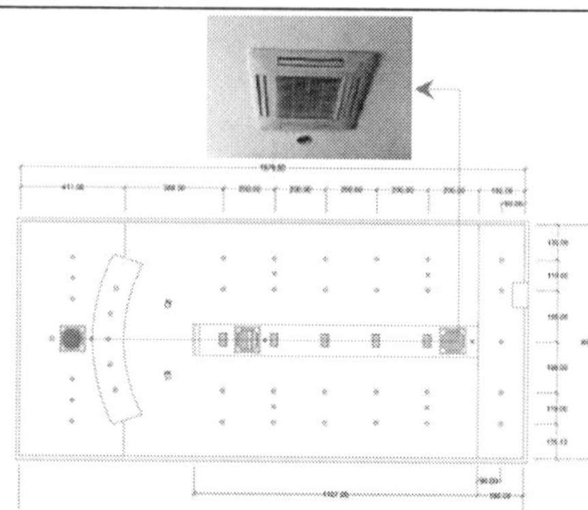
Audial	• Mic Sound (38 units) • Speakers (4 units) • Sound System (1 piece)	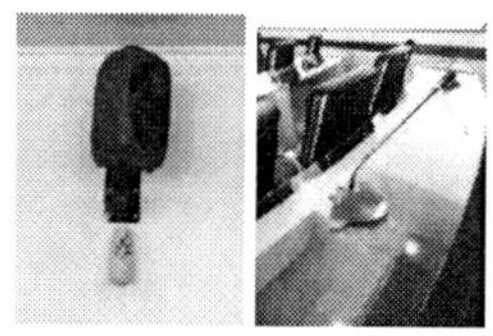
Visual	• TV monitor (1 piece) • LCD (2 pieces) • Lamp: downlight type (35 pieces), indirect type (4 pieces)	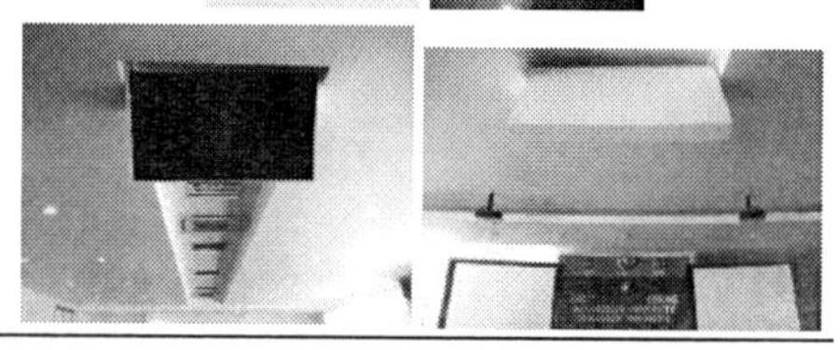

4.1.2. Functional Aspects. Table 4 shows the characteristics of functional elements in the Senate Room which include the space filler components, namely furniture: tables, chairs, and stage settings.

Table 4. Characteristics of functional elements

Components	Attribute	Pictures
Furniture	Table • Chairman meeting table (3) • Participants meeting table (32) • Signature table (1 piece) • Podium table (1 piece)	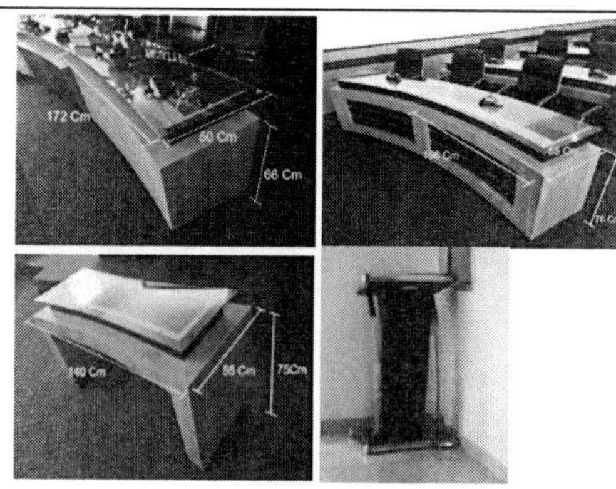

Components	Attribute	Pictures
	Chair • Chairman meeting chairs (5 units) • Participant meeting chairs (64 pieces)	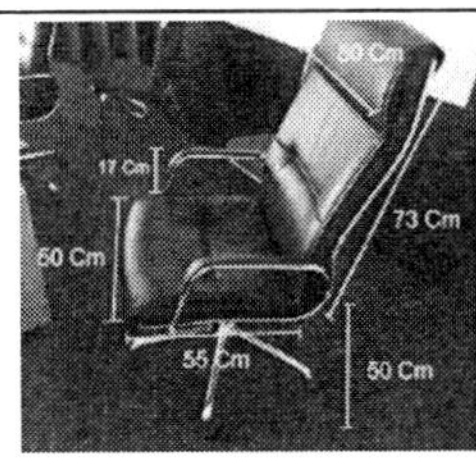 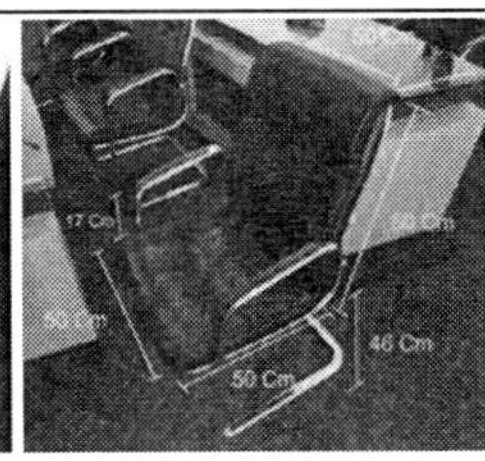
	Stage backdrop • LCD screen (2 pieces) • Background	 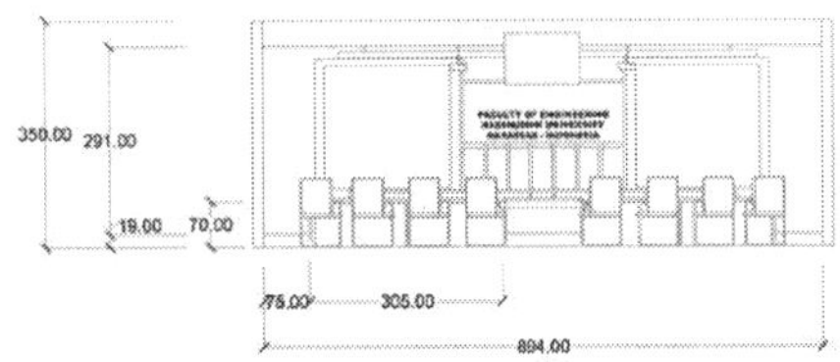

Table 5. Evaluation functional aspect

Criteria	Standard [4, 12]		Identified Problems
Furniture layout	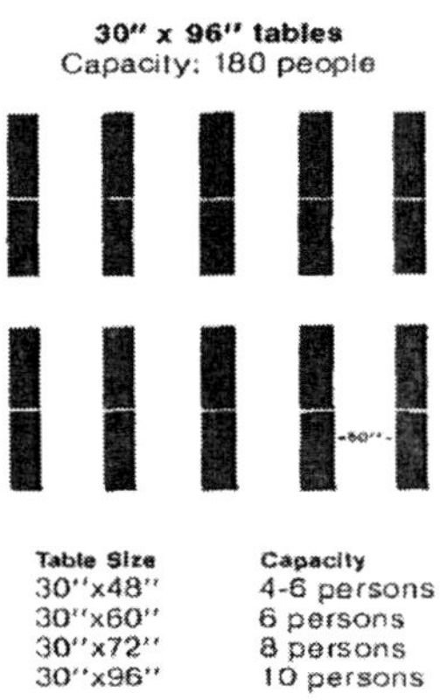	• Support comfortable eye contact and normal volume • Requires a range of space around it to be accessible and functional • Support circulation through the room and access to all furniture • The alignment mainly concerns the orientation of furniture items relative to each other and to the walls of the room	• Curved furniture layout affects the visual orientation of participants not focusing on the speaker • Location of furniture does not consider access to circulation from the entrance
Furniture design	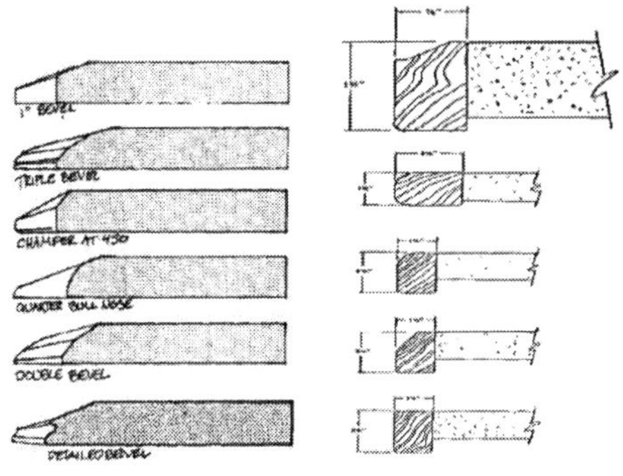	• Consider are the finish edges of glass and wood conference tables	• The conference table does not consider finishing table edges and user ergonomics, the design considers visual aesthetics more

Criteria	Standard [4, 12]	Identified Problems
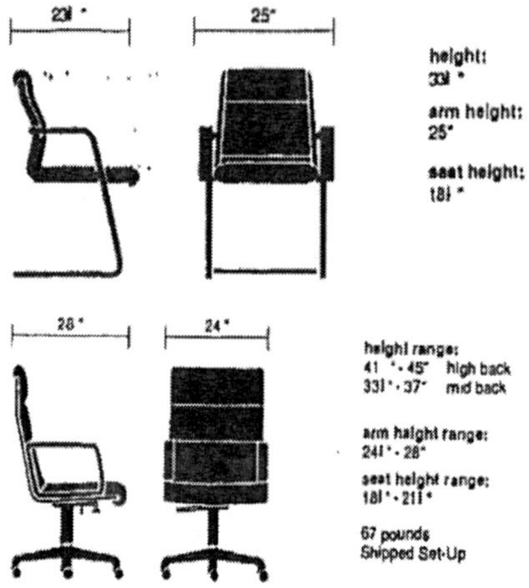	• Chair types should associated with certain general activities. • Ergonomic considerations are to be carefully reviewed in order to select the chair with appropriate attributes. • Eye height consideration in the sitting position is affected by the flexibility of the upholstery, height, and seat arrangement.	• The static type of meeting chair for participants tends to limit the movement of participants, and the height of the seat that cannot be set tends to make it difficult for participants to view the speaker.

4.3. Spatial Design Performance in Terms of Functional Aspects

From the spatial aspect there are three factors considered in the evaluation, namely; circulation, space layout, room size, and space density. The problems of users with various room features are shown in Table 6. These findings illustrate the drawbacks of feature conditions, such as limited participant visual problems and obstructed views of the speaker, unclear audials, ergonomics and improper furniture layouts, non-compact table designs and no electrical sockets are available, seats are static for participants, contrasting carpet colors tend to cause distraction. On the other hand, theater-type room layouts generally require a large enough space while the area of the Senate Room is 180 m^2. The capacity and dimensions of furniture are dominant and do not match the area of the room, causing space density and circulation problems.

Table 6. Spatial design and technical system

Features	Criteria [4, 12]	Suitability	
		appropriate	not appropriate
Circulation	• Sufficient space to walk comfortably to all parts of the room		√
Room layout	• The design and layout should therefore allow for different presentation styles, flexible seating, circulation within the space and entry/exiting that does not disrupt presentation.		√
Lighting	• Lighting should be zoned in such a way that front wall wash lighting fixtures can be dimmed to prevent light spill onto the screen during a presentation.	√	
	• Nominal min. 270 cm floor to ceiling height.	√	
Acoustic	• Acoustically treated walls, ceilings, floors, and window coverings provide absorptive qualities that minimize reverberation and improve voice communication in the space; should include carpeting with underlayment, acoustical ceiling tile (with an NRC of .7 to .9), and medium weight drapes or curtains.		√

Features	Criteria [4, 12]	Suitability	
		appropriate	not appropriate
	• Typical meeting rooms should have a lay-in acoustic tile ceiling with the specified tile having a minimum Noise Reduction Coefficient (NRC) rating of 0.65.		√
	• Walls separating meeting rooms should be a minimum construction of two layers of 5/8 in. (1.6 cm) gypsum board on one side of a metal stud and one layer of 5/8 in. (1.6 cm) gypsum board on the other.	√	
Fixtures, furnishings and special mountings	• Lecterns can either be stock items under the AV contract or custom-designed millwork.		√
	• AV equipment such as tape machines, video document camera, and computers can be built into walls, cabinets or credenzas for access by presenters.		√
	• Special mounting supports will be required for ceiling-mounted video projectors and any TV monitors suspended from the ceiling	√	
Natural Ventilation & Room Conditioning	• Natural ventilation not desirable.	√	
	• Manually openable windows are permissible.	√	
	• Mechanical cooling, Air Conditioning, $20^0 - 22^0$ C	√	
Audiovisual (AV) system	• Single image slides or dual side by side are most common.	√	
	• Slides can be front or rear-projected. The horizontal centerline of the lens is the horizontal centerline of the projected image.	√	
	• Video projectors can also project front or rear and can be suspended from the ceiling or mounted in a floor cradle in a portable configuration.		√
	• Large screen television and audiovisual systems are common features in the meeting rooms.	√	
	• The rear projection room should be 1/3 the size of the usable conference room space.		√
Audiosystem	• Audio systems required for audiovisual spaces may consist of two separate sound systems, one for speech reinforcement and one for amplification of program sound material.	√	
	• Speech reinforcement is accomplished through the use of microphones located at the lectern, head table, or even for the meeting participants in a larger space.	√	
	• Speech signals are mixed, processed, then amplified and delivered to flush-mounted ceiling loudspeakers.	√	
	• Uniform sound level throughout the audience area is measured with a pink noise signal, where 80 dBA is the target sound pressure level (SPL); with a tolerance of + or - 1.5 dBA.		√

Table 7. Evaluation functional aspect

Criteria	Standard [4, 12]	Suitability	
		appropriate	not appropriate
Room and Spaces Planning Requirements	• Room proportion to suit layout of furniture and desired occupancy		√
	• Design occupancy minimum 20 persons. Preferred 50 persons plus.	√	
	• Area to suit building occupants, allow approx.: - 0.8m^2 per staff member - 40 m^2 for 20 - 100 m^2 for 50	√	
	• One space a maximum of 80 m from any work-point.	√	
Crowding	• Larger objects carry more visual weight. The density of space is not dominated by furniture, so that the room feels relieved.		√
Visual	• Perception of layout as a visual composition, visual balance, alignment, and dominant points of emphasis.	√	
	• Place the average visual weight distribution at the center of the composition.	√	
	• Visual balance refers to the appearance of the arrangement from various points of view.	√	
	• The accent is desirable to have a dominant focal point in the interior, so that the eye can rest without competition for visual attention.		√
	• The focal point is often a protruding object such as a monitor screen.		√
Activities	• Furniture layout must supports human activities that occur in room.		√
	• Spaces are intended for comfortable human communication and ease of accommodation.		√
Ergonomic	• Distance needed around the object and for the exact distance and angle between the objects.		√
	• Objects should be positioned in pairs, such as chairs and tables of the same type and size.		√

Figure 3 shows the layout of furniture that tends not to consider the circulation of space, especially in the entrance area and the edge area of the room. Limited circulation space, causing the back table is not functioning.

The visual limitation towards the speaker, especially for participants who sit in the back, is due to the type of chair that is static-difficult to move and the suitability of the stage height to the eye height of the participant in a sitting position. Figures 4 and 5 show the visual area of the participants in the vertical and horizontal directions, where the podium and the screen becomes the main focus.

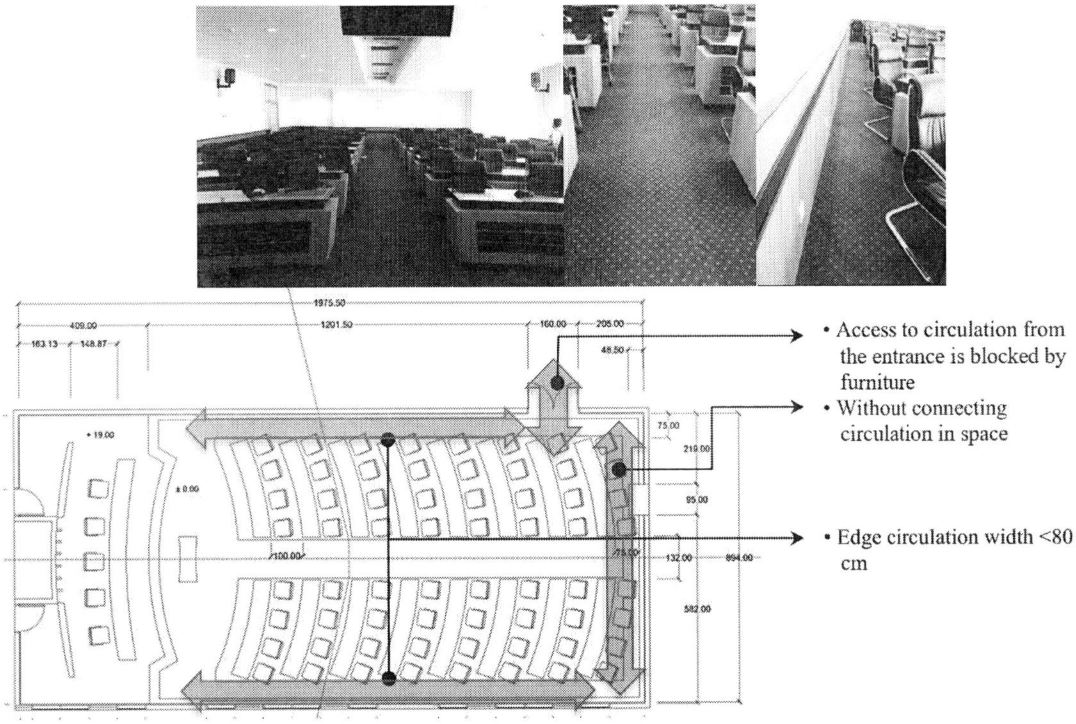

Figure 3. Circulation and room layout

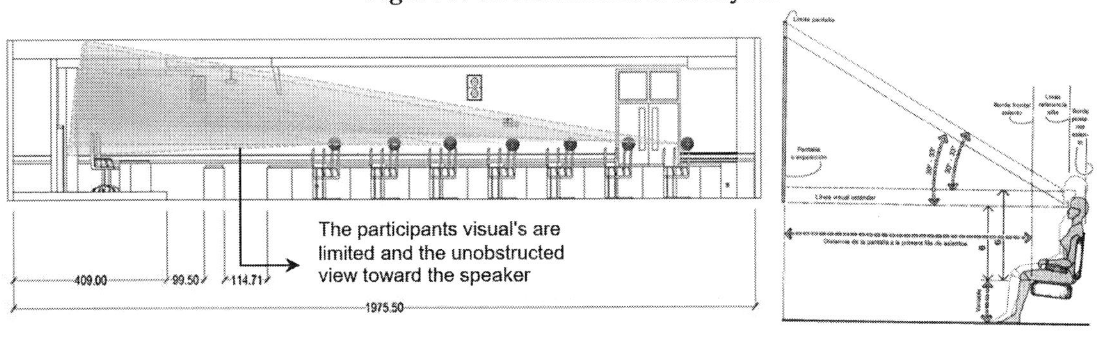

Figure 4. Visual area in a vertical plane [13]

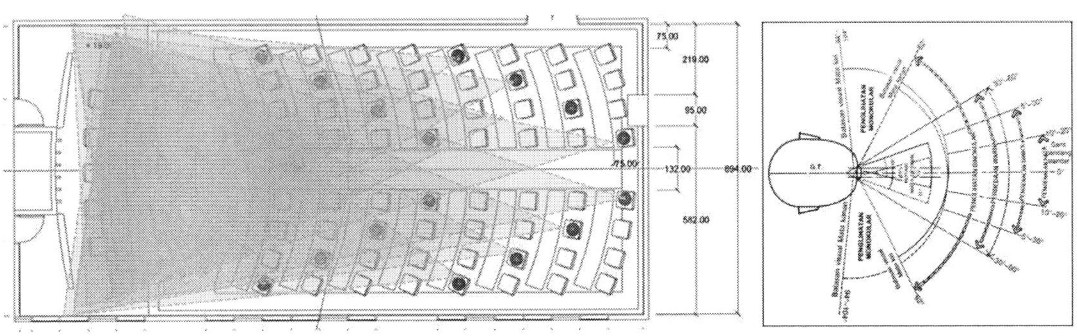

Figure 5. Visual area in a horizontal plane [13]

5. Conclusion

Characteristics of interior elements, both floors, walls, and ceilings can absorb sound well. Distraction due to color contrast tends to occur on the floor and sunlight glare from the window. Technical and functional aspects influencing the room so that the room becomes crowded, and less effectively accommodate the meeting activities. The results of this study are expected to contribute practical insights to architectural design, specifically related to the creation of meeting room interiors. Based on the results of the analysis and evaluation using POE, it is known that there are weaknesses of functional and technical aspects in the Senate Room, Faculty of Engineering, Unhas Gowa, namely:

- Functional aspects; furniture layout, proportion space, room circulation, furniture design, and visual space.
- Technical aspects; audials, acoustics, fixtures, furnishings and special mountings.

References

[1] Keputusan Menteri Kesehatan Republik Indonesia Nomor 288/Menkes/Sk/III/2003 tentang *Pedoman Penyehatan Sarana dan Bangunan Umum.*

[2] Hay R, et al. 2017 Post-occupancy evaluation in architecture: experiences and perspectives from UK practice *Journal of Building Research & Information* (Routledge).

[3] Haryadi 2010 *Arsitektur, Lingkungan dan Perilaku. Pengantar ke Teori, Metodologi dan Aplikasi* (Gadjah Mada University Press. Yogyakarta).

[4] De Chiara, J., Panero, J., and Zelnik, M. 2001 *Time-Saver Standards for Interior Design and Space Planning, 2nd Ed.* (Mcgraw-Hill Professional).

[5] Meeting Room. Retrieved from https://xwork.co/blog/kenali-5-jenis-seating-dalam-ruang-meeting-agar-tak-salah-pilih/, last accessed on April 10, 2019.

[6] Merrell & Schkufza, et al 2013 *Interactive Furniture Layout Using Interior Design Guideline* (Stanford University University of California, Berkeley).

[7] Panero, J., and Repetto, N. 1975 *Anatomy for Interior Designers, 3rd Ed.* (Whitney Library of Design).

[8] Deasy, C. M., and Lasswell, T. E. 1985 *Designing Places for People* (Watson-Guptill).

[9] Arnheim, R. 1974 *Art and Visual Perception* (University of California Press).

[10] Kubba, S. 2003 *Space Planning for Commercial and Residential Interiors* (Mcgraw-Hill Professional).

[11] Preiser 1988 *Post Occupancy Evaluation* (New York: Van Nostrand Company).

[12] Karlen, M. 2009 *Space Planning Basics, 3rd Ed.* (John Wiley & Sons).

[13] Panero, J., and Zelnik, M. 1979 *Human Dimension and Interior Space,* (Whitney Library of Design).

IOP Conf. Series: Materials Science and Engineering **875** (2020) 012010 doi:10.1088/1757-899X/875/1/012010

Characteristic of Student Informal Gathering Space (SIGS) Faculty of Engineering Hasanuddin University at Gowa Campus

A M Radja[1]*, R Wikantari[1], and A Harisah[1]

[1]Department of Architecture, Engineering Faculty, Hasanuddin University, Makassar, Indonesia

*Email: muftiradja@yahoo.com

Abstract. The purpose of this study are: 1) What are the characteristics of Student Informal Gathering Room (SIGS) in the Faculty of Engineering, and 2) What attributes are felt by students in the Student Informal Gathering Room (SIGS). To get the results of the research purposed applied qualitative research methods with purposive sampling techniques. Data collection is done by distributing questionnaires to students, in-depth interviews and mapping behavior with place-centered maps. The results of the study is the closeness of the gathering place with the main activities of students in the teaching and learning process which is a characteristic of SIGS found on the campus of engineering faculty at Gowa campus. While the environmental attributes felt by engineering students at SIGS are accessibility, comfort, sociality, visibility, and activities.

1. Introduction

Student informal gathering space (SIGS) is a space provided on campus that is used by students as a place to conduct activities outside of lecture activities. The space is very important because it can be used to shape student characters such as organizational activities, student activity units and social place for students [1]. The room is a space that is outside of the classroom, such as corridors, inter-building corridors, parks, canteens, libraries, and so on. Many activities carried out by students at SIGS, such as waiting for lecturers, chatting, doing assignments, eating and drinking, sleeping, playing games, and others [2].

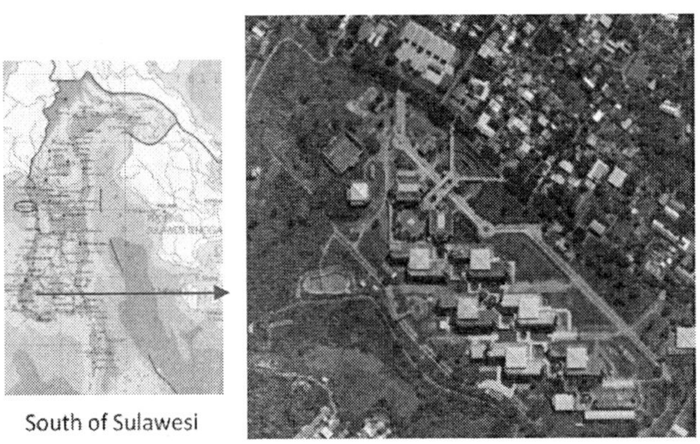

Figure 1. Location of campus of FoEHU

The Faculty of Engineering Hasanuddin University (FoEHU) is one of the faculties at Hasanuddin University which is located in the former of Gowa paper mill the KM 6 of Malino main road with an area of approximately 40 hectares (Figure 1). There are several buildings on the Gowa Engineering Campus, namely the Center of Technology (COT) Building, the Center of Science Activities (CSA), Classrooms (CR), AR buildings (Architecture and Regional and Urban Planning Department), Civil buildings (Civil Engineering and Environmental Engineering), Mechanical Building (Mechanical Engineering and Industrial Engineering), Electrical Building (Electrical Engineering and Information Engineering), Geological Building (Geological Engineering and Mining Engineering), Ship Building (Shipping Engineering, Shipping Engineering Systems, and Marine Engineering). Currently, additional campus facilities are being carried out, such as sports center and student center buildings which are still under construction.

The problem faced today with campus planning is the unavailability of public spaces or SIGS that can use by students as a gathering place to carry out various individual and social activities. There are 3 functions of campus, namely as an individual space as well as a public space for the exchange of information (scientific), a place to fulfill psychological needs (recreation) and biological functions (green open spaces) that are utilized by the academic community and the general public [3]. The SIGS must provide physical comfort, accessibility, and visibility [2]. In addition to the comfort in SIGS that students must feel, students must also feel some environmental attributes from the interaction process in a setting [4].

Initial observations showed that students in their spare time, they utilize the empty spaces around the buildings at Gowa campus as SIGS. The space like around the entrance of each building, a connecting corridor between buildings, corridor in the classrooms and hall in each building. The faculty built concrete chairs in the corner of the connecting corridor between buildings on the 1st floor, but not so many students gathered at the venue. The faculty also prepares concrete chairs in front of the CR and AR buildings, but it is not always used as a gathering place because it's hot by sunshine. The student gathering at the place when cool by the shadow of buildings and trees covering the area. The canteens between the buildings are also ineffective as a gathering space because it disrupts lecture activities and seems dirty [5]. Based on the problems above, the problem is: 1) What are the characteristics of SIGS in the Faculty of Engineering, and 2) What environmental attributes are students of Engineering faculty felt at the SIGS?.

2. Literature Review

Like public space [6, 7], SIGS is a space that can be accessed by all students. Research on behavior conducted by White in several public spaces in America, he noted several factors that influence the use of public space by people, namely: sitting space, sunlight, wind, vegetation and water, food, physical and direct visual access to the main road. Whereas the things that are not expected in public space are effective capacity, inner spaces, scale in small cities, and triangulation [8].

Based on its forming elements Rapoport explains that the settings can be distinguished namely [9]:

1) Fix components, which are basically fixed elements or changes that are rarely and slowly such as space, roads, pedestrians, and others.
2) Semi fix components, which are rather fixed elements, can change quite quickly and easily, such as trees, street furniture, street vendors.
3) Non-fix components, namely elements related to human behavior in using space.

The attributes that arise from interactions can be divided into 12 (twelve) namely [4]:

- Comfort is a state of the environment that provides a sense that is appropriate for the senses and anthropometrics accompanied by facilities that are appropriate to their activities. Anthropometrics are the proportions and dimensions of the human body as well as the physiological and ability characteristics associated with a variety of different human activities. Anthropometrics is also referred to as a human factor that dimensionally influences architectural design.

- Sociality is the level of a person's ability to carry out social relations in a setting. A level where humans can express themselves in the relationship of social behavior is directly related to the arrangement of seats and tables in a public space. The distance between individuals, non-verbal behavior such as body angle, eye contact, and facial expressions will indicate the quality of socialization.
- Visibility is the ability to see unobstructed visually on the intended object. Visibility is related to the distance felt by humans. But the perceived distance is not just a dimensional/geometric distance, but it involves a visual perception in which humans feel the presence or absence of obstacles to achieve the intended object.
- Accessibility is the ease of moving through and using the environment. The ease of movement in question is related to circulation (road) and visuals.
- Adaptability is the ability of the environment to be able to accommodate different behaviors that did not exist before.
- Sensory stimulation is the quality and intensity of stimulants as experiences experienced by the human senses.
- Control is the condition of an environment to create personality creates territory and limits a space.
- Activity is a feeling of intensity in behavior that continuously occurs in an environment.
- Crowdedness is a feeling of the level of density in an environment.
- Privacy is the ability to monitor the running of information that looks and sounds both from and in an environment. Privacy is the desire or tendency in a person to not be disturbed by his solitude.
- Meaning is the ability of an environment to present individual or cultural meanings to humans.
- Legibility is a convenience for someone to be able to recognize or understand key elements and relationships in an environment that causes the person to find a way or direction.

3. Method

This study applies qualitative research methods with a purposive sampling technique. The qualitative methods as a research procedure produce descriptive data in the form of written and oral words of people and their behavior observed [10-12]. The people referred to in this study are students of the Faculty of Engineering who carry out activities at SIGS on the Faculty of Engineering Gowa campus and beyond. Data collection is done by distributing questionnaires to students, in-depth interviews, and conducting behavior mapping.

4. Result and Discussion

FoEHU has not been planning space for public spaces used by students as SIGS, only spaces such as amphitheater which are located between the COT and CR buildings and empty spaces between the buildings. The empty space is then used as a canteen equipped with chairs and tables made of concrete. The parks in the electro and shipping buildings are not yet equipped with chairs and tables so the canteen manager prepares tents and temporary furniture. The design of the campus landscape is done by building a green open space located in front of the campus, but no students use for gathering place because it is hot and not maintained. There is concrete benches in the corner of the corridor on the 1st floor but fewer students gather because the space always hot by sunshine. Faculty and each department provided several sitting space like bench and student use as SIGS (Figure 2).

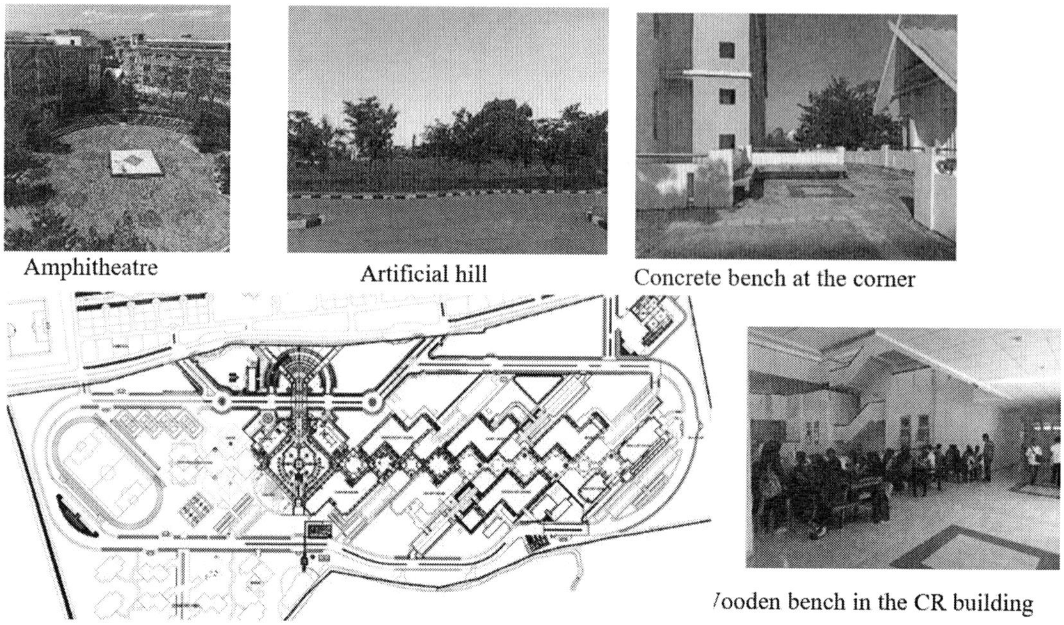

Figure 2. SIGS at the FoEHU

Based on the questionnaire, classrooms and laboratory buildings are places that are widely used for gathering places either before lectures (62%) or while waiting for the next lecture (33%). Students are in SIGS both before studying (33%), the time between lecture one to the next lecture (29%), and after lecture takes place before back to their homes (23%). SIGS are used for activities in these three places around the building entrance, indoor corridors, and canteen in front of the CR building or department/laboratory building. The choice of SIGS is because it is close to the classrooms and laboratories, making it easier for them when college starts. If they have time, they can go to the library or to their friend's boarding house located in front of the campus. Not so many students use SIGS after the lecture because many students go back home soon (60%) (Figure 3).

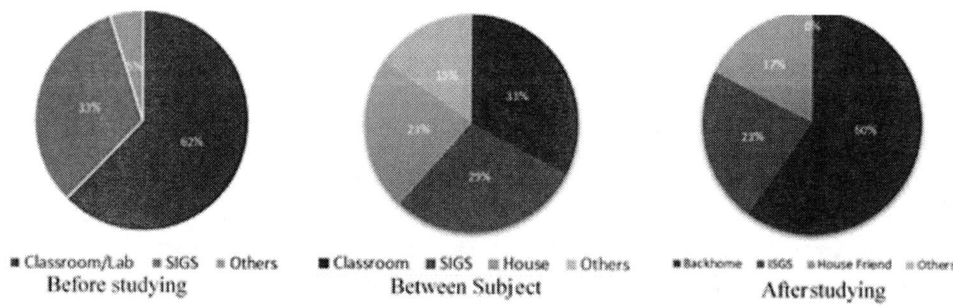

Figure 3. Time and place at the SIGS

The most commonly SIGS used by students are the department/laboratory building, CR building and park (canteen) and the corridor between the department buildings. The CSA building with some facilities like a library, computer center, display room, and around the entrance of the building becomes SIGS. As well as around the COT entrance and hall at 1st floor are often used by students as a gathering place. The corridor which is located around the amphitheater also becomes the SIGS for students before moving to the CR building (Figure 4).

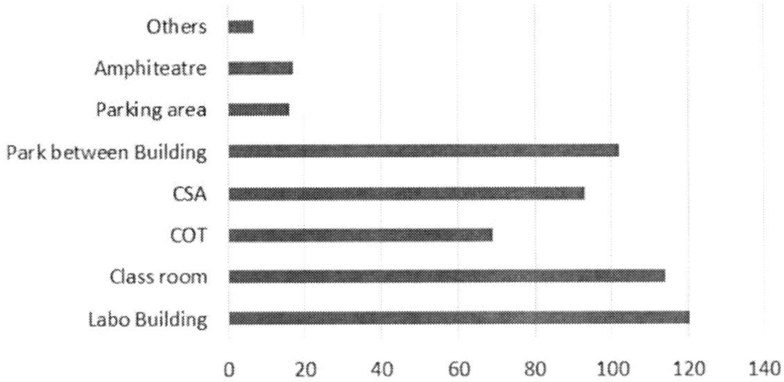

Figure 4. SIGS Commonly used by students

Some considerations of students choosing SIGS around it are the availability of chairs in classrooms, sitting space like concrete benches around the canteen and flower pots located along the ground floor corridors and around the building entrance. The availability of sitting space, protected from the sunshine, and feel the wind blows are the reason students in the SIGS. SIGS is a social space because many students always gather, which is the reason the student choose the space for gathering space. Space around the entrance of buildings, corridors in buildings are space where students can see the arrival of their lecturers and friends. Classrooms in the CR Building, libraries and showcase rooms at CSA have air conditioning, so students feel comfortable at the SIGS because they feel comfortable with the AC. Besides close to CR Building, library at CSA Building also students can study quietly without feeling disturbed by others. WIFI available in every building provides convenience for students to access information and lecture material. Boarding houses around the campus also become SIGS for students when waiting for college or after (Figure 5).

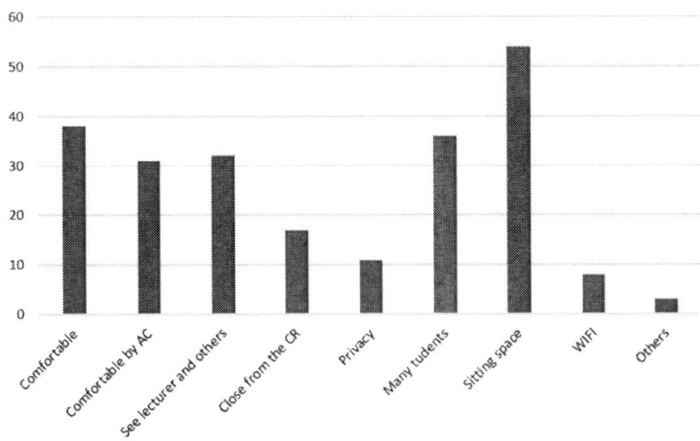

Figure 5. Reason to choose SIGS

The main activity carried out by students at SIGS is chatting with one another. In addition, many of them study and do assignments while searching for references on the internet. Eating and drinking are activities carried out by students in SIGS while waiting for the lectures in the classroom, in the canteen or in the resting time in the afternoon. Playing games by android or laptop during break time or waiting for the next lesson is also done by students at SIGS. Students also sleep in SIGS during break time, generally done in the student room or at the house/boarding house located around the campus. Other activities are just sitting without doing any activities or just looking at other student activities (Figure 6).

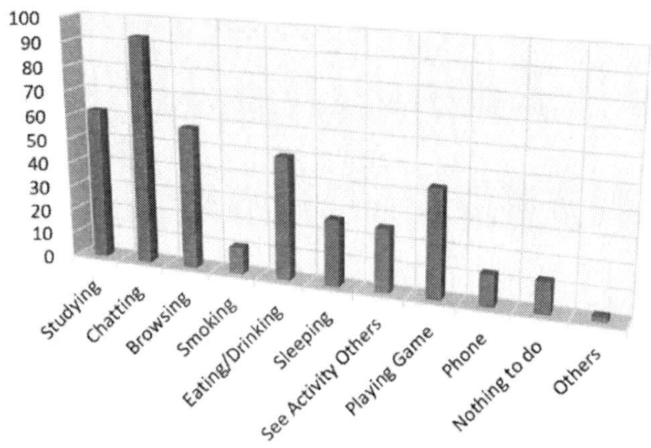

Figure 6. Activity on SIGS

5. Discussion

The main activity of students on campus is learning carried out in the classrooms (CR) and in the laboratory each department, so that moment is a great opportunity for students to meet in the building either before lecture or time between lecture. Because often used as a gathering place while carrying out various activities, the classroom has two functions, namely as a formal space for learning and informal space as SIGS. Students use classrooms as SIGS because these spaces provide several attributes to students [4].

Based on the favorite space used by students as SIGS according to Figure 3 above, the distribution of students on the Engineering campus is shown in figure 7. All students of each department use the space around their laboratory buildings, such as halls, entrances, and student rooms as SIGS. Outside the lecture hours, students use the space around the CR building, such as the corridor in the building, hall, around the entrance, and the corridor of the building as SIGS. They use the space as SIGS because they get the attribute of accessibility to CR building and laboratory building. The comfort attribute is felt by students by having a sitting space in the classroom, a wooden benches in the hall, and flowerpot wall along the corridor, and sitting on the floor. The comfort attribute is also felt with the presence of AC and protected from the sunshine. The SIGS gives an attribute of visibility to students to see the arrival of lecturers and other students. When students meet, they have a conversation and this scene make an attribute of sociality.

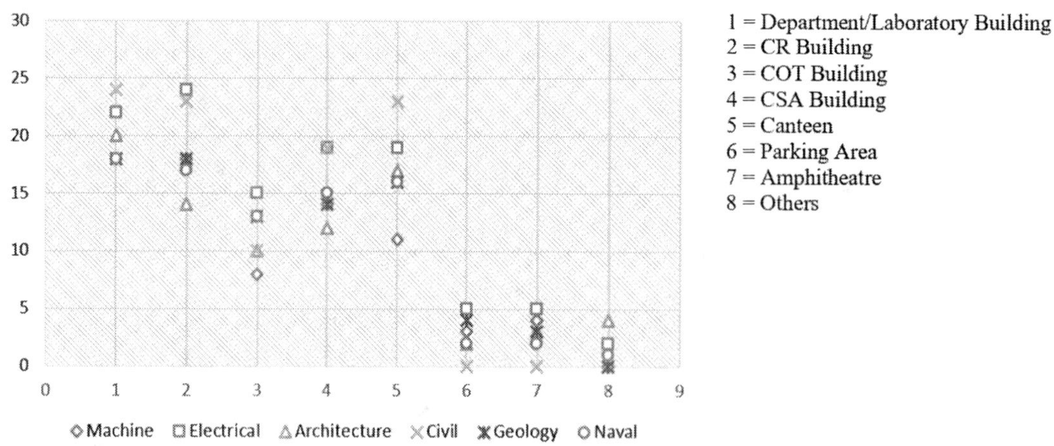

Figure 7. Distribution of students on SIGS

The space between buildings used as canteen is also used by students of all departments as SIGS, although not all canteens have permanent facilities such as roofs and furniture. Students get quick accessibility to the laboratory it is located between the laboratory buildings and located close to the entrance of the building. Students also get the visibility attribute, which is that they can easily see the lecturers and their friends because no prevent their view. Various kinds of activities can be carried out by students in the canteen, so students get the attributes of activities.

Rooms in the CSA building such as the library room, showcase room, and space in front of the elevator are the SIGS that is used by all students of the Engineering faculty to carry out various activities. In addition to getting the attribute of comfort because air-conditioned, students also get the privacy attribute in the absence of interference either sound disturbance or interference from other student activities. Besides being located in CSA which is located in the CR building, students get the attribute of accessibility. This room also has a chair that provides attributes of comfort in studying in accordance with the anthropometrics of the student body

Sitting space is one of the successes and gives life in public space [8] and there is no reason for people to be in public space except without a seat [13]. This is the main reason for students to be in SIGS, both chairs in the classroom, sitting space in the hall of the CR building and the laboratories of each department, concrete chairs in the canteen, and retaining walls in the ground of the plant's corridor, even student sitting on the floor. The siting space consists of 2 categories, namely primary seating in the form of a bench or chair and secondary seating in the form of a place where informally and spontaneously can rest while doing other activities [13, 14]. The sitting space provides student attributes of comfort (Figure 8).

Figure 8. Sitting space on SIGS

Conversation is a social activity as part of 3 types of outdoor activities that can occur at any time in public spaces [13, 15]. This is also done when students meet at RBIM while waiting for the lecturer, the main activity carried out is conversation. The conversation resulted in social attributes that were felt by students.

6. Conclusion

Based on the discussion and discussion above, it can be concluded as follows:

1) The closeness of the gathering place to the main activities of students in the teaching and learning process is a characteristic of SIGS found on the Engineering faculty at Gowa campus. The CR building as a place for students study, the laboratory building as a lecturer room, the administration room and the student practice room are the room for main students' activity, so there are many SIGS found in the building. The canteen which is located between the buildings is also the SIGS which is often used by students because it is close to the CR building and laboratory. The library and showcase room located in the CSA building are SIGS because they are located close to the CR building which makes it easy for students to go back to the classroom.
2) As a product of perception of the environment, the attributes perceived by students are accessibility, comfort, sociality, visibility, and activities.

Acknowledgments
Our gratitude to the Engineering faculty Hasanuddin University who provided research funding through LBE research in 2019, students of the Engineering faculty who took their time to be interviewed and filled in the questionnaires and students of Final Architecture assignment study period II / 2017 who helped circulate the questionnaire.

References
[1] Haristianti, Vika dkk. 2015 Kinerja Ruang Publik Kampus Ditinjau dari Faktor Attraction Studi Kasus: Lapangan Campus Center Timur ITB Prosiding Temu Ilmiah IPLBI. Universitas Sam Ratulangi, Hal. 95-100
[2] Widajanti, A. 2012 Kenyamanan Sebagai Salah Satu Kebutuhan Rang Berkumpjul Informal Mahasiswa. Studi Kasus: Universitas Mercu Buana Seminar Nasional-Menuju Arsitektur berEmpati Jurusan Arsitektur Universitas Kristen Petra. Hal. 547-554.
[3] Purnomo, Y. dkk. 2014 Konsep Ruang Terbukan Publik Mahasiswa Sebagai Penghubung Antar Unit di Universitas Tanjungpura (Jurnal Langkau Batang. Vol. 1 No. 1) Hal. 1-14.
[4] Weisman, J. 1981). Evaluating architectural legibility: Way-finding in the built environment. Environment and Behavior,13, 189-204.
[5] Radja, AM, Wikantari, Ria, dan Syam, Syahriana. (2017). Karakteristik Ruang Berkumpul Informal Mahasiswa (RBIM) Arsitektur dan PWK Fakultas Teknik Di Kampus Gowa Prosiding Saintek 3. COT. Fakultas Teknik, Universitas Hasanuddin.
[6] Carr. S. Etc. (1995). Public Space. Cambridge University Press. USA.
[7] Francis. M. (2003). Urban Open Space. Designing for User Needs. Landscape Architecture Foundation. USA.
[8] White.W.H. (1980). The Social Life of Small Urban Spaces. Project for Public Spaces. NewYork. NY.
[9] Setiawan, H.B, (1995). Arsitektur Lingkungan dan Perilaku. Dirjen Dikti Kemdikbud. Jakarta.
[10] Moleong, L.J. (2013). Metodologi Penelitian Kualitatif. Edisi Revisi. PT. Remaja Rosdakarya. Bandung.
[11] Sugiyono. (2014). Memahami Penelitian Kualitatif. CV. Alfabeta. Bandung.
[12] Denzin and Lincoln. (2009). Handbook of Qualitative Research. Pustaka Pelajar. Jogyakarta.
[13] Gehl, J. (1987). Life between Buildings; Using Public Space. New York. Van Nostrand Reinhold.
[14] Gehl, J. (2010), Cities for People, Washington - Covelo - London: Island Press.
[15] Radja, et al.(2012). Study on Bale Bale in Lae Lae Isle, Makassar, Indonesia. In Journal of Architecture, Planning and Environment (Transactions of the Architectural Institute of Japan, Vol. 77, p. 969-978, May 2012).

Post-Occupancy Evaluation on Infrastructure Service Performance of Mariso Low Cost Apartment (Rusunawa), Makassar

N Amri[1]*, Idawarni[1], E Syarif[1], S Amin[1], Y Siradjuddin[1], and R Hiromi[1]

[1]Architecture Department, Engineering Faculty, Hasanuddin University, Makassar, Indonesia

*Email: nurmaida.amri@gmail.com

Abstract. This study aims to measure the level of infrastructure service performance of Mariso low-cost apartment (Rusunawa), which is the key to the comfort of occupancy that ensure the sustainability of daily activities both physical and non-physical. Targets to achieve in this research are to formulate the infrastructure service performance according to the standard service of building infrastructure and facilities and as mandated in the standard of planning, building and evaluation of building post-occupancy. This method is used to determine the extent to which the perception of occupants responding to the results of a built environment after more than 10 (ten) years occupied. The applied research method was descriptive research method through the recording and infrastructure service performance analysis by linking between the comfort factor with the standard planning of Mariso low-cost apartment (Rusunawa). The results of the research are the recommendations on the suitability of the service standards/performance infrastructure of the building with the existing conditions of the building with attention to the variable availability of infrastructure that become the focus of research.

1. Introduction

Post-occupancy evaluation (POE) is an evaluation or diagnostic measurements to monitor the quality of the building, or post-occupancy evaluation is the systematic and accurate evaluation of a building after the complete construction of the building and occupied for some time. POE focuses on the occupants and the needs of occupants and therefore, they could provide basic information concerning the consequences of decision of past designs from the building performance result. This information will be a good basis for constructing better buildings in the future.

POE is an activity that originated from the concern with the occupied building performance. The basic approach to evaluate building performance is through the perspective of the building occupants [1]. Its main objective is to describe the relationship between building, management systems, and its occupants. The most commonly used POE scheme is through the assessment of occupant satisfaction [2]. The satisfaction derived from the comparison of the actual condition (performance) and the expected conditions (needs or interests) of residents. If the performance of the building does not meet the requirements of the occupants, dissatisfaction will occur. In addition to building performance, occupant's assessment may include other factors around the building, such as the management of facilities and the environment facilities. In this case, the occupancy system is the combination of physical and social aspects. Therefore, determining the scope of occupancy becomes very essential in identifying factors that affect satisfaction. The principle of EPH is to perform a systematic approach for improving the quality of a building which is responsive to the function desired and needed by the occupants [3].

Flats is a multi-blocks building built in an area that is structured functionally, and either horizontally or vertically. It consisted of several units that can be owned and utilized separately, primarily as an occupancy provided with collective unit, property and land [4]. A flats plan should be able to guarantee the occupancy sustainability by providing supporting facilities for environmental service for the flats. Complete utility of flats must fulfill the requirement of Government Regulation [5] regarding Technical Requirement of Flats Construction.

The aim of this study will focus on the interconnection between the standard apartment plan that is based on the standard low-cost apartment plan with the existing condition of current occupancy. Therefore, the results of the study could be formulated into recommendation focusing on the infrastructure and facility performance improvement sustaining the low-cost apartment. This research is important considering the absence of post-occupancy evaluation at Mariso low-cost apartment which has not yet reached a clear and thorough results on the infrastructure performance and supporting facilities of the flats, especially the description collected through direct recording/observation on the relevant condition.

2. Literature Review

2.1. The Philosophy and Definition of Apartment

The house or occupancy is one of the basic human needs that serve to support the implementation of the activities of human daily life. One of the main issues that metropolitan cities encounter is the unavailability of land while housing demand for urban population is on the hig level [6]. One alternative to solve this problem is to develop a model of vertical housing in the flats form. Flats construction is one of the strategic ideas in providing habitable occupancy, especially for low-income communities and serving as practical solutions in overcoming urban dense settlements [7].

Flats is a multi-blocks building built in an area that is structured functionally, and either horizontally or vertically. It consisted of several units that can be owned and utilized separately, primarily as an occupancy provided with collective unit, property and land [4].

2.2. The Requirement of Flats Infrastructure Environment

Environmental infrastructure is the basic physical environmental services which are required in the flats in order to sustain the flats properly including road, elevators, hallway, drainage, waste management system and clean water. Flats environment should be supported with the following facilities

a. The environmental infrastructure that serves as a liaison for the purposes of daily activities for the residents such as footpaths, vehicle and parking place;
b. Environmental infrastructure should consider the accessibility and harmony of the relationship in the daily activities and security when harmful situations occur, as well as the structure, size, and strength in accordance with the function and road use;
c. Clean water distribution, gas and electricity with all of the equipments such as water tank, water pump, gas tank and electrical substations;
d. Rain water sewerage connects the rain water from the flats to city water sewerage system;
e. Sewerage and / or septic that connects the waste water from the flats to the system sewage network of the city;
f. Trash, as garbage collector from the flats that will be dumped into the city dump, by considering the factors of transportation accessibility, cleanliness, health and elegance;
g. The water faucets to prevent and secure fire that can reach all places in the environment;
h. Vehicle parking and storage;
i. Telephone network and the communication equipment are in accordance with the purposes.

2.3. Flats Facilities

According to the Indonesian National Standard [8], environment Facilities are the supporting facilities that serve to the implementation and development of economic life, social and cultural rights, which encompasses commercial building or shopping center (the economic aspect), open space, education, health, religion, government facilities and public services, parks and cemetery (the location is outside of the flats or according to the spatial plan of the city).

a. Commercial facility is a supporting facility that enables economic activity and develop economic life of residents in form of building or business court as commercial service and workplace.
b. Educational facility is a supporting facility that enables students to optimally develop their knowledge, skills, and attitude with teaching-learning strategy based on the prevailing curriculum.
c. Health facility refers to the facility that sustains resident's health that functions to control the development and growth of the population.
d. Religious facilities are the facilities used to accommodate all the activities of worship and activity support.
e. Government facilities and public service facilities which can be used for the benefit of the public service, such as security guard service, meeting hall, neighborhood or community association office, police post, fire station, post office helpers, multipurpose buildings, administrative offices.

Open space planned with specific purpose that includes desired space quality and space functions. Therefore, it does not include open spaces as the remaining spaces or planned building groups. Facilities in open space each of the space and use of space outside of buildings, such as parks, roads, pedestarian, green lane, playing field, sports field and parking.

Environment facilities of flats must meet the following requirements:

1. Providing a sense of security, peaceful life, comfort and fit with local culture;
2. Fosteing a sense of belonging and to change habits that contradict with the life style in the flats;
3. Reducing the tendency to exploit or use the environment facilities for personal interests and specific groups;
4. Sustaining the most primary residential activity functions both the quantity and the types according to the existing environmental situation;
5. Accommodating functions related to the implementation and development aspects of economic and social culture.

2.4. Post-Occupancy Evaluation

Evaluation of post-occupancy is an evaluation to assess the success rate of a building in providing satisfaction of facilities to its residents. POE is performed to assess the suitability between building and built-environment with values and needs of residents, in addition to provide input in designing buildings with similar functions. POE is beneficial as short, intermediate, and long term reference and to provide improvements to increase residents' satisfaction on building and built-environment [9]. According to statement [3], Post-Occupancy Evaluation (POE) is defined as study or assessment on buildings' fulfilment in providing and supporting the residents, especially their values and needs. POE is performed to assess the suitability between building (built-environment) and values and needs of residents and as input in designing buildings with similar functions.

The concept of building performance is the basic philosophy and theoretical ground of POE that includes behavior, quality, and amenity aspects which are measured and evaluated [3]. The concept of building performance employed measurement, comparison, evaluation, and *feedback principles*. These are parts of systematic approach used to improve building environmental quality which includes variation of mechanism used to build buildings more responsive to the desirable functions and the needs of residents [3, 10]. POE is performed to assess the suitability between building and built-environment with values and needs of residents, in addition to provide input in designing buildings with similar functions. POE is beneficial due to its short, intermediate, and long term reference.

2.5. Investigation of Post-Occupancy Evaluation
The result of indicative POE influences the identification result of the main problem. Investigative POE includes various detailed and reliable topics. The main steps of investigative POE is identical with indicative POE, with higher effort, more time-saving, and the collected data and analysis technique are more perfect. Unlike indicative POE, where the criteria of evaluated buildings are based on the experience of evaluator, investigative POE employed a more objective and explicit research criteria. Evaluation criteria of investigative POE involves, at least, two activities: estimated reference compared to facilitation reference that is similar and already exist.

2.6. Diagnostic Post-Occupancy Evaluation
Using more sophisticated methods with more accurate result required several months to complete. The result is a comprehensive evaluation. Diagnostic POE follows diverse strategy methods, some of them are: questionnaires, surveys, and physical measurements where all approaches are adjusted with comparative evaluation on facilities with cross-sectionally similar types. Diagnostic POE is performed in a period of several months, one year, or more. The result and recommendation will be long-term oriented with objective not only to improve the main facility, but also building type reference given. The employed methodology is identical to the traditional method which focused on the use of scientific paradigm. Diagnostic POE is generally a large scale project that involves a wide range of variables. It is not infrequent that the effort performed is to improve the results that indicate relationship between variables. Therefore, Diagnostic POE employed both data collection and perfect analysis techniques which resulted on investigative and indicative POE [11]. The important part of diagnostic POE has been investigated, while its objective has correlation physically, environmentally, and behaviorally which give better experience on various relative form criteria. All the prerequisites demanded by diagnostic POE have considerable potential in creating accurate predictions on buildings shape and provide knowledge reference on types of building given through improvement on design criteria and literature guidelines. After being occupied, the possibility of performance change is present due to incompatibility of initial planning and building utilization, therefore it is not building performance evaluation. Building, in addition to physical requirements, needs to have functions based on the activity of its residents and subsequently, the interaction between building and residents exist [9, 12].

3. Research Methods
3.1. Material and Methods
This study employed descriptive qualitative method with objective to discover objective phenomenon based on observation and recording of physical and non-physical aspects of the research object. Descriptive qualitative analysis is employed to measure the comparison between the existing performance of infrastructures and supporting facilities of Mariso low-cost apartment (Rusunawa) with planning standard recommended by Ministry of Public Works in Government Regulation which regulate the minimum availability standard of infrastructures and facilities of low-cost apartment taken as research object.

The research location is administratively located in Lette, Mariso, Makassar, as shown in the following map:

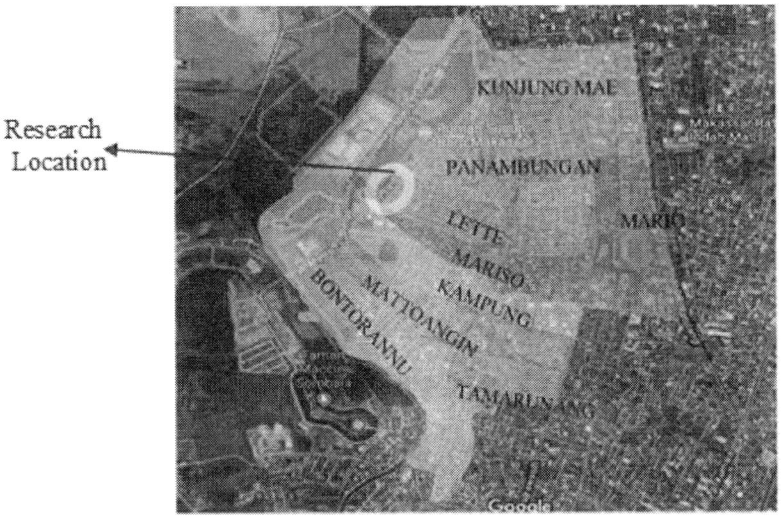

Figure 1. Research location [13]

3.2. Data Collection and Analysis Techniques
The data collection methods employed are observation and direct measurement for physical data, while non-physical data is recorded with direct interview on residents. The data analysis technique employed is qualitative descriptive with data grouping based on data categories which is further interpreted based on data categories, including comparison between planning standards and the availability of infrastructures and supporting facilities

4. Results and Discussion
This research is directed to post occupancy evaluation of infrastructures in Mariso low-cost apartment (Rusunawa) using references that regulate Ordinances of Environmental Facilities of Low-Cost Apartments, standards of environmental infrastructures of housing and settlement, and standards and technical regulations on environmental management of housing and settlement. In general, the research object (flats Mariso) is consisted of 6 five-story building blocks (1-4 floor for housing and ground floor for parking area). The size of house unit is 4x6 m with units per floor are 12 rooms. The total available room is 288 units with 1,152 total residents. The occupation of the residents is dominated by laborers and fishermen.

Figure 2. Mariso flats [13]

Following is the performance analysis of the flats infrastructures based on reference/standard infrastructures of settlement as mentioned above.

4.1. The Availability of Commercial Infrastructure

Commercial infrastructure is supporting infrastructure that enables economic activity and develop economic life of residents in form of building or business court as commercial service and workplace. Object of the study related to the performance of commercial infrastructures includes:

1. Stalls
2. Stores
3. Shopping center and business services

The standards of the commercial infrastructures above is presented in the Table 1.

Table 1. Commercial or work place facilities

Available Facilities	Minimal number of residents to be served (each unit of facilities)	Function	Location and maximum distance from the residential unit	Location and position at the building floor	Floor width	Land area (if it is a separate building)
Stalls	250 residents/50kk	Staple Food Seller	1. Located at the center of the environment 2. Accessible 3. Maximum 300 M radius	Located at the ground floor	18-36 M	$72M^2$ (with BCR 50%)
Stores	2500 residents	Selling daily necessities including clothing and food	Located at the center of the environment with access of 500 M radius	Located at the separated building	± 50 M^2	100 M^2 (with BCR 50%)
Shopping Center and Business Services	≥ 2500 residents	Selling clothing and food needs and services	Located at the center of the environment with access of 1000 M radius	Located at the separated building	± 600 M^2	1200 M^2 (with BCR 50%)

Source: National Indonesian Standards [14]

Total residents of 1,152 residents, the mandatory infrastructures in residential environment is stalls with an analysis as follows:

1 stall is reserved for 250 residents. Therefore, based on National Indonesian Standard [14] the number of stalls need to be prepared is 1,152/250=4.6 = 5 units. Recordings of the availability of commercial infrastructures in research site showed that only 1 unit of stalls is available. Therefore, to achieve the optimal service of commercial infrastructure, the addition of 4 units of stalls is needed.

4.2. The Availability of Educational Infrastructure

Educational infrastructure is supporting infrastructure that enables students to optimally develop their knowledge, skills, and attitude with teaching-learning strategy based on the prevailing curriculum. Educational infrastructure includes basic planning, design, and implementation of school buildings, in accordance with the decision of Director General for Secondary Education, Department of Education and Culture.

Object of the study related to the performance of educational infrastructures includes: (1) preschool; (2) elementary school; (3) junior high school; (4) senior high school. The standards of the educational infrastructures above is presented in Table 2.

Table 2. Educational facilities

Learning Psace Facilities	Minimal number of suitable resident	Function	Location	Distance	Number of Classrooms Needed	Floor Area Required	Land Area Required
Pre-Learning Level	1500 populatio n where children at age of 5-6 years accounted for 8%	Accommodate the implementatio n of pre-school education of 5 or 6-year old children	Within the neighborhood / integrated with playground within the neighborhood	Distance access of 500 M. Calculate d from the most distant unit and the highest floor is 500 M	Calculated based on the system of elementary school education of 5-6 years using the formula (1)	125 M^2 1.5 M^2/ Student	2,500 M^2
Elementar y School	1600 populatio n	Accommodate the implementatio n of primary school education	Not crossing the environment road and within the neighborhood	Reachable within the maximum radius of 1000 M from the farthest unit and the highest floor	Calculated with formula (2)	1.5 M^2/ Student	2,000 M^2
Junior High School	4800 populatio n	Accommodate the implementatio n of junior high school	Not located at the center of the environment, able to integrate with sports field and other educational facilities.	Maximum Radius of 100 M	Calculated with formula (3)	1.75 M^2/ siswa	9000 M^2
Senior High School	≥ 4800 populatio n	Accommodate the educational implementatio n of Senior High School	Able to integrate with sports field or other educational facilities and not located at the center of the environment	Maximum radius of 3 km from the unit	Calculated with formula (4)	1.75 M^2/individ ual	1. one-story senior high school for 12.500 M^2 2. two-story senior high school for 8000 M^2 3. three-story senior high school for 5000 M^2

Source: National Indonesian Standards [14]

Total residents of 1,152 residents, there is no obligation to build educational infrastructures in the research site. Nevertheless, educational problems of children on site can still be fulfilled with the nearest, existing educational infrastructures.

4.3. The Availability of Health Infrastructure

This infrastructure refers to infrastructure that support the health of the residents also functions to control the development and growth of the population. Object of the study related to the performance of health infrastructures includes: (1) integrated health post; (2) clinic; (3) Maternal and Children Health Center and maternity hospital; (4) community health center; (5) physician; (6) pharmacy. Furthermore, the standards of the health infrastructures above is presented in Table 3.

Table 3. Health facilities

Facilities	Minimum residents served	Functions	Distance	Minimum space requirements	Floor area required	Land area required
Integrated Health Post (Posyandu)	1000 residents	Provide health services to toddlers	Reachable within the maximum radius of 2000 M from the farthest unit and the highest floor	A room to accommodate health activities	30 M^2	60 M^2 (KDB 50%)
Clinic	1000 residents	Provide health services to residents	Reachable within the maximum radius of 400 M from the farthest unit and the highest floor	-	150 M^2	300 M^2 (KDB 50%)
Maternity and Children Health Center and maternity hospital	10.000 residents	Provide services to mothers prenatal, natal, and postnatal and provide services to children up to six years old	Reachable within the maximum radius of 100 M from the farthest unit and the highest floor	At least two rooms are available as checking and waiting	600 M^2	1200 M^2 (KDB 50%)
Community Health Center	30.000 residents	Provide comprehensive health services to residents that include child specialist, dental specialist, and provide services to children up to six years old	Reachable within the maximum radius of 1000 M from the farthest unit and the highest floor	At minimum, doctor examination room, dental examination room, and waiting room are available	350 M^2	-
Physician	5000 residents	Provide first general/specialist health services to residents	Reachable within the maximum radius of 1000 M from the farthest unit and the highest floor	An examination room and waiting room.	Minimum 18 m^2	-
Pharmacy	10.000 residents	Serve residents in drugs precurements	Reachable within the maximum radius of 1000 M from the farthest unit and the highest floor	Store room, drug formulation room, and waiting room.	Minimum 36 M	-

Source: National Indonesian Standards [14]

Total residents of 1,152 residents, the mandatory health infrastructures are integrated health post and clinic. Recordings of the availability of health infrastructures in research site showed that only health infrastructure available is 1 unit of integrated health post. Based on the standards of health infrastructure, this amount is unable to serve the needs of residents. However, based on the same standard, there is a lack of clinic in research site. Therefore, to achieve the optimal service of health infrastructure, the addition of 1 unit of clinic is needed.

4.4. The Availability of Religious Infrastructure

Religious infrastructure is any infrastructures used to accommodate all religious activities and supporting activities of residents in research site. Object of the study related to the performance of religious infrastructures includes: (1) prayer room; (2) local mosque; (3) community mosque; (4) subdistrict mosque; (5) other religions place of worships. With total residents of 1,152 residents, the mandatory religious infrastructures is prayer room. Furthermore, the standards of the health infrastructures above is presented in Table 4.

Table 4. Religious facilities

No.	Types of Facilities	Suitable Number of Population	Needs For Each Unit of Facillity		Standard (m^2/ individual)	Criteria	
			Floor area Min (m^2)	Land Area Min. (m^2)		Access Radius	Location and Completion
1.	Musholla/ Small Mosque	250	45	100 when built separatedly	0.36	100 m^2	Within the neighborhood Can be a part of the building other means
2.	Residential Mosque	2,500	300	600	0.24	1,000 m^2	Within the neighborhood Not crossing the highway Able to integrate with community centers.
3.	Urban Village Mosque	30,000	1,800	3,600	0.12		Reachable by public transportation
4.	Sub-district Mosque	120,000	3,600	5,400	0.03		Adjacent to the environment/ urban village. As a two-storey facility, CBR 40%
5.	Other religion's worship place	Depend on the kinship system / hierarchy of institutions	Depend on local customs	Depend on local customs	-	-	-

Source: National Indonesian Standards [14]

Field observation showed that in research site only 1 mosque available and usually functioned to perform congregate Friday prayer. Therefore, there is no obligation to build prayer rooms in the research site.

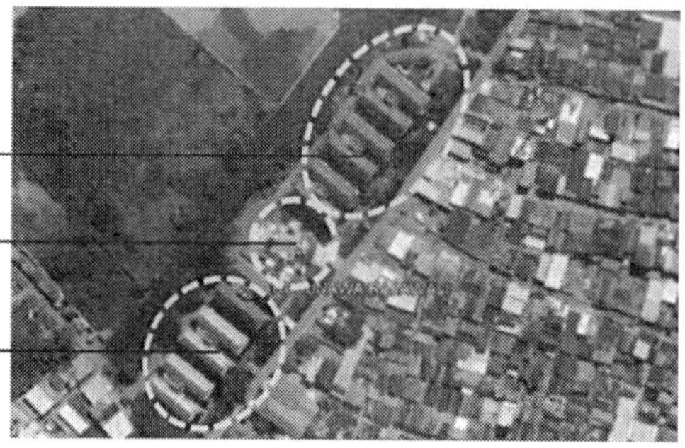

Figure 3. Location of mosque in Mariso Low-cost Apartment [14]

4.5. The Availability of Government and Public Services Infrastructures

Government and public services infrastructure are facilities used for the benefit of public services such as security post, meeting hall, neighborhood and community association offices, security post, police post, public telephone station, multipurpose building, open space, and mail box. The standards of the government and public services infrastructures above is presented in Table 5.

Table 5. Government and public services facilities

No.	Provided facilities	Maximum number of that can have access	The location and the maximum distance from occupancy unity	Location at the building floor	The minimum floor area	Minimum floor area (Separated building)
1.	Neighborhood Office	250 residents	At the center of flats environment	Can be located on the occupancy floor	18 M^2 – 36 M^2	-
2.	Hamlet office	1000 residents	Located at the center of the neighborhood and integrated with multipurpose room.	Can be located on the occupancy floor	36 M^2	-
3.	Security Post	200 residents	Located at the center of the neighborhood with a maximum reachable distance of 200 M	Can be located on the occupancy unit ground floor	4 M^2	6 M
4.	Police Post	2000 residents	Located in the front part or around the environment	Can be located on ground floor of the occupancy unit building	36 M^2	72 M
5.	Payphone	200 population	Located next to the other public services	At the ground floor	60 x 60 cm	-
6.	Multipurpose Building	10000 population	Located at the center of the neighborhood with a maximum reachable distance of 500 M	At the ground floor	250 M^2	500 M^2
7.	Open Space	200 population	Can be integrated or may utilize multipurpose room	At the ground floor	100 M^2	-
8.	Post Office Box	1000 population	In front of each residential building	At the ground floor	-	-

Source: National Indonesian Standards [14]

Based on the analysis of activities, with the total of residents of 1,152 residents, the mandatory government and public services infrastructures are neighborhood and community association offices, security post, public telephone station, multipurpose building, open space, and mail box. From field observation, it was found that the available public services are management office, security center, power house, and garbage bank. Whereas the public services facilities which are not available and need to be established are, neighborhood and community association office, public telephone station, multipurpose building, open space, and mail box.

Figure 4. Management office of Mariso Low-cost Apartment

Figure 5. Security post of Mariso Low-cost Apartment

Table 6. Standards of open spaces

No.	Types of Facilities	Suitable Number of Population	Land Requirement Min. (M^2)	Standard (m^2/ individual)	Reachable Radius (M)	Location Criteria and Completion
1.	Park/ Playground	250	250	1	100	At the center of neighborhood
2.	Park/ Playground	2,500	1,250	0.5	1,000	At the center of activities of the environment
3.	Park and Sports Field	30,000	9,000	0.3		Better if integrated with educational facilities
4.	Park and Sports Field	120,000	24,000	0.2		Located in the main road. Better if integrated with educational facilities
5.	Green line	-	-	15 M		Spread
6.	Cemetery/ Public Cemetery	120,000				Considering the reachable radius and the facilitated area.

Source: National Indonesian Standars [14]

4.6. The Availability of Open Space

The intended open space in this study is any open spaces planned with specific purpose that includes desired space quality and desired space functions. Therefore, it does not include open spaces as the remaining spaces or planned building groups. The standards of open spaces infrastructures for urban environment is presented in Table 6. Based on the analysis of activities, with the total of residents of 1,152 residents, the mandatory infrastructure in residential environment of Mariso Low-cost apartment is park/open field as social interaction facility between residents.

5. Conclusion

Mariso low-cost apartment is one of vertical housing intended for Low-Income Communities (LIC) in Makassar and is intended to fulfill the governmental duty of providing housing for citizens. Mariso low-cost apartment evaluated by Post-occupancy evaluation is to measure the performance of supporting its infrastructures of building with 10 years of occupancy. This evaluation is intended to evaluate the performance, in addition to evaluate the actual availability of infrastructures during evaluation period based on technical standards, both SNI and regulations according to the function of the building. During its operation, the decrease of performance is present in addition to other deficiency in terms of infrastructures availability based on the standards of infrastructures and facilities of urban environment. Thus, improvements and addition of supporting infrastructures are necessary to ensure the function of Mariso Low-cost apartment as housing that fulfill the sustainability of physical and non-physical activities of its residents.

References

[1] Thaddi, Z. R., and Admane, S. V. (2015). *Evaluation of factors for Post occupancy Satisfaction analysis of a Residential Building* – A review. Engineering and Scientific International Journal (ESIJ), 2(2), 53–59.

[2] Ning, Y., and Chen, J. (2016). Improving residential satisfaction of university dormitories through post-occupancy evaluation in China: *A socio-technical system approach. Sustainability (Switzerland)*, *8*(10), 1–7.

[3] Presiser, W. F. E, Rabinowitz, H.Z, and White, E.T. (1988). *Post-Occupancy Evaluation.* New York: Van Nostrand Reinhold Company.

[4] Law No. 20 of 2011 Regarding Condominium, Apartment or Flats. [in Bahasa]

[5] Government Regulation no. 4 of 1988 on Technical Requirement of Condominium Construction. [in Bahasa]

[6] Vitriana, A. (2018). Research Report on Post-Occupancy Evaluation of Rancakaek Transit Apartment, Bandung. [in Bahasa]

[7] Prasojo, W., and Frida, N. (2014). Analysis of Resident's Satisfaction Level on Apartment Occupancy In The Perspective of Building Quality in Surabaya Regency, Rekayasa Teknik Sipil, 3(1), 54–62. [in Bahasa]

[8] Indonesian National Standard No. 03-7031-2014 on Mechanism of Low-Cost Apartment Environmental Facility Planning. [in Bahasa]

[9] Suryadhi AB, 2005. Post-Occupancy Evaluation of Emergency Department at Tabanan General Hospital, Thesis, Universitas Gadjah Mada, Indonesia. [in Bahasa]

[10] Waani, J. O. (2015). Post-Occupancy Evaluation (POE): spatial behavior aspects of SLB YPAC Manado, *Media Matrasain*, *12*(3), 1–13. [in Bahasa]

[11] Budiarso, 2007., UR. Wisnu, ST, MT., Study Design of Post-Occupancy Evaluation Method of Sarijadi Bandung Flats, Fakultas Teknik Universitas Budi Luhur. [in Bahasa]

[12] Khair, N., Ali, H. M., Sipan, I., Juhari, N. H., and Daud, S. Z. (2015). Post occupancy evaluation of physical environment in public low-cost housing. *Jurnal Teknologi*, *75* (10), 155–162.

[13] Anonim. 2019. Mariso Flats. URL: https://www.google.com/maps/search/rumah+susun+mariso/.

[14] National Indonesian Standard (SNI)-03-1733-2004 on Planning of Residential Neighborhood in Urban Area Procedure.

Factors Affecting Thermal Comfort of Elementary Schools' Students in Makassar

B Hamzah[1]*, R Mulyadi[1], and M T Ishak[1]

[1]Department of Architecture, Faculty of Engineering, Universitas Hasanuddin, Kampus Fakultas Teknik Universitas Hasanuddin, Bontomarannu Gowa, Sulawesi Selatan, Indonesia 92171

*Email: baharsyah@unhas.ac.id

Abstract. The article aims to analyze the most affecting thermal environmental factors on students' thermal comfort in elementary schools. The questionnaire has been collected from 1,111 students. The one-minute interval of data measurements was carried out in the 33 classrooms at six primary schools in Makassar. Environmental data involved measurements of air temperature, globe temperature, air velocity, and relative humidity. Students wrote down their personal data and their response to the thermal environments of classrooms in the questionnaire sheets. The results show that the most affecting factor of thermal comfort of students is operative temperature (T_O). Based on the students' thermal sensation votes, the operative temperature ranges from 29.1 to 30.0 °C is recommended to provide a comfortable condition for students to learn more successfully in the naturally ventilated primary schools' classrooms. This result is slightly lower than the neutral temperature obtained from the regression between actual votes and the operative temperature, which was 30.2 °C. The results agree with the neutral temperature calculated using the revised Humphreys adaptive model which give the result of 29.4 °C. The second variable that has a significant effect on the students' thermal comfort is air velocity. The students felt comfortable at air velocity at about 0.25 m/s and also at air velocity more in the ranges of 1.00 to around 1.50 m/s.

1. Introduction

It is essential to provide good environmental conditions in the classrooms for students that enable them to learn more effective and successfully [1-3]. One of these environmental conditions is thermal comfort. ASHRAE standard 55 [4] defines thermal comfort as "the state of mind that expresses satisfaction with the surrounding environment." Fanger [5] describes three main factors determining thermal comfort of occupants. These factors are including thermal environments, individual, and other contributing factors. Environmental factors involve air temperature, air velocity, relative humidity, and radiant temperature. Individual factors involve clothing index and type of activity. Food and drink, acclimatization, body shape, subcutaneous fat, age and sex, and state of health are among contributing factors.

Several thermal comfort studies of students in the elementary schools have been carried out worldwide. The most updated studies in elementary schools [6-10] will be discussed. The studies analyzed different and broad aspects of thermal comfort in the elementary schools. Kim and de Dear [6] analyzed the thermal comfort expectations and behavior adaptation of students in secondary and elementary schools in Australia. Kim and de Dear found that among many other adaptive options, air conditioning is the most favorable adaptation method. Ali and Al-Hashlamun [7] assessed the thermal environment of indoors at different prototypical school buildings in Jordan. They compared the indoor thermal comfort between new and old schools. They found that the new school buildings provide a more satisfactory thermal environment than the old one. Jindal [8] presents a study of thermal comfort in naturally ventilated classrooms in India. Jindal found that there was a variation of comfortable classrooms between winter and summer session. Jiang et al. [9] explain a study on learning performance of students and thermal comfort of elementary schools in China. Jiang et al. [9] found that the environmental factors were strongly affecting the students' votes on overall indoor environmental quality (IEQ). Moreover, the learning performance of students is affected by all IEQ complaints and there was a good correlation between the number of complaints and the learning performance. Trebilcock et al. [10] discusses an adaptive thermal comfort taken from a field study in elementary schools in Chile. They found that socio-economic background of students affected the their thermal comfort votes.

In Indonesia, several studies assessing thermal comfort in educational buildings including primary schools [11-16] have also been carried out. The previous articles based on Makassar data have highlighted that even though the temperature of classrooms was hot, students in primary schools [13], secondary schools [12], and university [14] were mostly felt comfortable with around 80% of thermal acceptance. This phenomenon has led to another study that analyzed the application of adaptive model thermal comfort [15]. Using data collected from six elementary schools in Makassar, Hamzah et al [15] found a good agreement between the neutral temperature originated from actual votes and the one calculated from adaptive model [17-19].

This article aims to analyze the most affecting thermal environment factors that influence the students' thermal comfort in naturally ventilated classrooms in the elementary schools located in Makassar, Indonesia. The study is focused on the effect of each thermal environmental factors i.e. operative temperature, relative humidity, and air velocity on the students' thermal comfort. Besides, the integrated effect of the thermal environment variables is also analyzed and discussed in this article.

2. Research Method

2.1. Data Collection

Data have been collected from six nominated primary schools located in the city of Makassar, Indonesia. The data collection involved 33 classrooms and 1,111 students. The measurements and questionnaires data have been discussed in the previous conferences [13, 15]. The study used two protocols to collect data as follows:

- The objective surveys or measurements were carried out to record the thermal environment parameters and also the personal data of respondents. The one-minute interval of data measurements was carried out in the 33 classrooms at six primary schools in Makassar. The survey made use of one set of LSI TC, a multi logger of thermal comfort and four Hobo data-loggers. The LSI TC was manufactured by LSI-Lastem Italy. In this survey, The LSI TC connected with three sensors. Those sensors consist of one sensor (BST131) for measuring globe temperature, one sensor (BSU102) for measuring the air temperature and relative humidity, and one sensor (BSU121) for collecting wet bulb temperature. Besides, the measurements of thermal environments were also recorded by Hobo data-loggers, which was developed and manufactured by Onset in the United States. There were four standard Hobo loggers (Hobo-1) and two Hobo loggers (Hobo-2) with the external sensor were used in the surveys. Hobo-1 measured and collected air temperature and relative humidity. The external sensor attached in the Hobo-2 was a wire anemometer, which used to collect the velocity of air inside the classrooms. All sensors were placed in the six locations inside the classrooms at about 1 m above the floor level, following the same high as in Wong and Khoo [20]. The arrangement of instruments in a typical classroom is illustrated in Figure 1. The instruments used in this article has been explained in Hamzah et al. [13, 15]. Because of the number of LSI TC instruments are limited, two environmental parameters could not be measured at the six positions. The MRT was only measured and recorded at position A, and the air velocity at positions A and B. The position of instruments in the classrooms has been also explained in Hamzah et al. [13]. Students wrote their details of clothing data and indicated their position in the questionnaires. While the surveyors recorded the activity of students in the classrooms. The situation of typical measurement is shown in Figure 2.

- The subjective measurements or surveys were carried out to collect the students' responses on the thermal environment conditions. This was collected by questionnaire method. The questionnaire was developed and modified based on the questionnaire used in Wong and Khoo [14] in Singapore. This modified questionnaire has also been used in the previous study in Makassar, Indonesia [11-15]. The adapted questionnaire includes seven questions. Four questions recorded the students votes on the thermal aspects (thermal sensation, comfort level, thermal preference, and thermal acceptance). The following two questions asked the students' votes on the air velocity. The last question intended to get the students' votes on the humidity of classrooms. This questionnaire asked the TSV of respondents in seven scales following the ASHRAE standard protocol [4], namely: cold (-3), cool (-2), slightly cool (-1), neutral (0), slightly warm (+1), warm (+2), and hot (+3).

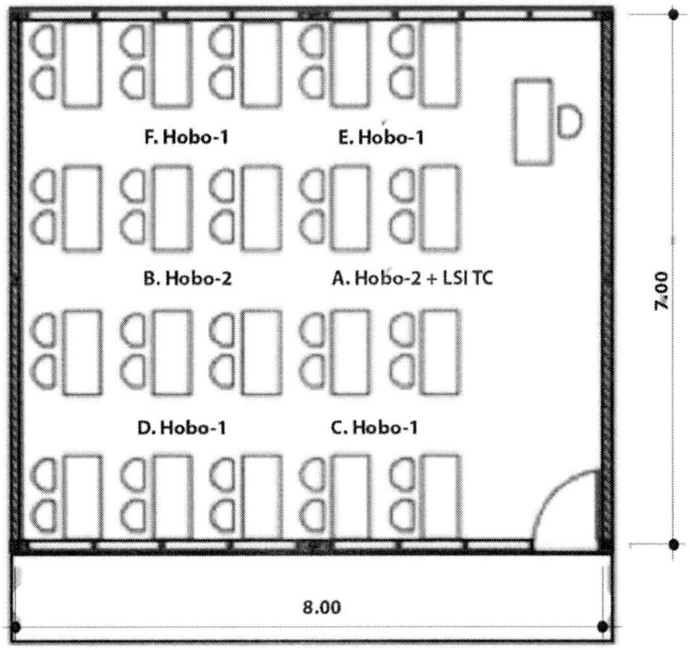

Figure 1. The placement of measurement instruments in a typical classroom

2.2. Data Analysis

Spreadsheet and statistical analysis software have been used in data analysis. The mean value of environmental variables has been calculated using the spreadsheet software Microsoft (MS) Excel. The same software has also been used to generate bar-chart graphics illustrating the percentage of TSV for each environmental variable (operative temperature, air velocity, and relative humidity). The scatter plot graph and linear regression analyses have been calculated using a statistical package for social science (SPSS) version 25. The regression illustrates the graphical of respondents' thermal sensation votes over the operative temperature (T_O).

Figure 2. The situation of measurement in a typical classroom

3. Results and Discussion

3.1. The Effect of Temperature on the Thermal Comfort of Students

The effect of temperature on the thermal comfort of students can be assessed based on operative temperature (T_o). Operative temperature is equal to the average between air temperature and radiant temperature.

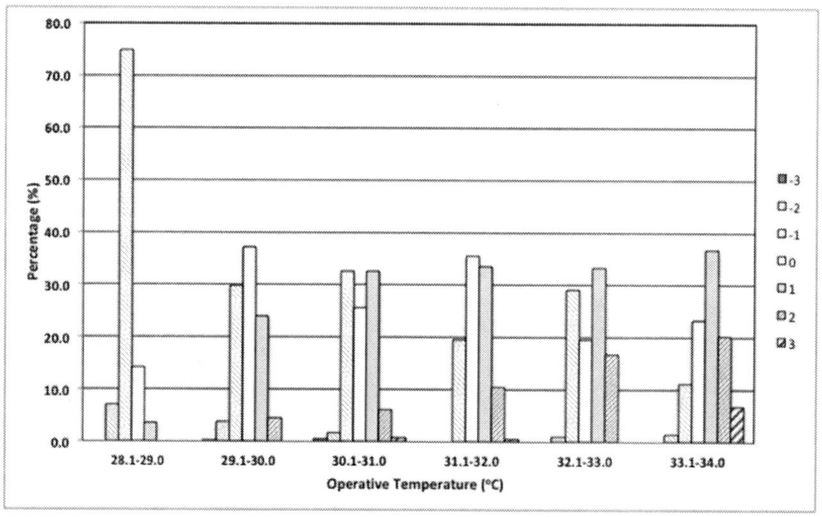

Figure 3. The percentage of thermal sensation votes (TSV) based on the operative temperature (T_O)

The distribution of students' thermal sensation votes (TSV) in each operative temperature group is presented in Figure 3. In the temperature range of 28.1-29.0 °C, most of the students voted in the slightly cool (-1). The highest percentage of students who votes neutral (0) is in the temperature ranges 29.1-30.0 °C, followed by 31.1-32.0 °C. In temperature ranges from 29.1 to 34.0 °C, the percentage of students votes neutral has fluctuated with a downward trend. On the contrary, the percentage of students who votes warm and hot regions (+1 to +3) is mostly increased. The highest percentage of student who felt slightly warm, warm and hot is the temperature range 33.1-34.0 °C. Based on this analysis, the temperature range of 29.1-30.0 °C is the most comfortable temperature for students in the naturally ventilated classrooms of elementary schools in Makassar. This figure is similar to the neutral temperature calculated from Humphreys' adaptive model [17] i.e. 29.4 °C as described in Hamzah et al. [15].

Figure 4 shows the scatterplot and regression between students' votes and the operative temperature (T_O). The formula of regression analysis between TSV and the operative temperature (T_O) is presented as follows:

$$TSV = 0.22T_o - 6.64 \qquad (1)$$

where,

TSV = thermal sensation vote

T_O = operative temperature

The neutral temperature (T_O) obtained from the regression between actual votes (TSV) and the operative temperature is 30.2 °C (T_O). This figure is quite close to the neutral temperature (i.e. 29.4 °C [15]) calculated using Humphreys adaptive model [17].

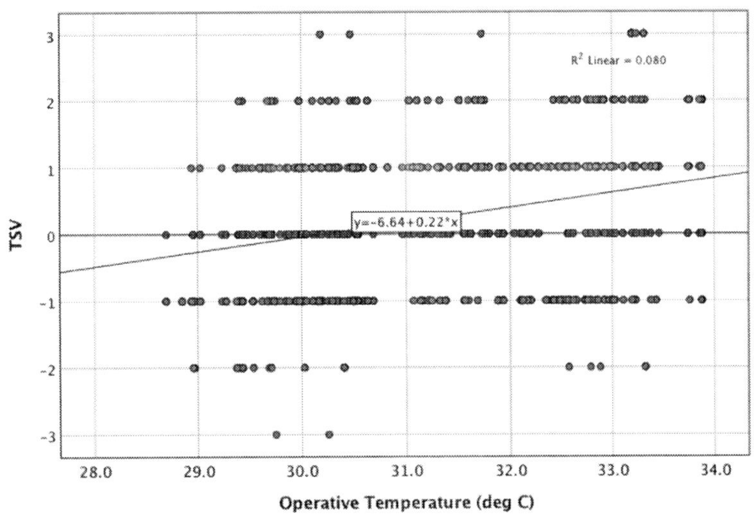

Figure 4. The scatterplot with regression equation between operative temperature and TSV

3.2. The Effect of Air Velocity on the Thermal Comfort of Students

Figure 5 shows the percentage of thermal sensation votes (TSV) based on groups of air velocity. The air velocity is ranging from less than 0.25 m/s to more than 1.25 m/s. In the <= 0.25 m/s group, the percentages of respondents who felt slightly cool (-1), neutral (0), and slightly warm (+1) are almost similar. This indicates that the air velocity of less than 0.25 m/s has very little impact on the thermal comfort of respondents. In the 0.26-0.50 and 0.51-0.75 groups, the percentage of students who felt slightly warm (+1) is higher than those who felt slightly warm and neutral (0). This indicates that in medium air velocity group students tend to feel warm. While in the last two groups (1.01-1.25 and >1.25 m/s), the percentage of students who felt slightly warm (+1) is lower than those who felt slightly cool (-1) and neutral (0). This indicates that in the group of air velocity of more than 1.00 m/s to around 1.25 m/s, the percentage of students felt comfortable increased from previous air velocity group.

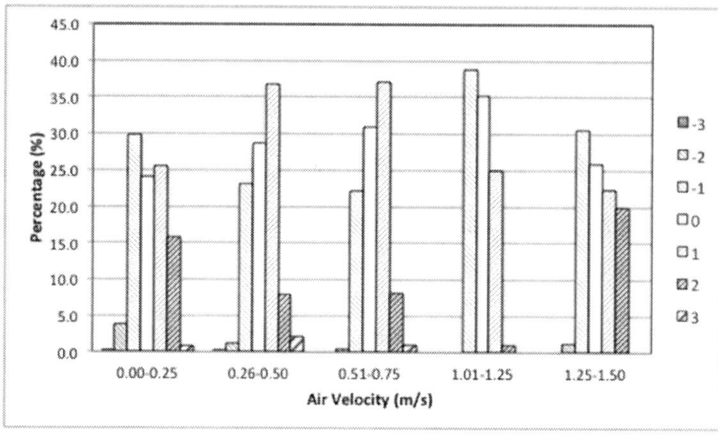

Figure 5. The percentage of thermal sensation votes (TSV) based on the air velocity

3.3. The Effect of Relative Humidity on the Thermal Comfort of Students

The effect of relative humidity experienced by students in the classrooms on their thermal comfort is presented in Figure 6. As seen in the figure, the highest percentage of students votes neutral (0) is in the humidity range in between 65.1 to 70.0%. In this group, about 35% of students were felt neutral. The highest percentage of students who felt either slightly warm (+1), warm (+2), or hot (+3) is the relative humidity range from 60.1 to 65.0%.

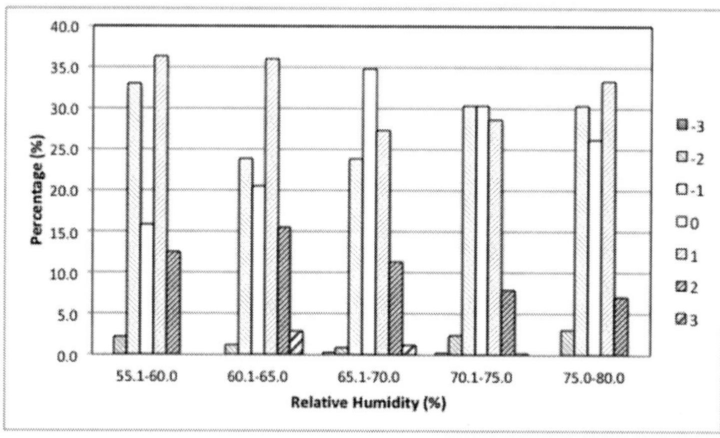

Figure 6. The percentage of votes based on the relative humidity (RH)

3.4. The Effect of an Integrated Thermal Environment on the Students' Thermal Comfort

Multiple regression analysis was used to assess the effect of an integrated thermal environment on the students' thermal comfort in the classrooms. With the F = 47.345 and Sig. = 0.000, ANOVA table confirms that the multiple regression is linear (Table 1). The test of coefficient as presented in Table 2 shows that all coefficients are significant. Therefore, the TSV can be predicted using this multiple linear regression formula as follows:

$$TSV = 0.385T_o + 0.049RH - 0.29AV - 15.006 \qquad (2)$$

where

TSV = thermal sensation vote
T_o = operative temperature
RH = relative humidity
AV = air velocity

Table 1. ANOVA of multiple regressions

Model		Sum of Squares	Degree of freedom	Mean Square	F	Sig.
1	Regression	144.008	3	48.003	47.345	0.000[b]
	Residual	1122.375	1107	1.014		
	Total	1266.383	1110			

Table 2. The coefficients of multiple regressions

		Unstandardized Coefficients		Standardized Coefficients		
Model		B	Std. Error	Beta	t	Sig.
1	(Constant)	-15.006	1.587		-9.455	0.000
	Operative Temperature (T_o)	0.385	0.035	0.495	11.08	0.000
	Relative Humidity (%)	0.049	0.009	0.243	5.541	0.000
	Air Velocity (m/s)	-0.29	0.081	-0.105	-3.583	0.000

Equation 2 shows that the dependent variable TSV has a positive correlation with the independent variable of the operative temperature and the relative humidity. This means that the TSV will increase when the operative temperature or the relative humidity increase. Based on the comparison between the coefficient of operative temperature and relative humidity, the operative temperature will affect the thermal comfort by about eight times the effect resulted from relative humidity. The coefficient of operative temperature will give an impact on the increase or reduction of one TSV if there is an increase or reduction of three °C. The same effect will be expected from an increase or a reduction of 24% of RH.

4. Conclusion

Based on the results and discussion, it is found that the most affecting factor of thermal comfort of students is operative temperature (air temperature and radiant temperature). Based on the students' thermal sensation votes, the operative temperature ranges from 29.1 to 30.0 °C is recommended to provide a comfortable condition for students to learn in the naturally ventilated primary schools. This result is slightly lower than the neutral temperature obtained from the regression between actual votes and the operative temperature. The neutral temperature is 30.2 °C. The results agree with the neutral temperature calculated using the revised Humphreys adaptive model. The second variable that has a significant effect on the students' thermal comfort is air velocity. The students felt comfortable at air velocity at about 0.25 m/s and also at air velocity more in the ranges of 1.00 to around 1.50 m/s.

Acknowledgment

The authors would like to express their appreciation to the Faculty of Engineering, Universitas Hasanuddin University for providing research funding under the Laboratory-Based Education (LBE) scheme with contract No. 6294/UN4.7.2/PL.00.00/2019 dated 12 April 2019. The authors would also like to thanks the students in the Building Science Laboratory who involved in this research during data collection and analyses.

References

[1] Pepler R D and Warner R E 1968 Temperature and learning: an experimental study *ASHRAE Transactions*, 74(1) 211-224.

[2] Bellia L, Boerstra A, Dijken F V, Ianniello E, Lopardo G, Minichiello F, Romagnoni P and Silva M G d 2010 *Indoor Environment and Energy Efficiency in Schools - Part 1*, ed. F.R.d.A. Alfano. Brussels: REHVA.
[3] Mendell M J and Heath G A 2005 Do indoor pollutants and thermal conditions in schools influence student performance? A critical review of the literature *Indoor Air*, **15** 27-52.
[4] ASHRAE 2013 *Thermal Environmental Conditions for Human Occupancy (ANSI/ASHRAE Standard 55-2013)*. Atlanta, US: ASHRAE.
[5] Fanger P O 1970 *Thermal comfort - Analysis and application in environmental engineering*. Copenhagen: Danish Technical Press.
[6] Kim J and de Dear R 2018 Thermal comfort expectations and adaptive behavioural characteristics of primary and secondary school students *Building and Environment*, **127** 13-22.
[7] Ali H H and Al-Hashlamun R 2019 Assessment of indoor thermal environment in different prototypical school buildings in Jordan *Alexandria Engineering Journal*, **58**(2) 699-711.
[8] Jindal A 2018 Thermal comfort study in naturally ventilated school classrooms in composite climate of India *Building and Environment*, **142** 34-46.
[9] Jiang J, Wang D, Liu Y, Xu Y and Liu J 2018 A study on pupils' learning performance and thermal comfort of primary schools in China *Building and Environment*, **134** 102-113.
[10] Trebilcock M, Soto-Muñoz J, Yañez M and Figueroa-San Martin R 2017 The right to comfort: A field study on adaptive thermal comfort in free-running primary schools in Chile *Building and Environment*, **114** 455-469.
[11] Hamzah B, Kusno A and Mulyadi R 2019 Design of energy efficient and thermally comfortable air-conditioned university classrooms in the tropics *International Journal of Sustainable Energy*, **38**(4) 382-397.
[12] Hamzah B, Gou Z, Mulyadi R and Amin S 2018 Thermal comfort analyses of secondary school students in the tropics *Buildings*, **8**(4).
[13] Hamzah B, Mulyadi R and Amin S 2017 *Thermal Comfort Analyses of Elementary School Students in the Tropical Region* in *51st International Conference of the Architectural Science Association (ANZAScA)* Wellington New Zealand: Department of Architecture, Victoria University of Wellington.
[14] Hamzah B, Ishak M T, Beddu S and Osman M Y 2016 Thermal comfort analyses of naturally ventilated university classrooms *Structural Survey*, **34**(4/5) 427-445.
[15] Hamzah B, Mulyadi R, Amin S and Kusno A 2020 Adaptive thermal comfort of naturally ventilated classrooms of elementary school in the tropics in *IOP Conf. Ser.: Earth & Environment Science* **402** 012021.
[16] Karyono T H, Heryanto S and Faridah I 2015 Air conditioning and the neutral temperature of the Indonesian university students *Architectural Science Review*, **58**(2) 174-183.
[17] Humphreys M A and Nicol J F 2000 Outdoor temperature and indoor thermal comfort: raising the precision of the relationship for the 1998 ASHRAE database of field studies *ASHRAE Transactions*, **106** 485-92.
[18] Humphreys M A 1981 *The dependence of comfortable temperatures upon indoor and outdoor climates*, in *Bioengineering, Thermal Physiology and Comfort*, Cena and Clark, Editors. Elsevier: Amsterdam. p. 229-250.
[19] Auliciems A 1989 *Human response to the environment*, in *Building Design and Human Performance*, N.C. Ruck, Editor. Van Nostrand Reinhold: New York.
[20] Wong N H and Khoo S S 2003 Thermal comfort in classrooms in the tropics *Energy and Buildings*, **35**(4) 337-351.

IOP Conf. Series: Materials Science and Engineering **875** (2020) 012013 doi:10.1088/1757-899X/875/1/012013

Relationship between Regional Development Level and Accessibility Index in Makassar Coastal Area

A Akil[1]*, A Yudono[1], S Wahyuni[1], R Ramadhani[1], and I Kamaruddin[1]

[1]Department of Regional and Urban Planning, Engineering Faculty, Hasanuddin University, Makassar, Indonesia

*Email: arifuddinak@yahoo.co.id

Abstract. Makassar population growth is also quite large, accompanied by the development of urban activities. The development was followed by city facilities and road. The connection between central and periphery area is a demand for community activities in an efficient, safe and comfortable manner. However, in general, the area has low-level accessibility, although the distance is relatively close. The study aims to examine the relationship between regional development level and accessibility level in Makassar coastal area. The accessibility index is analysed using the alpha index (α), while regional development level using LQ-analysis which focuses on residential, education, health, worship, and commercial. Furthermore the relationship between accessibility index and regional development using correlation analysis. The study results showed that the highest accessibility index was located in the city centre and the lowest in the urban periphery. It was also found that urban facilities tend to be higher in city centre than periphery direction. Low accessibility as a result of inequality in road infrastructure has implications for low performance of urban development. It shows that there is a connection between regional development and accessibility level. Therefore, it is recommended to policymakers to build access with high connectivity between the city centre and periphery area while paying attention to the availability of public open space.

1. Introduction

It has become the common view that a development good attention to the development of the society, social and environmental needs economic cost [1]. Urban space needs to have implications for the development of the built area. There are two aspects that affect urban development, namely physical aspects such as geographical location and non-physical aspects such as socio-economic potential of the region [2]. Development of urban areas uses a variety of resources, both natural resources, artificial and human resources optimally in the region. The development of urban area is a continuous process of the various results decision making inside or outside that effect an area. Factors that influence urban development are influenced by various locational aspects, urban systems, and agglomeration mechanism.

Inter-regional quality in the level of development, level of welfare, and rate of economic growth exists to some extent in almost every country [3]. Each region has different social, economic, cultural and physical conditions, so the concept of regional development should emphasize its potential. According to [4], regional development is an effort to spur socio-economic development, reduce disparities between regions and preserve the environmental sustainability of a region. Meanwhile, according to [5], one indicator of regional development is the ease of people obtaining the necessities of daily life. The life needs of the community can be met from various urban service facilities, especially social and economic facilities such as education, health, worship, government and public services, trade and industry, as well as sports facilities and green open spaces. This is relevant to the view of [6] that the built environment variable can be defined in four factors, including: population density factor, land use diversity factor, design factor (road quality, block size), and accessibility factor. In this case, in addition to the city service facilities, the availability of infrastructure and facilities is also needed. The infrastructure and facilities referred to include the completeness of the physical basis that enables urban areas to function. Some types of urban infrastructure, according to [7] include transportation, freshwater, greywater, drainage, waste, electricity, and telecommunication.

Between regional development and accessibility, there is very dynamic and complex interaction, involving various aspects of activities and interests. Changes in land use always influence road development and vice versa. The condition then creates a spatial relationship between the socio-economic and physical characteristics of the environment [8]. In line with that, [9] states that the pattern of change and the magnitude of movement and the choice of movement modes are functions of land-use change patterns. Likewise, any change in land use will certainly require an improvement in the transportation system. According to [10], accessibility is a concept that combines a geographic land use system with a transportation system that connects it. Furthermore according to [11] land use can identify urban activities in each zone. According to [11], each zone can be characterised by three measures, such as the type of activity, intensity of use, and accessibility between land use. The development of the area will quickly develop if supported by the fulfilment of facilities and good accessibility. While the development of urban infrastructure, especially road will facilitate the achievement of area.

Accessibility is one part of the analysis of the activities that interact with the transportation network system. One variable that can be stated whether the level of accessibility is high or low can be seen from the number of network systems available in the area. The more network systems available in the area, the easier accessibility is obtained and vice versa the lower the level of accessibility is obtained, the more difficult the area is reached from another region [12]. Besides, what determines the level of access is the pattern of land use management. Land use that has characteristics and is closely related to the function will cause high accessibility, and conversely, irregular land use planning will have an impact on inefficiency inaccessibility between regions. Thus the distance factor is not the only element that determines the level of accessibility. The fact, two adjacent zones cannot be said to have high accessibility if there is no connecting road between these zones. Likewise, on the contrary, two zones that are far apart cannot be fully said to have low level of accessibility, if between the zones there are adequate road infrastructure and fleet services [13]. According to [13], the level of regional accessibility can be determined based on several variables, namely availability of roads, the number of transportation means, the length and width of roads, and quality of the road. According to [14], accessibility conditions can be represented by road conditions, class of roads, and public transport lines. While, according to [15], accessibility is often associated with distance, travel time and travel costs. Based on the discussion above, the road within a region or inter-region connecting road is very influential in increasing the people flow and goods. If the accessibility of the area becomes greater, it will open up possibility of urbanisation process and regional development, in the development form of various regional function facilities such as housing, trade and services, government, and socio-cultural facilities. According to [12] and [16] are located in the focus of heavy traffic will experience rapid development. The development of an urban area and its population is largely determined by the availability of road facilities.

The literature review above is relevant to the regional development phenomena in Makassar city, especially in some suburban areas. Concerning its role as the main gateway to the eastern part of Indonesia, Makassar city is very potential to experience development due to high commercial and residential activities. This certainly requires roads that can access service centres such as housing, education, health, worship, and commercial facilities. The results of observations, especially in areas along the southern coast of Makassar, show that there is a tendency for roads in the city centre area (CBD Karebosi) to be more dominant than periphery area (Tanjung Bunga). In this case, the city

IOP Conf. Series: Materials Science and Engineering **875** (2020) 012013 doi:10.1088/1757-899X/875/1/012013

centre area represents a regional variable with a much higher development level than the periphery area. Besides, several periphery areas were also seen which although the distance was relatively close, but because of limited road, accessibility was low. This phenomenon is interesting to study further, especially related to the presence or absence of correlation between road construction level and regional development level.

One of the problems that occur in the connecting road of the city centre area with urban periphery of Makassar is congestion that occurs at some of road points due to inefficient accessibility between the two regions. The amount of mobility to the city centre area is not supported by an evenly distributed road system in several alternatives but tends to be concentrated. The road connecting the city centre area around Karebosi area with Tanjung Bunga area is relatively only served by one main road so that there is excess capacity at certain times. This research is classified as important to find out the accessibility level and the regional development level on connecting roads in the city centre and periphery to improve the economy of Makassar, and to improve quality of life safely, comfortable, and productive due to equitable distribution of urban infrastructure and facilities. The results of this study are expected to be basis for enriching norms, standards, guidelines, and criteria, related to increasing the accessibility and urban development area that are anticipatory and adaptive to the dynamics of science and technology development.

2. Methods

The object of study covers an area along Makassar coastal area that connects the downtown area (Karebosi) with the periphery area (Tanjung Bunga). The study area covered four districts, namely Ujung Pandang, Wajo, Mariso, and Tamalate, which included as many as sixteen sub-districts (see Figure 1). The research method design includes:

a. Conduct field observations by identifying road conditions along the coastal area that connects the city centre and periphery. The observed road includes the length and width of the road, the number of road, the number of nodes, and area of the delineation, which is contained in the GIS map. The data is analysed with alpha index analysis techniques to find the accessibility index (road connectivity) of the area to other regions.

b. Conduct field observations by identifying the area of several city facilities, such as education, worship, health, commercial, and residential. The data is analysed with LQ analysis techniques to determine the superior sector of the facility in terms of area.

c. Analysing the relationship between accessibility index variables and regional development variables using correlation analysis.

In addition, the research sample was also observed in the form of several urban facilities. The sampling technique is done by purposive sampling method. There are two types of alpha index (α), namely planar and non-planar graphs. A block is defined planar if the graph (path) is without intersecting segments. Whereas a block is defined as non-planar if the graph cannot be presented without intersecting segments. Road characteristics at the study site are classified as the alpha planar index. The Alpha Index formula is as follow [12]:

$$Alpha\ Index\ (\alpha) = \frac{m-t+s}{2t-s} \quad \text{.......} \quad (1)$$

where,

Alpha (α)	: Accessibility level
m	: Road (graph)
t	: Place point (node)
s	: Region (sub-graph)

To determine the level of regional development, Location Quotient (LQ) analysis techniques are used to understand the leading sectors that are driving growth. LQ measures the relative concentration or degree of specialisation of economic activity. This technique presents a relative comparison between the ability of an observed regional sector with the ability of a broader regional sector.

IOP Conf. Series: Materials Science and Engineering **875** (2020) 012013 doi:10.1088/1757-899X/875/1/012013

The LQ formula is as follows [17].

$$LQ = \frac{\frac{Xij}{Xi}}{\frac{Xj}{X.}} \quad \ldots\ldots\ldots\ldots\ldots\ldots\ldots\ldots \quad (2)$$

where,

Xij: Degree of 'j' activity in the 'i' region
Xi: Total activity in the 'i' region
Xj: Total of 'j' activity in all regions
X.: Degree of total area activity

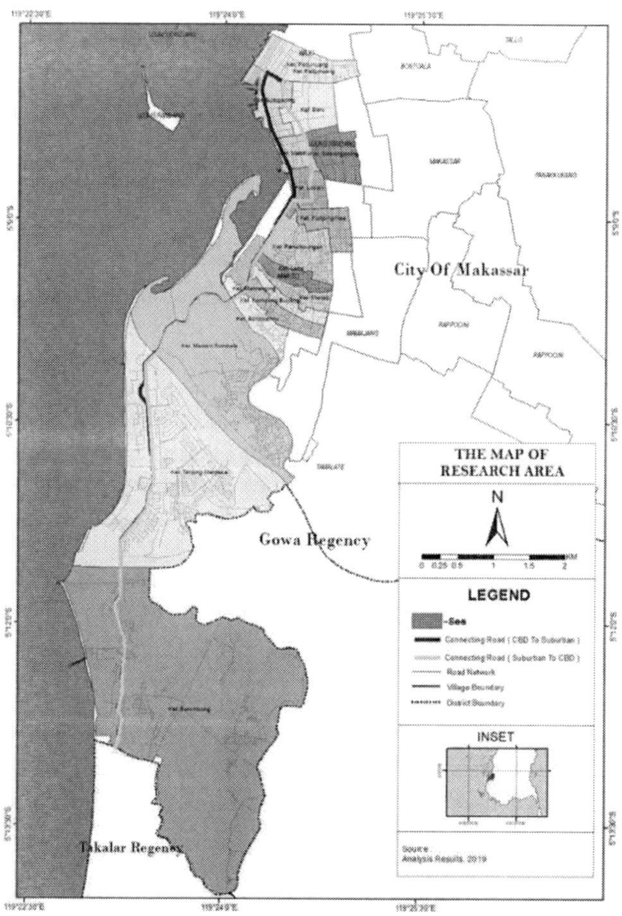

Figure 1. Map of research area

LQ> 1 indicates that the area has proper facilities so that the development of the region is classified as high. LQ = 1 indicates that the area has adequate facilities so that the development of the area is classified as moderate. LQ <1 indicates that the area has infeasible facilities so that the development of the region is classified as low [11]. To find out the relationship between regional development and accessibility, a correlation analysis (SPSS) was conducted. The correlation coefficient is stated as follows [18].

$$r = \frac{N\sum xy - \sum x \sum y}{\sqrt{N\sum x^2 - (\sum x)^2 N\sum y^2 - (\sum y)^2}} \quad \ldots\ldots (3)$$

where,
r = Correlation coefficient
x = Regional accessibility
y = Regional development
N = Number of the city area

Value of "*r*" range from -1 to 1. If *r* approaches 0 value, the relationship between the two variables is very weak (no relationship). If *r* approaches to 1 or -1 value, the relationship between the two variables is very strong. Variable '*X*' is road, while variable '*Y*' is number of facilities such as education, health, worship, commercial, and residential.

3. Results and Discussion

3.1. Analysis of Accessibility Levels in Regions between City Centre and Periphery Area

Accessibility level analysis aims to determine the connectivity of a graph (road or symbolized-s) to the area. Roads and number of nodes are used as variables in determining the level of accessibility in each sub-district. Roads and nodes are known through two stages, namely by using secondary data in the form of road segment attribute data. Furthermore, interpretation and digitation of GIS-based image maps are carried out by determining the number of road (symbolized-m) and the nodes number of road (symbolized-t) in each sub-district. The level of accessibility is calculated based on the number of roads and nodes compared to the area of each sub-district using the Alpha Index formula. The alpha index calculation to determine the accessibility of the area is done with the assumption that the area under study is flat. Accessibility index calculation is only based on the appearance of the results of the interpretation of satellite images as previously explained. The Alpha Index is relatively high if it approaches to 1. Based on the calculation that has been done using the formula, the Alpha Index is obtained as follows:

Table 1. Accessibility index values for each sub-district

No.	Sub-District	Number of Road (m)	Number of Node (t)	Sub-graph (s)	(m-t+s)	(2t-s)	Index Value α
1	Pattunuang	42	41	6	7	76	0.092
2	Bulogading	28	26	4	6	48	0.125
3	Baru	18	17	3	4	31	0.129
4	Maloku	24	24	4	4	44	0.091
5	Sawerigading	36	33	3	6	63	0.095
6	Losari	27	25	3	5	47	0.106
7	Kunjungmae	35	34	4	5	64	0.078
8	Panambungan	91	113	8	-14	218	-0.064
9	Lette	43	44	5	4	83	0.048
10	Mariso	52	51	5	6	97	0.062
11	Matoanging	40	38	4	6	72	0.083
12	Kampung Bayang	32	30	4	6	56	0.107
13	Bontomarannu	54	51	5	8	97	0.082
14	Maccini Sombala	169	204	9	-26	399	-0.065
15	Tanjung Merdeka	297	334	8	-29	660	-0.044
16	Barombong	180	195	13	-2	377	-0.005

If the Alpha Index value is close to 1, the value is getting higher, or the level of accessibility is high. Based on calculations that have been made using the alpha index formula in each sub-district, the alpha index value is found, based on the following formula.

$$Interval\ Class = \frac{Maximum\ Value - Minimum\ Value}{Number\ of\ Class\ Desired} \quad \ldots\ldots (4)$$

$$= \frac{0.129 - (-0.065)}{3} = 0.065$$

The interval from each index classification is 0.065. Then the interval class category is determined with the following values: High Class = 0.129 - 0.064; Medium class = 0.063 - (-0.001); and Low class = (-0.002) - (-0.066). For an explanation of the accumulation of index classification along the coastal area between the centre and the periphery area of Makassar, see Table 2.

Table 2. Level of region accessibility by sub-district

Sub-District	Number of Roads (m)	Number of Node (t)	Subgraf (s)	(m-t+s)	(2t-s)	Index Value	Interval Value	Level of Accessibility
Baru	18	17	3	4	31	0.129	0.065	High
Bulogading	28	26	4	6	48	0.125	0.065	High
Kampung Bayang	32	30	4	6	56	0.107	0.065	High
Losari	27	25	3	5	47	0.106	0.065	High
Sawerigading	36	33	3	6	63	0.095	0.065	High
Pattunuang	42	41	6	7	76	0.092	0.065	High
Maloku	24	24	4	4	44	0.091	0.065	High
Matoanging	40	38	4	6	72	0.083	0.065	High
Bontomarannu	54	51	5	8	97	0.082	0.065	High
Kunjungmae	35	34	4	5	64	0.078	0.065	High
Mariso	52	51	5	6	97	0.062	0.065	Medium
Lette	43	44	5	4	83	0.048	0.065	Medium
Panambungan	91	113	8	-14	218	-0.064	0.065	Low
Maccini Sombala	169	204	9	-26	399	-0.065	0.065	Low
Tanjung Merdeka	297	334	8	-29	660	-0.044	0.065	Low
Barombong	180	195	13	-2	377	-0.005	0.065	Low
Average Accessibility Index level on Research Sites						0.058		Medium

The analysis showed that the highest Alfa Index value was found in Baru Sub-District (0.129), and the smallest in the Maccini Sombala Sub-District (-0.065). There are ten sub-districts that receive an accessibility index score of the High category, all of which are located in the city centre. In these ten sub-districts, the number of road networks is greater than the node. This proves that the city centre has a more optimal road arrangement compared to the suburbs. The city centre has good connectivity between regions. The arrangement of the road network is used optimally in the city centre by connecting all the existing node points to facilitate the accessibility of each location of activity points.

Furthermore, there are as many as two sub-districts that get the accessibility index value of the Medium category, all of which are located in the suburbs. This is based on the number of roads that tends to be proportional to the number of existing nodes. Whereas the sub-district with low accessibility are four sub-districts, all of which are located in the periphery area. The low accessibility is due to the small number of road compared to the sub-district area. The small number of the road is also caused by a large amount of land that is not developed with dominant agrarian function of land.

The level of accessibility of the area traversed by the city centre connecting roads with Makassar peripherals is highest in the Baru Sub-District. The number of roads and nodes in Baru Sub-District is relatively small, but because the area in the region is relatively small, the accessibility is relatively high. Baru Sub-District, aside from being evenly distributed, is also caused by the arterial road, Jl. Ahmad Yani and Jl. Jend. Sudirman whereas the lowest level of accessibility is in Maccini Sombala. This is due to the uneven road nodes, which tend to be concentrated in only one area in a residential area. Overall, the level of accessibility on the downtown connecting road (Karebosi) with the periphery area (Tanjung Bunga) has index of 0.058 or is included in the classification with the accessibility level of the Medium category. For details, the results of calculating the level of accessibility can be seen in Figure 2.

3.2. Regional Development between the City Centre and the Periphery

Regional development is calculated by sub-district that is traversed by the connecting road from the downtown area to the periphery of Makassar using the Location Quotient (LQ) analysis. This method is used to determine the potential for economic activity as an indication of the basis sector by looking at the comparison between the ability of certain sectors in the study area. In order to be spatially analysed, the level of regional development is measured based on the area of each sub-district facility. The results of the analysis show that the LQ value of urban facilities in the coastal area of Makassar city along the road between the city centre (CBD Karebosi) and the periphery area (*Tanjung Bunga*) is classified as uneven. Some facilities tend to be distributed in one area or do not spread evenly within a region, so accessibility to reach these facilities is inefficient. Based on Table 3, the average LQ obtained for each Sub-district in the study location is 1.61. Furthermore, the LQ value is classified into three classes with the following categories:

LQ> 1, states that the facilities are proper and categorised as High level of development.
LQ = 1, stated that the facilities are adequate and categorised as moderate level of developmental.
LQ <1, stated that the facilities are infeasible and categorized as Low level of development.

In Table 3 can be seen the level of regional development by the sub-district in the study area.

Table 3. Level of regional development by sub-district

Sub-District	LQ Education	LQ Health	LQ Worship	LQ Commercial	LQ Residential	LQ Average	Level of Regional Development
Baru	0.97	1.31	1.85	1.21	1.17	1.30	High
Bulogading	1.28	5.33	0.43	1.04	17.08	5.03	High
Kampung Buyang	0.58	5.33	0.43	4.43	1.03	2.36	High
Losari	1.30	0.47	1.12	3.09	5.56	2.31	High
Sawerigading	1.74	2.72	0.53	1.39	1.44	1.57	High
Pattunuang	0.58	0.49	6.00	0.82	2.90	2.16	High
Maloku	1.28	2.96	0.43	0.80	0.94	1.28	High
Mattoanging	0.58	2.72	0.43	0.20	1.03	0.99	Low
Bontorannu	2.49	0.00	0.43	0.20	0.73	0.77	Low
Kungjungmae	0.58	3.91	1.02	3.76	0.70	1.99	High
Mariso	0.58	3.91	1.02	2.35	0.70	1.71	High
Lette	0.58	4.90	0.43	0.28	2.74	1.79	High
Panambungan	0.58	0.94	0.43	0.45	0.70	0.62	Low
Maccini Sombala	1.28	2.72	0.43	0.64	0.33	1.08	High
Tanjung Merdeka	0.78	0.00	0.43	0.56	0.56	0.47	Low
Barombong	0.58	0.00	0.43	0.46	0.51	0.40	Low
Average LQ on Research Sites						1.61	High

Table 3 shows that the highest level of regional development along the study site is the Bulogading Sub-District (LQ = 5.03) located in the downtown area. The region has adequate facilities, such as education, worship, sports and recreation, trade and services, and transportation. On the contrary, the lowest level development in the whole area of the study area is Barombong Sub-District (LQ = 0.40) which is located in the periphery area (southern part of Makassar). The low development in the region is influenced by the factor of considerable distance from the city centre. This area is classified as the lowest availability of urban facilities such as education, health and trade facilities, compared to other regions. It can be said that in general the development of the research area is quite high (LQ=1.61). Among them, there are eleven sub-districts with a high level of development in the region and five sub-districts with a level of development in the region that are classified as Low.

In general, regions that have a relatively high level of development are located in the city centre (north), except Mattoanging (LQ = 1.03) and Bontorannu (LQ = 0.73). This is caused by not many and not yet spread the number of roads built in the sub-district compared to the surrounding area. Likewise, in general, areas that have a relatively low level of development are located towards the outskirts of the city (south), except Maccini Sombala Sub-District (LQ = 1.08) classified as having a high level development. This is caused the area functions as an elite residential and tourism area where there are relatively large numbers of roads spread in the area compared to the surrounding area.

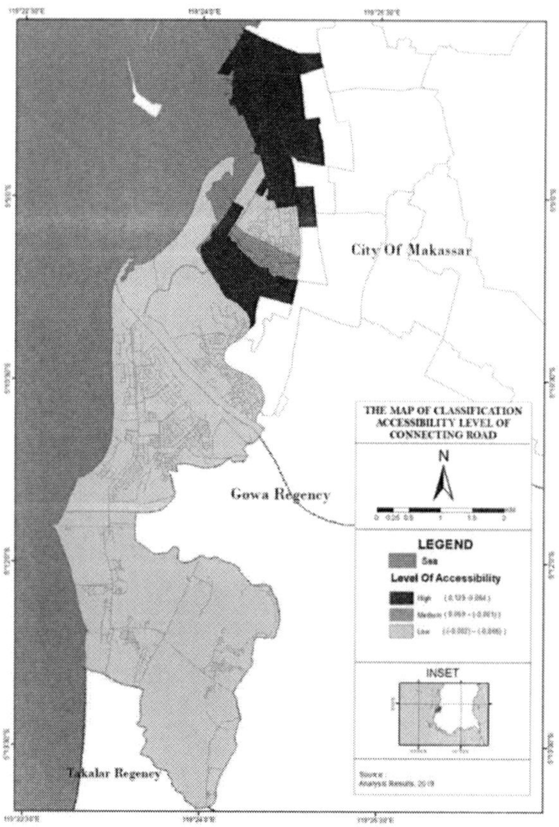

Figure 2. Map of classification of accessibility level in research area

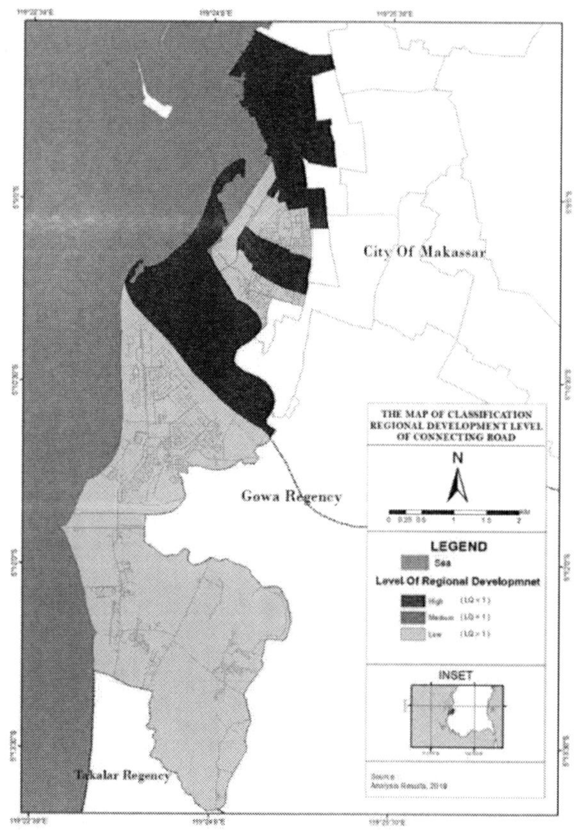

Figure 3. Map of classification of regional development in research area

3.3. The Relation between Accessibility and Regional Development in Makassar Coastal Area

The main objective of this study is to determine the relation (correlation) between accessibility (independent variables) with regional development (dependent variables) in several urban areas that are traversed by central city and connecting road in Makassar. Correlation analysis was performed using the SPSS for Windows program using the Spearman Rank Correlation analysis. This analysis aims to look for relationships or test the significance of the relationship between accessibility variable with road indicator and regional development variable with an indicator of urban facility availability. The accessibility value variable areas of the Alfa index value (see Table 1), while value of the regional development variable is obtained the LQ calculation results for each region (see Table 3). For clear results of the correlation analysis between accessibility level and regional development level on each area, can be seen in Table 4 below.

Tabel 4. Value of variables x and y in the research area

No	Sub-District	Accessibility (X)	Regional Development (Y)
1	Pattunuang	0.09	2.16
2	Baru	0.13	1.30
3	Losari	0.11	2.31
4	Maloku	0.09	1.28
5	Sawerigading	0.10	1.57
6	Bulogading	0.13	5.03
7	Barombong	-0.01	0.40
8	Maccini Sombala	-0.07	1.08
9	Tanjung Merdeka	-0.04	0.47
10	Bontorannu	0.08	0.77
11	Kampung Buyang	0.11	2.36
12	Kungjungmae	0.08	1.99
13	Lette	0.05	1.79
14	Mariso	0.06	1.71
15	Mattoanging	0.08	0.99
16	Panambungan	-0.06	0.62

Based on the Spearman Rank correlation rule, to measure the level of relationship between two variables can be seen based on the value of the relationship coefficient (r). The greater of "r" value means, the stronger of relationship, and conversely the lower of "r" value means, the weaker of relationship. To determine significance level of the relationship between variables using the criteria for the closeness of the relationship [19], as follows:

a. r - value < 0.2: Relation are very small
b. r - value $0.2 \leq X < 0.4$: Relation are small
c. r - value $0.4 \leq X < 0.7$: Relation are significant enough
d. r - value $0.7 \leq X < 0.9$: Relation are significant
e. r - value $0.9 \leq X < 1.0$: Relation are very significant
f. r - value 1.0: Perfect relation

Based on the results of the correlation that has been done, it is known the significance level of the relationship between variables as stated in Table 5.

Table 5. Relationship between accessibility levels and regional development

Correlations			
		Accessibility	Regional Development
Accessibility	Pearson Correlation	1	.596*
	Sig. (2-tailed)		0.015
	N	16	16
Regional Development	Pearson Correlation	.596*	1
	Sig. (2-tailed)	0.015	
	N	16	16

**. Correlation is significant at the 0.01 level (2-tailed).

Based on the results of the correlation analysis between accessibility and regional development in the coastal area that connects the city centre to the periphery area of Makassar shows "*r*" value (Pearson correlation) of 0.596 which has a fairly close and direct relationship, with a significance value of 0.015 (<0.05). This shows that if regional accessibility has increased, regional development will also experience the same thing. It can be concluded that there is a significant relationship between the accessibility level and the regional development level in the southern coastal area of Makassar city.

The connecting road between the downtown area (CBD Karebosi) and periphery area of the southern part of Makassar is one of the economic pathways that connect many economic activities both industrial and tourism sector. Based on interpretations on image map Makassar, it shows that at present there are more roads in the city centre and surrounding area than the periphery area. The lack of roads in Tanjung Bunga area has implications for Low accessibility of several urban facilities in general in the region as well as accessibility between the surrounding area. Good accessibility will facilitate community interaction inter-regions, resulting inequitable development. Fulfilment of road infrastructure will have an impact on regional development in a city or region.

4. Conclusion

The highest level of regional development along the southern coastal area of Makassar is Bulogading Sub-district, located in downtown area. The region has adequate facilities, such as educational, worship, sports and recreation, trade and services, and transportation. Conversely, the lowest level of regional development is Barombong Sub-District in the periphery area, which are relatively far from the city centre and availability of the facilities is still Low. However, in general, the development of the area in the research area is quite High. In this case there are eleven sub-districts with a high development level and five sub-districts that are classified as Low. The areas with high development lead to the centre of the city except Mattoanging and Bontorannu Sub-District, due to limited the number of roads compared to surrounding area. On the other hand, the Low development urban sub-district is located on the periphery area, except Maccini Sombala Sub-District, which has developed environment related to its function as a residential and tourism development area. Furthermore, the level of accessibility in general along the south of Makassar coast is classified as Moderate (α=0.058). Among them, there is ten regions with High accessibility, two regions with Medium level, and four regions classified as Low. Pearson correlation analysis results show that there is a significant direct relationship between the level of accessibility and regional development. Baru Sub-District, located in the city centre, is the region with the Highest accessibility value. While Maccini Sombala Sub-District which is located in the periphery area, has the Lowest accessibility value. This analysis explains that the accessibility level in the city centre is generally Higher than the periphery area. This is indicated by as many as twelve sub-districts that are relatively close to the city centre that has a High level of accessibility, and four sub-districts are relatively far from the city centre that has a Low level of accessibility. This fact corresponds to the regional development conditions of the sixteen regions studied and is relevant to the results of the correlation analysis. It can be concluded that there is a relationship between the accessibility level and regional development level in the coastal area of Makassar. In addition to accessibility, regional development is also influenced by other factors such as regional potential.

Acknowledgment

Authors wishing to acknowledge to the Dean of Engineering Faculty, Hasanuddin University has facilitating this paper.

References

[1] Azwar, S.A., Suganda, E., Tjiptoherijanto, P., Rahmayanti, H., 2013, Model of Sustainable Urban Infrastructure at Coastal Reclamation of North Jakarta, *Procedia Environmental Sciences*,Vol.17, pp 452-461.
[2] Sumadi, S., Franklin, P,J,C., Makainas, I., 2012. *The Relationship of Accessibility to the Level of Development of the District Area in Tomohon City. Manado*: Unsrat. [in Bahasa]
[3] Lipshitz,,Gabriel, 1993, The Main Approaches to Measuring Regional Development and Welfare, *Social Indicators Research*, Vol. 29, No. 2. pp 163-181.
[4] Riyadi and Bratakusumah, D.S., 2005, *Regional Development Planning*. Jakarta: PT Gramedia Pustaka Utama. [in Bahasa]
[5] Adisasmita, R., 2011. *Land Transportation Management*. Makassar: Graha Ilmu. [in Bahasa]
[6] Comendador, J., Lambar, M.E., Monzon, A., 2014, Urban built environment analysis: Evidence from a mobility survey in Madrid, Procedia - Social and Behavioral Sciences 160. pp 362-371.
[7] Yunus, H. S., 2000. *City Spatial Structure*. Yogyakarta: Pustaka Pelajar. [in Bahasa]
[8] Rudiarto, Iwan, 2010. Spatial Assessment of Rural Resources and Livelihood Development in Mountain Area of Java: A Case from Central Java – Indonesia. *Disertation* (unpublished), Universität Hohenheim, Jerman.
[9] Black, J.A.1981, *Urban Transport Planning: Theory and Practice*. London:Cromm Helm
[10] Basri, Hasan. 2012. *Measurement of Accessibility Index in Depok with Gravity Model*. Depok: Universitas Pancasila. [in Bahasa]
[11] Warpani, Suwardjoko. 1984, *City and Regional Analysis*. Bandung: ITB. [in Bahasa]
[12] Bintarto, R, et al. 1982. *Geographic Analysis Method*. Jakarta: LP3ES. [in Bahasa]
[13] Tamin, O.Z., 2003. *Transportation Planning and Modeling*. Publisher: Bandung Institute of Technology, Second Edition. [in Bahasa]
[14] Farida, U., 2013, The Effect of Accessibility on the Social Economic Characteristics of Rural Communities in the Bumijawa District, Tegal Regency, *Jurnal Wilayah dan Lingkungan*, Volume 1, No. 1, pp 49-66. [in Bahasa].
[15] Suthanaya, 2009. *Analysis of Accessibility of Public Transportation Passengers Towards the Center of Denpasar, Bali Province*. Gane Swara, Special edition, Vol. 3 No.3. [in Bahasa]
[16] Bintarto, R.et al, 1989, *Town-Village Interaction and Its Problems*. Jakarta: Ghalia Indonesia. [in Bahasa]
[17] Rustiadi, et al., 2009, *Regional Planning and Development*. Jakarta: Yayasan Obor Indonesia. [in Bahasa]
[18] Tika, Pabundu, 2005, *Geography Research Methods*, Jakarta: Bumi Pustaka, [in Bahasa]
[19] Guilford, J.P. 1956, *Fundamental Statistic in Psychology and Education*. 3rd Ed. New York: McGraw-Hill Book Company, Inc.

IOP Conf. Series: Materials Science and Engineering **875** (2020) 012014 doi:10.1088/1757-899X/875/1/012014

Value of Concrete Compressive Strength on 28 days with Variation of Candlenut Shell Applicated to Plate on a Non-destructive Test Using Hammer Test

Nasruddin[1]*, V Sampebulu'[1], and P Mushar[1]

[1]Laboratorium of Building Material, Structure and Construction, Architecture Department, Engineering Faculty, Hasanuddin University, Makassar, Indonesia

*Email: nas_junus@yahoo.com

Abstract. Concrete compressive strength is one of the parameters used to control the quality of a concrete. Concrete compressive strength testing method which is considered the highest level of reliability is a destructive test using a compressive testing machine. Testing requires quite high costs and a longer processing time. But sometimes testing must be done in the scene. Based on the case, a testing tool is needed to measure or determine the compressive strength of concrete without spoil, which is known as the non-destructive test. The tool commonly used for this test is the hammer test. The type of research is experimental. The purpose of this research is to get the compressive strength value of a concrete plate variation of 30% candlenut shells from the test results using a hammer test so as the results can be close to the results of testing using a compressive testing machine (destructive test). The research variable is the testing method (3 point of testing with angle 0°), the age of concrete plate curing, the method of curing (wet curing and dry curing). The number of samples of reinforced concrete slabs is 8 pieces measuring 50cm x 25 cm x 12cm. The research results are Compressive strength of concrete slabs of 30% candlenut shell variations using a Hammer test showed that the concrete strength reached 23-27 Mpa both in wet curing and dry curing or equivalent to the quality of K250 concrete at 28 days. The results also indicate higher values compared to normal concrete slabs.

1. Introduction

Many buildings are often found using reinforced concrete construction systems. Even some buildings have a long life like a monument or historic building. the more old of building, definitely, the strength of itself more decrease, for example the building with reinforced concrete construction. According to the theory, as we know the concrete reaches its maximum age at 28 days but more than it, the concrete is actually experiencing an increase in strength but not too significant. However, when the concrete reaches more than 50 years the strength of concrete will decrease. Concrete compressive strength is one of the parameters used to control the quality of a concrete. Concrete compressive strength testing method which is considered the highest level of reliability is a destructive test using a compressive testing machine. Testing requires quite high costs and a longer processing time. But sometimes testing must be done in the scene. Based on the case, a testing tool is needed to measure or determine the compressive strength of concrete without spoil, which is known as the non-destructive test. The tool commonly used for this test is the hammer test.

Based on the explanation, this article discusses about the purpose of research which is to get the compressive strength value of a concrete plate variation of 30% candlenut shells from the test results using a hammer test so as the results can be close to the results of testing using a compressive testing machine (destructive test).

2. Literature Review

According to [1], concrete can generally be divided into groups, namely: 11 - Concrete based on concrete class and quality that are: (a) Class I concrete. It is a class of concrete for non-structural works. For its implementation, this concrete does not require special expertise. Its quality control is limited to light supervision of the ingredients quality. As for the compressive strength, inspection is not required. Class I concrete is expressed as Bo; (b) Class II concrete. It is a class of concrete for general structural works. Its implementation requires sufficient expertise and must be carried out under the supervision of experts. Class II concrete is divided into B1, K125, K175, and K225 standard qualities. In the B1, quality control is only limited to the supervision of the ingredients quality. As for the compressive strength, inspection is not required. In the K125 and K175, there is a need to check the concrete compressive strength continuously from the results of the test specimens; (c) Class III concrete. It is a class of concrete for structural works higher than K225. Its implementation requires special expertise and must be carried out under the supervision of experts. It requires a concrete laboratory with complete equipment that is served by experts who can carry out continuous quality control of concrete.

Table 1. Concrete quality conversion from Kg/cm^2 to Mpa [2]

CONCRETE QUALITY K (kg/cm^2)	CONCRETE QUALITY fc (mpa)
K100	fc 8,3 mpa
K-150	fc12.35 mpa
K-175	fc 14.53 mpa
K-200	fc 16.60 mpa
K-250	fc 20.75 mpa
K-275	fc 22.83 mpa
K-300	fc 24.90 mpa
K-350	fc 29.05 mpa
K-400	fc 33.20 mpa
K-450	fc 37.35 mpa
K-500	fc 41.50 mpa

Based on Table 1 above, the concrete quality is classified into three quality classes, namely class I concrete, class II concrete, and class III concrete. Class I concrete, which are K-100, K125, K-150, K-175, and K-200, are used for non-structural work; Class II concrete, which are K-225, K-250, and K-275, are used for structural works such as floors, roads, foundations, floors, columns, etc.; and Class III concrete, which are K-325, K-350, K-375, K450, and K-500, are special concrete, for example for bridge beams and floors, runways, etc.

Concrete Hammer Test or Schmidt Hammer Test is an easy and simple testing method for estimate the quality of concrete.

2.1. How to Use the Hammer Test

Because the principle of work and how to use the tool were very easy, it was used generally to estimate the quality of concrete, especially for finished building structures. Furthermore, the quick test process that hammer test have can testing whole part of building structure for identicating uniformity of concrete. For noted, the concrete hammer test just become initial indication for grade and its grade uniformity because the tool only read the hardness of concrete on the surface layer (+4 cm) [3].

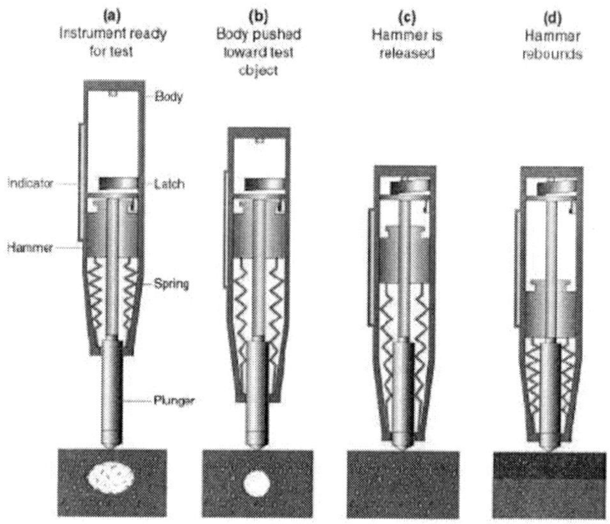

Figure 1. Principle of work concrete hammer-schimidt rebound hammer [4]

2.2. Advantages of the Hammer Test Method:

1. Cheap
2. Measurements can be made quickly
3. Easy to use
4. Do not damage the object

2.3. Disadvantages of the Hammer Test Method:

1. The test results are influenced by the surface flatness, concrete moisture, the properties and types of coarse aggregate, the degree of carbonization and the age of the concrete. Therefore it is important to remember that the concrete to be tested must be of the same type and condition.
2. It is difficult to calibrate the test results
3. The reliability level is low
4. Only provide information about concrete characteristics on the surface.

2.4. Collection Data:

1. Prepare a test schedule plan, prepare the equipment needed.
2. Finding data about the location of the construction detail, spatial layout and quality of construction materials during the building implementation.
3. Determine the test point.
   a. Test points for columns taken as many as 5 (five) points, each test point consisting of 8 (eight) firing points
   b. blocks are taken as many as 3 (three) test points each point consists of 5 (five) firing points
   c. floor plates taken as many as 5 (five) test points each consisting of 5 (five) firing points.

3. The Method

This research was conducted in several stages of work. It starts with the preparation of research tools and materials, material inspection, mixture planning, manufacturing of test specimens and testing of specimens. All work is carried out based on applicable regulations / standards with adjustments to existing laboratory conditions and facilities. The stages are as follows:

a. Preparation of tools and materials
b. Examination of material that aims to determine the characteristics of each concrete constituent material.
c. Manufacture of test specimens consisting of: Material preparation and Printing of test specimens
d. Test specimens using a hammer test on the test specimen plate
e. Processing test results and drawing conclusions.

The research variable is the testing method (3 point of testing with angle 0°), the age of concrete plate curing, the method of curing (wet curing and dry curing). The number of samples of reinforced concrete slabs is 8 pieces measuring 50cm x 25 cm x 12cm.

Figure 2. Type of hammer test that was used (CONTROL Model 58-C0181/N) [5]

4. Result and Discussion

4.1. Previous Research

In 2018, previous research had been carried out on cylindrical specimens and produced a concept of concrete material products with candlenut shell waste as a gravel substitution and as an innovative building material for self-help housing construction where the resulting cylindrical concrete showed that a variation of candlenut shells of 30% had concrete compressive strength is stronger and achieves the concrete quality of K200-K225. Then, from the results of these data, it becomes the basis for consideration to make a 30% candlenut shell variation by applying it to specimens with the function as a plate structure system.

(a) (b) (c)

Figure 3. (a) Candlenut shell material;(b) Normal concrete slump (0%);(c) 30% Concrete variation slump

Figure 4. Printing on mold plates that have been made

Figure 5. Plates treatment with wet and dry curing methods

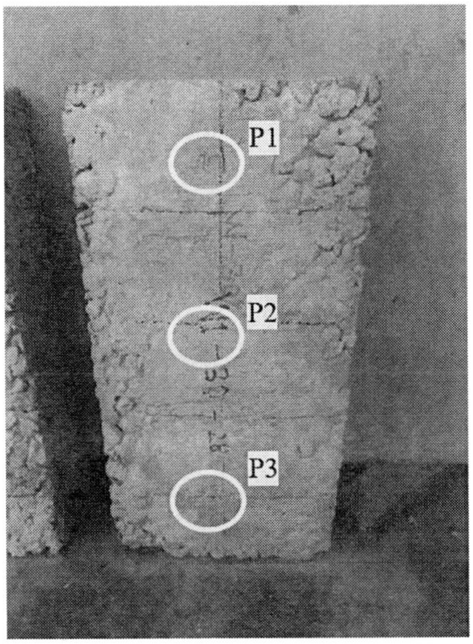

Figure 6. Shoot point of sample

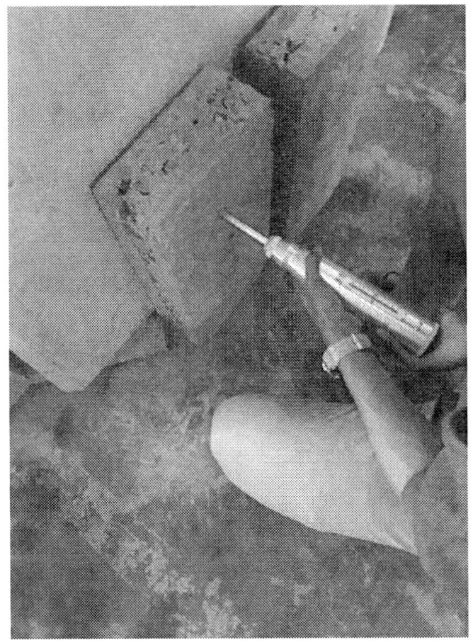

Figure 7. Testing process (angle 0°) by using hammer test

4.2. Variety

Based on Table 2, it can be seen that a concrete plate with a variation of candlenut shell has a lighter weight than a normal concrete variation of 0%. The difference in weight reaches 10-15%.

Table 2. Test object identification

Test Object	Weight (Kg)	Length (Cm)	Width (cm)	Height (cm)
D0-01	35.96	50	25	12
D0-02	35.17	50	25	12
W0-01	36.94	50	25	12
W0-02	36.81	50	25	12
D30-01	32.52	50	25	12
D30-02	30.94	50	25	12
W30-01	31.73	50	25	12
W30-02	30.61	50	25	12

Table 3. Non-destructive concrete compressive strength test results by using Hammer Test

	Concrete Compressive Strength (average of 8 samples) Mpa							
	D-1		D-2		W-1		W-2	
	0%	30%	0%	30%	0%	30%	0%	30%
P1	18	36	20	25	35	36	35	26
P2	25	10	27	10	15	22	14	20
P3	18	35	28	36	24	24	24	24
Average Value	20.3	27	25	23.7	24.7	27.3	24.3	23.3

D = Dry Curing, W = Wet Curing, P = Point (Shoot Point)

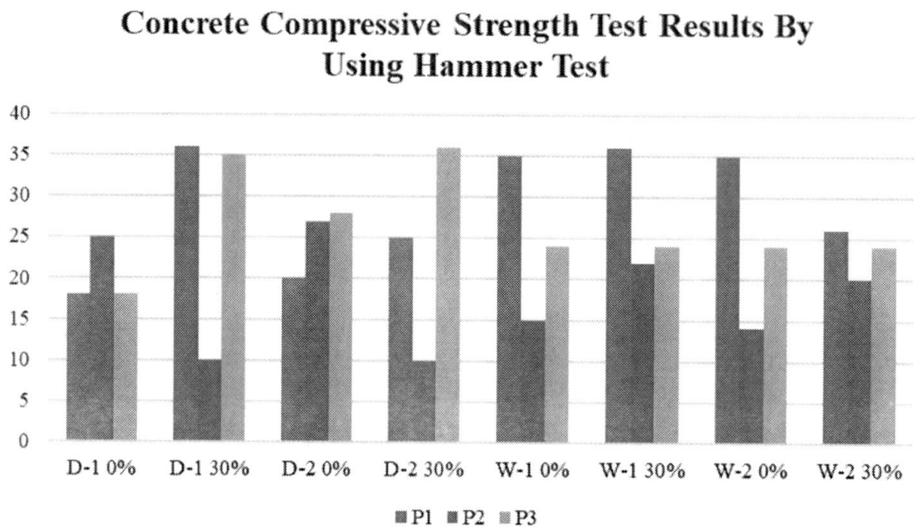

Figure 8. Chart of concrete compressive strength by using hammer test

Based on the results of the concrete plate testing in Table 2 and figure 6, it was found that the variation of 30% (candlenut shell variation) has a higher compressive strength value of the concrete compared to the concrete plate variation of 0% normal concrete, in the dry curing methods. Which is able to 23-27 Mpa equivalent to the quality K250-K300, while the variation of 0% has higher value in the wet curing and reaches 24 Mpa or equivalent to the quality K225-K250 (Table 1). Based on this, [1], concrete plates with a variation of 30% (both in the wet curing and dry curing) based on the class classification of concrete shows that the plates are included in class II concrete which is a class of concrete for general structural work and its implementation requires sufficient expertise. Hammer test is not an alternative method for testing the compressive strength of concrete, but as an indicator to assess the quality of concrete [6] is correct. In reality, the hammer test is used to determine whether the test object has a uniform or precise quality. If it is found that there is no uniformity in the reading of the rebound number in the hammer test, core samples from test specimens are taken for laboratory testing regarding their suitability and quality testing or compressive strength [7].

5. Conclusion
Compressive strength of concrete slabs of 30% candlenut shell variations using a Hammer test showed that the concrete strength reached 23-27 Mpa both in wet curing and dry curing or equivalent to the quality of K250 concrete at 28 days. The results also indicate higher values compared to normal concrete slabs.

References
[1] Mulyono, Tri, 2004, Teknologi Beton, Penerbit C.V Andi Offset, Yogyakarta.
[2] SNI 03-2847-2002 Tata Cara Perhitungan Struktur Beton untuk Bangunan Gedung.
[3] *CE 165. Concrete Material & Concrete Construction.*
[4] https://hesa.co.id/uji-kekuatan-beton-dengan-hammer-test/ (accessed in 2019).
[5] www.google.com. Type of hammer test that was used (CONTROL Model 58-C0181/N) (accessed in 2019).
[6] SNI 03-4430-1997 Metode Pengujian Elemen Struktur Beton dengan Alat Palu Beton.
[7] Darmawan Weka, et.al.2016. Media Komunikasi Teknik Sipil. Jurnal Ilmu Terapan Bidang Teknik Sipil. Volume 22, No. 1, Juli 2016.

Identification of Child-Friendly Integrated Public Spaces in Makassar City (Case Study: Losari Beach)

M Ali[1]*, S Aliah E[1], M Fachrul R[1], and Musdalifah[1]

[1]Waterfront Laboratory, Departement of Urban and Regional Planning, Engineering Faculty, Hasanuddin University, Makassar, Indonesia

*Email: mukti_ali93@yahoo.com

Abstract. The availability of urban public space supports the survival of the community. The importance of public space sometimes ruled out by the government. The problem that often occurs in urban areas is the transfer of the function of free land to built land. Child Friendly Integrated Public Space (CFIPS) is an open space as a play ground, for the growth and development of children. However, the number and quality of its benefits are still not proportional. In this study taking up waterfront public space in the city of Makassar, case study in the Losari Beach to satisfy the Child Friendly Integrated Public Space aspects, must adjust to the criteria set out in the CFIPS policy. This research was conducted to identify the availability of CFIPS in Makassar City, determine the condition of existing CFIPS and measure the level of visitor satisfaction through direct observation, questionnaires and visitor interviews. The result of this research is the Losari Beach as a friendly public space for children with activities with variable accessibility, safety, comfort, cleanliness and completeness of various facilities of the community views both for the child-friendly aspects.

1. Introduction

In the global era, children spend more time by themselves. Children do more activities with gadgets. With daily activities with gadgets children now no longer think about the importance of social interaction. On the other hand, children are the nation's next generation who need education and fulfillment of their rights to be able to grow and develop in accordance with their potential. Under these conditions, children need space as a place to do social activities and also need space for them to play. However, nowadays public spaces in big cities are increasingly decreasing. This is precisely what causes the lack of social interaction of children with the surrounding environment. The problem of open space in big cities is increasingly minimal due to changes in land use from a free land to be a built area [1]. The government and other stakeholders consider that open space as a place to play is not something important. Even some facts show that due to the development of the city there is a tendency to change the function of space, and the most frequently affected is the playground [2]. Currently Makassar City is facing a problem related to the lack of available land for public open space designation. The rise of developments that are more concerned with personal and group benefits without thinking about the negative impacts in the future still continue to this day. This phenomenon also has an impact on the needs of the community, especially children of unmet public spaces, where the public open space itself serves as a forum for social interaction. Lack of social interaction will affect the psychological health of children, which will have an impact on children's ignorance towards the environment and fellow humans.

One of the most open public spaces in Makassar is the Losari Beach Platform. In this area various activities can be carried out as the role of public open space itself, namely as a place for socializing and recreation. However, the Losari Beach Platform is still considered unable to meet the needs of children as a Child Friendly Integrated Public Space (CFIPS) when compared to the characteristics of friendly public spaces. The purpose of this study is to identify public perceptions about the level of satisfaction and how important the application of Integrated Child Friendly Public Spaces in Makassar City. Research Questions are (1) what is the condition of Public Spaces for children in Makassar City, and (2) How important is public space for children in Makassar City?.

2. Literature Review

2.1. Public Space

In accordance with the term used, the word "public", this space is a place for people to carry out various activities that are open access for all levels of society. The characteristics of public space are as follows:

- Communities are free to enter public spaces without any membership format (open access).
- People can come and go as they please.
- Public spaces remain open during and outside working hours.
- Has a creative character.
- Provides psychological comfort.
- Can be a political pulpit.
- Low profile.

Furthermore, there are 5 basic needs that must be met when citizens use public spaces, namely comfort, relaxation, passive engagement, active engagement, and adventure / diversity of features (discovery) [3].

2.2. Study of Child-Friendly Public Spaces

The aim of the Child Friendly Integrated Public Space (CFIPS) is basically to transform children's rights by realizing regions as Child-Friendly Cities. The strategic role of the CFIPS in fulfilling children's rights to the environment is based on the Guidelines for the Management of Integrated Child Friendly Public Spaces [4]. The objectives are expected to realize:

1. The need for open space to fulfill children's rights so that children can live, grow, develop and participate optimally in accordance with human dignity and dignity.
2. Providing infrastructure and means of partnership between the Regional Government and the community in fulfilling children's rights
3. Providing urban infrastructure and facilities as a City for Children

Policy evaluation is a mechanism intended to assess the effectiveness of public policies to be accountable to the public and to see the extent to which the objectives of a policy are achieved [5]. In a policy evaluation or policy, there are criteria in it. The strategic role of the CFIPS in fulfilling children's rights to the environment is based on the Child Friendly Integrated Public Space Management (CFIPS) Guidelines can be done by looking at the evaluation of the CFIPS policy. The goals are expected to be able to realize:

a. Efficiency: A policy can be said to be efficient if the net benefit (total benefit - total cost) is greater than zero or higher. The indicators used are the optimal allocation of funds in the implementation of the CFIPS Policy, the implementation of the socialization in the CFIPS Policy has the maximum facilities and infrastructure used in the implementation of an adequate policy. Then the value of efficiency in terms of CFIPS is measured by how influential the policy or facilities are provided to the object in this case the user.
b. Adequacy: Adequacy with regard to how far an alternative can meet a level of needs in question maximally. Adequacy in question is to state the extent of achieving the desired results in solving

problems. Adequacy criteria can be judged by the extent to which actions in the Policy can solve the problems that arise.

c. Equal distribution: Equity is closely related to legal and social rationality, where these criteria can be met if a policy allows the creation of a fair and fair distribution of results, to the resources that exist in society. Leveling criteria or similarity regarding the costs and benefits of equitable distribution to different groups. A policy may be effective, efficient and adequate but is rejected because it results in an uneven distribution of benefit costs.
d. Responsiveness: Responsiveness is related to responsiveness or how far the results of the policy satisfy the needs, preferences or values of the people (certain groups). Indicators of responsiveness are community satisfaction with the implementation of the CFIPS Policy, coordination with related parties as implementing the CFIPS activities, the implementation team receives reports and complaints about the implementation of the CFIPS Policy.
e. Accuracy: Accuracy is related to substantive rationality, because the question of accuracy is closely related to the substance of the expected goal. According to Willian N. Dunn the accuracy criteria, namely with regard to the question of whether the desired results (goals) are truly useful or valuable. Accuracy criteria question whether the support of the objectives of a policy is in accordance with the needs for public welfare.

3. Research Methods

3.1. Research Types and Location

This research is an identification research. In this study, the qualitative and descriptive methods are more dominant than the quantitative methods. Therefore, the main data mining techniques are through observation, questionnaires, interviews, and documentation. While the qualitative method is carried out by conducting a literary study obtained from various sources such as books, journals, and other scientific works.

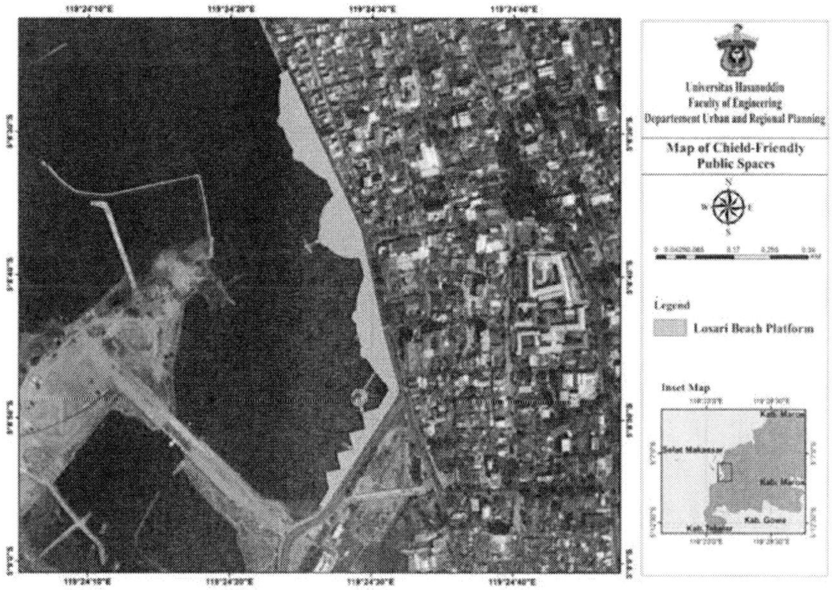

Figure 1. Losari Beach platform

The study was located in the Losari Beach Platform in Makassar City. Losari Beach Platform is located on the coast of Makassar City, a plaza and open space that is frequented by Makassar people and foreign tourists from various backgrounds and ages. In addition, in this study the Losari Beach

Platform is divided into 4 segments. The research variables in this study are the characteristics of visitors, facilities and infrastructure (facilities) of public open space, and visitor perceptions.

3.2. Data Collections Method
In this study using 4 data collection techniques, namely observation, questionnaires, interviews, and documentation. Data collection techniques carried out in accordance with the problems that will be identified in Makassar City, are as follows:

1. Observation is collecting data by direct survey to the field to examine the object to be examined. Field activities carried out are; (a) Reviewing the existing condition of the public space at Losari Beach Platform, (b) Reviewing facilities and infrastructure at the location to be investigated, namely Losari Beach Platform, and (c) Reviewing community activities especially children in the Losari Beach Platform.
2. Questionnaire, which is a data collection technique carried out by distributing questionnaires to visitors to the Losari Beach Platform to find out their level of satisfaction with the infrastructure available at that location.
3. Interview, this technique is carried out to find out whether the facilities and infrastructure of the Losari Beach Platform can be used well by the community especially children, measure the level of visitor satisfaction, as well as suggestions from visitors.
4. Documentation is data collection by taking pictures in accordance with observational activities carried out, namely pictures of existing conditions, facilities and infrastructure, as well as community activities, especially children in the study site.

3.3. Analysis Method
The analysis method used in this research is descriptive statistical analysis method. Statistical analysis itself is an analysis that emphasizes its analysis on numerical data which will then be processed into statistical information. The statistical information will then be described to draw conclusions.

As in this study, researchers analyzed how the assessment or assessment of child-friendly public spaces at the Losari Beach Platform. In analyzing, the researcher looked at the strengths and weaknesses according to the perceptions of visitors and who had visited the Losari Beach Platform using the Likert scale to determine the level of community satisfaction with the Losari Beach Platform. Furthermore, the results of the Likert Scale will produce a level of community satisfaction with the condition of the Losari Beach Platform as a Child Friendly Integrated Public Space. Following is the formula used to obtain a percentage of community satisfaction:

$$\% = \frac{n}{N} \times 100\%$$

Where, n is number of respondents' average scores, N is maximum number of scores.

The next step used to conclude the assessment or level of visitor satisfaction with the Losari Beach Platform as a Child Friendly Integrated Public is to determine the class interval criteria for the assessment of the level of child friendly public spaces [6] shown in Table 1.

Table 1. Class interval criteria level of assessment of child friendly public spaces

Percentage	Score Value
81% - 100%	Very Good
61% - 80%	Good
41% - 60%	Passably
21% - 40%	Bad
0% - 20%	Very Bad

4. Overview of Research Locations

Losari Beach which runs along the entertainment road located West of Makassar City. Administratively, the Losari Beach is belonging to Ujung Pandang District. The Ujung Pandang district consists of 10 sub-district with an area of 2.63 km^2 with 4 of them are in coastal areas and 6 other sub-district are non-coastal areas. Ujung Pandang District is bordered with Wajo District in the north, Mariso District in the south, in the East side with Makassar and Gowa Districts, and west side with the Makassar Strait.

Visitors can do various activities such as sports, enjoying the scenery, culinary tours, playing, or taking pictures in Losari Beach. Land use in the Losari Beach Platform area is settlements, trade and services, health, offices, and tourist areas. The following is a map of land use in the Losari Beach area.

5. Results

5.1. Characteristics of Losari Beach

Based on the results of the questionnaire that has been distributed, it is known that visitors in the Losari Beach area are more dominated by women compared to men on holidays and on weekdays with a percentage of 59%. This is also based on the reason visitors visit the Losari Beach Pavilion to bring their children to play in this place.

Figure 2. Land use around Losari Beach platform

Table 2. Comparison of the availability of CFIPS facility criteria

Standard Criteria	Existing
Play Facilities	Unfulfilled
Sports facilities	Fulfilled
Supporting Facilities	Fulfilled
Relaxing Facilities	Fulfilled
Accessibility	Fulfilled
Security	Fulfilled

IOP Conf. Series: Materials Science and Engineering **875** (2020) 012015 doi:10.1088/1757-899X/875/1/012015

In the morning, the dominant activity carried out by visitors is exercising, and in the afternoon visitors prefer to relax by enjoying the view in the form of sunset, and at night more dominated by culinary tourism activities. Meanwhile, based on the criteria for integrated child-friendly public space facilities, the criteria fulfilled for the Losari Beach Platform is written on Table 2.

The existing situation in Losari Beach Platform is carried out by direct observation and mapping methods carried out with the aim to see what facilities are available in this case the researchers divide the Losari Beach Platform into 4 segments with the aim of being able to see specifically the comparison of activities carried out by visitors along the research area. Based on observations and interviews, it can be seen that there are specific differences in segment 1, ie in this segment the differentiating activity is this segment specifically for the street vendor area. In segment 2 there is an office that is a differentiator from other segments, for segments 3 and 4 this is a segment that is an area for gymnastics, sports or event areas. For the different activities of segments 1-4, it can be seen from and to see the usual activities carried out by visitors at the Losari Beach Platform through the mapping made as follows.

Figure 3. Condition of segment 1

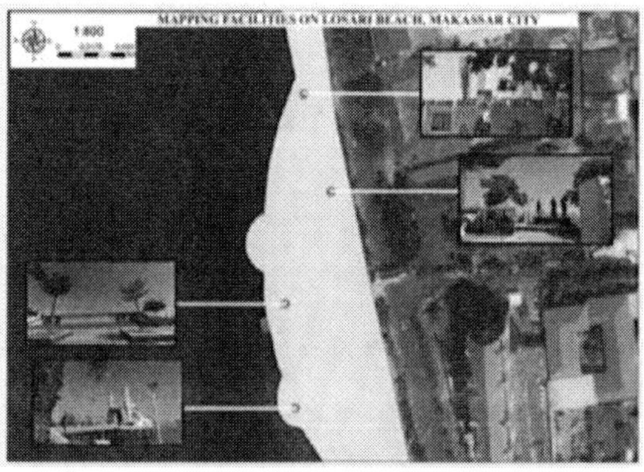

Figure 4. Condition of segment 2

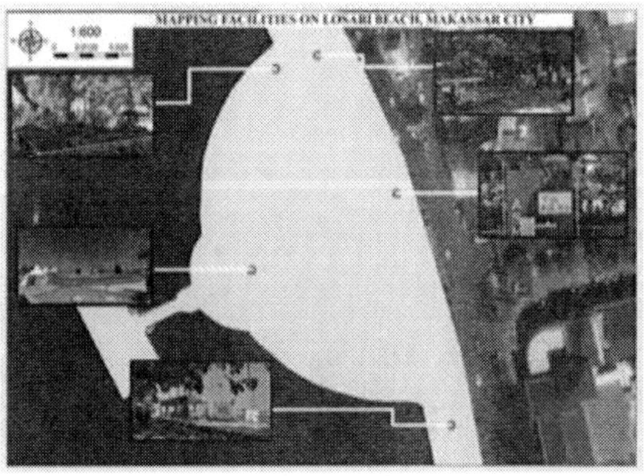

Figure 5. Condition of segment 3

Figure 6. Condition of segment 4

5.2. Visitor Perception

Based on observations made consisting of respondents as many as 107 respondents with more visitors or female respondents, as many as 61 people. The number of male visitors is 46 people. In addition, on another day, namely working days, the number of respondents was 92 respondents with more visitors or female respondents, namely 56 people. The number of male visitors is 36 people.

IOP Conf. Series: Materials Science and Engineering **875** (2020) 012015 doi:10.1088/1757-899X/875/1/012015

Based on the two previous statements, it can be concluded that the number of visitors / respondents who dominated were women. This is also very related to the reason they visit this CFIPS, in addition to needing entertainment to bring children to play in this CFIPS. It can be seen from the many mothers who bring their children both themselves and with other family members.

The daily activities of Losari Beach Public Space visitors only last for two periods of time, namely: in the morning, with the main activity being sports and supporting activities is the morning snack center, while in the afternoon until the evening the afternoon snacks are formed and enjoying the sunset panorama, making it the main destination of Makassar City. The peak activity in public spaces is on Sunday mornings.

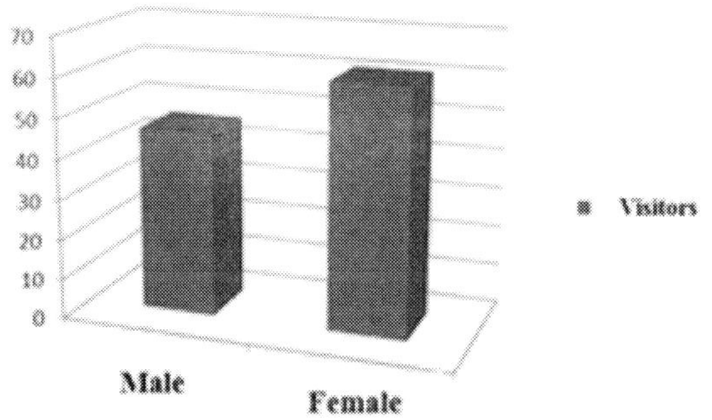

Figure 7. Graph of number of visitors on weekdays

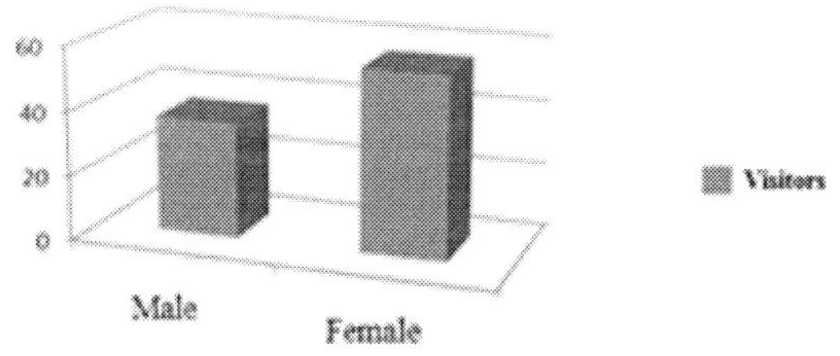

Figure 8. Graph of number of visitors on holidays

For more details, in this study, researchers divided 4 segments of the Losari Beach Platform area with specific public spaces of Losari Beach having a length of ± 950 meters with potential such as strategic locations of public spaces and located on the coast of Makassar City which has beautiful beaches. From the results of surveys and interviews that have been carried out for two days on holidays and weekdays, as many as 59% of the visitors to CFIPS Losari Beach are women and on average are teenagers with an age range of 13-25 years.

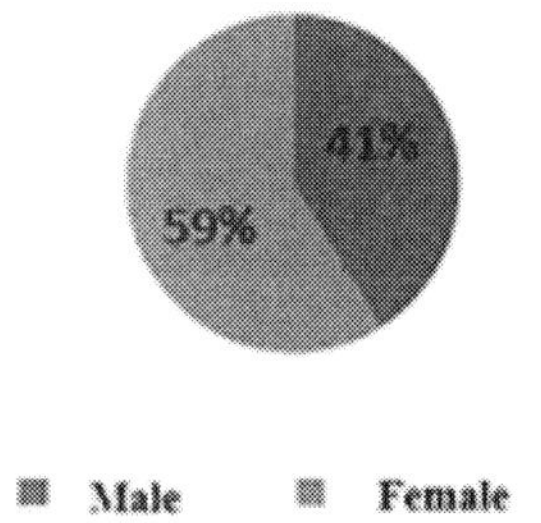

Figure 9. Percentage of number of visitors by gender

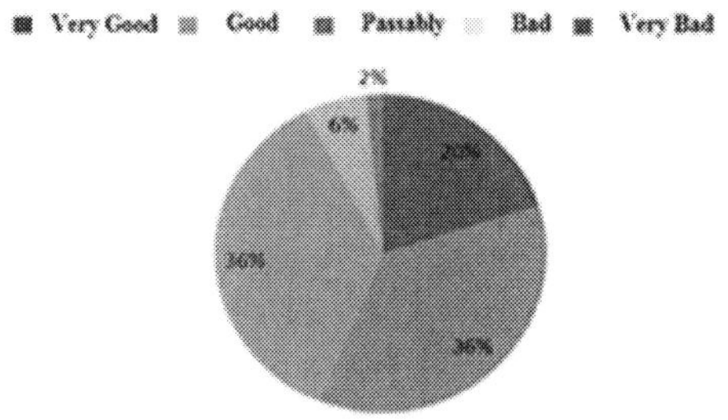

Figure 10. Percentage of level of community satisfaction with the Losari Beach platform

In general, the opinion of visitors to the Losari beach CFIPS is quite good. Can be seen based on the diagram below where the level of visitor satisfaction with the Losari Beach CFIPS as much as 56% of the number of respondents gave a positive response to the current condition of the Losari Beach CFIPS. This is influenced by the facilities available at this CFIPS which are already quite comfortable and also that this CFIPS can accommodate a variety of activities carried out by visitors, such as every afternoon the Anjungan Pantai Losari will be packed with teenagers because many community activities or performances are carried out in this CFIPS In addition, the existence of a Floating Mosque that makes it easy for visitors to worship and become one of the attractions in the Losari Beach CFIPS.

Based on respondents' answers to questions about facilities that support visitors' activities in the Losari Beach Public Space, the community considered it good for parking facilities, places of worship, accessibility, and comfort of other facilities to support the activities carried out. Based on the answers of respondents in this study, it can be analyzed respondents' statements on the perception of the Losari Beach Public Space perception as follows.

Table 3. Analysis of visitor ratings

No.	Questions	A n = 5	B n = 4	C n = 3	D n = 2	E n = 1	Total Score
1.	Parking lot	25	76	69	26	3	691
2.	Worship place	84	69	30	9	7	811
3.	Children's Playground	26	65	63	41	4	665
4.	Trash can	29	63	72	19	16	667
5.	Accessibility	26	82	64	22	5	699
6.	Toilet	13	37	43	63	43	511
7.	Rest area	31	88	61	13	6	722
8.	Other Comfort Facilities	21	56	100	16	6	667

Actual Score (ΣX) = 5433
Ideal Score = 5 x 199 x 8 = 7960

Percentage (%) $\sum X = \frac{\sum X \times 100\%}{\text{Ideal Score}}$ = 0.682 x 100% = 68.20% 5433

A: Very Good B: Good C: Passably D: Bad E: Very Bad

Based on the results of the analysis, the public perception of the Losari Beach Public Space is considered good, because the percentage of visitors is in the range of scores 61% - 68.20% bigger than 80%.

6. Conclusions

Based on the results of research conducted, it can be concluded that the community assessment of child-friendly public spaces at the Losari Beach Platform to support the Losari Beach Platform as a friendly public space for children with activities with variable accessibility, safety, comfort, cleanliness and completeness of facilities with various the view of the community is fairly "good" for the child-friendly aspect with the scoring method obtained results 61% - 68.20% bigger than 80%. The visitor / community perception identifies the importance of child-friendly public spaces as spaces that can accommodate the psychological needs of children to play, relax, exercise, learn, express, and explore. This is seen from the suggestions and public response to public spaces that have sports facilities, play areas, trees, park benches, and others.

Acknowledgment
This research was supported by the LBE FT-UH research fund. We thank the entire Faculty of Engineering, Universitas Hasanuddin, for providing the opportunity and financial assistance that greatly assisted this research.

Reference

[1] A. Dwiyanto., 2009. *Quantity and Quality of Green Open Spaces in Urban Settlements.* TEKNIK Vol. 30 No. 2 pp. 89-93. [in Bahasa]
[2] B. Saragih., 2004. *Concept Design for Children's Playground.* Jakarta: Bina Nusaantara University.
[3] E. Darmawan., 2003. *Theory and Study of Urban Public Spaces.* Semarang: Diponegoro University Publisher Agency.
[4] Indonesian Government. 2011. Minister Regulation (PERMEN) PP&PA Number 11 of 2011 concerning Child-Friendly Regency / City Development Policy.
[5] B. Winarno., 2002. *Public Policy: Theory and Process.* Media Presindo: Yogyakarta.
[6] A. Azis., 2019. *Child-Friendly Urban Green Open Space Assessment at Colliq Pujie Park in Barru District.* Thesis. Makassar: Universitas Hasanuddin.

IOP Conf. Series: Materials Science and Engineering **875** (2020) 012016 doi:10.1088/1757-899X/875/1/012016

Travel Behavior of the Engineering Students of UNHAS and UIN Alauddin in Meeting the Needs of Shopping Facilities

W W Osman[1]*, S Wunas[1], M Arifin[1], and G Lakatupa[1], and Misra[1]

[1]Laboratory of Housing and Settlement, Department of Urban and Regional Planning, Engineering Faculty, Hasanuddin University

*Email: w_wahidahosman@yahoo.com

Abstract. In the last ten years, there have been two developing universities in Gowa Regency (Faculty of Engineering, University of Hasanuddin and UIN Alauddin). Both universities are the centres of growth in housing and settlements, especially student rental houses. The development of housing has not yet been served with adequate housing/city infrastructure and facilities. The purpose of this study are: 1) to analyze the student travel behaviour in meeting their shopping needs (daily or monthly), 2) to explain the visit location of shopping facilities of those groups of student, 3) to arrange the resource development directions for environmental and sub-district development. The data were collected from field surveys, and online questionnaire using non probability sampling. The analytical method used were spatial and descriptive analysis. The results of analysis showed that the dominant student travel behaviour using a motorcycle, both for daily and periodic shopping needs. Travel behaviour always done together (two people). Behaviour movements, using motorcycles, both to the service centre in the environmental scale or district and city scale. The location of the visit is an environment scale service centre with a distance of ± 200 meters and a district or city scale service centre with a distance of > 5 km. Direction of development to support the dominant path of students using motorbikes within short reach is, provided a bicycle lane and pedestrian shade. In addition, the procurement of mass transportation to support long-distance travel to district and city-scale shopping centres. Development of sub-district service centres can be directed using the concept of mixed use and smart transportation (TOD), where the distance of the two groups of student housing is within a radius of ± 5.2 km.

1. Introduction

A campus or university is a small community that can trigger the growth of the surrounding urban area. At present the Faculty of Engineering of Unhas and UIN Alauddin campuses are centers of growth that encourage the construction of rental housing to meet the needs of students living around the campus. The construction of the campus in the Samata-Bontomarannu education area has a significant influence on the surrounding physical area such as an increase in building density and population. The education area also influences the social conditions of the community. Like the level of education, work, culture, community habits, and religion.

Along with the increase in students each year also raises a variety of student behavior, especially student travel behavior in meeting their needs. This is certainly a special concern for business people to understand the behavior of Unhas and UIN Alauddin Engineering students as one of the main samples in getting better understanding and planning to conduct business activities around Bontomarannu-Samata.

The purpose of the research: 1) analyze the behavior of the students of Unhas Engineering and UIN Alauddin in meeting the needs of shopping facilities; 2) explain the location of the visit of Unhas Engineering and UIN Alauddin students' shopping facilities in meeting their needs 3) arrange direction for the development of environmental infrastructure in meeting the shopping needs of Unhas Engineering and UIN Alauddin students according to the student's travel behavior.

2. Literature Review

2.1. Behavior

Behavior is an action that is directly involved in obtaining a need for goods and services, besides that the behavior involves a decision making process before doing something, behavior is a process (individual or group) in meeting their needs and desires [1]. Behavior is dynamic, meaning that behavior is always changing and moving all the time. This has implications for the study of consumer behavior [2]. Social behavior is a term used to describe the general behavior exhibited by individuals in society, which is basically in response to what is considered acceptable or unacceptable by one's peer group [3].

Behavior in selecting a product is based on or influenced by factors: a) cultural factors, b) social class, c) social group and reference factors, d) family, e) personal factors, f) location factors.

2.2. Travel Classification

Travel is a one-way movement from the home zone to the destination zone, including walking. Stopping by chance is not considered a destination of movement even if it is forced to make a change of route. Movement is often interpreted as the movement back and forth, in the science of transportation usually the analysis of the two must be separated. Five categories of residence-based movement objectives, namely:

a. Movement to work
b. Movement to school or university (movement with the aim of education)
c. Movement to shopping
d. Movement for social interests
e. Movement for recreational purposes

The purpose of the work and education movement, called the main movement goal which is a must to be carried out by everyone every day, while the other movement objectives are only optional and not routinely carried out [4].

2.3. The choice of Mode of Travel

The interaction between students and shopping locations requires students to decide how the interaction should be done, because interactions that occur between students and shopping locations will lead to travel, which in turn requires a choice of modes [5]. In choosing a mode of transportation there may be little choice or no choice at all. People who have a mode choice are called Captive to that mode. If there is more than one mode, the mode chosen will usually have the shortest, fastest and cheapest route, or a combination of the three. Decisions in modal choice are based on several factors such as time, distance, efficiency, cost, safety and comfort [1].

3. Research method

The research method is descriptive research, using qualitative and quantitative analysis with ArcGis 10.1 analysis tools. The research locations in Somba Opu and Bontomarannu Subdistricts were divided into 3 location segments, namely the distance of 500 m, 501-800 m, and 801-1200 m with the coordinates of the Unhas Technical Campus and UIN Alauddin.

The population in this study were all Unhas Engineering students and UIN Alauddin who were still actively studying and living around the study site. The research sample consisted of 184 students, consisting of 89 students from Unhas Engineering and 95 students from UIN Alauddin.

Research variables namely social community include: individual / group, community internal and external community and use of modes include the type of mode, vehicle ownership and costs. Visiting locations include: places visited, distance of shopping locations, length of travel, scale of service centers.

4. Findings and Discussion

4.1. The Analysis of Student Travel Behavior in Meeting the Needs Of Shopping Facilities

Tables 1 and 2 show the student's travel behavior (Unhas Engineering and UIN Alauddin) according to the location of residence based on the number of respondents to the social travel behavior of the student community according to the location of residence in terms of 2 factors namely internal and external factors. Internal factors that are intended in this study are friends of one study program, one boarding / rented building and fellow regional friends, while for external factors that are friends from different campuses / departments and with local residents.

Table 1. Unhas engineering student behavior according to the location of residence

No	Travel Behavior / Location of Residence	Number of Unhas Respondents												Total
		Internal Community						Eksternal Community						
		1		2		3-5		1		2		3-5		
		n	%	n	%	n	%	n	%	n	%	n	%	
1	500	9	10,1%	14	15,7%	0	0%	1	1,1%	1	1,1%	0	0%	25
2	501-800	18	20,2%	35	39,3%	2	2,2%	2	2,2%	2	2,2%	0	0%	59
3	801-1200	1	1,1%	3	3,4%	0	0%	0	0%	0	0%	0	0%	5
	Total	28	31,5%	53	59,5%	2	2,2%	3	3,4%	3	3,4%	0	0%	89

Tabel 2. UIN Alauddin student behavior according to the location of residence

No	Travel Behavior / Location of Residence	Number of UIN Alauddin Respondents												Total
		Internal Community						Eksternal Community						
		1		2		3-5		1		2		3-5		
		n	%	n	%	n	%	n	%	n	%	n	%	
1	500	4	4,2%	23	24,2%	9	9,5%	0	0%	3	3,2%	0	0%	39
2	501-800	9	9,5%	22	23,2%	8	8,4%	1	1%	2	2,1%	2	2,1%	44
3	801-1200	2	2,1%	6	6,3%	2	2,1%	0	0%	2	2,1%	0	0%	12
	Total	15	15,8%	51	53,7%	19	20%	1	1%	7	7,4%	2	2,1%	95

Based on table 1 and table 2, where the radius of 501-800 is the most dominant service radius of activity occurring by the perpetrators of the trip. Student behavior based on internal factors and external factors where the behavior of traveling with internal communities has a percentage of 61.7% of Hasanuddin University students and 41.1% of UIN Alauddin students is more dominant behavior occurs compared to the behavior of external communities. This is due to close kinship relations.

4.2. The used of Mode

4.2.1. Type of Mode. Type of mode is the type of vehicle used by students from the location of residence for travel divided into 5 types of modes, namely: walking, bicycle, motorcycle, car, and public transportation.

Based on Tables 3 and 4, the types of transportation modes used by students of Hasanuddin University and UIN Alauddin according to the location of residence are divided into five categories, where the motorized transportation mode is the most dominant mode used by students of both Hasanuddin University and UIN Alauddin by 88.8% and 81.1% to shopping facilities both living in the three segments and outside the study location.

The existence of public transportation is still rarely used by students who live in rental residences around the Technical University of Unhas and UIN Alauddin as modes, due to the large number of private motorbike ownership in Tables 6 and 7 and inadequate public transport routes to the location of student shopping visits at both study sites Table 5.

In this study shopping facilities can be reached on foot but most students prefer to use motorbikes (Tables 3 and 4). However, this behavior is dominant in Unhas Campus students, whereas for UIN Alauddin students most prefer to walk because in the 500 meter service radius they have shopping facilities in the form of supermarkets (shops).

Table 3. The Types of mode of transportation of Unhas engineering students according to the location of living for travel

Travel Behavior according to Residences	Individual						2 People						3-5 People						Total
	500		501-800		801-1200		500		501-800		801-1200		500		501-800		801-1200		
Type of Transportation	n	%	n	%	n	%	n	%	n	%	n	%	n	%	n	%	n	%	
Walk	0	0%	2	2,2%	0	0%	1	1,1%	3	3,4%	0	0%	0	0%	0	0%	0	0%	6
Bicycle	0	0%	0	0%	0	0%	0	0%	0	0%	0	0%	0	0%	0	0%	0	0%	0
Motorcycle	10	11,2 %	14	15,7 %	1	1,1 %	14	15,7 %	34	38,2 %	4	4,5%	0	0%	2	2,2 %	0	0%	79
Car	0	0%	4	4,5%	0	0%	0	0%	0	0%	0	0%	0	0%	0	0%	0	0%	4
Public transportation	0	0%	0	0%	0	0%	0	0%	0	0%	0	0%	0	0%	0	0%	0	0%	0
Total	10	11,2 %	20	22,4 %	1	1,1 %	15	16,8 %	37	41,6 %	4	4,5%	0	0%	2	2,2 %	0	0%	89

Table 4. The Types of mode of transportation of UIN Alauddin students according to the location of living for travel

Travel Behavior according to Residences	Individual						2 People						3-5 People						Total
	500		501-800		801-1200		500		501-800		801-1200		500		501-800		801-1200		
Type of Transportation	n	%	n	%	n	%	n	%	n	%	n	%	n	%	n	%	n	%	
Walk	0	0%	1	1%	1	1%	4	4,2%	5	5,3%	1	1%	1	1%	0	0%	1	1%	14
Bicycle	0	0%	0	0%	0	0%	0	0%	1	1%	0	0%	0	0%	0	0%	0	0%	1
Motorcycle	4	4,2%	9	9,5%	1	1%	22	23,1 %	16	16,8 %	7	7,4%	8	8,4%	9	9,5%	1	1%	77
Car	0	0%	0	0%	0	0%	0	0%	1	1%	0	0%	0	0%	1	1%	0	0%	2
Public transportation	0	0%	0	0%	0	0%	0	0%	1	1%	0	0%	0	0%	0	0%	0	0%	1
Total	4	4,2%	10	10,5 %	2	2,1%	26	27,4 %	24	25,3 %	8	8,4%	9	9,5%	10	10,5 %	2	2%	95

Table 5. The use of mode of students in general

Research Location	Shopping Area	Types of Mode	
		Private	Public Transportation
Engineering Faculty of Unhas	Panakukang Mall	Motorcycle and Car	Pete-pete from the Poros Malino St. to Alauddin
	Giant	Motorcycle and Car	Public transportation not available
	Ramadhan Stationery	Motorcycle	Public transportation not available
	Royal Mart	Motorcycle	Public transportation not available
	Balang-Balang Market	Motorcycle and walk	Pete-pete along the Poros Malino St.
	Afternoon Market	Motorcycle and walk	Pete-pete along the Poros Malino St.

Research Location	Shopping Area	Types of Mode	
		Private	Public Transportation
UIN Alauddin Samata	Mall Panakukang	Motorcycle dan mobil	Pete-pete from UIN Alauddin to Emmy Saelan field
	Giant	Motorcycle, car and public transportation	Pete-pete from UIN Alauddin to Emmy Saelan field
	Ramadhan Stationery	Motorcycle and walk	Pete-pete from UIN Alauddin to Emmy Saelan field
	Royal Mart	Motorcycle, Bicycle and walk	Pete-pete from UIN Alauddin to Emmy Saelan field
	Balang-Balang Market	Motorcycle	Public transportation not available
	Panakukang Mall	Motorcycle and Car	Pete-pete to Emmy Saelan field

4.2.2. The Ownership of Vehicle. Based on table 6 and table 7, the vehicle ownership of Unhas and UIN Alauddin students according to the location of residence is divided into five categories, where motorbike vehicles are the type of vehicle most commonly owned by Unhas and UIN Alauddin students as many as 67.5% and 60% this shows that motorcycle ownership has become one of the dominant modes used by students for access to shopping facilities both in segments one and segments two and three.

Table 6. The Ownership of vehicle of Unhas engineering student according to the location of residence

Travel Behavior according to Residences / Type of Transportation	Individual						2 People						3-5 People						Total
	500		501-800		801-1200		500		501-800		801-1200		500		501-800		801-1200		
	n	%	n	%	n	%	n	%	n	%	n	%	n	%	n	%	n	%	
Car and Motorcycle	1	1,1%	0	0%	0	0%	1	1,1%	2	2,2%	1	1,1%	0	0%	0	0%	0	0%	5
Motorcycle	7	7,9%	15	16,9%	1	1,1%	10	11,2%	23	25,9%	3	3,4%	0	0%	1	1,1%	0	0%	60
Car	0	0%	1	1,1%	0	0%	0	0%	0	0%	0	0%	0	0%	0	0%	0	0%	1
Bicycle	0	0%	1	1,1%	0	0%	0	%	0	0%	0	0%	0	0%	0	0%	0	0%	1
No Vehicle	2	2,2%	3	3,4%	0	0%	4	4,5%	12	13,5%	0	0%	0	0%	1	1,1%	0	0%	22
Total	10	11,2%	20	22,5%	1	1,1%	15	16,9%	37	41,6%	4	4,5%	0	0%	2	2,2%	0	0%	89

Table 7. The Ownership of vehicle of UIN Alauddin student according to the location of residence

Travel Behavior according to Residences / Type of Transportation	Individual						2 People						3-5 People						Total
	500		501-800		801-1200		500		501-800		801-1200		500		501-800		801-1200		
	n	%	n	%	n	%	n	%	n	%	n	%	n	%	n	%	n	%	
Car and Motorcycle	0	0%	0	0%	0	0%	2	2,1%	0	0%	2	2,1%	1	1%	3	3,1%	0	0%	7
Motorcycle	4	4,2%	7	7,4%	1	1%	17	17,9%	16	16,8%	2	2,1%	5	5,3%	5	5,3%	0	0%	58
Car	0	0%	1	1%	0	0%	0	0%	0	0%	0	0%	0	0%	0	0%	0	0%	1
Bicycle	0	0%	1	1%	1	1%	0	0%	0	0%	0	0%	0	0%	0	0%	0	0%	2
No Vehicle	0	0%	1	1%	0	0%	7	7,4%	8	8,4%	4	4,2%	3	3,1%	2	2,1%	2	2,1%	27
Total	4	4,2%	10	10,5%	2	2%	26	27,4%	24	25,3%	8	8,4%	9	9,5%	10	10,5%	2	2,1%	95

4.2.3. Lane. The lane that are passed by Unhas Engineering students and UIN Alauddin in traveling in meeting the needs of shopping facilities are as follows:

1. Afternoon Market
   - Faculty of Engineering of Unhas (STTP Gowa St. and Poros Malino St.).
   - UIN Alauddin Samata (Sultan Alauddin St., H.M Yasin Limpo St., Macanda St., Macanda II St., STTP Gowa St. and Poros Malino St.) and also (Sultan Alauddin St., Poros Malino St., STTP Gowa St. and Poros Malino St.).
2. Balang-Balang Market
   - Faculty of Engineering of Unhas (STTP Gowa St. and Poros Malino St.).
   - UIN Alauddin (Sultan Alauddin St., H.M Yasin Limpo St., Macanda St., Macanda II St., STTP Gowa St. and Poros Malino St.) and also (Sultan Alauddin St., Poros Malino St., STTP Gowa St. and Poros Malino St.).
3. Royal Mart
   - Faculty of Engineering of Unhas (Poros Malino St., STTP Gowa St., Poros Malino St., Sultan Alauddin St.) and (Poros Malino St., STTP Gowa St., Macanda II St., Macanda St., H.M Yasin Limpo St., Sultan Alauddin St.).
   - UIN Alauddin (Sultan Alauddin St.).
4. Ramadhan Stationery
   - Faculty of Engineering of Unhas (Poros Malino St., STTP Gowa St., Poros Malino St., Sultan Alauddin St.) and (Poros Malino St., STTP Gowa St., Macanda II St., Macanda St., H.M Yasin Limpo St., Sultan Alauddin St.).
   - UIN Alauddin (Sultan Alauddin St.).
5. Giant
   - Faculty of Engineering of Unhas (Poros Malino St., STTP Gowa St., Macanda II St., Macanda St., H.M Yasin Limpo St., Sultan Alauddin St., Tun Abdul Razak St.).
   - UIN Alauddin (Sultan Alauddin St., H.M Yasin Limpo St., Sultan Alauddin St., Tun Abdul Razak St.).
6. Panakukang Mall
   - Faculty of Engineering of Unhas (Poros Malino St., STTP Gowa St., Macanda II St., Macanda St., H.M Yasin Limpo St., Sultan Alauddin St., Tun Abdul Razak St., Aroepala St., Letjen Hertasning St., Adiyaksa St.).
   - UIN Alauddin (Sultan Alauddin St., H.M Yasin Limpo St., Sultan Alauddin St., Tun Abdul Razak St., Aroepala St., Letjen Hertasning St., Adiyaksa St.).

4.3. Analysis of the Location of Shopping Area for Unhas Engineering Students and UIN Alauddin in Meeting the Needs

The location of the University of Hasanuddin University and UIN Alauddin's student shopping visit with the travel distance and service center scale visited as follows:

4.3.1. The Location of Shopping Area. The dominant shopping locations were visited by Unhas Engineering students and UIN Alauddin doing shopping at a distance, namely Panakukang Mall and Giant. Next to the location of short-distance student shopping visits, namely in the Balang-Balang Market for Unhas Engineering Students, and Ramadhan Stationery for UIN Alauddin students.

4.3.2. Distance of the Location of Shopping Area. The distance of the shopping location to the two research locations, the closest distance that can be reached by UIN Alauddin students is 0.2 km to the shopping location, the Royal Mart. The closest distance that can be reached by Unhas students is 0.2 km to the shopping location, the Evening Market. Whereas the farthest distance reached by UIN Alauddin students was 6.6 km and the farthest distance reached by Unhas students was 10.3 km with the same shopping location, Mall Panakkukang.

4.3.3. Travels Time. Travel time of students from the rental residence to the location of the means of meeting the needs of shopping facilities according to the location of residence. For Unhas Engineering students, the biggest travel time is 15-30 minutes to get to shopping facilities by 51.7%, while for UIN Alauddin students 54.7% spend <15 minutes.

4.3.4. Scale of Service Centers. The service zone for student shopping locations is divided into three segments: first, the Environmental Service Center (PPL) is a settlement center that serves to serve inter-village scale activities; second, the Regional Service Center (PPK) is an urban area that functions to serve sub-district or several village scale activities; third, the Regional Activity Center (PKW), which is an urban area that functions as a center for industrial and service activities serving provincial or several regency scales. For the first zone, the scale of the Environmental Service Center (PPL) found in the two research sites, namely Balang-Balang Market and Royal Mart, is the location of the largest shopping visit visited by students on both campuses, then the Regional Service Center (PPK) is in Giant and Regional Activity Centers (PKW) are located in Panakukang Mall (MP).

4.4. Direction of Environmental Infrastructure in Meeting Needs
4.4.1. The planning of Pedestrian. In general, students make a dominant movement using motorized vehicles to shopping locations, so it can be directed towards the provision of shaded pedestrian paths, and bicycle lanes in order to encourage students to use modes of transportation that can improve health, safety, and environmentally friendly.

4.4.2. Provide the Mass Transit. In general, students travel to shopping locations at a distance (> 5 km) using a motorized transportation mode, then the procurement of mass transportation (pete-pete, BRT) is directed, to support student travel to the location of service centers at the sub-district and city scale.

4.4.3. Development of Service Center in District Scale. Unhas Engineering Students and UIN Alauddin made a shopping trip at three service centers, namely: environmental service center, sub-district service center, and city service center. To reduce the movement of student shopping at the scale of sub-district service centers and city service centers, directed towards the development of sub-scale service centers that can support both Unhas and UIN Alauddin student residential centers (± 5.2 km distance) can be directed by considering the concept of developing mixed land functions (mixeduse) and transportation friendly (TOD).

5. Conclusion

Students' trips to shopping locations generally behave in general terms, that is, having friends on the road, dominant on a small scale of 2 people (all students behave similarly in 3 segments of the research location). Travel companions are generally from the same study program or neighboring boarding. Movement behavior using motorcycle transportation modes, both in close proximity to the neighborhood shopping center or to the district and city shopping centers. His travel behavior to the environmental service center is via the environmental road (± 6.00m). Behavior of travel to the service center at the sub-district and city scale is via heavy motorways (Jn. Poros Malino, or Jln. Hertasning).

Behavior of student shopping visit locations has the same goal, namely at an environmental scale service center with a distance of achievement ± 200 meters (Pasar Balang and Ramadhan Stationery) and a service center at the sub-district and city scale with a distance of> 5 km (Giant Hertasning , and Panakukang Mall). Directions for the development of environmental infrastructure to support the movement path of students who are dominantly using motor vehicles to shopping locations, so that the direction of shaded pedestrian pathways and bicycle lanes can be directed to encourage students to use modes of transportation that can improve health and be environmentally friendly. In addition it is directed towards the procurement of mass transportation to support student trips to service centers at the sub-district and city scale. The development of district scale service centers that can support both

Unhas and UIN Alauddin student residential centers (± 5.2 km distance) can be directed by considering the concept of developing mixed use and transportation-friendly functions (TOD).

References

[1] Bahar, Erwin. (2019). *Strategi Pengembangan Ruang Terbuka Publik Berdasarkan Identitas Sosial Masyarakat Permukiman Sekitar Kampus Teknik Unhas Gowa*. Skripsi Teknik Perencanaan Wilayah dan Kota, Fakultas Teknik Universitas Hasanuddin.

[2] Dwi Hapsari, Meifinta dan Wisnu Pradoto. (2013). *Perkembangan Permukiman di Sekitar Lingkungan Kampus Undip Tembalang*. Jurnal Pembangunan Wilayah dan Kota. Biro Penerbit Planologi Undip. Volume 9 (4): 404-415 Desember 2013.

[3] Sabri, Ali. (2013). *Analisis Perilaku Perjalanan Mahasiswa dan Aksesibilitas Pada Perguruan Tinggi Di Makassar (Studi Kasus Fakultas Exact Unhas)*. Jurnal Teknik Sipil, Universitas Hasanuddin. Vol. 7.

[4] Iswahyudin, dkk. (2018). *Pola Pergerakan Penduduk Pekerja di Kecamatan Somba Opu Kabupaten Gowa Sebagai Daerah Hinterland Kota Makassar*. Skripsi Teknik Perencanaan Wilayah Kota, Fakultas Teknik Universitas Hasanuddin.

[5] Samang, Lawalenna, Sumarni Hamid Ali, dkk. (2013). Studi Kebutuhan Moda Transportasi Pengoperasian Kampus Fakultas Teknik Unhas Gowa. Jurnal Teknik Sipil, Fakultas Teknik Universitas Hasanuddin. Vol. 7.

IOP Conf. Series: Materials Science and Engineering **875** (2020) 012017 doi:10.1088/1757-899X/875/1/012017

The Effectiveness of Incinerator at the Integrated Waste Treatment Plant in the Campus of Engineering Faculty - Hasanuddin University

A T Lando[1]*, I Djamaludin[1], A N Arifin[2], N Oktorina[1], R Danah[1], and M F Sulistyowati[1]

[1]Environmental Engineering Department, Engineering Faculty, Hasanuddin University, Makassar, Indonesia
[2]Agro-Industrial Manufacture Engineering Department, ATI Makassar Polytechnic, Makassar, Indonesia

*Email: asiyanthit.lando@unhas.ac.id

Abstract. Waste is an endless environmental problem. The amount of waste production has increased due to the increasing population. Besides being produced in a city, waste generation is also produced by universities or campuses. One of them is the Gowa campus, the campus of Faculty of Engineering, Hasanuddin University. Current waste management is carried out by holding, collecting and burning in open areas without further management. It can cause environmental pollution and health problems. The objectives of this study were to determine the amount of campus waste generation, to analyze the needs of incinerators in campus, to determine incinerator tools, and to determine the advantages, disadvantages, efficiency and effectiveness of the incinerators in Gowa Campus. The results showed that the amount of waste generated at the Gowa campus was 0.25 liters/person/day and 0.12 kg/person/day. This incinerator requirements analysis uses the geometry method and the efficiency and effectiveness method. From the calculation, it is known that the use of incinerators for the next ten years is one unit. The disadvantage of using an incinerator is that the cost is very high. However, the advantage is that it can reduce waste generation and overcome environmental and health problems. The comparison between Double Chamber and Maxpell incinerator gave the results that the most effective and efficient is the Maxpell incinerator with an effectiveness of 97.5% and efficiency of 17.17%.

1. Introduction

One of the factors that cause the environment to be polluted is waste. Waste generation is directly proportional to the population, which is the increase in population, so the waste generation also increases. In addition to waste generation in a city, a place that also produces waste generation is a campus or university that has routine activities even on holidays. One campus that produces waste is the Gowa Campus, the campus of Engineering Faculty, Hasanuddin University located in Gowa Regency, South Sulawesi. Waste that is produced in the form of organic waste comes from the canteen or food stalls. Moreover, inorganic waste comes from teaching and learning activities and other activities. Waste generated from each building is then collected, transported, disposed of and burned in open fields without further handling. This poor management can cause environmental problems such as air pollution and health problems. One example of poor environmental management is the open dumping landfill which is indicated to have released dangerous gas. It should be immediately anticipated, and this dangerous gas can claim the lives of people who are in the closest radius of the landfill. People who breathe in methane gas every day may be able to experience damage to organs and cells of the body or even die if they continue to breathe in methane gas. Also, methane gas can explode at any time. Based on previous research [1, 2, 3] shows that the concentration and emissions of methane from the Tamangapa landfill in Makassar city are in a very varied range, from 12 - 425 parts per million (ppm) for methane concentrations and 2.44-18 Gg/year for methane emissions.

Based on these problems, waste needs to be appropriately managed, namely by burning waste using incinerator technology. However, the problem in applying incinerator is emission in the form of particulate matter (PM), Sulfur Dioxide (SO_2), Carbon Monoxide (CO), Carbon Dioxide (CO_2), Hydrogen Chloride (HCl), dioxin, furan and heavy metals. Therefore, incinerator that can be used must be equipped with a control system to meet the limits of particulate emissions and exhaust gases so that the smoke coming out of the incinerator is neutral smoke/gas. To find out the incinerator used is useful or not, then we need a method to determine the value of the efficiency and effectiveness of using an incinerator. Efficiency is a calculation to find the value of accuracy in carrying out incinerators as well as the benefits derived from the management of incinerators, while effectiveness is the effect or the results obtained from the incinerator technology.

The objectives of this research are: (1) to find out the amount of waste generated, (2) to analyze the need for incinerators, (3) to find out which incinerator tools can be used, it is adjusted to the amount of waste generated, and (4) to determine the strengths, weaknesses, efficiency, and effectiveness of incinerators that will be used in the Gowa campus.

According to the Law of the Republic of Indonesia No. 18 of 2008, Chapter One Article 1, waste is defined as the remnants of daily human activities and/or solid natural processes. Sources of waste come from campuses, markets, households, highways and public places. The types of waste generated are organic, inorganic, domestic, building waste and others [4]. In general, the most common types of incinerators applied are rotary kiln, multiple hearth, and fluidised bed incinerators. However, these types of incinerators are often applied for combustion of toxic and hazardous waste. The types of incinerators that can be used to burn domestic waste and as research material are as follows: Maxpell and Double Chamber Incinerator.

Maxpell incinerator is incinerators that do not use fuel, have a hydro process system that absorbs poisons and odours and converts carbon particles into magnets that can be captured by the split cell before exiting through the chimney. Moreover, Double chamber incinerator is a type of incinerator that uses diesel fuel. This incinerator has two furnaces, with temperatures of 300-600 °C for the first furnace and 600-1000 °C for the second furnace. This incinerator is equipped with a water scruber that serves to prevent air pollution by using water spray against gas coming out of the second combustion chamber before entering the chimney.

Figure 1. Maxpell Incinerator [5]

Figure 2. Double Chamber Incinerator [6]

2. Methods

The tools used in this study, in measuring the waste generation in Gowa Campus, were 60-litre garbage bag, 500-litre measuring box, meters, 50 kg scales, tarpaulins, stationery, and personal protective equipment (masks and gloves). The material used is waste generation from the Gowa campus. This study was conducted in four months consisting of preparation, experiment and report preparation. At the experiment stage, taking and measuring samples eight days. The experiment was carried out at the Gowa campus, Campus of Engineering Faculty, Hasanuddin University.

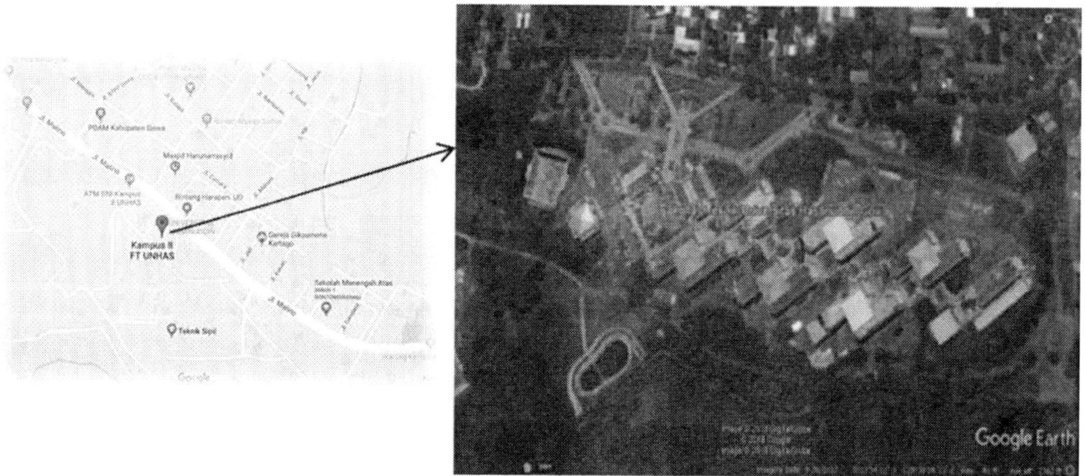

Source: Google Earth

Figure 3. Study site (Gowa Campus - Hasanuddin University)

The experiment began by distributing waste bags to the cleaning service on duty at each building. Waste originating from the cleaning service is then brought to the point of collection. The waste is then put into a 500-litre measuring box and compacted by dropping the box three times with a height of 20 cm. Then, measuring the area of the box and the height of the waste to get the waste volume. Next, sort waste according to the type of waste, weigh and record the weight of the waste. Add up the total weight of the waste from each type of waste to get the overall weight of the waste [7].

Figure 4. Measuring of waste generation in Gowa Campus

Data analysis was performed using the following equations: (1) waste volume = *area of the base of the measuring box x height of waste*, (2) *waste density = waste mass (kg)/waste volume (m^3)*. For the analysis of waste characteristics, sorting according to their characteristics and then calculated using the equation of *percentage characteristics (%) = total mass of waste (kg)/characteristic mass (kg) x 100%*. Next, calculate the waste generation per person per day = $v_1 + v_2 + v_3 + v_n/n$ (litre/person.day). Waste generation and the need of incinerator are directly proportional to the population, then the population and the rate of waste generation are projected using the Geometry method: $P_n = P_0 + (1 + r)n$ and $r = 1/t \ln P_t/P_0$. Where, P_0 = population in the starting year, P_n = Population in n year, 1 = constant number, r = population growth rate, n = total number of years from start to year n, and t = time period.

Furthermore, the calculation of efficiency and effectiveness of the use of incinerators is: (1) *efficiency = amount of input (cost)/amount of output (benefit) x 100%* and (2) *effectiveness = tangible result/target x 100%*. To find out the efficiency and effectiveness of using an incinerator can be seen from the criteria of financial efficiency and the interpretation of the effectiveness value as follows:

Table 1. Criteria of financial efficiency [8]

Value of Efficiency Ratio	Interpretation of Efficiency Value
> 100 %	Not Efficient
90 % - 100 %	Less Efficient
80 % - 90 %	Efficient Enough
60 % - 80 %	Efficient
< 60 %	Very Efficient

Table 2. Interpretation of effectivity value (%) [8]

Effectivity Value (%)	Interpretation of Effectivity Value
80 – 100	High
60 – 79.9	High enough
40- 59.9	A little low
20 – 39.9	Low
0.0 – 19.9	Very low

To analyze the costs and benefits of an activity, in this study is about the procurement of incinerators on the Gowa campus, several methods can be used [9]. Two of the methods are: (1) the Return Period method, used for valuing an investment project, based on the length of time the investment can be covered with cash inflows and (2) the Return of Investments method, used to return investment which is used to measure the percentage of benefits generated by the project compared to the costs incurred.

3. Results and Discussions

The total population in the Gowa campus is 4843 people, divided into 4479 students, 242 lecturers, 88 employees, and 34 cleaning services. The existing condition of waste management on the Gowa campus still adheres to the old paradigm of gathering, transporting, then dispose of it, as can be seen in Fig. 5. The waste collection that has been sorted is carried out in a container and transported at 07.00 AM and 4.00 PM. At the transportation stage, the waste is transported by the cleaning service officer, and then the waste is disposed of at the point of disposal then burned on open land without further management. This waste management model can pose a risk of air pollution and health problems.

Judging from the existing conditions of waste management at the Gowa campus, an integrated waste treatment plant is needed, as shown in Fig. 6. The integrated waste management paradigm is described based on Law No.18 of 2008 as follows: (1) waste sorting, (2) waste collection, (3) waste transportation, and (4) waste treatment.

Figure 5. Existing condition of waste management in Gowa campus

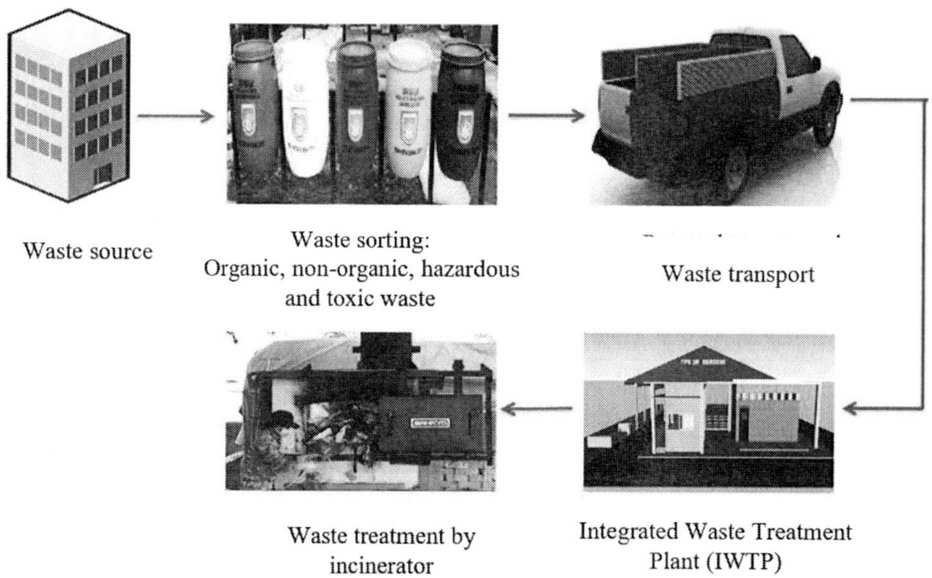

Figure 6. Integrated waste management

3.1. Integrated Waste Management

Waste sorting can be divided into organic, non-organic, hazardous and toxic waste, recycling, and others. Waste bins/containers are useful for distinguishing types of waste from sources of waste, making it easier to transport and manage the waste. Waste collection is conducted by using a vehicle that is modified by using a bulkhead to separate the waste according to the sorting that occurs at the waste source. The capacity of the transport vehicle is adjusted to the capacity of the waste generated each day. The collection time is equaled twice a day which is carried out at 7 am and 4 pm.

3.2. Waste Generation and Characteristics in Gowa Campus

Measurement of waste generation is divided into two: waste (1) mass and (2) volume measurements. The following is the result of measuring the weight and volume of waste generated from each building (waste source).

Table 3. Waste weight of Gowa campus

No.	Name of Building	Waste Weight (kg)								Waste Average Per Day
		Wed	Thu	Fri	Mon	Tue	Wed	Thu	Fri	
1	COT	2.46	2.78	2.16	2.5	2.06	2.14	1.92	1.76	2.22
2	CSA	9.16	10.9	8.19	11.49	9.82	9.2	11.02	7.66	9.68
3	Classroom	13.35	10.96	8.49	16.03	12.7	11.08	10.43	9.07	11.51
4	Civil	14.86	10.86	8.45	17.8	20.62	15.75	10.22	7.07	13.20
5	Architecture	2.66	3.8	2.31	2.46	3.63	2.85	2.12	1.6	2.68
6	Mechanical	0.81	2.42	1.02	6.3	2.96	1.27	2.72	1.29	4.18
7	Electrical	7.5	5.85	1.5	1.35	2.55	3.46	4.55	1.83	3.57
8	Geology	1.93	1.92	1.35	2.25	2.28	2.73	2.95	2.35	2.22
9	Naval	3.16	2.24	2.2	1.81	2.05	2.1	2.16	1.7	2.18
10	Ramsis	11.85	9.6	11.37	18.26	13.54	13.74	14.36	11.11	12.98
	Total	67.74	61.33	47.04	80.25	72.21	64.32	62.45	45.44	64.42

Table 4. Waste volume of Gowa campus

No.	Name of Building	Waste Volume (l)								Waste Average Per Day
		Wed	Thu	Fri	Mon	Tue	Wed	Thu	Fri	
1	COT	62.57	84.34	58.98	72.77	57.33	55.13	48.79	47.41	60.91
2	CSA	157.11	161.24	147.74	171.99	154.07	124.86	170.06	131.75	152.35
3	Classroom	224.08	189.63	180.53	235.38	215.54	210.85	189.08	186.05	203.89
4	Civil	201.21	172.82	134.78	308.70	319.17	261.29	167.03	120.17	210.65
5	Architecture	85.44	126.24	74.14	104.19	119.07	78.55	74.42	62.57	90.58
6	Mechanical	13.00	83.00	19.00	172.00	61.00	10.00	96.00	17.00	58.92
7	Electrical	88.20	66.15	49.61	64.22	74.42	61.74	57.88	65.60	65.98
8	Geology	52.37	54.57	51.54	60.91	60.64	72.77	75.80	63.39	61.50
9	Naval	90.96	67.53	62.29	64.77	68.91	68.08	60.09	63.12	68.22
10	Ramsis	248.06	159.86	238.97	349.49	271.49	278.93	100.60	229.87	234.66
	Total	1223.22	1165.07	1017.88	1604.14	1401.28	1222.15	1040.21	987.29	1207.65

Table 5. Waste composition/characteristics of Gowa campus

No.	Day	Waste Weight Based on Composition from All Buildings (kg)							
		PET	Non PET	Paper	Tissue	Card board	Styrofoam	Metal	Organic
1	Wed	14.21	10.06	6.76	4.54	6.89	1.62	0.77	22.48
2	Thu	13.76	10.34	6.68	4.69	7.68	1.98	0.35	15.85
3	Fri	10.63	7.58	4.91	3.58	5.22	1.47	0.10	13.55
4	Mon	18.14	10.53	9.18	6.93	10.27	3.00	1.15	21.05
5	Tue	14.47	11.27	9.30	7.20	7.94	2.23	0.45	19.35
6	Wed	13.19	12.80	8.33	4.00	6.29	1.85	0.62	17.24
7	Thu	13.27	13.82	5.21	4.06	6.46	2.32	0.77	16.54
8	Fri	9.46	6.71	4.97	3.70	3.99	1.60	0.21	14.80
	Total	107.13	83.11	55.34	38.7	54.74	16.07	4.42	140.86
Percentage (%)		21 %	17 %	11 %	8 %	11 %	3 %	1 %	28 %

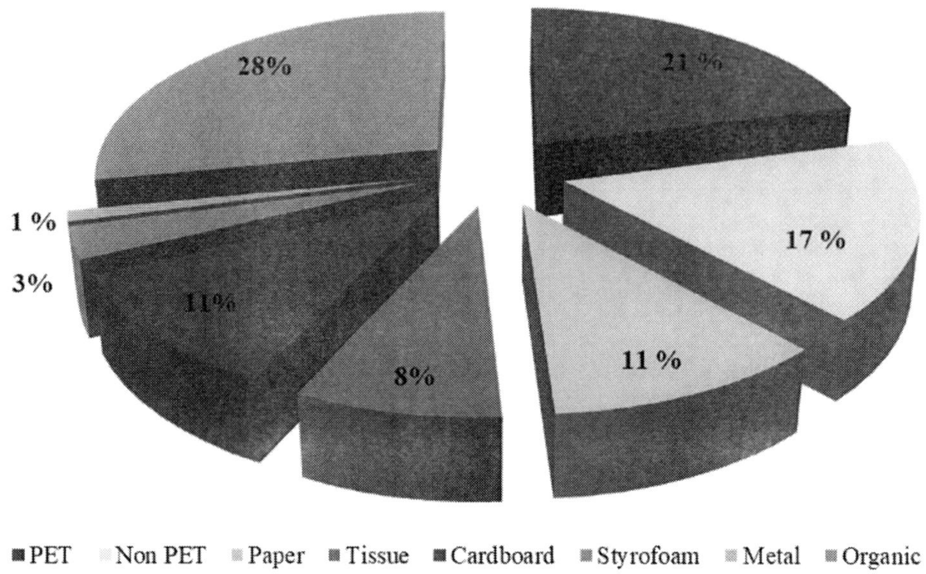

Figure 7. Percentage of waste composition of Gowa campus

Based on Tables 3 and 4, the results of waste generation per individual unit in Gowa campus is 0.013 kg/person/day and 0.25 litre/person/day. The data above is very important in determining the required land area of Integrated Waste Treatment Plant of Gowa Campus and the facilities contained therein, including incinerator. The tables also show the weight and volume of waste generation on the Gowa campus. From the whole building and the total population, the generation of waste generated every day is 64.42 kg/day and 1207.65 L/day. It can be seen the amount of waste that is burned every day with improper combustion techniques. It can cause significant problems for health and environmental pollution. Regarding these problems, incinerator technology is needed on Gowa Campus to reduce the amount of waste generated.

3.3. Projected Population and Waste Collection Rate in Gowa Campus

The results of the calculation of the projected population and waste generation on the Gowa campus are shown in Table 6.

Table 6. Total population and waste generation projection

No	Year	Total Population	Population Growth (r)	Waste Generation (L/people/day)	Waste Generation Rate (m^3/hari)	Total Waste Generation (m^3)
1	2018	4,843	0.0128	0.25	1.21	441.92
2	2019	4,905			1.23	447.58
3	2020	4,967			1.24	453.24
4	2021	5,031			1.26	459.08
5	2022	5,095			1.27	464.92
6	2023	5,161			1.29	470.94
7	2024	5,227			1.31	476.96
8	2025	5,294			1.32	483.08
9	2026	5,361			1.34	489.19
10	2027	5,430			1.36	495.49

The results of the projected population of the Gowa campus in 2027 were 5,430 people with the generated waste of 0.25 litre/person/day. Therefore, the waste generation on Gowa campus in 2027 is 5,430 people x 0.25 liters/person/day = 1.36 m^3/day, then the total waste production in 2027 is 1.36 m^3/day x 365 days = 495.49 m^3.

3.4. Analysis of the Needs and Effectiveness of Incinerators in Gowa Campus

Study and consideration in the selection of incinerator technology is a study to determine the effectiveness and efficiency of a waste burning technology. The main criteria considered for technology selection are the number of tools in reducing waste, the capacity for combustion, and the final results released by the incinerator. Errors in the selection of waste combustion technology can cause the failure of the entire waste treatment system that leads to health problems and environmental damage. Of the many companies making incinerator technology, two types of technology were taken for this study, namely the Double Chamber and Maxpell incinerator technology. Both of these machines were chosen to see how much effectiveness in reducing waste. The comparison of the specifications of these two technologies, it can be determined the amount of waste reduced from the residual ash of combustion and the time of operation carried out. Comparison of the effectiveness of the two incinerator technologies is as follows:

Table 7. Comparison of incinerator technology specifications

Comparison	*Double Chamber* Technology	*Maxpell* Technology
Combustion Capacity	0.075 m^3/hour	0.085 m^3/hour
Combustion Time	16 hour/day (8 times combustion process)	16 hour/day (8 times combustion process)
Combustion System	Double Chamber	Gasification
Combustion Temperature	> 900°C	600 - 1000°C
Smoke Neutralizer	Wet Scruber	Smoke Scruber
Fuel Usage	Diesel Fuel	No Fuel
Electricity	6.5 kWh	No Electricity
Combustion Ash	10 %	5 %
Area Needs	3 m x 3 m	3.5 m x 3.5 m

The results of the comparison of incinerator technology in the form of the effective value of Double Chamber and Maxpell incinerator technology can be seen in Figure 8. This figure shows the effectiveness value of the Double Chamber incinerator is 90% and 95% for Maxpell incinerator. Thus the technology that has a high value of effectiveness in reducing waste is the Maxpell incinerator. This effectivity is because the burning of waste is conducted only three times a week, not every day, in order to minimise damage to incinerators and further reduce the impact on air pollution.

Maxpell incinerator has greater effectiveness in reducing waste generation and is environmentally friendly because this incinerator does not use fuel compared to Double Chamber incinerators. Moreover, this incinerator has air pollution control devices (Hydro process) so that the air coming out of the chimney is safe for the environment.

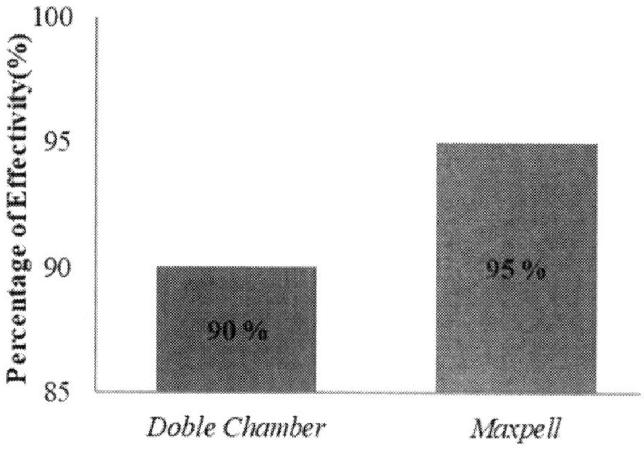

Figure 8. Percentage of effectivity of incinerator technology

Based on the results of projected waste generation, specifications, and calculation of the combustion rate of the Maxpell incinerator, the incinerator requirements in Gowa Campus can be analyzed, as shown in Table 8.

Table 8. Analysis of incinerator needs

No.	Year	Waste Generation Rate (m^3/hari)	Total of Waste Generation (m^3)	The Incinerator Needs (unit)
1	2018	1.21	441.92	1
2	2019	1.23	447.58	1
3	2020	1.24	453.24	1
4	2021	1.26	459.08	1
5	2022	1.27	464.92	1
6	2023	1.29	470.94	1
7	2024	1.31	476.96	1
8	2025	1.32	483.08	1
9	2026	1.34	489.19	1
10	2027	1.36	495.49	1

Based on the results of the calculation of the need for incinerators in the above table, it can be seen that the number of incinerators needed until 2027 is only one unit. Incinerator needs are known from the calculation of the rate of combustion. The volume of waste is divided by the time of combustion so that the resulting waste is burned for one hour is 0.075 m^3/hour with a total waste generation of 1.207 m^3/day. It can be assumed that the combustion capacity used is 0.085 m^3/hour or proportional to the amount of waste generated 1.36 m^3/day.

3.5. Analysis of Incinerator Efficiency in Gowa Campus

The two incinerators are then compared using the efficiency formula, the value of the retrieval period and the investment return method. The comparison can be seen in Figures 9, 10, and 11.

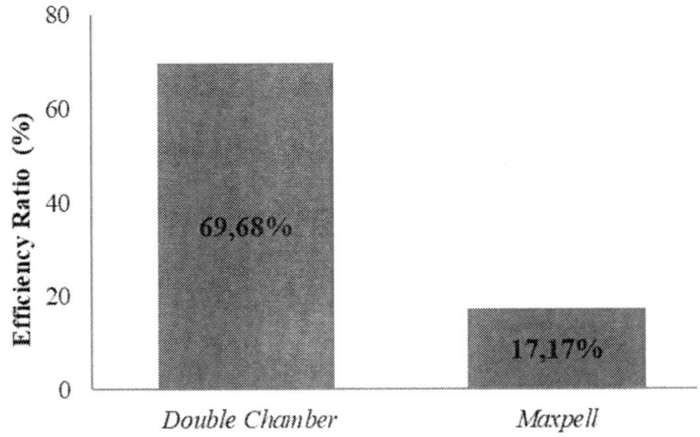

Figure 9. Value of efficiency ratio of incinerator technology

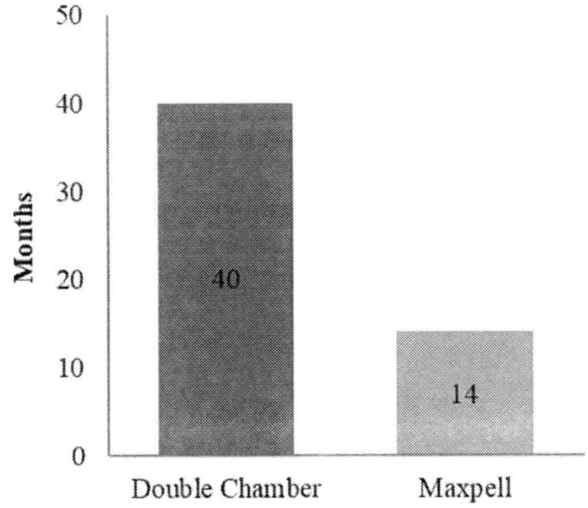

Figure 10. The value of retrieval period

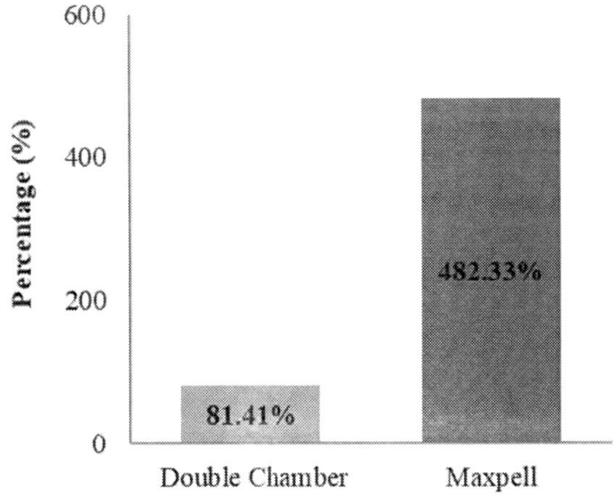

Figure 11. The return of investment (ROI)

Based on figures 9, 10, and 11 and the table of efficiency criteria for financial performance, the value of the efficiency ratio for Double Chamber incinerator is in the efficient range of 69.68%, and Maxpell incinerator is in the highly efficient criteria of 17.17%. The return period value for Double Chamber incinerator is 40 months and 14 months for Maxpell incinerator. While the Return of Investment (ROI) value in the Double Chamber incinerator is 81.41% and 482.33% in the Maxpell incinerator. The benefits of using incinerators can be increased, such as processing waste into electrical energy and so on, because it has a more significant advantage compared to the sales of waste ash produced. However, because the operational costs of incinerators are significant, the campus needs to work closely with other parties to ease the costs of procuring and operating incinerators on campus.

4. Conclusions

Based on the objectives of this study which are based on the results of research and data analysis, the following conclusions are obtained:

1) The total solid waste generation in the Gowa campus is 1207.65 litres/day and 64.42 kg/day, with individual units: 0.25 liters/person/day and 0.12 kg/person/day. The composition of waste is dominated by non-organic waste by 72% and organic waste by 28%.
2) The pattern of waste management on the Gowa campus is to collect, transport, and dispose of it. Therefore, it is necessary to properly process waste management, which is sorting, collecting, transporting and managing. At the management stage, the technology needed to reduce waste other than compost and enumeration is the incinerator. The advantage is that it can reduce waste quickly and has air pollution control devices so that it can reduce the impact of environmental and air pollution.

3) Based on the projection results, the population and the rate of waste generation in the Gowa campus in 2027 are 5,430 people and 495.4 m^3. Waste that is burned for one hour is 0.075 m^3/hour. It can be assumed that the combustion capacity used is 0.085 m^3/hour or proportional to the amount of garbage generated 1.36 m^3/day.
4) From the results of the study, it is known that the lack of incinerators is that they have extensive procurement, operational and maintenance costs while the excess of incinerators is that it can reduce waste generation and can overcome environmental problems and health problems. From the results of the calculation of the effectiveness and efficiency of the right incinerator used on the Gowa campus is the Maxpell incinerator with an effectiveness of 95% and an efficiency of 17.17%.

References

[1] Lando A T, Nakayama H, and Shimaoka T 2015 *Development of Laser Scanning Method for Estimation of Spatial Distribution of Methane Emission in A Landfill*. Memoirs of The Faculty of Engineering, Kyushu University, Vol. 75 (1), pp. 29-46, Kyushu University, Japan.

[2] Lando A T, Nakayama H, and Shimaoka T 2016 *Development of Scanning Method for Measurement of Spatial Distribution of Methane Emissions in Landfill Site*. Dissertation of Urban and Environmental Engineering, Kyushu University, Japan.

[3] Lando A T, Nakayama H, and Shimaoka T 2017 Application of Portable Gas Detector in Point and Scanning Method to Estimate Spatial Distribution of Methane Emission in Landfill. Waste Management, Vol. 59 (2017), pp. 255-265, Elsevier.

[4] Kodoatie, R.J 2003 *Manajemen dan Rekayasa Infrastruktur*. Pustaka Pelajar, Yogyakarta.

[5] Maxpell Technology 2015 *Incinerator Alat Pembakar Sampah Ramah Lingkungan*. Maxpell Technology [www.maxpelltecnology.com], Bandung. Accessed on June 24, 2018.

[6] AJM Tech Indonesia 2017 *Teknologi Inisinerator*. The Royal Town Regency, Surabaya.

[7] Standar Nasional Indonesia Nomor SNI-19-3964-1994 1994 *Metode Pengambilan dan Pengukuran Contoh Timbulan dan Komposisi Sampah Perkotaan*. Badan Standar Nasional (BSN), Indonesia.

[8] Marlina 2018 Analisis Efisiensi Digester Biogas Sebagai Energi Alternatif Skala Rumah Tangga (Studi Kasus Di Kecamatan Tompobulu Kabupaten Maros). Tesis Program Pascasarjana Universitas Hasanuddin, Makassar.

[9] Selin Aprilyani, 2015. *Analisis Biaya dan Manfaat.* Jakarta, Indonesia.

Soil-Compost Mixture as Alternative Material for Soil Cover Landfill

K Sari[1]*, I Ridwan[1], and N A Putry M[1]

[1]Department of Environmental Engineering, Engineering Faculty, Hasanuddin University, Makassar, Indonesia

*Email: kartikasari@unhas.ac.id

Abstract. One of typical problem in a developing country is the lack of controlling odors, vectors, fires, and litter in the waste landfill. It is known that this problem can be reduced by landfill. However to maintain cover material always available, it is uneasy, especially due to limited budget. This study intends to propose compost-soil mixture as an alternative material using residual soil cover from completed waste landfill. In order to investigate and assess the geotechnical properties of soil-compost mixture as an alternative material, several tests were conducted such as compaction, unconfined compression, and permeability tests. Compaction tests showed 0.4%, 0.8% and 1.2% of compost in soil mixtures results the values of dry density in the range between 1.297 g/cm^3 and 1.330 g/cm^3, which meet the minimum requirement according to Indiana environmental department USA. Shear strength of compost-soil mixture that derived from results of unconfined compression tests, reduced slightly for additional compost without curing time. However, the values increased sharply for 3 and 7 days of curing time. All the shear strength values of 0.88 kg/cm^3 to 34.76 kg/cm^3 passed the minimum requirement for cover material according to Environmental Protection Agency (EPA) and Environment Agency. In case of ability of the soil mixture to transmit water, permeability coefficient were 1.5 x 10^{-3} cm/sec, 7 x 10^{-4} cm/sec and 3 x 10^{-4} cm/sec for 0.4%, 0.8% and 1.2% of compost respectively. Referring to modern sanitary landfill and design, it was shown that, the soil mixtures meet minimum requirement for permeability coefficient. To sum up, compost-soil mixtures with composition tested can be considered as material alternative for soil cover in landfill.

1. Introduction

Inappropriate solid waste management is one of the global issues in the world especially in the case of environmental problem such as soil and ground water contamination, and air pollution [1]. One of the cause of exist problem is amount of waste and number of facilities available are not matched [2], [3], [4]. Another cause is improper structure of waste landfill. Many developing countries still apply open dumping and control open dumping rather than sanitary landfill. Waste disposal plays important role in this issue, since uncontrolled waste disposal can generate serious disease to the human, animal and ecosystem in that area due to heavy metals pollution in water, air and plant [5]. Burning waste openly also can produce CO, CO_2, SOx, NOx, PM10 and other pollutant in the atmosphere [6]. Moreover waste pickers can be exposed to disease while working in open dumping area [7].

Since several decades ago, sanitary landfill has been recommended by government of developed countries to control odors, fires and litter in the waste landfill [8]. Sanitary landfill is a waste landfill concept, where the waste was dumped, buried in the holes underground, compacted, and covered by material. One of the basic differences between system of conventional waste landfill and sanitary landfill is covering layer. Conventional system is normally designed without properly soil cover while sanitary landfill has regular daily cover [9]. However, providing cover materials every day is not easy, especially due to limited budget.

Tamangapa waste landfill is one of controlled dumping that exists in Indonesia regularly cover almost once a month. Some areas in Tamangapa landfill are inactive, but still can be reused as cover material. Another material that can be used to combine with soil is compost which made from degradation of leaves, trunks and some vegetation in that landfill. The purpose of this study is to evaluate geotechnical properties and possibility of soil-compost mixture use as an alternative cover material in Tamangapa landfill.

2. Methodology

2.1. Sampling Method

Soil sample was taken randomly in landfill of Tamangapa under disturbed condition. This landfill is located at Manggala Tamangapa with coordinate of 5°10'30'' LS and 119° 29' 36''LE, ±15 km from center of Makassar. The soil samples were covered by material of inactive landfill, at depth of 0-100 cm from surface. For compost, it was provided from the same place of soil sample.

2.2. Sample Preparation

The soil sample that collected from the landfill site by scoop was brought to Soil Mechanics laboratory in Civil Engineering Department UNHAS. Soil sampel was screened using 1 cm in diameter of sieve to separate soil from organic waste. Then soil was put in the wide plate to get air dried until dry air condition. Almost the same treatment was done for compost. Sample test was prepared in composition of soil with 0%, 0.2 %, 0.4 %, 0.8% and 1.2% of compost. Then several tests were conducted on these samples such as Atterberg limit tests, grain size analysis, specific gravity, compaction test, unconfined compression test and permeability test.

2.3. Laboratory Tests

Before conducting mechanical properties tests, physical characteristic tests were performed on soil samples, such as moisture content, liquid limit, plastic limit, shrinkage limit and specific gravity tests. For moisture content, the tests were performed according to ASTM D2216-10[10], while other tests such as liquid limit and plastic limit tests referred to ASTM D4318-17[11] and shrinkage limit tests were based on ASTM D4943[12]. For specific gravity, the test was done referred to ASTM D854-14[13]. It was found that soil sample contains 18.10% of water and specific gravity of 2.662 g/cm^3.

Another test performed on soil and compost samples was sieving analysis test, which referred to ASTM D 422-63[14]. This test was intended to determine the distribution of fine and coarse aggregate. Five hundred grams of soil samples were used for this test. This experiment used sieve no.4 (7.75mm), 10 (2 mm), 20 (0.84mm), 40 (0.42mm), 60 (0.25mm), 100 (0.15 mm), 200 (0.075), and they were shaken by shaker for 15 minutes.

Standard proctor compaction tests were performed in order to define maximum dry density and optimum moisture content of the soil-compost mixture. Following ASTM D558-96[15], moisture content of sample was increased gradually by pouring water in the sample. Then plot dry density and moisture content in the graph to find optimum moisture content and maximum dry density of the mixture sample.

Permeability test was conducted with falling head test method. The test aimed to find coefficient of permeability of soil-compost mixture. Experiments were done following ASTM D2434-68[16]. Unconfined compression test was conducted to determine unconfined compression strength values then they were derived into shear strength by following ASTM D2166-00 [17].

3. Results and Discussion

3.1. Laboratory Tests

3.1.1. Basic Properties. Several basic properties of soil and compost were defined from physical properties as mention in earlier. The tests were conducted to these materials separately. Results of the tests are shown in Table 1.

Table 1. Basic properties of soil and compost

Soil Properties	Value
Liquid limit (%)	73.62
Plastic limit (%)	34.29
Index Plastic (%)	39.33
Shrinkage limit (%)	16.14
Specific gravity (g/cm^3)	2.662
Water content of soil (%)	18.10
Water content of compost (%)	39.25

3.1.2. Sieve Analysis Tests. From sieve analysis tests, it found that 81.10 % of soil (> 50%) passed sieve no. 200, which consisted of 19.40 % of sand, 9.80 % of silt, and 71.30 % of clay. It was concluded that soil sample was clay soil with high plasticity. Furthermore, it was found also various size of compost particle that is 0.18 % of compost was bigger than 4.75 mm-diameter, 2.82 % was bigger than 2 mm and more than 81 % was smaller than 0.075 mm as seen in Table 2. According to SNI 19-7030-2004 [18] the sample was categorized as organic compost.

Table 2. Result of sieve analysis of compost

No. Sieve	Diameter (mm)	Passed (%)
4	4.75	99.81
10	2	97.17
20	0.84	91.71
40	0.42	87.15
60	0.25	84.55
100	0.15	82.89
200	0.075	81.83
PAN	-	0

3.1.3. Standard Proctor Compaction Tests of Soil-Compost Mixture. Compaction tests were conducted on four variant samples, 0%, 0.2%, 0.4%, 0.8% and 1.2% based on procedures were mention above. Dry density varied between 1.32 g/cm^3 and 1.3 g/cm^3 as shown in Table 3.

Table 3. Result of standard compaction tests

No.	Variation (%)	Optimum water content (ω_{opt}) %	Dry density ($\gamma_{dry\ max}$) gr/cm^3
1	0 %	33.77	1.316
2	0.4 %	33.92	1.330
3	0.8 %	33.82	1.297
4	1.2 %	33.79	1.300

IOP Conf. Series: Materials Science and Engineering **875** (2020) 012018 doi:10.1088/1757-899X/875/1/012018

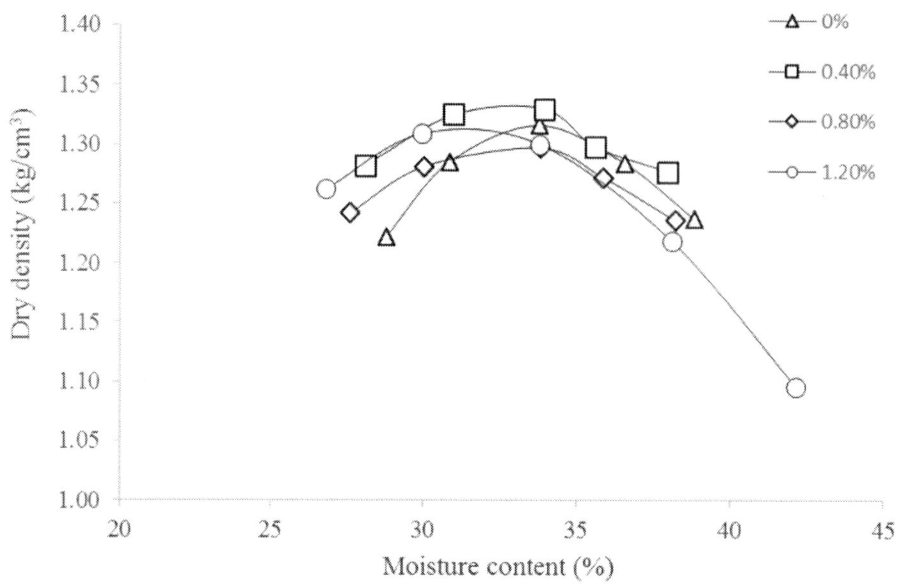

Figure 1. Result of compaction test in the four sample variants

3.1.4. Permeability Tests. Falling head test was conducted to determine permeability coefficient of the soil compost mixture. Similar to compaction test, four variants of samples were also tested on these tests. Result of the tests is shown in Figure 2. From result tests, it can be seen that permeability coefficient tend to reduce as percentage of compost increasing. It might be happened because the particles of compost filled in pore leading to reduction of empty spaces in soils.

3.1.5. Unconfined Compression Tests. As can be seen in the graph below, unconfined compression strength (qu) increased as percentage of compost in soil mixture increased. In case of variation of curing time, it found that qu of 3 and 7 days of curing time increased from 20 up to 40 times than 0 day. Similar trend was happened on shear strength in Figure 3.

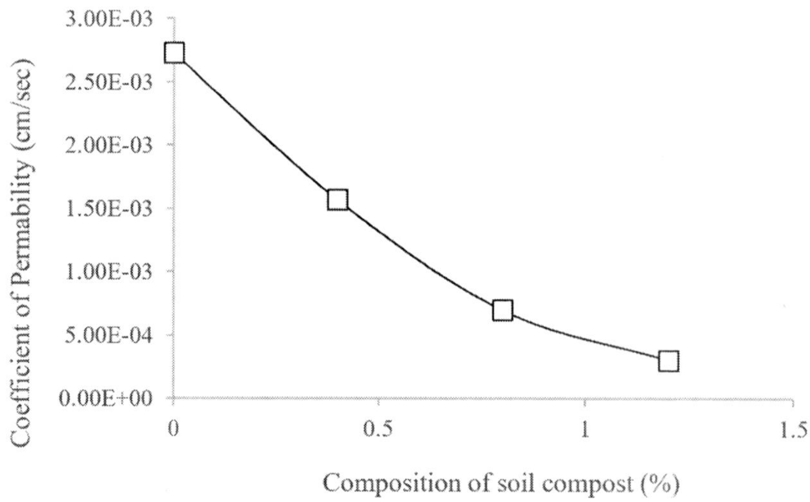

Figure 2. Coefficient of permeability versus composition of soil compost

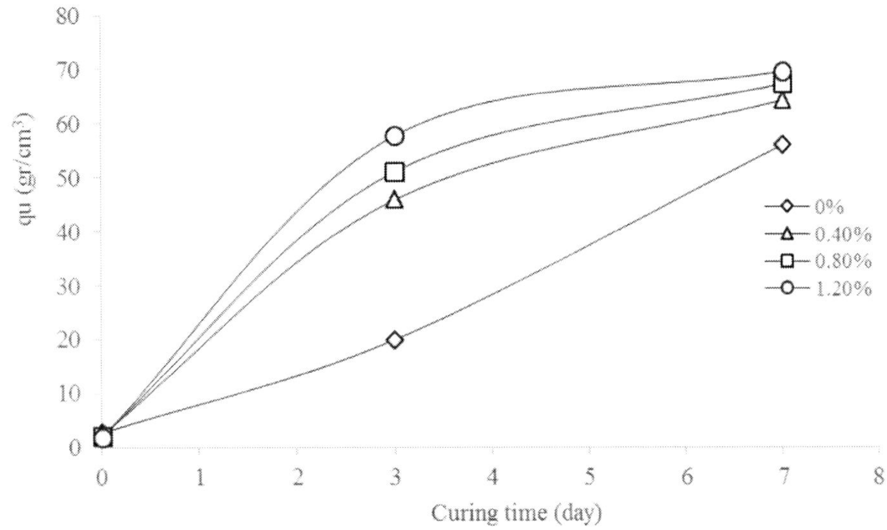

Figure 3. Unconfined compression strength of soil-compost

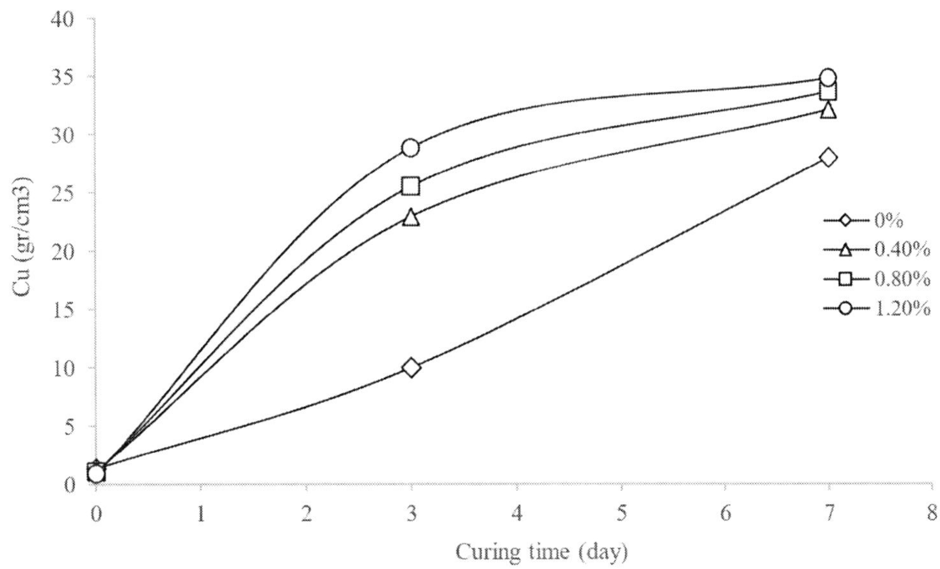

Figure 4. Shear strength of soil-compost

3.2. Test Results Analyses to the Alternative Daily Cover Requirement

Three type of composition of soil-compost mixture have been tested, 0%, 0.4%, 0.8% and 1.8% to assess possibility of soil-compost to become cover material alternative. In this case zero percent of compost was also tested just for comparison to other variants.

3.2.1. Liquid Limit and Plastic Limit. Since there was small part of sand in compost tested so value of liquid limit and plastic limit of soil-compost tested were not available but it will refer to liquid and plastic limit of soil cover tested. According to [19] and [4] liquid limit of cover material in landfills should be within the range of 20 % to 90%. Plasticity index should be in the range of 7% to 65%.

Table 1 shown that the liquid limit and plastic limit meet the requirements as alternative cover liner material.

3.2.2. Grain Size Distribution. Referring to [4], soil cover material should contain at least 10% of clay, which makes the soil tested suitable for using as cover material since the soil content 71.30% of clay

3.2.3. Bulk Density. Result of standard compaction test shown that all soil-compost tested meet requirement of cover material. As stated by Indiana department of environmental management USA 2013 [14], minimum bulk density of material should be at least 1 g/cm^3. Table 3 show the bulk density of all soil-compost tested were about 1.3 g/cm

3.2.4. Coefficient of Permeability. Referring to [20], alternative cover material should have permeability coefficient between 10^{-2} and 10^{-5} cm/sec. Result of falling head tests shown that permeability coefficient of all soil-compost tested were ranging 2.7 x 10^{-3} to 3.0 x 10^{-4} cm/sec means meet requirement.

3.2.5. Shear Strength. Derived from unconfined compression strength, value of shear strength of soil compost was defined. From these results, it shown that shear strength of sample tested meet minimum requirement which ranging from 10 kg/cm^3 to 67 kg/cm^3.

4. Conclusion

Based on study above it can be concluded that soil sample which taken from Tamangapa waste landfill and then mixed with compost can be used as alternative cover material. Value of liquid limit, plasticity index, clay fraction, bulk density, coefficient of permeability and shear strength meet minimum requirement of cover material in landfill area. Increasing composition of compost in sample tested tend to reduce coefficient of permeability. Similar tendency also for shear strength value, it tends to increase as percentage of compost was added. Even the value can be increased when the sample was cured about 3-7 days.

Acknowledgments

The authors would like to thank to Engineering Faculty of Hasanuddin University for providing financial support via LBE Development System.

References

[1] Bing, X.; Bloemhof, J.M.; Ramos, T.R.P.; Barbosa-Povoa, A.P.;Wong, C.Y.; van der Vorst, J.G.A.J. 2016, *Research challenges in municipal solid waste logistics management.* Waste Management (Amsterdam : Elsevier)., 48, 584–592.

[2] Gupta, N.; Yadav, K.K.; Kumar 2015, V. *A review on current status of municipal solid waste management in India.* (China: Journal Environment Science), 37, 206–217.

[3] Ferronato, N and Toretta, V., 2019, *Waste mismanagement in developing countries: a review of global issues*, (Italy : International journal of environmental research and public health).

[4] Sullivan, D., and Quigley, P., 2003, *Geotechnical Engineering &Environmental Aspects of Clay liners for landfill Project*, Ireland.

[5] Vitorino de Souza Melaré, A.; Montenegro González, S.; Faceli, K.; Casadei, V. 2017, *Technologies and decision support systems to aid solid-waste management*: A systematic review. (New York : Waste Management), 59, 567–584.

[6] Vongdala, N.; Tran, H.D.; Xuan, T.D.; Teschke, R.; Khanh, T.D. 2019, *Heavy metal accumulation in water, soil, and plants of municipal solid waste landfill in Vientiane, Laos.* (Italy : International Journal of Environment. Research and Public Health) 16, 22.

[7] Gutberlet, J.; Baeder, A.M. 2008, *Informal recycling and occupational health in Santo André, Brazil.* (USA : International Journal of Environment Health Research), 18, 1–15.
[8] Indiana Department of Environmental Management, *Use of alternative Daily Cover at Landfills*, office of land Quality, 2013.
[9] Damanhuri, E, and Tri, P, 2016, *Integrated Waste Management*, Publisher ITB, Indonesia.
[10] ASTM D2216-10 *Standard Test Methods for Laboratory Determination of Water (Moisture) Content of Soil and Rock by Mass*, 2014.
[11] ASTM D4318-17, 2017, *Standard Test Methods for Liquid Limit, Plastic Limit, and Plasticity Index of Soils*. ASTM International, West Chonshohocken, PA 2017.
[12] ASTM D4943, 2018, *Standard Test Method for Shrinkage Factors of Cohesive Soils by the Water Submersion*. ASTM International.
[13] ASTM D854-14, 2018, *Standard Test Methods for Specific Gravity of Soil Solids by Water Pycnometer* ASTM International, West Chonshohocken, PA.
[14] ASTM D 422-63, 2007, *Standard Test Method for Particle-Size Analysis of Soils*, ASTM International, West Chonshohocken, PA.
[15] ASTM D558-96, 1996, *Standard Proctor Compaction*, ASTM International, West Chonshohocken, PA.
[16] ASTM D2434-68, 2000, *Falling head test*, ASTM International, West Chonshohocken, PA.
[17] ASTM D2166-00, 2000, *Standard Test Method for Unconfined Compressive Strength of Cohesive Soil*. ASTM International, West Chonshohocken, PA.
[18] SNI 19-7030-2004: 2004, Compost Specification of domestic organic waste (In Bahasa)
[19] Benson, C., H., Zhai, H., and Wang, X., 1994, *Estimating Hydraulic Conductivity of Clay Liners*, Journal of Geotechnical Engineering ASCE, Vol. 120, No.2, pp 366-387.
[20] Qian, X., D., Shin, J., Y., Liu, X, D., 2019, *Modern Sanitary Landfill Design and Construction*, (Beijing : China construction industry press).

Mechanical Properties of Concrete Using Plastic Waste

R Irmawaty[1]*, H Parung[1], R Djamaluddin[1], A A Amiruddin[1], and M P Faturrahman[1]

[1]Department of Civil Engineering, Faculty of Engineering, Universitas Hasanuddin Jl. Poros Malino KM. 06 Bontomarannu, Gowa, 92171, Indonesia

*Email: rita_irmaway@yahoo.co.id

Abstract. This paper presents strength and behaviour of concrete by the addition of percentages recycled waste plastic (polyethylene terephthalate-PET). Almost 60 concrete samples with a cylindrical shape of 100 x 200 mm and a prism of 100 x 100 x 400 mm were prepare. Concrete strength (compressive, indirect tensile and flexural strength) were investigated throughout 7, 14, and 28 days using 10%, 15% and 20% of PET plastic aggregate as a substitute for coarse aggregate. It is found that when the percentages of PET plastic increase in concrete, the slump value and weight of concrete decrease significantly. The compressive strength, indirect tensile strength and modules of rupture are also found to be decreased with increasing the percentages of PET plastic. The addition PET plastic more than 10%, the strength dropped drastically more than 40%. The usage of waste plastic in concrete leads to a change in the modes of failure from brittle (rapid) failure to more ductile failure.

1. Introduction

Concrete is the most widely used construction material, obtained by mixing Portland cement, water, aggregates and additives to improve its performance. The use of waste and recycled materials in concrete mixes is the current trend in concrete development. Besides helping to reduce environmental pollution, by utilizing and treating solid waste generated by industrial waste and municipal waste, it also reduces the use of natural resources.

Plastic is one of the most significant innovations of materials in the 20th century. It is not surprising that plastic is widely used. Plastics have many advantages compared to other materials, such as plastic which has a low density, is insulating against electricity, has varying mechanical strength, limited temperature resistance, and chemical resistance varies. In addition, plastic is also lightweight, easy to design and production costs. But behind all the advantages, plastic waste creates problems for the environment. The nature of plastics as non-biodegradable materials causes plastic to take 100-500 years to fully decompose. The use of plastics has increased from time to time, causing the amount of plastic waste also continues to increase. According to the Minister of Environment and Forestry Siti Nurbaya, Indonesia is expected to produce 9 million tons of plastic waste in 2019 [1] and the total plastic waste in 2017 will be 65.8 million tons [2].

To overcome this, environmental experts and scientists from various disciplines have conducted various studies and actions. One of them is by recycling plastic waste. But this method is less effective. Only about 4% can be recycled, the rest is piled up in a landfill. Thus, it is necessary to handle plastic waste more efficiently, by utilizing plastic waste into artificial aggregates in concrete mixtures. The volume of concrete contains 65-80% aggregate and plays an important role in concrete properties such as workability, strength, dimensional stability and durability, therefore the use of plastic waste in concrete as a substitute for aggregate can reduce the amount of waste significantly.

In civil engineering construction, the use of shredded plastic has increased dramatically as a partial replacement of aggregates. Because it provides environmental benefits and economic value where the use of waste is increasing in concrete [3-4]. Previous studies have shown that partial replacement of aggregate with waste plastic can improve concrete performance such as resistance to abrasion and impact loads, ductility, shock absorption, and thermal conductivity [5]. Some studies also present that the addition of plastic waste to concrete mixes causes a decrease in mechanical properties such as compressive strength, tensile strength and flexural strength [6-7]. Other studies have shown that the addition of steel fibers in normal concrete increases mechanical properties (compressive strength, flexural strength, tensile strength) significantly [8-9].

This study aims to determine the mechanical properties of concrete using PET plastic bottle waste as an artificial coarse aggregate. To increase the strength of plastic waste concrete, 0.5% steel fiber is added to the weight of cement. The crack patterns and the distribution of chopped plastic and steel fibers in concrete will be discussed.

2. Experimental programs

Laboratory studies were carried out to evaluate compressive strength, indirect tensile strength and flexural strength of concrete containing shredded PET plastic and steel fibers.

2.1. Materials

Portland Composite Cement (PCC) that meets SNI 15-7064-2004 Indonesian cement production and is available in the market is used as a binding material. Crushed stone (aggregate maximum size of 30 mm and a fineness modulus 7.54) and river sand (fineness modulus 2.27) meet standards SNI 03-1968-1990 for coarse and fine aggregate. Source of aggregate from the Jeneberang river, Bili-bili. Table 1 presented the physical properties of aggregates.

Table 1. Aggregate characteristics

Characteristics		Crushed stone	Sand river	PET Plastic
Specific gravity	Dry oven	2.56	2.46	-
	Surface dry (SSD)	2.62	2.58	1.26
Water absorption, %		2.17	4.82	-

2.2. Specimens

Three standard cylindrical specimens of 100 x 200 mm are formed for each mix of compressive and indirect tensile test, while bending test used prism of 100 x 100 x 400 mm. There are 4 mix variations with 0%, 10%, 15% and 20% substitution of shredded PET plastic on the coarse aggregate volume and water cement ratio was 0.40, as presented in Table 2. Tests carried out were compressive test (7, 14 and 28 days), indirect tensile test, and bending test (28 days) using Universal Testing Machine (UTM). The specimen is cured in fresh water until the age of testing.

Table 2. Mix design of plastic concrete (PC) in kg/m^3

No.	Material	NC	PC-10	PC-15	PC-20
1	Water	182.69	182.69	182.69	182.69
2	Cement	462.50	462.50	462.50	462.50
3	River sand	613.49	613.49	613.49	613.49
4	Crushed stone	1021.63	919.47	868.39	817.33
5	Dramix 3D 80/60	2.31	2.31	2.31	2.31
6	PET Plastic	-	46.38	69.57	92.75
7	Superplasticizer	-	-	-	2.31

The PET plastic used in this study passed ¾" and retained on sieve No. 4. Physical appearance of PET plastic and type of testing are presented in Figures 1.

a. Plastic PET — b. Compressive and elastic modulus test — c. Splitting test — d. Bending test

Figure 1. Plastic PET and type of testing

3. Results and Discussion

3.1. Fresh concrete properties

Slump test was performed to determine the consistency of concrete mixtures, which can illustrate the workability of concrete to be stirred, transported, poured, and compacted without causing segregation of the constituent material. The consistency of concrete mix should be matched to the requirements for the finished product quality. Slump values decrease with increasing percentage of PET plastic in concrete. Concrete slump values for the addition of 0%, 10%, 15% and 20% PET plastics are 12, 9, 3 and 8 cm, respectively. The addition of 10% PET plastic still meets the target slump value of 8 ± 2 cm. However, when PET plastic is added more than 10%, the slump decreases drastically, causing the concrete is stiff and difficult to stir. The addition of superplasticizer (a water reducing-retarding admixtures type) to the substitution of 20% PET plastic could improve concrete workability, making it easier to compact.

Table 3 describes the reduction in weight of concrete volume to the increase in PET plastic substitution in concrete. The greater the volume of PET plastic added, the greater the weight reduction in concrete volume. However, the substitution of PET plastic up to 20%, it cannot be categorized as lightweight concrete with a weight between 1140 - 1840 kg/m^3 according to SNI 03-2847-2013.

Table 3. Weight of concrete

Name	Volume of PET Plastic (%)	Average weight of concrete (kg/m^3)	Reduction (%)
NC	0	2349.41	-
PC-10	10	2257.95	3.89
PC-15	15	2122.42	9.66
PC-20	20	2051.75	12.67

3.2. Hardened Concrete Properties

Compressive strength testing was carried out according to ASTM C39/C39M-01 at the age of 7, 14, and 28 days using cylinders of 100 mm x 200 mm. Compressive strength is the most important parameter in structural design. Like concrete in general, compressive strength increases with age for all variations of the addition of PET plastic as presented in Figure 2. This means that the hydration process in the cement paste runs well.

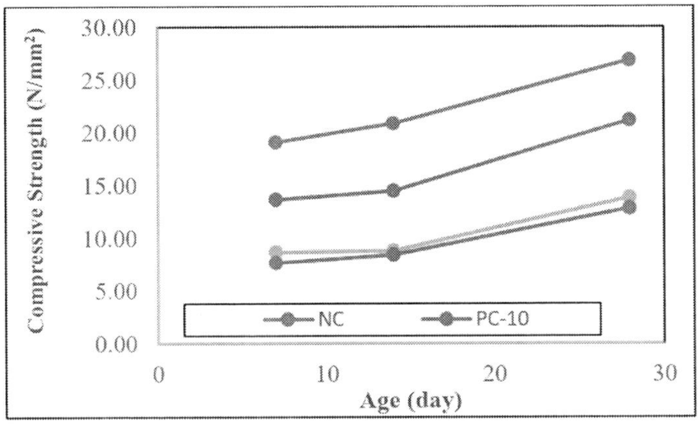

Figure 2. Compressive strength with variations in the addition of PET plastic

It is obvious from Table 4 that the compressive strength of concrete decreases with increasing the percentages of PET plastic. The addition PET plastic more than 10%, the compressive strength dropped drastically to 48%. The loose in strength attributed the plastic aggregate doesn't bond to the cement paste in the material in the same way that a crushed stone particle would. Therefore, substitution of PET plastic is limited to only 10% of the coarse aggregate volume.

Table 4. Compressive strength reduction

Name	Volume of PET Plastic (%)	Compressive strength (MPa)	Reduction (%)
NC	0	26.84	-
PC-10	10	21.14	21.23
PC-15	15	13.86	48.38
PC-20	20	12.84	52.15

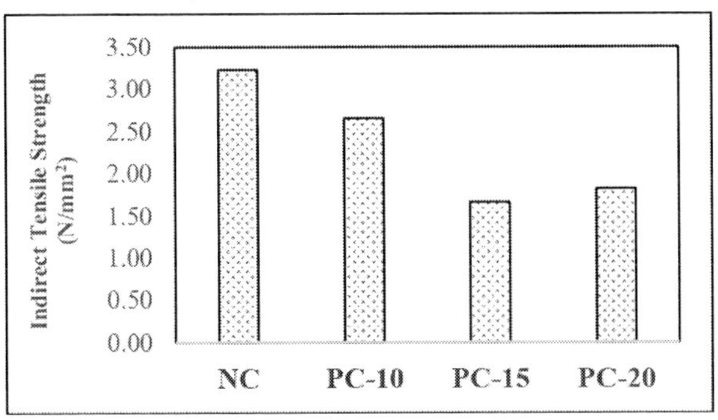

Figure 3. Indirect tensile strength with variations in the addition of PET plastic

Figure 3 indicates the test results of indirect tensile strength of hardened concrete, which was conducted according to ASTM C496-96 at 28-day. The test results showed a decrease in split tensile strength by increasing the volume of PET plastic in concrete. Addition of 10% PET plastic reduced split tensile strength by 17.8%. Then, the addition of 15% and 20% PET plastic, tensile strength dropped dramatically to 48.6% and 43.7% respectively. Therefore, it is not recommended to add more than 10% PET plastic. A large decrease in tensile strength of PN-15 compared to PN-20 due to the low consistency of PN-15 without the addition of superplasticizer, making it difficult to compact and produce less dense concrete. Visually observed the distribution of coarse aggregate, steel fibers and PET plastics on the surface of the specimens that have been split. Overall the specimen shows coarse aggregates, steel fibers and PET plastics evenly distributed, as shown in Figure 4.

Figure 4. Distribution of aggregate, steel fiber and PET plastic

Flexural strength testing was carried out according to ASTM C78-02T at 28 days old. It is clear from Figure 5 that when the percentage of added PET plastic increases, the bending strength on the contrary decreases. The greatest decrease in strength occurs in PN-15 due to low consistency without superplasticizer, making it difficult to compact and produce less dense concrete.

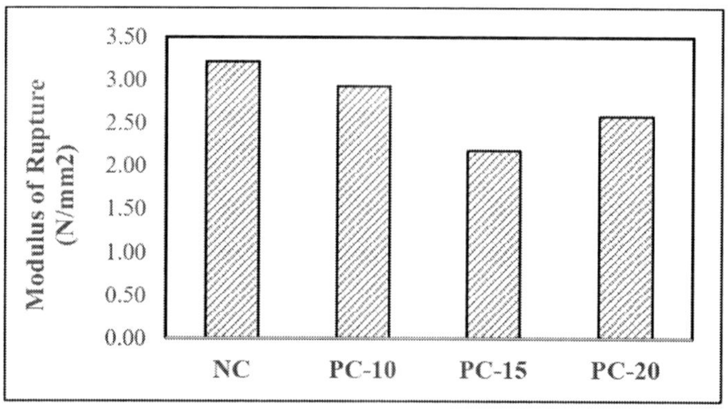

Figure 5. Bending strength with variations in the addition of PET plastic

Compressive test
Indirect tensile test
Bending test

Figure 6. Failure modes

IOP Conf. Series: Materials Science and Engineering **875** (2020) 012019 doi:10.1088/1757-899X/875/1/012019

The usage of PET plastic in concrete leads to a change in the modes of failure from brittle (rapid) failure to more ductile failure as shown in Figure 6. This indicates the bridge effect works well due to the use of steel fiber and chopped PET plastic with a size of about 20 mm.

4. Conclusion

From the results and discussion, it can be concluded as follow:

a. In general, the compressive strength, indirect tensile strength and modules of rupture are found to be decreased with increasing the percentages of PET plastic bottle. The addition PET plastic more than 10%, the compressive strength dropped drastically to 48%. Therefore, substitution of PET plastic is limited to 10% of the coarse aggregate volume.
b. The usage of waste plastic in concrete leads to a change in the modes of failure from brittle (rapid) failure to more ductile failure.

Acknowledgments

The author would like to thank the Faculty of Engineering, Hasanuddin University, for providing financial support to this research (LBE).

References

[1] E. Permana., 2019, Indonesia Produces 67 Million Tons of Trash in 2019, Available from: https://www.aa.com.tr/id/headline-hari/indonesia-hasilkan-67-juta-ton-sampah-pada-2019/1373712. Update : Jan 24, 2019. Accessed April 12, 2019. [in Bahasa]

[2] CNN Indonesia, 2019, Rules for Waste Reduction in the Business Sector, Available from: (https://www.cnnindonesia.com/ekonomi/20190311151320-92-376221/aturan-soal-pengurangan -sampah -di-sektor-bisnis-terbit-2019). Accessed April 12, 2019. [in Bahasa]

[3] D.Y. Osei., 2014, Experimental Investigation on Recycled Plastics as Aggregate in Concrete, *International Journal in Structure and Civil Engineering Research* 3(2), p. 168-74.

[4] F. Liu., Y. Yan., and C. Lan., 2015, *Performance of Recycled Plastic-Based Concrete Journal Materials and Civil.*

[5] A. Sofi., and B.R. Phanikumar., 2015, Effect of Pond Ash and Steel Fibre on Engineering Properties of Concrete, *AinShams Engineering Journal,* 7(1), pp 89-99.

[6] K.S. Rebeiz., J.W. Rosett., and A.P. Craft., 1996, Strength Properties of Polyester Mortar using PET and Fly Ash Wastes, *Journal Energy Engineering,* 122(1) 11052.

[7] P.S. Patil., J.R. Mali., G.V. Tapkire., and H.R. Kumavat., 2014, Innovative Techniques of Waste Plastic Used in Concrete Mixture, *International Journal Research Engineering Technology*, 3(9), pp 29-32.

[8] Y. Zhen., 2018, Mechanical Properties of Steel Fiber-Reinforced Concrete by Vibratory Mixing Technology, *Advances in Civil Engineering*, Volume 2018.

[9] J. Thomas., and A. Ramaswamy., 2007, Mechanical Properties of Steel Fiber-Reinforced Concrete, *Journal of Materials in Civil Engineering,* 19(5).

Investigation of Air Pollution Impact on Kinta River Water Quality at a Tropical Region

N A Mokhtar[1], W A Madhoun[2,6*], M Y D Alazaiza[3], A L Balogun[4], N K B E M Yahaya[5], and F C Ros[6]

[1,4]Universiti Teknologi PETRONAS, Malaysia
[2]Universiti Tun Hussein Onn Malaysia, 86400 Parit Raja, Johor, Malaysia
[3]Hasan Kalyoncu University, 27410 Gaziantep, Turkey;
[5]National Hydraulic Research Institute of Malaysia (NAHRIM), Malaysia
[6]Malaysia-Japan International Institute of Technology, Universiti Teknologi Malaysia Kuala Lumpur, 54100 Kuala Lumpur, Malaysia

*Email: wsah79@gmail.com

Abstract. Critical air quality levels lead to an unhealthy environment which disrupts physical activities and human health. Wet deposition of air pollutants might cause a high concentration of water pollution due to rain water washout of nitrate and particulate matter (PM). This study aimed to investigate the impacts of air pollutants deposition on river water quality in Malaysia. The methodology involved in the analysis of secondary data (January to December 2013) for air quality and river water quality using factor, correlation, and regression. Parameters of air quality were PM_{10}, Nitrate (NO_3), ozone (O_3) and temperature while water quality data were turbidity, Nitrate and PM_{10} (Ca, As, Hg, Cd, Cr, Pb, Zn, Cl, Fe, K, Mg, Na). The results show that there were positive correlations between air quality indicators and Kinta river water quality parameters. Correlation matrix shows that in terms of turbidity, air and water data were having 96% similarities. Regarding Nitrate concentrations, air and water records had only 30% of correlation matrix, which can be due to other sources of Nitrate which was agriculture activities near Kinta River. The factor analysis results showed that PM was the main contributor to river water quality particles with 94%.

1. Introduction

Air quality and river water quality are the most crucial pollution that need to be taken seriously. Both pollutants affect human health and cause fatality if they are excessive intake of polluted water and toxic air. Air pollution is the accumulation of substances in the atmosphere which then can be cleansed by wet and dry depositions [1]. Wet and dry depositions consist of soluble and insoluble particles which usually affect river water quality [2, 3]. In Malaysia, the data for air quality and river water quality sample were collected by sampling site and measurement of airborne particles [4].

The existing data from air quality station and river water quality station were used for analysis to determine the impacts on air pollution to river water quality. Prior to the water quality, the wet and dry depositions might be divided into three main sources: coastal, riverine and atmospheric inputs [5]. This means that the quantity of nitrate and PM_{10} from air particles can be found in the river water particles.

2. Materials and Methods

The scope of this study focused on the analysis of PM_{10}, nitrate, ozone, temperature, and turbidity from existing data situated at Ipoh, Perak.

Correlation analysis means a statistical measure that indicates the extent to which two or more variables fluctuate together. For this study, data from the existing air pollution and river water quality are correlated together based on its parameters. If the parameters for both data are majorly same with each other, therefore, it shows that there is a correlation between air quality and river water quality.

Factor analysis means a statistical data reduction which helps to obtain maximum common characteristic from the data to the factor score [6]. The number of iterations and the matrix show the relation between air quality and river water quality based on the parameter used by calculating the factor scored. For example, the data of PM_{10} for both air and river water quality will be analyzed and the factor pattern will show same value if there is a relationship between both data.

Regression model is a multivariate statistical method with a powerful pattern recognition capability. It is used for a simultaneous observation and analysis of more than one variable [7, 8]. It is used to reveal the variability that exist between a dependent and independent variable.

3. Results and discussions

3.1. Air Quality and River Water Quality Relationships under Weather Influence.

For PM_{10}, the highest value was during the dry season in June which was 56.4 μg/m³ and 64.4 μg/m³ as shown in Figure 1. The lower amount of precipitation during dry season causes less particulate matter in the atmosphere to be cleansed by rain. As the amount of particulate matter increases, PM_{10} were settle down as dry depositions and settled at the surface of river water and eventually soluble in water. Then, for the lowest value of PM_{10} was in April which was after wet season. The value of PM_{10} was 32.6 μg/m³ and 43.6 μg/m³ for air and river water quality respectively.

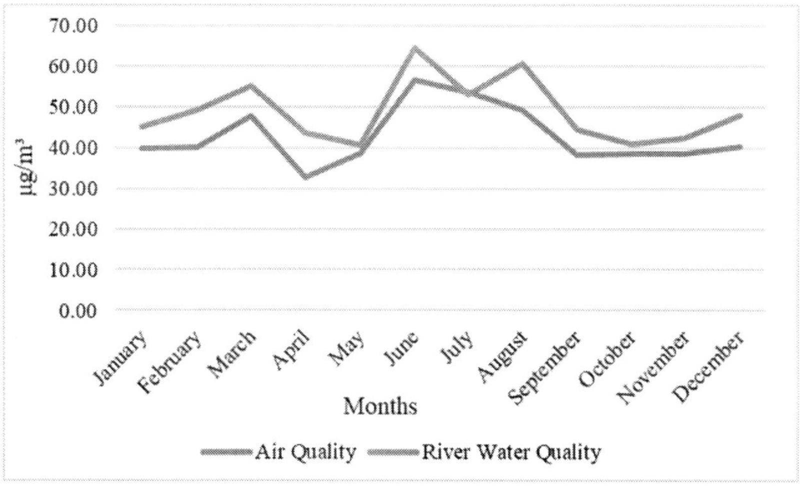

Figure 1. Correlation between air and water quality

As shown in Figure 2, Turbidity has highest value of 31.7 NTU in June which also considered as dry season. Hot weather affected the solubility of ozone in water and leads to inefficient purifying of river water. Ozone is known as a natural method to purify river water and can be done efficiently if the temperature of river water was at room temperature. It was a strong oxidation agent for water treatment from 100 years back [7]. Turbidity of river water increases when solubility of ozone decreases especially during dry season on June. In the month of December, the turbidity was the lowest due to lower temperature which was 31.2 NTU.

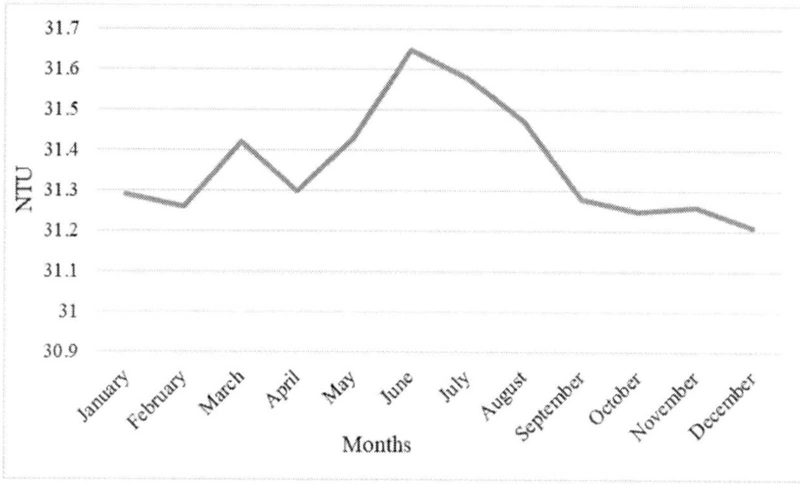

Figure 2. Monthly monitoring of turbidity in Kinta river

3.2. Correlation Analysis.

From both air quality and river water quality data in Table 1, the correlation matrix between air temperatures with turbidity was 0.96. The correlation matrix supported the factor analysis score for temperature and turbidity since temperature effect the solubility of ozone in water which then also effected the turbidity of water. There was a strong relationship between air temperature and turbidity due to concentration of dry and wet deposition.

Table 1. Correlation Matrix for Temperature, Ozone, pH value and Turbidity

Parameter	Air Temperature	Ozone	pH	Turbidity
Air Temperature	1.00	-	-	-
Ozone	0.37	1.00	-	-
pH	0.04	0.84	1.00	-
Turbidity	0.96	0.34	0.02	1.00

For ozone and pH of water, it stated that there is a relationship between them because the correlation matrix was 0.84. When solubility of ozone in water is higher, the pH value increases. As mentioned above, ozone is a natural method to purify water and by that, the pH value of water will not be too acidic and nearly neutral. Based on air quality data for July, the highest concentration of ozone was 0.022 DU. As ozone settle down into river water and as oxidizing agent, it helped in purify the river water. So, when the concentration of ozone was 0.022 DU, the pH value of Kinta river water was 7.02. meanwhile, when the ozone concentration was 0.01 DU in May, the pH value of river water was 6.04 which means it is acidic.

The correlation matrix between air quality and river water quality was 0.88 and that means air and river water quality has a strong relationship between them. This correlation matrix also supported the factor analysis score to show a strong relationship between air quality and river water quality. The indicators for PM_{10} in river water were Calcium (Ca), Arsenic (As), Mercury (Hg), Cadmium (Cd), Chromium (Cr), Lead (Pb), Zinc (Zn), Chlorine (Cl), Iron (Fe), Potassium (K), Magnesium (Mg) and Sodium (Na).

The correlation matrix for nitrate was 0.27. Atmospheric nitrate concentration between air quality and river water quality data are varied especially in this case where Kinta river located at the city of Ipoh, Perak. This is because the amount of nitrate dissolved in the river water also due to other factors like artificial fertilizers, disposal of wastes and the earth itself as organic matter contains nitrogen compounds [9]. The high differences between concentration of nitrate in air and river water are due to other sources that contribute to the concentration of nitrate as there was a previous study support this result. From the study of analysis of surface water pollution in the Kinta river using multivariate technique, there were many sources of nitrate and one of them are artificial fertilizers [6].

3.3. Factor analysis.

From Table 2, the factor scores of PM was 0.94, PM from atmosphere are possible to dissolve in water as rainwater cleanse the atmosphere and therefore settle in the nearest water catchment area [10]. Factor score for air quality and river quality were the same because it showed the communality between them. For initial communality, the factor score was 0.78 and final communality was 0.88. Thus, from the communality and specific variance after 11 iterations been done to the data, the final factor score was 0.94.

Table 2. Correlation matrix for nitrate

	F1	Initial communality	Final communality	Specific variance
Air Quality	0.94	0.78	0.88	0.12
River Water Quality	0.94	0.78	0.88	0.12

3.4. Regression Analysis.

The first two parameters for regression analysis are turbidity and temperature. Multiple R basically a correlation coefficient which shows how strong the linear relationship between multiple data. From the regression statistic for turbidity and temperature in Table 3, the multiple R is 0.96 which indicates that there is a positive correlation.

Table 3. Regression analysis for temperature and turbidity

Multiple R	0.96
R Square	0.92
Adjusted R Square	0.91
Standard Error	0.17

The linear lines for turbidity and temperature show only slight different with actual data and the R square is 0.91% which means the relationship is strong as turbidity and temperature can affect each other by slowing down the solubility of ozone dissolved in water as the temperature increases and result to high water turbidity. Constant concentrations of nitrate from river water quality station showed Kinta River were closer to agricultural area since the fertilizers recharged by runoff [7]. For PM regression analysis, the result for multiple R is 0.88 thus shows a positive relationship between air and river water quality data.

4. Conclusion

During wet season, the concentration of PM and turbidity decreased as they were washed out and diluted by high amount of precipitation. However, the concentration of nitrate increased. The runoff soil from nearby agricultural area flows into Kinta River and consequently increased the concentration of nitrate from fertilizers. Moreover, pH value of river water decreases when the concentration of ozone decreases during dry season. The factor analysis showed that the parameters that contribute to river water pollution were temperature, ozone and particulate matter. These parameters showed a strong correlation between air and river water quality indicators as the factor score between temperature and turbidity was 96.1% while the factor score for particulate matter was 94% similarities. Furthermore, the correlation analysis supported the factor analysis as the correlation matrix was almost 100% for temperature, turbidity, and PM. On the contrary, nitrate's result showed slight differences between air and river water quality, which can be explained by the influence of other sources such as agricultural activity which surround Kinta

River. The runoff from the soil that contained fertilizers increased the concentration of nitrate in Kinta River, especially during wet season.

References

[1] Pan, Y.P, Wang YS. (2015) Atmospheric wet and dry deposition of trace elements at 10 sites in Northern China. Atmospheric Chemistry and Physics 2015; 15, 951-972.

[2] Saeedi, M., Pajooheshfar S.P. (2012) Acid Rain Examination and Chemical Composition of Atmospheric Precipitation in Tehran, Iran. EnvironmentAsia, 5(1), 39-47.

[3] Mukherjee, S., Sarkar P, Roy A (2019). Dust Removal Capacity of Some Tree Species from Ambient Air in an Indian Metropolitan City, EnvironmentAsia, 12(1), 120-126.

[4] Ayers, G.P., Fook LS, Gillet RW, Kong CW, Peng LC. Atmospheric concentrations and deposition of oxidized sulfur and nitrogen species at Petaling Jaya, Malaysia 2000; 1993-1998. Tellus, 52B, 60-73.

[5] Sergey, M., Volodymyr M. (2012) Investigation of Atmospheric Wet and Dry Depositions to Marine Surface in Western Part of the Black Sea. Turkish Journal of Fisheries and Aquatic Science; 12, 497-505.

[6] Aziz, A., Baharuddin N, Juahir H, Osman R, Saim NA, Zain SM. Characterization of Spatial Patterns in River Water Quality Using Chemometric Techniques, Sains Malaysiana , 2014; vol 43, 1355-1362

[7] McMartin D.W., Stavrinides J., & Zanaic E. (2016). Field analysis of potable water quality and ozone efficiency in ozone assisted biological filtration systems for surface water treatment. *Water Research*, 104; 397-407.

[8] Isiyaka, H.A., Mustapha, A., Juahir, H. & Phil- Eze, P. (2019). Water quality modeling using artificial neural network and multivariate statistical. Model. Earth Syst. Environ. 5:583-593.

[9] World Health Organization (WHO). Nitrate and nitrite in drinking water (2011). WHO Guidelines for Drinking Water Quality. WHO Press, World Health Organization, 20 Avenue Appia, 1211 Geneva 27, Switzerland.

[10] Ebenstein, A, Fan M, He G, Greenstone M, Zhou M. (2017) New evidence on impact of sustained exposure to air pollution on life expectancy from China's Huai River Policy. PNAS, 114, 10384-10389. Harvard University, Cambridge, MA.

IOP Conf. Series: Materials Science and Engineering **875** (2020) 012021 doi:10.1088/1757-899X/875/1/012021

Appraisal of Indoor Microbial Pollutants at University Research Labs and Offices under Mediterranean Climate

E Y Abed[1], W A Madhoun[2,3]*, A A Elmanama[4], H Kim[5], X Xu[6], and F C Ros[2]

[1,4] Islamic University of Gaza, P.O.Box 108. Rimal St Gaza City, Palestine
[2] Malaysia-Japan International Institute of Technology, Universiti Teknologi Malaysia Kuala Lumpur, 54100 Kuala Lumpur, Malaysia, Malaysia
[3] Universiti Tun Hussein Onn Malaysia, 86400 Parit Raja, Johor, Malaysia
[5] University of Seoul, 163 Seoulsiripdaero, Dongdaemun-gu, Seoul 02504, Korea
[6] University of Windsor, 401 Sunset Avenue Windsor, Ontario N9B 3P4, Canada

*E-mail: wsah79@gmail.com

Abstract. Bacteria and fungi grow indoors when sufficient moisture is available. This study aims to determine the total viable bacterial count and fungi levels and to compare these levels among various science teaching laboratories and staff offices in Gaza city universities. A cross-sectional prospective study was conducted, sixty-five (65) air samples were collected from three local universities using Air Sampler. Auto Ranging Multimeter was used to record humidity and temperature. Air samples with counts of bacteria or fungi more than 500 Colony forming units "CFU"/m^3 were considered polluted according to the world health organization (WHO) standards. Among the 65 samples from the three universities, 48 sample (73.8%) had more than 500 colonies/100L. In term of fungal concentration, nine samples (13.8%) exceeded the WHO standards of 500 CFU/m^3. The highest percentage of bacterial load in air samples at the three universities (more than 500 CFU/m^3) was at IUG with 80.8% due to its old existence, more labs and lab activities. The lowest was at AQU with 61.5% due to less lab and lab activities. The highest percentage of fungal load in air samples at the three universities also were at IUG with 19.2 % and the lowest was at AQU with 7.7%.

1. Introduction

Quality of indoor air is an important element that influences the well-being of humans as they breathe 10m^3 air every day and generally spend 80–95% of the time indoors [1, 2]. Indoor environments contain a complex mixture of live and dead micro-organisms, fragments, toxins, allergens, volatile microbial organic compounds and other chemicals [3]. Bacteria, fungi and virus may form biological air contaminants and their distribution may vary according to the environment conditions. Heating, ventilation and air conditioning systems may also be microbial sources [4].

There is a growing evidence that exposure to biological agents in the indoor environment can have adverse health effects. Generally biological hazards to human arise from exposure to high concentrations of bio-aerosols; there are three main groups of diseases; infectious diseases, respiratory diseases and cancer [5]. A recent indoor air quality report by the World Health Organization (WHO) provided sufficient epidemiological evidence that inhabitants of damp or moldy buildings are at increased risk of respiratory symptoms, respiratory infections and exacerbations of asthma [6]. Moreover, the quality of indoor environments affects comfort and productivity [7].

The presence of many biological agents in the indoor environment is due to dampness and inadequate ventilation where excess moisture may increase the chemical emissions from building materials and floor covers [3]. Micro-organisms including fungi and bacteria are important factors influencing indoor air quality. The variety of species may alter with nutrient source, water availability and temperature. Both fungi and bacteria numbers are seasonally influenced; where they have higher levels in summer and autumn [8, 6]. Several studies were conducted previously and shows the trends of indoor biological pollutants at different environments, while this research investigated the university indoor environment at a Mediterranean climate and examined the influence of weather parameters.

This research was carried out to determine the total viable bacterial count and fungi levels as well as to investigate the relationship between pollutants levels, humidity and temperature in the studied universities labs and offices.

2. Materials and Methods

2.1. Study Location

The study was conducted to assess bacterial and fungal counts in the indoor air of research labs and faculty offices of the selected universities (The Islamic University of Gaza "IUG", Al Azhar University "AZU" and AL Aqssa University "AQU") in Gaza City, Palestine. These universities are serving students' population 63000 and 4000 staff.

2.2 Sampling

Samples were collected from the labs and offices of IUG, AZU and AQU over one month (April-May 2013). Samples were collected from different labs at three selected universities. Twenty-six samples were collected from IUG, 26 samples from AZU and 13 samples from AQU. Samples from fifty-five labs of biology, chemistry, environment science, medical sciences and physics were used for the study. More so, 10 faculty offices were tested in this research.

2.3 Materials

Culture media used for counting of bacteria and fungi were as follows:
A - Dichloran Rose Bengal Chloramphenicol (DRBC) agar (Oxoide- UK) DRBC Agar was used for the enumeration of yeasts and molds, colonies of molds and yeasts should be apparent within 5 days of incubation. Colonies of yeast appear pink due to the uptake of rose Bengal.

2.4 Statistical Analysis

Chi square test was performed using SPSS 14.0 to examine the relationship between fungi and bacteria as criterion dependent variable, humidity and temperature as criterion independent variables. One-way analysis of variance (ANOVA) was performed to compare colonies count by concentration.

3. Result and Discussion

A total of sixty-five air samples were collected from various locations at the selected three local universities and they are shown in Table 1. Twenty-six samples were collected from IUG and AZU respectively while 13 samples were selected from AQU. The samples were classified to two groups: Labs and faculty offices. At IUG, 22 labs and 4 faculty offices were selected for sampling while, at AQU, 10 labs and 3 offices were sampled. The total of sample size in all the universities was 65 samples (55 labs and 10 faculty offices).

3.1 Bacteria Counts

The average of bacterial counts for all sample population (n= 65) was 926 colony forming units "CFU"/m3 with a range of 160- 2420. This value exceeded the WHO standards of 500 CFU/m^3 for indoor bacteria concentrations. These values are consistent with the levels reported by Karwowska [9] with average bacteria in laboratory rooms in China ranged from 1600 – 2000 CFU/m^3 [10].

The highest mean of colonies of bacteria was at IUG with 1048 CFU/m^3 and the lowest was at AQU with 815 CFU/m^3 and an extreme bacterial count of 2420 CFU/m^3 was detected at a faculty office at

Table 1. Total bacterial count at the studied locations

Locations	N	Mean CFU/m³	Std. Dev	Minimum CFU/m³	Maximum CFU/m³
University Level					
IUG	26	1048.08	624.481	190	2320
AZU	26	858.46	515.796	200	2000
AQU	13	815.38	630.933	160	2420
Total	65	925.69	584.201	160	2420
Labs Level					
Environment and Earth Science	14	1000	724.2	190	2420
Biology	14	694.3	288.2	320	1300
Chemistry	14	920.7	498.9	380	1920
Medical Sciences	13	1176.9	729.6	160	2280
Physics	10	826	534.7	240	1880
Total	65	925.7	584.2	160	2420
Labs vs Faculty offices					
Labs	55	884.55	561.284	160	2320
Faculty offices	10	1152.00	685.173	190	2420
Total	65	925.69	584.201	160	2420

AQU. The highest percentage of samples that exceeds the standards was at IUG which recorded 81% and this could be due to the crowdedness at the old university buildings which were built in the 80's. AQU had the lowest number of samples that exceeded the WHO standards with 62% and this could be due to less number of labs and less crowdedness. These differences were not statistically significant where the p value was 0.643.

The results show that the highest mean of bacterial counts was at Medical Sciences lab which has 1177 CFU/m^3 and the lowest count was at Biology lab with 694 CFU/m^3.These results agree with Bholah and Subratty [11] who found that the number of indoor bacterial bio-aerosol concentrations ranged between 3 and 1110 CFU/m^3.

3.2 Fungi

The average count for all samples was 313 CFU/m^3 with a range of 20- 1320 CFU/m^3. At the university level, the values are higher at IUG (348 CFU/m^3), slightly lower at AZU (317 CFU/m^3), and much lower at AQU (238 CFU/m^3). The higher concentrations at IUG labs may resulted due to labs locations at the basement and the poor ventilation. ANOVA results showed insignificant differences between universities with p-value of 0.884. When averaged by type of labs, the highest concentration was observed at Environmental and Earth Science Department Labs (430 CFU/m^3) and the lowest was at Physics Department Labs of 245 CFU/m^3.

Those average values are acceptable because they meet the WHO standards of 500 CFU/m^3 for indoor fungi. These results are consistent with that by Ross et al., [12]. They found fungal concentrations of 338 CFU/m^3 and 312 CFU/m^3 at a shopping center and a company respectively. While lower levels of 194 CFU/m^3 and 178 CFU/m^3 were observed at an auditorium and a hospital respectively. Karwowska [9] determined fungi in a lab rooms ranged between 160 and 780 CFU/m^3. Other studies have reported lower indoor fungal counts. For example, Bholahand and Subratty [11] reported concentrations ranged 0-196 CFU/m^3 at office buildings. Omoigberale et al. [13] observed fungal population from 3 to 45 CFU/m^3 in hospitals' indoor environments.

IOP Conf. Series: Materials Science and Engineering **875** (2020) 012021 doi:10.1088/1757-899X/875/1/012021

However, 6 out of 55 labs and 3 out of 10 offices tested exceeded the standards. Two labs at IUG and another lab at AZU had fungal counts over 1000 CFU/m^3. Further investigation of fungi sources in those places and corresponding clean-up measures are warranted. As shown in Table 2, the highest mean fungal concentration was detected at faculty offices (405 CFU/m^3) while the labs had an overall mean of 299 CFU/m^3. However, the difference in the mean values of office and labs are statistically significant due to large variability within each group. Also, 30% of the samples collected from the faculty offices exceeded the WHO standard of 500 CFU/m^3, while only 11% of the samples collected from labs did not fulfill with the same standard. The high level of Fungal in offices could be due to higher level of humidity (67%) and temperature (24.5°C) which exceeded the ASHARE Standards.

Table 2. Total fungal count at the monitored locations.

Locations	N	Mean CFU/m^3	Std. Dev	Minimum CFU/m^3	Maximum CFU/m^3
University Level					
IUG	26	347.69	263.747	60	1120
AZU	26	316.54	254.919	60	1320
AQU	13	238.46	186.451	20	660
Total	65	313.38	246.369	20	1320
Labs Level					
Environmental and Earth Science	14	430	423.1	100	1320
Biology	14	268.6	146.8	60	480
Chemistry	14	311.4	182.4	100	680
Medical Sciences	13	290.8	188.08	20	600
Physics	10	245	127.9	140	540
Total	65	313.4	246.3	20	1320
Labs vs Faculty offices					
Labs	55	296.7	216.1	20	1120
Faculty offices	10	405	374.8	100	1320
Total	65	313.38	246.3	20	1320

3.3 Influence of Temperature and Humidity

The overall average of temperature was 24.3°C and this range acceptable for indoors environment. The temperature at labs level with the w that the highest temperature was at Physics Labs with 25.6°C and the lowest was at Environment Science labs with 22.9°C. In terms of temperature, the relationship between labs was statistically significant where p value was 0.001. The difference of temperature between labs and faculty offices, where the mean of temperature at faculty offices was 24.5°C while it was 24.2°C at the labs. The difference between labs and faculty offices has no statistical significance where p value was 0.681.

As shown in Table 3 also, the highest level of humidity was at AQU with 66.8% and the lowest level was at IUG with 65.1%. There is statistical significantt relationship between IUG and AQU where the p value was 0.047.

Regarding the humidity on the labs level, it can be noticed from Table 3 that the highest humidity level was at Environment Science labs with 66.7% while the lowest level was at Chemistry labs with 64.9%. Moreover, there was a statistical significance between the levels of humidity in the different labs where p value was 0.037, which means that the humidity varies from one lab to the other due to differences in size, occupancy or layout. Humidity level at the faculty offices was greater than the levels at the labs, where humidity was 67.06% at the offices and 65.6% at the labs. The relationship between the two groups was statistically significant, where the p value was 0.032.

Table 3. Humidity trends

Location	N	Mean (%)	Std. Deviation	Minimum	Maximum
Universities Level					
IUG	26	65.1	2.05	61.3	68.7
AZU	26	66	2.1	62.4	69.4
AQU	13	66.8	0.73	65.9	68.9
Total	65	65.8	1.9	61.3	69.4
Labs Level					
Environmental and Earth Science	14	66.7	1.5	63.7	68.7
Biology	14	65	1.1	63.4	66.5
Chemistry	14	64.9	1.6	62.4	67.1
Medical Laboratory Sciences	13	66.5	3.2	61.3	69.4
Physics	10	66.1	1.4	64.2	68.9
Total	65	65.8	1.9	61.3	69.4

4. Conclusion

It can be concluded that the highest mean of bacterial count was detected at the Medical Sciences labs with an average of 1177 CFU/m^3 and the lowest was at Biology labs with an average of 94.3 CFU/m^3. In term of bacterial counts, chemistry labs were considered the most polluted where 86% of the samples exceeded the WHO standards of 500 CFU/m^3 while physics labs were the least polluted with 60% of the samples. Environment Science labs were considered to be the most polluted with fungi and the lowest was at biology labs with 0%. Faculty offices had the highest counts of bacteria with an average of 1152 CFU/m^3 and 405 CFU/m^3 of fungal counts. The overall humidity level was 65.8% and this is higher than the ASHARE Standards of 40 %- 60%. The highest humidity percentage was at faculty offices with a level of 67.1%. In order to improve the indoor environment quality, it is recommended to perform a periodical monitoring of indoor quality and laboratory disinfection along with good ventilation. Separation of labs from faculty offices, organize awareness programs for staff and students on indoor pollution.

References

[1] Cabral J., (2010). Can we use indoor fungi as bioindicators of indoor air quality? Historical perspectives and open questions. Science of the Total Environment. 408, 4285 – 4295.
[2] Mohd Dzulkifli, S. N., Abdullah, A. H., Yong, L. Y., Suhaimi Shamsuri, M. M., Daud, Z., (2018). Classification of Potential Risk Factors through HIRARC Method in Assessing Indoor Environment of Museums. International Journal of Integrated Engineering, Vol. 10 No. 8, p. 43-55.
[3] World Health Organization, (2007). WHO Guidelines for Indoor Air Quality –Dampness and Mould.
[4] Grigorevski A., Lima R., Silva L., Linares R., and Coelho R., (2006). Occurrence of actinomycetes in indoor air in Rio de Janeiro, Brazil.Building and Environment, 41, pp: 1540–1543.
[5] Srikanth P., Sudharsanam S. and Steinberg R., (2008). BBio-aerosols in Indoor Environment: Composition, Health Effects and Analysis. Indian Journal of Medical Microbiology 26, pp: 302-312.
[6] McMahon S., Hope J., Thrasher J., Rea W., Vinitsky A. and Gray M., (2012). Common Toxins in Our Homes, Schools and Workplaces.
[7] Samet J. and Spengler J., (2003). Indoor Environments and Health: Moving Into the 21st Century. American Journal of Public Health 93(9), pp: 1489-1493.
[8] Douwes, J. Thorne, P. Pearce, N. Heederik, D. (2003). Review: Bioaerosol Health Effects and Exposure Assessment: Progress and Prospects. 47 (3), 187–200.

[9] Karwowska E., (2003). Microbiological Air Contamination in Some Educational Settings. Polish Journal of Environment Studies, 12(2), pp: 181-185.

[10] Kalwasinka A., Burkowska A. and Wilk L., (2012). Microbial air contamination in indoor environment of a university library. Annals of Agriculture and Environmental Medicine 19, pp: 25-29.

[11] Bholah R. and SubrattyAH., (2002). Indoor biological contaminants and symptoms of sick building syndrome in office buildings in Mauritius. International Journal of Environmental Health Research vol. 12.

[12] Ross C., Roberto J., Inez T., et al., (2004). Studies on Fungal and Bacterial Population of Air-conditioned Environments. Brazilian Archive of Biology and Technology 47, pp: 827-835.

[13] Omoigberale M., Amengialue O. and Iyamu M., (2014). Microbiological assessment of Hospital Indoor Air Quality in Ekpoma, Edo State, Nigeria.Global Research Journal of Microbiology 4, pp:1 –5.

IOP Conf. Series: Materials Science and Engineering **875** (2020) 012022 doi:10.1088/1757-899X/875/1/012022

Impact Assessment of Traffic Emission on the Respiratory System of Non-Smoking Traffic Policemen in Palestine

W A Madhoun[1,2,*], E Salem[3], A Eljedi[3] , H A Isiyaka[4], and F C Ros[2]

[1]Universiti Tun Hussein Onn Malaysia, 86400 Parit Raja, Johor, Malaysia
[2]Malaysia-Japan International Institute of Technology, Universiti Teknologi Malaysia Kuala Lumpur, 54100 Kuala Lumpur, Malaysia
[3]The Islamic University of Gaza, Gaza, Palestine
[4]Kano University of Science and Technology, Wudil, Nigeria

*Email: wsah79@gmail.com

Abstract. This study was carried out to investigate the effects of prolonged exposure of non-smoking traffic police men to vehicular emission within the northern Gaza Governorate-Palestine. Samples of atmospheric pollutants such as PM_{10}, CO and CO_2 were collected alongside respiratory data involving lung function, chest and respiratory status and direct physical examination. This was done in order to understand the level of association between respiratory status such as effect on breathing, upper respiratory tract irritation, frequent cough and eye irritation. Chi-square statistical techniques and descriptive statistics were applied in the datasets. The result shows that there is a strong association between air pollution and respiratory disease with a p-value 0.027 at significance of alfa 0.05. The statistical result shows that about 20 non-smoking traffic policemen have upper respiratory tract infection, 15 with frequent cough symptom, 6 with shortness of breathing and 25 with eye irritation. This study is an indication that vehicular emission within the study area is a primary source of pollution that possesses a risk in exposing non-smoking policemen to series of respiratory diseases. Government agencies and other stakeholders should provide preventive measures in either reducing the number, concentration and sizes of pollutants emitted or provide protective measures to reduce the level of exposure of traffic policemen and other habitants.

1. Introduction

Quality human health is very paramount in any societal development. This is because the physical and mental stability of individuals and their ability to discharge their duties in any organization can be retarded when exposed to poor air quality condition. Air pollution is a well- recognized public health problem associated with a range of adverse health outcomes, ranging from premature mortality to sub clinical respiratory symptoms [1]. According to WHO [2] an estimate of 2.4 million people dies each year from causes directly attributed to air pollution. The USA for instance, spends over $ 200 million annually on routine ambient air monitoring programs [3]. The most dangerous atmospheric air pollutant is PM_{10} that negatively affects human health because it can easily diffuse into the gas-exchange region of the lung when inhaled. This can cause lung cancer, asthma as well as respiratory and cardiovascular diseases [4]. CO reduces the body's ability to deliver oxygen to the organs and tissue because it bonds with haemoglobin more easily compared with oxygen [4].

However, pregnant women and infants are more affected by CO. This is because CO affects the amount of oxygen fed to the foetus through the mother as well as affects infants negatively before and after birth [5]. Ozone is formed when Nitrogen Oxide and Volatile Organic Compounds in heat and sunlight react. O_3 is an extremely reactive compound that reduce the functionality of the lung, damage tissues, accelerates the level of decay and discoloration of material's [1]. Ozone is a greenhouse gas that causes global warming by trapping infrared radiation emitted by the earth surface [6].

Air pollutants found in urban area represent a mixture of primary particles emitted from various sources and secondary particles from aerosols formed by chemical reactions [7]. Air Quality is a measure of the concentrations of gaseous pollutants and sizes or number of Particulate Matter emitted by human-induced activities or natural processes [2]. Air is polluted when particulate toxic elements emitted by natural or anthropogenic sources exceed the normal threshold required for a normal life [8]. Air pollution can also be seen as an excess in the concentration of atmospheric gasses and aerosol particles induced by anthropogenic activities sufficiently high to damage plants, animals other life forms, ecosystem structures or works of art [9].

According to [10] air is polluted mostly by anthropogenic induced pollutants and non-point source pollutant in the environment. These non-point sources pollutants are mainly generated from sources that cannot be easily spotted out which affects the environment directly or indirectly. This has directly and indirectly accelerated the rate of morbidity, mortality and environmental degradation in both short and long-term exposures [11]. Despite the effort put in place by government and other stakeholders to set an emission standard and technically improve the emission control pattern, the expanding industrialization, urbanization and increase traffic volumes especially in places along the Gaza-strip have increased the level of air pollution.

Traffic policemen who work in the busy traffic signal areas for years are exposed to the risk of air traffic pollution. This is because vehicles and motorcycles discharge dangerous air pollutants that can negatively impact policemen when inhaled, and expose them to serious health predicaments. According to [12] the largest contributor to PM_{10} emission in most urban centres is vehicle emission especially from gasoline and diesel fuel combustion. References [13-16] recognised traffic emission as the major contributor to air pollution that can pose a serious health challenge. Air is a major risk factor for the development of chronic lung disorders such as Chronic Obstructive Pulmonary Disease (COPD), which may enhance the risk of lung cancer and also exacerbate bronchial asthma in sensitive individuals [17]. A report by Ministry of Planning and International Cooperation, Palestine stated that there are over 70,500 motor vehicles that cluster within a short distance and are dominated with old and out-dated. More so, pollutants are discharged from coal operation power stations in Israel factories that are transported to the Gaza strip by wind. In addition, soil dust is re-suspended due to poor and haphazard road construction within the study area.

2. Materials and Methods

2.1. Study Area

Gaza Strip is a territory of historical Palestine on the eastern coast of the Mediterranean Sea that shares borders with Egypt on the southwest and occupied Palestine on the east and north [18]. It is 41 kilometres long, and from 6 to 12 kilometres wide, with a total area of 365 square kilometres. The population of Gaza Strip is about 1.7 million people with a yearly growth rate of about 3.2%. Gaza Strip has the seventh highest population growth rate in the world [18]. Gaza strip has a special regional importance in Palestine due to the fact that it constitutes the only linkage between Egypt and the Palestine [18, 19].

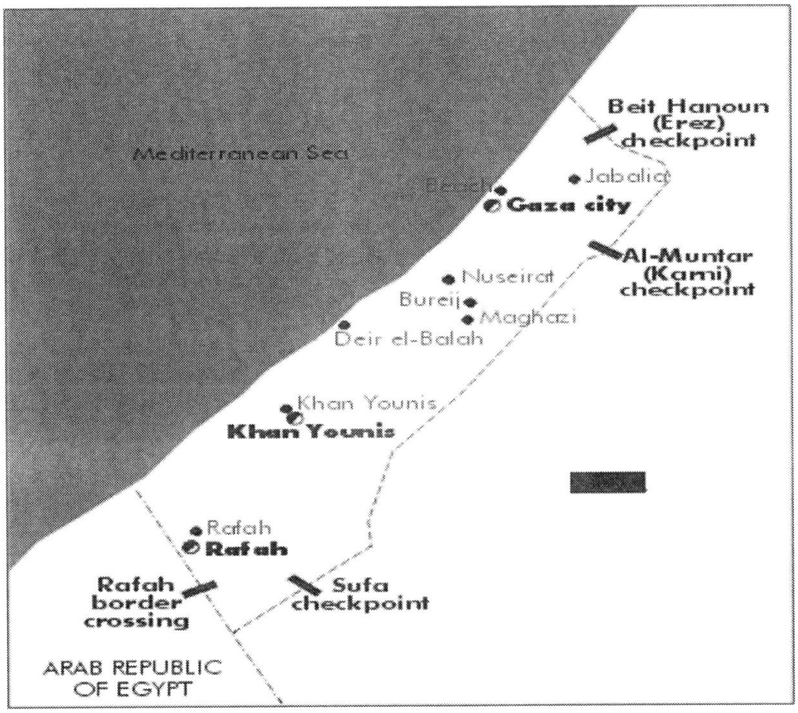

Figure 1. Map of study area showing sampling points

2.2. Sampling Data and Equipments
The sampling strategy comprises of healthy non-smoking traffic policemen within the Gaza-strip with no any history of drug abuse and other respiratory diseases. Forty-one road stationed traffic policemen assigned to control the flow of vehicles in roads and areas with high, medium and low traffic intensity in crossroads and parking areas were selected for the study.

2.2.1. Respiratory Data and Instruments. Spirometers are non-invasive diagnostic instruments for screening and basic testing of pulmonary function. Offering essential diagnostic insight into the type and extent of lung function impairment, spirometer tests can be performed fast at fairly low cost. The data comprises of chest and respiratory condition of the sampled population, physical examination by thoracist, forced vital capacity, forced expiratory volume and peak expiratory flow rate.

2.2.2. Quality Data Instruments. Kanomax Handheld IAQ Monitor Model 2211 was used to collect air quality data such as CO, CO2 PM. This instrument is very simple, flexible and can capture accurately air pollution data in a location.

3. Method of Data Analysis
Chi-Square statistical test was applied in this study to see the extent to which the observed variables are associated with the expected in a contingency table. It shows how categorically variables are related. This technique was introduced to observe if there is any association between air pollutants and respiratory disease.

Table 1 describes the result of Pearson Chi-Square Tests with a p value = 0.027 at degree of freedom of 0.05. Consequently, the general assumption here is that, if the p-value is less than the degree of freedom, and then there is a strong association between the observed and expected. In this study, it can be concluded that there is a significant relationship between the level of air pollution and the effects on respiratory system of traffic police men within the northern region of the Gaza-strip. This is an indication that traffic policemen responsible for traffic regulation in the study area are exposed to high risk of respiratory challenge.

Table 1. Result of Pearson Chi-Square test

	Value	df	Asymp. Sig. (2-sided)	Exact Sig. (2-sided)	Exact Sig. (1-sided)
Pearson Chi-Square	4.917[a]	1	.027		
Continuity Correction[b]	3.867	1	.049		
Likelihood Ratio	5.014	1	.025		
Fisher's Exact Test				.044	.024
Linear-by-Linear Association	4.843	1	.028		
N of Valid Cases	41				

This result is an indication that in most urban areas, vehicular emission is predominant and significantly contributes to air quality problems [13, 15]. This is due to the traffic congestions especially during the early hours of the morning when people are rushing to work and, in the evening, when people close from work. The particles emitted from the vehicular exhaust especially are held in upper respiratory tract and accumulates in the lung and produces respiratory abnormalities. Hence, PM_{10} are of great concern in air pollution studies. The effects of air pollution include breathing and respiratory problems, aggravation of existing respiratory and cardiovascular disease, alterations in the body defence systems against foreign materials, damage to lung tissue, carcinogenesis and premature death [19]. Furthermore, Table 2 describes the physical symptoms indicating the possibilities of traffic policemen casualties within the study area.

Table 2. Physical symptoms occurrence

S/n	Symptoms	Sample	Prevalence of symptoms	
			Yes	No
1	Eyes irritation	Traffic Police	25	16
2	Frequent coughing	Traffic Police	15	26
3	Upper respiratory tract irritation	Traffic Police	20	21
4	Shortness of breathing	Traffic Police	06	35

However, the result shows that 15 traffic policemen have frequent coughs that indicate a symptom of respiratory infection while 16 have no any trace of such health predicaments. More so, 20 traffic police men have symptom of upper respiratory tract infection, while 6 have short breathing capability. 25 out of the sampled population are also exposed to frequent eye irritation.

4. Conclusion

This study was done in order to understand the level of association between respiratory status such as breathing, upper respiratory tract irritation, frequent cough and eye irritation with the level of air pollution. Chi-square statistical techniques and descriptive statistics were applied in the datasets. The result shows that there is a strong association between air pollution and respiratory disease with a p-value 0.027 at significance of alfa 0.05. The statistical result shows that about 20 non-smoking traffic policemen have upper respiratory tract infection, 15 with frequent cough symptom, 6 with shortness of breathing and 25 with eye irritation. This study is an indication that vehicular emission within the study area is a primary source of pollution that possesses a risk in exposing non-smoking policemen to series of respiratory diseases. Government agencies and other stakeholders should provide preventive measures in either reducing the number, concentration and sizes of pollutants emitted or provide protective measures to reduce the level of exposure of traffic policemen and other habitants.

IOP Conf. Series: Materials Science and Engineering **875** (2020) 012022 doi:10.1088/1757-899X/875/1/012022

References

[1] Currie, J., Neidell, M., & Schmieder, J. F. (2009). Air pollution and infant health: Lessons from New Jersey. Journal of health economics, 28(3), 688-703.

[2] World Health Organization (2002). World Health Report: Reducing Risk, Promoting Healthy Life. Geneva, Switzerland.

[3] Scheffe, R. D., Solomon, P. A., Husar, R., Hanley, T., Schmidt, M., Koerber, M., & Valentinetti, R. (2009). The national ambient air monitoring strategy: Rethinking the role of national networks. Journal of the Air & Waste Management Association, 59(5), 579-590

[4] MacNee, W. and Donaldson, K. (2003). Mechanism of lung injury caused by PM10 and ultrafine particles withspecial reference to COPD. European Respiratory Journal 21(40): 47s-51s.

[5] Ballester, F., Llop, S., Estarlich, M., Esplugues, A., Rebagliato, M. and Iñiguez, C. (2010). Preterm birth and exposure to air pollutants during pregnancy. Environmental Research 110(8):778-785.

[6] Moustris, K. P., Larissi, I. K., Nastos, P. T., Koukouletsos, K. V. and Paliatsos, A. G. (2013). Development and Application of Artificial Neural Network Modeling in Forecasting PM10 Levels in a Mediterranean City.

[7] Salma, I., Chi, X., & Maenhaut, W. (2004). Elemental and organic carbon in urban canyon and background environments in Budapest, Hungary. Atmospheric Environment,38(1),27-36.

[8] Bernstein, J. A., Alexis, N., Barnes, C., Bernstein, I. L., Nel, A., Peden, D., & Williams, P. B. (2004). Health effects of air pollution. Journal of Allergy and Clinical Immunology, 114(5), 1116-1123.

[9] Jacobson, M. Z. (2002). Atmospheric pollution: history, science, and regulation. Cambridge University Press

[10] Razak, M. I. M., Ahmad, I., Bujang, I., Talib, A. H., & Ibrahim, Z. 2013. Economics of Air Pollution in Malaysia. Int J Humanit Soc Sci. 3(13): 173-177.

[11] Cohen, A. J., Ross Anderson, H., Ostro, B., Pandey, K. D., Krzyzanowski, M., Künzli, N., & Smith, K. (2005). The global burden of disease due to outdoor air pollution. Journal of Toxicology and Environmental Health, Part A, 68(13-14), 1301-1307.

[12] Kumar, K., & Thakur, G. S. M. (2012). Advanced applications of neural networks and artificial intelligence: A review. International Journal of Information Technology and Computer Science (IJITCS), 4(6), 57.

[13] Latif, M. T., Azmi, S. Z., Noor, A. D. M., Ismail, A. S., Johny, Z., Idrus, S., Mokhtar, M. B. (2011). The impact of urban growth on regional air quality surrounding the Langat River Basin, Malaysia. The Environmentalist, 31(3), 315-324.

[14] Dominick, D., Juahir, H., Latif, M. T., Zain, S. M., & Aris, A. Z. (2012). Spatial assessment of air quality patterns in Malaysia using multivariate analysis. Atmospheric Environment, 60, 172-181.

[15] Azid, A., Juahir, H., Toriman, M. E., Kamarudin, M. K. A., Saudi, A. S. M., Hasnam, C. N. C., Abdul Aziz, N.A., Azaman, F., Latif, M. T., Zainuddin, S. F. M., Osman, M. R. and Yamin, M. (2014). Prediction of the Level of Air Pollution Using Principal Component Analysis and Artificial Neural Network Techniques: a Case Study in Malaysia. Water, Air, & Soil Pollution, 225(8): 2063 – 2077.

[16] Isiyaka H. A., Juahir, H., Toriman, M. E.,Azid, A., Gasim, B.M Kamarudin, M.K. (2014) Assessment of the spatial variation and source apportionment of air pollution based on chemometric techniques: a case study in the Peninsular Malaysia Jurnal Technologi 77:1 33-44.

[17] MPIC (2012). Ministry of Planning and International Cooperation-Palestine.

[18] National Institute of Health, National Heart, Lung and Blood Institute (1995) Global initiatives for asthma., a global strategy for asthma management and prevention. NHLBI/WHO Workshop Report 20

[19] Cotes JE., 1978- Lung function Assessment and application in medicine. Blackwell Scientific Publication, 4th edition, Melbourne.

A Study of Assessment and Mapping of Carbon Footprints to Campus Activities in Hasanuddin University Faculty of Engineering

R Zakaria[1]*, S H Aly[1], M Hustim[1], and A D M Oja[1]

[1]Department of Environmental Engineering, Engineering Faculty, Hasanuddin University, Makassar, Indonesia

*Email: rasdianazakaria@unhas.ac.id

Abstract. The Hasanuddin University Faculty of Engineering in Gowa is the largest engineering campus in South Sulawesi. The lecture activities are very complex which are not only teaching and learning but also households and offices activities. Therefore, it is important to know the mapping of existing carbon footprint on the campus due to the complexity of the campus activities. The purpose of this study is to determine the level and mapping of the carbon footprint on campus. The method used is to halve emissions into primary and secondary carbon footprint emissions. The results of this study indicate that the greatest emissions is generated from the secondary carbon emissions which is electricity consumption amounting to 1,315,429.63 $KgCO_2$/Year, and the lowest emissions is resulted from the use of paper at 49.39 $kgCO_2$/Year. For the buildings the highest CO_2 emissions is generated from the Architecture Department building at 289,176.26 $KgCO_2$/year and the lowest CO_2 emissions is generated from the Naval B Department building at 20,599.20 $KgCO_2$/year. The mitigation scenario in reducing CO_2 emissions effectively is by implementing Earth Hour for 1 hour every month which can reduce electricity consumption by 582,541.78kWh/Year and CO_2 emissions at 164,428.70 $KgCO_2$/year. The effective scenario to decrease waste is by implementing 3R with 2% waste reduction which is 167,332.79 Kg/Year and emissions CO_2 are 4,772.41 $KgCO_2$/Year.

1. Introduction

The carbon footprint is a general term used to describe the total amount of GHG emissions from responsible individual or organization. Each person contributes to GHG, thus everyone has a carbon debt that must be paid [1]. The more the human activity, the more energy is used so that the larger the carbon footprint. The carbon footprint refers to all human activities that cause carbon emissions and left trace in the earth. Environmental changes due to increase of the carbon footprint are caused by human activities, so the environmental recovery must be carried on by humans. Therefore, it is necessary to study human contribution to the carbon footprint of our daily activities. The goal is that humans can limit the amount of the carbon footprint produced and thus can help restore the living environment [2].

The carbon footprint is often used as a reference to measure how much GHG emissions generated by activity or by a process of production of goods or services. Carbon footprint is defined as a measurement of human activity which impacts the environment determined by how many carbon emissions (by-product) resulted in increased GHG, usually counted in unit ton of CO_2 [3]. All activities such as electrical consumption (lights, kitchen utensils, razors, and electronic devices), daily garbage (organic waste, papers, bottles of mineral water (drinking water) and transportations (cars and motorcycles) can produce carbon dioxide (CO_2) [4].

Makassar, a metropolitan city located in South Sulawesi Province has a population of 1,469,601 inhabitants. The high number of people with a variety of activities will certainly affect the greenhouse gas emissions generated. According to the National Development Planning Agency (BAPPENAS), Makassar produces the biggest gas greenhouse despite being the most developed city in Indonesia, in terms of economy and infrastructure sectors. There is an increase of greenhouse gas emissions, which is more than quadrupled in the year 2020 [5]. The presence of universities in Makassar may contribute to the amount of greenhouse gas emissions, which has impact to the environment either direct or indirect.

Activities that occur in the campus are extremely complex such as learning activities, administrations, usage of laboratories, and others [6]. But it is not yet known how the amount of greenhouse gas emissions produced in the educational institutions and the spread of emissions at universities. Educational activities, similar to office and household activities have the potential producing greenhouse gas emissions [7]. For example, Hasanuddin University Faculty of Engineering has buildings that have activities, such as the use of LPG, electricity, vehicles, and garbage. This research could be a reference for Gowa Campus of Hasanuddin University to become an eco-green campus.

Based on this background, the research was conducted to determine the emissions of CO_2 from the campus activities where the study area is Gowa Campus of Hasanuddin University Faculty of Engineering. Therefore, the authors take the title "A Study of assessment and mapping of carbon footprints to campus activities in Hasanuddin University Faculty of Engineering"

2. Methodology

This research was conducted at the campus of the Hasanuddin University Faculty of Engineering in Jl. Malino Km 6 Bontomarannu, Gowa and was held on 4 until 13 March 2019. This campus has 10 buildings in total. This study was conducted for seven days of research in which two days to do a collection of electronic equipment in every building, the use of paper and usage Petroleum Liquefied Gas (LPG), waste generation, and five days to calculate the amount of traffic volume at 07:00 to 18:00 pm. This research is a descriptive study which requires data to be obtained directly or indirectly. The primary data in this study were obtained from direct data collection and interviews in the field. The primary data are as follows:

- Use of electronic tools such as lights, air conditioners, mobile phones, laptop, workstations, refrigerators, fans, LCD monitors, photocopy machines, printers, water dispensers, rice cookers, TVs, electric stoves, and servers.
- Papers use by staffs
- Use of Liquefied Petroleum Gas (LPG)
- Calculation of vehicles such as motorcycles, cars, and trucks passing the street entrance of the campus gates to the parking.
- Mixed waste generation from buildings.

The secondary data were obtained from the study of literatures in the form of journals and previous researches, as well as image location through google earth, as shown in Figure 1.

Figure 1. Hasanuddin University Faculty of Engineering

2.1. Carbon Footprint Mapping Using Surfer 12.0

To find out the pattern of carbon footprint spread that occurs in the campus area, carbon footprint mapping were created using the Surfer 12.0 program. This program was used to create the visual of carbon footprint spreads from traffic activities around the school after calculating the noise level at each point of observation where the distribution is classified in color.

3. Results and Discussion

Based on the research, there are three parts of results and discussion, which are the amount of carbon emission for each building, the amount of carbon emission for every activity, and the mitigation to reduce the carbon emission.

3.1. The Amount of Carbon Emission for Each Building

In this section, CO_2 emissions are CO_2 emissions generated from each building which divided by primary and secondary. The calculation of carbon emission in every building is shown in Table 1, the histogram of the emissions is shown in Figure 2, and the map of the emissions in Figure 3.

Table 1. Calculation of carbon emission in every building

No.	Location	Emission CO_2		Total Emission CO_2 ($KgCO_2$)
		Emisson Primary	Emission Secondary	
1	Department of Naval A	78,138.22	112,610.95	190,749.17
2	Department of Naval B	163.64	20,435.56	20,599.20
3	Department of Geology	130,735.06	132,102.92	262,837.98
4	Department of Electrical	56,354.93	197,799.93	254,154.86
5	Department of Mechanical	112,050.03	166,933.56	278,983.59
6	Department of Civil	132,318.32	154,432.66	286,750.98
7	Department of Architecture	110,561.49	178,614.77	289,176.26
8	Classrooms	1,730.56	133,478.05	135,208.61
9	Center of Scientific Activities	38,255.91	134,523.12	172,779.03
10	Center of Technology	56,151,83	84,547.49	140,699.32

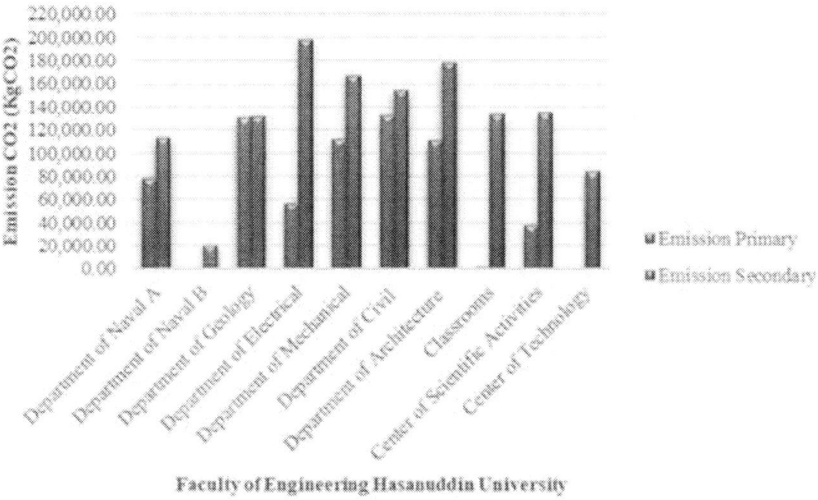

Figure 2. The histogram of CO2 emission in every building

Figure 3. The map of CO2 emission in every building

Table 1, Figure 2, and Figure 3 show that the largest CO_2 emissions are generated from Architecture Department at 289,176.26 $KgCO_2$/Year/Building having primary emission 110,561.49 $KgCO_2$/Year/Building and secondary emission 178,614.77 $KgCO_2$/Year/Building, due to the uses of the electronic devices found in Architecture Department, such as computers and lights.

3.2. The Amount of Carbon Emission for Every Activity

In this section, CO_2 emissions are CO_2 resulting from campus activities which then classified into primary and secondary. The calculation of emissions in the Hasanuddin University Faculty of Engineering for every activity can be seen in Table 2, the histogram of the emissions can be seen in Figure 4, and diagram of CO_2 emissions can be seen in Figure 5.

Table 2. Recapitulation result of total CO_2 emissions for every activity

No.	Activity	Type of Emission CO_2	Emission CO_2 ($KgCO_2$)		
			Day	Month	Year
1	Consumption of LPG	Primary	3,382.90	67,657.92	708,716.71
2	Waste Generation		175.98	703.94	178,095.59
3	Vehicles		3.57	92.95	743.57
4	Electricity Consumption	Secondary	3,653.97	109,619.14	1,315,429.63
5	Usage Paper		0.16	4.12	49.39
	Total Emission CO_2 ($KgCO_2$)		7,216.58	178,078.06	2,203,034.88

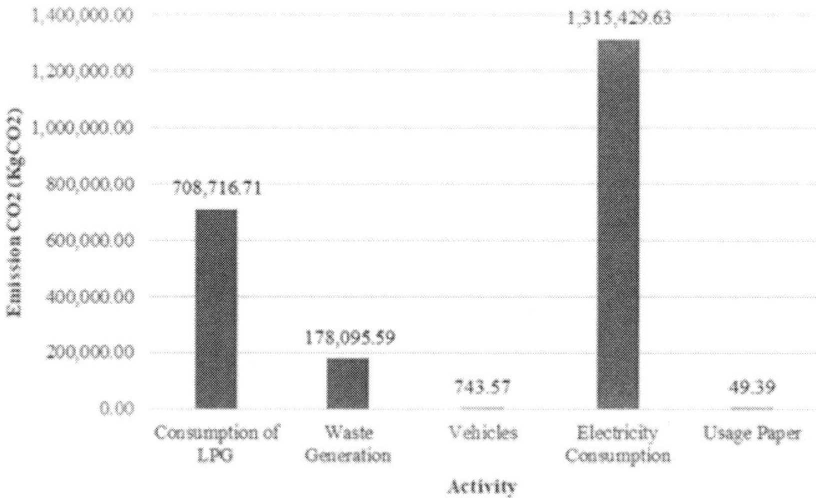

Figure 4. The histogram of CO_2 emissions total per activity/year

From Table 2 and Figure 4 the largest CO_2 emission was on electricity usage at 3,653.97 $KgCO_2$/Day, 109,619.14 $KgCO_2$/Month, and 1,315,429.63 $KgCO_2$/Year, due to the large number of electronic devices found in each buildings. This is reinforced by previous research stating that the factor which has most influences on CO_2 emissions is electric power, because teaching and learning activities require electronic devices to take place [8]. Based on the BAPPENAS emissions baseline (2014) for the energy-based sector and using LEAP software, stated that in 2020, the production of greenhouse gas emissions in the province of South Sulawesi are more than quadrupled to as much as 15,000,000 tons of CO_2eq [5] , for 2019 the emissions was 14,000,000 tons of CO_2eq so that the Hasanuddin University Faculty of Engineering in Gowa campus contributed 1,315 tons of CO_2eq or equivalent to 0.0075% of the total greenhouse gas emissions in South Sulawesi. Because of this high increase, mitigation of gas emissions from campus activities was conducted.

3.3. The Mitigation to Reduce the Carbon Emission

From the results of the research, the largest emissions is resulted from the use of electricity and also from the waste generated even though it is not the biggest contributor but also we can anticipate an increase in emissions so that the mitigation scenarios can be prepared. There are six scenarios we can use. For the electricity usage there are 3 scenarios, electricity shut down until 30 minutes, 1 hour, and 2 hours, while for waste generation, we use 3R waste manufacturing scenarios in order to reduce waste to 2%, 5%, and 10%. For mitigation scenario, an effective campus activity can be done is to shutdown electricity for 1 hour and a reduction in waste generation to 2 %, details are as follows:

3.3.1. One Hour Electricity Shutdown Scenario. This scenario is based on earth hour activity [9]. The scenario to shutdown the electricity for 1 hour. Below is the electricity consumption after 1 hour shutdown in Gowa campus Hasanuddin University Faculty of Engineering.

Table 3. The electricity consumption before and after 1 hour shutdown

No.	Location	Total (kWh)		Reduction
		Before	After	%
1	Department of Naval A	418,605.12	366,279.48	1.07
2	Department of Naval B	75,968.64	66,472.56	0.19
3	Department of Geology	491,077.44	429,692.76	1.26
4	Department of Electrical	741,052.80	648,421.20	1.89
5	Department of Mechanical	620,559.36	542,989.44	1.59
6	Department of Civil	535,772.16	468,800.64	1.37
7	Department of Architecture	644,676.48	564,091.92	1.65
8	Classrooms	548,029.44	479,525.76	1.4
9	Center of Scientific Activities	500,051.52	437,545.08	1.28
10	Center of Technology	84,541.32	73,973.66	0.8
	Total kWh	4,660,334.28	4,077,792.5	12.5

From the results above we can see that electricity usage is 4,660,334.28 kWh/Year. The total usage after Earth Hour, electricity usage is 4,077,792.5 kWh/Year, we can see electricity consumption was reduced by the scenario and the total reduce was 582,541.78kWh/Year or equivalent to 12.5%. And for the emission, we can see the following Table 4 below:

Table 4. The results of CO_2 emissions before and after 1 hour shutdown

No.	Location	Emission CO_2 ($KgCO_2$)		Reduction
		Before	After	%
1	Department of Naval A	112,604.78	98,529.18	0.29
2	Department of Naval B	20,435.56	17,881.12	0.05
3	Department of Geology	132,099.83	115,587.35	0.34
4	Department of Electrical	199,343.20	174,425.30	0.51
5	Department of Mechanical	166,930.47	146,064.16	0.43
6	Department of Civil	144,122.71	126,107.37	0.37
7	Department of Architecture	173,417.97	151,740.73	0.44
8	Classrooms	147,419.92	128,992.43	0.38
9	Center of Scientific Activities	134,513.86	117,699.63	0.34
10	Center of Technology	84,541.32	73,973.66	0.22
	Total Emission CO_2 ($KgCO_2$)	1,315,429.63	1,151,000.92	3.36

From the results above we can see that CO_2 emissions was at 1,151,000.92 $KgCO_2$/year. The electricity usage before Earth Hour has total CO_2 emissions of 1,315,429.63 $KgCO_2$/year, we can see that with this activity the CO_2 emissions can be reduced to as much as 164,428.70 $KgCO_2$/year or equivalent to 3.36%. The graphic consumption of electricity before and after 1 hour shutdown is shown in the following Figure 5.

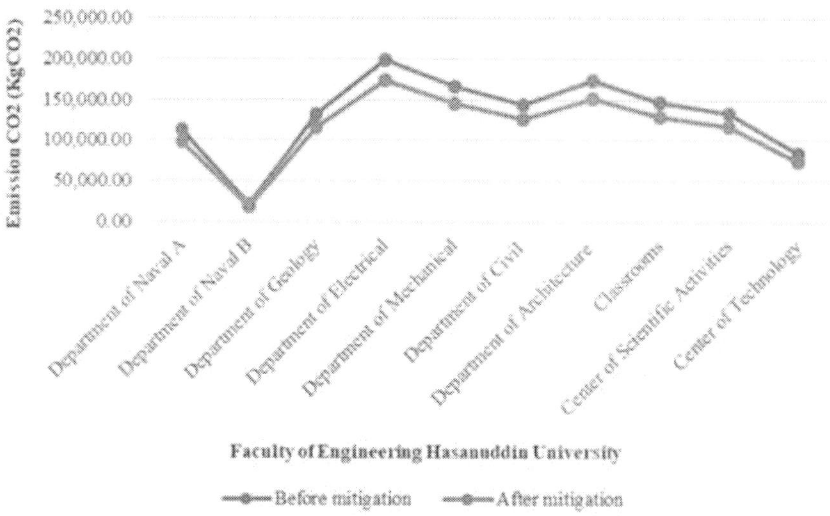

Figure 5. CO_2 emission before and after 1 hour shutdown

3.3.2. Reduction of Waste Generation to 2 % Scenario. According to the ministry of environment and forestry (2017), waste management is up to 80% and waste reduction has reached 20% in the last 5 years [10]. The percentage of organic waste component is 60%, plastic 14%, paper 9%, metal 4.3%, rubber 5.5%, fabric 3.5%, Glass 1.7%, and others 2.4% [11]. The decrease is due to the application of 3R Principle or Reduce, Reuse, Recycle. The following is the Table of the results of the application of garbage collection after the implementation of the 3R program:

Table 5. The generated waste before and after reduce to 2%

No.	Location	Waste Generation (Kg/ Year)	
		Before	After
1	Department of Naval A	776.6	761.068
2	Department of Naval B	383.24	375.5752
3	Department of Geology	1,155.88	1,132.76
4	Department of Electrical	1,257.96	1,232.80
5	Department of Mechanical	970.64	951,22.2
6	Department of Civil	4,863.76	4,766.48
7	Department of Architecture	1,445.84	1,416.92
8	Classrooms	4,052.84	3,971.78
9	Center of Scientific Activities	2,445.08	2,396.18
10	Center of Technology	782.32	766,673.6
	Total (Kg/ Year)	18,134.16	17,771.48

From the results we can see that waste generation is 17,771,48 Kg/Year. The total waste before 3R program is 18,134.16 Kg/Year, with this scenarion the waste generation can be reduced as much as 167,332.79 Kg/Year. The CO_2 emissions that can be reduced can be seen in the following Table:

Table 6. The CO_2 emissions before and after implementation of the 3R program reduce to 2%

No.	Location	Waste Generation (Kg/ Year)	
		Before	After
1	Department of Naval A	7,626.99	3,688.52
2	Department of Naval B	3,763.80	3,688.52
3	Department of Geology	11,351.90	11,124.86
4	Department of Electrical	12,354.43	12,107.34
5	Department of Mechanical	9,532.66	9,342.00
6	Department of Civil	47,766.99	46,811.65
7	Department of Architecture	11,624.16	13,915.60
8	Classrooms	39,802.94	39,006.88
9	Center of Scientific Activities	24,013.13	23,532.87
10	Center of Technology	7,683.16	7,529.50
	Total ($KgCO_2$ / Tahun)	175,520.16	170,747.75

From these results we can see that the CO_2 emissions from waste generation are 170,747.75 $KgCO_2$/Year. The waste generation prior to 3R program is at 175,520.16 $KgCO_2$/Year, we can see a reduction of CO_2 emission of 4,772.41 $KgCO_2$/Year.

4. Conclusion

- The measurement of CO_2 emissions generated by the Gowa Campus, Hasanuddin University Faculty of Engineering are 2,032,682.58 $KgCO_2$/Year, 178,078.06 $KgCO_2$/Month, and 7,216.59 $KgCO_2$/Day. For Primary CO_2 emissions are equal to 717,203.56 $KgCO_2$/Year, 68,454.80 $KgCO_2$/Month, and 3,563.45 $KgCO_2$/Day, and for secondary CO_2 emissions are 1,315,479.02 $KgCO_2$/Year, 109,623.26 $KgCO_2$/Month, and 3,654.13 $KgCO_2$/Day.
- From the mapping of CO_2 emissions can be seen that the highest usage of building/year is the Architecture Department, with the total of CO_2 emissions are 289,176.26 $KgCO_2$/Year/Building, with primary CO_2 emissions are 110,561.49 $KgCO_2$/Year/Building, and secondary CO_2 emissions are 178,614.77 $KgCO_2$/Year/Building.
- Mitigation scenarios that can be used to reduce the biggest CO_2 emissions are by implementing Earth Hour every 1 hour a day for every month which can reduce electricity consumption at 582,541.78kWh/Year and CO_2 emissions at 164,428.70 $KgCO_2$/year. The second scenario is to reduce waste generation by implementing the 3R which can reduce waste up to 2% in the amount of 167,332.79 Kg/Year and emissions CO_2 in the amount 4,772.41 $KgCO_2$/Year.

Acknowledgement
This research was funded by Faculty of Engineering, Hasanuddin University in Laboratory Based Education (LBE) Research Program.

References

[1] Aulia, Hiqma, Aryo, dan Jecky. 2019. Studi Carbon Footprint untuk Menghitung Emisi CO2 Primer dari Aktivitas Rumah Tangga di Wilayah Pengembangan I Kota Pekanbaru Riau : Universitas Riau.
[2] Astari, Ratih Gita. 2012. Studi Jejak Karbon dari Aktivitas Permukiman di Kecamatan Pademangan Kotamadya Jakarta Utara. Jakarta : Universitas Indonesia.
[3] Intergovernmental Panel On Climate Change. 2006. IPCC Guidelines For National Greenhouse Gas Inventories, Volume 2, Energy.

[4] Masyiah, Nurul Rani. 2016.Analisis Besaran Emisi CO2 pada Kawasan Perumahan di Kota Makassar Berbasis Quantum GIS. Makassar : Universitas Hasanuddin.
[5] Badan Perencanaan Pembangunan Nasional. 2014. Pedoman Teknis Perhitungan Baseline Emisi Gas Rumah Kaca Sektor Berbasis Energi. Jakarta : Jakarta.
[6] Putri, Dian Ananda. 2017. Kajian Jejak Karbon dari Aktivitas Kampus Fakultas Ekonomika dan Bisnis Universitas Diponegoro. Semarang : Universitas Diponegoro.
[7] Nur Azizah, Tasha. 2017. Kajian Jejak Karbon dan Pemetaannya dari Aktivitas Kampus di Fakultas Ilmu Budaya Universitas Diponegoro. Semarang : Universitas Diponegoro.
[8] Aroonsrimorakot, S., Y uwaree, C., Arunlertaree, C., Hutajareorn, R., Buadit, T. 2013.Carbon Footprint of Faculty of Environment and Resource Studies Mahidol University Salaya Campus Thailand.Journal of Cleaner. Universitas Mahidol : Thailand.
[9] Peraturan Presiden Republik Indonesia No. 61 Tahun 2011 tentang Rencana Aksi Nasional Penurunan Emisi Gas Rumah Kaca.
[10] Kementrian Lingkungan Hidup. 2017. Sistem Informasi Pengelolaan Sampah Nasional. Jakarta : Jakarta.
[11] Syakinah, Siti. 2019. Studi Perencanaan Teknis Tempat Pengelolaan Sampah Terpadu 3-R (TPST-3R) Kampus Fakultas Teknik Universitas Hasanuddin. Gowa: Universitas Hasanuddin.

IOP Conf. Series: Materials Science and Engineering **875** (2020) 012024 doi:10.1088/1757-899X/875/1/012024

Model Microsimulation Roundabout Utilities in Makassar

M Pasra[1]*, S Hamid[1], A Faisal[1], and H Yatmar[1]

[1]Department of Civil Engineering, Engineering Faculty, Hasanuddin University, Makassar, Indonesia

*Email: mubapasra@gmail.com

Abstract. The vehicle growth is one of the problems affecting traffic congestion. The circulation of entry vehicles and circulating vehicles in confluence made the conflict point. Avoidance the conflict is the need to increase road performance. The roundabout of the road is one of the utilities to increase the performances of traffic and reduce congestion. The prevention step is to maximize the utilities to control and manage the traffic. The roundabout dispersed and controlled the traffic flow. According to Makassar roundabout observation, the queuing and delay of vehicles recently occurs. This study aims to an analysis of the roundabout using the microsimulation Vissim program. The model uses the geometric condition of the road and volume of the vehicle that produce the models based on real condition and validation. This model performed the visual condition and calculated the capacity and shows the emission of the operational vehicle, and the general performances are summarized for operation performances at roundabout confluence sections. The result indicated that the roundabout is still useful to reduce the potential of queuing and delay the time of vehicles, and it needs supervision periodically for preventing the traffic from the potential congestion, especially in developing countries. The research result is useful as a reference for improving the capacity of roundabouts.

1. Introduction

Transportation is the movement of people or goods from one place to another by using a vehicle driven by man or machine. Along with population growth, the travel movement is also increasing. Vehicle growth is not matched by adequate infrastructure would be problematic, particularly for transportation. The congestion problem often occurs in individual road sections at peak hours (rush hour), for it is one of the ways to overcome transport problems is to maximize the functions of facilities and transport infrastructure.

Makassar City is one of the metropolitan cities in eastern Indonesia. Data motor vehicles in 2016, as many as 1,425,151 units increased from 2014, the number of vehicles ranging from 1,252,755 units. The growth rate of vehicles in 2015 for two wheels in Makassar is 6.29%, 10.20% of passenger cars, buses, and trucks 0.74% 7.55%. The long road of Makassar in 2016 was 2977.50 km with 66.80% good condition, the condition being 13.46%, 14.87% lightly damaged condition and severely damaged 4.87% (the city of Makassar in Figures, 2017), The growth of motor vehicles in the city of Makassar reach 7% per year, while the growth of road <1% per year [1].

Congestion often occurs on roads and intersections, especially at peak hours. Traffic congestion often occurs at the intersection of demand impact of traffic density are long queues and delays by vehicle [2]. This condition requires treatment that traffic congestion is not continuous. The handling facility at the intersection can be a roundabout and traffic lights [3]. The research explains that the handling of the plot intersection, there are two types of treatment with the handling of traffic lights and no traffic lights [4].

There are several roundabouts in the city of Makassar, among others, the roundabout Tugu Mandiri, Mandai Simpang Lima roundabout and the roundabout Samata Gowa. This roundabout performance will be analyzed Vissim software using model-based microsimulation. This model shows the visual conditions such as field conditions. In this study, an analysis of the problems with using simulation software in order to produce a valid analysis and in accordance with conditions on the ground.

2. Literature Review

2.1. Characteristics of Macro & Micro Traffic

Parameters needed to define macro traffic directly related to the traffic system are [5]:
- Volume
- Speed
- Density

Traffic microscopic approach examines some of the critical parameters which influence the response to the vehicle itself in traffic on the highway while the parameters - parameters include [6]:
- *Spacing* and headway
- *Lane Occupancy*
- *Clearance* and Gap

2.2. Intersection and Roundabout

According to the Ministry of Transportation Directorate, the junction is a node on the network where the streets meet and intersect the vehicle trajectory. Intersection created to reduce potential conflicts between vehicles (including pedestrians) while providing maximum comfort and ease of movement of the vehicle [7]. The traffic circle (roundabout) can be regarded as a particular case of the canalization of the islands in the middle can act as a divider, and a steering controller for the traffic system rotates in one direction [8].

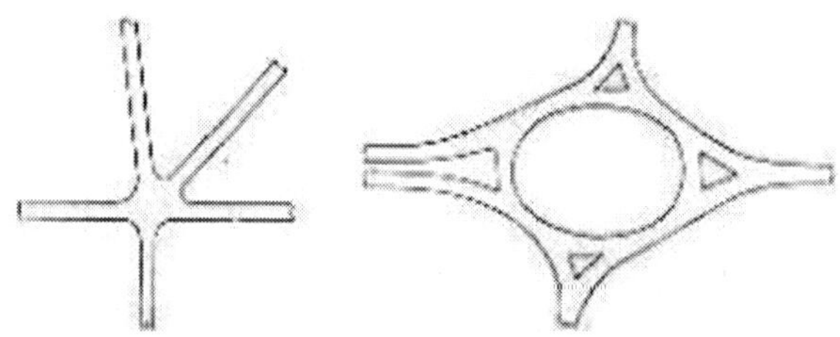

(a) (b)

Figure 1. Examples junctions' piece and roundabout

2.3. Traffic Management

Purpose of implementation of traffic management are [9]:
1) Getting the level of efficiency of traffic movement overall with the level of accessibility (comfort size) high by balancing demand with the movement of existing support facilities.
2) Increasing the level of safety of the user that can be accepted by all parties and improve the level of safety as possible.

According to the Regulation of the Minister of Transportation, one of the traffic planning is the determination of the level of service desired [10]. As for the level of service at the intersection of priority, "stop" can be seen in the Table 1 as follows:

Table 1. Income multiplier of transportation sector

Level of Services	The average delay is stopped (seconds per vehicle)
A	< 5
B	5-10
C	11-20
D	21-30
E	31-45
F	> 45

2.4. Microsimulation Model

The concept of a simulation model is very often used in traffic in transportation planning an activity exceptionally dynamic and very spacious. Basically, the simulation models are grouped in three dimensions [11]:

a) Deterministic Simulation Model with Stochastic Simulation Model
b) Continuous Simulation Model with Discrete Simulation Model
c) Static Simulation Model with Dynamic Simulation Model

Micro-simulation model of a concept or system of analysis which is widely used today because with this concept helps users define and evaluate the best parameters to be used in the context of specific issues-based computerization [12].

Car Following Model is a model that is used to control the behavior of the driver, or the driver of the other riders who were on the same track of this model was developed.

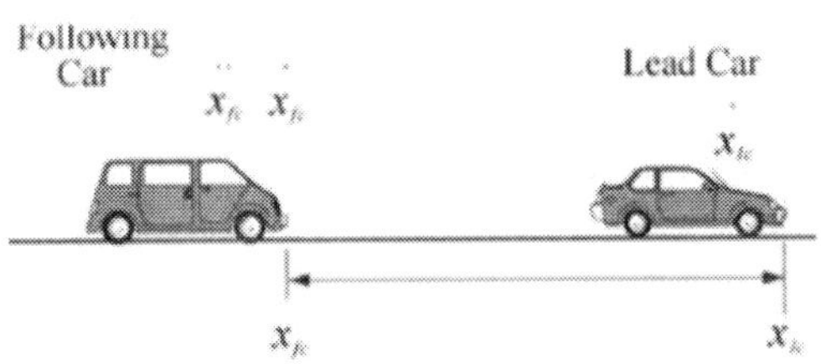

Figure 2. Illustration of car following model

2.5. Applications PTV VISSIM

PTV Vissim software is used for simulation of microscopic traffic flow leading developed by PTV Planing Transport [13]. VisSim tool traffic micro-simulation used for planning and traffic modeling for urban areas in the countryside want any good for any analysis of the flow of vehicles or pedestrians flows and have the ability to simulate various types of traffic modes simultaneously [14]. Vissim can be used for several cases, among others:

- Building roads and intersections
- Planning and development
- Mass transport planning

3. Research Methodology

The method used in this research is to conduct a series of surveys of traffic especially, the survey vehicle volume and speed.

3.1. Place of Research

Location of the research carried out on three roundabouts in Makassar namely, Tugu Mandiri, Mandai and roundabout in Samata.

3.2. Research Framework Flow

Before doing research, then made steps groove implementation of research activities to be run in a systematic and targeted achievement of the objectives of the study. The first step that needs to be done is a preliminary study that consists of background, problem formulation, and purpose of the study, then examined in the literature review and a variety of basic theory. The flow can be seen in Figure 3.

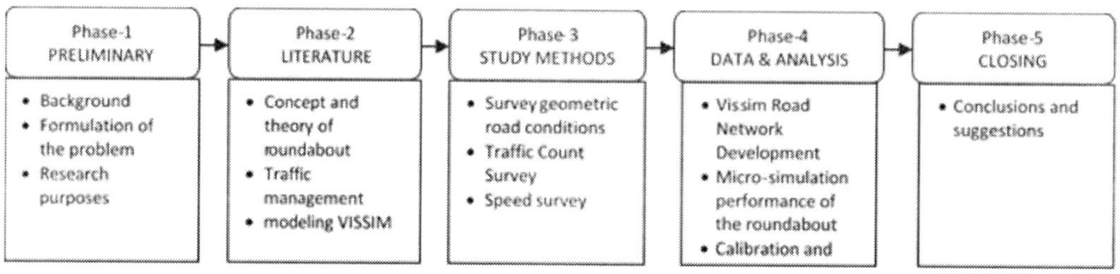

Figure 3. Research framework flow

3.3. Procedure Research (Survey)

The survey was conducted based on the needs of the data as input to the application. The survey is divided into three categories:

1) Survey traffic count is traffic volume by recording the number of vehicles on the roundabout.
2) Speed survey conducted using a speed gun test by directing a tool to the type of vehicle to be measured speed.
3) Survey geometric conditions are conducted by recording the geometric road conditions such as lane width, lane, shoulder, etc.

The third survey was conducted on peak hour traffic volume in the morning and afternoon on weekdays and holidays. Other data supporting the research is secondary data such as the number of vehicles the city of Makassar, the population, and more.

4. Result and Discussion

4.1. Volume of Traffic Roundabout Intersections Tugu Mandiri

The volume of traffic flow on each - each closers intersection shows the flow of vehicles that will cross the intersection. Here's the trend of vehicle volume of closer crossroads of observation area (Ujung Pandang-Nusantara-Sulawesi-Bundaran-Balai Kota- Slamet Riyadi).

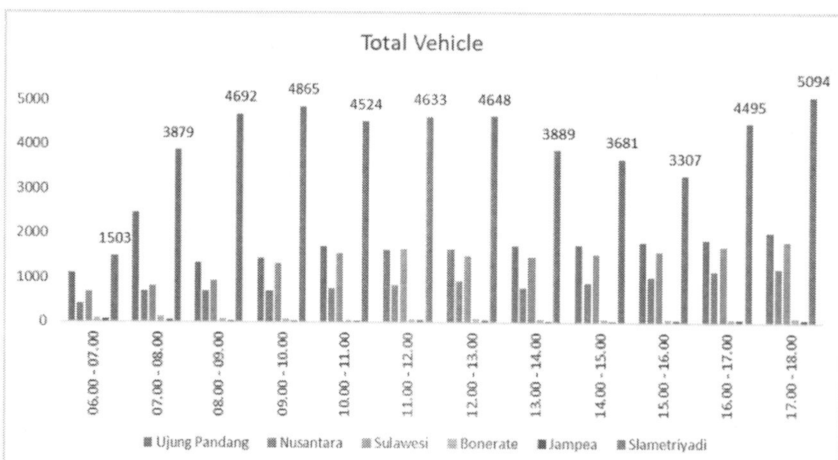

Figure 4. The trend of vehicle volume closers roundabout intersections

4.2. Profile Speed Vehicle Traffic Flow at Intersection

The results of the survey of vehicle speeds in Simpang roundabout Tugu Mandiri as shown visually in in intersection shows in figure 5, figure 6, and figure 7 bellow.

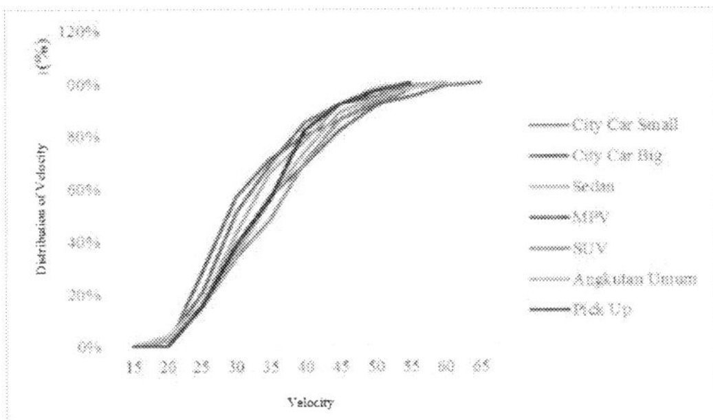

Figure 5. Vehicle speed of Light Vehicle (LV)

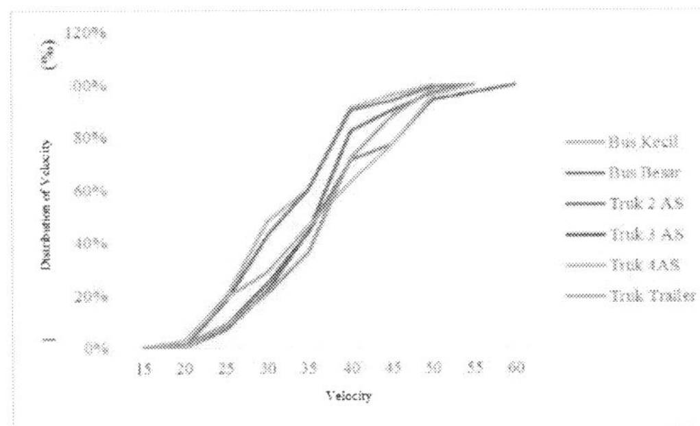

Figure 6. Vehicle speed of Heavy Vehicle (HV)

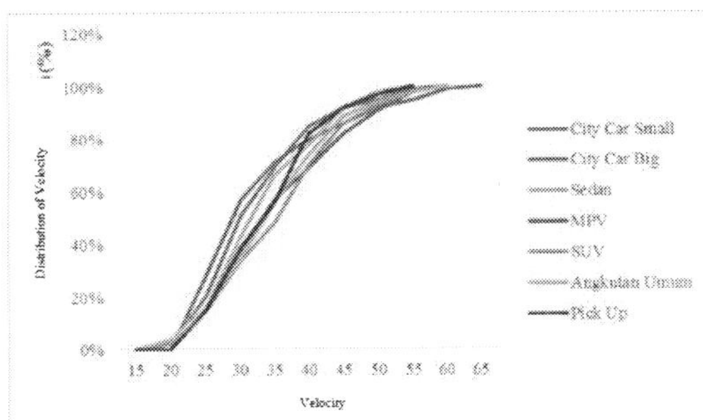

Figure 7. Vehicle speed of MOTORCYCLE

The above figures show that the frequency of the phenomenon of the vehicle speed at the intersection tends distributed normally, the graph shows that the frequency of the velocity distribution by type of vehicle is almost identical except for the type of trucks and buses that have relatively little frequency.

4.3. Calibration Model Micro - Simulation

Based on the parameters that exist in VISSIM been several parameters by different traffic conditions than do the calibration of the volume of vehicles in the period of peak hour (peak hour) that the peak hours in the morning, afternoon, and evening. The value of the calibration and calibration test results are accepted using GEH test. Then, the model is used to make the simulation of the roundabout, especially in Tugu Mandiri intersection. The visualizations are shown in the following figure 8.

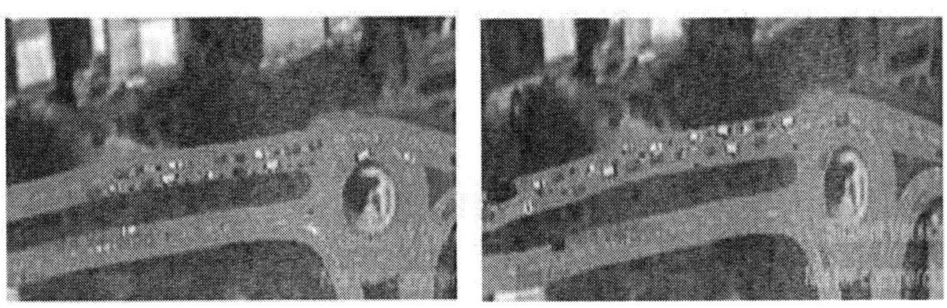

Figure 8. The 3D visualization roundabout of Tugu Mandiri intersection

4.4. Validation Results Model Micro - Simulation

The analysis used is the chi-square test. Where would seem the probability value of the chi-square test. The validation results also show that the model is accepted.

4.5. Analysis of Micro - Simulation Traffic Existing

Analysis of microsimulation by the results of the value - the value of the parameters used to analyze the performance of the traffic junction roundabout Tugu Mandiri, long queues, and delays of vehicles for this research.

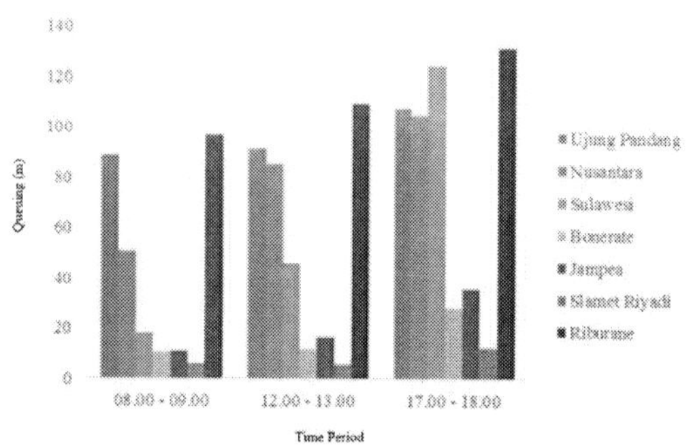

Figure 9. Vehicle queue length

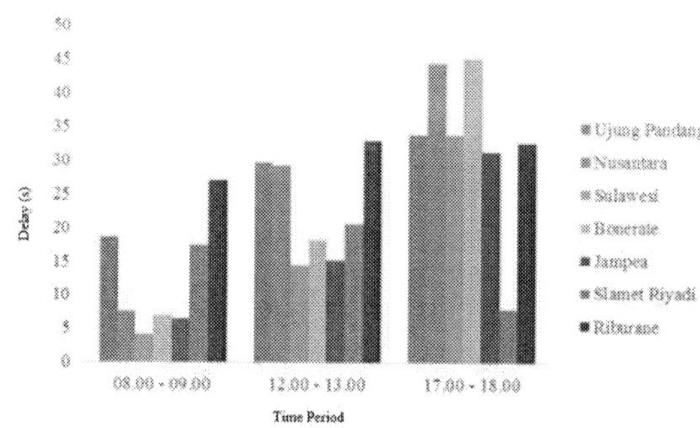

Figure 10. Vehicle delay

The traffic performance based on simulation results is shown in figure 9 and figure 10. The figure shows that the phenomenon of queuing and delays vehicles in each leg intersection is quite different, especially the long queues. For the long queues occurs at Jalan Ujung Pandang, which is longer than another side with the highest value at 38.12m, this has occurred at 12:00 to 13:00. Otherwise, the traffic in Jalan Nusantara with value is around 8.89 m, which is closers to Road Bonerate and occurs at 15:08 am. As for the delay of vehicles show the opposite phenomenon was a delay at the intersection of Jalan Ujung Pandang foot taller by about 6-15 seconds compared closers Jalan Slamet Riyadi west just 5-9 seconds.

4.6. Improving Simulation Performance (Alternative of U-Turn on Jl. Riburane)

The alternative movement of a traffic intersection is made by the addition of U-Turn at the base of Riburane road, so vehicles initially have to pass through the roundabout into a U-turn ot the U-Turn. The comparative visualization results after optimization models with field conditions at each closer are shown in figure 11 and figure 12 below.

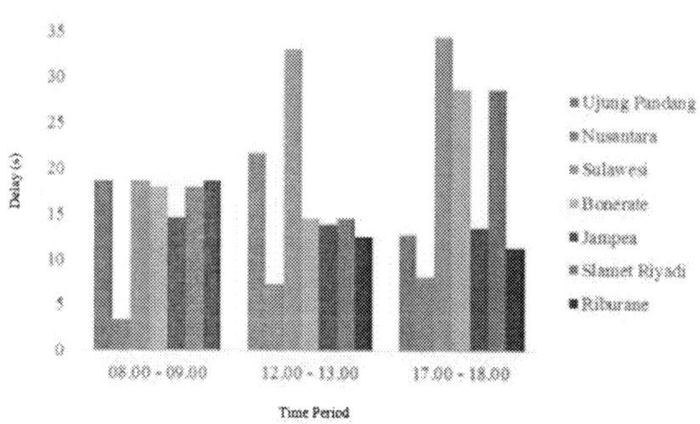

Figure 11. Vehicle queue length

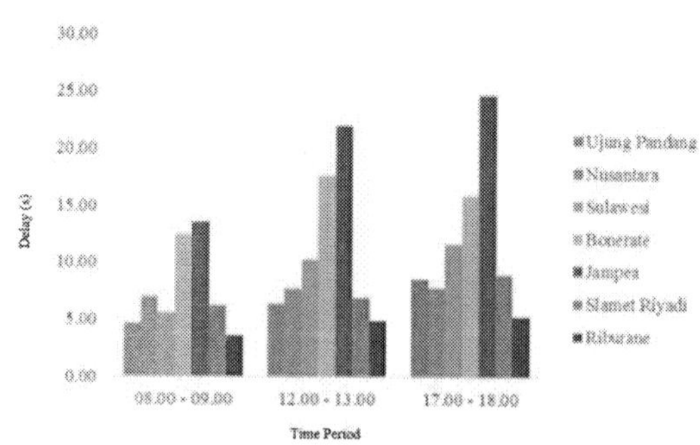

Figure 12. Vehicle delay

Figures 11 and 12 show that the optimum form of an adverse performance by adding a U-Turn at the base of the road Riburane. The results of the long queues and the maximum delay occur on Jalan Ujung Pandang 21.68 m; 8:56 second, Jalan Nusantara as much as 18,02 m; 7.76 seconds, Jalan Sulawesi, 34.46; 11.63 seconds, Jalan Bonerate 28.66 m; 17.60 seconds, Jalan Jampea 14.63 m; 24.66 seconds, Jalan Slamet Riyadi 28.66 m; Jalan Riburane 8.89 seconds and 18.68 m; 5:28 seconds. They were giving the U-Turn in the Road Riburane a positive impact on Jalan Ujung Pandang, namely a reduction in long queues at 31.41%, 12.15%, and Jalan Jalan Nusantara Riburane 53.57%. On the other hand, the lead of increasing of long queues occurs at Jalan Sulawesi of 13.65%, Jalan Bonerate 13:38%, 18.72% and Jampea Jalan Slamet Riyadi of 0.05%. After optimization was obtained, then it was improving the performance of road networks. Figure 14 shows the visualization of simulations results.

Figure 13. Improving work simulation results

5. Conclusions

After the analysis and validation model, connecting the conclusions and linked directly as follows:

a. Model microsimulation of traffic conditions at the intersection of Jalan Nusantara - Jalan Ujung Pandang - Road Riburane using Vissim software through the calibration and validation of simulation models by using volume and long queues of vehicles in the field.
b. Traffic performance of existing conditions at the Roundabout Simpang Tugu Mandiri obtained value of the queue length, and the maximum delay occurs are Jalan Ujung Pandang 38.12 m at 14:32 seconds; Jalan Nusantara 8.99 m and 8:38 seconds; Sulawesi street 31.67 m and 8:06 seconds; Jalan Bonertate 15.53 m and 18:46 seconds; Jalan Jampea 16.50m and 10:13 seconds; Jalan Slamet Riyadi 28.78 m and 9:04 seconds; Riburane street 34.63 m and 5:30 seconds.
c. Giving U-Turn in the Road Riburane leave a positive impact on Jalan Ujung Pandang, namely a reduction in long queues at 31.41%, 12.15% and Jalan Jalan Nusantara Riburane 53.57%, but lead to increased long queues at Jalan Sulawesi amounted to 13.65%, Roads 13:38 Bonerate%, 18.72% and Jampea Jalan Slamet Riyadi of 0.05%. Overall the positive impact is resulting from the provision of U-Turn at broad Leh Riburane road. So in the case of a roundabout intersection Tugu Mandiri, to produce optimum performance, sufficient intersections are wearing alternative addition of U-Turn in the Road Riburane for generating traffic performance is better than the other phases of the movement.

Acknowledgements
The work was supported by the Faculty of Engineering with Laboratory Based Education Research Grant. We would thankfully to Department Civil Engineering and all lecturer that contributes to this research project.

References

[1] Departemen Perhubungan. Dinas Perhubungan Kota Makassar. 2017
[2] Morlok, E. K. 1998. Pengantar Teknik dan Perencanaan Transportasi. Erlangga. Jakarta
[3] Oglesby, C. H., Hicks, R.G. 1982. Highway Engineering, Fourth Edition,Edisi Indonesia,1993, Teknik Jalan Raya edisi ke empat, jilid I. Penerbit Erlangga. Jakarta.
[4] MI Ramli, H Yatmar. 2020. Delay Distribution Estimation at A Signalized Intersection. IOP Conference Series: Earth and Environmental Science 419 (1), 012090.
[5] Law, A. M., and Kelton, D. M. 1991. Simulation Modeling and Analysis, Second Edition. McGraw-Hill Inc. New York, USA.
[6] Hobbs, F.D. 1979. Traffic Planning and Engineering, Second edition, edisi Indonesia, 1995, Perencanaan dan Teknik Lalu Lintas, Edisi kedua. Gadjah Mada University Press. Yogyakarta.
[7] Alamsyah, Ansyori. 2003. *Rekayasa Jalan Raya*. Universitas Muhammadiyah Malang. Malang
[8] Banks, J. H. 2002. Introduction to Transportation Engineering. 2nd ed. McGraw-Hill. New York.
[9] MI Ramli, H Yatmar. 2019. An Optimum Route Analysis of Fire Rescue According to Fire Station Location (Case Study: Residence Area in Makassar's Central Business District). IOP Conference Series: Earth and Environmental Science 235 (1), 012072.
[10] Alik Ansyori Alamsyah. 2005. Rekayasa Lalu-lintas. Universitas Muhammadiyah. Malang.
[11] Direktorat Jenderal Perhubungan Darat. 1996. Pedoman Teknis Perekayasaan Tempat Perhentian Kendaraan Penumpang Umum. Jakarta.
[12] MI Ramli, H Yatmar. 2019. Study On Speeding Behaviour of Driver's Motor Vehicles to Traffic Accident Risk in National Road in Makassar City. Journal of Indonesia Road Safety 2 (3), 159-170.
[13] O'Flaherty, C. A. 1997. Transportation Planning and Traffic Engineering. Hodder Headline Group. London.
[14] C. Jotin Khisty & B. Kent Lall. 2005. Dasar-dasar Rekayasa Transportasi. Jilid I. Erlangga. Jakarta

IOP Conf. Series: Materials Science and Engineering **875** (2020) 012025 doi:10.1088/1757-899X/875/1/012025

Phytoremediation of Fe and Mn Metal in Acid Mine Drainage Using *Typha Angustifolia*

R Ibrahim[1,2]*, A Zubair[1,2], M Selintung[1,2], and A R Dwirachma[1]

[1] Department of Environmental Engineering Department, Engineering Faculty, Hasanuddin University, Makassar, Indonesia
[2] Laboratory of Water Treatment, Department of Environmental Engineering Department, Engineering Faculty, Hasanuddin University, Makassar, Indonesia

*Email: ibrahim_roslinda@yahoo.com

Abstract. Coal mining activities in Indonesia are quite high and generally use the open-pit method. The problem that arises due to open mining is the presence of acid mine drainage (AMD) containing heavy metals such as Fe and Mn. The existence of heavy metals can cause water pollution and cause the death of aquatic organisms. Active processing methods that are widely used to process AMD have shortcomings, including high costs and can produce large amounts of sludge. One of the passive processing methods, namely phytoremediation method using constructed wetland system, is an alternative processing method which has advantages such as easy design, low cost of financing, and does not require expert operations, but has a fairly good ability to reduce pollutants, including heavy metals. The purpose of this study was to examine the effectiveness of a constructed wetland system in reducing Iron metal (Fe) and manganese metal (Mn) in AMD. This study was designed in a laboratory-scale research experiment by flowing AMD into a reactor that has regulated discharge and growing media. In addition to these two variables, the success of the constructed wetland system in this study was also influenced by the presence of *Typha angustifolia* which functions as a phytoremediation agent in reducing the concentration of Fe and Mn metals. The analysis showed that the best treatment in this study was Q1M2 which had the highest removal efficiency of Fe and Mn respectively 94.35% and 85.21%.

1. Introduction

Indonesia is one of the countries with quite high coal mining activities in the world. In general, coal mining uses the open-pit mine method. The problem caused by open mining is acid mine drainage (AMD). AMD is water that is formed due to oxidation of exposed or exposed sulphide minerals in the air in the presence of water catalysed by iron oxide and sulphur bacteria [1].

AMD has a low pH and contains dangerous heavy metals such as Fe, Al, Mn, Cu, Zn, Cd, Pb, As and usually also contains high sulphate, so it is a source of environmental contamination [2-5]. The existence of heavy metals can cause water pollution so that the water contained in the vicinity of the mining site cannot be consumed and supports people's lives. Besides, it can also cause the death of aquatic organisms. The content of heavy metals that accumulate in seawater and sediments will enter the food chain system and affect the life of the organism [6].

There are two methods for processing AMD, which are active treatment technology and passive treatment technology. One of the active processing technologies that are widely used is neutralisation which serves to neutralise the pH of AMD so that the iron removal process in water can run well.

Active processing generally uses chemicals containing lime, in the form of $CaCO_3$, $Ca(OH)_2$, CaO or the addition of caustic soda (NaOH) and ammonia (NH_3). The weakness of this method is that it is expensive and can cause sludge to be drained regularly [7].

Passive treatment method with phytoremediation method using constructed wetland system is an alternative treatment to reduce the concentration of AMD heavy metals. Phytoremediation is a technique that uses plants to reduce or reduce levels of pollutants in the environment so that they are no longer dangerous [8]. Categories in phytoremediation consist of phytoextraction, *phytofiltration*, *phytostabilisation*, phytovolatilization and phytodegradation depending on the mechanism of remediation [9]. Phytoextraction involves the use of plants to remove contaminants in the soil. *Pytofiltration* is the process of removing metals from water that is polluted by roots or saplings of plants. Whereas *phytostabilisation* involves roots to absorb pollutants from the soil and store them in the rhizosphere, and reduce the spread of pollutants. *Phytovolatization* involves the use of plants to remove pollutants through the evaporation process in leaf foliage, such as Se and Hg pollutants. *Pyhtodegradation* is the use of plants to associate with microorganisms in reducing levels of pollutants [10]. The use of plants in the constructed wetland system contributes to the increase in the content of organic matter through substances produced by secretions and decomposition of plant residues, helps stabilise the substrate, helps maintain microbial populations, and provides aesthetic quality for wetlands [11].

This study was designed to analyse the removal efficiency of Fe and Mn metals through a constructed wetland system using Typha angustifolia, which was adjusted to East Kalimantan Regional Regulation No. 02 of 2011 concerning Water Quality Standards from the Coal Industry [12].

2. Methods

This research is an experimental pilot project by conducting a research experiment to see the effect of the studied variables on the content of ferrous metals and acid mine manganese before and after being constructed wetland.

2.1. Research Design

This study consisted of two independent variables, namely flowrate and growth media composition. The plant used is a type of aquatic plant that is Typha angustifolia with the number of plants meeting the reactor with a spacing of 15 cm (30 plant clumps). Water discharge is divided into two variations namely Q1 = 0.00125 l/s with HRT 5 days and Q2 = 0.00200 l/s with HRT 3 days. The composition of the planting media also consists of two variations namely M1 = consisting of gravel; limestone; compost and soil with a height of 5 cm: 15 cm: 30 cm and M2 = consist of the same planting medium with a height of 10 cm :20 cm :20 cm.

2.2. Research Implementation

2.2.1. Reactor Preparation. This study uses a constructed wetland reactor made of box-shaped fibre with dimensions of 100 cm x 100 cm x 100 cm. Reactor provided as many as four pieces that are adjusted to the specified treatment variations. The treatment variations are presented in the following Table 1:

Table 1. Treatment variations

Water discharge	Growing media composition	
	M1	M2
Q1	Q1M1	Q1M2
Q2	Q2M1	Q2M2

2.2.2. Acclimatisation of Typha Angustifolia. Acclimatisation of plants to the environment has only been carried out for 15 days after planting. After the plant adaptation period ends, the process of plant acclimation to the water to be treated is carried out using groundwater with a characteristic pH of 6.94, Fe content of 0.05 mg/L and Mn content of 0.7 mg/L. The ability of plant adaptation is characterised by the visual appearance of fresh / not withered leaves.

2.2.3. Research Experiment. The acid mine water used in this study is collected in a water tank, then flowed into each of the available reactors. Filling wastewater up to the reactor height, then flowing out of the reactor through a tap and collected in a plastic container. The discharge of wastewater flowed from the water tank to the reactor is regulated so that it is balanced with the discharge of wastewater coming out of the reactor. Then the water sample is processed and placed in a plastic bottle. Laboratory analysis of wastewater parameters according to SNI standards.

3. Result and Discussion

3.1. Characteristics of Acid Mine Drainage

The characteristics of acid mine drainage used in this study can be determined based on the water quality parameters listed in Table 2.

Table 2. Characteristics of acid mine drainage

Parameter	Unit	Value	Quality standards
pH	-	2.9	6 - 9
Iron (Fe)	mg/L	11.32	7
Manganese (Mn)	mg/L	16.9	4
TSS	mg/L	326	300

Data from the laboratory analysis results of AMD sample parameters indicate that the pH, Fe, Mn and TSS values exceed the required standards based on East Kalimantan Regional Regulation No. 02 of 2011 concerning Water Quality Standards from the Coal Industry.

3.2. pH Value of Acid Mine Drainage

PH measurements for all effluent reactors constructed wetland carried out five times. The measurement result data is in Table 3. Then the data is compared with the required quality standards.

Table 3. Effluent constructed wetlands pH value

Treatment variations	Observation time (day)					Quality standards
	2	6	10	14	18	
Q1M1	6.82	7.10	6.55	.6.70	7.56	6 - 9
Q1M2	6.86	7.04	6.58	6.58	7.16	
Q2M1	3.47	3.48	3.21	3.97	3.05	
Q2M2	3.42	6.04	4.82	6.08	5.40	

The data presented in Table 3 shows that all effluents of reactors constructed wetland with different treatments have increased pH. However, variations in the treatment of Q1M1 and Q1M2 showed an increase in pH that was higher than the treatments of Q2M1 and Q2M2 and had met the required quality standards of 6-9. The Q1 treatment with HRT of 5 days was seen to have a positive influence on the increase in pH of AMD because the contact time with lime was longer than the Q2 treatment. Water discharge affects AMD retention times in artificial swamp systems [13]. The use of lime in each reactor is intended as an active treatment to increase the pH due to an increase in alkalinity in both the limestone layer and the organic / compost layer. In anaerobic wetland systems with the composition of growing media such as compost and soil added with active lime can stimulate the growth of sulphate reducing bacteria to increase the alkalinity that can increase the pH of AMD [14, 15].

3.3. The concentration of Iron Metals (Fe) Effluent Constructed Wetland

Data from laboratory analysis results for the Fe concentration of water samples through the constructed wetland reactor carried out five times are presented in the following Table 4.

Table 4. The concentration of iron metals (Fe) effluent constructed wetland

Treatment variations (mg/L)	Observation time (day)					Highest removal efficiency (%)	Quality standards
	2	6	10	14	18		
Q1M1	7.92	4.18	4.16	5.64	0.75	93.37	7
Q1M2	2.25	4.23	3.14	3.16	0.64	94.35	
Q2M1	11.25	4.49	11.52	5.02	11	60.34	
Q2M2	3.60	4.86	9.81	3.26	6.48	68.20	

The data presented in Table 4 and Figure 1 shows that all constructed wetland reactors with different treatment variations can reduce the concentration of Fe contained in acid mine drainage. Wetlands function to absorb and bind heavy metals and slowly precipitate them as sedimentary deposits to become part of the geological cycle [16]. The reactor with Q1M2 treatment has a higher ability than other treatments because it is proven that the concentration of Fe effluent is lower, ranging from 0.64 to 4.23 mg/L with the highest removal efficiency of 94.35%. The level of Fe effluent produced has met the required quality standards of 7 mg/L.

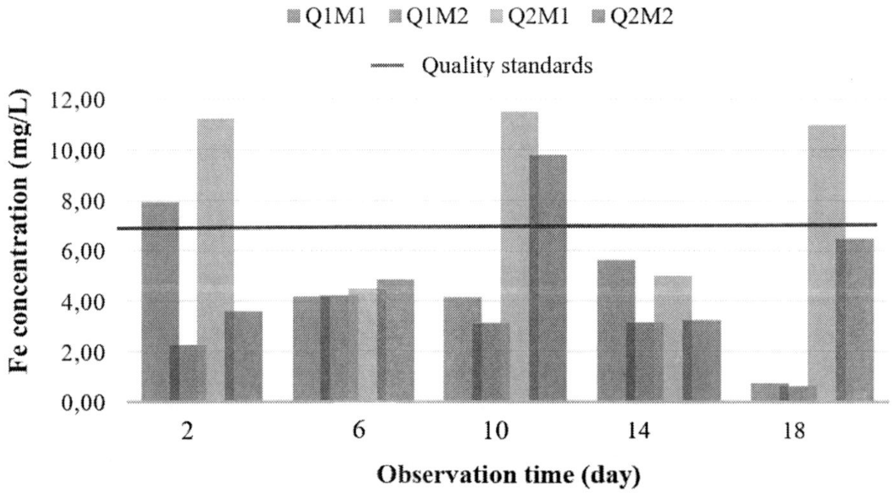

Figure 1. The concentration of iron metals (Fe) effluent constructed wetland

The decrease in Fe effluent concentration after passing through the constructed wetland reactor with Q1M2 treatment was due to the longer HRT and the composition of the planting medium with higher gravel and limestone layers than other treatments. HRT 5 days gives sufficient time for contact between AMD and the planting media and plants to occur in the process of reducing Fe metals. The process of reducing metal Fe occurs because of the activity of microorganisms that live on plant roots that produce oxygen so that aerobic conditions that support the decomposition of metals are created.

Plants also have a role in the process of reducing Fe metals. Fe metal is absorbed by plants because this metal element is a micronutrient needed in the process of forming proteins, as a catalyst for the formation of chlorophyll. The element Fe is important in the synthesis of chlorophyll which is a place for plant photosynthesis [17]. Some plants identified can be used in the phytoremediation process because of their ability to accumulate heavy metals, one of which is Typha angustifolia [18]. Plants that can accumulate heavy metals in their organ tissues are known as hyperaccumulators. The properties possessed by hyperaccumulator plants include fast-growing, being able to consume large amounts of water in a short time, being able to remediate more than one pollutant, and high tolerance to pollutants [19].

Another factor that influences the high efficiency of the reactor removal by Q1M2 treatment is the height of the limestone media. The presence of a sufficient amount of limestone causes an increase in pH value to the neutral pH range. At neutral pH and the presence of sufficient dissolved oxygen, soluble ferrous ions can be ionized into ferries and subsequently form ferrous hydroxide deposits which are difficult to dissolve, in the form of crystals (precipitates) which are usually brownish yellow, due to acidic and aerobic conditions the ferrous form soluble in water [20].

3.4. The concentration of Manganese Metals (Mn) Effluent Constructed Wetland

Laboratory analysis data for the Mn concentration of water samples through the constructed wetland reactor conducted five times are presented in the following Table 5.

Table 5. The concentration of manganese metals (Mn) effluent constructed wetland

Treatment variation (mg/L)	Observation time (day)					Highest removal efficiency (%)	Quality standards
	2	6	10	14	18		
Q1M1	7.70	6.70	6.90	6.90	5.90	65.09	4
Q1M2	2.50	8.30	6.80	6.90	5.90	85.21	
Q2M1	13.40	9.20	17.00	3.26	6.48	80.71	
Q2M2	8.60	6.80	11.10	7.40	7.30	59.76	

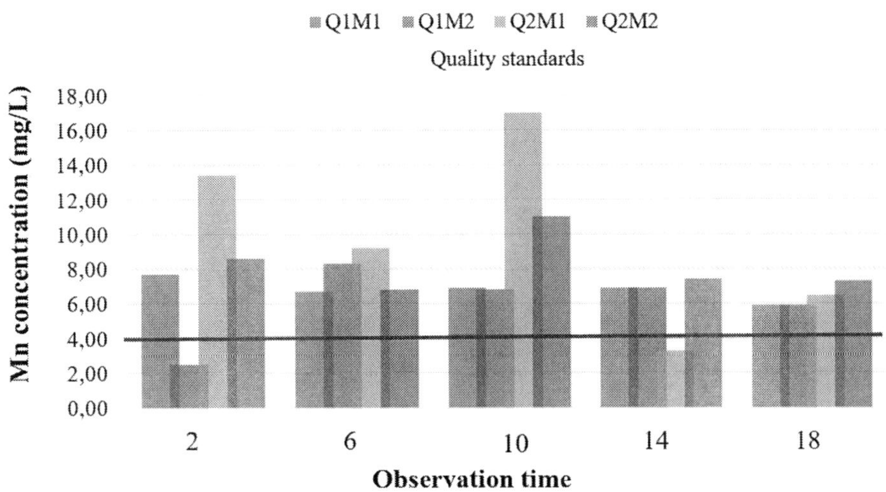

Figure 2. The concentration of manganese metals (Mn) effluent constructed wetland

The data presented in Table 3 and Figure 2 show that all reactors constructed wetland with different treatment variations can reduce the concentration of Mn in acid mine drainage. However, the reactor with Q1M2 treatment has a higher ability than other treatments because it is proven because the Mn concentration of the effluent is lower which ranges from 2.50 - 8.30 mg/L with the highest removal efficiency of 85.21%. The resulting effluent concentration has met the required quality standard of 4 mg/L, but only at one observation time. Likewise, the reactor with Q2M1 treatment, there is only one time of observation that shows the ability to reduce levels of Mn in acid mine drainage to meet quality standards. The effluent of the reactor with the treatment of Q1M1 and Q2M2 on each observation does not show the ability to reduce the levels of Mn metals that meet the required quality standards.

The decrease in Mn effluent levels after passing through the constructed wetland reactor is due to HRT and the composition of the planting media. Similar to the reduction process of Fe metals in acid mine drainage, the reduction of Mn metals also occurs because of the activity of microorganisms that live on plant roots and absorption by plant roots. Several types of microbes can oxidise and reduce manganese, among others, Leptothrix pseudoochraceae, Arthrobacter siderocapsulatus and Metallogonium significantly play a role in the oxidation process of Mn divalent ions and Mn-oxide reduction [21]. Mn metal is one of the microelements needed by plants. Mn is an essential element in photosynthesis reactions, especially the process of breaking down water fractions into free hydrogen ions and oxygen [22]. Mn functions as an activator in the oxidoreduction system in the photosynthetic electron transport system [23].

Mn metal processing is not as easy as Fe metal processing. In Mn processing, preliminary treatment needs to be carried out before entering in a constructed wetland. AMD with acidity level > 4 can be channelled into the settling pond first to reduce the heavy metal content. Besides, the optimisation of reduction of Mn metal is done by aeration.

4. Conclusion

All reactors constructed wetland with different treatment variations showed the ability to remove Fe and Mn metals. The highest removal efficiency of Fe and Mn metals was shown by the Q1M2 reactor at 94.35% and 85.21%, respectively and had met the required quality standards.

Acknowledgement

Thanks to the Faculty of Engineering Hasanuddin University for funding the 2019 LBE research activities.

References

[1] Lottermoser, B.G., 2010. Mine Wastes, Characterization, Treatment and Environmental Impacts. 3rd edition. London (GB) : Springer.

[2] Achterberg, E.P., Herzl, V.M.C., Braungardt, C.B., Millward, G.E., 2003. *Metal behaviour in an estuary polluted by acid mine drainage: the role of particulate matter. Environ*. Vol.121, pp. 283–292.

[3] Blodau, C., 2006. *A review of acidity generation and consumption in acidic coal mine lakes and their watersheds. Science of the Total Environment,* Vol. 369, pp. 307–332.

[4] Elisa, M., Gomes, P., and Favas, J.C., 2006. *Mineralogical controls on mine drainage of the abandoned Ervedosa tin mine in north-eastern Portugal. Applied Geochemistry*. Vol. 21, pp. 1322– 1334.

[5] Jeremy, D., Atkin, S., Beale, G., and Alexdaner, G., 2004. *Development of the Sleeper Pit Lake. Mine Water dan the Environment*, Vol. 23, pp. 2–11.

[6] Meador, J. P., Ernest, D. W., and Kogley, A. N., 2005. *Science of the Total Environmental.* 339:189 – 205.

[7] Johnson, D. B., and Hallberg, K. B., 2005. *Acid Mine Drainage Remediation Option: A Review, science of The Total Environment,* vol.338, pp. 3 – 14.

[8] Salt, D.E., Smith, R.D., Raskin, L., 1998. Phytoremediation. Ann. Rev. Plant Phys. Plant Mol. Biol., 49(1):643-668.

[9] Lone, M.I., He, Z., Stoffella, P.J., Yang, X., 2008., *Phytoremediation of Heavy Metal Polluted Soils and Water: Progress and Perspective., Journal of Zhejiang University SCIENCE B.*, Vol. 9, No. 3, pp. 210-220.

[10] Garbisu, C., and Alkorta, I., 2001. *Phytoextraction: A cost-effective plant-based technology for the removal of metals from the environment. Biores. Technol*, Vol. 77, No. 3, pp. 229-236.
[11] Younger, P. L., 2000. The Adopted and Adaptation of Passive Treatment Technologies for mine Watersbin The United kingdoms. Mine Water and Environment.
[12] East Kalimantan Regional Regulation Number 02 of 2011 concerning Water Quality Standards from the Coal Industry.
[13] Sim, C. H., 2003. The Use of Constructed Wetlands for Wastewater Treatment. Wetland International – Malaysia Office. Selangor.
[14] Chang, I. S., Shin, P. K., Kim, B. H., 2000. *Biological treatment of acid mine drainage under sulphate-reducing conditions with solid waste materials as substrate. Water Research*, Vol. 34, pp. 1269-1277.
[15] Gibert, O., De Pablo, J., Cortina, J. L., Ayora, C., 2004. *Chemical characterisation of natural organic substrates for biological mitigation of acid mine drainage. Water Research,* Vol. 38, pp. 4186-4196.
[16] Sheoran, A.S., Sheoran, V., 2006. Heavy metal removal mechanism of acid mine drainage in wetlands: A critical review. Minerals Engineering, Vol.19, pp. 105–116.
[17] Emmons, R. D., 2000. Turfgrass Science and Management 3rd. Delmar Thomson Learning. The USA. pp. 395.
[18] Muhammad D, Chen F, Zhao J, Zhang G, and Wu F., 2009. *Comparison of EDTA and Citric Acid-Enhanced Phytoextraction of Heavy Metals in Artificially Metal Contaminated Soil by Typha angustifolia. Int J Phytoremediation*. Vol.11, No. 6, pp. 558-74.
[19] Morel J. L., Echevarria G, and Goncharova N., 2006. Phytoremediation of Metal- Contaminated Soils. Netherland: Springer.
[20] Joko, T., 2010. Unit Produksi dalam Sistem Penyediaan Air Minum. *Graha Ilmu*, Yogyakarta.
[21] Ghiorse, W.C., 1978. Microbial Reduction of Manganese and Iron in Biology of Anaerobic Microorganisms edited by Zehnder, A.J.B. A Wiley Interscience Publ. John Wiley & Sons. New York.
[22] Hassett, J.J. dan Banwart, W.L., 1992. Soils and Their Environment. *Prentice-Hall*. New Jersey.
[23] Jones, Jr., Wolf, B. dan Mills, H.A., 1991. Plant Analysis Handbook. *Micro-Macro Publ. Inc*. Georgia.

Experimental Study of Wave Reflection in Breakwater Overtopping Catcher Model

M P Hatta[1], A I D Puspita[1*], M A Thaha[1], R Karamma[1], S Pongmanda[1], A S Mustari[1], and M Ibrahim[1]

[1]Civil Engineering Department, Hasanuddin University, Makassar, Indonesia

*Email: ildhaandi@gmail.com

Abstract. Breakwater overtopping catcher model is a breakwater or shore protection model, which can accommodate or catch wave overtopping. This breakwater model is different from most other conventional breakwater models, at the back of this breakwater model, an opening is made to form a reservoir. This reservoir serves to collect overtopping wave, which is then collected and released back to the sea through an energy conversion process. How many waves overflow at the peak of the breakwater model is strongly influenced by several factors, one of which is the wave characteristics. This research was carried out experimentally to determine the characteristics of the waves, especially the influence of wave steepness to the wave reflection coefficient. The experiment was carried out in a wave flume with regular wave generation. Variations of research carried out on variations in the model that is slope and freeboard and wave variations that is period and wave height. The results of the study show that the wave reflection coefficient decrease with the increase of wave steepness, it is also seen that steep slopes produce the largest wave reflection coefficient.

1. Introduction

The breakwater is a coastal protection building that works by dissipating wave energy. Whereas sea waves are one of the renewable energy sources that are very environmentally friendly. These ocean waves energies have the potential to be harnessed, rather than being destroyed. Ocean waves as a source of renewable energy are always available and even abundant in island countries like Indonesia. This energy source should not be destroyed by a breakwater but should be utilized.

The use of ocean waves for energy conversion has been known for a long time as the wave energy converter (WEC). Wave energy converter is a device to convert ocean wave energy into electrical energy. At present, the number of WEC devices has been very large, and even some of the devices have already been applied, but the device which is then termed as a single stand wave energy converter requires a very expensive cost when applied. Therefore, the researcher considers an idea to reduce the high cost of the single stand WEC with the concept of integrating the breakwater with the wave energy converter.

The integration of WEC devices with other marine facilities has become common especially for near-shore applications. The main reason is triggered by better economic viability through cost-sharing on construction, installation, maintenance, and operation as aligned with the main issue faced especially by the current stand-alone WEC device [1].

The breakwater overtopping catcher model is a breakwater model that not only functions as a coastal protector but also functions as a wave energy converter. This breakwater model is different from most other conventional breakwater models, at the back of this breakwater model, an opening is made to form a reservoir. This reservoir serves to collect overtopping wave, which is then collected and released back to the sea through an energy conversion process.

This paper aims to determine the characteristics of the waves, especially the influence of wave steepness to the wave reflection coefficient on the breakwater overtopping catcher model.

1.1 Wave Reflection Coefficient

Wave reflection is one form of wave deformation and is one of the important parameters in coastal or breakwater protection studies. When a wave comes and hits an obstacle or coastal structure, some of the wave energy will be reflected, some will be transmitted, and some will be destroyed (dissipation). To know how much a coastal structure material reflects waves, it is usually known from how much its wave reflection coefficient. The wave reflection coefficient can be calculated using the following equation [2, 3].

$$Cr = \frac{Hr}{Hi} = \sqrt{\frac{Er}{Ei}} \quad (1)$$

The formula shows that the reflection coefficient is the ratio between the height of the reflected wave to the height of the incident wave, where the height of the reflection wave is symbolized by *Hr* and the height of the incident wave is symbolized by *Hi*. The reflection wave coefficient can also be the root of the comparison of the reflection wave energy to the incident wave energy.

1.2. Wave Steepness

The wave steepness is also one of the most important wave characteristics that can give an idea of an incoming wave that will hit a coastal structure. An incoming wave has a height and a wavelength, and the ratio between the height and the wavelength (s_0 = Hm_o/L_0) is known as the wave steepness. In general, when the wave steepness is about s0=0.01 that it can indicate that it is a typical swell sea and when the wave steepness is about $s_0 = 0.04$ to 0.06 that it can indicate that it is a typical wind sea. Long-period waves are often be associated with the Swell sea, where the main parameter that affects overtopping is the period. Definition of wave steepness shown in Figure 1.

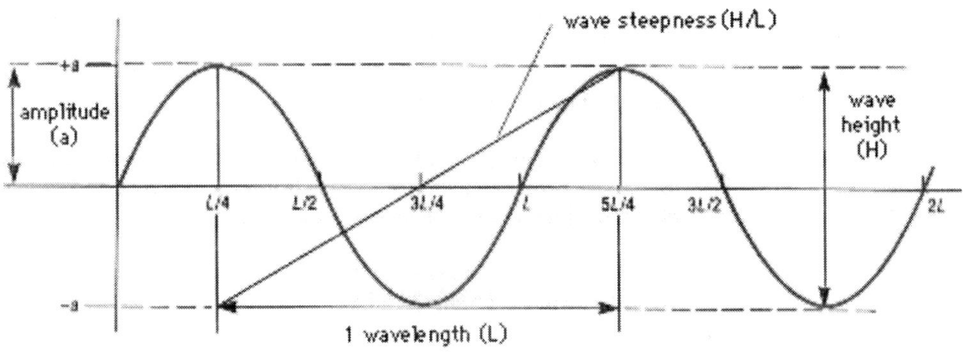

Figure 1. Definition of wave steepness [4]

Also, wind seas may become seas with low wave steepness if the waves break on a gentle foreshore. By wave breaking the wave period initially does not change much, but the wave height decreases. This leads to a lower wave steepness. A low wave steepness on relatively deep water means swell waves, but for depth limited locations it often means broken waves on a (gentle) foreshore [5].

2. Experimental Study

2.1. Experimental Facility and Equipment

This research is experimental research conducted in the hydraulics laboratory of the Civil Engineering Department Faculty of Engineering Universitas Hasanuddin. This experimental study carries out physical modeling where the scale used is 1:20 length scale. Breakwater building models are tested in a wave channel which has a length of 15 m and a width of 30 cm, the effective depth of the wave channel is 45 cm. The generator of the wave is a flap type. Wave making flaps create wave motion. The top of the flap is connected to the drive plate using a stroke and the bottom of the flap is a hinge. Flap movement is a rotational motion that is controlled through a rotary motion of the drive plate. This flap movement/flutter that raises the waves. The wave flume photo is shown in Figure 2.

Figure 2. Wave flume photos used in the study

The wave generators on the wave channel generated wave height in the range of 2-12 cm and wave periods in the range of 0.6 to 6 seconds. The wave generators generated the regular wave. By adjusting the stroke in several variations to change the magnitude of the flap deviation, the wave height can be varied. And by adjusting the rotational speed of the drive plate, the wave period (T) can be varied.

2.2. Physical Model

The breakwater overtopping catcher model is a breakwater model that not only functions as a coastal protector but also functions as a wave energy converter. This breakwater model is different from most other conventional breakwater models, at the back of this breakwater model, an opening is made to form a reservoir. This reservoir serves to collect overtopping wave, which is then collected and released back to the sea through an energy conversion process. The breakwater overtopping catcher model is in 1:20 scale length. The freeboard of the breakwater overtopping catcher model is varied with 3 variation which is 12.5 cm, 10 cm, 7.5 cm, and the slope of the breakwater overtopping catcher model is varied with 3 variation which is 0.5, 0.4, and 0.3 or 1:2, 1:2.5 and 1:3 respectively. The freeboard and slope variations (model parameters) are shown in Figure 3.

IOP Conf. Series: Materials Science and Engineering **875** (2020) 012026 doi:10.1088/1757-899X/875/1/012026

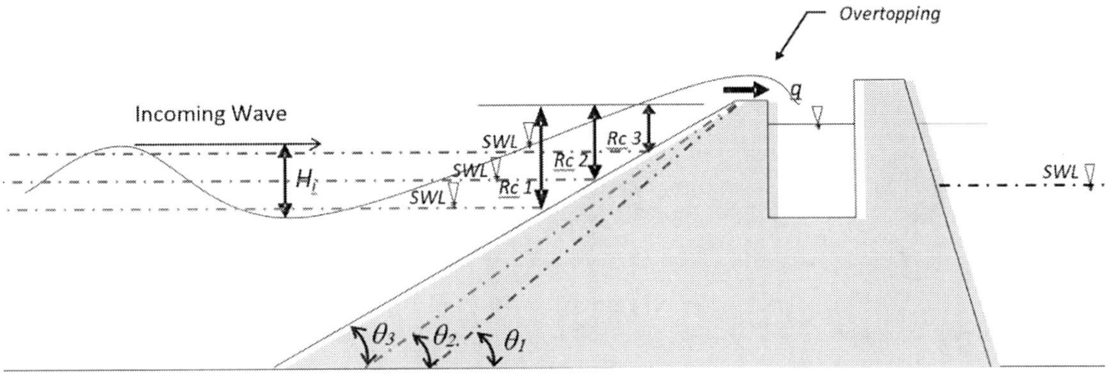

Figure 3. Illustration of slope variations and freeboard variations of the breakwater overtopping catcher model

The incoming wave height (Hi) varied with three variations which are 6 cm, 7 cm, and 8 cm. The wave period (T) which is directly related to the wavelength (L) also varied with seven variations which are 0.6 s, 0.7 s, 0.8 s, 0.9 s, 1.1 s, 1.2 s, and 1.3 s. The wave period and the incoming wave height (wave parameters) are presented in Table 1.

Table 1. The variation of wave parameters

Incoming wave Parameters	Wave parameters variation
Wave Height (cm)	6 - 8
Wave Period (s)	0.6 – 1.3

3. Result

The experimental study carried out is summarised in Table 2. The test carried out with a smooth impermeable slope formed the reference case.

Table 2. Test condition summary

Model	Tan θ	Rc	Hi	T
1	0.5 (slope 1 : 2)	12.5 cm	3 variation	7 variation
		10 cm	3 variation	7 variation
		7.5 cm	3 variation	7 variation
2	0.4 (slope 1 : 2.5)	12.5 cm	3 variation	7 variation
		10 cm	3 variation	7 variation
		7.5 cm	3 variation	7 variation
3	0.3 (slope 1 : 3)	12.5 cm	3 variation	7 variation
		10 cm	3 variation	7 variation
		7.5 cm	3 variation	7 variation

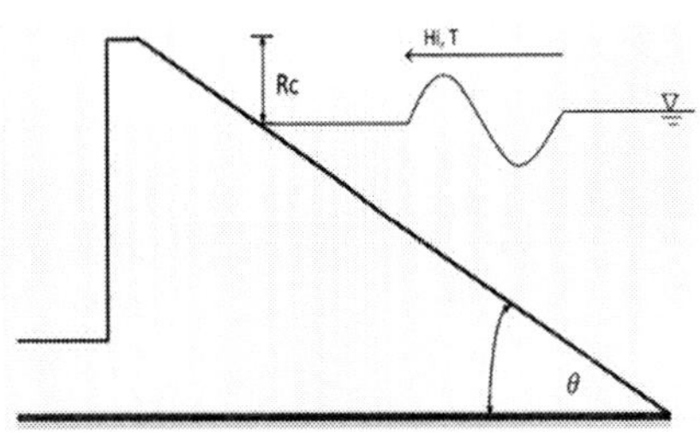

Figures 5 (a)-(c) shown the resulting graph of the relationship between wave steepness and the wave reflection coefficient, with slope variations and freeboard variations on the breakwater overtopping catcher model.

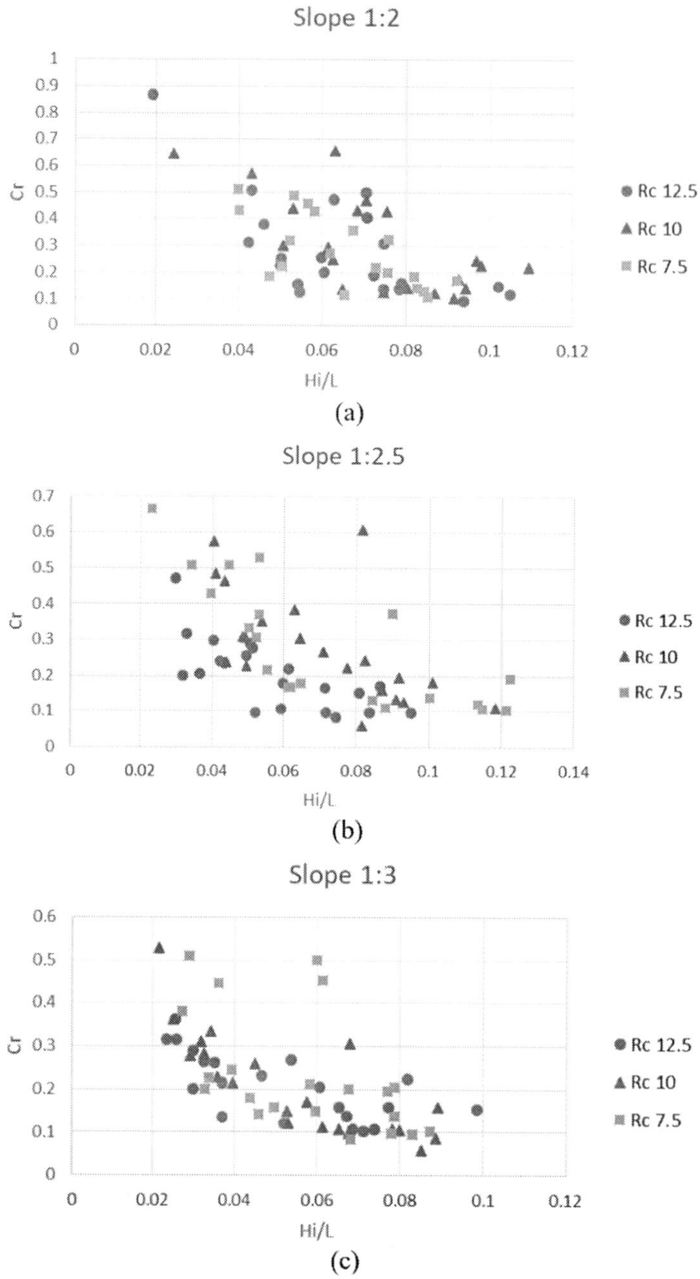

Figure 4. (a) Relationship between wave steepness and wave reflection coefficient for Rc 12.5 cm, Rc 10 cm, and Rc 7.5 cm for slope 1:2; (b) Relationship between wave steepness and wave reflection coefficient for Rc 12.5 cm, Rc 10 cm, dan Rc 7.5 cm for slope 1:2.5; (c) Relationship between wave steepness and wave reflection coefficient for Rc 12.5 cm, Rc 10 cm, dan Rc 7.5 cm for slope 1:3

From the chart in Figure a (slope 1 : 2), it appears that the wave steepness is in the range of values of 0.02 to 0.12 and the resulting reflection coefficient is in the range of values of 0.1 to 0.9. From the chart in Figure b (slope 1 :2.5), it appears that the wave steepness is in the range of values of 0.02 to 0.12 and the resulting reflection coefficient is in the range of values of 0.1 to 0.7. From the chart in

Figure c (slope 1 : 3), it appears that the wave steepness is in the range of values of 0.02 to 0.1 and the resulting reflection coefficient is in the range of values of 0.1 to 0.6.

4. Conclusions

A few conclusions can be drawn from figure 5 (a, b, c). Firstly, in all freeboard heights condition shown that large wave steepness value, results in a small wave reflection coefficient value, or it can be said that, wave reflection coefficient decrease with the increase of wave steepness. Secondly, it appears that the steeper the slope of the model, the larger the wave reflection coefficient produced. It can be seen that slope 1 : 2 produces the highest wave reflection coefficient values.

References

[1] M A Mustapa, O Y K 2017 Wave energy device and breakwater integration: a review *Renewable and Sustainable Energy Reviews* 77 43-58.
[2] A I D Puspita, M S Pallu, M A Thaha, F Maricar 2018 *International Seminar of Infrastructure Development* (Manado).
[3] Triatmodjo B 1999 *Teknik Pantai* (Yogyakarta: Beta Offset).
[4] http://www.marine.tmd.go.th/marinemet_html/lect18.html 2019 August 2 *Motion in the Sea -- Waves.*
[5] The Eurotop Team 2018 *EuroTop Wave Overtopping of Sea Defences and Related Structures Assessment Manual* (Hamburg: Boyens Offset).

IOP Conf. Series: Materials Science and Engineering **875** (2020) 012027 doi:10.1088/1757-899X/875/1/012027

An Estimation of Origin-Destination Matrices for a Public Transport Network in Makassar using Macrosimulation Visum

M I Ramli[1]*, D Runtulalo[1], H Yatmar[1], and A Mangessi[1]

[1]Department of Civil Engineering, Engineering Faculty, Hasanuddin University, Makassar, Indonesia

*Email: isran2609@gmail.com

Abstract. The origin-destination (O-D) matrix is fundamental data for transportation planning. It is a primary process for modelling approaches the public transport. Generally, the observation survey of O-D volumes takes much cost, time, and effort. Therefore, establishing an inexpensive and simple observation method is expected to be able to carry out the continuous observation of passengers O-D volumes. Defined that many problems in transport planning and management tasks require an origin-destination (O-D) matrix to represent the travels pattern. The need for low-cost methods to estimate current and future O-D matrix is even more valuable in developing countries because of the rapid changes in population, economic activity, and land use. In this study, the O-D matrix use for modelling trip distribution of passenger in the current and future situation and plotting the model with Visum as a macro simulation program. This study focuses on the O-D volumes on a line of local public transport service. The on-board observation is carrying out, and the frequency of service is observed. Then, the passengers O-D matrix is estimated. These results must be useful to improve the operation of the service of public transport.

1. Introduction

Public transportation plays an important role in urban transportation system. It has a potential to mitigate traffic congestion by reducing private car users. Furthermore, it is necessary to secure urban mobility for citizens in transportation, such as employee, students and others. An effective operation is expected to be carried out based on a fine management. It can be done on condition that the information of passengers demand can be obtained, because it must be fitted to the demand of passengers. So, the effective management of public transport requires the current demand. On the other hand, the demand of passengers of public transportation is changed in the lapse of time. Therefore, not only obtaining the current demand but also must be required to keep the operation be an effective one. Generally, observation survey of O-D volumes take much cost, time and effort. Therefore, establishing an inexpensive and simple observation method is expected to be able to carrying out the continuous observation of passengers O-D volumes.

A high-quality Origin–Destination (OD) matrix is a fundamental prerequisite for any serious transport system analysis. However, it is not always easy to obtain it because OD surveys are expensive and difficult to implement. This is particularly relevant in large cities with congested networks, where detailed zonification and time disaggregation require large sample sizes and complicated survey methods [1].

Furthermore, the researcher explain that models of transport demand have been used for many years to synthesize O-D matrix in study areas. A typical example is the gravity model that had a functional form, plus the appropriate values for the parameter involved, is employed to produce acceptable matrix representing trip making behavior for many trip purpose and time periods. In this study, using gravity model based models for estimating ODM [2].

This study focuses on the O-D volumes on a line of local transport service. Local transport is one of a public transportation similar as a bus service called *local transport*. Its vehicle size is small and the capacity of the vehicle is around 10 people. It is served on pre-determined route and passengers can get on and off at any place on the line. Instead of using a general O-D observation, passengers O-D estimation method using simple onboard observation is established. In the onboard observation, only the number of getting on and off passengers at every stopping point. These data would recorded in every stopping point and tracking the location by GPS. The counted numbers are aggregated by unit of pre-determined road sections ("zones" thereafter). Then, zone to zone passengers O-D volumes are estimated by estimation method of gravity model and plotting the zone of catching area of passenger with *Visum* program. Finally, the whole passengers O-D volumes are estimated by expanding the obtained O-D considering the sampling rate of onboard observation.

The proposed estimation method is applied to a local transport line served in Makassar, Indonesia. The onboard observation is carrying out and the frequency of local transport service is observed. Then, the passengers O-D matrix are estimated. These results must be useful to improve the operation of local transport service.

2. The study Methods

The study implement of priority policy to develop public transport in many countries, the public transport has become the major way for the passenger transport in city. Especially about the location of the origin-destination, the influence of the bus station on the urban transport is very important, which has become the bottleneck section in urban transport [3].

On the other hand, for the developing countries, the local transports that doesn't have bus station are existed. These situation increase the likelihood of the bottleneck, the public transport can stop in everywhere and time to boarding and alighting the passengers. Thus, better understanding of the origin-destination of public transport based on demand. This study conducted with observation survey by boarding and alighting (on-off) passengers in Makassar. The data of passengers took in holiday and workday on totally 60 samples of public transport through B1 line (Hasanuddin University-Cenderawasih). The estimation of O-D matrix would be established using the Gravity model with the production-attraction-constraint (PACGR). Generally, observation survey of O-D volumes takes much cost, time and effort. In this study, the passengers of O-D volumes carrying out by a simple observation. The simple observation is inexpensive in terms of man power, low cost, organizational and management requirements, in which it can be collected without disrupting travelers. In this study, proposed an inexpensive and simple observation method is expected to be able to carrying out the O-D passengers volumes.

The OD research defined that the synthesis model famous and commonly is gravity (GR), because it simple and easy to understand [4, 5]. This method using the gravity concept whereas the production and attractive trips have the relationship to origin zones, such us population and the value of O-D matrix relate to accessibility (constraint) as a distance, time, also cost function [6].

Eventually, for transportation purpose, the gravity (GR) stated as:

$$T_{ij} \approx O_i \cdot D_j \cdot f_{ij} \tag{1}$$

where:

Tij = the trips between origin (i) to destination (d)
Oi and Dj = quadratic distance between two zones
f_{ij} = deterrence function

Oi and *Dj* stated the total trips origin from zone *i* and ends in the zone *j*, f_{ij} here is the deterrence function. Therefore, the sum of O-D matrix cell based on 'row' produce the total trips from each zone, whereas the sum based on a column produces total trips towards each zone [7]. The new equation become as follows:

$$T_{ij} \approx A_i \cdot O_i \cdot B_j \cdot D_j \cdot f_{ij} \quad (2)$$

This equation *Ai* and *Bj* are getting by iteration and with simply checked for *Tij* in the equation (1) already fulfill the constrained, which then the value has to use again to count the *Ai*. This process had been iterating until the value of *Ai* and *Bj* generating a certain value (converge). Some sample probably has different iteration depend on the value of *Oi* and *Dj* [8]. Then finally converge value would use for counting each value of O-D and established the matrix. Generally define the value of *Ai* and *Bj* will be calculated iteratively. This calculation can be started by assuming the value of *Ai* or *Bj*. In order to starting the iteration for the calculation the value of *Ai* will be assumed as one for all zones (*Ai* =1). The first value of *Ai* could be assumed any numbers bigger than zero. This assumption value just affect to the number of iteration to achieve the converge value. The bigger difference between the first and last number make more iteration required to achieve the converge value. The same way also applied in case the iteration started by *Bj* value. Bellows is an example to get the converge value of *Ai* and *Bj* on morning sample for more specific value in each cell matrix. The first step is determining the value of Ai as one in the first iteration. Then, this value will be used to calculate the value of B_j.

The O-D matrix is required for directional traffic assignment. In this matrix, the rows and columns represent the origin and destination zone respectively (Table 1). The converge value with getting *Ai* and *Bj* for every i and j, also for deterrence function, then all of matrix cell of sample one as the example bellows.

$$t_{11} = A_1 \cdot O_1 \cdot B_1 \cdot D_1 \cdot C_{ij} - \alpha$$

$$\cdot$$

$$\cdot$$

$$t_{n-n}$$

In the on-off data, total passengers from origin and destination input is create on the following table

Table 1. The trips from origin and destination zone

Zone		Destination							Sum
		1	...	...	j	...	...	n	O_i
	1	t_{1-1}	...	...	t_{1-j}	...	...	t_{1-n}	O_1
	:	:			:				
	:	:			:				
Origin	i	t_{i-1}	...	...	t_{i-j}	...	...	t_{i-1}	O_i
	:	:			:				
	:	:			:				
	n	t_{n-1}	...	...	t_{n-1}	...	...	t_{n-n}	O_n
Sum D_j		D_1	...	...	D_j	...	...	D_n	T

3. Result and Discussion

3.1. The Onboard Observation Data

The onboard observation bring out the public transport travel time and the number of passengers of pete-pete sampling. Every sample has departure and arrived time when started and finished one destination A – B and B – A. The pete-pete sample are classified in interval times. For the total travel time and sample of pete-pete are shown in table bellows.

Table 2. Travel time of samples

Workday	Line			
	A-B		B-A	
	Departure	Arrived	Departure	Arrived
Morning	6:30	7:41	7:48	8:34
	6:35	7:29	7:29	8:29
	6:44	7:40	7:40	8:35
	6:30	7:38	7:43	8:34
	6:31	7:26	7:34	8:26
	6:32	7:26	7:28	8:22
	8:40	9:27	7:54	10:42
	8:42	9:42	9:42	10:42
	8:36	9:43	9:45	10:42
	8:44	10:17	10:18	11:05
	8:41	9:36	9:37	10:35
	8:28	9:32	9:33	10:27
Noon	11:42	12:46	12:48	13:47
	11:49	12:49	12:49	13:51
	11:44	13:01	13:03	14:27
	11:08	12:07	12:10	13:01
	11:48	12:51	12:53	13:51
	11:28	12:32	12:33	13:27
	14:40	15:25	15:26	17:05
	14:10	15:15	15:15	16:20
	14:34	15:54	15:54	17:04
	14:12	15:26	15:27	16:20
	14:06	15:08	15:09	16:09
	14:09	15:18	15:20	16:25
Afrer-noon	17:18	18:55	18:57	20:05
	17:24	18:39	18:40	19:59
	17:05	18:25	18:26	19:38
	17:01	18:00	18:03	19:36
	16:19	18:00	18:00	19:08
	17:27	19:01	19:01	20:17

Holiday	Line			
	A-B		B-A	
	Departure	Arrived	Departure	Arrived
Morning	6:46	7:34	7:35	8:25
	6:45	7:33	7:34	8:25
	6:36	7:30	7:34	8:22
	6:50	7:31	7:35	8:25
	6:36	7:15	7:16	8:12
	6:54	8:24	8:28	9:06
	8:39	9:11	9:18	10:06
	8:28	9:21	9:21	10:14
	8:22	8:58	8:58	10:51
	8:42	9:25	9:25	9:58
	8:36	9:34	9:34	10:36
	9:08	10:20	10:20	11:10
Noon	11:35	12:12	12:18	12:55
	11:26	12:19	12:19	13:13
	10:52	11:42	11:44	12:30
	11:08	11:49	11:56	12:37
	10:37	11:24	11:31	12:17
	11:12	12:06	12:08	12:54
	13:21	14:13	14:21	14:54
	14:20	15:10	15:10	16:09
	14:10	14:47	14:47	15:36
	14:17	15:07	15:13	16:06
	14:17	15:06	15:03	15:50
	12:56	15:03	15:03	15:58
After-noon	16:17	17:09	17:13	18:01
	16:13	17:04	17:05	18:01
	17:10	17:47	17:47	18:36
	16:10	16:49	16:53	18:33
	16:59	17:50	17:50	18:47
	16:01	17:42	17:43	18:42

The tables are the data result from passenger record and marking by GPS for totally samples. The tables record the passengers and stopping point are analyzed to determine the position of each passengers of sampling in line. These created by Visum software to read the GPS track and waypoint. The total passengers separately more details in zones. Established the number of passengers based on marking stopping point based on zone is illustrated in figures below.

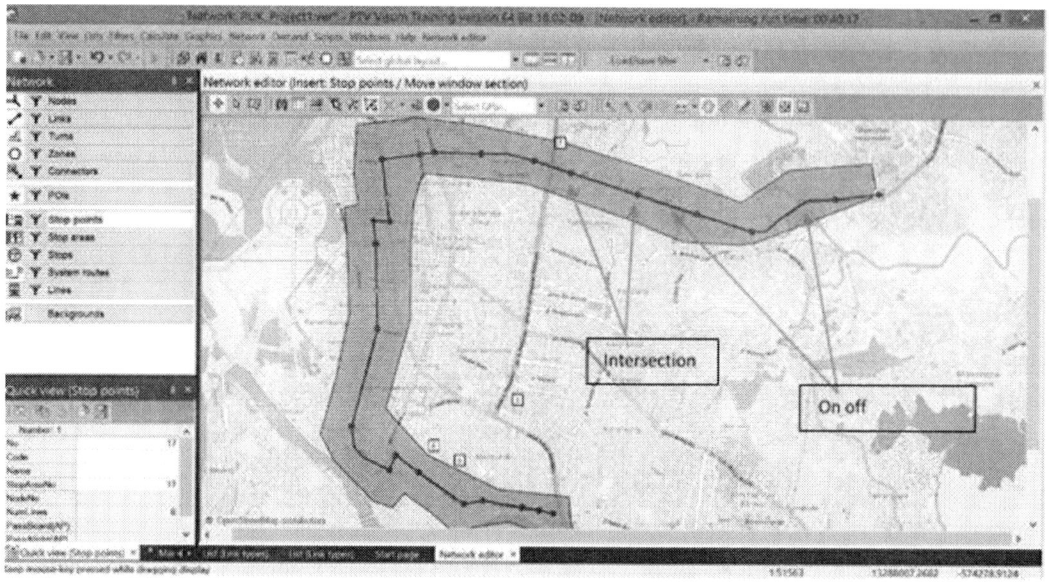

Figure 1. The figure of catchment area of passengers' by Visum

3.2. The Traffic Counts

According to counts data, the total operating pete-pete of B1 line are obtained. In the table bellows show the total vehicles by time interval.

Table 3. Pete-pete counts

Time	Vehicles Zone (A-B)	Vehicles Zone (B-A)
06:00 – 07:00	45	22
07:01 – 08:00	45	68
08:01 – 09:00	58	40
09:01 – 10:00	51	45
10:01 – 11:00	42	44
11:01 – 12:00	51	35
12:01 – 13:00	39	38
13:01 – 14:00	39	41
14:01 – 15:00	36	56
15:01 – 16:00	48	42
16:01 – 17:00	44	37
17:01 – 18:00	47	45

a. workday

Time	Vehicles Zone (A-B)	Vehicles Zone (B-A)
06:00 – 07:00	13	16
07:01 – 08:00	42	34
08:01 – 09:00	41	35
09:01 – 10:00	39	40
10:01 – 11:00	44	37
11:01 – 12:00	31	37
12:01 – 13:00	37	33
13:01 – 14:00	33	32
13:01 – 14:00	38	31
15:01 – 16:00	32	28
16:01 – 17:00	39	28
17:01 – 18:00	38	45

b. holiday

The pete-pete passing through observation area has counted. For total unit that registered administratively of B1 Line (sample observed) is 146 units.

3.3. The O-D Passengers Volume

The O-D matrix have been established for all sample in morning, noon, and afternoon. Onboard data observation produces the cell value for *Oi* and *Dj*. Then, using gravity model method to represent the O-D matrix. The total value in cell not only give information about the O-D volume of passenger but also the total passengers each zone for totally sample.The result of estimation shows the highest O-D volumes as the pattern according to zone for totally sample in interval time, morning, noon, and afternoon that shown on the following figures. The total passenger each zone would be one of figure to evaluate the service of pete-pete and realize the fine management of public transport in Makassar. The histogram bellows explain information for total passenger on zone of B1 line.

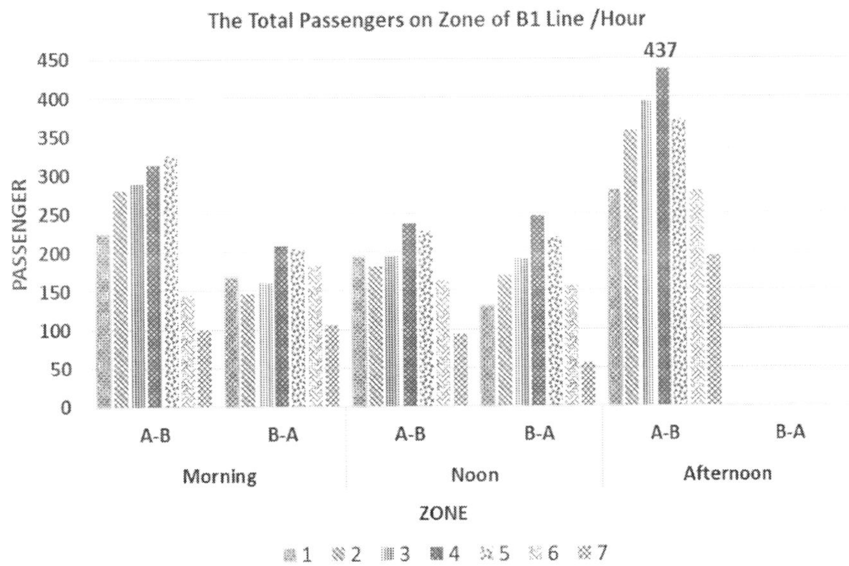

Figure 2. The total passenger on zone of B1 line (A-B) and (B-A) on workday

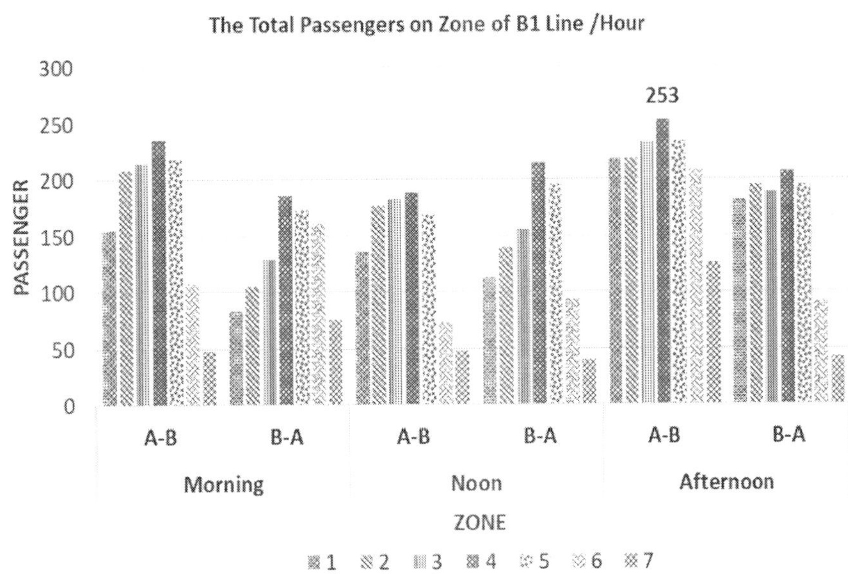

Figure 3. The total passenger on zone of B1 line (A-B) and (B-A) on holiday

4. Conclusion

In this study, the O-D matrix of Makassar has estimated in a couple of days represented in workday and holiday at morning, noon, afternoon. As the result we can see the O-D passenger's volumes of sampling tend to the OD general of B1 line. Both of OD passengers' volumes sampling and general have similarly trend to values as follows:

- OD Passengers Volume Pattern (workday): in the morning, the OD pattern on zone 1-5 and return direction on zone 7-1, in the noon, the OD pattern on zone 1-7 and return direction on zone 6-1, in the afternoon, the OD pattern on zone 1-7. The highest passengers is 437 / hour. The demand of pete-pete is around 45 units.
- The OD Passengers Volume Pattern (holiday): in the morning, the O-D pattern on zone 1-5 and return direction on zone 7-1, in the noon, the O-D pattern on zone 1-5 and return direction on zone 5-1, in the afternoon, the OD pattern on zone 1-7 and return direction on zone 5-1. The highest passengers is 253 / hour. The demand of pete-pete is around 26 units.

Compare to general O-D survey that need more time, effort, and resource such us observer was identify an origin and destination location of each passenger. The simple observation in this study just use general equipment such GPS with marking the stopping point of passengers and determined the location in zones compare with pete-pete counting was give same result about the total O-D of public transportation and total passengers in observation line..

Acknowledgments
Authors wishing to acknowledge assistance or encouragement from Faculty of Engineering Universitas Hasanuddin especially for Laboratory Based Education (LBE) Research Grant for funding and also for Civil Engineering Department, Faculty of Engineering that support and contribute in the research.

References
[1] Marcela A.Munizagaa, Carolina Palmab 2012 Estimation of a disaggregate multimodal public transport Origin–Destination matrix from passive smartcard data from Santiago, Chile. Transportation Research Part C: Emerging Technologies Volume 24, October 2012, Pages 9-18.
[2] HU Junhui et al., 2011. Urban Mixed Traffic Flow Considering the Influence by Origin-Destination of Public Transportation. Journal of Transportation Systems Engineering and Information Technology Volume 11.
[3] Tamin O.Z., Willumsen L.G., 2011. Transport demand model estimation from traffic counts. Transportation 16: 3 – 26. Kluwer Academic Publishers. Netherland.
[4] Tamin O.Z., Suyuti R., 2007. The Study of Developing Accuracy of O-D Matrix from Traffic Count Data in Equilibrium. Journal Science and Technology of ITB. Vol. 39 A, No. 1 & 2, 2007, 23-39. Bandung.
[5] Tamin, Ofyar Z., 2005. Integrated Public and Road Transport Network System for Bandung Metropolitan Area (Indonesia). Proceeding of the Eastern Asia Society for Transportation Studies, Vol. 5 pp 1281-1300.
[6] Ali N., Ramli I., 2007. Optimization of Public Transport (Mini Bus) in Makassar with Break Event Point. Simposium X FSTPT, Tarumanegara University. Jakarta.
[7] M. D. S Iqbal et al., 2014. Development of origin-destination matrix using mobile phone call Data. Transportation Research Part C 40 (2014) 63-74.
[8] O. Feldman et al., 2012. Alternative Gravity Modelling Approaches for Trip Matrix Synthesis. Association for European Transport and Contributors.

IOP Conf. Series: Materials Science and Engineering **875** (2020) 012028 doi:10.1088/1757-899X/875/1/012028

The Airport Development due to the Demand of Passengers (Case: Mutiara Palu, Indonesia)

S A Adisasmitha[1]*, H Yatmar[1], S Rauf[1], and M Hustim[1]

[1] Department of Civil Engineering, Engineering Faculty, Hasanuddin University, Makassar, Indonesia

*E-mail: adjiadisasmitha@gmail.com

Abstract. The influence of service quality has limited attention to the airport study. The services are according to the demand that directly influences the development of the airport. This study investigates the development of the airport. The increasing of passengers used aircraft for the mode choice of transportation made the polemic. The quality of service with the growing demand of passengers could satisfaction the passenger to use the airport facilities and made a flight. The rising trend of passengers has to cover the quality of service performance with the expansion of the airport. The demand analysis is conducted to predict and optimize the services of the airport facilities According to the population and daily flight information. Demand shows the forecast of passengers use the semi-average method, least square method, and the quadratic trend method. The result confirms that the trend of passengers is attracted to choose a cheap flight as a priority than the safety factor. The population affected by the demand for a flight. Otherwise, the growth of domestic products is related to increasing flight demand. Finally, the result shows the factors that affected the demand of passengers to travel with aircraft influenced the expansion in developing the airport facilities.

1. Introduction

The airport is one of the transport infrastructures is very important for Indonesia as an archipelago. The growing demand for flight business has an important role as one of the transportation services. In Indonesia, the function of the airport is a strategical function cause the travel time will be short, fast, and effective, especially to the long travel distance and between the other island.

The development of the flight in transportation services is aiming to fulfill the demand or air transport services, which is increasing by the year. The fast transportation, safety, and comfort become an important thing to support economic growth so that airport as an indicator of developing city.

Palu, as the central city of Central Sulawesi Province, is located in the valley plain and bay. The area consists of five elements like mountain, valley, river, bay, and sea. The total area of Palu reaches 395,06 kilometers with a total population that according to the Statistical of Palu Government in 2015, around 367.342. Palu was also known as a transit city for water transportation that comes from the outer of Celebes Island [1].

The strategical location and support by the facilities of transportation: port, airport, and road make Palu as the busy city. The condition is making the other problems, especially for airport transportation services, while the quality service of the airport is the main part of air services.

Palu Mutiara Airport is located in the south of the Palu region. The airport has become the main gate of Palu City. According to the statistical of government, the total activity of airport records every year, especially for 2015 to 2016, the increase of passengers are more than 100.000 people [2]. The trend of increasing passengers is not followed by the facilities and the quality, so that makes a new polemic. Further, the development of Palu Airport has become an international airport that has a special constraint in order to perform a good quality service and comfortable for totally customers and users of the airport. The masterplan of the airport is established gradually, continuously, comprehensives, and well-integrated.

According to the tren of passengers of the Mutiara Palu airport, the condition certainly concerns to the quality of service and affect the analysis of the developing airport is important. The demand analysis of passengers of the airport according to the income and the population as the initial information that supports the planning of the improvement and developing the quality of services. The services not only from the airport side but also comes from the assessments of the airline that operated in Mutiara airport as the preference for continuous improvement.

2. Study Literature

2.1. Air Transport

Transportation has an important role in developing countries. The transport facilities are used to growth finance, infrastructure, movement, others. Totally transportation is affecting social life. The transportation has provided effective services to open the isolation area, service the village, support to the rural area.

The transportation itself is an activity to transport people, stuff, and others that come from origin to destination. This is the similarity with the function of an airport to transporter people or cargo from origin to destination with a plane. The plane as air transportation has a characteristic, fast speed, and reachable to place that services by the other transportation mode [3].

2.2. Transportation Demand

Demand transportation services are determined by the transport of goods and passengers carried. Demand by consumers is very closely related to the consideration of service (service) arising from the use of transport services. The quality of transport services has to be provided effectively and efficiently. Transportation services are continuous/available, secure, safe, adequacy, frequency, regular, responsible, acceptable cost, and comfortable [4].

Transport services need to be highly qualitative and having different characteristics as a function of time, travel destinations, frequency, type of cargo transported, and others. Transportation services that do not fit the needs of the movement led to the transportation system useless [5]. Economically, an inefficient transportation system or transportation problems is a big waste [6].

Six factors that affect demand for air traffic is on the following description. First, the economic factors with elements that gross domestic income and exchange rates. Gross domestic income is attributed to how income levels, while the exchange rate affects the amount of travel. Second, the demographic factors such as population or the number of inhabitants of a region and urbanization. Third, the supply factors that are indicators of aviation service providers. As for factors including supply are the cost of transport (passengers per km), size of the aircraft, fuel costs, technology (navigation, communication and so on), management (network airlines, new routes, and so on), capacity constraints, and the ticket. Fourth, economic regulation, such as airline and airport privatization, free trade, open skies, foreign ownership, and tax regulation. Fifth, the regulatory environment. An example is the imposition of a carbon tax, which provides an additional impact on the cost of the trip. Sixth, cargo transportation, which related to the logistics needs of air transport [7].

2.3. The influence of Income and Demand

The per capita income is the amount of the average income of the population in a country. Per capita income show that construction has been implemented by the government have made it, how big this success, and as a result, incurred by the increase [8]. The GDP (Gross Domestic Product) is defined as the total value added generated by all business units within a region or the total value of final goods and services produced by all economic units in a region. The value of GDP divided by the population in the region resulted in income per capita [9].

The analysis of the development of airports based on passenger demand can use the value of the GDP of a region and forecasting the population growth rate. There are two general approaches used in forecasting, as follows [10]:

- Quantitative forecasting, using various mathematical models that use historical data and or causal variables to forecast demand.
- The qualitative forecasting or subjective forecasting, utilizing important factors such as intuition, personal experience, and the value system of decision-making.

2.4. Mutiara Airport of Palu

Recently, the Palu city becomes the special economic zones (SEZ) in eastern Indonesia. The government of Palu already provides all instruments to support the special status, and for the total area of 1,520 hectares in the district of North Palu is prepared, which includes the Village Pantoloan, Baiya, and Bambara. The land area of 1,520 hectares will be divided into an industrial area of 700 hectares, residential area (500 hectares), the area of education and research (100 hectares), commercial area (100 hectares), sports area (50 hectares), warehouse area (50 hectares), plantation and garden area (20 hectares) [11].

Mutiara Palu Airport was located at an altitude of 86 meters (282 ft) above sea level, has two paved runways and concrete surfaces designation 15R / 33L with 2,500 x 45 meters and 15L / 33R with 3,450 x 60 meters. The airport has an extensive terminal building that covers 15,196 square meters [12]. The operation of the airport is facilitated by various support systems. The airline that services the flight are Batik Air, Citilink, Garuda Indonesia, Kalstar Aviation, Sriwijaya Air, Susi Air, Wings Air, and Xpress Air. Central Sulawesi government is trying to develop Mutiara airport into an international airport that given perspective with the high-interest air transport according to the population growth demand [13]. Furthermore, the analysis of Palu Mutiara airport development is essential as an input and as a preference.

3. Research methodology

The method used in this research is to interview the respondents who have a passenger that exists in the airport terminal building that would travel using the Mutiara Palu airport. The analysis of the study will use a forecasting method and the trend with linear regression, the average value, and moderate value to forecast the passenger demand with the variable value of GDP and population growth.

3.1. Location & Framework

The location of the research is at the airport terminal building. Previously, the study conducted the flow of research, including the preparation, the literature study, the method, the data analysis, and the following flow (Figure.1).

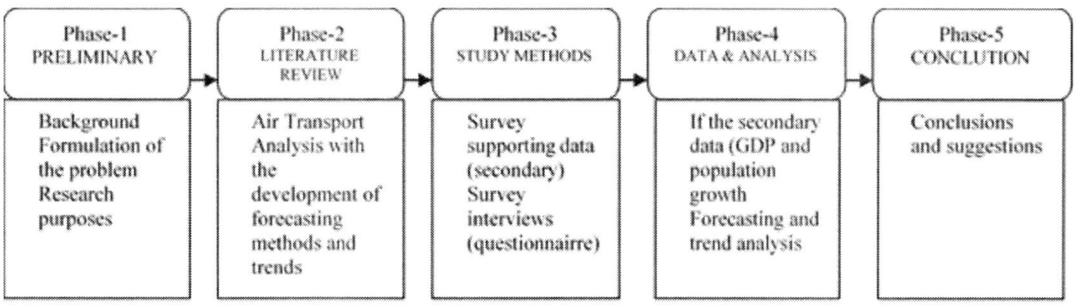

Figure 1. Flow framework research

3.2. The survey method

The interview was conducted with respondents passengers who were in the waiting room, arrival, and the terminal building. This interview was conducted by asking the respondents available time to interviewed about their opinion du to the airline services. Personal information has also recorded the preference through airfare, destination, time travel, and the quality of services. Other preferences as secondary data are received from official government websites such as location map, geographical conditions, the total number of operational airlines, the total passengers, and related information as like as domestic income and growing population.

4. Result and Discussion

The total respondents are 1006 passengers who have travel and use the Mutiara Palu airport services. The preferences of passengers are applied to knows the potential demand and the relation of gross domestic income and population due to the demand of airlines.

4.1. The Preferences of airlines selection

The preferences of passengers' airlines through the services on airlines that fly from origin and destination to Palu city. The result is shown in figure 2. The total respondents' preferences according to the services provided by the airlines such as services quality, safety, security, facility, timeliness, airfare cost, others. The respondent's willingness priority services to improve, and each airline is various. As an example, Transnusa is focused on airfare costs and timeliness, Lion air is on the airfare cost and safety, Batik is on the secure and airfare cost, Sriwijaya is on the airfare cost and facility, Garuda is on the security and safety, and for the Wings is on the airfare cost. The majority respondent's willingness priority is tending to airfare cost, followed by security and safety.

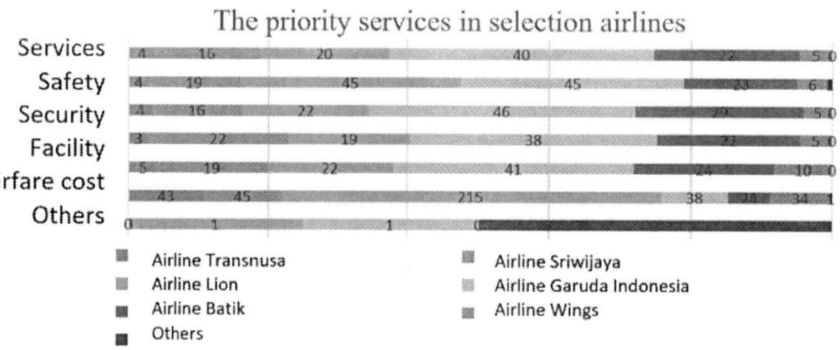

Figure 2. Respondents priority services in select the airlines

4.2. Respondents comparison priority in selected airlines

The whole airline has been assessed by the respondents. The total airlines are evaluated according to the providing services. The following figure 3 shows the percentage of respondent's priority services for the total airline's services.

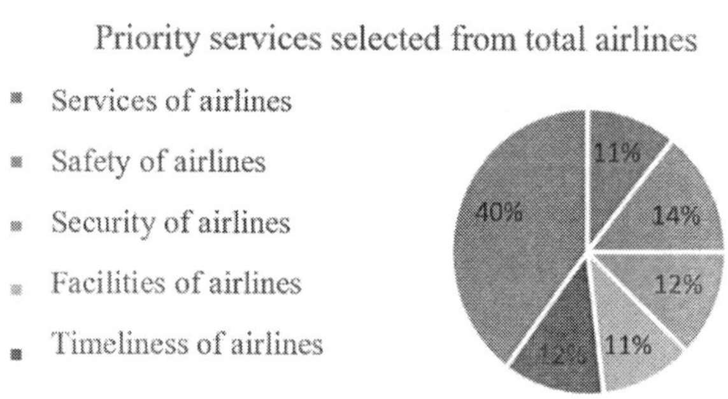

Figure 3. Respondents comparison chart priority selection airlines

Based on this, the comparison of priorities of respondents in the election of the airline in figure 3 shows that as much as 107 respondents (11%) priority the services, 143 respondents (14%) select the safety level, 122 respondents (12%) select the security level, 109 respondents (11%) choose the facility, 121 respondents (12%) selected the departure and arrival time, and the majority respondents stated that the airfare cost is suitable to the finance condition of respondents.

4.3. Regression result

The influence of the variable of services to the gross domestic regional and the population is calculated using linear regression. The multiple linear regression results are shown on the following table.1.

Table 1. Multiple linear regression results

Variables	Coefficient	Standard Error	t-statistics	Probability
C	3558534.355	1006452.493	3.535710208	0.071516923
X1	-9.349842469	2.878541705	-3.248117771	0.083135306
X2	0.054201215	0.009090084	5.962674744	0.026992994
R-Square	0.952144713			
Adjusted R Square	0.904289427			
Standard Error	51244.89701			
F-statistic	19.89633284			
Prob (F-Statistics)	0.047855287			

Note:
C : constants / intercept
X1 : Total population
X2 : GDP

Based on Table 1, the regression model as follows:

$$Y = (-9.349842469X1) + 0.054201215X2 + 3558534.355 \tag{1}$$

4.4. Significant of F-Test

The result of F-test shows the values for 5.79 (α = 5%) and F = (K; n - k = F (2,5) = 5, 79. According F-statistics/ F-count equal to 964.153. Compare the-test and F-table, then the F-count is obtained, which is greater than F-table (19.89633284 > 5.79). Therefore, the independent variables simultaneously influence the dependent variable.

4.5. Significant of T-Test

The t-test usually shows the influence of the independent variables partially that explains the variations in the dependent variable. The influence of the independent variables on the dependent variable can be seen in Table 2 below.

Table 2. Partial regression results (T-test)

	Coefficients	Standard Error	t-Stat	P-value
Intercept	3558534.355	1006452.493	3.535710208	0.071516923
Population (X1)	-9.349842469	2.878541705	-3.248117771	0.083135306
GDP (X2)	0.054201215	0.009090084	5.962674744	0.026992994

According to the Table 2 result shows that several independent variables are significantly assuming the value of t-count > t-table. The t-table obtained with 2.77645 (α = 5%) and t (α / 2; n - k - 1) = t

(0,025; 4) = 2.77645. Compare the value of the t-test and t-table, the t count of population variables (X1) is smaller than t-table (-3.248117771 > 2.77645) and the value t count the GDP variable (X2) by (5.962674744 > 2.77645). Therefore, it shows that the independent variable (X2) partially affects the dependent variable.

4.6. The forecast of Airlines according to GDP value

The forecast of airlines use the growth rate method is shown in figure 4. The equations are used the growth rate with 13%, otherwise, while using linear regression is lightly straight with X value in equations that also as a total GDP in forecast year with the yellow line. The blue line is a moderate value compare to the linear regression and trend data.

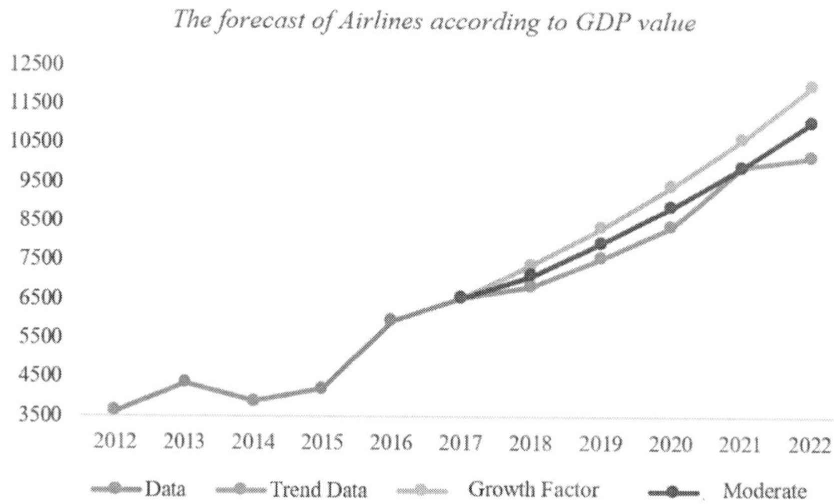

Figure 4. The airlines' forecast

5. Conclusions

According to the results and discussion before, it can be concluded:

a. The result shows that 40% of respondents to prioritize prices in the offer of the airline, 14% prioritize the safety level of the airline, 12% prioritize security levels, 12% prioritize the punctuality of departure and arrival, 11% prioritize services that are provided, and 11% prioritize facilities given airline.
b. The population has not significantly influence, and there is no linear relationship between the number of populations to the demand of passengers at the airport Mutiara according to the t-count value that smaller than t-table (-3.248117771 < 2, 77 645).
c. The GDP has significant influence and has a linear relationship between the GDP of the passenger demand on the airport, with the t-test calculations more than t-table (5962674744> 2.77645). The influence the GDP on passenger demand comes from the regression coefficient valued with 0.054201215, which is means that the increase of the GDP will increase in passenger demand at the airport.

Reference

[1] SA Adisasmita. 2015. Perencanaan Transportasi Publik. Graha Ilmu: Yogyakarta.
[2] Badan Pusat Statistik (BPS). 2016. Palu Dalam Angka. BPS Palu 2016.
[3] SA Adisasmita. 2012. Level of Service Analysis and Airport Terminal Development. International Journal of Engineering & Technology Vol 12.
[4] MI Ramli, H Yatmar. 2020. Delay Distribution Estimation at A Signalized Intersection. IOP Conference Series: Earth and Environmental Science 419 (1), 012090.

[5] MI Ramli, H Yatmar. 2019. An Optimum Route Analysis of Fire Rescue According to Fire Station Location (Case Study: Residence Area in Makassar's Central Business District). IOP Conference Series: Earth and Environmental Science 235 (1), 012072.
[6] MI Ramli, H Yatmar. 2019. Study On Speeding Behaviour of Driver's Motor Vehicles to Traffic Accident Risk in national Road in Makassar City. Journal of Indonesia Road Safety 2 (3), 159-170.
[7] SA Adisasmita. 2013. Mega City and Mega Airport. Graha Ilmu: Yogyakarta.
[8] SA Adisasmita. 2014. Tatanan Bandar Udara Nasional. Graha Ilmu: Yogyakarta.
[9] SA Adisasmita. 2011. Transportation and Regional Planning. Graha Ilmu: Yogyakarta.
[10] SA Adisasmita. 2007. The Design of hasanuddin International Airport Based on Eco-Airport System. Graha Ilmu: Yogyakarta.
[11] SA Adisasmita. 2005. Air travel Demand Forecasting and Passenger Flow Distributions at Airport Terminal Building. University of Newcastle: Australia.
[12] SA Adisasmita, S Hamzah, MI Ramli, et all. 2017. Pengembangan Infrastruktur Bandar Udara Menuju Bandar Udara Masa Depan dengan Konsep Airport City. IPTEK Journal of Proceedings Series 3 (5).
[13] RU Latief, S Pallu, SA Adisasmita, SH Aly. 2014. Risk Response Preference on Public-Private Partnership (PPP) in Indonesia Airport Infrastructure Development. International Journal of Application on Innovation in Engineering & Management Vol 3 (8) 120-124.

IOP Conf. Series: Materials Science and Engineering **875** (2020) 012029 doi:10.1088/1757-899X/875/1/012029

Experimental Study of Natural Materials Utilization as a Stabilization Agent on Soft Soil

T Harianto[1], N Marfu'ah[1]*, I Rauf[1], and T B C Leatemia[1]

[1]Department of Civil Engineering, Faculty of Engineering, Hasanuddin University, Makassar, South Sulawesi, Indonesia

*Email: nurulmarfuah.nm@gmail.com

Abstract. Chemical soil stabilization is known as the most economical method compared to others. The development of stabilization materials has led to studies of the utilization of natural materials and waste materials to replace cement and lime which has become the most popular binder agent ever known. This research utilizes natural minerals such as overboulder, zeolite and Palm Oil Fuel Ash (POFA) as its binder agents. The amount of natural material varies at intervals 3% - 15% by the weight of soil. Additional of natural materials is done to increase bearing capacity as a subbase layer, which is based on FHWA standards CBR minimum is 20%. California Bearing Ratio (CBR) tests were carried out to see the effect of each material for increase bearing capacity of soft soil. Curing periode is varies at 7 days, 14 days and 28 days. The results shows, by adding natural minerals such as overboulder, zeolite, and POFA increasing the CBR value up to 5 times, 6 times and 5 times compared to untreated soil and close to the CBR value of cement stabilized soil. Referring to FHWA standards, we can conclude that natural materials can be used for stabilization materials on subbase layer.

1. Introduction

In the world of construction, soil is a very important materials in supporting durability of a construction building. Reference [1] states that soil is construction materials which is available directly in the field, so this will be very economical if it can be used directly as a subbase materials, specifically at the locations that difficult to accessing sirtu. However, the soil must have a high bearing capacity. In order to gain a high bearing capacity, stabilization is carried out. The most popular method for increasing the bearing capacity of soil is the chemical stabilization method. This method is usually done by adding pozzolanic materials like cement and lime [2]. But, utilization of cement are avoided because the production process are considered not environmentally friendly [3]. Therefore, finding alternative materials become the needs of sustainable developments. As technological advancement in material measurement, the development of chemical material in research of stabilization leads to the utilization of wastes and other natural materials such as the utilization of fly ash [4], waste of Buton asphalt [5], fiber waste [6], and utilization of bacteria [7].

2. Materials and Method

2.1. Materials

Materials in this research divided into two kinds: soil and the stabilization material. The stabilization material that used in this research is natural materials such as overboulder, zeolite and burning palm oil shells which is known as palm oil fuel ash (POFA). We will explain each materials as follows:

2.1.1. Soil

Soil samples used in this study was taken around the Engineering Faculty, Gowa Regency. For untreated soil, physical and mechanical characteristics were tested to determine the initial condition of the soils before addition of materials. Testing method of these samples refers to the ASTM laboratory testing standards (American Standards for Testing and Materials), while USCS (Unified Soil Classification System) and AASHTO (American Association of State Highway and Transportation Officials) were used as its standard classification system.

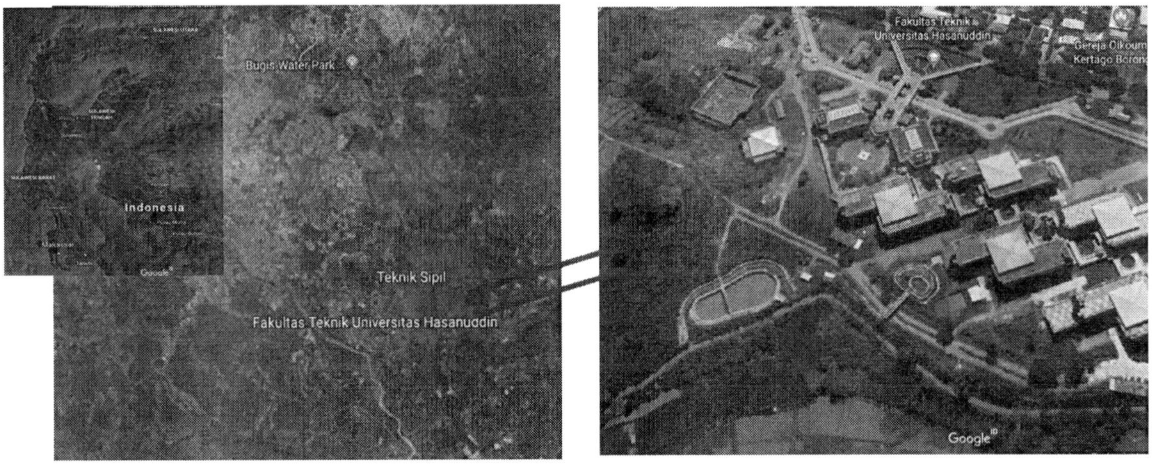

Figure 1. Soft soil location

2.1.2. Palm Oil Fuel Ash

This is often called as ashes produced from palm oil solid waste at temperatures around 800°C-1,000° C at a steam power plant in a palm oil mill. The palm oil industry produces solid waste such as fiber, shells and empty plants. The extraction process of 100 tons of fresh tan and fruit will produce 20 tons of shells, 7 tons of fiber, and tons of empty tons [8]. POFA can be used as pozzolan, which is a fine material containing silica and alumina which can react and form cement material. POFA contain high silicon dioxide and have the potential to be used as a cement alternative. POFA is a promising pozzolanic material and is widely available in all parts of the world .Utilization of the right POFA can reduce the use of cement and reduce the volume of waste so that it is very beneficial for environmental sustainability [9].

2.1.3. Overboulder

This is commonly known as overburden materials or buton asphalt cover. This material is rarely used. However, it is known that overboulder has a deposit almost 800 million tons in Buton Island. [10] stated that, according to X-Ray Diffraction (XRD) data mineral that contained in OB in the form: CaO of 79.64%, SiO_2 of 9.63%, so based on these results OB has the potential as a pozzolanic material that can be used as a stabilization material on soft soils.

2.1.4. Zeolite

Zeolite has several properties including: easy release of water due to heating, but also easy to bind water molecules in moist air. Because of its nature, zeolites are widely used as drying agents in industrial fields. But lately, zeolite known containing a lot of silica which potentially to used as pozzolanic material like overboulder and POFA. In this study, all the stabilization materials that used is the fraction that passed no. 200 sieve.

This is caused by the reactivation of stabilization material which is strongly influenced by the degree of fineness of the grains [11]. In other words, the higher the specific surface area of the stabilization material, the faster the reaction between soil and stabilization material.

2.2. Research Methodology

In Table 1 the amount of stabilization materials is presented. The percentage of material used are POFA 5% and 10% with a curing period of 7 days and 28 days, zeolite 3% with a curing period of 14 days and 28 days and overboulder 5%, 10% and 15% with a curing period of 7 days, 14 days and 28 days. Curing aims to maximize the performance that occurs between its stabilization material and soil.

Table 1. Mix design

No.	Specimens	Percentage (%)	Curing Time (days)
1	Untreated Soil	0	0
2	POFA	5 and 10	0, 7 and 28
3	Zeolite	3	0, 14 and 28
4	Overboulder	5, 10 and 15	0, 7, 14 and 28

The specimens making and testing for California Bearing Ratio is referred to ASTM standard D1883 - 16. Where the samples are made with dimensions of 15.2 cm x 17.8 cm and compacted with 56 layers of bedding. This sample was made using a standard pounder weighing 5 lbs. The CBR sample then cured with time varies from 7 days, 14 days and 28 days.

3. Results and Discussion

3.1. Soil Classification

Based on the laboratory results, (Unified Soil Classification System) standard clay material used in this study included in the classification of CH or inorganic clay and fertile soils with high plasticity. Based on the AASHTO (American Association of State Highway and Transportation Officials) standards, the soil used in this study is classified as A-7-6.

Table 2. Characteristics of Soil

No	Characteristics	Unit	Result
	Physical Properties		
1	Specific Gravity (Gs)	-	2.71
2	Initial Water Content (Wn)	%	12,17
3	Sieve Analysis		
4	a. Sand	%	31,60
	b. Silt	%	19,28
	c. Clay	%	49,12
5	Atterberg Limits		
	a. Liquid Limit (LL)	%	58,37
	b. Plastic Limit (PL)	%	29,08
	c. Plasticity Index (PI)	%	29,29
	Standard Proctor Test		20,16
6	a. Dry density (γd max)	gr/cm^3	1,41
	b. Optimum Water Content (wopt)	%	25,28
	Mechanical Properties		
	CBR Unsoaked	%	6,84

3.2. Effect of Mineral Additions on CBR Value

3.2.1. CBR Value Due to Additional of Palm Oil Fuel Ash

Figure 2 shows the effect of adding POFA on CBR value by curing period 7 days and 28 days.

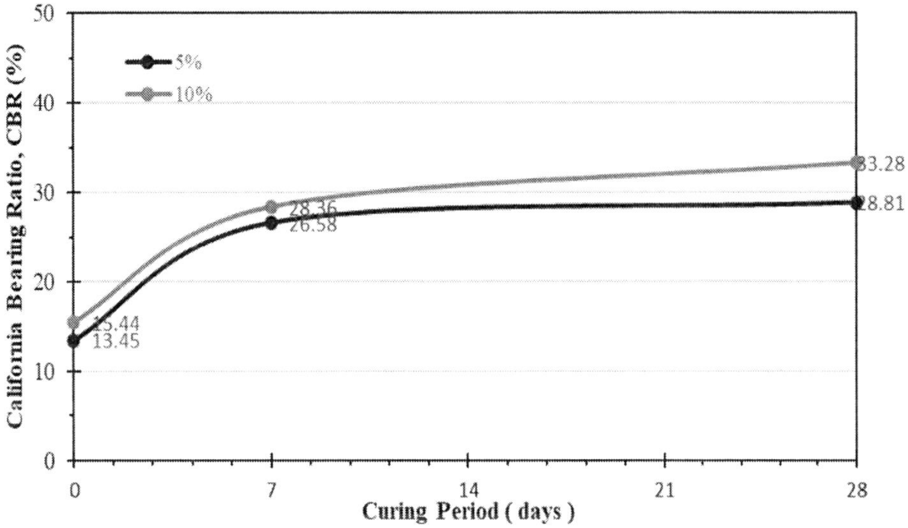

Figure 2. CBR value due to additional of POFA

Based on datas shown, additional of palm oil fuel ash increase CBR value. The peak was obtained at 28 days by increasing CBR value 4 times at 5% POFA and 5 times at 10% POFA comparing to untreated soil which is the CBR value is 6,84% . From the figure, we can conclude, additional of 10% POFA is better than 5% POFA.

3.2.2. CBR Value Due to Additional of Overboulder

Figure 3 shows the effect of Overboulder addition to CBR value in curing period 7 days, 14 days and 28 days [10]. Based on datas shown, additional of overboulder increase CBR value. The peak of CBR value was obtained in the 28 days curing period by increasing the CBR value each: 3 times greater at 5% addition, 4 times greater at 10% addition and 5 times greater at 15% addition compared to CBR value of untreated soils. From Figure 3, we can conclude, 15% overboulder is the most effective percentage.

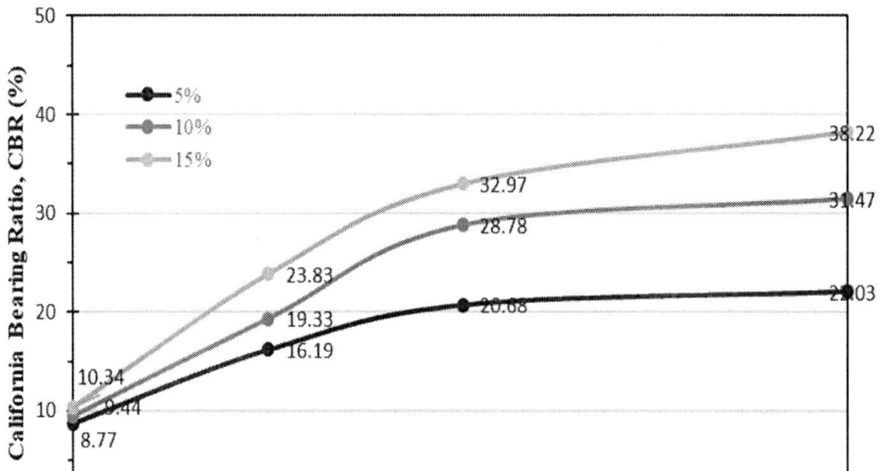

Figure 3. CBR Value Due to Additional of 5%,10% and 15% Overboulder

3.2.3. CBR Value Due to Additional of Zeolite

Based on the data attached to Figure 4, it can be seen that the CBR value increases with increasing curing period. The addition of zeolite by 3% can increase the CBR value. It can be seen, the peak increase in CBR value is obtained during the 28 day curing period with a value 6 times greater than the CBR value of untreated soil.

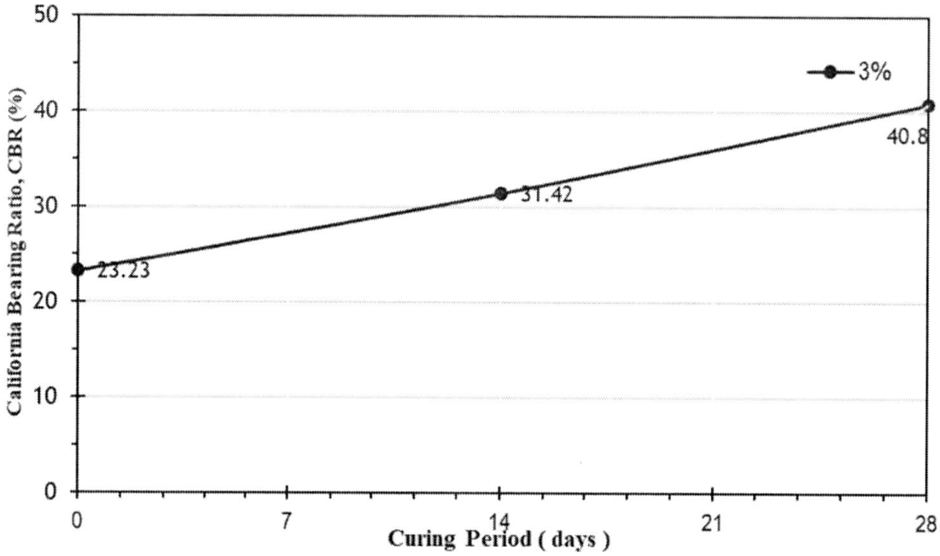

Figure 4. CBR Value Due to Additional of 3% Zeolite

3.3. Comparison of the Materials Effect on CBR Value

Below is shown a comparison of zeolite, POFA and Overboulder datas on its optimum content in this research (3% zeolite, 20% POFA and 15% Overboulder). This comparison aims to determine the effectiveness of all the stabilization material for increasing CBR Value.

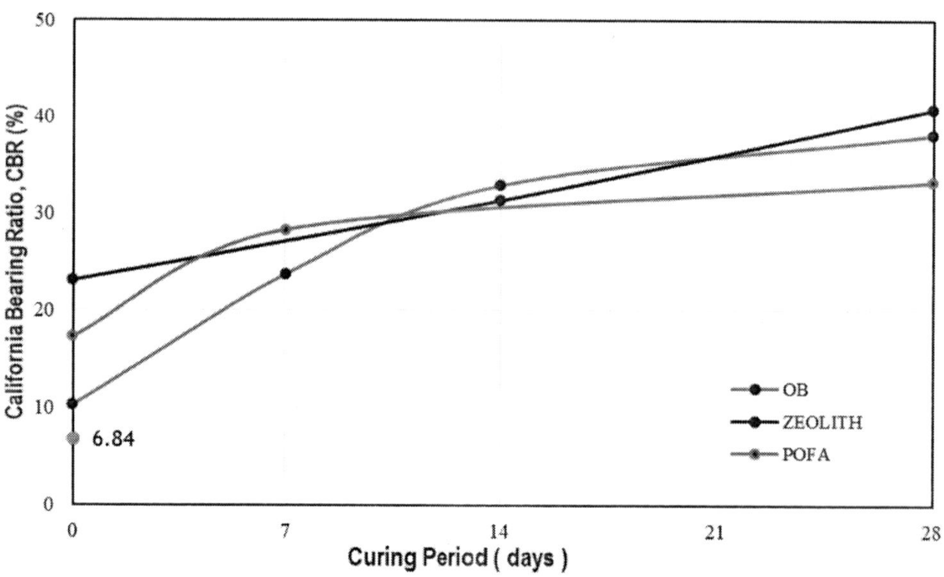

Figure 5. Comparison of 3% Zeolite, 10% POFA and 15%Overboulder

As we can see,at data without curing (0 day), the CBR value of 15% Overboulder addition has the lowest value at 10,34%, then followed by 10% POFA at 15,44 %, and the highest obtained by 3% zeolite addition which the value is 23,23%. After cured 28 days , the trend of this stabilization material is slightly different from 0 days. The lowest CBR value is shown by 10% POFA addition where the CBR value is 33,28%, then followed by 15% Overboulder addition with a CBR value 38,22%, and the highest obtained by 3% zeolite addition with CBR 40,8%.

By comparing data at 28 days, zeolite obtained the highest CBR value among all materials. Where the CBR value increased 5 times in the addition of Overboulder, 6 times in the addition of zeolite and 5 times in the addition of POFA compared to CBR value of untreated soil which is 6.84%. Zeolite is consistently shows a significant increase in CBR values more than POFA and Overboulder, even without curing (0 days), additional of zeolite has greatly increased the CBR value almost 4 times from untreated soil.

Federation of Highway Association Standard requires a minimum CBR value for the subbase layer is 20%. The results shows that in general the average value meets the requirements for utilize as subbase layer except, the addition of overboulder and POFA without curing.

4. Conclusion

1. The addition of all natural materials increases CBR value and as we can see the longer the curing period, the higher CBR value that can be obtained
2. Based on FHWA standards about CBR minimum for subbase layer, zeolite, POFA and overboulder can be used as a stabilization material.
3. From the following materials, zeolite has the best performance compared to POFA and Overboulder.

References

[1] Bowles, J., 1986. *Physical and Geotechnical Properties of Soils (In Bahasa).* Jakarta: Erlangga.

[2] B M Das. 2000. *Chemical and Mechanical Stabilization. Transportation in the New Era Millenium.*

[3] R Cahyaputra and A Pratiwi. 2013. *Green Strategy Sebagai Jawaban "Now and the Future" Semen Indonesia*. Surabaya.

[4] S Nurdin, L Samang, J Patanduk, T Harianto. 2016. *Performance of Fly Ash on Soft Soil Stabilization with Natural Fiber Reinforcement as Landfill Cover Layer.* Inersia. [In Bahasa]

[5] T. *Harianto*, S H Nur, I Rauf and N Marfu'ah. 2019. *Study of CBR Value of Lightweight Geomaterial (Soil-EPS) Stabilized by Waste of Buton Asphalt (In Bahasa).* Prosiding Seminar Nasional Teknik Sipil 2019. Surakarta.

[6] T Harianto, S Hayashi, Y J Du and D Suetsugu. 2008. *Effects of Fiber Additives in the Desiccation Crack Behavior of the Compacted Akaboku Soil as A Material for Landfill Cover Barrier.* J. Water Air Soil Pollution. pp 141-149.

[7] Hasriana, L Samang, T Harianto, M N Djide. 2018. *Bearing capacity improvement of soft soil subgrade layer with Bio Stabilized Bacillus Subtilis*. MATEC .

[8] T E Tay.G Liu.V B C Tan. X S Sun. D C Pham. 2008 *Progressive Failure Analysis of Composites.* J.Composite Materials. 42(18): 1921-1966.

[9] W Tangchirapat. C Jaturapitakkul. P Chindaprasirt.2009. *Use of Palm Oil Fuel Ash as a Supplementary Cementitious Material For Producing High-Strength Concrete.* J. Construction and Building Materials, 23(7): 2641-2646.

[10] N Dhani. L Samang. T Harianto. A R Djamaluddin.2018. *Characteristics Study of Over Boulder Asbuton as Pozzalanic Material for Soft Soil Stabilization.* International Seminar On Infrastucture Development. Manado.

[11] M Janz & S.E.Johansson. 2002. *The Function of Different Binding Agents in Deep Stabilization.* Swedia: Swedish Deep Stabilization Research Centre.

Type and Quantity of the Residual Construction Material of Low-rise Building Construction in Makassar

M Asad A[1]*, Rusdi U L[1], Rosmariani A[1], Suharman H[1], Evi A[1], and Yulio O[1]

[1]Civil Engineering Department, Engineering Faculty, Hasanuddin University, Makassar, Indonesia

*Email: muh.asad@yahoo.co.id

Abstract. Waste material is considered as residual material which is not used during construction projects. These materials are unable to be reused, returned, recycled, or salvaged which have significant impact to environment. The residual material that occurs is caused by inaccurate planning and inefficient use of material in the project. This research is intended to identify quantity of residual material in the construction of low-rise buildings. Data was collected from 25 housing and shophouse projects. It then analysed using quantitative analysis to determine the type and quantity of the construction waste material. The result shows that the highest waste material for housing is sand while for shophouse projects is bricks.

1. Introduction

The use of materials in the construction industry can cause residual material. The rest of this material is part of unused materials in the implementation of construction projects and is not a part of the building. So that the more remaining material that occurs, the more inefficient use of material in the project.

Alternatives to overcoming and minimizing the rest of the construction materials, as well as improving company performance began to be felt necessary to find. Minimization of the huge remaining material will help the contracting company to increase profits as much as possible. Several studies focusing on rearrange the way to deal with residue by using recycle and reusable method. Moreover, it also focuses on the impact of burning waste material the construction process.

The method of recycling is still difficult to implement in Indonesia, because the trash cans have not been sorted according to the type of waste. Therefore, the garbage put together in one shelter. In addition, it requires sophisticated technology applications that require high costs, and the results of recycling have not been studied to be utilized. This method of combustion will adversely affect air pollution and the environment. In the method of reusing remaining material, it is usually limited to materials that are not part of the building structure (non-consumable material), such as formwork and scaffolding [1].

This research is related to the handling of the remaining material that might be done in Indonesia, namely through material management to minimize the remaining material that occurs in the field, this is due to cost considerations. technology that is still simple, and at the same time is environmentally conscious.

2. Literature Review

Material is an important component in determining the cost of a project. More than half of the project cost is absorbed by the uses of material. During the construction stage, material usage in the field often results in significant number of residual material. So, the efforts to minimize the remaining material are important to apply. Materials used in construction can be classified in two major parts, they are:

a. Consumable materials, is a material that will eventually become part and physical structure of the building, for example: cement, sand, gravel, brick, reinforcing iron, steel, and others.
b. Non-consumable materials, is a supporting material in the construction process and is not a physical part of the building after finishing, for example: scaffolding, formwork and temporary retaining walls.

The flow of construction materials starts from delivery to the location, construction process, until the last position will end in one of the four positions below, namely:

a. Physical structure of the construction
b. Excess material (leftover)
c. Reuse in the same project
d. Waste

The rest of the core construction materials will continue to grow in accordance with the progress of development carried out, in addition to affecting project costs will also cause new problems that can disrupt the project environment and surroundings. Control of the quantity of the remaining material can be done in several ways including:

a. Looking for ways to reuse the remaining material.
b. Recycle the remaining material into useful items
c. Destroy the remaining material by burning
d. Look for ways to reduce the remaining material that arises [2].

Preventing the remaining material by means of prevention is the best way, because it is felt to be more economical, lower and safer than other solutions. Expenditures to control the remaining material from the outset will be more profitable than expenses incurred due to the arising of the remaining material. Therefore the magnitude of the quantity of the remaining material that occurs is very closely related to material management.

Preventing the remaining material by means of prevention is the best way, because it is felt to be more economical, lower and safer than other solutions. Expenditures to control the remaining material from the outset will be more profitable than expenses incurred due to the arising of the remaining material. Therefore, the magnitude of the quantity of the remaining material that occurs is very closely related to material management [3].

The remaining material that arises can be categorized into two parts, namely:

a. Demolition waste is the remaining material arising from the demolition or destruction of old buildings.
b. Construction waste is the remnants of construction materials originating from the construction or renovation of privately owned, commercial buildings and other structures. The rest of the material is in the form of rubbish consisting of concrete, brick, stucco, wood, pipes and electrical components [4].

Construction Waste can be classified into two categories based on its type: direct waste and indirect waste. Direct waste is the remaining material that arises in the project because it is damaged and cannot be used anymore which consists of:

a. Transport and delivery waste (remaining transportation & shipping)
All remaining material that occurs when transporting material in the work location, including dismantling and placement in storage such as throwing / throwing cement, ceramics when moved.
b. Site storage waste (remaining storage)
The residual material occurs due to the accumulation / storage of material in unsafe places, especially for sand and broken stone material. or in a humid environment, especially for cement material.

c. Conversion waste (residual deformation)
The residual material occurs due to cutting material with an uneconomic form such as concrete steel materials, ceramics, etc.
d. Fixing waste (remaining installation)
Material that is scattered damaged or wasted during use in the field such as sand, cement, coal, etc.
e. Waste cutting
The rest of the material is produced due to cutting of materials such as piles, concrete steel, brick, ceramics, concrete iron, etc.
f. Application and waste residue
The remaining material that occurs such as mortar that fell / scattered at the time of execution or mortar is left behind and has hardened at the end of the work.
g. Criminal waste (residual due to criminal acts)
The residual material occurs due to theft or acts of destruction at the project site.
h. Wrong use waste (remaining usage errors)
Use the type or quality of material that is not in accordance with the specifications in the contract, the board of directors will instruct the contractor to replace the material in accordance with the contract, thereby causing the remaining material in the field.
i. Waste management (residual error management)
The occurrence of material residue is due to wrong decision making or doubt in making decisions, this is due to weak project organization, or lack of supervision [5].

Indirect waste is the residual material that occurs in the form of a cost loss, there is an excess use of the material and planned volume, and there is no physical residual material in the field. This indirect waste can be divided into three types, namely:
a. Waste substitution (the remainder of the turnover)
The remaining material that occurs due to its use deviates from its original purpose, thus causing the loss of costs that can be caused by three reasons;
 1. Too much material is purchased
 2. Damaged material
 3. Increasing certain material needs

b. Production waste (the rest of the production)
The remaining material caused by excessive use of material and the contractor is not entitled to claim the excess volume because the basis of payment is based on the volume of the contract, for example uneven masonry walls cause excessive use of mortar due to thick plastering.
c. Negligence waste (residual due to negligence)
The rest of the material that occurs due to an error at the site (site error), so that the contractor uses more material than specified, for example: excavation of the foundation that is too wide or deep caused by error / carelessness of workers, resulting in excess use of the volume of concrete at the time of foundation casting [6].

Waste prevention should be the main consideration. It is the avoiding of using much materials than needed which create less waste. It can be done by in several ways. For instance, selling the timber that has been used as concrete formwork. However, when it is not possible to prevent waste, thus waste should be managed wisely. Construction waste should be managed wisely using the following approach:
a. Reusing, where using the same material again with the same usage without turning it to any other form. For example, using the same multiplex several times for concrete formwork.
b. Recycling, where turning particular materials into other useful forms when directly reusing is not possible. Various components of construction and demolition waste can be recycled. For example, steel have the highest recycling percentage over the materials recovered from construction and demolition sites.

c. Disposal is the last choice should be considered, as it means removing, sometimes materials are unusable after first usage. Therefore, in this case burning and landfill is included.

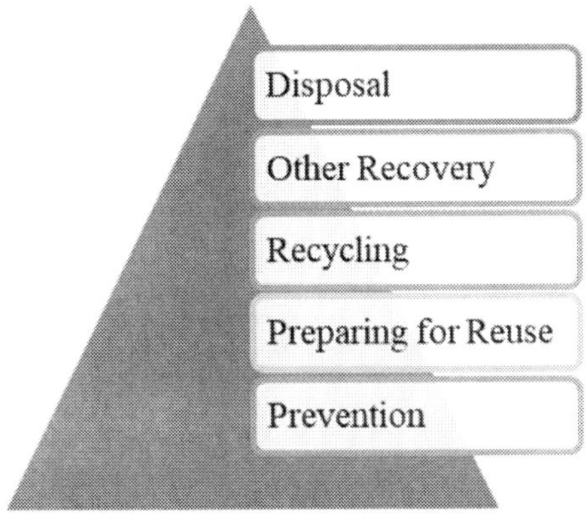

Figure 1. Waste hierarchy summary [7]

3. Research Methodology

An outline scheme of the work program of this study is describe with the following methods:

1. Preliminary Study in presenting ideas or problems to be discussed, namely the study of the remaining material in the construction of low-rise buildings. Conducting literature studies that support this research. Formulate the problem that will be examined and discussed in this study.
2. Research variables are material requirements, material storage, material transfer, material usage, and material handling
3. Data retrieval in the assessment of material requirements index is done by inventorying secondary data on construction projects that have been. As a guide, the questionnaire is used as a measurement tool. Inventory secondary data from interviews to contractors, consultants, planners, government agencies or other randomly chosen construction implementers.
4. Processing data using descriptive and inferential statistics to determine the index of material requirements, worker productivity and tool use. Estimated average waste cost is calculated using the following equation

$$\text{Estimated average waste cost} = \text{Estimated waste quantity} \times \text{material price} \qquad (1)$$

4. Result and Discussion

Determination of the type of material to be quantified based on the concept of the Pareto Law in which obtained 7 types of material for residential houses and shop houses are sand, brick, wood, concrete steel, tile, ceramic and cement. From the above results, a quantitative analysis of the remaining material is carried out by calculating the volume of ready-made material in the field reduced by the volume of design material based on project plan drawings and the Bill of Quantity (BOQ), then reduced by the remaining material in the field which can still be used if there is. The results obtained are for the highest residential dwellings with sand with an average remaining construction material ranging from 13.6% while for the highest shop houses are bricks with an average remaining construction material ranging from 12.08% as shown in Table 1.

The main causes of the residual construction material in this project are:

a. Changes from the initial design
b. Orders must be placed with a certain volume in one order or can't be done in small quantities.
c. Damage from tiles and ceramics during material delivery.
d. The sand material sent is not in the volume that should be in one truck
e. The material is not according to specifications
f. Damage in handling material during the construction process

Table 1. Percentage of remaining material volume

Type of Material	Average amount percentage	Estimated average waste cost per project
Residential		
Brick	15.92%	Rp. 254,000 - Rp. 635,000
Sand	17.96%	Rp. 1,142,000 - Rp. 1,713,000
Tile	10.92%	Rp. 214,000 - Rp. 877,400
Wood	14.34%	Rp. 924,500 - Rp. 1.571.650
Concrete Steel	6.94%	Rp. 84,000 - Rp. 210.000
Ceramics	16.34%	Rp. 156,200 - Rp. 296.780
Cement	12.94%	Rp. 124,500 - Rp. 311.250
Others	4.64%	Rp. 243,000 - Rp. 510.300
Shops House		
Brick	17.08%	Rp. 164,000 - Rp. 410.000
Sand	13.14%	Rp. 1,520,000 - Rp. 2.280.000
Tile	13.14%	Rp. 183,000 - Rp. 750.300
Wood	11.47%	Rp. 1,328,000 - Rp. 2.257.600
Concrete Steel	8.45%	Rp. 79,000 - Rp. 197.500
Ceramics	14.45%	Rp. 213,000 – Rp. 404.700
Cement	16.24%	Rp. 111,000 - Rp. 277.500
Others	6.03%	Rp. 315,000 - Rp. 661.500

For the lowest material the rest is concrete steel. The low residual of this material is due to the relatively high price so that construction implementers continue to use scraps of residual steel concrete in construction even though it does not meet the criteria of propriety of a reinforced concrete structure.

5. Conclusion

The remaining dominant construction materials in the form of sand and bricks where for residential projects respectively 17.96% and for shop-house projects 17.04%. Pure sand and bricks have no impact on the environment, where the rest of this type of construction material can still be reused again.

Acknowledgement

The author would like to express great gratitude to the Faculty of Engineering, Hasanuddin University. This work is supported by the Laboratory Based Education (LBE) research fund.

References

[1] Ekanayake, Lawrence Lesly, 2000, Construction Material Waste Source Evaluation, Proceedings: Strategies for a Sustainable Built Environment, Pretoria, 23-25 August.

[2] Lamahayu, Andre Julius, & Widjaja, Katarina Raninda, 2008. Penelitian awal penanganan kontraktor terhadap direct waste material pada proyek konstruksi di Surabaya. www.getcited.org/pub/103495511.

[3] Bossink, B.A.G.,and Brouwers, H. J. H., 1996, Construction waste: Quantification and source evaluation, Journal of Construction Engineering and Management, Vol 122, No 1, pp. 55–60.

[4] Gavilan, R. M., and Bernold, L. E., 1994, Source evaluation of solid waste in Building construction, Journal of Construction Engineering and Management, September 1994. pp. 536 – 552.

[5] Farida, R. 2013. Analisa Sisa Material Konstruksi Dan Penangannya Pada Proyek Gedung Pendidikan Profesi Guru Univeristas Negeri Surabaya, Universitas Sebelas Maret, Surakarta.
[6] Asnudin, A & Labombang, M. 2012. Faktor-Faktor Penyebab Terjadinya Waste Material Pada Proyek Konstruksi Gedung (Studi Kasus: Pembangunan Gedung Rektorat Blok B dan Rumah Sakit Pendidikan Universitas Tadulako). Jurnal Konferensi Nasional Teknik Sipil 8, Institut Teknologi Nasional, Bandung.
[7] Curran, H. (2014, May 9). Draft construction environmental management plan. Retrieved from http://infrastructure.planninginspectorate.gov.uk/wp-content/ipc/uploads/projects/EN020001/2.

IOP Conf. Series: Materials Science and Engineering **875** (2020) 012031 doi:10.1088/1757-899X/875/1/012031

Analysis of Sediment Distribution at the Intake Structure

B Bakri[1]*, S Pallu[1], R Lopa[1], F Maricar[1], A Sumakin[1], M F Maricar[1], and Ridwan[1]

[1]Department of Civil Engineering, Engineering Faculty, Hasanuddin University, Makassar, Indonesia

*Email: bambangbakri@gmail.com

Abstract. As a Maritim country, most of Indonesia's population lives near the coast or estuary. The main problem felt by the community is sanitation, clean water. The utilization of groundwater is very limited in number because of the influence of seawater intrusion. On the other hand, excess freshwater from upstream is very abundant, especially around the estuary. However, the morphological condition of the river downstream or bay has a significant depth and dimension, so it requires expensive exploitation costs in using the water. One of the solutions that have been done currently to use the water as raw water for clean water is by building free intake around the estuary. Along with river utilization, problems have been found regarding intake capacity and sediment entering the intake beyond the expectation. As a result, it gives an excessive burden on the water treatment plant. This research is an experimental study conducted in the laboratory that aims to determine the distribution of sediment in free intake building. The aims of the research end up to get the most effective intake placement elevation in terms of the capacity and the amount of sediment entering the intake. As a result of this study, it can cope with the high sedimentation rates in the free intake building. The results showed that the minimum discharge to 0.5 lt/second in the intake high (h1) was 0.04 gr, while in the intake high (h3) was 0.01 gr. For the maximum discharge to 0.7 l/sec in the intake (h1) was 1.7 gr while in the intake high (h3) was 0.02 gr. The amount of sediment that comes out through the intake will be directly proportional to the fluid flow in the channel with the same slope and elevation. However, the amount of sediment that comes out through the intake will inversely proportional to the intake elevation, discharge, and the same slope.

1. Introduction

As a maritime country, most of Indonesia's population lives near the coast or estuary. The main problem felt by the community is sanitation, especially clean water. The utilization of groundwater is very limited in number because of the influence of seawater intrusion. On the other hand, excess freshwater from upstream is very abundant, especially around the estuary. However, the morphological condition of the river downstream or estuary has considerable depth and dimension, so it requires expensive exploitation costs in using the water. One of the solutions that have been done currently to use the water as raw water for clean water is by building free intake around the estuary [1].

The existence of this intake is very helpful in supplying raw water for clean water needs, especially in coastal areas. Along with the use of tenure, it was found that the existing intake capacity is far below the role of the plan. Variables affecting the function of intake are the shape and velocity of the flow in the river, river profile, and the placement of the intake pipe from the river to the intake [2]. Therefore, the application of the theory of hydraulics to the intake design can result in an appropriate real condition, and thus sufficiently rigorous for practical design [3, 4]. The previous study obtained the most effective intake placement in terms of the intake capacity (elevation of 0.25 h) [5]. However, the current intake problem is not only a lack of intake capacity that has been planed but the rate of sediment entering the intake that exceeds the previous planed before. It is one of the problems that has been facing now. As a result of the high rate of sediment entering the intake, the burden of drinking water treatment plant is also increasing. It affects the high cost of water treatment, and the quality of clean water that reaches the community is also reduced due to the high rate of sedimentation.

Based on the mentioned above, intake design in the current condition applies the principle of hydraulic flow due to the absence of previous studies conducted. So, the authors consider the need for in-depth research on the analysis of sediment distribution at the intake due to the influence of intake discharge and elevation [6].

2. Methodology

The principle of flow at the intake in a river can be said to follow the law of flow at the hole. A hole is an opening in the wall or bottom of a tank where liquid flows through it. The shape of the tunnel is a rectangular, triangular, or circular. The upstream side of the hole can be sharp or rounded. Because of its ease of manufacture, it also commonly used in raw water intake because water is generally used in closed channels or circular pipes. According to its size, the hole is above the water surface in the tank, and such an opening called spillway. The spillway also functions as a flow measurement tool and is widely used in irrigation networks. A large size spillway is called a weir. The function of weir not only measures the discharge/debit but also to increase the water level in irrigation and raw water network. Overview hydraulic reservoirs are the same as a spillway. Spillways are usually made of plates, while weirs are made of concrete or masonry. The upstream depth is measured from the hole axis with the height of energy (head) H. In the flow through the tunnel or the spillway, the high energy can be constant or change due to an outflow. If the high energy is consistent, then the flow is steady, whereas if the high energy changes, then the flow is unsteady. Based on the principle of Bernoulli's application, the velocity or discharge is influenced by the location and hole's elevation from the surface of the water and the cross-sectional area of the hole [7]. The equation is illustrated below.

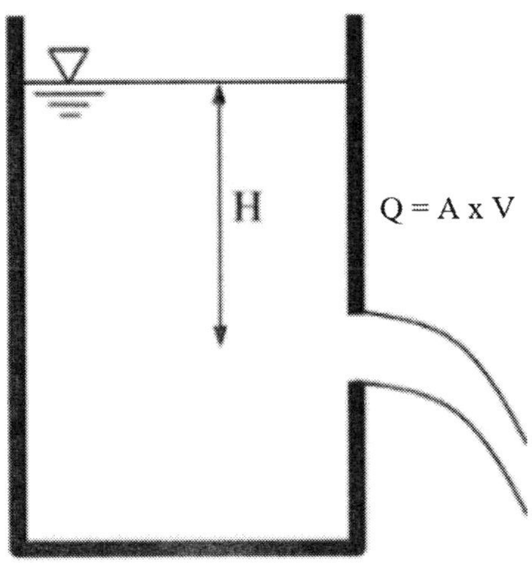

Figure 1. The principle of flow in a tank or hole

Based on Bernoulli flow principle above, the magnitude of flow velocity can be written as in the following equation:

$$V = \sqrt{2gH} \tag{1}$$

so that the amount of water flowing through the intake is:

$$Q = A \times \sqrt{2gH} \tag{2}$$

where:

Q = Flow discharge at the intake (m^3/s)
V = velocity of bursts of water at the hole or intake (m/s)
H = height of water above the hole (m)
g = Acceleration of gravity (m/s^2)
A = Cross-sectional area of the hole (m^2)

The flow in the tank is influenced by water height above the hole, while other flow parameters also affect the amount of water flow at the intake. The benchmark includes water flow in the river, the slope, and even the distance between weirs and intake. The flow discharge is affected by the water level. As height as the level of the surface water, the higher the discharge occurs. The increase of downstream surface water is followed by the rise of the release through the rectangular spillway. In other words that the changes in square spillway flow discharge are also accompanied by the changes in the surface water level at the upstream. A water level with flowrate, better known as the curve of release curved. The magnitude of the increase in water level downstream is not significant to changes in discharge flow. On the other hand, the flow velocity is not the same along the river. It depends on the shape, the roughness of the river canal and river patterns. The most significant velocity is located in the middle and upper part of the deepest part of the river, which is far from the frictional drag on the walls and bottom [8]. The maximum velocity in the channel usually occurs below the free surface as deep as 0.05 to 0.25 times its depth. The closer the edge, the deeper the velocity reaches the maximum [9]. The velocity distribution in the channel cross-section also depends on other factors, such as the unusual shape of the cross-section, channel roughness, and bending. Wide, swift, and shallow currents or very slippery channels, the maximum velocity often occurs on a free surface. The presence of intake in the river does not significantly affect changes in the velocity distribution in the river [5]. It is shown that the highest velocity remains located around the center of the channel and tends to fall towards the surface of the water, as shown in the following figure:

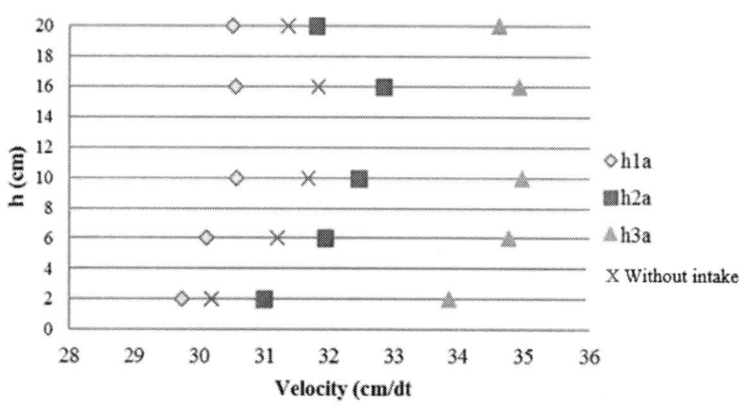

Figure 2. Distribution velocity without intake and with intake

On the other hand, the sediment transport capacity on the cross-section of the river is the magnitude of the sediment passing through the cross-section within a particular time unit. Scouring, sedimentation, and experiencing balanced transport needs to know the quantity of sediment transported in the process. The river is called balance condition when the capacity the sediment comes through the cross-sectional river is equal to the capacity sediment goes out the river. Sedimentation occurs when the sediment capacity comes in more significant than balanced sediment capacity in units of time. However, the crushing condition occurs when the capacity of sediment comes in smaller than stable sediment capacity in units of time. Riverbed material that floats in the stream consists of fine grains of sand floating above the riverbed. As a result, it is always pushed up by the turbulence of the flow. Sediment particles still float in the river flow if the tide is turbulent. If the flow is luminaire, the sediment concentration will decrease and eventually settle, just as if the river flow did not flow.

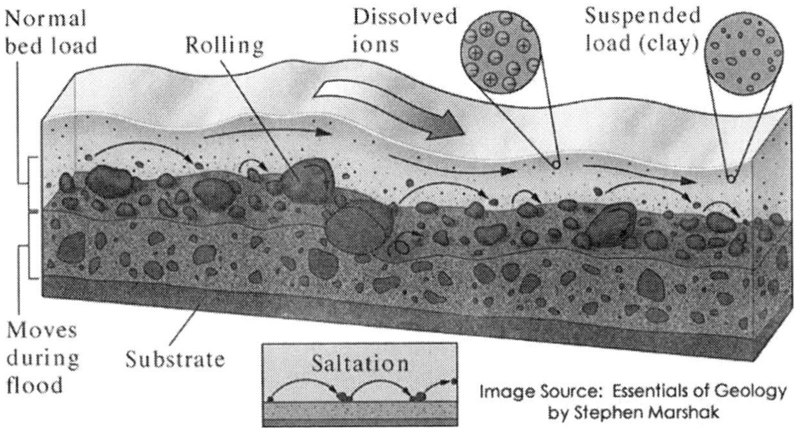

Figure 3. The movement of the sediment

This research was conducted at the Hydraulics Laboratory of the Department of Civil Engineering, Hasanuddin University. The scale model is used To describe the conditions in the field or prototypes with the model. The model is not distorted with a scale of 1:10. The flow in the river or channel uses a flume with a flume length is 9 m and a width of 0.08 m, as well as a flume height of 0.2 m. The weir or overflow height used in the flume is 0.15 m. The channel wall is installed with an intake pipe with a diameter of 7 mm to simulate the flow of water entering the intake. It is placed in 3 variations of water level, namely: h1 = 0.25H, h2 = 0.5H and h3 = 0.75H from the surface water. The distance of the intake to the weir varied: L1 = 1 m, L2 = 2 m, and L3 = 3 m. The slope of the channel is also varied 3, namely the slope of I1 = 0.17%, I2 = 0.25% and I3 = 0.33%. There are also 3 variations of water discharge in the channel, namely Q1 = 0.5 l / s, Q2 = 0.75 l / s and Q3 = 1 l / s.

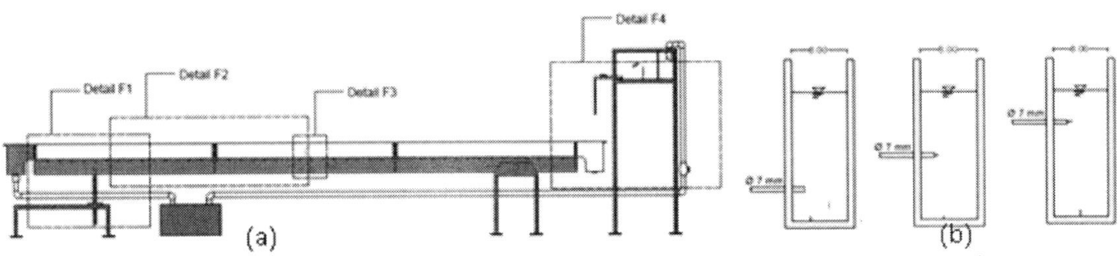

Figure 4. Research equipment a) Flume b) Elevation of intake placement

While the sedimentary characteristics used in this study are the aggregates that pass in sieve # 4 and # No.10, which are 100%. Then, the sedimentation characteristic used is 98% that pass in sieve #No.20, followed by 67.4% that pass in sieve No. 40, and filter No.60 passed 12.4%. Filter No.100 showed 3.8% passes while in filter No.200, it is 0.8%. In this test, it was found that the sample used passed filter # No. 4. This shows that the characteristics of the sample used were fine-grained, as shown in the figure below:

IOP Conf. Series: Materials Science and Engineering **875** (2020) 012031 doi:10.1088/1757-899X/875/1/012031

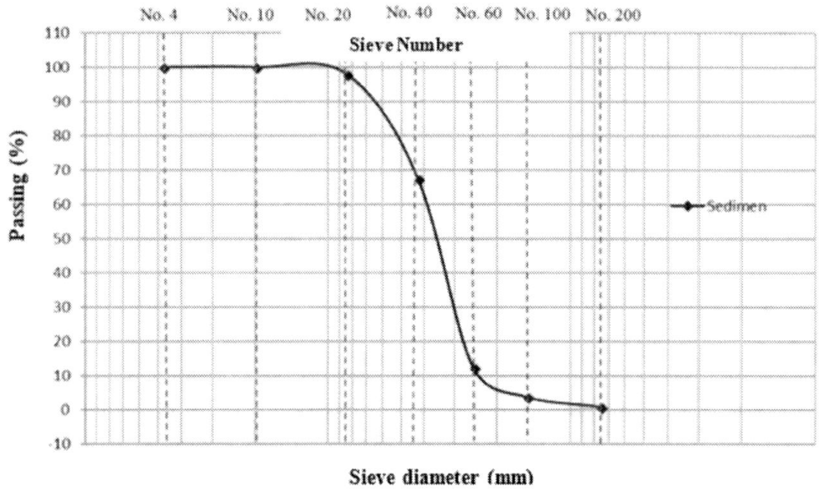

Figure 5. Sediment sample

3. Results and Discussions

3.1. Effect of Discharge on Sediment at the Intake

Based on research results, the amount of deposit that comes out due to the influence of discharge on the intake can be seen in the following:

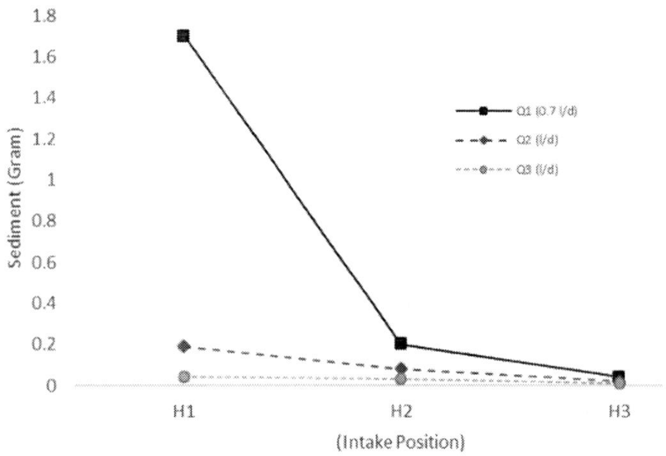

Figure 6. Comparison between sediment and water debit

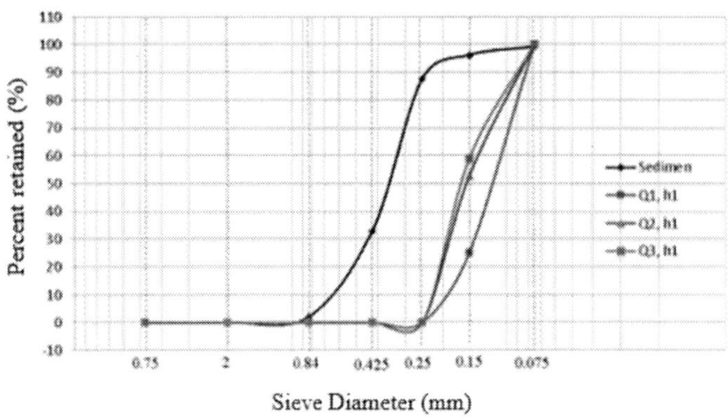

Figure 7. Sediment distribution due to changes in discharge (h1)

Figure 6 shows the relationship between sediment emitted through the intake and discharge (Q) entering the flume with slope (I) = 0.17% and elevation height (h) intake = 0.25 cm. In this experiment the three types of discharge are Q1 = 0.5 l / sec Q2 = 0.6 l / sec and Q3 = 0.7 l / sec. The value obtained from each discharge is Q1 = 0.04 gram or 5.71% and in Q2, and it increases to 0.19 gram or 27.14%, and finally, in Q3, the sediment obtained is 1.7 grams. It can be seen in the comparison graph between Q1, Q2 is 0.15 grams, and 21.43%, while in Q2, Q3 is 0.51 grams or 91.18%, respectively. It shows that with the increase in discharge in the channel, the amount of sediment that comes out at the intake will be even greater.

Based on Figure 7, it can be seen that the sediment entering the intake in conditions Q1h1 is No.4, 10, and 60 all pass 100% while the condition filter No.100 that passes through is 75%. In the Q2h1 condition in filters, No.4, 10, 40, and 60 all passed 100%, while in filter No.100, that passed is 48%. In the Q3h1 condition in filters, No.4, 10, 40, and 60 all passed 100%, while in filter No.100, that passed is 40%. It shows that the increase in discharge causes more and more diameter of the sediment to pass through the No.100 filter. Increased discharge from Q2-Q1 shows a rise of 16.7%, while an increase in the diameter of the sediment that passed the No.100 filter is 56.25%. Similarly, the increase in discharge from Q2-Q3 shows a rise of 40%, while the increase in the diameter of the sediment that passed the No.100 filter is 87.5%.

Figure 6 shows the relationship between sediment released through intake and discharge (Q) with elevation height (h2) intake = 0.53 cm. The value obtained from each discharge is Q1 = 0.03 gram or 15% and in Q2 increases to 0.08 gram or 40%. Finally, in Q3, the sediment obtained is 0.2 gram, it can be seen in the comparison chart between Q1, Q2 is 0.05 gram or 25% while in Q2, Q3 is 0.12 gram or 60%. It shows that with the increase in discharge in the channel, the amount of sediment that comes out at the intake will be even higher.

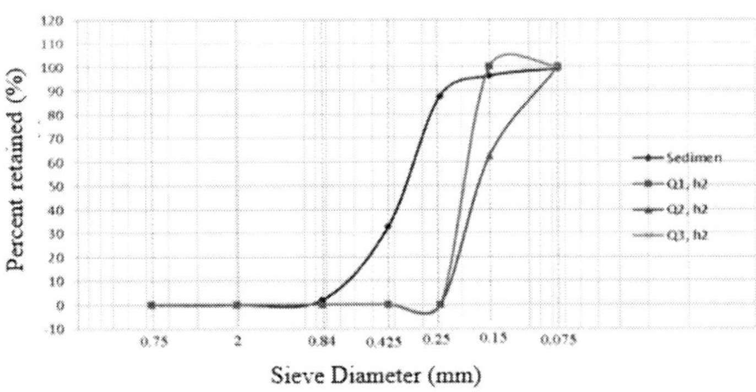

Figure 8. Sediment distribution due to changes in discharge (h2)

Figure 8 shows that the sediment that enters the intake at Q1 conditions in filters No.4, 10, 40, and 60, all passes 100%, while at filter No.100 to the pan that passes is 0%. At condition Q2 on filter No.4, 10, 40, and 60 all passed 100%, while in filter # No.100, that passed is 37.5%. In condition Q3 in filter No.4, 10, 40, and 60 all passed 100%, while in filter No.100 to the pan, that escapes is 0%. It indicates that sediment entering the intake is fine-grained sediment, while the percent held in this condition can be seen that sediment coming to the intake under conditions Q1, h2, I and Q3, h2, I in filter No. 4, 10, 40 and 60 are retained by 0%. In filters, No.100 and 200 are retained by 100%, while in conditions Q2, h1, I in filters No.4, 10, 40, and 60 are retained by 0%. In filter No. 100, it was maintained as much as 62.5% and in filter No. 200 stuck by 100%.

Figure 6 also shows the relationship between sediment emitted through intake and discharge (Q) entering the flume with elevation height (h) intake = 0.75 h. The value obtained from each discharge is Q1 = 0.01 gr or 25%, and in Q2, it increases to 0.02 gr or 50%. Finally, in Q3, the sediment obtained is 0.04 gr. It can be seen in the comparison chart between Q1, Q2 is 0.01 grams or 25% whereas, in Q2, Q3 is 0.02 grams or 50%. It shows that with the increase in discharge in the channel, the amount of sediment that comes out at the intake will be even higher.

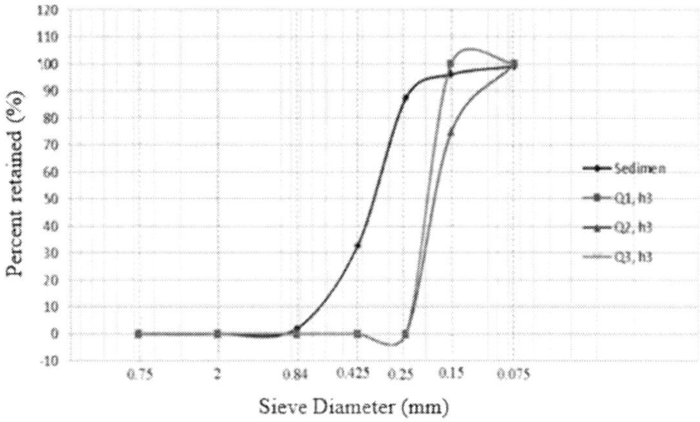

Figure 9. Sediment distribution due to changes in discharge (h3)

Based on the graph above, it can be seen that the sediment that enters the intake under Q1 conditions at the same elevation and slope, in filters No.4, 10, 40, and 60 all pass 100%, then hold in the No.100 filter to 0% pan. In the Q2 conditions in filters No.4, 10, 40, and 60 all passed 100%, then in filter No.100 that passed is 25%. In the Q3 conditions in filters No.4, 10, 40, and 60 all passed 100%, then in filters No.100 and 200, they are held 100%. It indicates that the sediment that enters the intake is fine-grained. The percent retained in this condition can be seen that the sediment that enters the intake under conditions Q1, h3, and Q3, h2 on the filters No.4, 10, 40, and 60 are retained as much 0%. In filter No.100 and No.200, it is retained by 100%. It is in Q2 conditions, h1 in filter No.4, 10, 40, and 60 are retained by 0%, while in filter No.100, it is retained by 75%, and on sieve No.200, it is held 100%.

3.2. Effect of Intake Elevation on Sediment at the Intake

Based on the results of research conducted in the laboratory that the elevation of the intake building influences the amount of sediment that comes out through the intake pipe. This can be seen in the following picture below:

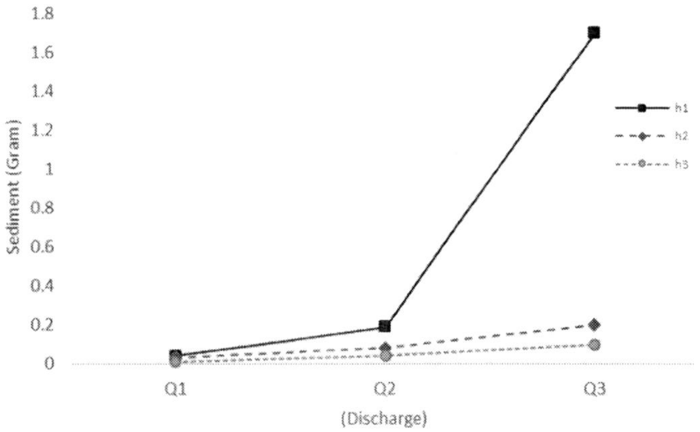

Figure 10. Comparison between sediment and water debit

A comparison between different sediments (grams) and elevation under Q1 conditions can be analyzed graphically. Figure 10 shows the relationship between sediments coming out through the intake based on intake and discharge elevations (Q) entering the flume at Q1 = 0.5 l / sec. This experiment was carried out with three types of elevation namely h1 = 0.25cm, h2 = 0.5 cm and h3 = 0.75 cm. The value obtained from each elevation is h1 = 0.04 gram, and on h2 decreased to 0.03 gram or 75%. Finally, on h3, the sediment obtained is 0.01 gram or 25%. It can be seen in the comparison chart between h1-h2 is 0.01 gram or 25%, whereas in h2 - h3 is 0.02 gram or 50%. It shows that with the increase in intake elevation in the channel, the amount of sediment that comes out in the intake pipe will be smaller.

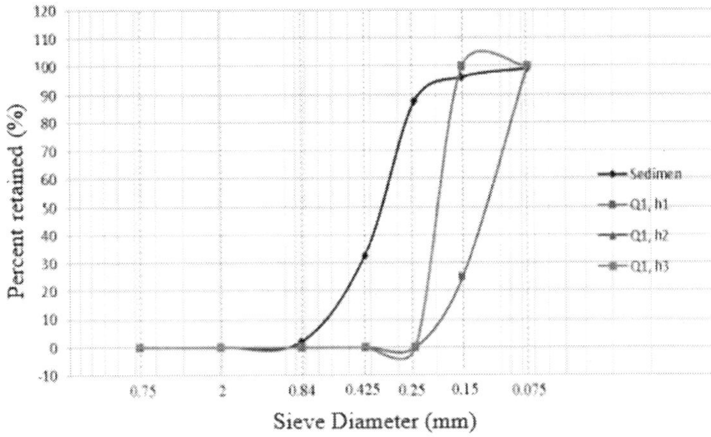

Figure 11. Comparison graph between sediment and intake elevation (Q1 = 0.5 l / sec)

Based on the picture above, it can be seen that the sediment is entering the intake in conditions Q1h1 and Q1h2 in filters No.4, 10, 40, and 60 are retained as much as 0%, while in filter No. 100 is retained as much as 25%. In filter, No.200 is retained as much as 100%. Meanwhile, in the Q3h1 condition is in filters No.4, 10, 40, and 60, then in filter No.100 it was retained by 52.63%, and in filter 200, it was held 100%. It indicates that the sediment that enters the intake is fine-grained sediment.

Figure 10 shows the relationship between sediment emitted through intake based on intake and discharge elevation (Q) entering the flume at Q2 = 0.6 l / s. The value obtained from each elevation is h1 = 0.19 gr, and at h2 decreases to 0.08 gram or 42.10%. Finally in the h3 sediment obtained was 0.04 gram or 21.05%. The ratio between h1-h2 was 0.11 gram or 57.9%, whereas in h2 - h3 it was 0.04 gram or 21.05%. It shows that with the increase in intake elevation in the channel, the amount of sediment that comes out in the intake pipe will be smaller.

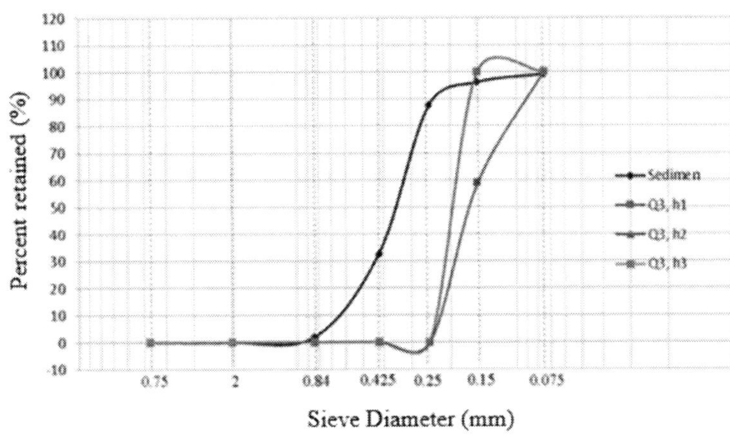

Figure 12. Sediment distribution due to changes in elevation (Q3)

Based on the graph above, it can be seen that the distribution of sediment entering the intake, Q3h1 and Q3h2 in sieves No.4, 10, 40 and 60, is retained by 0%, then in filter No. 100 it is maintained by 60%, while Q3h3 is retained by 0% in sieve No. 4, 10, 40, 60, and 100. It indicates that the intake elevation affects the diameter of the sediment entering the intake. The higher the intake elevation from the riverbed channel, the smaller the diameter of the sediment that enters the intake. A comparison of the elevation of the intake elevation from h1-h2 shows an increase in the height of 100% while the increase in sediment distribution with the diameter held in the No.100 filter is the same or 0%. Likewise, the rise in intake elevation from h1-h3 showed an increase in the height of 200%, while the increase in the distribution of sediment with a diameter held in the No.100 filter was only 66.7%.

4. Conclusion

Based on the results of the analysis shows that changes in intake discharge and elevation affect the distribution of sediment that enters the intake as summarized as follows:

- The amount of sediment coming out through the intake is affected by the intake discharge and elevation. The lower the elevation intake, the more sediment that will come out due to the movement of sediment that occurs under the channel that flows faster,
- The amount of sediment coming out through the intake will be inversely proportional to the intake elevation (h) with the same flow and slope channel. The minimum amount of sediment that comes out through the intake is at the highest intake elevation, h = 0.75 cm.
- The amount of sediment coming out through the intake is so significant compared to the discharge and other elevations of the intake. It is due to the pattern of sediment movement and discharges pressure at the highest Q3 h1 conditions and closest to the bottom of the channel
- Placement of intake elevation shows changes in sediment distribution, the closer the intake is to the bottom of the canal or river, the more varied the sediment that enters the intake. Whereas fine-grained diameter sediments will dominate the further the input from the bottom of the channel/river, the sediment entering the intake.

References

[1] Ahn S.H., Xiao Y., Wang Z., Zhou X. and Luo Y. 2017. Numerical Prediction on the Effect of Free Surface Vortex on Intake Flow Characteristic for Tidal Power Station. Journal of Renewable Energy. 101:617-6128.

[2] Devi K. and Khatua K.K. 2016. Prediction of Depth Averaged Velocity and Boundary Shear Distribution of a Compound Channel Based on the Mixing Layer Theory. Journal of Flow Measurement and Instrumentation. 50:147-157.

[3] B. Bakri, Y. Arai, T. Inakazu, A. Koizumi, S. Pallu, and H. Yoda (2015). A multi-step genetic algorithm model for ensuring cost-effectiveness and adequate water pressure in a trunk/limb mains pipe system. Journal of Water Supply: Research and Technology—AQUA. Vol. 64.2, 2015, pp 176-185.

[4] Kodoatie. 2001. Hidrolika Terapan pada Saluran Terbuka, Andi, Yogyakarta. Page. 215.

[5] B. Bakri, M. Akbar, S. Pallu, R. Lopa, M. Ihsan, and Y Arai. 2017. Study Of Flow Velocity Distribution On Free Intake Structure And Its Influence To Intake Capacity. Journal of Engineering and Applied Sciences.

[6] Wang J., Xiang S., Fu S., Cao P., Yang J. and He J. Experimental Investigation on the Dynamic Responses of a Free-Hanging Water Intake Riser Under Vessel Motion. Journal of Marine Structures. 50:1-19.

[7] Brandt M.J., Johnson K., Elphinston A.J., and Ratnayaka D.D. 2017. Twort's Water Supply. Seven Edition. Elsevier. Pages 205-223

[8] Kumbhakar M, and Ghoshal K. 2016. Two Dimensional Velocity Distribution in Open Channels Using Renyi Entropy. Journal of Physica A: Statistical Mechanics and its Applications. 450:546-559.

[9] Lu Y., Tong Z., Glass D.H., Easson W.J., and Ye M. 2016. Experimental and Numerical Study of Particle Velocity Distribution in the Vertical Pipe After a 90° elbow. Journal of Powder Technology. In Press.

IOP Conf. Series: Materials Science and Engineering **875** (2020) 012032 doi:10.1088/1757-899X/875/1/012032

Preliminary Study on Compressive Strength of Porous Asphalt Containing Modified Buton Asphalt, Waste Plastic and Limestone Powder

M W Tjaronge[1]*, Fakhruddin[1], M A Caronge[1], M H Zaifullah[1], and M F Rahmat[1]

[1]Departement of Civil Engineering, Faculty of Engineering, Universitas Hasanuddin, Indonesia.

*Email: tjaronge@yahoo.co.jp

Abstract. Porous asphalt is a surface layer or a wear layer is one of the asphalt material developed to overcome the problem of inundation during the rainy season making it suitable for use in Indonesia which has high rainfall intensity. Furthermore, it is necessary to optimize the use of a number of natural resources in Indonesia such as natural rock asphalt in Buton Island (BRA) and limestone which is available in many places. In addition, reducing of plastic waste is attracts the attention of a number of countries including Indonesia. This paper is part of a series of investigate study related to the behaviour of porous asphalt mixture made with modified Buton asphalt (MBA), waste plastic and lime stone powder. The compaction results showed good adhesion can unite the waste Polyethylene terephthalate (PET), limestone powder and coarse aggregate particles. There is no visible negative effect of waste PET and limestone powder in porous asphalt mixture against compressive strength.

1. Introduction

Indonesia has high rainfall intensity. Often when it rains heavily, the rain floods the road a few moments before it flows into the drainage drain. Inundation that occurs on the road body causes several problems for motorists. Some of these include splashing water that causes view obstructions and reduced tire grip on the road causing easy slip. Porous asphalt is a surface layer or a wear layer is one of the asphalt material developed to overcome the problem of inundation during the rainy season. In contrast to the dense asphalt mixture which has a micro porosity of about 4-6%, porous asphalt mixture is formed by most of the coarse aggregates bound by bitumen, where fine aggregate or filler is added in small amounts. Because the mixture is made using mostly coarse aggregates with small amount of filler or fine aggregates, coarse aggregate skeletons can be formed and bond with bitumen where interconnected porosity with diameters of about 5 mm occurred. Porosity volume is around 10-25%. Interconnected porosities have a function to flow water from the upper part of the surface layer through the porous asphalt body to the waterproof layer below. Thus the water will be reduced quickly, thereby reducing inundation on the surface. Coarse aggregates well arranged on porous asphalt will increase the friction between the tire and the road surface so that the vehicle does not slip easily [1-3].

Many places in Indonesia are covered by limestone hills and mountains. As a result of the weather and temperature that often changes, the upper parts of some layers of limestone are easily peeled off due to scouring and erosion to form powder that pass sieve no.4 to sieve no.200. This study used limestone powder that have been eroded and peeled from the top part of the limestone layer. The limestone powder was grinded and filtered through the No. 200 sieve to obtain limestone powder which was used as filler in this study.

Nowadays the use of plastics including PET plastic continues to increase. PET plastic is widely used, one of which is as an ingredient for making soft drink plastic bottles. This resulting in the amount of plastic including PET plastic that is not used and becomes waste continues to increase. Plastic materials that are wasted are not easily decomposed and remain in their intact form for a long time in nature, causing many problems to the environment.

On the other hand, the nature of plastics including PET plastic that is not easily biodegradable shows the nature of its good resistance to environmental influences. Good resistance of waste PET is expected to have a positive effect when PET waste is used with other materials. Based on existing research literatures, PET in natural state is semi-crystalline resins [4-6], and has a glass transition temperature (Tg) around 70°C [7-9].

One positive effort to overcome the problem of waste plastic is to use it as a construction material. In recent years there have been several studies using waste PET as an added material in the manufacture of asphalt concrete mixtures. The use of waste PET as added material produced concrete asphalt mixtures which have better physical properties. This is caused when heating waste PET that has been softened turned into a material with properties of crystalline and semi-crystalline so that when mixed with bitumen causes the dense asphalt concrete mix is stiffer with better physical properties than dense asphalt concrete mix without waste PET [10, 11].

Natural rock asphalt is formed from hydrocarbon material that has been strongly integrated with sediment deposits. In some areas in southern Buton, Buton Island, Indonesia are covered with natural rock asphalt deposits. The content of Buton Rock Asphalt (BRA) is about 30% bitumen and 70% mineral, most of the bitumen characteristics of BRA resemble petroleum bitumen characteristics. In recent decades these natural asphalt rocks have been successfully processed into several forms. Concrete asphalt mixtures with physical characteristics that meet the requirements for concrete asphalt mixtures can be achieved by using BRA products as a partial replacement for petroleum bitumen [12, 13].

In recent years, a number of natural rock asphalt processing plants have succeeded in refining the bitumen content so that it is separated from minerals. Bitumen distillated from natural rock asphalt is mixed with petroleum bitumen to produce modified Buton asphalt (MBA). The use of the MBA is expected to further reduce dependence on petroleum bitumen as a binding material in making asphalt concrete mixtures.

The research presented is one part of a series of effort to increase the use of processed Buton natural asphalt products. This research is an initial attempt to provide an overview of the use of MBA, waste PET and limestone powder to make the basic design of a mixture of porous asphalt. Compressive strength is a test of physical characteristics used to assess the preliminary mixture of porous asphalt containing MBA, PET waste and limestone powder.

Figure 1 shows a surface layer including a layer made with porous asphalt when receiving vehicle loads. At the top part of each layer, compressive stress will arise and at the bottom part, tensile stress will occur. Therefore, it is deemed necessary to assess the ability of a surface layer including the porous asphalt layer to accept loads that will cause compressive stress.

Figure 1. Surface layer including a layer made with porous asphalt when receiving vehicle loads

2. Material and Procedures

2.1. Modified Buton Asphalt (MBA)

This study uses an MBA produced by a national company that processes rock asphalt. The MBA used in this study can be widely obtained in the market. Table 1 shows the physical properties of the MBA. Most of the physical properties of the MBA meet the requirements of the bitumen used to make concrete asphalt mixtures.

Table 1. Physical properties of Modified Buton Asphalt (MBA)

No	Properties	Value	Unit
1	Penetration at 25°C	78.6	0.1 mm
2	Softening Point	52.0	°C
3	Ductility	114.0	cm
4	Flash Point	280	°C
5	Density	1.12	
6	Loss on Heating TFOT	0.3	% wt
7	Penetration after loss on heating	86	%
8	Viscosity 135 Cst (Temp. mixing)	1826	°C

Table 2. Physical properties of coarse aggregate and filler

No.	Properties	Crushed stone	Limestone powder filler
1	Specific gravity		
	a. Bulk/dry	2.62	2.55
	b. Surface saturated dry	2.68	2.61
	c. Apparent	2.77	2.71
2	Abrasion (%)	25.72	-
3	Unit weight (kg/m^3)		
	a. Loss	1.30	1.08
	b. Compact	1.49	1.32
4	Flakiness (%)	10.23	-
5	Sand equivalent (%)	-	85.71
6	Absorption (%)	2.07	2.35

2.2. Coarse Aggregate and Filler

River stone that have been processed into crushed stone was used as coarse aggregate in this study. The largest size of coarse aggregate passed the 20 mm sieve. The water absorption, particle size distribution, density and fineness modulus of the aggregates met the specification of of material for road pavement. The physical properties of the coarse aggregate and filler are given in Table 2.

2.3. Sample Preparation

According to the selected materials characteristics, the optimum MBA was 6.0% of the total aggregate weight. Specimens of porous asphalt mixtures were prepared using different ratios of waste PET (0–1.5%) and lime powder filler content was observed to be 10% of the total aggregate weight. The mixtures matrix is shown in Table 3. Impact compaction of Marshall Hammer was applied with 50 blows to each face of the specimens. Porous asphalt mixture was prepared in amount to produce specimen weighting 1,600 g. The specimens were mixed and compacted at temperature of 135–140°C.

Table 3. Porous asphalt mixture composition (1,600 grams)

No	Description	Unit	Weight in the mixture			
1	Waste plastic PET	%	0	0.5	1.0	1.5
2	weight	gr	0	7.52	15.04	22.56
3	MBA weight	gr	96.00	96.00	96.00	96.0
	Number of sieve					
4	3/4"	gr	0.00	0.00	0.00	0.00
5	1/2"	gr	38.27	37,02	35,76	34,51
6	3/8"	gr	638,60	657,35	636,09	634,84
7	No. 4	gr	601.53	600,28	599,02	597,77
8	No. 8	gr	75.20	73,95	72,69	71,44
9	No. 200	gr	98.21	96,96	95,70	94,45
10	Pan (Lime)	gr	52.19	50,94	49,68	48,43
11	Total aggregate weight (4+5+6+7+8+9+10)	gr	1504.00	1504	1504	1504
12	Test object weight (3+11)	gr	1,600.00	1,600.00	1,600.00	1,600.00

Figure 2. Equipment of compressive strength test

2.4. Compressive Strength Test

Figure 2 shows the compressive strength test. Axial strain was derived from two linear variable differential transducers (LVDTs). Compressive strength was conducting according to SNI 03-6758-2002 [14].

3. Result and Discussion

3.1. Visual Observations

Visual observations at the time of mixing showed that waste PET that has reached the softening point can be fused with limestone powder and MAB. No accumulation of the softened waste PET or limestone powder at one point. Homogeneous mixing of materials (coarse aggregates, limestone powder, waste PET and MAB) results coarse aggregates well covered by MAB. Likewise, the results of visual inspection on the test specimens that have been compacted, no visible buildup of PET waste that has been softening or limestone powder coagulation in one place. The compaction results showed good adhesion can unite the waste PET, limestone powder and coarse aggregate particles. Connected porosities were formed firmly from coarse aggregate particles which are well covered by MAB, PET waste and limestone powder.

3.2. Compressive Strength and Peak Strain

The compressive strength test results are shown in Table 4. The compressive strength of porous asphalt specimens that did not contain waste PET was 0.83 N/mm^2, while specimens containing waste PET of 0.5%, 1.0% and 1.5% were 1.05, 1.19, and 1.28 N/mm^2, respectively. The magnitude of the increase in compressive strength of specimens containing PET waste by 0.5%, 1.0% and 1.5% was 26.50, 43.37, and 54.21%, respectively, compared to specimens without waste PET. Compressive strength increased as the percentage of waste PET added rose.

The peak strain at peak compressive stress are shown in Table 5.The peak strain that correlated with compressive strength of porous asphalt specimens without waste PET was 0.06 mm/mm, while specimens containing waste PET of 0.5%, 1.0% and 1.5% were 0.04, 0.05, and 0.04 mm/mm, respectively. Referring to Table 4, it can be noted that waste PET did not have a significant effect on peak strain of the porous asphalt mixture prepared with MBA.

Table 4. Compressive strength test results

Waste PET content (%)	Compressive strength (N/mm^2)	Increament (%)
0	0.83	100
0.5	1.05	26.50
1.0	1.19	43.37
1.5	1.28	54.21

Table 5. Peak strain

Waste PET content (%)	Peak strain (mm/mm)
0	0.06
0.5	0.04
1.0	0.05
1.5	0.04

4. Concluding Remarks

The research described in this paper has four basic objectives, namely developing the use of MAB which is a derivative of BRA, reducing waste plastic by utilizing waste PET as an additive, strengthening the use of limestone powder which is widely available in Indonesia. All of the mentioned materials can be used to make parous asphalt mixture that had the adequate ability to bear the compressive load.

Acknowledgments

Authors wishing to acknowledge assistance or encouragement from Civil Engineering Department, Faculty of Engineering that support and contribute in the research.

References

[1] REAM – SP 5/2008, Spesification for Porous Asphalt, Road Engineering Assosiation of Malaysia.

[2] J. S. Chen., S. F. Chem., and M. C. Liao., 2015, Laboratory and field evaluation of porous asphalt concrete", Asian Transport Studies vol. 3 Issue 3, pp. 298–311.

[3] D. H. Shen., C. M. Wu., and J. C. Du., 2008, Performance evaluation of porous asphalt with granulated synthetic lightweight aggregate, Construction and Building Materials Vol. 22, 2008, pp. 902–910.

[4] D. Raabe., N. Chen., 2004, Recrystalization in Deformed and Heat PET Polymer Sheets. Mater Sci Forum; 467-470:551-6.

[5] R.J. Seyler., 1997, Semi-Crystalline Polymers. Two Phases or Three? An Overview and Perspective. J Therm Anal; Vol. 49, pp. 491-498.

[6] O. Gueguen., S. Ahzi., A. Makradi., S. Belouttar., 2010, A New Three-Phase Model to Estimate The Effective Elastic Properties of Semi-Crystalline Polymers: Application to PET. Mech Mater; 41:1-10.

[7] Z.D. Cheng., editor. 2007, Handbook of thermal analysis and caloimetry. Gallagher Patrick K, series editor. Applications to polymers and plastics, vol. 3. Elsevier.

[8] M.Y. Keating., B. Malone., and W. D. Saunders., 2002, Annealing effect on semi-crystalline materials in creep behavior. J Therm Anal Calorim; 69:37–52.

[9] J.M. Li., C. Liu., H.C. Qiao., L.Y. Zhu, G. Chen., and X. D. Dai., 2008, Hot embossing/bonding of a poly(ethylene terephthalate) (PET) microfluidic chip. J Micromech Microeng;18-015008:10pp.

[10] Ahmadinia E.; Zargar M.; Karim M. R.; Abdelaziz M.; and Shafigh P. 2011. Using Waste Plastic Bottles as Additive for Stone Mastic Asphalt. Material and Design 32 (2011) 4844–4849.

[11] E. Ahmadinia., M. Zargar., M.R. Karim., M. Abdelaziz., and E. Ahmadinia., 2012. Performance Evaluation of Utilization of Waste Polyethylene Terephthalate (PET) in Stone Mastic Asphalt. Construction and Building Materials Vol. 36 pp. 984–989.

[12] M.W. Tjaronge., and R. Irmawaty., 2012. Influence of Water Immersion on Physical Properties of Porous Asphalt Containing Liquid Asbuton as Bituminous Asphalt Binder. Third International Conference on Sustainable Construction Materials and Technologies http://www.claisse.info/Proceedings.htm.

[13] Directorate General of Highways Department of Public Works., 2006. Construction and Building Guidelines No: 001 - 01 / BM / 2006, Utilization of Buton Asphalt: General.

[14] Standard National of Indonesia., 2002, Standard Test Method of Compressive Strength Test of Asphalt Mixture. SNI 03-6758-2002.

Prototype of Electrical Generator Device from Piezoelectric with Serial Configuration - Based on Mechanical Pressure

Yusran[1]*, A A Madia[1], and Z Zainuddin[1]

[1]Department of Electrical Engineering, Engineering Faculty, Hasanuddin University, Makassar, Indonesia

*Email: yusranibnu@yahoo.com

Abstract. This study discussed about of prototype of electrical generator device from piezoelectric with serial configuration - based on mechanical pressure. A prototype of a piezoelectric-based power plant that is able to convert mechanical energy (footing) or object load on piezoelectric into electrical energy was the aim of this research. The piezolectric generation system can be applied to serve limited loads and can be a support system for a distributed generation system. The method of research was device design and experimental testing. The experimental testing was carried out by varying pressures on piezoelectric to produce the desired output voltage. In testing, the pressure was generated by two different objects, a human footing and a object load with varying weight. The average voltage produced on human footing was 9.736 V for 60 kg body weight. The average voltage produced on object loads was 1.33 V for 5 kg object weight. This research has succeeded in making a prototype of a power plant using piezoelectric.

1. Introduction

Electrical energy can be converted from various sources of energy, both non-renewable or renewable. The energy source can be heat energy, motion energy, sun energy and other forms of energy. Electricity can be used to supply both high and low energy consumption equipment. A number of electronic equipment consume low energy which is generally supported by batteries.

This study discusses the prototype of alternative power plants. Electrical power is generated through the conversion of kinetic energy generated from human movements in the form of mechanical pressure through a piezoelectric harvesting system [1]. This prototype can be used to produce electricity for the low-power electronic instruments consumption.

Specifically, the purpose of this research is to make a prototype of a piezoelectric-based power plant that is able to convert mechanical energy (footing) or object load on piezoelectric into electrical energy. In this study, the magnitude of voltage generated by the prototype is also measured. The piezolectric generation system can be applied to serve limited loads and can be a support system for a distributed generation community [2-4].

The research with aim to design energy hybrid harvesting system was carried out in 2011 by Wei C K and Ramasamy G [5]. The system is combination of mechanical and piezoelectric harvesting system. The another research by Pisharody H G in 2011 [6]. Piezoelectric transducers were planted on a mat. The paper studied about a number factor that influence the system design for piezoelectric energy harvesting was conducted by Ali W G and Nagib G in 2012 [7].

IOP Conf. Series: Materials Science and Engineering **875** (2020) 012033 doi:10.1088/1757-899X/875/1/012033

2. Piezoelectric System

Piezoelectric material is a material that when exposed to voltage or mechanical strain will generate an electric field. On the other hand, if the material is exposed to an electric field, the material will experience mechanical strain or stress. There are two types of piezolectric materials based on the manufacturing process, namely natural and artificial. Quartz, Berlinite, Tourmaline and Rossel Salt are examples of natural materials. Meanwhile, Barium Titanate (BaTiO3), Lead Zirconium Titanate (PZT), Lead Titanate (PbTiO3) are artificial materials.Piezoelectric can be formed by several different types of material which can be classified into three parts [1].

- Crystal material, for example: Quartz (SiO2), Gallium Orthophosphate (GaPO4)
- Ceramic material, for example: Barium Titanate (BaTiO3), Lead Zirconate Titanate (PZT)
- Polymeric material, for example: Polyvinylidene Diflouride (PVDF)

The displacement of the electric charge density in piezoelectric material can be expressed through Equation (1).

$$D = \varepsilon E \tag{1}$$

where:

D = transfer of electric charge density
ε = permittivity
E = electric field strength

while the equation for finding a strain based on Hooke's Law is stated by Equation (2).

$$S = sT \tag{2}$$

where:

S = strain
s = compliance
T = pressure

These two equations, when combined will produce Equations (3) and (4).

$$\{S\} = \left[s^E\right]\{T\} + \left[d^E\right]\{E\} \tag{3}$$

$$\{D\} = [d]\{T\} + \left[\varepsilon^t\right]\{E\} \tag{4}$$

where

[d] = matrix for direct piezoelectric effects
$[d^t]$ = matrix for the opposite piezoelectric effect
power E = zero electric field, or constant
power T = zero field pressure, or constant
power t = transposition of a matrix

3. The prototype

The initial step taken was to design a design that will be applied in a prototype. Figure 1 shows the electrical circuit design of the prototype. While. Figure. 2 shows the physical design of the prototype.

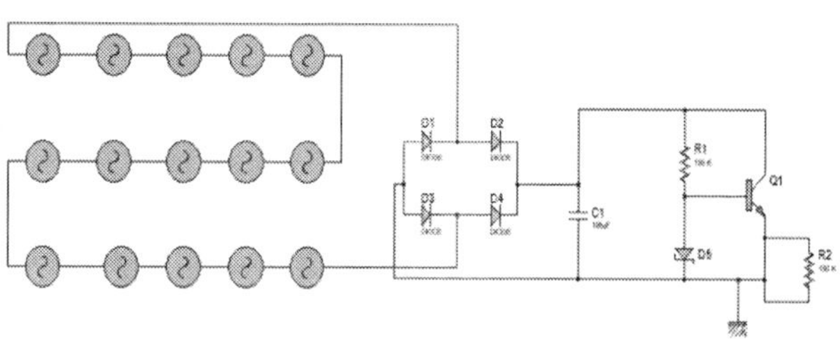

Figure 1. Electrical circuits design of prototype

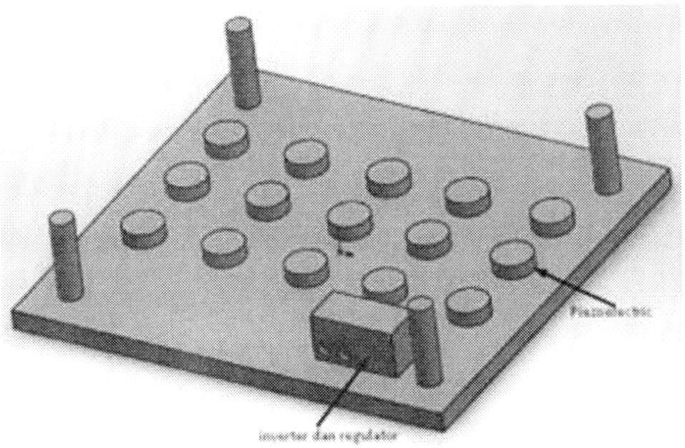

Figure 2. Prototype physical design

The second step was the process of assembling and making prototypes. The assembly and manufacturing process conducted through two stages. The first stage was the process of making a prototype frame that serves as a place for laying piezoelectric and other electronic components. The design of the framework on the prototype as shown, to protect connecting cables and diodes. It also serves as a place for piezoelectricity that will function to change mechanical pressure into electrical energy.

Figure 3. Prototype display

The second stage was the process of piezoelectric assembly and the electronic components in the prototype. Electronic components contained in the prototype include piezoelectric, diode, resistor, capacitor. In addition, there was a converter circuit that will convert the AC output voltage into DC and a cable (connector) that will connect all the electronic components. The physical appearance of the prototype is shown in Figure 3.

4. Results and Discussion

The type of piezoelectric material was used is PZT type ceramics. The series circuit testing was carried out using 15 piezoelectric pieces with 3 experiment sets. The process of measuring the voltage generated by the prototype was carried out using a digital multimeter. Human body weight at intervals of 20 kg to 60 kg with functions to produce mechanical stress that is correlated with the output voltage. In addition, object loads were also used at intervals of 1 kg to 5 kg.

4.1. Testing with Human Footing

The output voltage generated by the prototype in a series circuit with human footing is shown in Table 1.

Table 1. Testing result with human footing

No	Mass (kg)	Voltage (V)			
		Exp. 1	Exp. 2	Exp. 3	Average
1	20	4.0	3.8	3.8	3.9
2	35	6.0	6.2	6.0	6.1
3	42	6.7	6.5	6.5	6.6
4	49	7.7	7.6	7.6	7.6
5	55	9.9	9.5	9.8	9.7
6	60	10.6	10.3	10.2	10.4

The results in Table 1 show that testing with a weight of 20 kg produces an average voltage of 3.9 volts. The last test with a body weight of 60 kg produced an average voltage of 10.4 volts. The data generated by 3 experiments set on prototype series circuits for human footing shows a difference that is not too large. Although there are some data that experienced a decrease in output voltage, it does not show such a large difference. The difference in the three sets of experiments above can be seen through Figure 4.

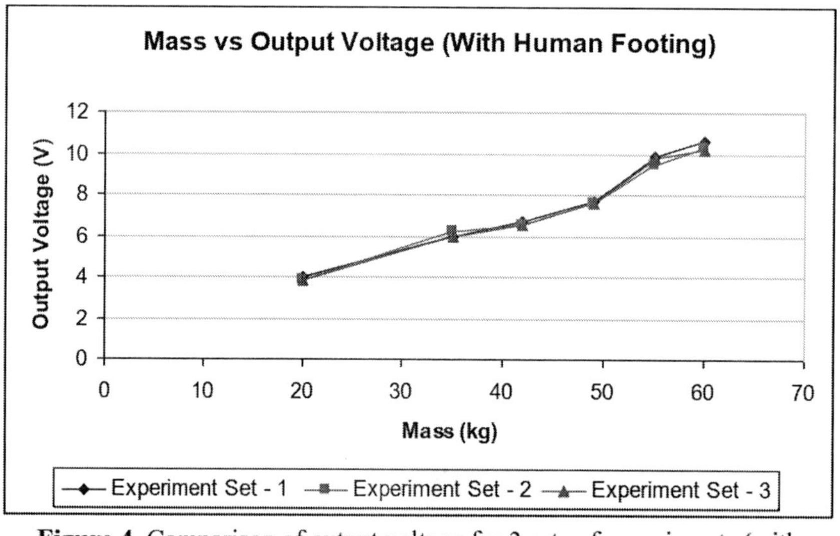

Figure 4. Comparison of output voltage for 3 sets of experiments (with human footing)

4.2. Testing with Object Loads

The output voltage generated by the prototype in a series circuit with an object load is shown in Table 2.

Table 2. Testing result with object load

No	Mass (kg)	Voltage (V)			
		Exp. 1	Exp. 2	Exp. 3	Average
1	1	0.1	0.09	0.09	0.093
2	2	0.4	0.6	0.6	0.53
3	3	0.6	0.8	0.8	0.73
4	4	0.8	0.9	0.9	0.87
5	5	1.2	1.4	1.4	1.33

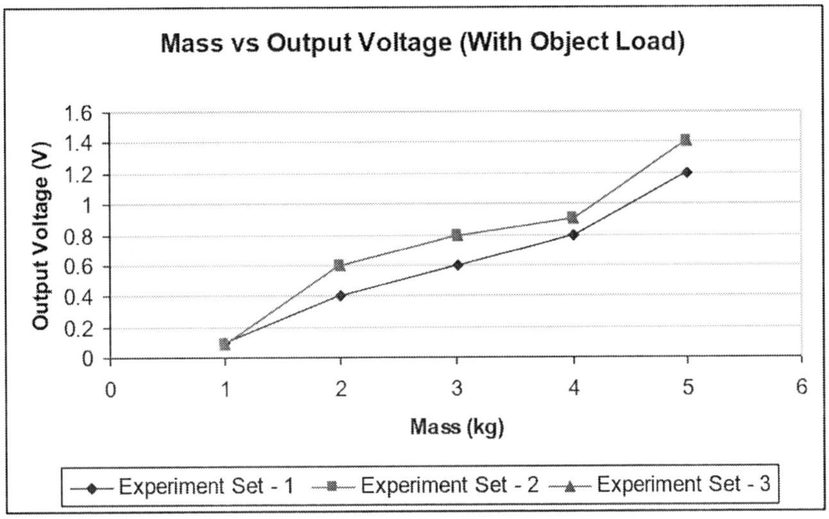

Figure 5. Comparison of output voltage for 3 sets of experiments (with object loads)

The results in Table 2 show that a mass of 1 kg produced an average voltage of 0.093 volts. The last test by using a mass of 5 kg produced an average voltage of 1.33 volts. The resulting output voltage increases when given an increase in load (mass) even though the output produced is not as large as that produced by a series of circuits on a human footing. The difference in the three sets of experiments above can be seen through Figure 5.

5. Conclusion

This research has succeeded in making a prototype of a power plant using piezoelectric. This prototype is able to convert mechanical energy derived from human footing and object load pressure into electrical energy. The largest average voltage generated in a series circuit for a footing load of people weighing 60 kg was 10.4 volts. As for the load object with a weight of 5 kg produced an average voltage of 1.33 volts

References

[1] Ebrahimi F (ed) 2013 *Piezoelectric materials and devices-practice and aplications* InTech Rijeka Croatia.
[2] Yuli Asmi Rahman, Salama Manjang, Yusran, Amil Ahmad Ilham 2018 Distributed generation's integration planning involving growth load models by means of genetic algorithm *Archives of Electrical Engineeering* 67(3) p. 667-682.
[3] Yusran 2014 Electrical network power quality improvement through distributed generation optimum placement based on breeder genetic algorithm method *Proc. The 4th Makassar International Conference on Electrical Engineering and Informatics* (Makassar) p. 20-22.
[4] Yusran 2019 Electromagnetic field impact on 150 kV Raha-Baubau transmission line *IOP Conf. Ser.: Earth Environ. Sci.* 235 012107.
[5] Wei C K and Ramasamy G 2011 A hybrid energy harvesting system for small battery powered applications Proc. *IEEE Conference on Sustainable Utilization and Development in Engineering and Technology (STUDENT)* Semenyih p. 165-170.
[6] Pisharody H G 2011 An optimal design for piezoelectric energy harvesting system *Proc.* ISGT 2011-India, Kollam, Kerala, pp. 244-247.
[7] Ali W G and Nagib G 2012 Design considerations for piezoelectric energy harvesting systems *Proc.* International Conference on Engineering and Technology (ICET) Cairo p. 1-6.

IOP Conf. Series: Materials Science and Engineering **875** (2020) 012034 doi:10.1088/1757-899X/875/1/012034

Potential of Heart Disease Detection Based on Iridology

C Yohannes*[1], I Nurtanio[1], and K C Halim[1]

[1]Department of Informatics, Faculty of Engineering, Universitas Hasanuddin, Makassar Indonesia

*Email: christoforus@unhas.ac.id

Abstract. Every year the number of deaths caused by heart disease continues to increase. This is mostly caused by misdiagnosis or misinterpretation of the heart disease symptoms. Therefore, people need to be aware of this disease by maintaining a healthy lifestyle and conducting regular checks on the potential for heart disease. This early examination can be done using the iridology method, namely by analyzing the iris of the eye. This paper presents the implementation of computer vision and machine learning of iridology to detect the potential for heart disease. This system uses Canny edge detection and Principal Component Analysis (PCA) to extract features in the iris region of the eye, and Backpropagation Algorithm of Artificial Neural Networks to create the predictive model. There are 110 data used in this system, consisting of 55 eye images from subjects suffering from heart disease and 55 images of normal subjects. The data is divided into 88 data for training and 22 data for testing. The proposed system produced accuracy up to 95.45% for the test data using the sigma 0.3 for canny edge detection, 50 principal components, 50 hidden neurons, and 0.01 for the error limits.

1. Introduction

Iridology is a method that can show health problems in various organs in the body by analyzing the colored areas in the eye, or also called iris. Iridology was popularized by Dr. Ignatz Van Peczely in 1881 in his book "Discoveries in the Field of Natural Science and Medicine". This method has been practiced for hundreds of years in various forms, including by utilizing digital imaging to determine the patient's health condition [1].

Cardiovascular and heart disease has become the main cause of about 17 million deaths in the world each year. WHO on its publication "The Atlas of Heart Disease and Stroke" stated that by 2020, heart disease and stroke will become the leading cause of both death and disability worldwide. They also recommend several precautions against the risk of this disease, one of which is by conducting periodic health checks [2]. Therefor this paper introduces a system for early detection of potential heart disease based on iridology.

This paper presents a system that uses Canny edge detection and Principal Component Analysis (PCA) as a method for extracting cardiac features in the iris. Canny edge detection is used to detect features in the form of blemishes in the heart region of the iris, while PCA is used to extract the strongest features found in the Canny. The results of the PCA are then used as input to the backpropagation neural network.

Research related to the detection of potential heart disease based on iridology has previously been conducted. The research uses the Canny edge detection and PCA to extract features. While the method used for classification is Support Vector Machine. This research achieved 100% accuracy for the training process, and 80% for the testing process [3].

Other research using artificial neural networks has also been carried out for cases of identification of autonomic nerve damage based on iridology. This research uses PCA for feature extraction and backpropagation neural network to do the classification. The data used in this study amounted to 80, consisting of 40 data from subjects suffering nerve damage and 40 data from normal subjects. This research produced the highest disease recognition rate of 83.33% [4].

2. Research Method

2.1. Iridology

Iridology is a method of medical examination popularized by Dr. Ignatz Van Peczely in 1881. The method of iridology works by analyzing the colored parts of the eye or commonly called Iris. Iris is the most complex network that can be seen from outside of the body. Iris is connected to hundreds of thousands of nerve endings, microscopic blood vessels, muscles, and other tissues. In other words, the iris is connected to all organs and tissues of the body through the brain and nervous system so that through the iris can be known the health conditions of various organs in the body.

Based on the iridology chart developed by Dr. Bernard Jensen, the heart organ can be clearly identified in the left eye with a position in the direction of 2 to 3 o'clock. Figure 1 shows the mapping of organs based on iridology chart by Dr. Bernard Jensen.

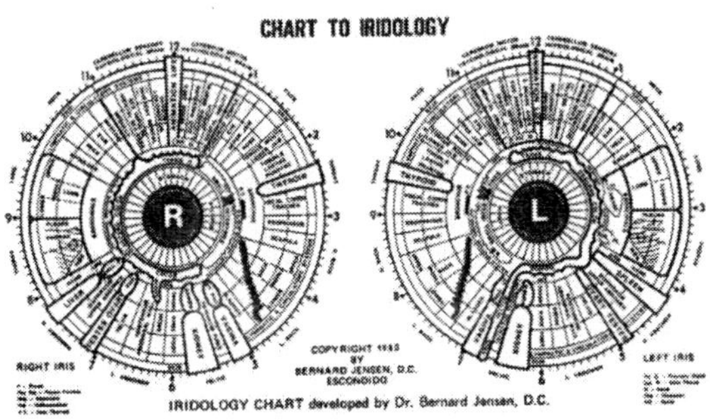

Figure 1. Iridology Chart by Dr. Bernard Jensen [5]

2.2. Dataset

In this study, 110 eye images consisting of 55 normal images and 55 images with heart disease were used. Data was obtained from one clinic that provides iridology service and was taken from a variety of different people in the same lighting conditions. The image used is in the .JPG format with a size of 500 × 500 pixels.

The first step in data preparation is to do manual isolation of the iris portion of the eye image for all existing images. After that, the pupil region is masked into black to facilitate the Region of Interest (ROI) selection process.

In this study, the data is divided into train and test data. For the training data, 88 data were used, consisting of 44 normal data and 44 data with heart disease. As for the test data, 22 were used consisting of 11 normal data and 11 data with heart disease.

2.3. Image Pre-processing

The initial stage in this system is data preprocessing. Preprocessing conducted in this study included converting RGB image into grayscale, histogram equalizing, smoothing, and pupil masking. Converting from RGB to grayscale is used to reduce the system computing load from three color channels to one channel. Histogram equalization method used in this study is Contrast Limited Adaptive Histogram Equalization (CLAHE). Histogram equalization is used to increase the contrast between the iris background and the iris fiber features. While smoothing is used to reduce noise arising from increased contrast in the previous stage. Smoothing method used in this study is a simple median filtering method. Figure 2 shows the results of the preprocessing performed.

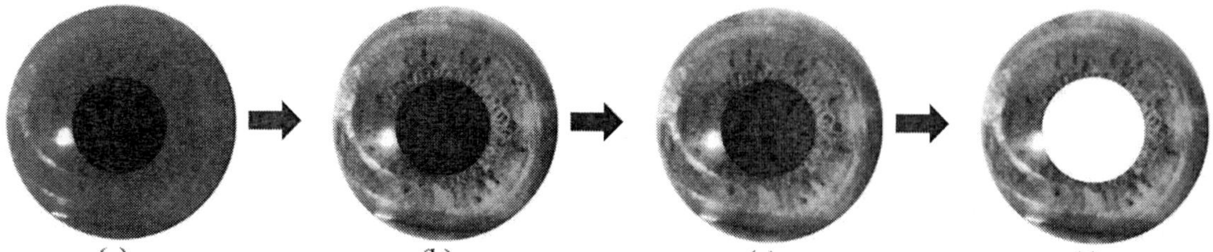

Figure 2. (a) Grayscale Image, (b) Histogram Equalized Image, (c) Smoothened Image, (d) Masked Pupil Image

2.4. Region of Interest (ROI) Selection

ROI is an area that contains features that will be used to recognize the condition of the heart in the iris. Based on the information of iridologists, the ROI position of the heart in the iris image is in the direction of 2 to 3 o'clock, so the system will only take the region as input. The ROI selection process is carried out automatically by the system.

The initial stage of choosing ROI is the process of transforming iris images after preprocessing into a linear image. The transformation used in this process is based on warpPolar function on [6]. This process will change the initial image of a circular iris into a linear form as shown in Figure 3(a). After that, the ROI of the heart can then be taken from a linear image. For the cardiac ROI, the portion of iris to be taken is only 50% of the iris band starting from the end of the pupil to the middle of the iris band as in Figure 3(b). The results of the ROI selection are then resized to a size of 50 × 50 pixels. The results of the ROI selection can be seen in Figure 3(c).

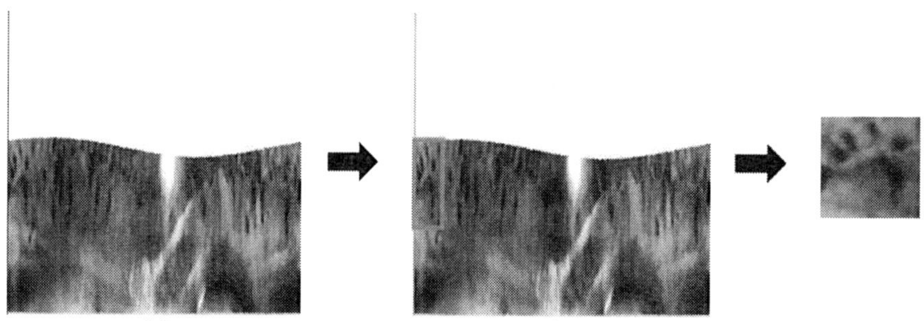

Figure 3. (a) Transformed Image, (b) Target ROI, (c) Resized ROI

2.5. Canny Edge Detection and Principal Component Analysis (PCA) for Feature Extraction

Canny edge detection is used to extract edges from iris fibers in ROI. The Canny edge detection method used has two main parameters namely *low threshold* and *high threshold* where the determination of these parameters is done automatically using a sigma value. Calculation of low threshold and high threshold values is done with the following Equation (1) based on [7]:

$$\left.\begin{aligned} low &= \max(0, (1.0 - sigma) \times median \\ high &= \min(255, (1.0 + sigma) \times median \end{aligned}\right\} \quad (1)$$

In this study an experiment was conducted on five sigma Canny values, namely 0.1, 0.2, 0.3, 0.4, and 0.5. For each sigma value, different accuracy will be obtained in the system prediction results. The best accuracy is obtained at sigma value 0.3. The results of Canny edge detection are shown in the following Figure 4.

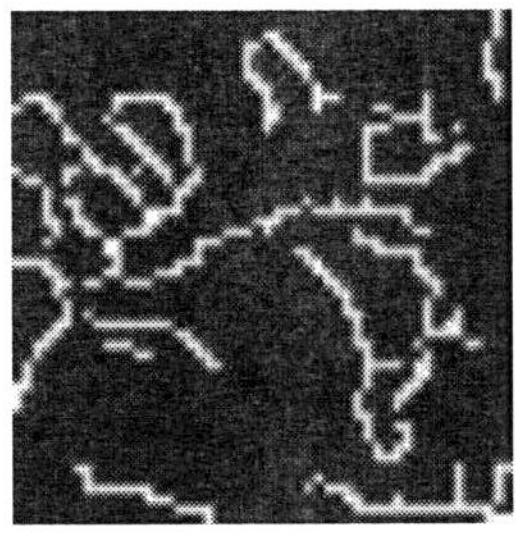

Figure 4. Canny Image for Sigma 0.3

All previous Canny edge detection images are then flattened to produce image array with a size of 1 × 2500 pixels, then combined into a dataframe. PCA is then used to retrieve the most important features or also called Principal Components (PC) contained in the data frame. The PC determination is done by determining the eigen value and eigenvector of each feature column. Some features with the highest eigen value can be taken as a PC to be used as input in neural networks. This PC selection process can also reduce the computational burden because it only takes a portion of all the features available. Broadly speaking, the PCA process is explained as follows according to [8].

The first stage in PCA is to standardize the range of values of data variables so that each variable has the same contribution to the analysis. The standardization process is done by subtracting the value of the variable with the average value of the variable as in Equation (2).

$$z = value - mean \tag{2}$$

The next step is to calculate the covariance matrix to find out whether there is a relationship between variables in the data frame. This is also done to find out whether there is information redundancy in the dataframe so that it can be removed. This calculation is done for each dimension contained in the dataframe as shown in the Equation (3).

$$C(X,Y) = \frac{\sum_{i=1}^{n}(X_i - \bar{X})(Y_i - \bar{Y})}{(n-1)} \tag{3}$$

Next is to calculate the eigen vector and eigen value of the covariance matrix to determine the PC. Determination of the features that become PCs is done by sorting the eigen vector based on the eigen value starting from the largest to the smallest. The greater the eigen value, the more information contained in the component.

After sorting the eigen vector, we can then choose the number of PCs that will be used as input in the neural network. This can be done by using the entire eigen vector or by removing a portion of the eigen vector that has a small eigen value. Discarding the eigen vector with a small eigen value can reduce the burden and system computing time because the amount of data to be processed is getting smaller.

After determining the eigen vector to be used, the feature reconstruction is carried out by multiplying the transposed standardized data by transposed feature vector as in Equation (4).

$$Final\ Set = FeatVector^{T} \times StandarizedOriDataset^{T} \tag{4}$$

In this study, several experiments were carried out for the number of PCs including 10, 20, 30, 40, and 50 where the best accuracy was obtained on the 50 PCs.

2.6. Artificial Neural Network Backpropagation for Classification

Backpropagation algorithm is one algorithm that is widely applied to solve machine learning problems. This algorithm has an architecture consisting of several layers including the input layer, hidden layer, and output layer. Where hidden layer can consist of one or more hidden layers and each hidden layer can consist of more than two neurons. The architecture of backpropagation artificial neural networks is shown in Figure 5.

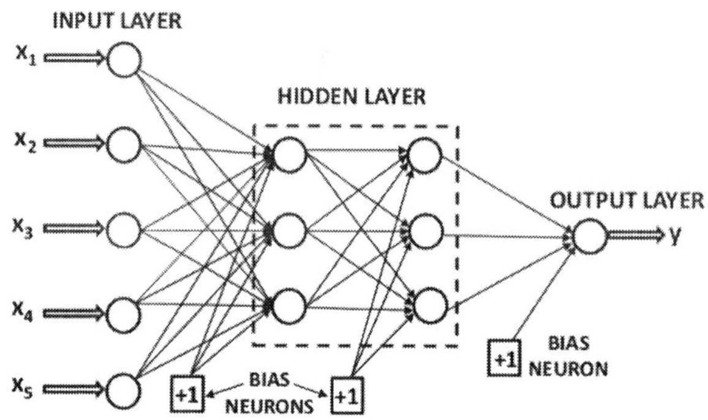

Figure 5. ANN Backpropagation Architecture

Backpropagation algorithm is divided into two stages, namely the training phase and the testing phase. The training phase aims to obtain the right weight values to be used in calculations so that they can produce accurate predictions. The training phase is divided into two processes, namely *feed forward* and *backpropagation*, where the feed forward process will be tasked to produce an output value, while backpropagation is to improve the initial weight value based on the error value obtained during the feed forward process. For the testing phase, the process that occurs is only the feed forward process that takes place using the final weight value obtained at the training stage [9].

The backpropagation algorithm in this study is used as a machine learning model that will produce predictions from iris images. The resulting prediction will be the presence or absence of potential heart problems experienced by the patient. In this case, the system output will be coded into 1 and 0, respectively representing potential and normal.

The architecture used in this study consists of an input layer, a hidden layer, and an output layer, where the number of neurons in the input layer follows the number of PCs used, and the number of neurons in the output layer is two for each class. As for the number of neurons in the hidden layer, an experiment is performed on several values including 10, 20, 30, 40, and 50. In addition to these parameters, this model also limits the number of iterations by 20,000 epochs, and uses error limits of 0.1 and 0.01. In this study, the best accuracy was obtained in model with the number of hidden neurons of 40 and 50 and at an error limit of 0.01.

3. Result and Analysis

The system was tested on machine with Intel(R) Core(TM) i7-5500U CPU @ 2.4GHz, 8GB of RAM, and operating on Windows 10 64-bit. The backpropagation model uses the same initial weight value for each experiment.

3.1. Results comparison for each Canny's sigma value

The best result from system testing are obtained at sigma value of 0.3 for the number of PCs as much as 50, 50 hidden neurons, and the error limit of 0.01 with an accuracy score of 95.455%. The accuracy comparison table between sigma values is shown in Table 1.

Table 1. Results comparison for each Canny's sigma value

Sigma Value	Classification Accuracy
0.1	72.727%
0.2	77.273%
0.3	95.455%
0.4	77.273%
0.5	77.273%

3.2. Results comparison for each Principal Component size

Determination of PC size greatly affects the performance of the backpropagation network model created. This is because the size of the PC also acts as the number of neurons contained in the input layer, so that it can affect the learning and testing of the system. The best accuracy of the system is obtained at the number of PC 50 for sigma Canny 0.3, 50 hidden neurons, and error limit 0.01. A comparison table of results for each PC size is shown in the following Table 2.

Table 2. Results comparison for each PC size

Principal Components	Classification Accuracy
10	68.182%
20	77.273%
30	59.091%
40	86.364%
50	95.455%

3.3. Results comparison for each number of hidden neuron

The number of hidden neurons can also affect the performance and results of the backpropagation network. The number of hidden neurons will affect the number of weights that connect between neurons at each layer so that it can affect the calculation results.

Table 3. Results comparison for each number of *hidden neurons*

Number of Hidden Neurons	Classification Accuracy
10	77.273%
20	81.818%
30	77.273%
40	95.455%
50	95.455%

In this study, the best accuracy result was obtained on the number of hidden neurons as much as 40 and 50 for sigma Canny 0.3, PC size 50, and error limit 0.01. The comparison table of the accuracy results for each number of hidden neurons is shown in Table 3.

3.4. Results comparison for each maximum error limit value

The maximum error limit value is used to limit the learning process carried out during training. Determination of error limits is done to prevent overfit so that the model can make better predictions against data that has never been encountered before. In this study, an experiment was conducted on two error limit values, 0.1 and 0.01, where the best accuracy was obtained at the error limit 0.01 with a sigma value of 0.3, PC size 50, and the number of hidden neurons 50. Accuracy comparison tables for each error limit value can be seen in Table 4.

Table 4. Results comparison for each maximum error limit value

Error Limit Value	Classification Accuracy
0.1	90.909%
0.01	95.455%

4. Conclusion

Detection of potential heart disease based on iridology can be done using Canny edge detection and Principal Component Analysis (PCA) for feature extraction, and Backpropagation Artificial Neural Networks as classifiers. The data used in this paper amounted to 110 data which were divided into 88 training data and 22 test data. The best accuracy was obtained in experiment using sigma 0.3, PC size of 50, hidden neurons by 50, and error limits 0.01, where the resulting accuracy was equal to 95.45%. Accuracy achieved successfully passed the accuracy of previous studies related to the detection of potential heart disease based on iridology.

Acknowledgment

This research is supported by Lab Based Education Grant 2019, Faculty of Engineering, Hasanuddin University.

References

[1] "10 Conditions That Iridology Can Detect | Better Living Wellness Clinic," 20 Februari 2017. [Online]. Available: https://www.betterliving.co/2017/02/20/10-conditions-that-iridology-can-detect/.

[2] "WHO publishes definitive atlas on global heart disease and stroke epidemic," World Health Organization, 23 September 2004. [Online]. Available: https://www.who.int/mediacentre/news/releases/2004/pr68/en/.

[3] L. I. Permatasari, A. Novianty and T. W. Purboyo, "Deteksi Gangguan Organ Jantung Menggunakan Komputerisasi Iridologi Dengan Metode Klasifikasi SVM (Detection of Heart Organ Disorders Using Computerized Iridology With SVM Classification Method)," 2016.

[4] E. S. Wicaksono, I. Santoso, A. A. Zahra and R. R. Isnanto, "Identifikasi Kerusakan Saraf Autonomik Melalui Citra Iris Mata Menggunakan Ekstraksi Ciri Analisis Komponen Utama (PCA) dan Jaringan Saraf Tiruan Perambatan Balik (Identification of Autonomic Nerve Damage through Iris Eye Images Using Feature Extraction Analysis of Main Components (PCA) and Back Propagation Neural Networks)," 2017.

[5] B. Jensen, Iridology Simplified: An Introduction to the Science of Iridology and Its Relation to Nutrition, Book Pub Co, 2011.

[6] "OpenCV Geometric Image Transformations," [Online]. Available: https://docs.opencv.org/3.4/da/d54/group__imgproc__transform.html#ga49481ab24fdaa0ffa4d3e63d14c0d5e4.

[7] A. Rosebrock, "Zero-parameter, automatic Canny edge detection with Python and OpenCV," 6 April 2015. [Online]. Available: https://www.pyimagesearch.com/2015/04/06/zero-parameter-automatic-canny-edge-detection-with-python-and-opencv/.

[8] L. I Smith, A Tutorial on Principal Components Analysis, 2002.

[9] C. C. Aggarwal, Neural Networks and Deep Learning: A Textbook, Springer International Publishing, 2018.

IOP Conf. Series: Materials Science and Engineering **875** (2020) 012035 doi:10.1088/1757-899X/875/1/012035

Allocation of Wind Power Plant Construction Using Adaptive Neuro-Fuzzy Inference System

H Jasman[1]*, S M Said[1], and I Nurtanio[2]

[1]Electrical Engineering Department, Engineering Faculty, Hasanuddin University, Makassar, Indonesia
[2]Informatic Engineering Department, Engineering Faculty, Hasanuddin University, Makassar, Indonesia

*Email: henra.jasman.te10@gmail.com

Abstract. This research aims to determine the feasibility of a location to build Wind Power Plant (WPP). One of the challenges in wind energy development is to find area or location with high suitability for optimizing wind farm to allow the installation of as many wind turbines as possible within a limited area. To construct as many wind turbines as possible, the Adaptive Neuro Fuzzy Inference System (ANFIS) method is adopted using 7 parameters i.e wind potential, elevation, slope, type of land use, soil condition, access to roads, and populated areas and 2178 rules of fuzzy logic. The research was conducted in Kelara District, Jeneponto Regency, South Sulawesi, Indonesia. The most optimal result is obtained in the village of Tolo with a percentage of feasibility of 91% and error 0.48287. The output of this system can determine the feasibility of a location of wind turbine installation and can be used as a reference for determining wind farms in Indonesia and other countries.

1. Introduction

The increase in carbon dioxide (CO2) emissions caused by the consumption of fossil fuels has attracted worldwide attention because almost all of the world's gas emissions come from coal 44%, oil 36%, and natural gas 20%. One way to overcome the increase in CO2 emissions is the use of renewable energy sources to reduce dependence on fuels [1].

Renewable energy is energy made from sources that will never run out, or that can be replenished. The use of renewable energy sources is a form of commitment from countries in the world to reduce carbon emissions and limit changes in global average temperature. One of the possible renewables is wind. The wind is everywhere, easily available, easily converted, and does not require high operational costs [2].

One of the challenges in developing wind energy is to find areas or locations with high compatibility because building wind power plants requires a long concept and research [3]. Therefore, it is necessary to make a location mapping to determine the potential of an area based on the eligibility criteria for installing wind turbine sites. The more wind fields installed, the higher the amount of wind energy lost due to the of the turbine-to-turbine interaction. Optimizing wind farms are needed to achieve the highest efficiency of the wind and allow the installation of as many wind turbines as possible within a limited area. Therefore, it is necessary to optimize the installation of wind turbine locations at wind power plants [4].

Previously, research on location determination used a combination of Boolean logic, fuzzy logic, and Weighted Product methods to determine suitable locations for the development of wind power plants in Palladam Taluk, Tamil Nadu. The location selection process uses ten criteria, namely wind source, slope, hospital distance, river distance and water source, air force base and settlement distance, distance from the road and transmission, and distance from the forest. The results showed that 2.99% of the study area was classified as the most suitable for the development of wind power plants, 10.2% were found to be quite suitable, 0.535% was the least suitable, and 86.267% were permanently unsuitable [5].

Research conducted by Said et al shows that Kelera Sub-District, Jeneponto Regency has a lot of potential location suitable for building a Wind Power Plant. The research succeeded in getting a feasibility value according to the criteria needed by an area with the help of Geographic Information System (GIS), which was then calculated by the weighting or scoring method. However, the result obtained by weighting or scoring methods are still done manually to determine the feasibility of an area and the determination of wind turbine installation points [6].

Based on the description from previous studies, the authors propose a method that can optimize the determination of locations for the construction of wind power plants using Fuzzy Logic and Adaptive Neuro-Fuzzy Inference System (ANFIS) with the help of GIS. The output of this system can determine the feasibility of the location of wind turbine installation. Therefore, it can be used as a reference for determining wind farms in Indonesia and other countries.

2. Literature Review

2.1. Fuzzy Logic

A very important aspect of fuzzy logic is the use of fuzzy if-then rules. Although in the field of artificial intelligence systems that rely on rules have long existed, but the system has a deficiency in connecting the gap between fuzzy antecedents and fuzzy consequences. Fuzzy logic removes these shortcomings by adopting calculus practices for fuzzy rules. The ability to strike a balance between important factors and accuracy has made fuzzy logic an interesting field of research. Humans have practiced this balance for a long time. Therefore, fuzzy logic is simultaneously new and old.

Fuzzy systems have two main characteristics to ensure increased performance for specific applications. The two characteristics are as follows:

- Fuzzy systems are suitable for uncertain or estimated analyzes. In particular, fuzzy systems are suitable for applications that have complex mathematical derivations.
- Fuzzy logic can make decisions that have approximate values below partial or temporary information.

Instead of giving a right or wrong answer, fuzzy logic tries to integrate the decision-making process followed by humans. Therefore, fuzzy logic provides an approximate but efficient method for explaining system performance, which is usually very difficult to accurately illustrate. Thus, fuzzy logic can be applied to intelligent decision-making systems in unpredictable and dynamic environments [7].

2.2. Adaptive Neuro-Fuzzy Inference System

Adaptive neuro-fuzzy inference system (ANFIS) is a hybrid rule-based learning system that is a combination of artificial neural networks and fuzzy logic for logical calculations and analysis. The ANFIS structure, as shown in Figure 1, consists of nodes and directional links that connect nodes.

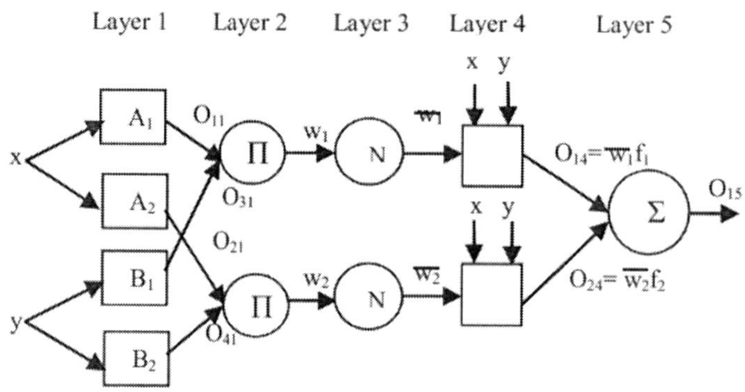

Figure 1. ANFIS architecture

These nodes may be adaptive when the output of the node depends on the parameters of the node ie, the output of the node depends on the output of the previous layer. The purpose of directional links is to connect nodes of various layers and provide a path for signal flow without weight requirements for modifying signals.

ANFIS structure consists of five layers. Each layer has a fixed or adaptive node, connected by a directional link. Layer 1 has four nodes namely A1, A2, B1 and B2. Each node is a fuzzy set which output gives ownership level of input to the fuzzy set. Outputs Ok1 (k = 1 to 4), the output of node k from layer 1 is given by equations (1) to (4).

$$O_{11} = \mu_{A1}(x) \tag{1}$$

$$O_{21} = \mu_{A2}(x) \tag{2}$$

$$O_{31} = \mu_{B1}(x) \tag{3}$$

$$O_{41} = \mu_{B2}(x) \tag{4}$$

Layer 2 consists of a fixed node that performs the multiplication function for the incoming signal. Equations (5) and (6) give the output of layer 2.

$$\omega_1 = \mu_{A1}(x) \times \mu_{B1}(y) \tag{5}$$

$$\omega_1 = \mu_{A2}(x) \times \mu_{B2}(y) \tag{6}$$

Node on layer 3 used for normalizing layer 2 output using equations (7) and (8).

$$\varpi_1 = \omega_1 / (\omega_1 + \omega_2) \tag{7}$$

$$\varpi_2 = \omega_2 / (\omega_1 + \omega_2) \tag{8}$$

The output of the layer 4 adaptive nodes depends on the consequence parameters of the node. The output is obtained using equations (9) and (10).

$$O_{14} = \varpi_1 f_1 \tag{9}$$

$$O_{24} = \varpi_2 f_2 \tag{10}$$

Where ϖ_1 and ϖ_2 are normalized outputs from layer three, while f_1 and f_2 are determined by the Sugeno fuzzy system, using equations (11) and (12).

$$f_1 = p_1 x + q_1 y + r_1 \tag{11}$$

$$f_2 = p_2 x + q_2 y + r_2 \tag{12}$$

where, p, q, and r are the system parameter sets.

The function of layer 5 is to do the sum of the incoming O14 and O24 signals to give the total output (O_{15}) calculated by equation (13) [8].

$$O_{15} = \left(\omega_1 f_1 + \omega_2 f_2\right) / \left(\omega_1 + \omega_2\right) \tag{13}$$

3. Proposed System

The study area for the development of WPP in Indonesia is focused on Jeneponto, as shown in Figure. 2. Jeneponto, which has eleven districts, is one of the districts in the province of South Sulawesi. It located in the northern boundary of Gowa and Takalar Regencies, the eastern border with Bantaeng Regency, the southern border with the Flores Sea, and the western border with Takalar District with an area of 749.79 km^2. Average wind speed is generally above 5 m/s. Thus, the area in Jeneponto has enough potential to develop WPP and interesting to be observed in more detail.

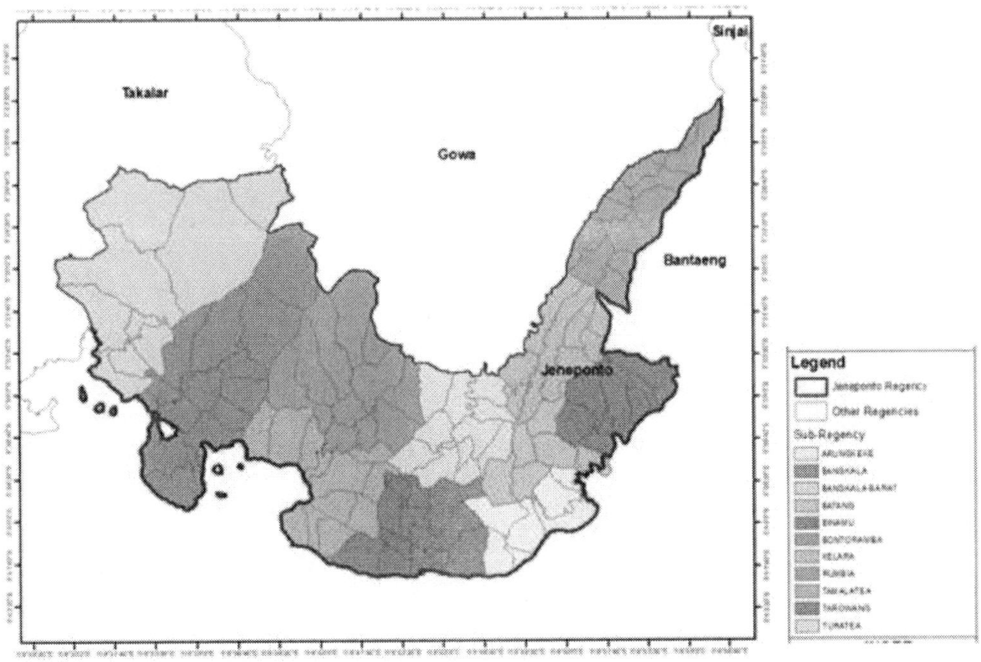

Figure 2. Location map of Jeneponto regency

Feasibility testing is done by the Fuzzy logic method shown in the flowchart in Figure 3.

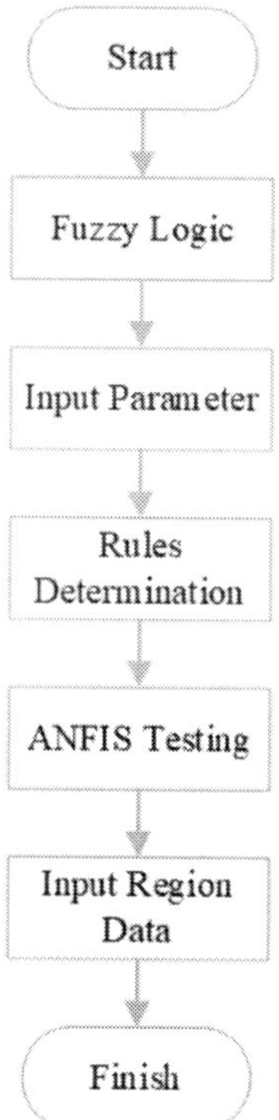

Figure 3. Flowchart of the research

The first step is input parameters, where the parameters used in this study are wind speed, elevation, slope, forest, land, road, and settlement with values, wind potential, elevation, slope, type of land use, soil conditions, access, and area populated.

Wind speed parameter is classified into 5 categories, namely very low (<2 m/s), low (2-3 m/s), medium (3-4 m/s), high (4-5 m/s), and very high (> 5 m/s). Elevation parameter (above sea level (ASL)), is categorized as high (> 100 m), medium (50-100 m), low (0-0 m). The slope is divided into five categories, which are very good (0–8°), good (9–15°), neutral (16–25°), bad (25–40°), and very bad (> 40°). Land use is divided into five categories, very good (non-forest (NFA)), good (protection forest (PF)), neutral (production forest (PRF)), poor (limited production forest (LPR)), and very poor (hunting park (HP)). Soil type is divided into very good (hap-ustalfs), good (haplustult), neutral (ustropepts), poor (ustipsamments), and very poor (dystrandepts).

As for the distance to the road (access), the road is classified into very suitable (> 300 m, fine asphalt), suitable (250-300 m, asphalt), neutral (150-250 m, fine soil), unsuitable (50- 150 m, non-smooth land), and very unsuitable (0–50 m, rocky land). Meanwhile for populated areas, it is divided into very suitable (> 300 m, very not crowded), suitable (250-300 m, not crowded), neutral (150-250 m, medium), unsuitable (50-150 m, busy), and very unsuitable (0–50 m, very crowded) [6].

After giving criteria, the next step is to set rules with all possible combinations of a total of 2178 combinations shown in Table 1 with the combination rules as follows:

- If the A1 wind combined with the others, it will produce a low value unless A1 is combined with C2 C3 C4 C5 C6 C7, then it will be of an average value.
- If the wind B1 combined with the others, it will produce an average value. But if B1 combined with B2 B3 B4 B5 B6 B7, it will be of high value.
- If the C1 wind combined with the others, it will produce high values unless it combined with C2 C3 C4 C5 C6 C7, then it will be very high.

Table 1. Rules table

Wind Speed	Elevation	Slope	Land Use	Soil Type	Access	Populated Area	Result
A1	A2	A3	A4	A5	A6	A7	Low < 30%
B1	B2	B3	B4	B5	B6	B7	Medium 30 - 60 %
C1	C2	C3	C4	C5	C6	C7	High 60 - 90%
							Very High 90 - 100 %

Code A represents the low parameter condition, code B represents the medium parameter condition, and code C represents the high parameter condition. The results have 4 conditions, namely low if below 30%, medium between 30 - 60%, high between 60 - 90%, very high above 90%. The number 1 represents wind speed parameter, number 2 represents elevation parameter, number 3 represents slope parameter, number 4 represents land use parameter, number 5 represents soil type parameter, number 6 represents access parameter, and number 7 represents populated area parameter.

Furthermore, these rules are tested using the ANFIS method by taking samples to test errors based on the rules that have been made and determine the suitability of the rules with parameters to get the feasibility of a location. The rules of all the parameters that have been set are:

- Wind area between 0 to 8
- The elevation from sea level is 0 to 200
- Slope type 0 to 50
- Forest or land use 0 to 100
- Soil types 0 to 100
- Access roads from 0 to 350
- Populated area from 0 to 350

After that, the minimum to a maximum range of each parameter is divided by 100 to equalize the conditions so that there are 100 consecutive samples of each parameter for the ANFIS test data, then the data are trained as shown in Figure 4.

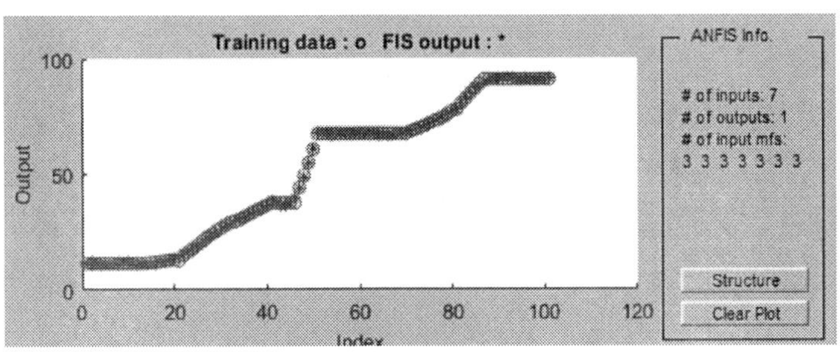

Figure 4. Training chart with ANFIS testing

Figure 4 illustrates the simulation of the data provided by the method, where blue color is the rule of testing data while red color is a simulation designed by the ANFIS method to mimic the obtained data. Furthermore, the two data trained with the results of the training test in Figure 5 shows that the rules made by the ANFIS method approach get an error value of 0.4287 in the second iteration as in Figure 3.

```
ANFIS info:
    Number of nodes: 4426
    Number of linear parameters: 2187
    Number of nonlinear parameters: 63
    Total number of parameters: 2250
    Number of training data pairs: 101
    Number of checking data pairs: 0
    Number of fuzzy rules: 2187

Warning: number of data is smaller than number of modifiable parameters

Start training ANFIS ...

   1      0.490977
   2      0.48287

Designated epoch number reached --> ANFIS training completed at epoch 2.
```

Figure 5. Training result with ANFIS testing

Then, the testing is done by trying each region data with the input of each village, according to the data taken. Village selection is focused on Kelara sub-district because the area is considered as a potential WPP development area [6].

The assignment of values for land/forest use conditions and soil types is only sorted according to the highest to lowest order and is given a range between 0 to 100 as shown in Tables 2 and 3.

Tabel 2. Soil type parameter

Criteria	Description Slope	Value
Very Poor	Dystrandepts (Andisol)	0-20
Poor	Ustipsamment (Entisol)	20-40
Neutral	Ustropepts (Inceptisol)	40-60
Good	Haplustults (Aluvial)	60-80
Very Good	Haplustalfs (Alfisolf)	80-100

Tabel 3. Land use parameter

Criteria	Land Use	Value
Very Poor	Hunting Park	0-20
Poor	Limited Production Forest	20-40
Neutral	Production Forest	40-60
Good	Protection Forest	60-80
Very Good	Non-Forest	80-100

The data of each area are shown in Table 4

Table 4. Data of each village

Village	Wind Speed	Elevation	Slope	Land Use	Soil Type	Access	Populated Area
Tolo	7.39	170	4	90	90	301	301
Tolo Barat	7.39	145	4	90	90	301	275
Tolo Selatan	7.39	150	4	90	90	301	275
Tolo Timur	7.39	255	12	50	80	301	275
Tolo Utara	7.39	365	32.5	70	80	301	275

Village	Wind Speed	Elevation	Slope	Land Use	Soil Type	Access	Populated Area
Bonto Lebang	7.39	170	12	50	80	301	275
Samataring	7.39	215	32.5	50	80	301	301
Bonto Nompo	7.39	240	32.5	50	80	301	75
Gantarang	7.39	180	32.5	50	80	301	75
Tombolo	7.39	275	32.5	50	80	301	75

Data are obtained by direct acquisition, such as wind data are obtained by measuring wind conditions for 2 months using anemometer measurement, soil type data by taking soil samples, road access data by observing the state of the village area, land use data by observing forests in the area, data settlements by directly observing the state of the number of houses in each village while data on sea level and slope obtained from the Geographic Information System (GIS) software shown in Figures 6 and 7.

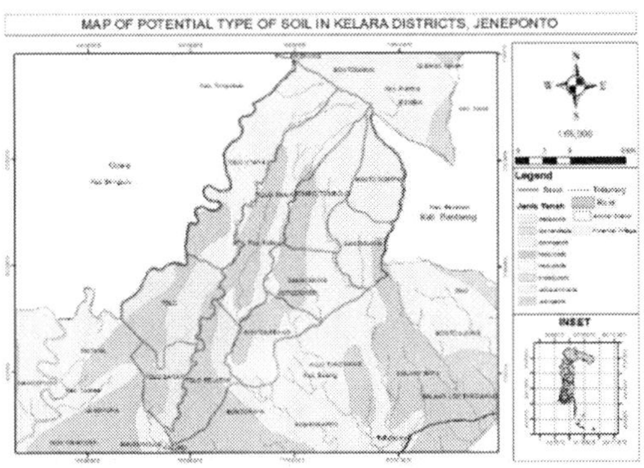

Figure 6. Map of potential type of soil in Kelara

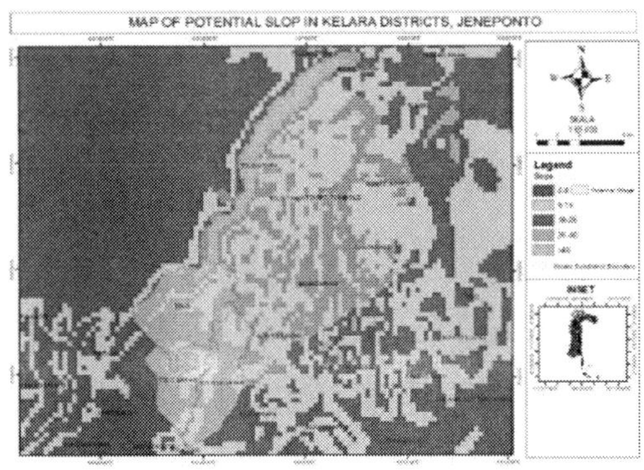

Figure 7. Map of potential slop in Kelara

4. Result and Discussion

The testing carried out in each village in the Kelara sub-district of Jeneponto using the mentioned parameter values produced the results as shown in Table 5.

Table 5. Test result

Village	Feasibility (%)
Tolo	91
Tolo Barat	78,4
Tolo Selatan	78,4
Tolo Timur	69,3
Tolo Utara	60,8
Bonto Lebang	65
Samataring	68,4
Bonto Nompo	55
Gantarang	55
Tombolo	55

Table 5 shows that the percentage of the feasibility values obtained is different, the highest value in the village of Tolo with a percentage value of 91%, then the villages of West and South Tolo get the second-highest value with a percentage of 78.4%, in the third place in the village of East Tolo with a percentage of 69.3%, in order fourth in the village of Samataring with a percentage of 68.4%, ranked fifth in the village of North Tolo with a percentage of 60.8%, and then in the last three positions with a percentage of 55% are the villages of Bonto Nompo, Gantarang and Tombolo. This allows that the feasibility of a location determined with the highest value.

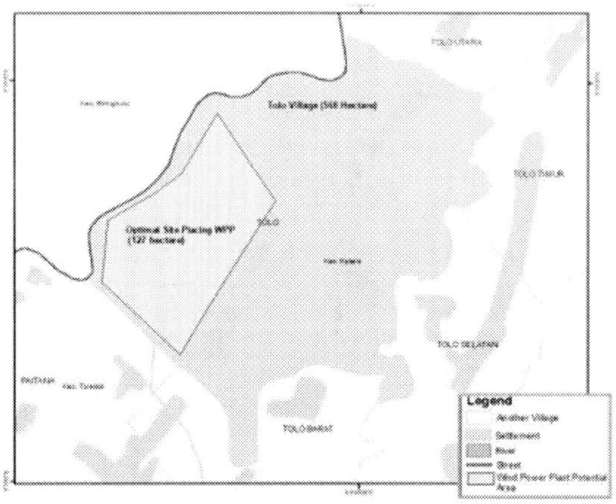

Figure 8. Map of Tolo village

The Tolo village area, as shown in Figure 8, is the best location to place the WPP with the highest percentage value using the ANFIS method. The area of Tolo village is 568 hectares. Based on the existing conditions, the potential location for placing wind turbines in Tolo is around 127 hectares for the optimal location of WPP.

The proposed method can be used to help determine the feasibility of a location of the wind turbine installation. Utilization of wind energy for WPP in Tolo Village, Jeneponto Regency, could be one of the solutions for the Indonesian government to increase the use of electricity generation by using renewable energy sources. Hence, the use of alternative energy to meet electricity needs in South Sulawesi in addition to the Sidrap WPP, and WPP in Kayuloe village, Jeneponto district in the next few years can be increased.

5. Conclusion

This study presents an analysis of location suitability for WPP placement using the Fuzzy Logic method and the Adaptive Neuro-Fuzzy Inference System (ANFIS) to find the feasibility value using seven parameters and 2178 rule sets. The obtained results are the Tolo village gets a percentage of the feasibility of 91% with an error of 0.48287. The proposed system can be used to determine locations in other areas by looking for values by adjusting the parameters that exist in that region.

References

[1] M. D. C. Koengkan and J. A. Fuinhas, "The Impact of Renewable Energy Consumption on Carbon Dioxide Emissions - The Case of South American Countries," *Rev. Bras. Energ. Renov.*, vol. 7, no. 2, Mar. 2018.

[2] M. M. Rifadil, E. Purwanto, A. Jaya, and G. Prabowo, "Wind Power Generation Systems Using Vertical Axis Windmills for Residential Loads," *14th Semin. Intell. Technol. Its Appl.*, p. 5, May 2013.

[3] A. Miller and R. Li, "A Geospatial Approach for Prioritizing Wind Farm Development in Northeast Nebraska, USA," *ISPRS Int. J. Geo-Inf.*, vol. 3, no. 3, pp. 968–979, Jul. 2014.

IOP Conf. Series: Materials Science and Engineering **875** (2020) 012035 doi:10.1088/1757-899X/875/1/012035

[4] A. M. Abdelsalam and M. A. El-Shorbagy, "Optimization of wind turbines siting in a wind farm using genetic algorithm based local search," *Renew. Energy*, vol. 123, pp. 748–755, Aug. 2018.
[5] L. Iyappan and P. Kasinatha Pandian, "Geoprocessing model for identifying potential wind farm locations," *IET Renew. Power Gener.*, vol. 10, no. 9, pp. 1287–1297, Oct. 2016.
[6] S. M. Said, Y. S. Akil, and M. H. Muzakir, "GIS approach for wind power plant development in South Sulawesi, Indonesia: A location suitability analysis," presented at the The 4th International Conference on Industrial, Mechanical, Electrical, and Chemical Engineering, Surakarta, Indonesia, 2019, p. 030085.
[7] R. Hossain, A. Maung, A. Than Oo, and A. Shawkat Ali, "A hybrid machine learning using Mamdani type fuzzy inference system (FIS) for solar power prediction," *Ann. Fuzzy Sets Fuzzy Log. Fuzzy Syst.*, May 2013.
[8] S. A. Khan, M. D. Equbal, and T. Islam, "ANFIS based identification and location of paper insulation faults of an oil immersed transformer," in *2014 6th IEEE Power India International Conference (PIICON)*, 2014, pp. 1–6.

IOP Conf. Series: Materials Science and Engineering **875** (2020) 012036 doi:10.1088/1757-899X/875/1/012036

Design and Implementation Monitoring and Booking Systems for Smart Parking at Engineering Faculty Campus

Muh Anshar[1]*, R S Sadjad[1], Dewiani[1], M Hanan[1], R Prayudha[1], and M Abry[1]

[1]Department of Electrical, Faculty of Engineering, Universitas Hasanuddin, Indonesia

*Email: muh.anshar@gmail.com

Abstract. Parking our vehicles sometimes become troublesome and situation gets worse when the user is required to park their cars at the unknown parking availability. Not to mention, this situation also occurs within the new Campus of Engineering Faculty of Universitas Hasanuddin. This campus has a motto to be the centre of technology, so it is something unfortunate that the parking management does not incorporate technology to support the parking system. This paper emphases on designing smart parking model which is suitable to be adopted within this new campus. Technology based Internet of Things is used to modernised the parking system with advanced features of monitoring and booking mechanisms. These features are accessible through smart phones which increases the usability of the proposed smart parking system. The experiments show that the parking management system can function accurate and effectively in enhancing the quality of parking system, with the two-mentioned features of the proposed smart parking system, which are monitoring and booking capabilities.

1. Introduction

The increasing use of vehicles and community mobilization from one place to another parallel with the community's need for land or parking spaces. In Makassar city there are at least 717 parking points spread across 14 sub-districts [1]. In terms of government regulations, particularly in the Makassar city area, South Sulawesi, local government regulations have been issued to regulate the Makassar City roadside parking system [2], and in general have been stipulated in regional regulations regarding the Makassar City regional spatial plan [3].

However, the increasing population of vehicle users, especially car users, is becoming a problem for the campus community today. The limited available parking space in the campus area and the lack of information regarding the availability of parking space creates a mismatch between vehicle users and the available parking space capacity. At the Engineering Faculty of Hasanuddin University itself, the parking information system used is still not able to inform the availability of parking lots to vehicle users. Drivers who want to park their vehicles must first check directly the availability of their parking locations. The impact is, in addition to the time loss experienced by the driver of the vehicle, this can also be a trigger for congestion on the streets around the parking area [4, 5].

This problem should be solved by creating an online parking lot monitoring information system. With the development of IoT technology and the use of smartphone technology, technological devices can monitor the availability of parking lots and then provide that information to vehicle users in real-time, so users can find out the available parking locations and make parking reservations before arriving at the parking location [6]. Therefore, in this paper, the author will try to make the design and implementation of an online parking reservation system on car parking lots at the Faculty of Engineering, Universitas Hasanuddin.

2. Technology Peripherals

The following descriptions provide details of peripherals to be used to develop the smart parking hardware design.

2.1. Sensor Unit

In order to detect the presence of objects, in this case, vehicles, a device is used and one of the commonly used to perform this functionality is ultrasonic sensors.

A technology that is often used for object detection applications and object height measurement is ultrasonic technology. Ultrasonic sensors work by utilizing the principle of sound waves. Ultrasonic terminology means waves with frequencies above the frequency of human hearing where these sensors use a type of frequency that cannot be heard by humans. This is important for sensor performance because certain frequencies are rarely generated for other functions which are useful for avoiding interference effects with other sound sources. Ultrasonic sensors transmit sound waves in sequence or series which will be reflected back by the object to be detected. Mileage is obtained by calculating the travel time of sound waves since they are transmitted until they are reflected back. The microprocessor can then calculate the level, the content and even open the channel flow. This mechanism is also used to detect the presence and profile of an object in a particular application.

Figure 1. Physical look ultrasonic sensor

However, this technology has several limitations, for example the shape of the object's surface detection, distance, size and angle can affect the accuracy of the reading. The ideal surface for this sensor application is a smooth and hard surface. This type of surface is able to reflect a large number of signals rather than soft or uneven surfaces. Grain and flour ingredients will reduce the operating distance from the sensor and lower measurement accuracy. Ultrasonic sensors require an unobstructed air column to be detected. In addition, the distance from the object can also affect the accuracy of the ultrasonic signal. If the object is to be measured or detected far away, it requires better reflection characteristic ability than closer objects [7]. Examples of ultrasonic sensor shapes can be shown in Figure 1 and a brief illustration of the sensor working system shown in Figure 2.

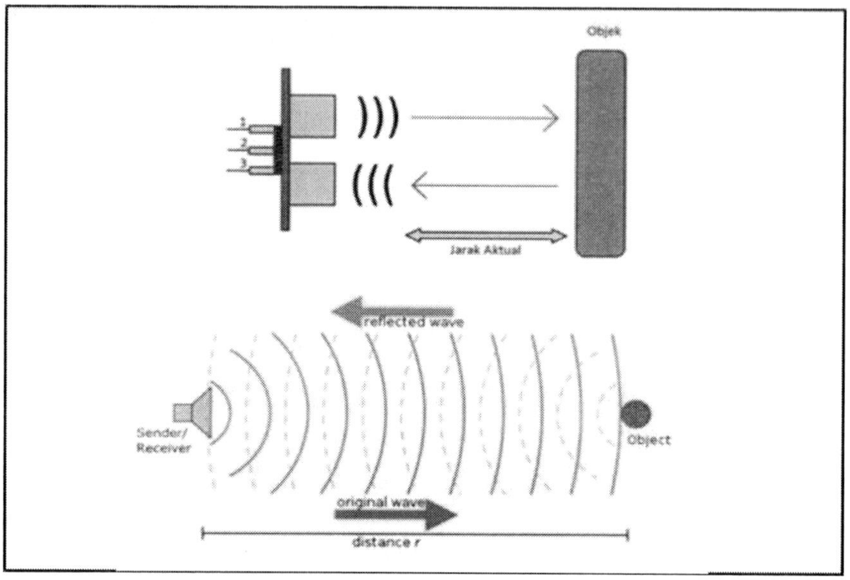

Figure 2. How Ultrasonic sensor works

2.2. Control Unit

Currently, the integrated system technology is increasingly dominating the market and microcontroller technology is a technology that is widely adopted, both among academics and industry. Some of the developments in this microcontroller system have evolved into user-friendly systems, among which the most popular are Arduino and Raspberry-based systems.

2.2.1. Arduino Unit. Arduino technology is a microcontroller with small dimensions that is equipped with a universal serial bus (USB) plug to connect computers and several connection sockets that can be connected by cables to electronic components such as motors, relays, light sensors, laser diodes, loudspeakers, microphones and so on. This system can be supplied with a voltage source via a computer USB, or from a 9-volt battery voltage source or from another DC voltage source. This technology can be controlled via a computer or programmed with a computer and then released and then the system can function alone or independently. The Ardunio Board is designed open source so that everyone can access and make boards that are compatible with Arduino [8]. Examples of Arduino board shapes can be seen in Figure 3.

Figure 3. Arduino board

2.2.2. Raspberry Pi. The Raspberry Pi device has various interfaces for installing hardware sensor devices. For example, by using the I2C bus for simple sensors (light, temperature, motion, sound) and USB for more complex sensors (Wifi). By using a Linux system, the use of daemons to collect realtime sensor readings and cache data in local memory. This data is periodically stored to the server via a secure TCP / IP connection. Each board currently requires two physical connections, for wired ethernet and DC power supply via an electrical transformer [9]. Figure 4 shows an example of a Raspberry-based board.

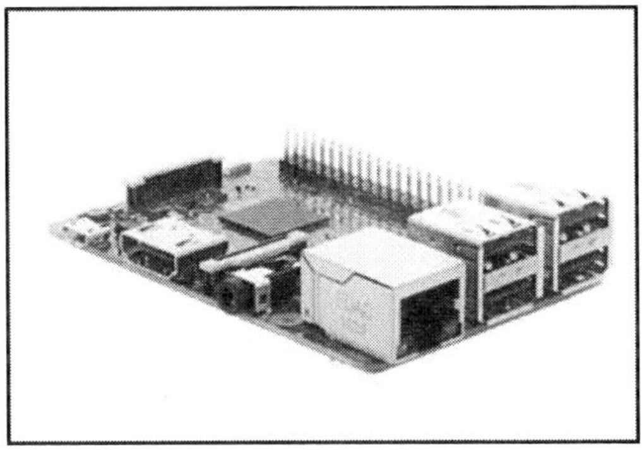

Figure 4. Raspberry Pi model

2.3. Communication Unit

One of the fast-growing devices used in Internet of Things is the LoRa technology (Long Range) which is a long range with low power wireless device. This device used a spread spectrum modulation technique originated from chirp spread spectrum or CSS technology. LoRa technology with the LoRaWAN protocol has enabled IoT utilisation which tackle some challenges facing out current technology, which are energy management, infrastructure efficiency, disaster related applications, etc [10,11].

It is mentioned in [12] that LoRa is the creation of a long-range communication link through physical layer or via wireless modulation. As the LoRa is based on CSS, it allows to maintain low power characteristic as those wireless systems that utilises frequency shifting (FSK) modulation and at the same time, it can increase significantly its communication range. Figure 5 shows an example of LoRa board available on the market.

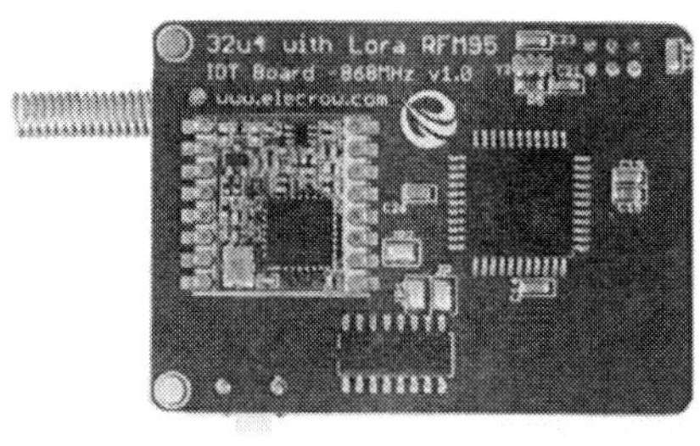

Figure 5. LoRa board

3. Smart Parking Diagram Design

Figure 6 below shows the overall diagram design of the Smart Parking System to be implemented in the parking management system at the Campus of Faculty of Engineering UNHAS.

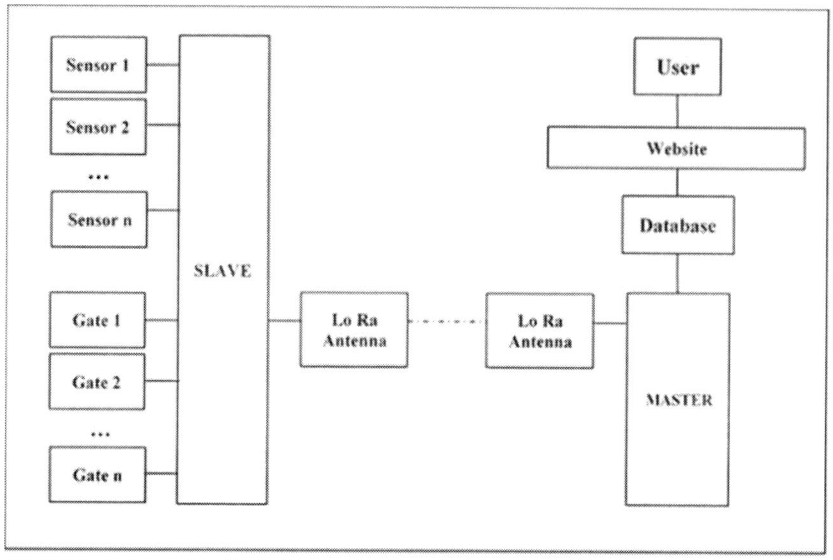

Figure 6. Smart parking block diagram

It can be seen from Figure 6 that the overall system is divided into three main sections, the Slave, the Master, and the Server/Database sections. The Slave part is a part of the system that is directly related to the parking lot to be detected. The main task of the Slave is to detect the presence of vehicles at the parking point and control the activity of the gate / indicator located on the parking lot. The Master part is in charge of acquiring and processing data, then to be sent to the database located on the server. The system offers two features, which are monitoring and booking features, that are available through a dedicated website and are accessible via personal computer or smart gadgets. Between the Master and Slave communication is done using wireless radio frequency technology, so that the Master and slave can be placed in separated distance.

Figure 7 is a picture of the data flow in the system. The sensor that has read the condition of the parking point then sends data to the Slave. The slave then collects data from each sensor and sends it to the Master. The Master part that has received the data then processes the data and then stores it. Data that has been stored by the Master will be forwarded to the database located on the server. The process of sending data from slave to Master is sequential process. Thus, the slave will only send the parking conditions to the Master periodically when all sensors have successfully sent the detection results to the slave.

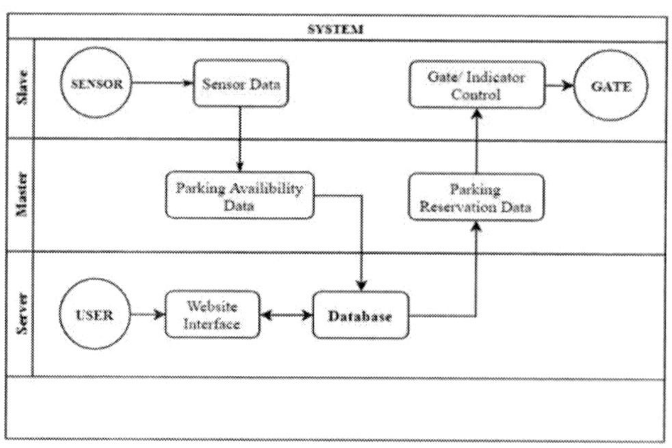

Figure 7. Flow data of the smart parking system

The Master not only receives data on parking conditions from the slave, the Master also receives data from the database about the points that have been reserved by the user. This data is then processed in accordance with the availability of parking lots that would be reserved. If parking is available, the Master will send an order to the slave to activate the gate / indicator at the reserved parking point.

4. Experiments and Results Analysis

There are several elements of the smart parking system and most of them are viewable from the webpage.

4.1. Webpage

The front webpage in which users can access the monitoring and booking features is shown in Figure 8 below.

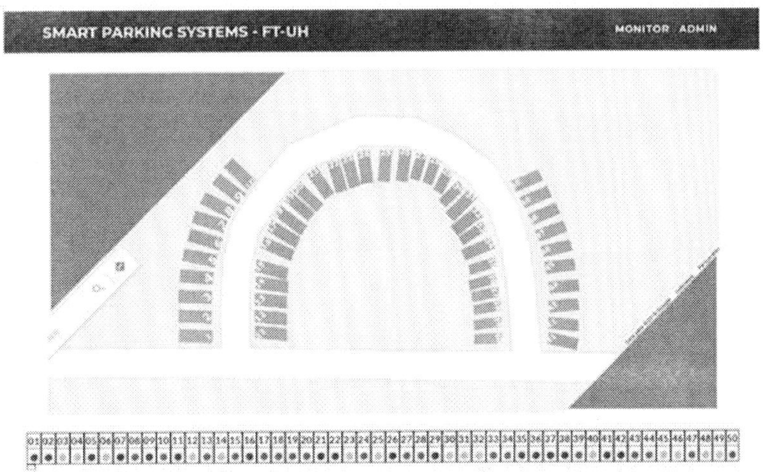

Figure 8. Front page of the webpage smart parking

Figure 8 is the front display of the website page of the online parking reservation system interface. Through the web page, users can find out the availability of each existing parking slot and latter to make a reservation at the desired parking lot. The picture shows the conditions of the indicators that exist on the website. The green circle indicates that the parking lot is empty, the red circle indicates that the parking lot is filled, while the yellow circle indicates that the parking lot has been reserved. Steps to access the features of the smart parking can be described as the following.

- Open web browser (either pc, laptops or smart phones
- Type **smartparking.unhas.ac.id**
- Check availability of slots by the colour of the indicators (available slot 18 to slot 33, others for future development)
- Click on the available slot (green indicator) by first activating GPS

Figures 9 and 10 show the booking appearances whether its is a successful booking or unsuccessful reservation.

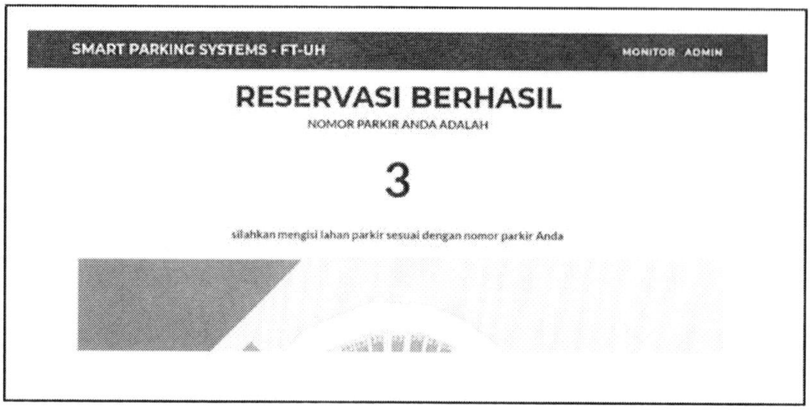

Figure 9. Successful reservation

Figure 9 shows the notification window when the users successfully book their parking spot. The page shows the position of the parking spot and the number of the parking slot. Figure 10 depicts the notification window when the user fails to book their parking spot. The page also shows the reason why the user unable to reserve the parking spot. To successfully reserve the parking spot, the system must know the location of the user and calculate the distance of the user to the parking point. So, the user needs to turn on the GPS on their smartphone and allow the system to know the location from the user internet browser. This procedure needed to avoid user reserve the parking spot from very distant places and making another user unable to reserve a parking spot.

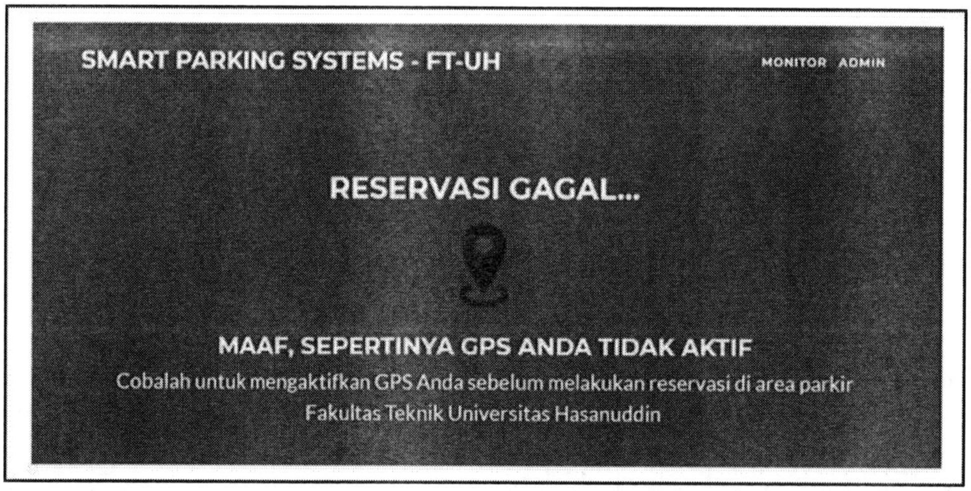

Figure 10. Unsuccessful booking

4.2. Measurements Delay Data

One set of data is collected from the delay time from the sensor reading detects a vehicle until the changing in the associated indicator colour (Table 1). It can be seen from Table 1 that there are some time delays occurs when sending data to the server or when retrieving data from the server with only three parking points. The longest delay in the table is 11.14 seconds while the shortest delay is 2.51 seconds. There are several factors contributes to the delay, one of them is the internet speed during the reservation or the number of the sensor used in the system.

Table 1. Delay Time of the Booking System

Cycle No.	Delay (s)	
	Sensor to Database	**Database to Gate**
1	5.19	5.41
2	2.82	7.87
3	3.72	2.51
4	5.63	1.54
5	3.01	11.14
6	3.29	3.42
7	10.48	4.78

The effect of the number of sensors used on the delay that occurs in the slave of the system is shown in Table 2. Each sensor takes approximately 68 ms to read the conditions of each parking point. The more sensors being installed, the longer delays generated for establishing data communication of the server (master)-slave. Hence, it will accumulate the delay of the overall existing sequential system.

Table 2. Delay by Slave (sensors being Utilised)

Number of Sensors	Delay (ms)
1	68
2	134
3	201
5	337
10	674
20	1350

4.3. Measurements Distance Data

Table 3 depicts the distance of the user to the server that could affect the success of a reservation. In the system that has been made, a maximum distance limit of reservation system is set in order to avoid users making reservations from places that are too far from the parking point.

Table 3. Distance access

No	Distance (m)	Result
1	210	Succes
2	400	Succes
3	850	Fail

It can be seen that the booking feature of the smart parking system is at maximum 400 meters and beyond that no reservation is allowed, however, only the monitoring feature that can be accessed by the users.

5. Conclusions

The system we design can work appropriately for handling parking spot detection and online reservation. Time requires to make a reservation is considerably fast with 11.14 seconds of delay time.

For user convenience, the booking feature is only accessible within the radius of 400 meters from the parking lots. In the future, there are some improvements to be made as to increase the usability of the smart parking system and to integrate with a complete opening-closing gate mechanism. There are still some delays in the process of sending and receiving data which needs to be tackled. Also, the system used today is still a sequential system in which the system will update data periodically even though there are no changes in the system. This also makes the system less responsive when changes occur in the system.

Acknowledgments
Authors wishing to acknowledge assistance from all members of the Research Group Cognitive Social Robotics and Advanced Artificial Intelligence Research Centre and also the Faculty of Engineering UNHAS for providing the research fund under the scheme of LBE, so that this research work can be carried out.

References
[1] Foundation of Indonesia Legal Aid – Makssar Legal Aid, 2015. Broadcast on Fixing Parking Management in the City of Makassar. http://lbhmakassar.org/wp-content/uploads/2015/08/press-relase-JUKIR.pdf, accessed online on 13 March 2017.
[2] Meyer of Makassar City, 2006. Local Regulation for the City of Makassar, No 17, 2006 about Parking Management on Public Road Side in the City of Makassar (in *Bahasa*).
[3] Meyer of Makassar City, 2015. Local Regulation for the City of Makassar, No 6, 2015 about Spatial Planning of the City of Makassar2005-2015 (in *Bahasa*).
[4] B. M. Kumar Gandhi, and M. K. Rao. May 2016. A Prototype for IoT based Car Parking Management System for Smart Cities. Indian Journal of Science and Technology, vol. 9, no. 17.
[5] N. Moses, and Y. D. Chincholkar. June 2016. Smart Parking System for Monitoring Vacant Parking. International Journal of Advanced Research in Computer and Communication Engineering, vol. 5, no. 6, pp. 717-720.
[6] Y. Zheng, and S. Rajasegarar. Apr 2015. Parking Availability Prediction for Sensor-enabled Car Parks in Smart Cities," in 2015 IEEE Tenth International Conference on Intelligent Sensors, Sensor Networks and Information Processing (ISSNIP), Singapore.
[7] G. Arun Francis, M. Arulselvan, et al. April 2019. Object Detection Using Ultrasonic Sensor. International Journal of Innovative Technology and Exploring Engineering (IJITEE) ISSN: 2278-3075, Volume-8 Issue-6S.
[8] B. Yusuf Abdullahi. 2014. The Working Principle Of An Arduino. 978-14799-4106-3/14, © 2014 IEEE
[9] N. Anand, and P. Vikram. 2015. Raspberry Pi-A Small, Powerful, Cost Effective and Efficient Form Factor Computer: A Review. International Journal of Advanced Research in Computer Science and Software Engineering (IJARCSSE). 5. 720-737.
[10] Semtech. "What is LoRa." Retrieved 1 July, 2019, https://www.semtech.com/lora/what-is-lora.
[11] Augustin, A., Yi, J., Clausen, T. and Townsley, W., 2016. A study of LoRa: Long range & low power networks for the internet of things. *Sensors*, *16*(9), p.1466.
[12] Alliance, L. (2015). "LoRAWAN." Technical Marketing Workgroup 1.0. from https://www.tuv.com/media/corporate/products_1/electronic_components_and_lasers/TUeV_Rheinland_Overview_LoRa_and_LoRaWANtmp.pdf.

IOP Conf. Series: Materials Science and Engineering **875** (2020) 012037 doi:10.1088/1757-899X/875/1/012037

Performance Evaluation of Mini Single Board Computer in Hadoop Big Data Cluster

Adnan[1]*, Z Tahir[1], C Yohannes[1], and Ariel[1]

[1]Department of Informatics, Faculty of Engineering, Hasanuddin University, Makassar, Indonesia

*Email: adnan@unhas.ac.id

Abstract. Hadoop is a Java-based and open source software framework that serves to process big size data in a distributed manner. Hadoop uses a framework for applications and programming called MapReduce. Along with the development of Hadoop, big data research using Hadoop is running using a single board computer (SBC). SBC is a mini computer which has a specification similar to a conventional computer but has lower specifications and power usage. However, the use of storage used on a single board computer has constraints, namely low bandwidth and limited local storage, therefore NAS is used because it has large storage and bandwidth speeds is upto1 Gbps. It uses as storage that is carefully used on Hadoop. Based on testing by running Hadoop in a single node and multi nodes using Micro SD storage and NAS does have a significant difference. When running Hadoop on a single node the difference in data access speed generated by Micro SD and NAS has a difference of 2 seconds. On multi nodes (clusters) the ratio is 4 seconds. The difference in NAS bandwidth that is 2 times faster than Micro SD show a significant difference if it is run on Hadoop.

1. Introduction

Evolution of Information and communication technology nowadays is mainly in unstructured data. Researchers faced to the rapid growth of data. The rapid growth of data lead to the trend called Big data. Big data is a term that describe those both structured and unstructured data sized extremely large. Big data is difficult to store, collected, managed as well as analyzsed using legacy database. It is because those data growth rapidly and continously. International Data Corporation estimated that the size of data in digital world is around 0.18 zettabytes. IDC also predicted big data will be 10 times fold every five years [1].

Parallel computing is a kind of computing process in which a large scale of of computation is performed simultaneously by multiple computing resources. Nowadays the popular parallel computing platform for which to process a large amount of data distributedly is Hadoop. Hadoop constitute a framework software developed based on Java. Dough cutting, the same person who developed Apache Lucene, was the first person to introduce Hadoop at the first time. The Hadoop project started with Apache Nutch project as a part of Apache Lucene [2].

Among the many ways used to analyze big data, Hadoop is the one framework used because of its Hadoop Distributed File Syatems as well as its method called MapReduce. Mapreduce make Hadoop is possible to process big data in parallel and distributed manner on hundreds or even thousands of computer.

In this research, we make use of Banana Pi M3. Banana Pi M3 is a single board computer that has component similar to the desktop computer. Banana Pi M3 has better specifications. They supported by eight core processors but lower price. However, Banana Pi is simlar to other type mini SBC. They use Micro SD as main storage. Unfortunately, Micro SD storage is known to slower and smaller capacity than HDD. Consequently we propose to leverage Network Attached Storage (NAS) as a alternative. We propose to relocate HDFS from Micro SD to Network Attached Storage. We suggest that MapReduce application may have better performance when we make use of NAS compared to default storage Micro SD. NAS is kown to have large capacity and low overhead [3].

This paper presents our experimental result on evaluating the performance of MapReduce aplication on SBC equipped with Network Attached Storage and Micro SD. We compared the result on both using Network Attached Storage and Micro SD.

2. Related Works and Tools

2.1. Hadoop

Hadoop constitutes an open source software framework developed on top the Java programming language. Hadoop functions to process extreeme size of data on cluster. Cluster is a kind of distributed memory parallel computer build from multiple computer connected using high speed network interconnection [1]. According to [4], Hadoop MapReduce is able to process data of very large size. The size of data could be in petabyte (10^{12}) order on top large scale cluster (upto thousand nodes of computer). Originally, Dough Cutting developed Hadoop as part of Nutch project belong to Apache Foundation.

Hadoop consists of common Hadoop which is usefull in providing access to Hadoop Distributed File System. This common hadoop constains packages needed by JAR files, scripts needed for strarting Hadoop, and documentatins for the jobs which are completed by Hadoop.

According [1], the core of the Hadoop is consist of :

1. Hadoop Distributed Files System (HDFS)

   This is a file system used by Hadoop to distribute data being processed by multiple nodes

2. Mapreduce framework

   Thiss a programming framework work for distributed processing using Hadoop.

Figure 1 shows the core a Hadoop which is consist of HDFS and MapReduce. Fig. 1(a) shows HDFS components which consist of NameNode and DataNode. Further, Secondary node is a backup for the NameNode. Fig. 1(b) shows that in MapReduce there is jobTracker as well as TaskTracker.

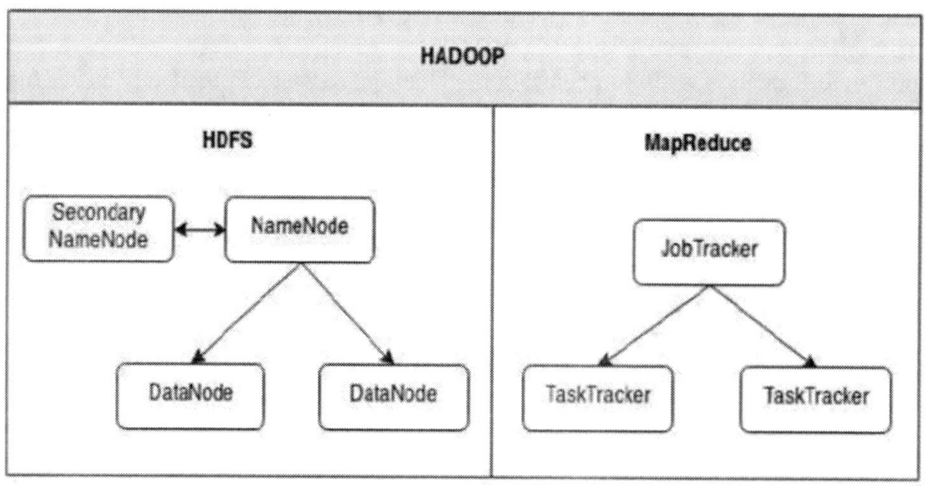

(a) (b)

Figure 1. Two main component of Hadoop. (a) HDFS (b) MapReduce

Hadoop on a small cluster could consist of one node of master and some nodes of slaves. The master node is consistof NameNode and JobTracker, whereas the Slave nodes are consist of some DataNodes and TaskTracker. Hadoop required JRE version 1.6 or newer version. When starting and also stopping the Hadoop sistem, SSH connections are needed among the nodes of the Hadoop cluster [5].

2.2. Hadoop Distributed File System

Hadoop Distributed File System (HDFS) is Java-based file system that is distributed for Apache Hadoop [1]. As a distributed file system, HDFS useful for handling large amounts of data which is stored and distributed in many connected computers which is commonly called a cluster. Distributed file systems on Hadoop can be assumed as a file system that stores data not in one local Hard Disk Drive (HDD) or other media storages , but the data is fragmented (files are distributed in block form with a size of 64MB – this size of block if configurable however), and distributedly stored in multiple nodes of cluster.
HDFS stores data in a way such that HDFS split the data into chunk of data of size 64MB (default). The chunk of the data are stored scattered in each nodes of cluster. These pieces of data in HDFS are called blocks. The block size in each file is not necessarly to be fixed to 64 MB, where the block size can be adjusted to the user's preference. Although the data is stored scattered to several nodes, but from the user's perspective, the data still looks like we access files on one computer. Files that are physically spread across multiple computers can be treated like treating files on one computer [2].
As a distributed file system, HDFS has main components such as NameNode, DataNode, and Secondary NameNode [1]. The architecture of the three components can be seen in Figure 2.

2.3. MapReduce

MapReduce is a framework for applications and programming that was introduced by Google and used to do a job of distributed computing that is run on a cluster [6]. MapReduce consists of the concept of map and reduce functions that are commonly used on functional programming [5].

One program that uses the MapReduce concept that has been provided by Hadoop is WordCount. WordCount is a benchmark program that aims to count words in a plaintext file. The MapReduce process in WordCount is divided into 2 stages: the process of mapping and reducing.

2.4. Hibench

HiBench is a big data suite benchmark that favor evaluating a variety of large data frameworks in terms of speed, throughput and utilization of system resources. It contains a set of Hadoop, Spark and streaming workloads, including Sort, WordCount, TeraSort, Sleep, SQL, PageRank, indexing of Nutch, Bayes, Kmeans, NWeight and DFSIO, etc. It also contains several streaming workloads for Spark Streaming, Flink, Storm and Gearpump [7].

Basically, the purpose of TeraSort is to sort the data of 1TB (or another amount of data that is preferred) enabled as possible. This is a benchmark that complements the HDFS and MapReduce layer testing of the Hadoop cluster. There are 3 steps that are used to carry out benchmarking benchmarking, namely:

1. Generate random data using Teragen
2. Sort the generated data using Terasort program
3. Validate sorted data using Teravalidate

2.5. Performance Analysis Tools

Performance analysis tool is a flexible performance analysis framework designed for Linux operating systems. PAT collects system-level performance metrics including CPU, DISK and NETWORK usage. And generate files in the form of pdf or Excel and provide a visual display of the results of the presentation of the data that has been collected.

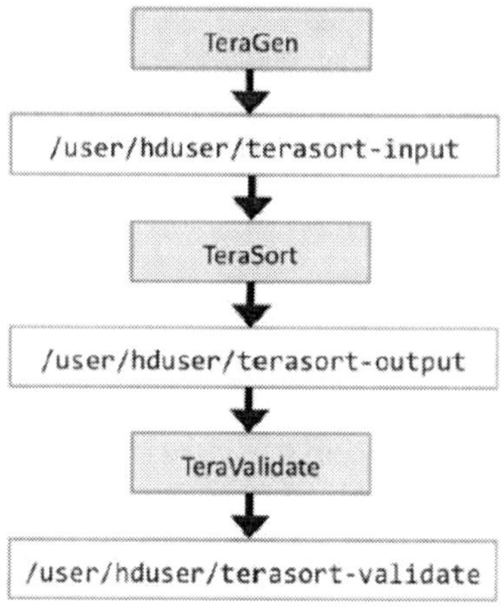

Figure 2. Basic work flow of Terasort benchmark using Hadoop MapReduce

3. Experimental Configuration

3.1. Hardware and Software Platform

In our works, we make use both hardware and software as follows : We make use six Banana Pi M3 equipped with two types of storages. In the first scenario, Banana Pi M3 used 32 GB Micro SD class 10 80 MB/sec. In the 2nd scenario, Banana PI M3 connected to a Network attached storage via gbe network using gigabit ethernet switch. The Banana PI M3's specification described as in Table 1, while the storage used in both scenario as present as in Table 2.

Table 1. Banana Pi M3 specification

Component	Specification
CPU	Octa-core A83T ARM Cortex A7 2 GHz
RAM	2 GB
Network interface	Gigabit ethernet and WiFi

Table 2. Storage specification

Storage type	Capacity and link speed
Micro SD	32 GB / 80 MB/sec
NAS	2TB /1000 Mbps

3.2. Scenario of Experiments

3.2.1. First Scenario

The first scenario intends to analyze performance of I/O (storage access) using single node. In this scenario, Hadoop MapReduce ran using single node of Banana Pi M3. We measured the performance of MapReduce using Micro SD and NAS as HDFS storage. We compared the execution time for both experiments while using two type of the storage. For both experiments in this scenario, PAT is installed on NAS, whereas the data being sorted are located in hdfs (Micro SD or NAS respectively). Performance analazyer tools (PAT) is to obtain performance data such as user CPU time, iowait, system time etc. We make the size of data being sorted varied from 320 MB up to 2 GB.

3.2.2. Second Scenario

The second scenario we intend to analyze the performance of I/O (storage access) using multiple nodes. In this scenario, Hadoop MapReduce bencmark, that is terasort, ran using multiple node of Banana Pi M3. We measured the performance of MapReduce using Micro SD and NAS as HDFS storage. We compared the execution time for both experiments while using two type of the storage as the base of HDFS (it means that data being sorted are located on files of HDFS). For both experiments in this scenario, PAT is installed on NAS. Performance analyzer tools (PAT) is to obtain performance data such as CPU time, iowait, system time etc. We make the size of data being sorted varied from 1 GB up to 4 GB.

4. Experimental Results

4.1. The first Scenario

The first scenario aims to determine the performance of Micro SD and NAS data access speeds on Hadoop. The first scenario is to run Hadoop in a single node using Micro SD and NAS storage. The file sizes used in the first scenario are 320MB, 500MB, 1GB, and 2GB. Experiments in the first scenario carried out 10 attempts.

The results of the measurement of data access speed in the first scenario by running Hadoop in a single node (cluster) using Micro SD and NAS storage can be seen in the Table 3.

Table 3. Execution time of Terasort with different type of storage

Storage type	File size/execution time			
	320 MB	500 MB	1 GB	2 GB
Micro SD	287.5 secs	423.76 secs	852.88 secs	1584.75 secs
Network Attached Storage	284.96 secs	423.57 secs	816.15 secs	1583.75 sec

4.2. Second Scenario

The first scenario aims to obtain the performance measure of Micro SD and NAS data access speeds on Terasort of Hadoop MapReduce. The first scenario is to run Terasort on multi-node (cluster) using Micro SD and NAS storage. The file size used in the second scenario is 1 GB, 2 GB, 3 GB, and 4 GB. Experiments in the second scenario carried out 10 attempts. The results of the measurement of data access speed in the second scenario by running Hadoop Terasort in a multi node (cluster) using Micro SD and NAS storage can be seen in Table 4.

Table 4. Execution time of Terasort with different type of storage

Storage type	File size/execution time			
	1 GB	2 GB	3 GB	4 GB
Micro SD	49.07 secs	56.18 secs	59.63 secs	67.79 secs
Network Attached Storage	46.50 secs	50.52 secs	52.83 secs	60.13 sec

Based on the results of experiments conducted in the second scenario, it can be seen that the difference in data access speed on Hadoop which is run in a multi node using NAS has a significant difference from Micro SD. The tables show that the speed produced by Micro SD and NAS have a significant difference, where the difference in the speed of time produced by the two storage is only 3 seconds different in each file tested, the insignificant difference.

4.3. CPU Utilization

This subsection reports experimental result on using PAT to obtain CPU utilization while running MapReduce benchmark. Data presented is obtained from experiments while using NAS and Micro SD as HDFS media.

4.3.1. Single Node

Table 5. Single node

Componen of CPU Utilization	Micro SD as HDFS media	Network Attached Storage as HDFS media
	Portion (%)	
User time	11.20	12.88
System	4.03	5.16
Nice	0.00	0.00
I/O wait	8.25	0.79
Steal	0.00	0.00
Idle	76.50	81.15

4.3.2. Multiple Nodes

Table 6. Multiple nodes

Component of CPU Utilization	Micro SD as HDFS media	Network Attached Storage as HDFS media
	Portion (%)	
User time	17.04	22.88
System	8.28	10.89
Nice	0.00	0.00
I/O wait	41.34	44.47
Steal	0.00	0.00
Idle	33.32	21.74

From the tables, using NAS is better than Micro SD as by using NAS could improve user time and reduces idle. Iowait increase slightly because of network contention by multiple nodes. This is must be a contention as in single nodes Iowait in using NAS is lower than in Micro SD.

5. Conclusion

In the test running Hadoop in a single node did not show a significant difference of 2 seconds. While in multi-node using Micro SD and NAS storage there is a significant difference of 7 seconds (10%). This difference is caused by all processors spending more time on user code and reducing idle even though iowait is slightly increased. This increase in iowait is very small compared to the large number of processors added, from 4 to 24.

Acknowledgment
This research is supported by LBE Grant 2019, Faculty of Engineering, Hasanuddin University.

References

[1] Lam Chuck, "Hadoop in Action" Stamford : Manning Publication. 2011.
[2] P. Kusumanegara, Analisis Performan Kecepatan MapReduc pada Hadoop Menggunakan TCP packet Flow Analysis. 2014.
[3] J. Webster, "Understanding Hadoop Technology and Storage," [online].
[4] White Colin, MapReduce and the Data Scientist. BI Research, 2012.
[5] Magang Industri, "Definisi Cloud Computing Meruvia.org Cloud Computing," 2013.
[6] Apache, "Apache TM Hadoop" [online] : https://hadoop.apache.org.
[7] Intel Hadoop, "PAT, Performance Analysis Tools". [online]. Available https://github.com/Intel-hadoop/.

IOP Conf. Series: Materials Science and Engineering **875** (2020) 012038 doi:10.1088/1757-899X/875/1/012038

Partial Discharge Analysis on Crossed-linked Polyethylene (XLPE) Cable 150 KV

M Salama[1]*, K Ikhlas[1], and S M Wahyu[1]

[1]Electrical Department, Engineering Faculty, Hasanuddin University, Makassar, Indonesia

*Email: salamamanjang@gmail.com

Abstract. Partial discharge is the flashover of sparks that occur in an insulating part. as a result of the high potential difference in the insulation. Partial discharge can occur in solid insulation materials. It is play an important role in the electrical power system to separate two or more conductor of electric voltage. So, between these conductors do not occur flashover. The material was chose as the insulation material for high voltage equipment is polymer material, it is because in addition to having a high breakdown voltage. Polyethylene is one of the materials included in the polymer group, generally better known as plastic because the material is thermoplastic. There are several methodologies before conduct this research, that includes literature studies. Literature Studies is a research method by finding material from various libraries on topics that will be discussed in this research. Do Partial Discharge testing. And then, record the data obtained. From the experiments that have been done, the results show that a good Partial Discharge value is below 10 pC, if the cable has more than doubled from the normal value, the cable have been damaged or have been consumed by the age of the cable.

1. Introduction

1.1. Partial Discharge

Partial discharge is the flashover of sparks that occur in an insulating part (in a deep cavity or on a surface) as a result of the high potential difference in the insulation. This phenomenon arises due to many factors including the quality of the dielectric material, the cavities in the dielectric material, as well as the damage or imperfections in the process.

Partial discharge (PD) phenomenon if it occurs continuously it will cause overheating in certain areas which will damage the insulating material and lead to system failure. Before all of this happens it is very important to do initial detection and identification to find the cause of electrical discharge that can reduce the quality of the dielectric material from a system that uses it. Some of these causes can be identified from the responses that arise in the electrical discharge test. In this test, the variety of responses that can be investigated with discharge detection devices for each type, shape or pattern of electrical discharge on the oscilloscope and its relationship with variations in voltage testing and voltage application within a certain time [1].

PD testing is related to quality and quantity values. The quality value is analyzed from the tendency of the data obtained from the characteristics of the material in a particular test. While the quantity value is the nominal value of PD which has the dimension of pico coulomb (pC). Both of these values must meet the standards on testing so that the quality of the equipment can only be assessed. In detecting Partial Discharge (PD) occurs at a certain voltage level called the inception voltage and does not occur at a certain voltage level called the Extinction voltage. If the voltage at the applied source is sinusoidal, the threshold and extinct stresses occur as described in Figure 1. Also, there are a lot of interference signals detected on the detector as noise, can cause misperceptions and reduced measurement sensitivity in testing. The correct interpretation of the characteristics of PD in the equipment being tested depends on the experience of the researcher in conducting the test.

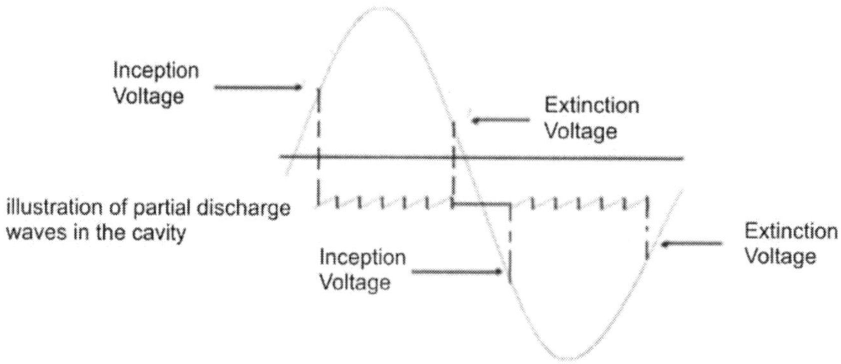

Figure 1. Inception voltage and extinction voltage in sinusoidal waves

PD is a consequence of local electrical stress concentrations in the insulation or on the insulation surface. PD occurs under less than 1 μs, with various occurrence phenomena (internal, surface, corona, and treeing). The figure above shows a model that illustrates the appearance of cavities in partial discharges in a cable insulation system, voids are modeled as Cc capacitors with lower permittivity than insulation permittances [2].

1.2. Types of Partial Discharge

Partial discharge appears in electrical equipment, including cable insulation, Transformers (winding, oil insulation), Generators (Stator, Motor), Gas Insulated Substation, and so forth. Internal discharge is PD which is located inside the insulation (both solid and liquid insulation). The form of internal discharge varies, among others: shaped like a sphere/space, long - both perpendicular or intersecting with the electric field. Corona discharge is not related to insulation systems like other types of partial discharges, corona appears on the gas media because there is no homogeneous field between high voltage conductors.

Channel discharge or treeing arises because of internal discharges that occur continuously in the dielectric discharges that arise continuously cause erosion of the material which results in the formation of treeing patterns. Treeing can occur due to the presence of metallic particles (due to imperfect insulation manufacturing processes) or voids inside the insulation, delamination in the insulation, and rough structure on the inside of the conductor [3].

1.3. Properties of Electronic Materials

Dielectric is a material that has a very small or even almost no current conductivity. The dielectric material can be in the form of solid, liquid and gas. Unlike a conductor, in dielectric materials, there are no conduction electrons which are free to move throughout the material by the influence of the electric field. The electric field will not produce charge movements in the dielectric material. It is this property that causes the dielectric material to be a good insulator. In a dielectric material, all electrons are firmly bound to their nuclei so that a solid body structure (lattices) is formed, or in the case of a liquid or gas, the positive and negative parts are bound together so that each mass flow is not a transfer of charge. Therefore, if a dielectric is given an electric charge, this charge will stay localized in the area where the charge was placed.

According to Gauss's law, the magnitude of the electric flux density that penetrates each closed surface is equal to the total charge covered by that surface can be calculated by the following equation [4].

$$B = \frac{Q}{A} = \mu H \tag{1}$$

where:

B = Flux density (Wb m^2 or Tesla)
Q = Big load on Coulomb
A = Cross-sectional area m^2
H = Electromagnetic field (A/m)
μ = Permeability (Henry/m)

The intensity of the magnetic field is affected by the permeability of the material and flux. So μ can be calculated by the equation

$$\mu = \mu_o \ \mu_r \tag{2}$$

where

μo = Empty space permeability
μr = Material permeability

The use of dielectric material as isolation between electrodes affects the capacitance. This capacitance effect will have different characteristics depending on the permeability of the dielectric material. The capacitance effect has the same properties as two plates parallel to a permeability of material flowed by a voltage source. This capacitance is very closely related to the intensity of the magnetic field (H) and the flux density (B) and the permeability of the empty space (μ_o) and the permeability of the material (μ_r). According to Gauss's law, the magnitude of the electric flux density that penetrates each closed surface is equal to the total charge covered by that surface can be calculated by the equation [4, 5].

1.4. The Mechanism of Partial Discharge (PD)
The manufacturing process of the insulator is expected to provide an average distribution of electrical stress from the voltage electrode. This is very difficult to achieve because in every manufacture of insulation materials still produce cavities in it. One of the mechanisms of PD is caused by a gap or cavity in the insulating material It was illustrated that when V is the applied voltage across the dielectric, the voltage V_c across the cavity will be given as [6]:

$$V_c = C\ C_b \left(C_b + C_c\right) \tag{3}$$

This voltage may cause void breakdown depending on the instantaneous value of V_c and dielectric strength of air inside the void, since stress in the void (E_c) may become much higher than the average stress in the dielectric. The energy released in this discharge will cause deterioration of the dielectric.

In solid insulation materials, the cavity contained in the insulation material is usually filled by air or gas which has a permeability of material lower than the surroundings. The mechanism of the occurrence of PD can be explained in more depth by using the illustration as shown in Figure 1, in the cavity contained in this insulating material, there is a partial capacitance effect.

The capacitance effect that occurs has a lower material strength, thus causing a greater field intensity in the cavity. The intensity of this large field can cause arcs of fire. This arc indicates the load jump in the cavity. Furthermore, the arc will be muffled and start charging until it finds another cavity to release it again. The phenomenon of short discharge and the length of time it occurs repeatedly is referred to as partial discharge. If it happens continuously it will be able to damage the insulating material [2, 7].

2. Experimental Method

2.1. Object Tester

In this research, there are two types of cables that we used, which is the new 150 KV Crossed-linked Polyethylene (XLPE) cable and the old of XLPE cable, the length of each cables is 1 meter.

2.2. Test Setup

This is the test setup that are used to this Partial Discharge Research.

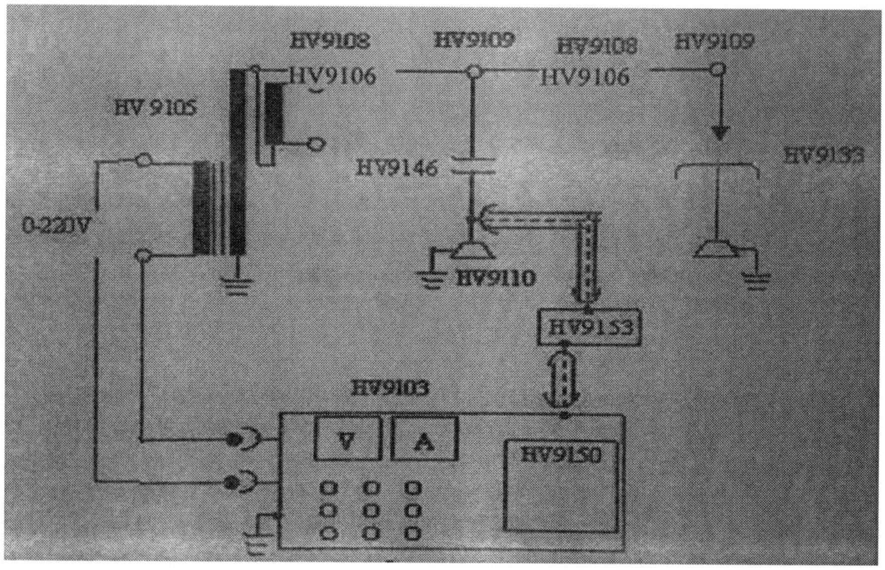

Figure 2. Test setup of PD measurement

The equipment to be used are: HV9105 (HV Test Transformer, HV9103 (Control Desk), HV9146 (Coupling Capacitor), HV9150 (AC Peak Voltmeter), HV9106 (HV Flexible Connector), HV9107 (Earthing Rod), and (HV9153) Partial Discharge Meter.

2.3. Partial Discharge Testing

Testing procedures that must be considered before carrying out Partial Discharge testing on the High Voltage test transformer are:

1. Preparation of tools and materials. In this preparation, we need to prepare partial Discharge test equipment including Coupling Capacitor, Transformer, Regulators, Resistors 100 M Ohm, Partial Discharge Meters, Jumpers, and Cable Grip. And, for the test material is the old of 150 KV XLPE cable and the new 150 KV XLPE cable
2. String up the Partial Discharge circuit in the test transformer by the experiment circuit guideline, after completing the stringing, apply voltage to the XLPE cable with a large 20 KV, 40 KV, 60 KV, and 80 KV. Apply each voltage in each side which are left side, right side and middle side of the cable and measure the Partial Discharge using Partial Discharge Meter and make sure the jumper is tightly connected to the grip of the cable.
3. Data Recording. In this test, data collection was carried out 4 times on each side, which is the right side, the left side and the middle, after each time the data was taken, the partial discharge meter must always be calibrated, so the data that taken will always be accurate.
4. Report: Entered data in the report is the result of data retrieval on each side of the cable, the amount of data obtained is as much as 12 for each cable.

3. Result and Discussion

3.1. Partial Discharge Test Results on the New XLPE Cable 150 KV

The data in the table below is the Partial Discharge test data on the new 150 KV XLPE cable.

Table 1. Partial discharge data on the new XLPE cable 150 KV

Voltage (V)	Partial Discharge Value (pC)		
	Left	Middle	Right
20	6.13	3.43	7.15
40	7.83	4.07	7.83
60	8.06	4.19	8.34
80	8.94	3.52	9.80

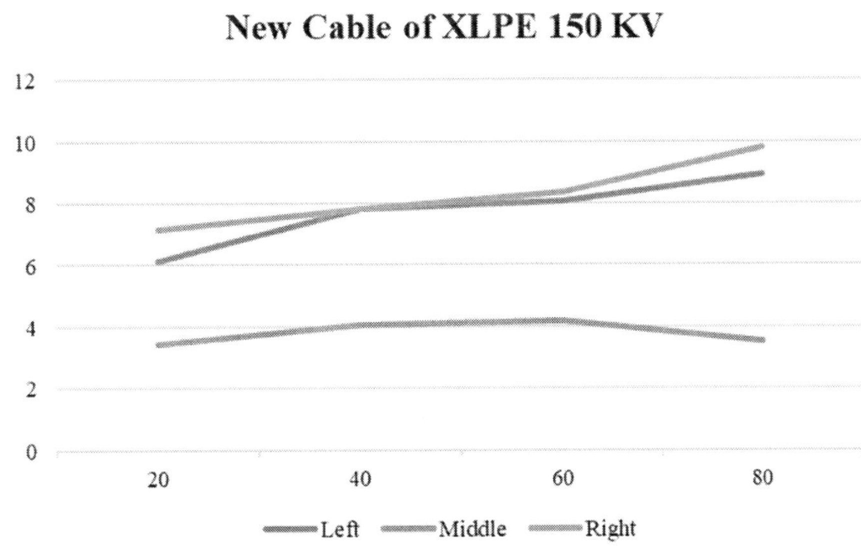

Figure 3. Partial discharge value on XLPE cable 150 KV

Table I and Figure 3 above show that on each side of the cable, showing different values, and when given a different voltage. From the picture above shows that the cable used is the new cable because it can be seen from the value of the Partial Discharge, each side shows below the maximum value of the IEC standard value 60270: 2000 which is 10 pC (pico Coulomb).

In terms of insulation, XLPE has a better work resistance but the price is more expensive compared to other types of synthetic insulation. XLPE has the best characteristics. But in general, synthetic insulation has advantages over paper insulation, that is, it is cleaner, the way to connect is easier, and the working temperature is higher

Partial discharge (PD) is the flashover of sparks, which results from imperfections or inhomogeneous insulation materials. The standard value of partial discharges that do not hurt an object of insulating material is 10 pC, and the scale of partial discharges is up to 10000 pC.

IOP Conf. Series: Materials Science and Engineering **875** (2020) 012038 doi:10.1088/1757-899X/875/1/012038

3.2. Partial Discharge Test Results on the Old XLPE Cable 150 KV

Table 2. Table of partial discharge data on XLPE cable 150 KV

Voltage (V)	Partial Discharge Value (pC)		
	Left	Middle	Right
20	63.51	16.87	79.15
40	77.84	22.77	83.48
60	82.78	89.51	88.84
80	89.51	89.51	89.51

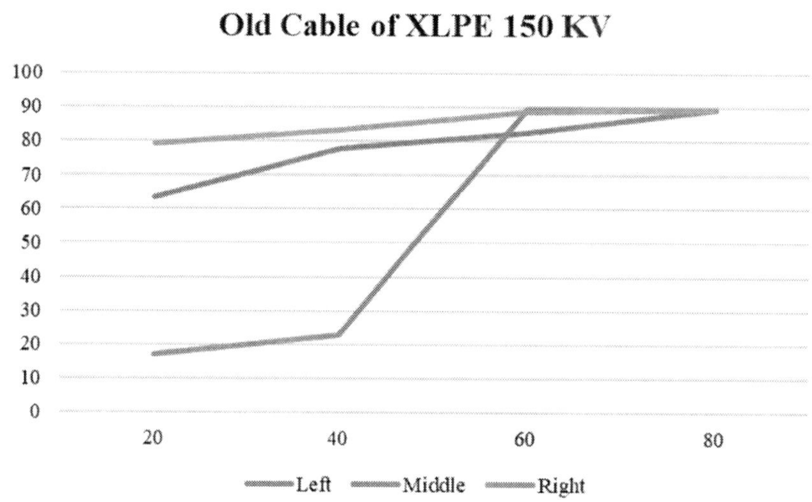

Figure 4. Partial Discharge value on XLPE cable 150 KV

From the data above, it can be seen that the resistance of the XLPE cable has reached its limit, it can be seen that there have been several points that have reached its maximum value, it is showing that the cable has been classified as a damaged cable because it has passed the standard value of Partial Discharge which is 10 pC.

There are several factors of underground cable damage that affect the value of each side of the cable, which are:

a. the level of soil moisture, when the cable has been aging, the cable surface can be easily penetrated by moisture causing corrosion around the cable.
b. ground movement causes the cable to shift, thus issue making it quickly damaged.

This factor plays a role in the destruction of underground cables and will damage to the cable so that Partial Discharge is very easy to penetrate in cable construction. If it is continuously, there will be a lot of energy losses that occur resulting in a decrease in efficiency in transmitting the electricity power.

4. Conclusion

The conclusions that can be drawn are:

1. Partial Discharge on the cable is influenced by the amount of the voltage that applied to the test object and is influenced by the physical condition of the cable and the quality of cable insulation.
2. The more partial discharge the cable produces, the more fully discharge are obtained.
3. Partial Discharge will easily penetrate to the core of the cable and will impact to the result of the amount of energy.

Acknowledgment
The authors would like to take this opportunity to thank the Ministry of Research, Technology and Higher Education of the Republic of Indonesia for its financial support this research. Also thanks to the head of the high voltage laboratory for supporting the testing equipment.

References
[1] Malik, H. M, A. A. Al-Araini, M.I. Qureshi, *Electrical Insulation in Power System*, Marcel Dekker, Inc, 1997.
[2] Arismunandar A. And Kawaharu, Susumu. 1979. *Teknik Tenaga Listrik Jilid II : Saluran transmisi.* Jakarta. PT Pradnya Paramitha.
[3] Black Burn, J. Lewis. 1987. *Protective Relaying Principles and Application, Electrical Engineering and Elektronics.* Boca Raton Florida, Marcel Dekker Inc.
[4] Kraus/Fleisch, *Electromagnetics With Application*, McGraw-Hill International Editions, Electrical Engineering Series, Fifth Edition, 1999.
[5] Graneau, Peter 1979. *Underground Power Transmission the Science,* Technology and Economics of High Voltage. California. University of California.
[6] Vogelsang, R, Fruth, B, Frohlic, K. 2005. *Detection of Electrical Tree Propagation by Partial Discharge Measurements" European Transactions on Electrical Power.* Switzerland. John Wiley & Sons, Ltd.
[7] N. Ra, T. Kawashima, N. Hozumi, Y. Murakami, U. Khayam, and S. Suwarno, "Simulation of Partial Discharge Induced Electromagnetic Wave in Power Transformer", *EPI International Journal of Engineering*, vol. 1, no. 1, pp. pp. 54-59, May 2018.

Implementation of Clustering and Similarity Analysis for Detecting Content Similarity in Student Final Projects

A A Ilham[1]*, A Bustamin[1], I Aswad[1], and F Armin[1]

[1]Department of Informatics, Faculty of Engineering, Hasanuddin University, Makassar, Indonesia

*Email: amil@unhas.ac.id

Abstract. To finish study, students are requested to submit final projects. In some universities, the final projects are not necessary to be submitted for publication. The final project reports are stored in a local database. As the number of final projects is growing in the local database, similar contents may exist among the documents. The commercial tools cannot be used to detect the content similarity since the documents are not published. This paper proposed a system to detect content similarity in documents that are stored in a local database. Considering the number of stored documents, this similar content detection system implements two step processes. First, clustering documents to find most related documents. Second, finding content similarity among the selected documents. The experiment results show that the system is successfully clustering documents and detecting content similarity by implementing TF-IDF and Cosine Similarity algorithms. This system is limited to proceed documents that are written in Bahasa.

1. Introduction

Similar contents among documents may indicated the documents contain plagiarism. Plagiarism is defined as taking of essays, opinions and so on from others and makes them appear like their own essays and opinions. The documents that have little content, the examination can be done manually without the help of the system. However, on documents that have thousands of lines and pages, it is certainly not possible. Many commercial tools exist, but they cannot be used to examine documents that are stored in a local database.

A number of algorithms have been used to detect content similarity in documents such as string-based detection technique, Vector Space Models (VSM), syntax and semantic based detection technique, structural-based detection technique and citation based detection technique [1]. The Rabin-Karp Algorithm has been used in [2]. The hash value generated from this algorithm is the benchmark of similarity between the documents being tested. However, this research still needs to be adjusted when choosing the K-gram value to be used in the tokenization process. The weakness of this algorithm is that the system cannot know the order in which documents appear in this case the time of publication. This algorithm only focuses on determining the similarity between documents being compared. The comparison of the level of document similarity using several methods such as K-Means Clustering, K-Nearest Neighbor and Shingling Algorithm is done in [3]. The purpose of this study is to compare two vectors of text documents that measure the degree of similarity. The best level of accuracy generated by the K-Nearest Neighbor algorithm is equal to 95%.

The Rabin-Karp algorithm is also applied in the Indonesian document similarity detection system. In addition to Rabin-Karp, the Confix-Stripping method is used to maximize the steps for searching basic words in text processing. The system works well in detecting document similarities with an average processing time of 0.0123 seconds and an average similarity level of documents tested 89.1967%. Overall the level of accuracy shown by the system is 0.7 [4].

A number of commercial tools have been developed such as PlagAware, PlagScan, Check for Plagiarism, iThenticate, PlagiarismDetection.org, Academic Plagiarism, The Plagiarism Checker, Urkund, Docoloc and Turnitin [5]. PlagAware provides two main features namely tracing content theft and proof of authorship [6]. PlagScan provides a number of features such as database checking, Internet checking, publications checking, synonym and sentence structure checking [7]. CheckForPlagiarism.net implements fingerprint and document sources to protect documents form plagiarism. It provides features such as Internet checking, publications checking, synonym and sentence structure checking, and multiple document comparison [8].

2. Research Method

Considering the number of stored documents, our system implements two step processes. First, clustering documents to find 10 most related documents. Second, finding content similarity among the selected documents. The system design can be seen in Figure 1.

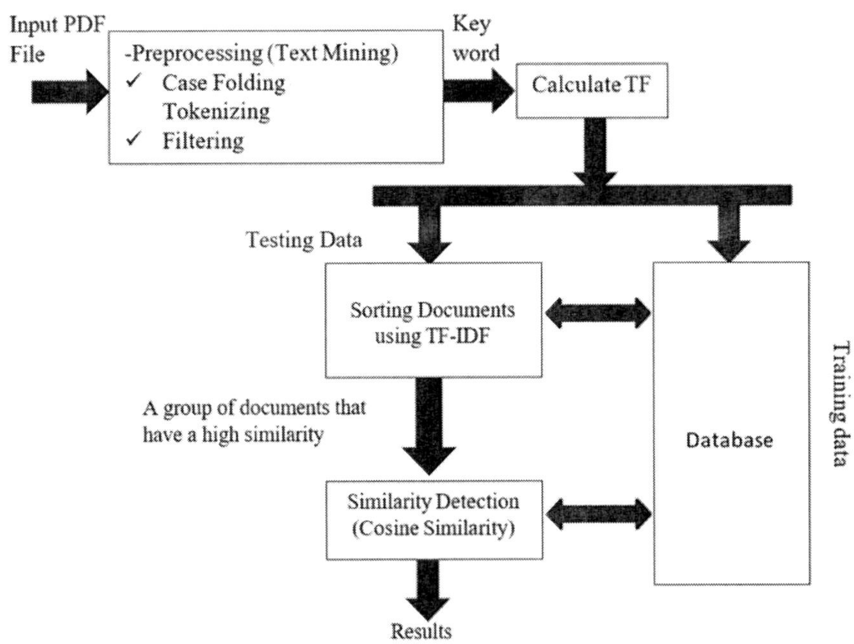

Figure 1. System design

2.1. Preprocessing

Preprocessing is the initial stage of text mining where the text data will be cleaned so that the text becomes more structured before entering the next stage for further processing. Continuous character sets (text) must be broken into more meaningful. This can be done on several different levels. Text preprocessing stages in this study consist of tokenizing, case folding, filtering, and stopword.

2.2. TF-IDF Weighting

Term Frequency (TF) is a simple weighting where the importance of a word is assumed to be proportional to the number of times the word appears in a document. Inverse Document Frequency (IDF) is a weighting that measures how important a word in a document is when viewed globally in all documents. The TF x IDF weighting value will be high if the TF value is large and the word observed is not found in many documents. The TF value is calculated using the following Equation (1):

$$TF(d,t) = f(d,t) \tag{1}$$

where *f(d,t)* is the number of times the word *t* appears in document *d*. IDF considers the frequency of words in all documents. IDF weighting assumes that the weight of a word will be large if the word often appears in a document but not many documents contain the word. IDF values are calculated using the following Equation (2):

$$IDF(t) = log\left(\frac{N}{df(t)}\right) \tag{2}$$

where *df(t)* is the number of documents that have the word *t*. The results of previous studies show that TF x IDF weighting can improve performance better. The TF x IDF value is calculated using the following function [9].

$$TFIDF(d,t) = TF(d,t)x\ IDF(t) \tag{3}$$

At Term Frequency (TF), there are several types of formulas used, i.e. [10]:

a. Binary TF: pays attention to whether or not a word exists in the document. If it exists, a value of one is given, if not, a zero value is given.
b. Raw TF: the TF value is based on the number of occurrences of a word in the document. For example, if it appears five times the word will be worth five.
c. TF logarithmic: used to avoid the dominance of documents that contain few words in the query but have a high frequency.
d. TF normalization: using a comparison between the frequencies of a word with the total number of words in a document.

2.3. Cosine Similarity

Cosine similarity is a method used to calculate the level of similarity between two objects. Cosine similarity is used to measure the closeness between two vectors. Cosine similarity is the result of the dot product of the two vectors that is normalized by dividing by Euclidean Distance between the two vectors. Equation (4) obtained is as follows [11]:

$$sim\left(d_j, q\right) = \frac{\sum_{i=1}^{N} w_{i,j} w_{i,q}}{\sqrt{\sum_{i=1}^{N} w^2{}_{i,j}}\sqrt{\sum_{i=1}^{N} w^2{}_{i,q}}} \tag{4}$$

where:
d_j = document *j*
q = query document
$\sum_{i=1}^{N} w_{i,j}$ = total weight of the word *i* in document *j*
$\sum_{i=1}^{N} w_{i,q}$ = the weights number of word *i* in the query

The TF-IDF weighting process can be seen in Figure 2. The TF-IDF method combines two concepts for weight calculation, namely the frequency of occurrence of a word in a particular document and the inverse frequency of the document containing the word. The frequency with which words appear in a given document indicates how important that word is in the document. The frequency of documents containing the word indicates how common the word is. So, the weight of the relationship between a word and a document will be high if the frequency of the word is high in the document and the overall frequency of the document containing the word is low in the document.

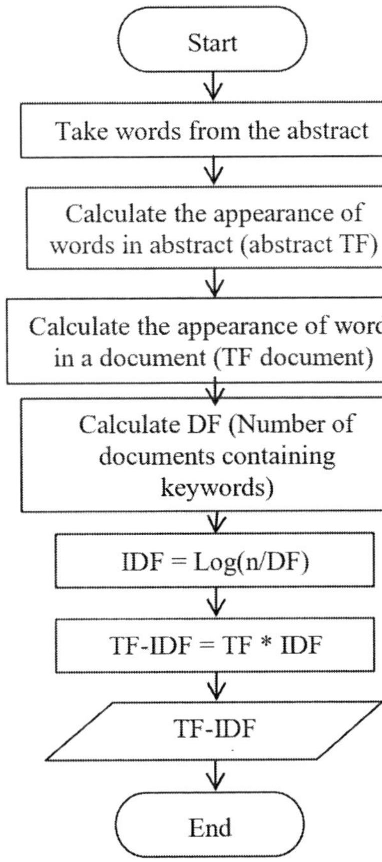

Figure 2. Weighting process

3. Implementation and Result

To illustrate how the system works, we compared a test document (Q) with four other documents (D2, D3, D4 and D5) as shown in Table 1. The test documents are written in Bahasa.

We perform preprocessing to all documents including tokenizing, case folding and filtering. In case folding, the contents of the document are converted into lowercase. Filtering process eliminates punctuation and words that are considered not important in Bahasa such as word "*di*", "*dan*", "*untuk*", etc. The output of the preprocessing can be seen in Table 2.

Table 1. Test document

Doc Name	Content
Q	*Fenomena plagiarisme yang lebih spesifik sering terjadi di dunia akademis, khususnya dilakukan oleh mahasiswa dalam menyelesaikan tugas kuliah maupun tugas akhir karena tersedianya fasilitas untuk menyalin suatu teks dan menaruh salinan teks tersebut dari satu dokumen ke dokumen lainnya.*
D2	*Perkembangan teknologi memiliki dampak yang sangat signifikan dalam kehidupan sehari-hari, mulai dari kegiatan yang sederhana hingga kegiatan yang membutuhkan tingkat ketelitian yang tinggi. Kegiatan yang umum dilakukan oleh sebuah instansi adalah kegiatan pengarsipan dokumen, baik dokumen dalam bentuk fisik maupun elektronik.*
D3	*Kegiatan menjiplak tugas sering dilakukan mahasiswa yang merupakan tindakan plagiat, banyak tugas yang terkumpul dan waktu yang terbatas membuat dosen sulit untuk memeriksa tugas satu per satu. Oleh karena itu, diperlukan suatu aplikasi yang dapat mendeteksian kemiripan dokumen teks.*
D4	*Paper ini mendiskusikan tentang deteksi plagiasi dengan menggunakan metode string matching algoritma rabin-karp. Metode diimplementasikan dalam aplikasi berbasis web untuk mendeteksi plagiasi dengan cara menguji teks (huruf) yang ada pada dokumen abstraksi dari kar ya skripsi ataupun jurnal mahasiswa.*
D5	*Kesamaan Dokumen dapat digunakan untuk menjadi petunjuk dan contoh mencari informasi yang sama. Kemampuan mencari kesamaan ini dapat mengurangi waktu. Untuk menggambarkan tingkat kesamaan antara dokumen dapat diukur oleh Metode Cosine Similarity. Berdasarkan tingkat kesamaan dokumen dapat diklasifikasikan dengan menggunakan Algoritma Single Pass Clustering.*

Table 2. Preprocessing results

Doc Name	Content
Q	*Fenomena plagiarisme spesifik dunia akademis mahasiswa menyelesaikan tugas kuliah tugas tersedianya fasilitas menyalin teks menaruh salinan teks dokumen dokumen alasan penulis mencoba membangun sistem deteksi plagiarisme dokumen bahasa indonesia metode vector space model*
D2	*Perkembangan teknologi memiliki dampak signifikan kehidupan seharihari kegiatan sederhana kegiatan membutuhkan tingkat ketelitian kegiatan instansi kegiatan pengarsipan dokumen dokumen bentuk fisik elektronik*
D3	*Kegiatan menjiplak tugas mahasiswa tindakan plagiat tugas terkumpul terbatas dosen sulit memeriksa tugas aplikasi mendeteksian kemiripan dokumen teks*
D4	*Paper mendiskusikan deteksi plagiasi metode string matching algoritma rabinkarp metode diimplementasikan aplikasi berbasis web mendeteksi plagiasi menguji teks huruf dokumen abstraksi karya skripsi jurnal mahasiswa*
D5	*Kesamaan dokumen petunjuk contoh mencari informasi kemampuan mencari kesamaan mengurangi menggambarkan tingkat kesamaan dokumen diukur metode cosine similarity berdasarkan tingkat kesamaan dokumen diklasifikasikan algoritma single pass clustering*

After the preprocessing, TF-DF is calculated by counting the appearance of each term in *Q* in other documents. The TF and DF values can be seen in Table 3.

Table 3. TF and DF results

Term	TF				DF
	D2	D3	D4	D5	
Fenomena	0	0	0	0	0
plagiarisme	0	0	0	0	0
spesifik	0	0	0	0	0
dunia	0	0	0	0	0
akademis	0	0	0	0	0
mahasiswa	0	1	1	0	2
menyelesaikan	0	0	0	0	0
tugas	0	3	0	0	1
kuliah	0	0	0	0	0
tersedianya	0	0	0	0	0
fasilitas	0	0	0	0	0
menyalin	0	0	0	0	0
teks	0	1	1	0	2
menaruh	0	0	0	0	0
salinan	0	0	0	0	0
dokumen	2	1	1	3	4
alasan	0	0	0	0	0
penulis	0	0	0	0	0
mencoba	0	0	0	0	0
membangun	0	0	0	0	0
sistem	0	0	0	0	0
deteksi	0	0	1	0	1
bahasa	0	0	0	0	0
indonesia	0	0	0	0	0
metode	0	0	2	1	2
vector	0	0	0	0	0
space	0	0	0	0	0

TF-DF values are used to calculate IDF using log(n/df) formula. TF x IDF values as can be seen in Table 4.

Table 4. IDF and TF-IDF calculation results

Term	IDF = log (n/df)	TF x IDF			
		D2	D3	D4	D5
fenomena	0	0	0	0	0
plagiarisme	0	0	0	0	0
spesifik	0	0	0	0	0
dunia	0	0	0	0	0
akademis	0	0	0	0	0
mahasiswa	0.301	0	0.301	0.301	0
menyelesaikan	0	0	0	0	0
tugas	0.602	0	1.806	0	0
kuliah	0	0	0	0	0
tersedianya	0	0	0	0	0
fasilitas	0	0	0	0	0
menyalin	0	0	0	0	0
teks	0.301	0	0.301	0.301	0
menaruh	0	0	0	0	0
salinan	0	0	0	0	0

Term	IDF = log (n/df)	TF x IDF			
		D2	D3	D4	D5
dokumen	0	0	0	0	0
alasan	0	0	0	0	0
penulis	0	0	0	0	0
mencoba	0	0	0	0	0
membangun	0	0	0	0	0
sistem	0	0	0	0	0
deteksi	0.602	0	0	0.602	0
bahasa	0	0	0	0	0
indonesia	0	0	0	0	0
metode	0.301	0	0	0.602	0.301
vector	0	0	0	0	0
space	0	0	0	0	0
Document Weight Value		0	2.408	1.806	0.301

As can be seen in Table 4, document *D3* has the highest weighting value among other documents. This shows that document *D3* has a high similarity with document *Q*. We implement Cosine Similarity algorithm to measure the level of similarity between document *Q* and *D3*, as shown in Table 5.

Table 5. Cosine similarity value

Term	nQ	nD3	(nQ x nD3)	$(nQ)^2$	$(nD3)^2$
fenomena	1	0	0	1	0
plagiarisme	2	0	0	4	0
spesifik	1	0	0	1	0
dunia	1	0	0	1	0
akademis	1	0	0	1	0
mahasiswa	1	1	1	1	1
menyelesaikan	1	0	0	1	0
tugas	2	3	6	4	9
kuliah	1	0	0	1	0
tersedianya	1	0	0	1	0
fasilitas	1	0	0	1	0
menyalin	1	0	0	1	0
teks	2	1	2	4	1
menaruh	1	0	0	1	0
salinan	1	0	0	1	0
dokumen	3	1	3	9	1
alasan	1	0	0	1	0
penulis	1	0	0	1	0
mencoba	1	0	0	1	0
membangun	1	0	0	1	0
sistem	1	0	0	1	0
deteksi	1	0	0	1	0
bahasa	1	0	0	1	0
indonesia	1	0	0	1	0

Term	nQ	nD3	(nQ x nD3)	$(nQ)^2$	$(nD3)^2$
metode	1	0	0	1	0
vector	1	0	0	1	0
space	1	0	0	1	0
model	1	0	0	1	0
kegiatan	0	1	0	0	1
menjiplak	0	1	0	0	1
tindakan	0	1	0	0	1
plagiat	0	1	0	0	1
terkumpul	0	1	0	0	1
terbatas	0	1	0	0	1
dosen	0	1	0	0	1
sulit	0	1	0	0	1
memeriksa	0	1	0	0	1
aplikasi	0	1	0	0	1
mendeteksian	0	1	0	0	1
kemiripan	0	1	0	0	1
SUM			12	45	24

nQ: the number of term occurrences in document *Q*
nD3: the number of term occurrences in document *D3*

Based on the Cosine Similarity formula in Equation (4), the result is:

$$Cosine\ Similarity = \frac{12}{\sqrt{45} \times \sqrt{24}}$$

$$= \frac{12}{6.708 \times 4.899}$$

$$= 0.365$$

The level of similarity between document *Q* and document *D3* is 0.365.

We developed a web-based interface and implemented the algorithms to find content similarity among documents in local database. The system can recognize the order in which documents appear, in this case the time of publication. Only documents that have older publication time are compared to the tested document. The web-based interface provides Add Document page, Administration page, Sorting Document page and Similarity page. Figure 3 shows Add Document page which is used to add a new document to be tested. Administration page as shown in Figure 4 is used to check similarity or delete a document. Figure 5 shows Sorting Document page which clustering most related document, and Figure 6 shows Similarity page which compare the tested document with the most related documents. Our functional testing shows that the system is able get the input, process the document and display the results as expected.

IOP Conf. Series: Materials Science and Engineering **875** (2020) 012039 doi:10.1088/1757-899X/875/1/012039

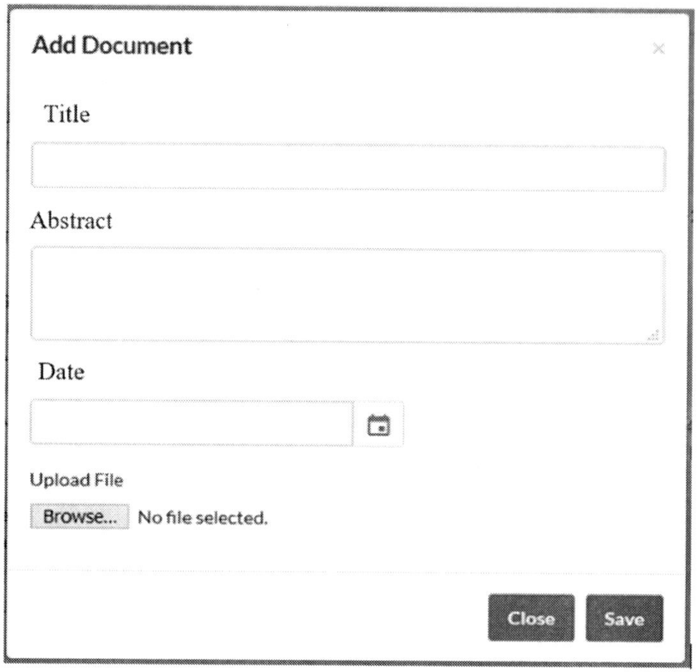

Figure 3. Add document page

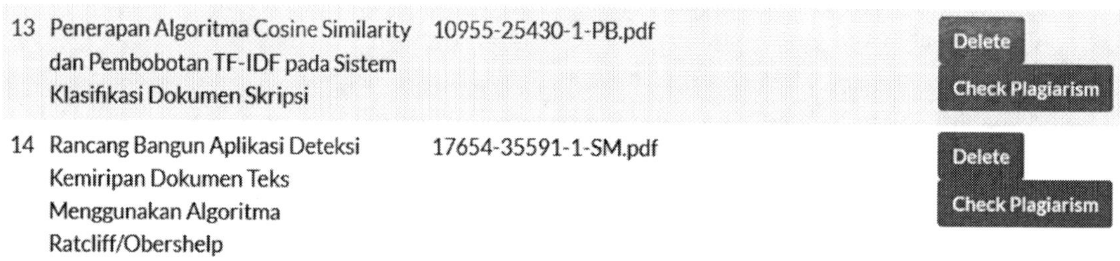

Figure 4. Administration page

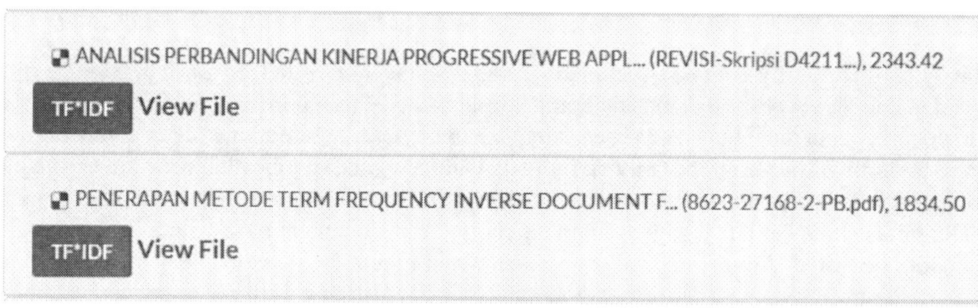

Figure 5. Sorting Document page

#	Source	Current	Value
1	sistem library yang berbasis website akan menyimpan data digital yang berukuran besar banyak sehingga membutuhkan kapasitas penyimpanan yang besar pengelolaan data yang terstruktur	permasalahan yang dihadapi pada text mining adalah jumlah data yang besar dimensi yang tinggi data struktur yang terus berubah serta data noise	0.6076436202502
2	tampilan beranda untuk anggota gambar	tampilan beranda user gambar	0.67082039324994

Figure 6. Similarity page

4. Conclusion

A system to detect content similarity in documents that were stored in local database was built in this study. Considering the number of stored documents, the system implemented two step processes by clustering documents to find most related documents, and then finding content similarity among the selected documents. The system was able to recognize the order in which documents appeared, in this case the time of publication. Only documents that had older publication time were compared to the tested document. The functional testing showed that the system was able get the input, processed the document and displayed the results as expected. The system was successfully clustering documents and detecting content similarity by implementing TF-IDF and Cosine Similarity algorithms. Future research can be developed using other weighting methods for large-scale documents.

Acknowledgment

This research is funded by Lab Based Education Grant 2019, Faculty of Engineering, Hasanuddin University.

References

[1] K. Vani and D. Gupta, "Study on Extrinsic Text Plagiarism Detection Techniques and Tools," *J. Eng. Sci. Technol. Rev.*, vol. 9, no. 3, pp. 99–105, 2016.

[2] R. E. Putri and A. P. Siahaan 2017 Examination of Document Similarity using Rabin-Karp Algorithm, *International Journal of Recent Trends in Engineering and Research,* vol. 03, no. 08, pp. 196-201.

[3] M. Oghbaie and M. M. Zanjireh 2018 Pairwise document similarity measure based on present term set *Journal of Big Data,* vol. 5, no. 52, pp. 1-23.

[4] D. D. Sinaga and S. Hansun 2018 Indonesian Text Document Similarity Detection System using Rabin-Karp and Confix-Stripping Algorithms, *International Journal of Innovative Computing, Information and Control,* vol. 14, no. 5, pp. 1893-1903.

[5] A. M. El Tahir Ali, H. M. Dahwa Abdulla, and V. Snášel, "Overview and comparison of plagiarism detection tools," *CEUR Workshop Proc.*, vol. 706, pp. 161–172, 2011.

[6] "PlagAware." [Online]. Available: https://www.plagaware.com/. [Accessed: 09-Apr-2019].

[7] "PlagScan: Online Plagiarism Checking." [Online]. Available: https://www.plagscan.com. [Accessed: 09-Apr-2019].

[8] "CheckForPlagiarism.net." [Online]. Available: https://www.checkforplagiarism.net/. [Accessed: 09-Apr-2019].

[9] R. Nabil, L. Amine, A. Issam, B. L. EL Habib and L. El Houssine 2015 A New Approach that improves TF-IDF Weighting Measure, *International Journal of Information and Communication Technology,* vol. 5, no. 10, pp. 1-10.

[10] C. Breitinger, B. Gipp and S. Langer 2015 Research-paper recommender systems: a literature survey, *International Journal on Digital Libraries,* vol. 17, no. 4.

[11] C. Manning, P. Raghavan and H. Schutze 2008 Introduction to Information Retrieval, Cambridge: Cambridge University Press.

IOP Conf. Series: Materials Science and Engineering **875** (2020) 012040 doi:10.1088/1757-899X/875/1/012040

Interactive Learning Media Based on Android for Computer Programming Course

A D Achmad[1]*, A Achmad[1], and Z M Putra[2]

[1]Department of Electrical Engineering, Universitas Hasanuddin, Gowa, Sulawesi Selatan, Indonesia 92171
[2]Department of Electrical Engineering, Universitas Fajar, Makassar, Sulawesi Selatan, Indonesia 90231

*Email: andini.achmad@unhas.ac.id

Abstract. Computer Programming is one of the basic compulsory subjects for the electrical engineering field. This subject needs continuous practice so that it can be mastered well. This study aims to develop an interactive learning media based on Android so that students can practice anywhere. The interactive learning media for computer programming course takes in explanation of operator c++, input-output statement, program structure, data type, condition, repetition, string, and array, complemented with evaluation facility using the cosine similarity method so that the automated essay scoring feature is available. Interactive features provided in this learning media are the auto complete syntax and function in the code editor. For the evaluation used the black box testing method to test the functionally of the android application and the distribution of questionnaire to determine the response of students. The responses of student have proven that the level of practically of this interactive learning media is 87.5%.

1. Introduction

The development of technology in education is very fast. The demand for quality learning outcomes is a correlation of technological consequences. Ranging from conventional learning models to computerized learning models. The presence of computerized media will be very helpful in teaching and learning [1]. In the university lecturers are expected to be able to transfer their knowledge to students well.

Android-based learning media is one of the most popular information technology media in the world of education this time. The transfer of knowledge can be carried out well through this media because of the ease of interacting between lecturers and students. The interaction is that the lecturer gives some learning materials complemented with evaluation in the form of student essay questions and answers based on the skills they have. The essay in question is an essay in Indonesian which has many variations of affixes. The Automatic Essay Score (AES) feature was added to the e-learning function as a final grade student so that the lecturer could streamline time and energy to check student essay answers [2].

Sixties years last century, researches conducted research on AES [3]. It loses about 30 billion pounds per year just because the teachers in Britain spending their 30% time for scoring student's answers about 30% [4]. By implementing AES, it's hoped that teaching staff will not spend much time examining the work of their students.

IOP Conf. Series: Materials Science and Engineering **875** (2020) 012040 doi:10.1088/1757-899X/875/1/012040

There are many kinds of method that support AES. The previous research [5] reveals that Cosine Similarity Method has a better server performance so that preferred to be implemented in e-learning automatic scoring system. In this research, to support the AES feature Cosine Similarity method is used in this design of the learning media.

The Cosine Similarity method [6] is a method used to measure the similarity between two different text or documents by measuring the cosine of the angle between the document representation vectors. The cosine value is between 0 and 1 where the greater value of cosine, the greater its similarity between the two sentences or documents that being compared. This method aims to equalize the two sentences in a document that will be compared, then the results of the comparison will be used as similarity values.

Cosine similarity is a measure of similarity more commonly used in information retrieval and is a measure of the angle between the document vector D_a (point (a_x, b_x)) and D_b (point (a_y, b_y)). Each vector represents each word in each document (text) that is compared and forms a triangle, so that the cosine law can be applied to state that:

$$\cos C = a^2 + b^2 - c^2 \Big/ 2ab \tag{1}$$

where

$$a^2 = a_x^{\ 2} + a_y^{\ 2}; b^2 = b_x^{\ 2} + b_y^{\ 2} \tag{2}$$

and

$$c^2 = \left(b_x - a_x\right)^2 + \left(a_y - b_y\right)^2 \tag{3}$$

Substituting the above equation values for a, b and c, so that they are obtained

$$\cos C = \left[a_x b_x + a_y b_y\right] / \left[\left[ax2 + ay2\right]^{1/2} \times \left[b_x^{\ 2} + b_y^{\ 2}\right]^{1/2}\right]$$

when two documents are identical, the angle is zero degrees (0°) and the similarity is one (1); and when the two documents are not identical at all, the angle is 90 degrees (90°) and the same is (0) [7].

Computer Programming is one of the basic courses with 2 credits found in the curriculum of Electrical Engineering Study Program at Universitas Hasanuddin. This course is intended to provide knowledge and understanding of how to make computer programs using programming languages to solve problems in various fields of life. The programming language used in this course is C++. Currently the lecture process is more often done in the classroom so that students are limited in doing programming activities directly, while in order to understand the material all students should immediately be able to practice it. To overcome this, an interactive learning media for computer programming course is created which is complemented with evaluation/assignment facility. This application is based on Android but also can be accessed via website.

2. Research Method

The development of this research is in accordance with the R&D method (Research and Development). R&D is a research method used to produce certain products and test the effectiveness of these methods. In education, R&D is a research method used to develop or validate products used in education and learning [8]. This research used R&D with the following steps:

IOP Conf. Series: Materials Science and Engineering **875** (2020) 012040 doi:10.1088/1757-899X/875/1/012040

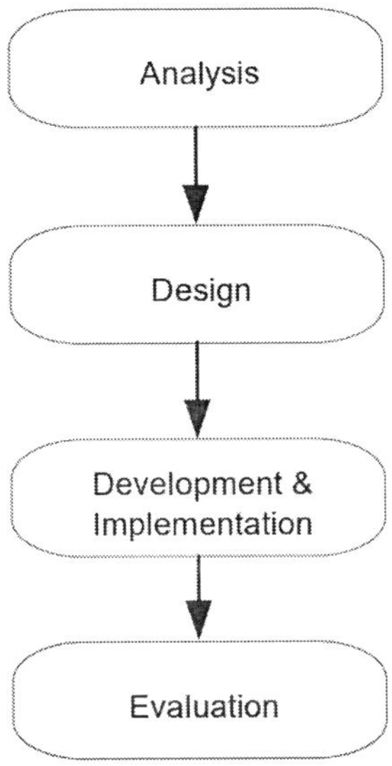

Figure 1. Stages of development and research

- *Analysis* – covering problem, need, and task identification.
- *Design* – covering data design for material and problem needs, navigation design, user interface design, and algorithm.
- *Develop* – covering making learning media using Android Studio, HeidiSQL, and Rational Rose.
- *Implementation* – in the form of a learning media—android application, that can be accessed using computer and android smartphone.
- *Evaluate* – user response assessments were conducted by students.

The research site was conducted at the Computer and Network Laboratory of the Department of Electrical Engineering, Faculty of Engineering, Universitas Hasanuddin at the Gowa Campus. For the evaluation used the black box testing method to test the functionality of the android application and the distribution of questionnaires to determine the response of students. Data collection tools used by using a questionnaire instrument. Context diagram for interactive media learning shown in Figure 2.

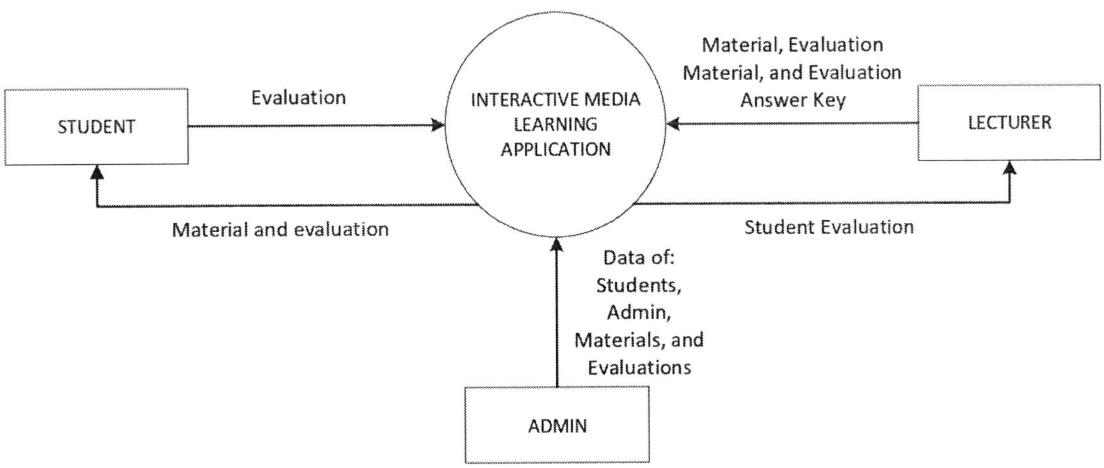

Figure 2. Interactive media learning for computer programming course context diagram

3. Results and Discussion

From the results of the system design that has been carried out, an interactive learning media is produced with an automatic assessment feature for essay answers and auto complete syntax and function in the code editor. The main parts of this research are as follows.

3.1. Lecturer Page

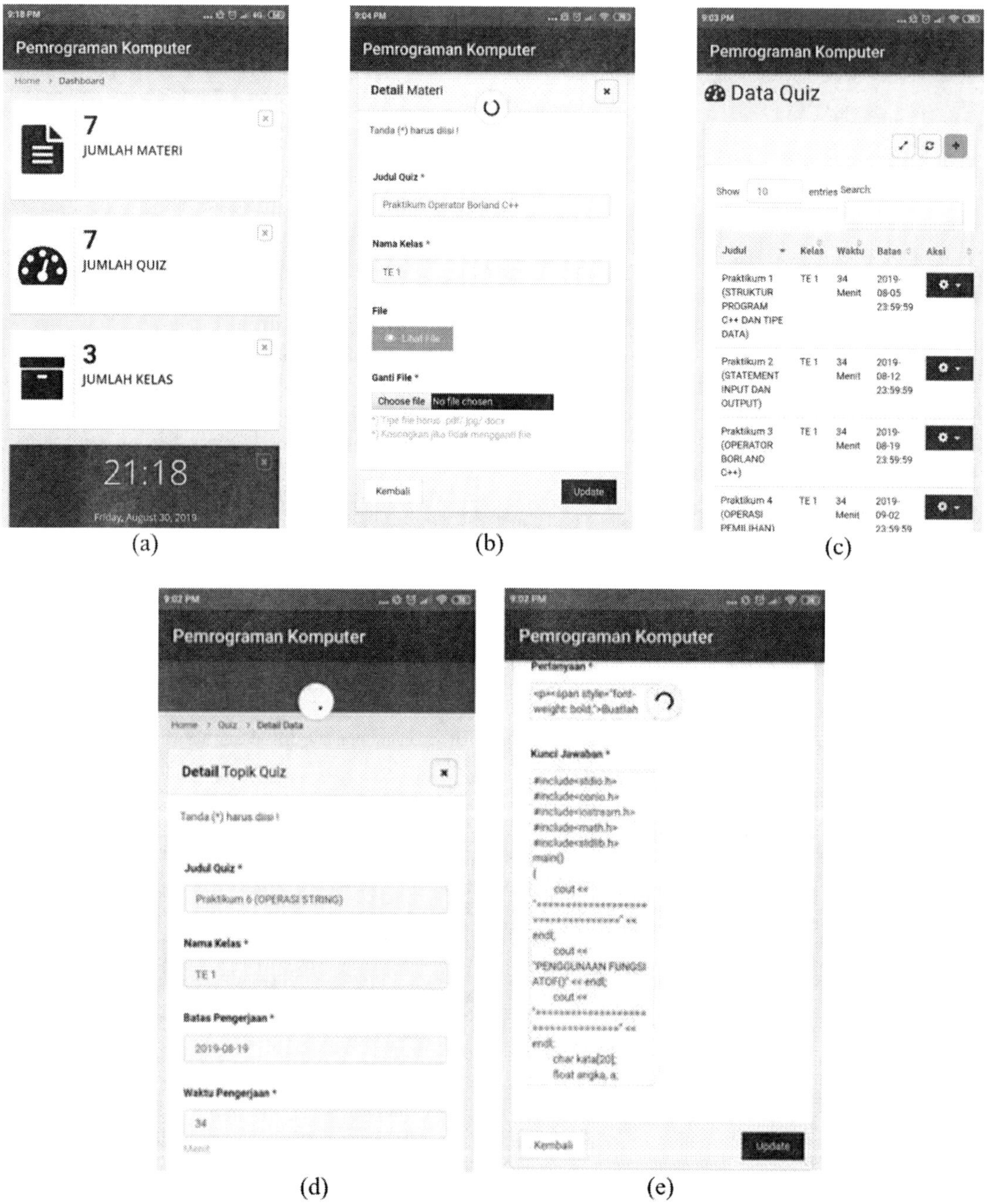

Figure 3. Lecturer result pages (a) Lecture menu page (b) Lecture learning material page (c) List evaluation/quiz page (d) Create evaluation/quiz page (e) Detail create evaluation/quiz page

3.2. Student Page

Figure 4. Student result pages (a) Student menu page (b) Student learning material page (c) List evaluation/quiz page (d) Quiz page, display problem (e) Quiz page, display answer editor

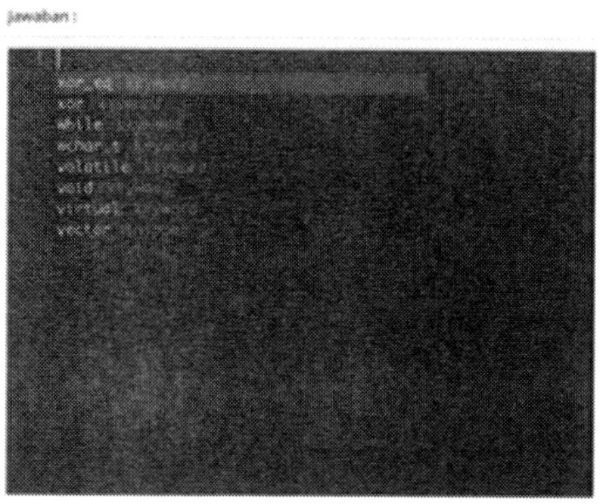

Figure 5. Interactive features, auto complete syntax and function in the code editor

Not all students have laptop at the time of learning as well as its prices relatively expensive. But most students use mobile phones based on Android. After the implementation and testing the application by filling out the questionnaire, the results show that 87.5% agree that the development of an interactive learning media based on Android is practical and can help students to understand the learning material.

4. Conclusions

Based on the research results obtained, it can be concluded that the interactive learning media based on Android can be used as a practical learning media to help students to understand learning material more which is equipped with interactive feture; auto complete syntax and function in the code editor. This application can also assist lecturers in directly monitoring the results of student evaluation during lecture time because it equipped with AES feature using cosine similarity method.

References

[1] Bhavna K. Pradeep C. and Manmohan S., 2013. *Comparative Analysis Study of E-Learning and Traditional Learning in Technical Institution* (India). International Conference on Communication Systems and Network Technologies.

[2] Astutik S. Cahyani A. D. and Sophan M. K. 2012, *Sistem Penilaian Esai Otomatis Pada E-Learning dengan Algoritma Winnowing. Vol 12 no 2* (Indonesia) Jurnal Informatika P 47-52.

[3] T. Kakkonen and E. Sutinen. 2004. *Automatic Assessment of the Content of Essays Based on Course Materials.* ITRE. 2nd International Conference Information Technology: Research and Education, pp. 126-130.

[4] S. Ghosh and S.S. Fatima. 2008. *Design of an Automated Essay Grading (AEG) System in Indian Context.* TENCON. IEEE Region 10 Conference, pp.1-6.

[5] E. S. Pramukantoro and M. A. Fauzi. 2016. *Comparative analysis of string similarity and corpus-based similarity for automatic essay scoring system on e-learning gamification.* International Conference on Advanced Computer Science and Information Systems (ICACSIS), Malang, 2016, pp. 149-155.

[6] Fauzi, Mochammad & Lieharyani, Djoko & Setiawan, Budi Darma & Pramukantoro, Eko. 2017. *Automatic Essay Scoring System Using N-Gram and Cosine Similarity for Gamification Based E-Learning.* Proceedings of the International Cnference on Advances in Image Processing (ICAIP 2017). Bangkok, Thailand. pp. 151-155. doi: 10.1145/3133264.3133303.

[7] Konchady M., 2006. *Text Mining Application Programming* (Boston: Charles River Media)

[8] Hanafi. 2017. *Konsep Penelitian R&D dalam Bidang Pendidikan.* Saintifika Islamica: Jurnal Kajian Keislaman, Volume 4. No. 2 Juli-Desember 2017. ISSN: 2507-053X, pp 129-150.

IOP Conf. Series: Materials Science and Engineering **875** (2020) 012041 doi:10.1088/1757-899X/875/1/012041

Effects of 500 kV EHV Transmission Lines on Location Determination and Capacity of Shunt Reactor in South Sulawesi Systems

I Kitta[1]*, A Suyuti[1], Z Muslimin[1], S Manjang[1], W Santoso[1], and M T Hidayat[1]

[1]Department of Electrical Engineering, Engineering Faculty, Hasanuddin University, Makassar, Indonesia

*Email: ikhlaskitta@gmail.com

Abstract. Electricity in South Sulawesi system is so extensive that it is necessary to construct a 500 kV Extra High Voltage (EHV) to avoid instability and waste of electricity during the distribution of electricity. need. Therefore, this paper describes the implementation of research on optimizing the electrical energy network system from renewable energy power plants in the form of the application of extra high voltage transmission. This research was carried out as an effort to implement a 500 kV extra high voltage transmission in South Sulawesi. The length of the 500 kV EHV to be installed includes an extra-long transmission, which is 365 kms which will produce more voltage on the network during light loads. So, it is necessary to make efforts to improve the electrical system to meet applicable standards, including the addition of the shunt reactor. The purpose of this study is to obtain conclusions about the effect of the shunt reactor on the South Sulawesi transmission system after the 500 kV EHV is applied. The research that has been carried out aims to obtain the optimal location and capacity of a shunt reactor. The results of this study will be very useful for electricity stakeholders, especially transmission companies in South Sulawesi.

1. Introduction

Current conditions in the South Sulawesi system, the distribution of electric power is done through a 150 kV high voltage (HV) transmission lines and 275 kV extra high voltage (EHV) transmission lines [1]. The power transmission system needs to be provided with a backbone so that the system is still sufficient to meet the development needs of South Sulawesi's load for the next 40 years. The system that is very suitable as a backbone is a 500 KV transmission system. Another factor driving the need for a 500 kV system in the South Sulawesi System is the presence of Wind Turbine Power Plant in Sidrap with a capacity of 70 MW and Jeneponto with a capacity of 60 MW (South Sulawesi) which has an impact on system operation due to its intermittent energy source characteristics. In addition to intermittent generating factors, another factor influencing the operation of the South Sulawesi system is load characteristics. The government encourages the smelter industry to grow in the South Sulawesi region and this will have an impact on the operation of the electricity system because of the load characteristics of the smelter in the form of an electric arc furnace so it must be able to be anticipated.

For future development, it is necessary to study the application of 500 kV EHV on the South Sulawesi system. From several studies conducted by researchers that the South Sulawesi system is currently unable to anticipate the development of such a burden. In 2019 the Electric Power Supply Business Plan was issued as a master plan document for the development of Indonesia's electricity system, including the plan to construct a 500 kV on South Sulawesi transmission system for the 2019-2028 planning period of 365 kmc [2].

The plan for the development of the 500 kV EHV by the Government of Indonesia is very encouraging. However, before the 500 kV system is implemented in South Sulawesi, a study needs to be done on the need for a 500 kV voltage level transmission system to get the condition of the system if 500 kV has been applied. One thing that needs to be studied is the use of shunt reactors that function to control the voltage increase in the 500 kV transmission system [3]. In an EHV transmission line system especially 500 kV, changing the voltage level becomes the most important problem. So the purpose of this paper is to determine the location and size of the shunt reactor capacity to increase the voltage level in the South Sulawesi transmission system.

2. Research Method

2.1. Test System Description

The South Sulawesi electricity interconnection system is supplied by large-scale power plants located at Jeneponto 400 MW and Sengkang 192 MW. The large-scale power output voltage after going through a step-up transformer is a 150 kV system. While the 190 MW Poso voltage of 275 kV. In Palopo substation there is an interconnection between the 150 kV system and the 275 kV system. South Sulawesi's electricity load is mostly in the southern area, namely in Makassar City. The total installed capacity of the plant in South Sulawesi is 1367 MW. The peak load is 950 MW. The number of existing substations (SS) in South Sulawesi is 33 units with a total capacity of 1888 MVA [4]. The length of the 500 kV transmission line to be built is 365 kmc, using a 4 x Zebra ACSR conductor [5]. Single line diagram of South Sulawesi's system is shown in Figure 1.

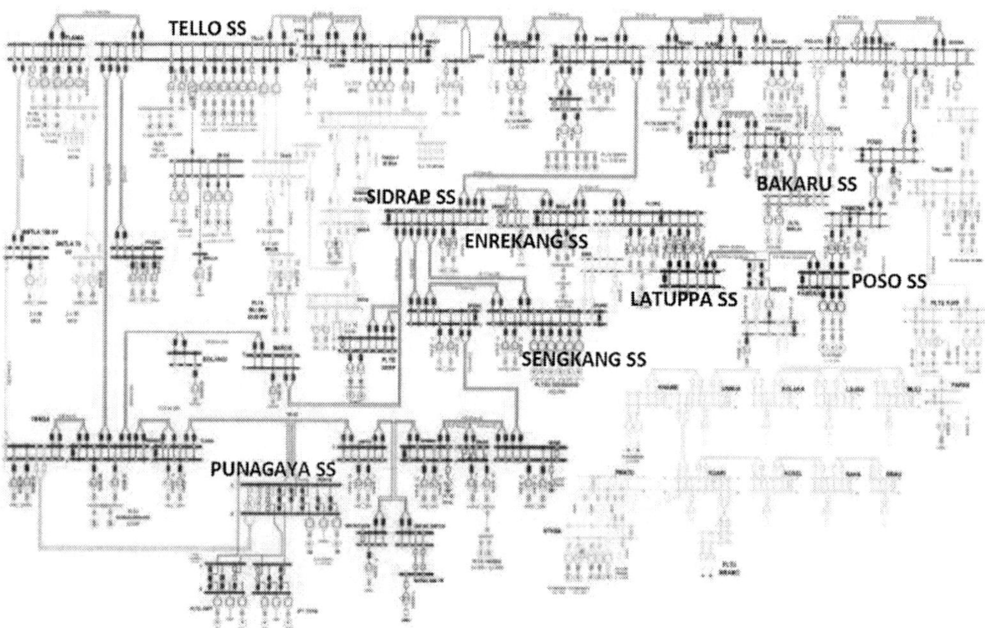

Figure 1. Single line diagram of South Sulawesi system [5]

2.2. Shun Reactor

The shunt reactor is a coil that is installed in parallel with the system so that it can absorb capacitive reactive power from the electric power system by the installed capacity of the shunt reactor. Shunt reactors are used in systems that have large charging currents during light / low loads. Shunt reactors can be installed directly on the EHV bus which functions for secondary voltage absorption during light/low loads.

2.3. Simulation Procedure

The work steps undertaken to carry out this research are: 1) Survey of technical data from government offices dealing with electricity, namely: PLN South Sulawesi (Indonesian Republic Electricity Company), as well as supporting data on writing from relevant agencies and reference books in libraries related to the subject matter. 2) Forming a model based on the case that might exist based on references and available data. 3) Calculation of parameters from existing data for 500 kV transmission line system of South Sulawesi. Existing data entered into the calculation model to complete the calculation of: shunt reactor inductance to increase the electric length, and determine the compensation of the shunt reactor to increase the voltage level. And 4) Getting results in the form of conclusions from calculations that have been done. In shunt reactor compensation in the South Sulawesi transmission system is done only in the substations associated with a 500 kV transmission system. As a result, the installation of shunt reactor compensation for shunt reactor use simulation was installed at 150 kV Punagaya substation, 150 kV Sidrap substation, 150 kV Bakaru substation, 275 kV Latuppa substation, and 500 kV Sidrap substation. To do the simulation, the system is connected to the shunt reactor, where the shunt reactor is installed in one of the substations, then the other substations are not given a shunt reactor. Each simulation is performed shunt record installations ranging from 400 MVAr to 600 MVAr with the addition of MVAr, which is 50 MVAr. The voltage will be observed on all five buses. The increase and decrease in voltage on each bus are observed to exceed the operating voltage limit. The voltage that occurs must be within the limit of the voltage that is still permissible, that is, it cannot be more than 105% of the working voltage and must not be less than 95% of the nominal voltage by following applicable standards.

Analysis of the effect of the shunt reactor will use power flow calculations using the Newton Raphson method [6]. This method is used to get the power flow in each network as well as the voltage of each voltage at each busbar / substation. The voltage profile of the power system operation by the reactive power balance in the South Sulawesi system can be expressed by the following Equation (1):

$$\sum Q_{gi} - \sum Q_{cl} = \sum Q_{ok} + \sum Q_L \tag{1}$$

where, ΣQ_L = reactive power loss of the system, ΣQ_{ok} = reactive power load on the load bus k, ΣQ_{cl} = injection of reactive power from other sources, and ΣQ_{gi} = reactive power generated by the generator.

3. Result and Discussion

3.1. Calculation Result

Table 1 shows the magnitude of the voltage at 5 (five) substations. The value of this voltage arises due to the installation of a shunt reactor on the 275 kV Latuppa substation. Table 2 is a result table of the installation of the shunt reactor in 150 kV Punagaya substation. Table 3 also shows the operational voltages of several substations in South Sulawesi, where these results are the effects of installing a reactor shunt on a 150 kV Sidrap substation. Table 4 shows the simulation results of placing the shunt adhesive on the 500 kV Sidrap substation.

Table 1. Placement of the shunt reactor on the 275 kV Latuppa substation

Shunt	Bus Voltage (kV)				
Reactor (MVAr)	Latuppa 275	Punagaya 150	Sidrap 150	Bakaru 150	Sidrap 500
400	Not convergent				
450	Not convergent				
500	272	154	160	154	512
550	267	153	159	153	508
600	262	153	159	152	504
650	255	151	157	151	500

Table 2. Placement of shunt reactor on the 150kV Punagaya substation

Shunt Reactor (MVAr)	Bus Voltage (kV)				
	Latuppa 275	Punagaya 150	Sidrap 150	Bakaru 150	Sidrap 500
400	Not convergent				
450	320	148	161	156	519
500	314	144	158	153	508
550	311	142	156	151	502

Table 3. Placement of the shunt reactor on the 150kV Sidrap substation

Shunt Reactor (MVAr)	Bus Voltage (kV)				
	Latuppa 275	Punagaya 150	Sidrap 150	Bakaru 150	Sidrap 500
400	Not convergent				
450	314	154	149	154	496
500	311	152	148	153	489
550	308	151	146	148	483
600	305	150	144	151	476

Table 4. Placement of a shunt reactor on a 500kV Sidrap substation

Shunt Reactor (MVAr)	Bus Voltage (kV)				
	Latuppa 275	Punagaya 150	Sidrap 150	Bakaru 150	Sidrap 500
400	313	152	159	152	504
450	309	151	157	150	496
500	304	149	155	149	489
550	301	148	154	147	483
600	297	146	153	145	476
650	295	145	152	144	471
700	293	144	151	143	467
750	291	143	150	142	462

3.2. Discussion

Table 1 above shows the application or use of a shunt reactor at 275 kV Latuppa substation located in Palopo. If the operating limit of the substation is based on a voltage limit of ± 5%, then at a capacity position between 600 - 650 MVAr, the voltage at each substation is within the operating limit. The voltage on the Latuppa substation is 255 kV, Punagaya substation is 155 kV, Sidrap150 substation is 157 kV, Bakaru substation is 151 kV, and Sidrap500 substation is 500 kV. Another case with another scenario, where the placement of the shunt reactor at 150 kV Punagaya substation, 150 kV Sidrap substation, 150 kV Enrekang substation, and 500 kV Sidrap substation produces operating stresses on all substations outside the standard operating voltage limits.

The results of this study show the effect of the reactor shunt on the South Sulawesi electricity system, where the reactor shunt decreases the stress on all substations / busbars. The effect of the reactor shunt on the distribution of electrical energy from power plants in the South Sulawesi system through a long transmission line by using AC EHV at low loads causes a large load on the line at each node. The charge flow is an electric current, a current caused by charging. Discharging a channel due to alternating voltage is called a charging current, the charging current always flows in the transmission line even though the network is open. The charging current due to the influence of network capacitance will cause reactive power. The reactive power generated by the 500 KV transmission line is approximately 1 kmc = 1 MVAr. If the 3 (three) phase network in South Sulawesi's 500 kV system has a length of 365 kms, the reactive power generated = 365 MVAr. To compensate for such a large reactive power, it is necessary to install an inductor in the form of a Shunt Reactor greater than 365 MVAr generated by the 500 kV transmission line.

South Sulawesi's 500 kV EHV transmission lines is a long network so it requires compensation equipment to control the working voltage at each node along the network which serves to reduce the dielectric length of the 500 kV transmission to increase the capacity of the power line. At present, the 500 kV long network compensation tool is using a Shunt Reactor which is connected to the 500 kV busbar side, 275 kV side, and the 150 kV side. When the load is full due to the device effect which causes the capacitive reactive power of the network is relatively large, it can be absorbed by the load so that there is no over-voltage, but at low load conditions ie on holidays / large MVAr the network cannot be absorbed by the load causing a voltage the substation exceeds the operational allowable working voltage from the nominal voltage (+5% to -10%).

4. Conclusion

The conclusions of this study are:

- Shunt reactors are needed when light loads occur, while the excess power system must be absorbed beforehand.
- The most ideal shunt reactor position is at 275 kV Latuppa substation, and the shunt reactor capacity is 600 - 650 MVAr.

Acknowledgments

The authors gratefully acknowledge Faculty of Engineering through the LBE scheme for financial support of this research.

References

[1] Kitta I, Manjang S, Rachmaniar I and Maricar F 2019 Tropical climate effects on corona power losses on 275 kV transmission lines in the South Sulawesi system *Przegląd Elektrotechniczny*, **95** : 1/2019.

[2] MER Ministry 2019 PLN Electric Power Supply Business Plan for 2019 – 2028 *Kementerian Energi dan Sumber Daya Mineral Republik Indonesia* Jakarta.

[3] Pramana PAA and Munir BS 2016 Modelling and Simulation of Electrical Resonance in EHV Transmission Line Case Study in West Java Region 500 kV System *MATEC Web of Conferences*, ACPEE 2016.

[4] Kitta I, Manjang S, Sahali IR and Maricar F 2019 Insertion of 275 kV Transmission Line for Improving the Voltage Profile and Efficiency of Electrical Power System IOP *Conference Series: Materials Science and Engineering,* 676 **(2019)**: 012006.

[5] PLN *Perusahaan Listrik Negara* Indonesian Republic Electricity Company.

[6] Saadat H 1999 *Power System Analysis* (New-York: McGraw-Hill).

IOP Conf. Series: Materials Science and Engineering **875** (2020) 012042 doi:10.1088/1757-899X/875/1/012042

Smart City Application: Community Survey System of Urban Comfort Level

E Warni[1]*, Z Zainuddin[1], and M F Amperiento[1]

[1]Department of Informatics, Engineering Faculty, Hasanuddin University, Makassar, Indonesia

*Email: elly@unhas.ac.id

Abstract. Based on the projected increase in urban population in Indonesia released by the Ministry of PUPR and Bappenas in 2017 that there was a significant increase in urbanization, which is estimated that in 2015-2035 the population in urban areas will exceed the population in rural areas (66.6% of the population urban and 33.4% rural population). Increasing the number of the urban population needs to get proper handling from the government. If not, it will trigger the emergence of various urban problems ranging from the physical city to the problems that will have an impact on the reduced level of comfort in life in the region. Based on this, it is necessary for the participation of city dwellers in playing the role of planners, actors, and supervisors of city development. This research work develops a framework which uses fuzzy Inference System to calculate urban comfort level, starting from a questionnaire-based survey. There are seven input variables used to produce an output of urban comfort level. Based on the stages of using the Mamdani Model Fuzzy Inference Logic System starting from the formation of the fuzzy set, the application of the impulse function, the composition of the rules to the process of affirmation (defuzzification), it can be proven that there is a correlation between the input variables so that it can determine the output of urban comfort level results. The outcomes help the city government in making public policy on city development.

1. Introduction

The development of cities has historically been seen as causes and solutions for social, economic, political and cultural improvement. However, in the development of the city, the environment changes. Environmental changes are influenced by the level and type of industrialization, the quality of housing, accessibility to green spaces and raising concerns about transportation [1]. Environmental damage is a problem of the city caused by the city development that affects large-scale urbanization in the city.

The increasing population continues to make the city services will be increasingly ineffective unless the city can provide the necessary service facilities for the community as a whole who live in the city. Today many cities throughout the world still cannot serve the people who live in it. This is because the city cannot provide infrastructure service facilities to accommodate daily community activities in the city. So many urban people no longer feel comfortable living in cities, due to overcrowding that makes city spaces more narrow, congestion, and environmental damage.

From the problems of the city above, the city community needs a city that is livable for them or called Livable City. Livable City is a keyword in urban planning because it can solve various city problems that disturb the comfort of the city. By increasing the quality of life that people living in cities are related to their ability to access infrastructure (transportation, communication, water, and sanitation), food, clean air, affordable housing, employment and space, and green parks. The Livable City concept

infrastructure in sustainable city representation [2]. In the context of sustainability is the ability to maintain the quality of life required by the city community

Conditions in the future will be even more uncomfortable if there are no bold, creative, and progressive actions from city leaders to take and implement bold city development policies, likewise, with the residents. The community must understand, understand and carry out their obligations as a good city citizen, not just being a city community but becoming a city citizen (citizen) who helps realize the comfort of the city. City residents must also play a role as planners, actors, and supervisors of city development. In this decade, citizen participation in the process of urban development planning has become a strategic matter in changing and formulating public policies on urban development. The city government and community must unite in collaboration to realize the livable city of Indonesia's Future.

Fuzzy logic is one method for analyzing systems that contain uncertainty. Fuzzy Mamdani is one method that is very flexible and has a tolerance on existing data. Fuzzy Mamdani has the advantage of being, more intuitive, accepted by many parties, more suitable input received from humans, not machines. Based on fuzzy logic, a Mamdani fuzzy model will be produced that can analyze the comfort level of the city.

2. Literature Review

2.1. Definition of Livable City

Nowadays, many urban people are complaining about the inconvenience of their neighborhood. These inconveniences can be found in problems ranging from traffic jams, poor maintenance of public facilities, and environmental cleanliness issues. In these conditions, every community wants a city that is comfortable and deserves to be inhabited or Livable City. "A Livable City is a city where I can have a healthy life and where I have the chance for easy mobility - by foot, by bicycle, by public transportation, and even by car where there is no other choice ... Livable City is a city for all people. That means that the Livable City should be attractive, worthwhile, safe for our children, for our older people, not only for the people who earn money there and then go and live outside in the suburbs and the surrounding communities. For the children and older people, it is especially important to have easy access to areas with green, where they have a place to play and meet each other, and talk with each other. Livable City is a city for all [3] ".

A livable city or Livable City is where people can live comfortably and quietly in a city. A livable city is a city that can accommodate all the activities of the city community and is safe for all people [4]. The concept of Livable City is used to realize that the idea of development as an improvement in the quality of life requires both physical and social habitat for its realization [5].

In realizing the concept of Livable City, it must be supported by sustainable cities, so that urban space planning can be realized according to plan. In the context of sustainability is the ability to maintain the quality of life that is needed by current and future urban communities. "Therefore a Livable City is also a 'sustainable city': a city that satisfies the needs of the present inhabitants without reducing the capacity of the future generation to satisfy their needs" [6]. Livable City is a city where public spaces are the center of social life and focus throughout the community [6]. The concept of Livable City is used to realize that the idea of development as an improvement in the quality of life requires both physical and social habitat for its realization [7].

"... there are those social groups for whom a Livable City is one where those elements have been preserved or renewed which have always been an integral part of people-friendly places. These are, as Peter Smithson once beautifully said 'relationships between streets and buildings, and buildings between themselves, and trees, and seasons of the year, and ornamentation, and events and other people [8].

The Livable City concept is also very much related to the environment. Livable City must be sustainable with the ecological system and the comfort of life for the city community. Ecological recovery can improve the environment in Livable City and sustainability. Livable City must create and maintain a clean environment. "The coin of livability has two faces. Livelihood is one of them. Ecological sustainability is the other. Livelihood means jobs close enough to decent housing with wages commensurate with rents and access to services that make for a healthful habitat. Livelihoods must also be sustainable. If the quest for jobs and housing is solved in ways that progressively and irreparably degrade the environment of the city, then the livelihood problem is not being solved. Ecological degradation buys livelihood at the expense of quality of life, with citizens forced to trade green space and breathable water for wages. To be livable, a city must put both sides of the coin together, providing livelihoods for its citizens, ordinary as well as affluent, in ways that preserve the quality of the environment [9].

Understanding Livable City from the perspective of people is a livable city where urban people can find work, serve basic needs including clean water and sanitation, have access to proper education and health, live in a safe community and a clean environment. It can be said that Livable City is a description of a comfortable environment and atmosphere of the city as a place to live and as a place for activities seen from various aspects, both physical aspects (urban facilities, infrastructure, spatial planning, etc.) as well as non-physical aspects (social relations, economic activity, etc.).

2.2. Livable City Principles

In realizing a livable city, Livable City must have basic principles. This basic principle must be owned by cities that want to make the city a livable and comfortable city for the people of the city. The following are the basic principles for realizing Livable City:

1. The basic principles for Livable City [8]
   a. Availability of various basic needs of urban communities (decent housing, clean water, electricity).
   b. Availability of various public and social facilities (public transportation, city parks, worship facilities/health/ worship services).
   c. Availability of public spaces and places for socializing and interacting.
   d. Security, free from fear.
   e. Supports economic, social and cultural functions.
   f. Environmental sanitation and physical beauty.
2. In Livable City, can be said to rest on 4 (four) pillars [10]
   a. Improve the system of life opportunities for the welfare of society.
   b. Provision of employment.
   c. A safe and clean environment for health, well-being and for sustaining economic growth.
   d. Good governance.

2.3. Livable City Case Study in Indonesia

To find out the perceptions of city residents regarding the comfort level of big cities in Indonesia, the Indonesian Planning Experts Association (IAP) conducted a 2011 Indonesia Most Livable City Index (MLCI) research. This activity is an annual index that shows the comfort level of city residents to live, live and activities in a city in terms of various aspects of urban areas. This index is produced with the following approaches: "Snapshot, Simple and Actual" conducted in 15 major cities in Indonesia, namely Yogyakarta, Denpasar, Makassar Makassar, Surabaya, Semarang, Banjarmasin, Batam, Jayapura, Bandung, Palembang, Palangkaraya, Jakarta, Pontianak, and Medan. The indicator criteria used by IAP in conducting this research consisted of 26 indicators which were grouped into several main criteria [9], namely:

1. The Spatial Aspect (Urban Planning, Green Open Space),
2. Environmental Aspects (Cleanliness, Pollution),
3. Transportation Aspects (Roads, Transportation),
4. The Health Facilities Aspect,
5. Educational Facility Aspects,
6. Infrastructure Aspects - Utilities (Electricity, Water, Telecommunications),
7. Economic Aspects (Field of Work, Location of Work),
8. Security Aspects,
9. Social Aspects (Culture, Citizen Interaction).

2.4. Fuzzy Logic
The concept of fuzzy logic was introduced by Prof. Lotfi Astor Zadeh in 1962. Fuzzy logic is a methodology of problem-solving control systems that is suitable for implementation on systems, ranging from simple systems, small systems, PC networks, multi-channel or workstations based on data acquisition, and control systems. In fuzzy logic it allows membership values to be between 0 and 1 [11].

According to Kusumadewi and Purnomo said, "Fuzzy logic is one of the components forming soft computing". Prof. Lotfi A. Zaedah first introduced fuzzy logic in 1965. The basis of fuzzy logic is fuzzy set theory. In fuzzy set theory, the role of the degree of membership as a determinant of the existence of elements in a set is very important [10].

Sutojo et al. said "Fuzzy logic is a methodology of problem-solving control systems, which is suitable for implementation on systems, ranging from simple systems, small systems, embedded systems, PC networks, multi-channel or workstations based on data acquisition, and control systems. This methodology is applied to hardware, software, or a combination of both [12].

2.5. Mamdani Model Fuzzy Inference System Logic
The Fuzzy Inference System Mamdani model is often also known as the max-min method. Ebrahim Mamdani introduced this method in 1975. The Mamdani method is often used in applications because of its simple structure, which uses MIN-MAX or MAX-PRODUCT operations [10]. To get the output, four stages are needed, namely:

a. Fuzzy set formation.
   In the Mamdani Method, both input and output variables are divided into one or more fuzzy sets.
b. Application function implications.
   In the Mamdani Method, the implication function used is Min.
c. Rule Composition
   Unlike monotonous reasoning, if the system consists of several rules, then the inference is obtained from the collection and correlation between rules. There are 3 methods used in conducting fuzzy system inferences, namely: max, additive and probabilistic OR (probe)
d. Affirmation (defuzzification)
   The input of the defuzzification process is a fuzzy set obtained from the composition of fuzzy rules, while the resulting output is a number in the fuzzy set domain. So if given a fuzzy set within a certain range, then a certain crips value can be taken as output.

3. Result and Discussion
3.1. Fuzzy Set Formation (Fuzzification)
The process of determining the assessment of city comfort levels has seven input variables and one output variable. Input variables consist of physical, environment, economic, transportation, facilities, infrastructure and social. The output variable to be processed is the city comfort variable.

Table 1. Fuzzy input and output variables

Function Name	Variable	Universe Talks	Description
	Physical city	[1 10]	Rating Points
	environmental	[1 10]	Rating Points
	Quality	[1 10]	Rating Points
Input	Transportation	[1 10]	Rating Points
	Facilities	[1 10]	Rating Points
	Infrastructure	[1 10]	Rating Points
	Economy	[1 10]	Rating Points
	Social	[1 10]	Rating Points
Output	urban comfort level	[1 10]	Rating result

In the process of producing values from the output variables, four stages are needed, namely the formation of fuzzy sets (fuzzification), the application of implication functions, the composition of rules to the affirmation process (defuzzification). Details of the data can be seen in Table 1.

Table 2. Fuzzy association

Variable Name	Fuzzy Association	Domain	Membership function
Physical city	Not good	[1 4]	Left shoulder
	Pretty good	[3 6]	trapezoidal
	Well	[5 8]	Trapezoidal
	Very good	[7 10]	Right shoulder
Environmental quality	Not good	[1 4]	Left shoulder
	Pretty good	[3 6]	trapezoidal
	Well	[5 8]	Trapezoidal
	Very good	[7 10]	Right shoulder
Transportation	Not good	[1 4]	Left shoulder
	Pretty good	[3 6]	trapezoidal
	Well	[5 8]	Trapezoidal
	Very good	[7 10]	Right shoulder
Facilities	Not good	[1 4]	Left shoulder
	Pretty good	[3 6]	trapezoidal
	Well	[5 8]	Trapezoidal
	Very good	[7 10]	Right shoulder
Economy	Not good	[1 4]	Left shoulder
	Pretty good	[3 6]	trapezoidal
	Well	[5 8]	Trapezoidal
	Very good	[7 10]	Right shoulder
Social	Not good	[1 4]	Left shoulder
	Pretty good	[3 6]	trapezoidal
	Well	[5 8]	Trapezoidal
	Very good	[7 10]	Right shoulder
Comfort Urban Level	Not good	[1 4]	Left shoulder
	Pretty good	[3 6]	trapezoidal
	Well	[5 8]	Trapezoidal
	Very good	[7 10]	Right shoulder

The graph of representation of the degree of membership of the fuzzy set of town comfort ratings is as follows:

a. Graphic representation of the degree of membership of physical city variables

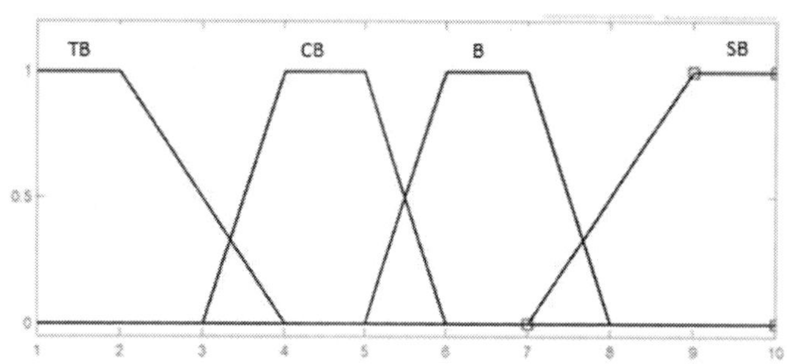

Figure 1. Degree of membership in physical city variables

IOP Conf. Series: Materials Science and Engineering **875** (2020) 012042 doi:10.1088/1757-899X/875/1/012042

b. Graphic representation of the degree of membership of environmental quality variables

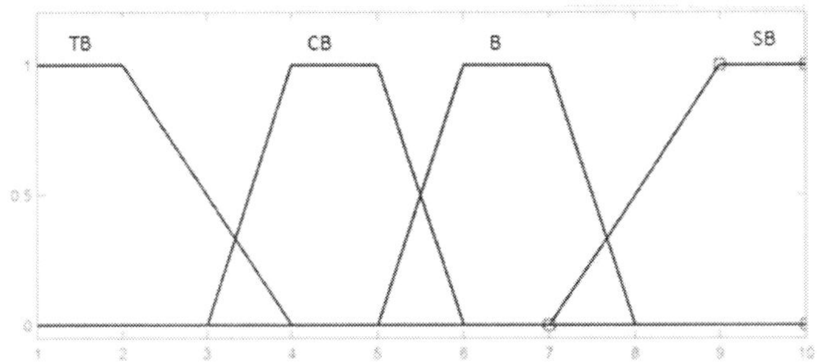

Figure 2. Degree of membership of environmental quality variables

c. Graphic representation of the degree of membership of transportation variable

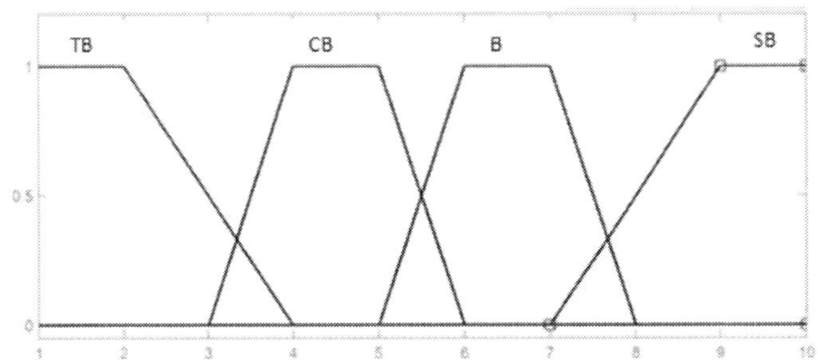

Figure 3. Variable transportation membership degree

d. Graphic representation of the degree of membership of variable facilities

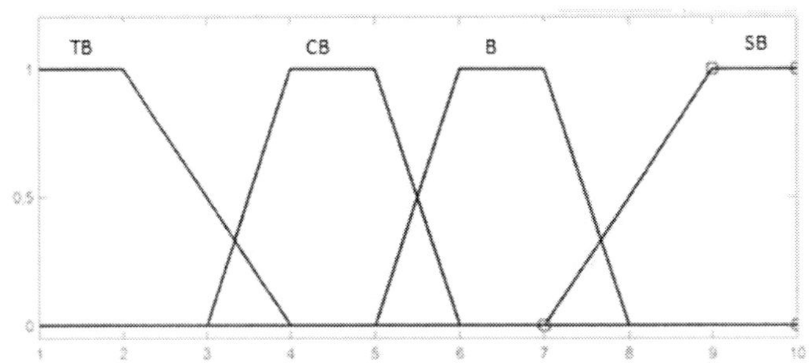

Figure 4. Degree of membership of variable facilities

e. Graphic representation of the degree of membership of the infrastructure variable

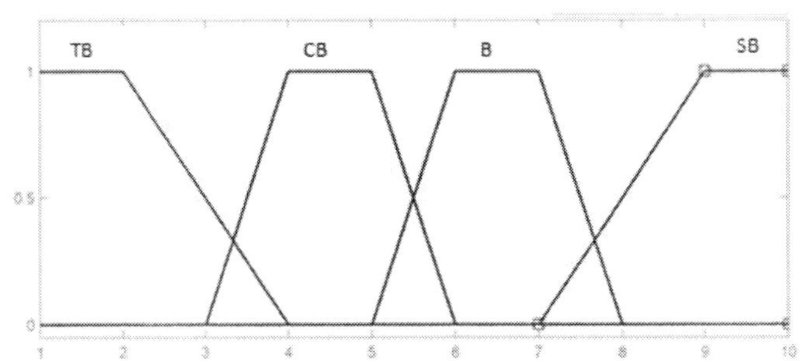

Figure 5. Variable infrastructure membership degrees

f. Graph representation of the degree of membership of economic variables

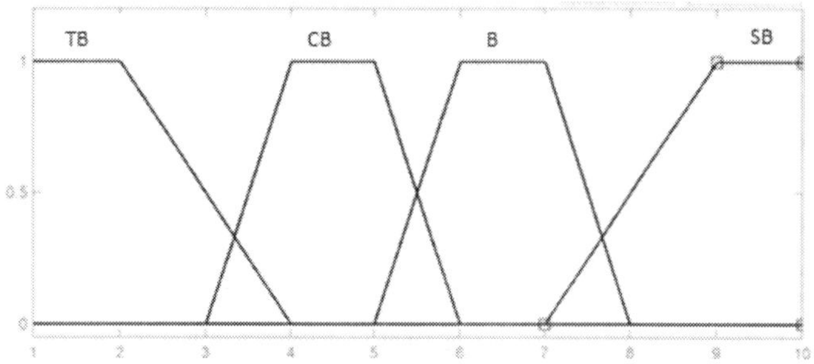

Figure 6. Degree of membership of economic variables

g. Graph representation of the degree of membership of social variables

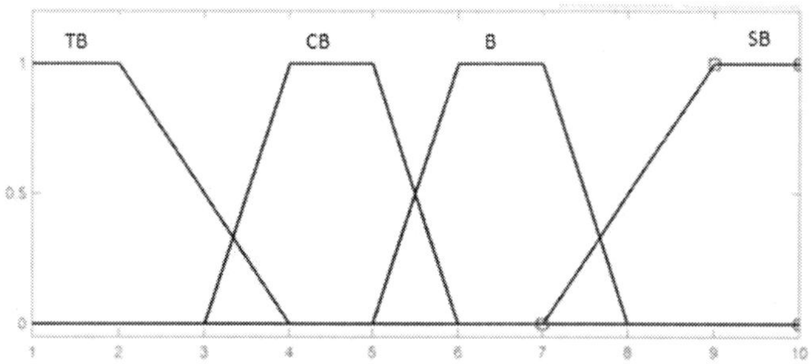

Figure 7. Degrees of membership of social variables

h. Graphic representation of the degree of membership in urban comfort level variables

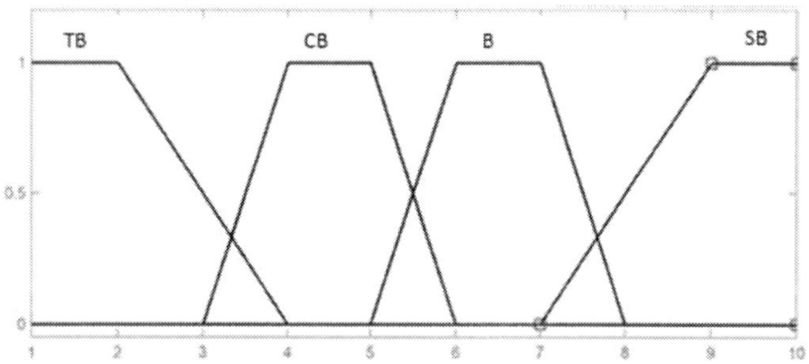

Figure 8. Degrees of membership in urban comfort level variables

3.2. Application function Implications

Rules can be formed to express the relationship between input variables and output variables. Every rule is an implication. The operator used to connect the two inputs is AND, and the one that maps between input variables and output variables is IF-THEN. The rules include:

Rule 1: if (physical city is not good) and (environmental quality is not good) and (transportation is not good) and (facilities are not good) and (infrastructure is not good) and (economy is not good) and (social is not good) then (urban comfort level is not comfortable)

Rule 2: if (physical city is not good) and (environmental quality is not quality) and (transportation is not satisfactory) and (facilities are not satisfactory) and (infrastructure is not good) and (economy is not good) and (social is Pretty Good) then (urban comfort level is Uncomfortable)

Rule 3: if (physical city is not good) and (environmental quality is not quality) and (transportation is not satisfactory) and (facilities are not satisfactory) and (infrastructure is not good) and (economy is not good) and (social is OK) then (urban comfort level is Not comfortable)

Rule 4: if (physical city is not good) and (environmental quality is not good) and (transportation is not good) and (facilities are not good) and (infrastructure is not good) and (economy is good enough) and (social is not good) then (urban comfort level is not comfortable)

Rule 5: if (physical city is not good) and (environmental quality is not good) and (transportation is not good) and (facilities are not good) and (infrastructure is not good) and (economy is good enough) and (social is Pretty Good) then (urban comfort level is Uncomfortable)

Rule 6: if (physical city is good enough) and (environmental quality is good enough) and (transportation is good enough) and (facilities are good enough) and (infrastructure is good enough) and (economy is good enough) and (social is Pretty Good) then (urban comfort level is Comfortable enough)

Rule 7: if (the physical city is Good) and (environmental quality is Good) and (transportation is Good) and (facilities is Good) and (infrastructure is Good) and (economy is Good) and (social is Good) then (urban comfort level is comfortable)

etc……

3.3. Composition of Fuzzy rules

In the Mamdani method, the composition between functions implies using MAX by taking the maximum value from the rule output, then combining each fuzzy area.

3.4. Defuzzification Process

The process of defuzzification is to convert the fuzzy output to a firm value based on a predetermined membership function. The defuzzification process uses the Centroid Method.

4. Conclusion

The result shows that using the fuzzy Mamdani model can explain the rules of connectedness between input variables namely physical city, environment, economic, transportation, facilities, infrastructure, and social aspects to produce an output of the city comfort level. This research has proven the correlation of these variables in determining the results of the city comfort level.

References

[1] Hahlweg, D., 1997. "The City as a Family" In Lennard, S. H., S von Ungern Sternberg, H. L. Lennard, eds. Making Cities Livable. International Making Cities Livable Conferences. California, USA: Gondolier Press.

[2] Wheeler, Stephen M. 2004. Planning For Sustainability, Creating Livable, Equitable, And Ecological Communities. New York: Routledge.

[3] Salzano, E. 1997. "Seven Aims for the Livable City" in Lennard, S. H., S von Ungern-Sternberg, H. L. Lennard, eds. Making Cities Livable. International Making Cities Livable Conferences. California, USA: Gondolier Press.

[4] Evans, Peter. 2002. Livable Cities? Urban Struggles for Livelihood and Sustainability. California: University of California Press.

[5] Palej, A. 2000. Architecture for, by and with Children: A Way to Teach Livable City. Presentasi paper Dalam International Making Cities Livable Conference, Vienna.

[6] Ard-Am, O., Douglass, M., Evans, P., Gille, Z., Hsin-Huang, Kim, M. H. M. E. K. I. K., . . . Pezzoli, K. (2002). Livable Cities? Urban Struggles For Livelihood And Sustainability (P. Evans Ed.). London: University Of California Press.

[7] Hahlweg, D., 1997. "The City as a Family" In Lennard, S. H., S von Ungern Sternberg, H. L. Lennard, eds. Making Cities Livable. International Making Cities Livable Conferences. California, USA: Gondolier Press.

[8] Douglass, Mike. 2002. "From Global Intercity Competition to Cooperation for Livable Cities and Economic Resilience in Pacific Asia". Environment and Urbanization, Vol. 14, April: pp. 53-68.
[9] IAP. 2011. Indonesia Most Livable City Index 2011. Jakarta: Ikatan Ahli Perencana Indonesia.
[10] Kusumadewi, S dan Purnomo, H, 2004. Aplikasi Logika Fuzzy untuk Mendukung Keputusan. Yogyakarta: Graha Ilmu.
[11] Rosenfeld, A., 1975, Fuzzy Graphs, Zadeh, L.A, Fu, K.S., dan Shimura, M. (Editor), Fuzzy Sets and Their Applications to Cognitive and Decision Processes, Academic Press, New York, 77-95.
[12] Sutojo, T., Mulyanto, Edy., Suhartono, Vincent. 2011. Kecerdasan Buatan. Yogyakarta: C.V. Andi Offset.

IOP Conf. Series: Materials Science and Engineering **875** (2020) 012043 doi:10.1088/1757-899X/875/1/012043

Comparison of Voltage Stability Index Before and After Wind Turbine Penetrated to Sulseltrabar Interconnection Power System Using Modal Analysis Method

I C Gunadin[1]*, Z Muslimin[1], A M Ilyas[1,2], and A Siswanto[1,3]

[1]Electrical Department of Hasanuddin University, Indonesia
[2]Khairun University, Indonesia
[3]University of 17 Agustus 1945 Cirebon, Indonesia

*Email: indarcg@mail.com

Abstract. Voltage stability problems and renewable energy penetration in power systems are the main issues in the electric power system at this time. This paper aims to examine the condition of voltage stability at the time before and after the entry of wind power plants in the sulseltrabar interconnection system. There are two wind power-generating units that will be studied, namely the PLTB Sidrap with a capacity of 75 MW and the PLTB Jeneponto with a capacity of 72 MW. The simulation using Matlab program. From the simulation results, it can be seen that Sulseltrabar's voltage stability interconnection system at the time of the entry of PLTB Sidrap and PLTB Jeneponto, has improved stability, this can be seen from the eigenvalue and bus participation factor values. From the minimum eigenvalues it can be seen that the penetration of PLTB Jeneponto is slightly better than PLTB Sidrap in improving voltage stability. The condition that has the greatest influence on improving stability is when the two wind power plant enter to the Sulseltrabar interconnection power system.

1. Introduction

The problem of voltage stability becomes a major problem in the electric power system. As the load continues to increase, the problem of voltage stability will also increase [1]. There are many cases of blackouts in the electric power system caused by voltage stability problems [2]. For the case of a total blackout on September 30, 2007, at 09.14.24, the system load was 270 MW off with the total electrical energy not distributed at 860.42 MWh with the extinction time ranging from 4 to 7 hours. If the assumed price of electricity at that time was Rp 580 / kWh, the total loss of PT.PLN at the time of the total blackout was Rp 499,043,600.-. On the consumer side, industries in the South Sulawesi region also suffered losses with hampered their production process. Then the fundamental question arises, whether the total blackout phenomenon due to voltage instability problems can be detected early and how to anticipate so that black-out can be prevented. This study seeks to provide an initial indication (stability index) of system operating conditions, stability limits and scenarios to improve system conditions.

The construction of the Bayu Power Plant (PLTB) in Jeneponto Regency with a capacity of 160 MW and in Sidrap amounting to 75 MW, is certain to begin immediately, after the South Sulawesi Provincial Government signed an MoU with PT Indo Wind Power Holdings, a subsidiary of Asia Grand Capital. The inclusion of the PLTB plant will increase the availability of electricity supply in the South Sulawesi interconnection system and is expected to be able to drive industrial growth and spur rapid economic growth in the South Sulawesi, West Sulawesi and Southeast Sulawesi regions. However, the installation of wind turbine will also cause other negative impacts including ecological impacts, noise impacts, impacts from the electricity side and others, it should also be considered before PLTB penetration [3].

This study aims to analyze the impact of the entry of PLTB on the Sulseltrabar interconnection system which is more focused on the study of the impact of the voltage stability of the electric power system. Does the PLTB penetration cause a significant effect on the Sulseltrabar electricity system and how to minimize the negative effects caused.

To monitor the stability of the system, there are several methods that are often used, including; PV and QV curves, Continuation Power Flow, L-Index. This research uses the Capital Analysis method by observing the eigen value and bus participation factor (BPF). Each bus will be sorted according to the value of its participation in unstable conditions [4].

2. Voltage Stability

In electric power systems which are operated in stress conditions, voltage stability becomes very important. In the planning and operation of power systems, stress stability analysis focuses on two aspects including:

1. Prediction, how close are the unstable voltage condition systems?
2. The mechanism, when voltage instability occurs, what is a factor?

Predictions provide a measure of voltage safety, while the mechanism will provide important information that is useful in the operation strategy in preventing voltage instability. The problem of voltage drop is a serious problem in the electrical system in many countries and this occurs in the system of interconnecting electricity [5].

There are several factors that can cause voltage drops including :

- Lack of reactive resources
- Heavy loading of active power
- Protection relay does not work well

Marison and Kundur have succeeded in finding a method to predict the occurrence of a voltage drop in an electric power system, this method is known as Modal Analysis. This method is based on the mechanism of calculating the smallest eigenvalue and eigenvector of the Jacobian matrix system. Eigen value has a correlation with changes in voltage values and changes in reactive power [6].

The stability of the Electric Power System can be evaluated by checking the state of the existing eigenvalue. If all eigenvalues are positive then the system is said to be stable, while the system is said to be unstable if the eigenvalue is negative. Eigenvalue of the Jacobian matrix which is zero indicates that the system is within the limits of instability [7].

The potential for voltage drop can be predicted through an evaluation of the minimum eigenvalue. The magnitude of the minimum eigenvalues indicates how close the system has fallen. By using the participation factor, the weakest bus or node can be determined.

The Q-V curve is the most commonly used method for describing voltage stability. This curve shows the sensitivity and changes in bus voltage to changes in reactive power injection and the stability limit or distance from each bus to a voltage drop condition.

The system has a stable voltage under operating conditions if on every bus in the system the magnitude of the bus voltage increases with increasing reactive power injection on the same bus. The system has an unstable voltage if at least one bus from the system has a reduced voltage magnitude when increasing reactive power injection on the same bus. In other words, the system is stable if the sensitivity of the Q-V curve is positive for each bus and unstable if the Q-V curve has a negative sensitivity on one of the buses [8].

From the results of previous research which is a collaboration of the Department of Electrical Engineering with PT. PLN Unit AP2B South Sulawesi Region, the results show that the transient stability conditions in the South Sulawesi system are very weak. The value of Critical Clearing Time

(CCT) on some buses indicating the value is below the value of 120 ms which is the standard time limit from relay work to open the Circuit Breaker (CB) [9].

From the interference data, it can be concluded that the disturbance trend in the sulseltrabar system has increased from year to year, this indicates that there is a stability problem in the interconnection system in South Sulawesi. The disturbance statistics can be seen in Figure 1.

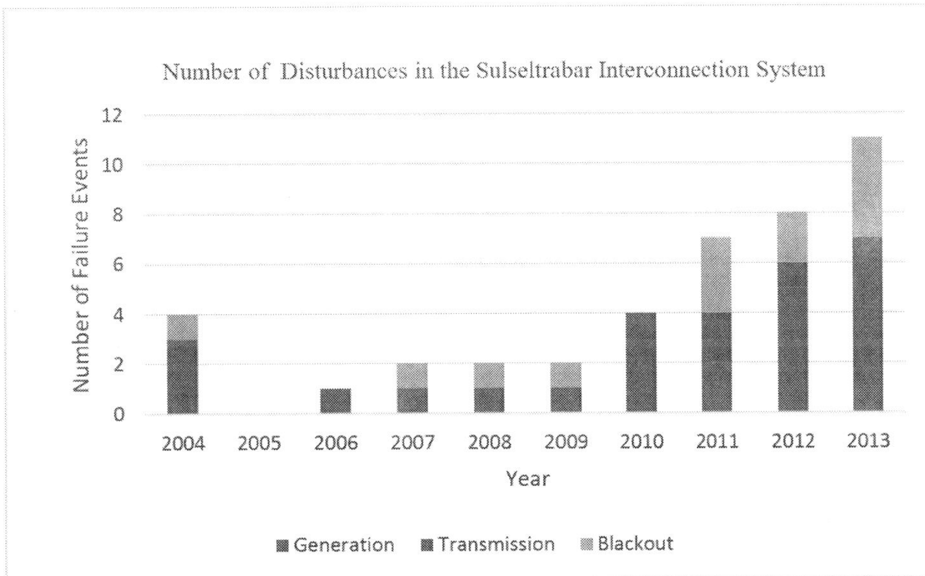

Figure 1. Statistics of the Southern Sulawesi electricity disruption

3. Determining the Value of Eigen Value

From Newton Raphson's Power Flow Equation obtained :

$$\begin{bmatrix}\Delta P\\ \Delta Q\end{bmatrix} = \begin{bmatrix}J_{P\theta} & J_{PV}\\ J_{Q\theta} & J_{QV}\end{bmatrix}\begin{bmatrix}\Delta\theta\\ \Delta V\end{bmatrix} \quad (1)$$

where :

ΔP = Changes in real power on the bus
ΔQ = Changes in reactive power on the bus
$\Delta\Theta$ = Changes in the voltage angle of the bus
ΔV = Changes in voltage magnitude on the bus

Conventional power flow models are used to analyze system voltage stability. Voltage stability is affected by changes in active power (P) and changes in reactive power (Q) even though at the operating point P is kept constant, so that voltage stability only considers the relationship between increasing the value of Q and decreasing the value of voltage V [10].

It is assumed that $\Delta P = 0$ is obtained:

$$\Delta P = 0 = J_{P\Theta}\ \Delta\Theta + J_{PV}\ \Delta V$$

$$\Delta\ \Theta = -J_{P\Theta}^{\ -1} J_{PV}\Delta V$$

$$\Delta Q = J_{Q\Theta}\,\Delta\Theta + J_{QV}\Delta V$$

Substituting Equations (2) and (3) obtains the following equation.

$$\Delta Q = J_R \, \Delta V \quad (2)$$

where :

$$J_R = J_{QV} - J_{Q\Theta} J_{P\Theta}^{-1} J_{PV}$$

J_R is the Jacobian reduction matrix obtained from the system of equations (2). Then the following can be written:

$$\Delta V = J_R^{-1} \, \Delta Q \quad (3)$$

From equation 3 it is found that the J_R matrix represents a linear relationship between the change in voltage and the change in reactive power injection on a bus. The eigenvalues and eigenvectors of the J_R matrix are used to analyze the system's voltage stability. The voltage instability can be identified from the form of the J_R eigen value matrix [11].

Analysis of the results of J_R eigenvalues is as follows:

$$J_R = \Phi \Lambda \acute{\Gamma} \quad (4)$$

where :

Φ = right eigenvector of the J_R matrix
$\acute{\Gamma}$ = left eigenvector of the J_R matrix
Λ = diagonal eigenvalues of the J_R matrix

by changing J_R to J_R^{-1} we get:

$$J_R^{-1} = \Phi \, \Lambda^{-1} \, \acute{\Gamma} \quad (5)$$

with : $\Phi\acute{\Gamma} = 1$

Equations (3) and (5) are substituted to obtain:

$$\Delta V = \Phi \Lambda^{-1} \acute{\Gamma} \Delta Q \quad (6)$$

$$\Delta V = \sum \frac{\Phi \acute{\Gamma}}{\lambda} \Delta Q \quad (7)$$

where :

λ is eigen value, provided that:

1. If $\lambda = 0$, the voltage will fall because the change in reactive power will cause the voltage change to be infinite
2. If $\lambda > 0$, the system voltage is stable
3. If $\lambda < 0$, the system voltage is unstable

The minimum value of an Eigen value is a value indicating closeness to the instability condition.

4. Methodology of Research

The South Sulawesi interconnection system was used as the case tested in this study. The following is shown in figure 2 single line diagram of the southern Sulawesi electricity interconnection system.

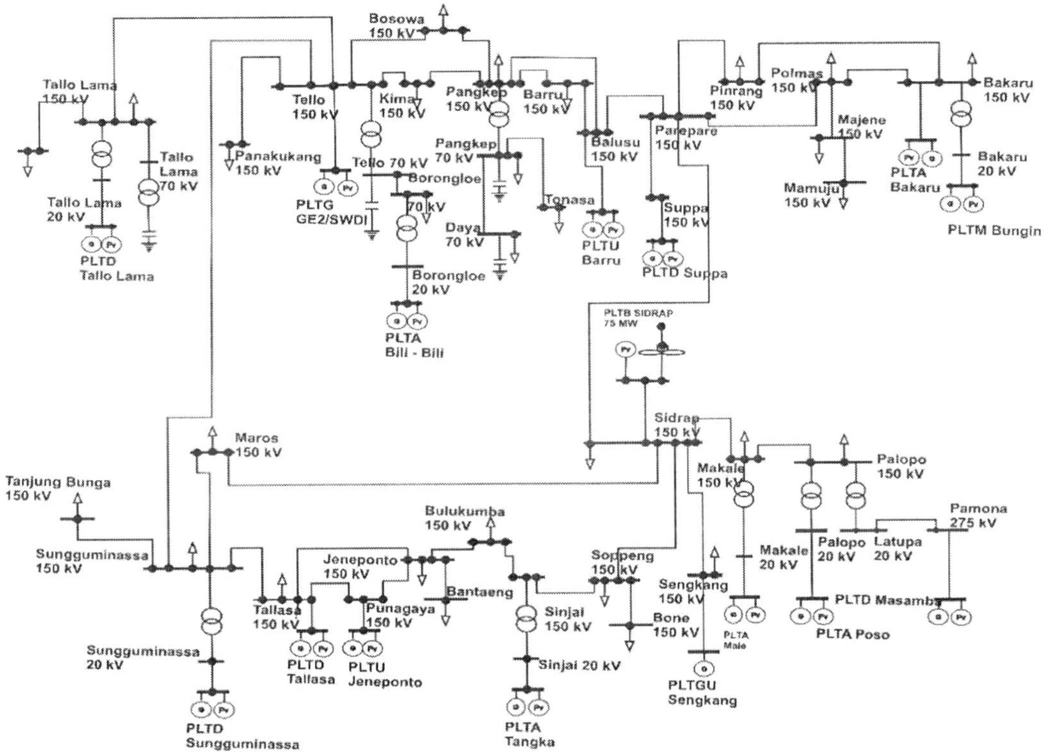

Figure 2. Single Line Diagram of Southern Sulawesi Electricity Interconnection System

In this study, the peak load data on the South Sulawesi interconnection system is 964 MW. The South Sulawesi interconnection system consists of 44 buses with a nominal voltage of 150 kV. Table 1 shows the loading and generation in the southern Sulawesi electricity system.

In this research several case studies will be conducted. There are 4 cases that will be tested in this study:

1. *Case-1* : Without Wind Turbine Penetration
2. *Case-2* : PLTB-Sidrap 75 MW Penetrated to System
3. *Case-3* : PLTB-Jeneponto 72 MW Penetrated to System
4. *Case-4* : PLTB Sidrap & Jeneponto 147 MW Penetrated to System

Furthermore, each case will see the minimum eigenvalue of J_R. The voltage stability level of the 4 cases can be compared by looking at the eigenvalue. The weakest bus can also be determined by looking at the value of the bus participation factor.

Table 1. Generating data and load data on the South Sulawesi interconnection system

No Bus	Bus Name	Voltage (pu)	Angle Degree	Load		Generator	
				MW	Mvar	MW	Mvar
1	Bakaru	1.030	0.000	3.50	0.20	47.225	79.568
2	Polmas	0.997	-1.366	17.10	4.10	0.000	0.000
3	Majene	0.988	-2.718	23.30	3.70	0.000	0.000
4	Mamuju	0.981	-3.274	9.60	4.80	0.000	0.000
5	Pinrang	1.000	-0.132	24.40	6.20	14.300	-49.721
6	Parepare	0.996	-0.870	18.70	4.70	0.000	0.000

7	Suppa	1.000	0.443	0.00	0.00	31.100	-2.244
8	Sidrap	0.974	4.734	26.50	10.30	0.000	0.000
9	PLTU Barru	1.000	-3.614	0.00	0.00	60.400	63.501
10	Barru	0.953	-12.058	10.10	2.40	0.000	0.000
11	Pangkep150	0.944	-14.570	22.10	8.00	0.000	0.000
12	Pangkep70	0.947	-15.928	0.00	0.00	0.000	0.000
13	Tonasa	0.918	-17.736	18.90	10.60	0.000	0.000
14	Bosowa	0.941	-15.264	33.10	15.40	0.000	0.000
15	Kima	0.953	-17.866	18.00	5.80	0.000	0.000
16	Tello150	1.000	-16.016	63.30	18.30	21.000	90.758
17	Panakukang	0.980	-18.572	68.30	17.70	0.000	0.000
18	Tello70	0.981	-20.521	0.00	-20.00	0.000	0.000
19	Borongloe	1.000	-21.317	11.40	0.00	5.200	13.564
20	Mandai	0.951	-19.782	24.30	2.60	0.000	0.000
21	Daya	0.957	-22.003	45.50	2.80	0.000	0.000
22	Tello30	1.000	-16.025	0.00	0.00	0.000	0.000
23	Barawaja	1.001	-16.034	0.00	0.00	0.000	0.000
24	Tellolama150	1.000	-16.880	19.70	4.70	12.600	18.115
25	Tellolama70	0.975	-19.285	0.00	0.00	0.000	0.000
26	Bontoala	0.958	-19.480	26.50	7.70	0.000	0.000
27	Sungguminasa	1.000	-9.356	15.70	3.60	20.000	129.111
28	Tanjung Bunga	0.968	-12.192	55.20	16.70	0.000	0.000
29	Tallasa	1.000	20.743	20.60	4.70	79.000	28.459
30	Maros	0.959	0.729	18.60	5.50	0.000	0.000
31	Punagaya	1.000	30.519	0.00	0.00	196.100	-0.573
32	Jeneponto	0.983	19.439	17.40	3.40	0.000	0.000
33	Bulukumba	0.989	12.671	27.10	6.50	0.000	0.000
34	Sinjai	1.000	11.106	21.90	4.60	4.000	69.558
35	Bone	0.990	10.533	32.10	8.20	0.000	0.000
36	Soppeng	0.975	10.142	14.10	3.40	0.000	0.000
37	Sengkang	1.000	14.142	28.40	11.50	265.200	-34.481
38	Makale	1.000	6.236	11.90	1.50	8.200	69.807
39	Palopo	1.000	39.200	49.20	0.00	4.000	128.970
40	Latuppa	0.972	62.540	0.00	0.00	0.000	0.000
41	PLTA poso	1.000	106.778	0.00	0.00	195.000	55.638
42	Pamona 275	0.993	105.315	0.00	0.00	0.000	0.000
43	Pamona150	0.989	104.738	4.90	0.50	0.000	0.000
44	Poso	0.986	103.885	11.00	1.80	0.000	0.000

5. Result and Discussion

5.1. Case-1

The first step, carried out a study load flow. For the case, it is found that the voltage value is within the allowable limit, which is between 0.9 p.u to 1.05 p.u. The voltage on bus 21 has the lowest value of 0.906 p.u and the Bakaru bus has the highest voltage of 1.03 p.u. Figure 3 shows the stress profiles on all buses in the South Sulawesi interconnection system.

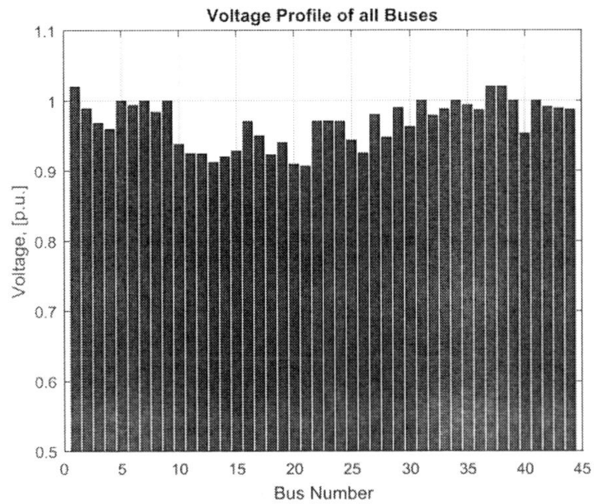

Figure 3. Voltage Profile of All Bus in in the South Sulawesi Interconnection System

After the load flow study, the eigen value is derived from the Jacobian Reduction matrix. For case-1, the 4 lowest eigenvalues are obtained:

Minimum Eigenvalue is $\lambda_{(17)} = 1.847$
Minimum Eigenvalue is $\lambda_{(26)} = 2.673$
Minimum Eigenvalue is $\lambda_{(20)} = 3.229$
Minimum Eigenvalue is $\lambda_{(24)} = 3.877$

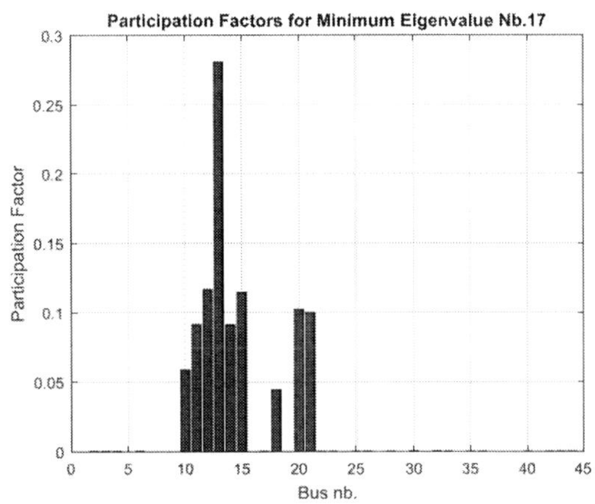

Figure 4. Bus Participation Factor for Case-1

With the lowest eigenvalue (λ_{-17}), the value of bus participation is obtained. The weakest bus is bus 13 (Tonasa Bus) as shown in Figure 4.

5.2 *Case-2*

Penetration of the Sidrap PLTB of 75 MW affects the stability of the system voltage. This can be observed from changes in eigenvalues. The 4 lowest eigenvalues at the time the PLTB Sidrap enters the system are:

Minimum Eigenvalue is $\lambda_{(18)}$ =1.856
Minimum Eigenvalue is $\lambda_{(26)}$ =2.673
Minimum Eigenvalue is $\lambda_{(20)}$ =3.221
Minimum Eigenvalue is $\lambda_{(24)}$ =3.884

The weakest bus is bus 13 (Tonasa Bus) as shown in Figure 5.

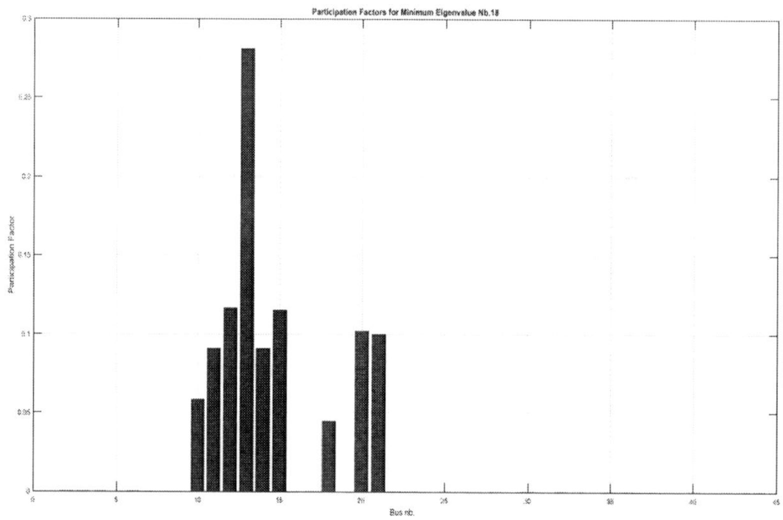

Figure 5. Bus participation factor for Case-2

5.3 Case-3
Penetration of the 72 MW PLTB Jeneponto also affects the stability of the system voltage. The effect of Jeneponto PLTB is better than Sidrap PLTB. This can be observed from changes in eigenvalues. The lowest eigenvalue when PLTB Jeneponto enters the system is higher than when the Sidrap PLTB enters the system

The 4 lowest eigenvalues when PLTB Jeneponto enters the system are:
Minimum Eigenvalue is $\lambda_{(19)}$ =1.860
Minimum Eigenvalue is $\lambda_{(26)}$ =2.673
Minimum Eigenvalue is $\lambda_{(20)}$ =3.222
Minimum Eigenvalue is $\lambda_{(24)}$ =3.888

5.4 *Case-4*
Penetration of PLTB Jeneponto and PLTB Sidrap of 147 MW greatly affects the stability of the system voltage. This can be observed from changes in eigenvalues. The lowest eigenvalue when the Jeneponto PLTB and PLTB SIdrap enter the system is higher compared to when one of the Wind Turbine enters the system.

The 4 lowest eigenvalues when Jeneponto PLTB and Sidrap PLTB enter the system are:
Minimum Eigenvalue is $\lambda_{(19)}$ =1.868
Minimum Eigenvalue is $\lambda_{(26)}$ =2.673
Minimum Eigenvalue is $\lambda_{(20)}$ =3.207
Minimum Eigenvalue is $\lambda_{(24)}$ =3.894

With the lowest eigenvalue (λ_{-19}), the value of bus participation is obtained. The weakest bus is bus 13 (Tonasa Bus) as shown in Figure 6.

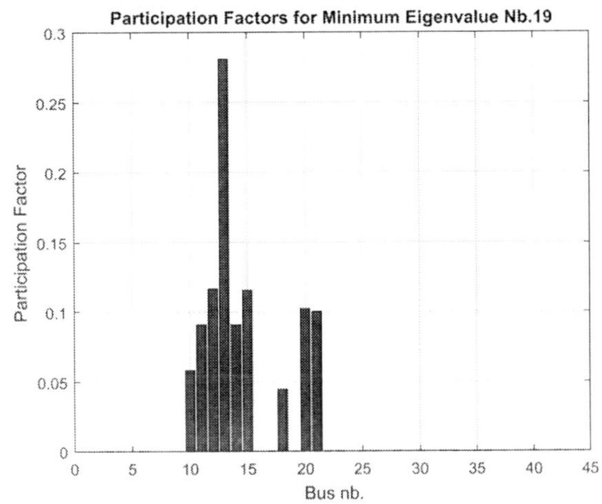

Figure 6. Bus Participation Factor for Case-4

From the simulation results for all cases the comparison of eigen value values for all cases can be seen in the table 2.

Table 2. Comparison of eigenvalues for all cases

No.	Case-1	Case-2	Case-3	Case-4
1	139.7688	140.1985	140.3649	140.7502
2	118.0498	115.4475	117.7507	115.0995
3	80.4251	95.9785	79.9885	95.9785
4	95.9785	58.6243	95.9785	58.5212
5	57.1805	53.8286	57.0366	54.032
6	53.6695	70.3481	53.8896	70.3481
7	70.3481	42.3584	70.3481	42.173
8	41.8707	40.68	41.7274	39.2901
9	40.6068	39.116	39.1684	36.4224
10	38.9803	31.118	36.3293	31.2149
11	31.042	30.6867	31.1473	30.3857
12	30.2834	28.5001	29.9962	28.5816
13	24.8471	19.7055	20.0097	19.6262
14	20.0664	27.9791	24.9226	27.9791
15	16.671	24.9015	16.6089	24.9712
16	17.9308	17.9389	27.9791	14.0541
17	14.0541	8.7551	8.768	8.7982
18	8.7213	14.0541	1.8598	1.8683
19	1.8463	1.8561	14.0541	3.2074
20	3.2019	3.2202	3.1921	9.8038
21	5.4029	9.8038	5.4227	5.4355
22	27.9791	5.4172	9.8038	23.0073
23	9.8038	23.0073	23.0073	3.8941

24	23.0073	3.8844	3.8874	4.9826
25	3.8765	4.9826	4.9826	2.6726
26	4.9826	2.6726	2.6726	4.1785
27	2.6726	4.1785	4.1785	3.9273
28	4.1785	3.9273	3.9273	
29	3.9273			

Table 3 shows the minimum values of eigenvalues for all cases. It can be seen that the lowest minimum eigenvalues occur in case-1 while the highest minimum eigenvalues occur in case-4. When the wind turbine has not entered the system, the minimum eigenvalue is 1.847, while when the Sidrap PLTB enters the eigenvalue system it becomes 1.856. The inclusion of Tolo PLTB makes the stability of the system better with an indication of the eigen value of 1.860.
These results indicate that when Wind Power enters the sulseltrabar electrical system, the steady state stability of the system gets better.

Table 3. Comparison of the lowest eigenvalues for all cases

Case	Minimum Eigen Value
1	Minimum Eigenvalue is $\lambda_{(17)}$ =1.847
2	Minimum Eigenvalue is $\lambda_{(18)}$ =1.856
3	Minimum Eigenvalue is $\lambda_{(19)}$ =1.860
4	Minimum Eigenvalue is $\lambda_{(19)}$ =1.868

6. Conclusion

From the simulation results it can be seen that the inclusion of wind turbine in the Sulseltrabar electrical system has the effect of improving the stability of the steady state system, this can be seen from the minimum eigenvalues. At the time before penetration of the wind turbine eigenvalues of 1.847. After the penetration of the wind turbine the eigenvalue increased to 1.868. An improvement in steady state stability in the Sulseltrabar system has taken place.

References

[1] Savu C.Savulescu. Real-Time Stability Assessment in Modern Power System Control Center. *IEEE Press. Wiley*. 2009
[2] I. C. Gunadin, A. Soeprijanto, and O. Penangsang, "*Steady state stability assessment using extreme learning machine based on modal analysis*," Int. Review of Electrical Engineering, vol. 7, no. 3, pp. 4532-4537, June 2012.
[3] Indar Chaerah Gunadin, Eka Sanjaya Putra Az, Yusri Syam Akil, and Steven Humena, *"The Impact of the Injection of Wind Power Plant on the Steady State Condition and the Dynamics of Sulselbar Power System"*, International Journal of Electrical and Electronic Engineering & Telecommunications Vol. 8, No. 6, November 2019
[4] Ching-Yin Lee, ShaoHong Tsai, Yuan-Kang Wu. A New Approach to The Assessment os Steady-State Voltage Stability Margin Using P-Q-V Curve, *Electrical Power and Energy System*. 2010
[5] Gao, B., Kundur, P., Morisson, G.K, "Voltage Stability Evaluation Using Modal Analysis", IEEE Transactions on Power Systems, vol. 7, no. 4, pp. 1423-1543, (November 1992).
[6] Kundur,P., "Power System Stability and Control", Mc-Graw Hill, New-York, 1994, pp 960-971.
[7] I. C. Gunadin, A. Soeprijanto, and O. Penangsang, "*Real power generation scheduling to improve steady state stability limit in the Java-Bali 500 kV interconnection power*

system,"World Academy of Science, Engineering and Technology, vol. 4, no. 12, pp. 1771-1775, December 2010.

[8] M.Moghavveni, M.O. Faruque, Power Systm Security and Voltage Collapse: A Line Outage Based Indicator for Prediction, *Electrical Power and Energy System*, 1999; 21; pp 455-461

[9] I. C. Gunadin, S. M. Said, and M. Irsan, "*Determination of stability index of electrical power system using Rei-Dimo methods*", Journal of Theoretical and Applied Information Technology, vol. 90, no. 1, pp. 161-167, August 2016.

[10] M.H. Haque. A Fast Method for Determining The Voltage Stability Limit of a Power System. *Electrical Power and Energy System*, 1995; 32; pp.35-43

[11] Dr. EnemuohF. O., Dr. Onuegbu J. C. and Dr.AnaziaE. A., *Modal Based Analysis and Evaluation of Voltage Stability of Bulk Power System*, International Journal of Engineering Research and Development, Volume6, Issue 12(May2013), pp. 71-79

Control Design of Automatic Voltage Regulator to Improve the Voltage Stability at Sengkang Power Plant, South Sulawesi Indonesia

I C Gunadin[1]*, Z Muslimin[1], A M Ilyas[1,2], and A Siswanto[1,3]

[1]Electrical Engineering Department, Faculty of Engineering, Hasanuddin University, Indonesia
[2]Electrical Engineering Department, Faculty of Engineering, Khairun University, Indonesia
[3]Electrical Engineering Department, Faculty of Engineering, University of 17 Agustus 1945 Cirebon, Indonesia

*Email: indarcg@gmail.com

Abstract. The change in voltage at the generator terminal is highly correlated with the change in reactive power at the center load. There are several reactive power sources in the power system including those from field amplifiers that are controlled by equipment called Automatic Voltage Regulators (*AVR*). Changes in reactive power in the system will change the voltage profile including the voltage at the generator terminal. To maintain the voltage level on the generator terminal, the *AVR* must be adjusted properly. Determination of the value of the amplifier gain on the *AVR* largely determines the performance of the *AVR* in stabilizing the voltage. Wind Turbine penetration greatly affects the system voltage. This research is to see the impact of voltage changes due to wind turbine penetration in sulseltrabar interconnection systems. Determination of *AVR* gain is generated from the Routh-Hurwitz equation. Furthermore, in this study PID control will be applied to increase the *AVR* output response. Determination of the correct *Kp*, *Ki* and *Kd* constants is obtained with the help of a MATLAB simulation model. The simulation shows that using a *PID* controller, the overshoot and settling times obtained are better. The maximum exceeded voltage is 1 per unit and the settling time is 10.15 seconds.

1. Introduction

Voltage stability in the power system is highly correlated with changes in reactive power in the system. The voltage collapse phenomenon is strongly influenced by systems that lack reactive power and control devices fail in the power system [1-3]. Reactive power comes from generator excitation as well as from reactive power compensation equipment such as capacitors and Flexible Alternating Current Transmission Systems (*FACTS*) devices [4-5].

An *AVR* is designed to maintain the terminal voltage of a generator. By adjusting the amplification current in the excitation, the terminal voltage can be controlled. When the system load increases, the voltage at the generator terminal will drop. By increasing the existing supply current, the generator voltage can be increased to a predetermined value. When generators work together in an interconnection system, the AVR on each generator will receive a command signal according to the change in voltage that occurs on the generator bus [6].

In the sulseltrabar interconnection system, wind turbine has entered the system. There are two wind turbine plants namely Sidrap Wind Turbine with a capacity of 75 MW and Tolo Wind Turbine with a capacity of 72 MW. Penetration of the wind turbine in the system will cause problems including the problem of voltage stability [7]. Intermittency of wind turbine does not affect the stability of the voltage so that the performance of the AVR to stabilize the voltage is needed. To improve AVR performance, a control method is needed. In this study PID (Proportional, Integral, Derivative) control is used as a control device [8].

2. Voltage Stability

In electric power systems which are operated in stress conditions, voltage stability becomes very important [9]. In the planning and operation of power systems, stress stability analysis focuses on two aspects including:

1. Prediction, how close are the unstable voltage condition systems.
2. The mechanism, when voltage instability occurs.

Predictions provide a measure of voltage safety, while the mechanism will provide important information that is useful in the operation strategy in preventing voltage instability [10]. The problem of voltage drop is a serious problem in the electrical system in many countries and this occurs in the system of interconnecting electricity. There are several factors that can cause voltage drops including: load increase, lack of excitation, loss of compensation equipment such as capacitors.

2.1. Voltage Settings

In practice, the reactive power supplied by the generator to the load continues to change. This causes the output voltage (voltage on the generator terminal) to be changed to guarantee the generator. it remains stable in compensating for the power requirements at that load. To regulate the size of the output voltage (V_t) that is by adjusting the size of the induced electromotive force (E) generated, by changing the amount of rotor flux (φ) where the magnitude of the rotor flux depends on the magnitude of the reinforcement current so that it can be said that to change the output voltage can be done by change the reinforcement current. If the output voltage exceeds the proper voltage, the gain current must be reduced so that the voltage drops again. Conversely, if the output voltage drops, the amplifying current must be increased so that the voltage returns to normal. In this strengthening current regulation system, there are two conditions that may occur, namely:

a. Under Excitation

Under excitation condition is a generator condition where the gain is less (less strengthened) so that the current will supply to the system. This can be reviewed according to the GGM / flux of the anchor reaction. When the generator is less amplified, it must provide an overtaking current to the system because the overtaking current will produce an anchor flux which will strengthen the rotor flux. The vector notation for under excitation conditions is as follows:

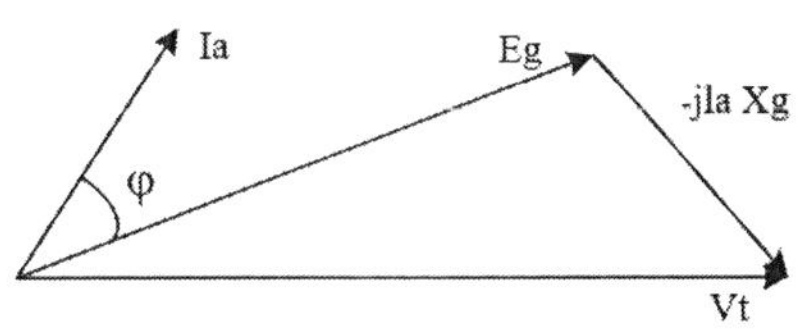

Figure 1. Vector diagram of under excitation conditions

b. Over Excitation

Over excitation condition is a generator condition where the gain is too large (too strengthened) so that it will supply the lagging current to the system. When the generator is too amplified, it must provide a lagging current to the system because the lagging current will produce an anchor flux that will fight the rotor flux to reduce too much gain. The vector notation for over excitation conditions is as follows:

IOP Conf. Series: Materials Science and Engineering **875** (2020) 012044 doi:10.1088/1757-899X/875/1/012044

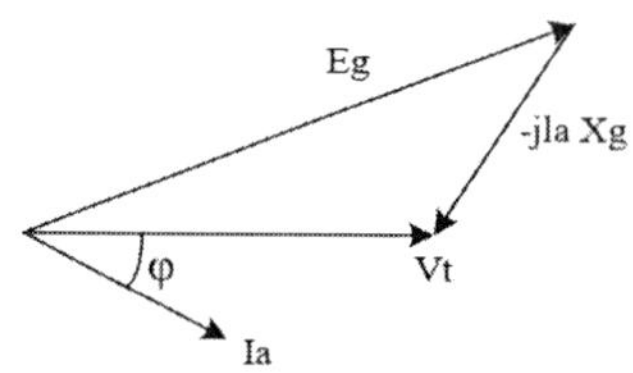

Figure 2. Vector diagram of over excitation conditions

2.2. *AVR (Automatic Voltage Regulator)*

AVR (Automatic Voltage Regulator) is a device mounted on a generator that can work automatically regulating the voltage or amplitude of the waves produced by the generator to remain stable. *AVR* works in regulating the generator output voltage by controlling the gain current of the generator [11].

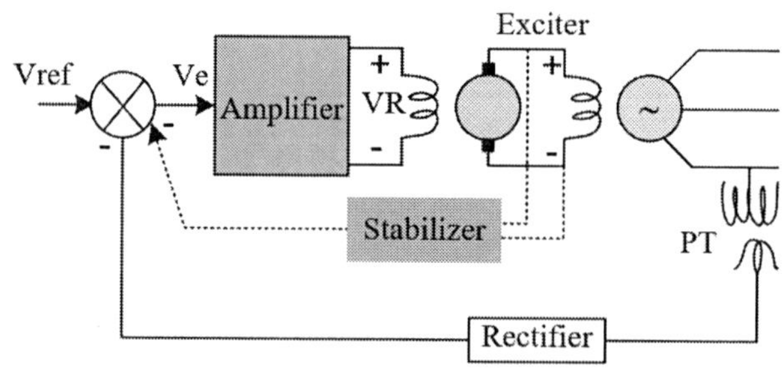

Figure 3. AVR Model

2.1.1. *Amplifier*

The reinforcement of the excitation system can be in the form of magnetic gain, loop gain, or electronic gain. Amplifiers can be represented as K_A with T_A time constants, which in the mathematical model are like the following equation:

$$\frac{V_R(S)}{V_e(S)} = \frac{K_A}{1+T_A S} \tag{1}$$

The value of the time constant (T_A) is very small, ranging from 0.02 to 0.1 seconds.

2.1.2. *Exciter*

Excitation commonly used in a generator there are several types ranging from using DC generators to modern types using *SCR* as a rectifier to produce DC power. Excitation can be represented as K_E with T_E time constant, which in its mathematical model is like the following equation:

$$\frac{V_F(S)}{V_R(S)} = \frac{K_E}{1+T_E S} \tag{2}$$

2.1.3. *Generator*

The terminal voltage of a generator is very dependent on the load. In a linear form, the relation of the terminal voltage function to the field voltage can be represented as K_G with the T_G time constant and the transfer function as follows:

$$\frac{V_t(S)}{V_F(S)} = \frac{K_G}{1+T_G S} \tag{3}$$

2.1.4. Sensor

The sensor consists of a voltage transformer (*VT*) and a rectifier. The sensor can be represented as K_R with the T_R time constant and its transfer function as follows:

$$\frac{V_S(S)}{V_t(S)} = \frac{K_R}{1+T_R S} \tag{4}$$

The working principles of the *AVR* are as follows:

- The generator output voltage is initially reduced by using PT (Potential Transformer) or the voltage transformer then rectified.
- The results of the rectification are then compared with the reference voltage (V_{ref}).
- If a difference occurs, the *AVR* will instruct the amplifier to increase or decrease the DC generator gain current so that the output voltage of the generator also changes.
- If the DC generator output voltage changes, the synchronous generator gain current also changes, consequently the generator output voltage returns stable [12].

The block diagram of an *AVR* is as shown below:

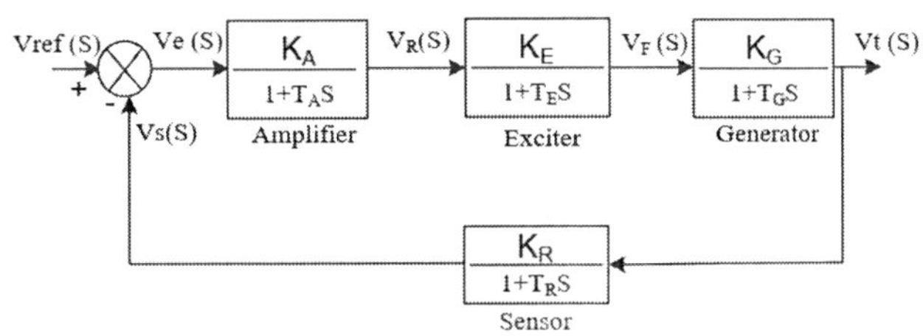

Figure 4. Block diagram of *AVR*

3. Proportional Integral Derivative

The *PID* controller (Proportional Integral Derivated) actually consists of three types of controllers that are mutually combined namely the *P* (Proportional) controller, the *I* controller (Integral), and the *D* controller (Derivative). Each has certain parameters that must be set to operate properly, called constants.

A. P Controller (Proportional)

Proportional controller has an output that is proportional to the size of the error signal/error. In simple terms it can be said that the output of the Proportional controller is the multiplication of proportional constants (*Kp*) with the input signal. A change in the input signal causes the system to directly change its output by its multiplying constant.

Experimentally, the use of Proportional controllers must pay attention to the following conditions:

1. If the K_p value is small, the Proportional controller is only able to make small error corrections, so that it will produce a slow system response.
2. If the K_p value is increased, the system response shows the faster it reaches its steady state.
3. However, if the K_p value is enlarged until it reaches an excessive price, the system will work unstable or the system response will oscillate.

B. Controller I (Integral)

The Integral Controller functions to produce a system response that has a zero steady state error. If the error signal does not change, the output will remain as before the input change occurred. Experimentally, the use of an Integral controller has the following characteristics:

IOP Conf. Series: Materials Science and Engineering **875** (2020) 012044 doi:10.1088/1757-899X/875/1/012044

1. The controller output requires a certain amount of time, so it tends to slow down the response.
2. If the error signal is zero, the controller output will remain at the previous value.
3. If the error signal is not zero then the output will show an increase or decrease which is affected by the magnitude of the error signal and the value of the Integral constant (K_i).
4. Large value Integral Constants (K_i) will accelerate the loss of offset, but will result in an increase in oscillation of the controller output signal.

C. *Controller D (Derivative / Differential)*

A sudden change in input from the Derivative controller will result in very fast and large changes. Experimentally, the use of Derivative controllers has the following characteristics:

1. This controller will not produce output if there is no change in the input (in the form of an error signal).
2. If the error signal changes with time, the output produced by the controller depends on the value of the Derivative constant (K_d) and the rate of change of the error simyal.
3. Derivative controller has a character to overtake, so that this controller can produce a significant correction before the error gets bigger, so it is used to speed up the initial response of a system, although it does not minimize errors in its steady state.

It can be said that each of the *P, I,* and *D* controllers has advantages and disadvantages. So that if the three are combined in parallel into a Proportional plus Integral plus Derivative controller or commonly called a *PID* controller, they will overlap. The controlling elements *P, I,* and D respectively aim to accelerate the reaction of a system, eliminate offsets and produce large initial changes.

4. Research Data and Results

The following is the *AVR* Parameter data derived from the manual book of the generator which is used as the object of research. In this research, the Sengkang sector generator will be used as a tested case.

Table 1. Generator excitation parameter data in the South Sulawesi system

No.	Generator	K_A (p.u)	T_A (Seconds)	*VA* max	*VA* min
1	PLTA Bakaru	400	0.04	0.71	-0.71
2	PLTA Teppo (Pinrang)	1	0.02	1	-1
3	PLTD Tallasa	1	0.02	1	-1
4	PLTD Suppa	1	0.02	1	-1
5	PLTU Barru	1	0.02	1	-1
6	PLTU Tallo	100	0.04	1	-1
7	PLTD Agreko (T Lama)	100	0.04	1	-1
8	PLTD Sungguminasa	10	0.02	18.3	-18.3
9	PLTU Arena (Jeneponto)	10	0.02	18.3	-18.3
10	PLTA POSO	400	0.04	0.71	-0.71
11	PLTA Tmanipi (Sinjai)	4	0.02	5.99	-5.99
12	PLTGU Sengkang	300	0.04	1	-1
13	PLTD Malea (Toraja)	10	0.02	18.3	-18.3
14	PLTD Palopo	4	0.02	5.99	-5.99
15	PLTA Bili-Bili	4	0.02	5.99	-5.99

Table 2. Generator constant data on Sengkang Power Plant

Gain	Time Constant
$K_A = 400$	$T_A = 0.04$ *second*
$K_E = 1$	$T_E = 0.5$ *second*
$K_G = 1$	$T_G = 1$ *second*
$K_R = 1$	$T_R = 0.025$ *second*

From the *AVR* parameter data, if it is made in the form of a block *AVR* diagram will be:

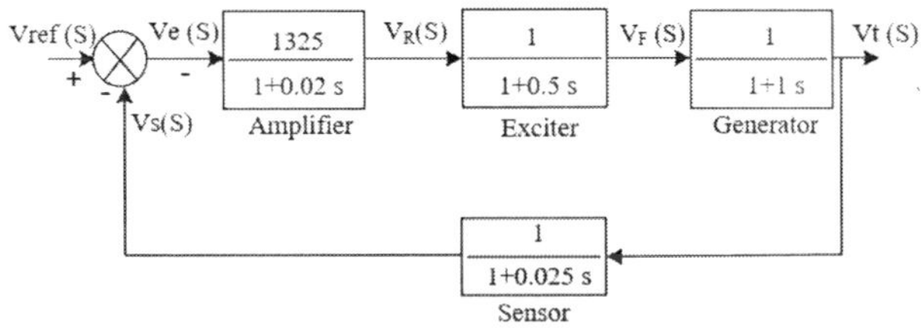

Figure 5. Block diagram of research results

where the K_A parameter value is variable, but the train constant used on the machine is now 1325. The simplified form of the *AVR* block diagram above is as follows:

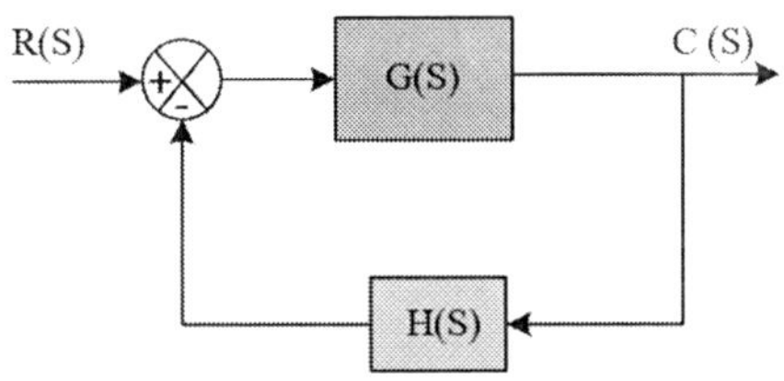

Figure 6. Simplification block diagram

so the equation for the function of the closed loop :

$$\frac{C(s)}{R(s)} = \frac{G(s)}{1+G(s)H(s)} \tag{5}$$

from the data obtained then made in the form of an open loop function G(s)H(s) to be

$$G(s)H(s) = \frac{K_A K_E K_G K_R}{(1+T_A s)(1+T_E s)(1+T_G s)(1+T_R s)}$$

$$G(s)H(s) = \frac{K_A.1.1.1}{(1+0.002\ s)(1+0.5\ s)(1+s)(1+0.025\ s)}$$

then the mathematical model simply becomes

$$G(s)H(s) = \frac{4000\ K_A}{S^4+93\ s^3+2272\ S^2+6180_s+4000} \tag{6}$$

the equation 6 is added with constant 1 (one) so that the system characteristic equation is obtained as follows:

IOP Conf. Series: Materials Science and Engineering **875** (2020) 012044 doi:10.1088/1757-899X/875/1/012044

$$1 + G(s)H(s) = 1 + \frac{4000\,K_A}{S^4 + 93\,S^3 + 2272\,S^2 + 6180\,s + 4000}$$

$$1 + G(S)H(S) = \frac{S^4 + 93\,S^3 + 2272\,S^2 + 6180\,s + 4000 + 4000\,K_A}{S^4 + 93\,S^3 + 2272\,S^2 + 6180\,s + 4000} \tag{7}$$

substitution Equation 7:

$$\frac{C(s)}{R(s)} = \frac{G\,(S)}{\frac{S^4 + 93\,S^3 + 2272\,S^2 + 6180\,s + 4000 + 4000\,K_A}{S^4 + 93\,S^3 + 2272\,S^2 + 6180\,s + 4000}}$$

$$= \frac{G(S)(S^4 + 93\,S^3 + 2272\,S^2 + 6180\,s + 400)}{S^4 + 93\,S^3 + 2272\,S^2 + 61800\,s + 4000 + 4000\,K_A}$$

According to the Routh stability criteria:

$$S^4 \quad +93\,S^2 + \quad 6180\,S + 4000 + 4000K_A = 0$$

S^4	1	2272	$4000 + 400K_A$
S^3	93	6180	0
S^2	A_1	A_2	
S^1	B_1		
S^0	C_1		

where:

$$A_1 = \frac{(93x2272) - (1X6180)}{93} = 2205.548$$

$$A_2 = \frac{\left(93x(400 + 4000\,K_A)\right) - (1x0)}{93} = 4000 + 4000K_A$$

$$B_2 = \frac{(2205.548x6180) - (93x(4000 + 4000K_A))}{2205.548} = 6011.334 - 168.66\,K_A$$

$$C_1 = \frac{\left((6011.334 - 168.66\,K_A)X(4000 + 4000K_A)\right) - (2205.508x))}{(6011.334 - 168.66\,K_A)}$$

$$4000 + 4000K_A$$

To get a range of K_A values where the condition of a system is stable if column 1 is positive (> 0):

$B_1 = 6011.334 - 168.66\,K_A > 0$

$- 168.66\,K_A > - 6011.334$

$- K_A > - 35.642$

$K_A < 35.642$

and

$C_1 = 4000 + 4000\,K_A > 0$

$4000\,K_A > - 4000$

$K_A > - 1$

So the range of values for K_A:

$-1 < K_A < 35.642$

From the results of calculations using the Routh stability criteria then the value is simulated using MATLAB as follows:

A. Without a *PID* controller

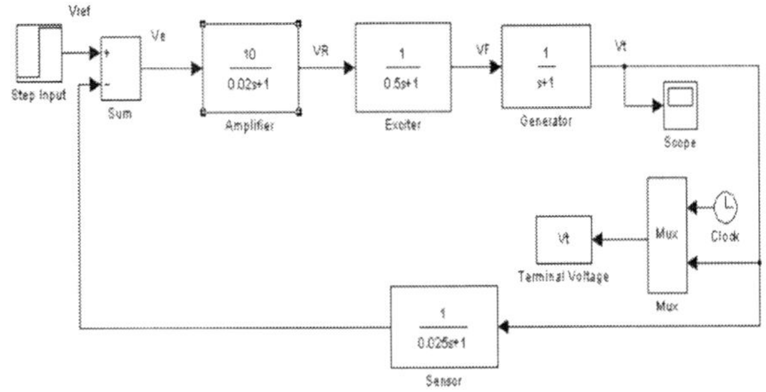

Figure 7. MATLAB Simulation Diagram Block without *PID* controller

For gain $K_A = 22$

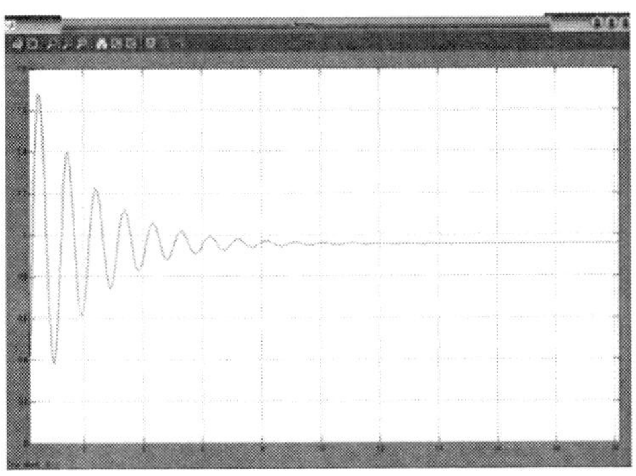

Figure 8. MATLAB simulation results for $K_A = 22$ without a *PID* controller

B. With a *PID* controller

Add the PID controller to the AVR simulation block diagram with the settings as shown below:

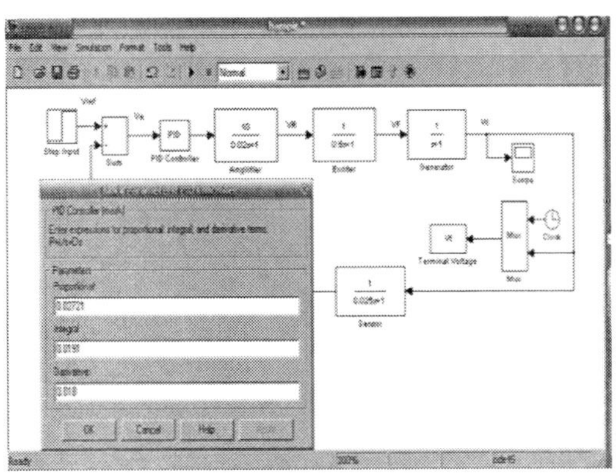

Figure 9. MATLAB simulation model with *PID* controller

For $K_A = 22$

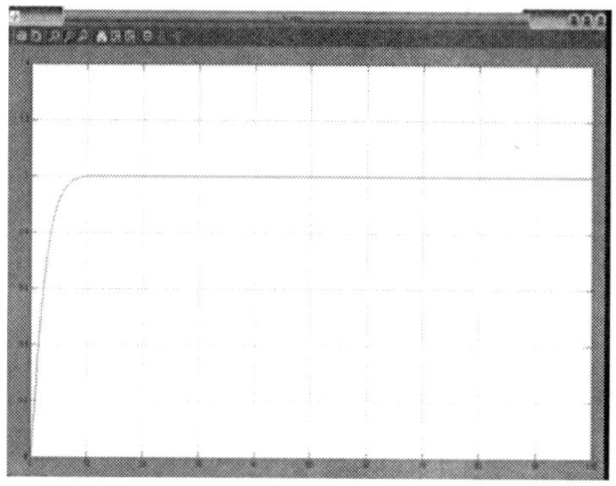

Figure 10. Simulation results for $K_A = 22$ with a *PID* controller

The simulation results obtained in the parameter settings $Kp = 0.0272$, $Ki = 0.0191$, and $Kd = 0.018$, in full as the following in Table 3:

Table 3. MATLAB simulation results

No	*KA* constant	No *PID* Control		With *PID* Control	
		Maximum overshoot (p.u)	Time (seconds)	Overshoot maksimum (p.u)	Time (seconds)
1	1	0.52	3.95	1	563.9
2	10	1.3	7.25	1	31.95
3	20	1.6	10.21	1	12.5
4	22	1.68	14.5	1	10.15
5	30	1.7	32.5	1.01	13.26
6	36	Stability is not achieved	Stability is not achieved	1.025	11.85
7	1325	Unable to run	Unable to run	1.2	8.9

Table 3 shows the simulation results without *PID* control and with *PID* control on generators in the Sengkang sector. By increasing the gain K_A value, it is seen that the overshoot and settling time values are getting bigger. For gain $K_A = 36$, for an uncontrolled AVR system, the *AVR* system is unstable. The next time the gain is increased again until it reaches $K_A = 1325$, the computational program will not converge so that the simulation cannot run. In contrast, when the *AVR* is mounted on the *PID* control, when the K_A gain = 36, the *AVR* is still stable with an output voltage having an overshoot of 1.025 per unit and a settling time of 11.85 seconds

5. Conclusions

Based on the research we have done, we can conclude the following:

1. Determination of the value of the amplifier constant (K_A) on an AVR (Automatic voltage Regulator) should refer to the value limit obtained from the calculation results with the *Routh* stability criteria method, which is in the range $-1 < K_A < 35{,}642$. Giving the best amplifier constant (K_A) is the setting $K_A = 1$ (without the *PID* controller), where the setting is obtained the fastest stable time with the lowest level of oscillation (transient).
2. The addition of PID controllers (Proportional, Integral, and Derivative) will be able to reduce or even eliminate oscillations (transients) that occur in the generator with the right K_p, Ki, and Kd parameter

settings namely in the parameter settings $K_p = 0.0272$, $Ki = 0.0191$, and $Kd = 0.018$ with a K_A value of 22, where the setting does not experience the slightest transient and the fastest stable time.

References

[1] Gunadin, I.C., A. Soeprijanto, and O. Penangsang, *Real Power Generation Scheduling to Improve Steady State Stability Limit in the Java-Bali 500 kV Interconnection Power System.* World Acad. Sci. Eng. Technol, 2010. **72**: p. 1-5.
[2] Indar Chaerah Gunadin, Sri Mawar Said, Muhammad Irsan, *Determination of stability index of electrical power system using REI-Dimo methods,* Journal of Theoretical and Applied Information Technology vol. 90, 15 August 2016
[3] Ansar Suyuti, Indar Chaerah Gunadin,, Nuhayati, PID implementation on real time 3-phase induction motor controlling and monitoring, Journal of Theoretical and Applied Information Technology vol. 89, 31 July 2016
[4] Gao, B., G.K. Morison, and P. Kundur, *Voltage stability evaluation using modal analysis.* IEEE Transactions on Power Systems, 1992. **7**(4): p. 1529-1542.
[5] Djari, M.A., L. Benasla, and W. Rahmouni. *Voltage stability assessment using the V-Q sensitivity and modal analyses methods*. in *2017 5th International Conference on Electrical Engineering - Boumerdes (ICEE-B)*. 2017.
[6] Gunadin, I., A. Soeprijanto, and O. Penangsang, *Steady state stability assessment using extreme learning machine based on modal analysis.* International Review of Electrical Engineering, 2012. **7**: p. 4532-4537.
[7] Gunadin, I., et al., *Determination of Steady State Stability Margin Using Extreme Learning Machine.* WSEAS Transactions on Power Systems, 2012. **7**.
[8] Goh, H.H., et al. *Validation of steady-state stability evaluation exerting with Dimo's approximation*. in *4th IET Clean Energy and Technology Conference (CEAT 2016)*. 2016.
[9] *Indar Chaerah Gunadin, Muhammad Abdillah, Adi Soeprijanto, Ontoseno Penangsang, Steady-State Stability Assessment Using Neural Network Based on Network Equivalent,* Telkomnika vol. 9, December 2011
[10] Indar Chaerah Gunadin, E.S.P.A., Yusri Syam Akil, and Steven Humena, *The Impact of the Injection of Wind Power Plant on the Steady State Condition and the Dynamics of SULSELBAR Power System.* Int. J. Elec. & Elecn. Eng. & Telcomm., 2019. .
[11] Siswanto, A., et al., *Stability improvement of wind turbine penetrated using power system stabilizer (PSS) on South Sulawesi transmission system.* AIP Conference Proceedings, 2018. 1941(1): p. 020036.
[12] Kundur, P., *Power System Stability and Control,* McGraw-Hill Inc. USA, 1994.

IOP Conf. Series: Materials Science and Engineering **875** (2020) 012045 doi:10.1088/1757-899X/875/1/012045

Analysis of Transmission Line Stability for Sulselbar Interconnection System with the Penetration of Renewable Energy to Prevent Voltage Collapse

I C Gunadin[1]*, A Suyuti[1], A M Ilyas[1,2], and A Siswanto[1,3]

[1]Electrical Engineering Department, Universitas Hasanuddin, Indonesia
[2]Universitas Khairun, Indonesia
[3]Universitas 17 Agustus 1945 Cirebon, Indonesia

*Email: indar@eng.unhas.ac.id

Abstract. There are three main requirements to meet the criteria both in the electric power system, namely reliability, quality, and stability. Stability is the ability of the system to return to work normally after experiencing interference. System stability analysis is a guide for operators to detect or prevent voltage drops and power outages. This research discusses the stability of the 150 KV electrical system in South Sulawesi before and after interconnection with the Sidrap and Jeneponto Wind Power Plant. The method used is the Line Stability Index (Lmn Index) to view the voltage profile. Index Lmn coding is run and simulated using Matlab. Validation is done by IEEE 14 bus simulation data, and comparing the results of the analysis before and after penetration with PLTB. The study was conducted on the Sulselbar 150 KV system before penetration with PLTB, namely 44 buses, 14 generators, and 52 lines / networks. The highest Lmn index result is the network from bus 38 to bus 39 that is equal to 0.0292. Second is bus 15 to bus 16 which is 0.0291, and third is bus 40 to 42 that is 0.0273. After penetration with PLTB Sidrap and Jeneponto, the number of generators will be 16 generators. The highest Lmn index results are from bus 38 to 39 that is 0.0292, then bus 15 to bus 16 that is 0.0290 ranks second, and bus 40 to 42 equals 0.0272 or keeps the third order. After the sidrap load bus is converted to a bus generator, the Lmn index before in reactive power supplied to the load by a generator is a certain thing, this condition must be maintained so that the plant continues to work in stable conditions.

1. Introduction

Voltage stability analysis illustrates the ability of the system to maintain its voltage value under normal conditions or after a disturbance has occurred. System instability caused by interruptions, load increases, and changes in system configuration can also affect the voltage stability of the electric power system. Voltage instability occurs when the voltage value on the receiver / consumer side will drop from its normal limit, in that case it can cause a condition called voltage collapse. Voltage will drop at the lowest point, so it can cause a blackout of the system.

Calculation of voltage stability is required when planning or operating an electric power system. If there is an unintentional change in system configuration such as contingency, it is very likely that the system's voltage stability will not be the same as normal conditions. An example is the loss of a channel or generator can result in a decrease in the ability of voltage stability in a system.

There are several methods for analyzing voltage stability, including the P-V curve, discussing the relationship between power and voltage, and the Fast Voltage Stability Index (FVSI). In this study using the Line Stability Index (Lmn Index) to determine the stability index of the 150 KV transmission system in Sulselbar.

2. Literature

Voltage stability analysis can be interpreted as the ability of the power system to maintain the value of the system voltage, under normal conditions and in a state of interference [1, 2]. There are two types of voltage stability based on the simulation time, namely, static voltage stability, and dynamic voltage stability. Dynamic analysis is used to study transient stability by paying attention to load dynamics, and generators. And static analysis uses algebraic equations which are computationally easier than dynamic analysis. Static analysis is more ideal for the study of voltage stability limits in cases before contingency, and after a system contingency. Static voltage stability analysis is based on curve calculations, or on the singular Jacobian matrix on power flow [3].

2.1. Power System Stability

The stability of a power system is a complex dynamic system, and consists of linear, nonlinear subsystems that are constantly experiencing internal, and external interference. Power system stability can be defined as the ability of the power system to remain in equilibrium conditions under normal operating conditions and to regain equilibrium conditions that can be received after a disturbance [4].

2.1.1. Rotor Angle Stability

The stability of the rotor angle of a power system is the ability of interconnected synchronous machines of the power system network to stay in line with each other ie in synchronous conditions [4]. Analysis of the stability of the rotor angle involves the study of electro-mechanical oscillations. A fundamental factor of rotor angle stability is the way in which the synchronous engine power output varies with the rotor oscillation. This can be either steady state stability or transient stability.

2.1.2. Frequency Stability

Frequency stability analysis refers to the ability of the power system to maintain a stable frequency after system interference. This results in a significant imbalance between generation and load. This depends on the network's ability to maintain or restore the balance between the system generator, and the load, with accidental minimum load losses. In general, the problem of frequency stability is related to insufficient equipment response, or inadequate generator reserves [4]. Frequency stability can be short-term (ranging from a fraction of a second) or a long-term phenomenon.

2.1.3. Voltage Stability

Voltage stability is related to the ability of the power system to maintain an acceptable voltage level in all buses in the system under normal operating conditions, and after a breakdown [4]. This involves all voltage levels on each bus under different load conditions to determine the stability limits, and margins. Based on the size of the disturbance, voltage stability can be classified into the following two subcategories:

a. Large fault voltage stability refers to the ability of the system to maintain a constant voltage after a large fault such as a system error, loss of generation, or circuit contingency.
b. Small fault voltage stability refers to the ability of the system to maintain a fixed voltage when experiencing a small disturbance such as a gradual change in the system load.

2.2. L-Index (L)

L-index based on power flow solutions was developed by Kassel [5]. This *L*-Index measures the voltage instability and is suitable for a constant type of power load. The values range from 0 to 1. The *L*-index formulation is as follows.

$$L = \max_{j\in\alpha L}\{Lj\} = \max_{j\in\alpha L}\left\{\left|1 - \frac{\sum_{i\in\alpha G} F_{ji}{*}\, V_i}{V_j}\right|\right\} \quad (1)$$

$$F_{ji} = \left|F_{ji}\right| \angle\, \theta_{ji}$$

where L is the load / consumer area, and G is the generator / generator area, L_j is the local indicator that determines the busbar from which the collapse might originate [5]. $[F]$ is calculated using the formula $[F] = [F_{LL}]^{-1}[Y_{LG}]$, where $[Y_{LL}]$ and $[Y_{LG}]$ computed using a matrix Y-bus. Voltage V_i and V_j is voltage at the i bus, and j [6].

2.3. Line Stability Index (Lmn-Index)

The stability index (*Lmn*) is derived based on the concept of power transmission lines in one network. Moghavvemi and Omar [7] lowered the network stability index to evaluate the stability between two buses in an interconnected system as shown in the following figure.

IOP Conf. Series: Materials Science and Engineering **875** (2020) 012045 doi:10.1088/1757-899X/875/1/012045

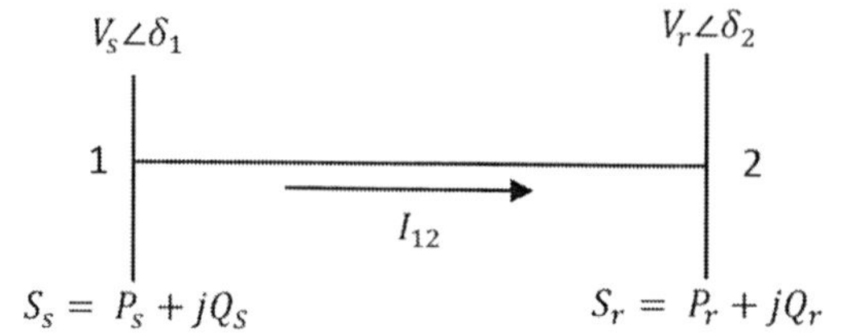

Figure 1. Tipical One-Line diagram of transmission line

where, V_s, P_s, and Q_s are the sending / generating voltage, real power and reactive power, respectively. V_r, P_r and Q_r are the receiver / load tip voltage, real power, and reactive power, respectively. δ1 is the phase voltage of the generator, and δ_2 is the phase angle of the load at the load, I_{12} is the line current and θ is the angle of the transmission line.

Power flow through the transmission line using the representation of the pie model (π) for a two-bus system uses a quadratic equation of voltage set greater than or equal to 0 (zero). If the value is less than 0, then the root will be imaginary indicating that there is instability in the system. The formula for the Lmn index is as follows.

$$Lmn = \frac{4\ X\ Q_r}{|V_s|^2\ sin^2\ (\theta - \delta)} \leq 1 \quad (2)$$

The *Lmn*-index is also directly related to reactive power, and not directly related to active power through the voltage phase angle δ. Networks in the system are said to be close to instability when the *Lmn*-index is close to one (1). On the other hand, if the Lmn-index value is less than 1, the system is said to be stable [7].

3. Materials and Methods

The Electricity System in Sulserabar consists of 44 buses and 14 conventional generator buses and 2 new renewable energy buses, namely PLTB located in Sidrap and Jeneponto Regencies. The bus generators in question are: Bakaru as slack buses and bus generators namely Pinrang, Suppa, Sidrap (PLTB), Barru, Tello, Borongloe, Tellolama, Sungguminasa, Tallasa, Punagaya, Sinjai, Sengkang, Makale, Palopo, Poso, the rest as much as 28 a bus is a load bus [8]. The results of the analysis of the transmission stability of the 150 KV system were compared before and after penetration with PLTB.

The transmission stability analysis method used is the Lmn Index. The generator data, and the single line system of South Sulawesi can be seen in Table 1 and Figure 2.

Table 1. The Data of Generator Sulselbar System

No	Bus Code	Voltage Mag.	Angle Degree	Load		Generator			Injected	
				MW	Mvar	MW	Mvar	Qmin	Qmax	Mvar
1	1	1.0300	0.0000	3.5000	0.2000	63.0000	3.1000	0.0000	0.0000	0.0000
2	0	0.9984	0.0000	17.1000	4.1000	0.0000	0.0000	0.0000	0.0000	0.0000
3	0	0.9900	0.0000	23.3000	3.7000	0.0000	0.0000	0.0000	0.0000	0.0000
4	0	0.9827	0.0000	9.6000	4.8000	0.0000	0.0000	0.0000	0.0000	0.0000
5	2	1.0000	0.0000	24.4000	6.2000	14.3000	0.8000	0.0000	0.0000	0.0000
6	0	0.9977	0.0000	18.7000	4.7000	0.0000	0.0000	0.0000	0.0000	0.0000
7	2	1.0000	0.0000	0.0000	0.0000	31.1000	8.2000	0.0000	0.0000	0.0000
8	2	0.9847	0.0000	26.5000	10.3000	75.000	0.0000	0.0000	0.0000	0.0000
9	2	1.0000	0.0000	0.0000	0.0000	60.4000	4.8000	0.0000	0.0000	0.0000
10	0	0.9387	0.0000	10.1000	2.4000	0.0000	0.0000	0.0000	0.0000	0.0000
11	0	0.9255	0.0000	22.1000	8.0000	0.0000	0.0000	0.0000	0.0000	0.0000
12	0	0.9253	0.0000	0.0000	0.0000	0.0000	0.0000	0.0000	0.0000	0.0000
13	0	0.8954	0.0000	18.9000	10.6000	0.0000	0.0000	0.0000	0.0000	0.0000
14	0	0.9208	0.0000	33.1000	15.4000	0.0000	0.0000	0.0000	0.0000	0.0000
15	0	0.9286	0.0000	18.0000	5.8000	0.0000	0.0000	0.0000	0.0000	0.0000
16	2	0.9700	0.0000	63.3000	18.3000	21.0000	7.9000	0.0000	0.0000	0.0000
17	0	0.9497	0.0000	68.3000	17.7000	0.0000	0.0000	0.0000	0.0000	0.0000
18	0	0.9400	0.0000	0.0000	-20.0000	0.0000	0.0000	0.0000	0.0000	0.0000
19	2	0.9400	0.0000	11.4000	0.0000	5.2000	0.2000	0.0000	0.0000	0.0000
20	0	0.9189	0.0000	24.3000	2.6000	0.0000	0.0000	0.0000	0.0000	0.0000
21	0	0.9205	0.0000	45.5000	2.8000	0.0000	0.0000	0.0000	0.0000	0.0000
22	0	0.9704	0.0000	0.0000	0.0000	0.0000	0.0000	0.0000	0.0000	0.0000
23	0	0.9710	0.0000	0.0000	0.0000	0.0000	0.0000	0.0000	0.0000	0.0000

No	Bus	Voltage	Angle	Load		Generator			Injected	
	Code	Mag.	Degree	**MW**	**Mvar**	**MW**	**Mvar**	**Qmin**	**Qmax**	**Mvar**
24	2	0.9700	0.0000	19.7000	4.7000	12.6000	0.0000	0.0000	0.0000	0.0000
25	0	0.9443	0.0000	0.0000	0.0000	0.0000	0.0000	0.0000	0.0000	0.0000
26	0	0.9260	0.0000	26.5000	7.7000	0.0000	0.0000	0.0000	0.0000	0.0000
27	2	0.9800	0.0000	15.7000	3.6000	20.0000	5.9000	0.0000	0.0000	0.0000
28	0	0.9474	0.0000	55.2000	16.7000	0.0000	0.0000	0.0000	0.0000	0.0000
29	2	0.9900	0.0000	20.6000	4.7000	79.0000	39.1000	0.0000	0.0000	0.0000
30	0	0.9632	0.0000	18.6000	5.5000	0.0000	0.0000	0.0000	0.0000	0.0000
31	2	1.0000	0.0000	0.0000	0.0000	196.1000	38.6000	0.0000	0.0000	0.0000
32	2	0.9786	0.0000	70.0000	12.5000	72.0000	0.0000	0.0000	0.0000	0.0000
33	0	0.9881	0.0000	27.1000	6.5000	0.0000	0.0000	0.0000	0.0000	0.0000
34	2	1.0000	0.0000	21.9000	4.6000	4.0000	0.5000	0.0000	0.0000	0.0000
35	0	0.9937	0.0000	32.1000	8.2000	0.0000	0.0000	0.0000	0.0000	0.0000
36	0	0.9869	0.0000	14.1000	3.4000	0.0000	0.0000	0.0000	0.0000	0.0000
37	2	1.0200	0.0000	28.4000	11.5000	265.2000	7.9000	0.0000	0.0000	0.0000
38	2	1.0200	0.0000	11.9000	1.5000	8.2000	2.1000	0.0000	0.0000	0.0000
39	2	1.0000	0.0000	49.2000	0.0000	4.0000	2.0000	0.0000	0.0000	0.0000
40	0	0.9528	0.0000	0.0000	0.0000	0.0000	0.0000	0.0000	0.0000	0.0000
41	2	1.0000	0.0000	0.0000	0.0000	195.0000	27.2000	0.0000	0.0000	0.0000
42	0	0.9913	0.0000	0.0000	0.0000	0.0000	0.0000	0.0000	0.0000	0.0000
43	0	0.9897	0.0000	4.9000	0.5000	0.0000	0.0000	0.0000	0.0000	0.0000
44	0	0.9881	0.0000	11.0000	1.8000	0.0000	0.0000	0.0000	0.0000	0.0000

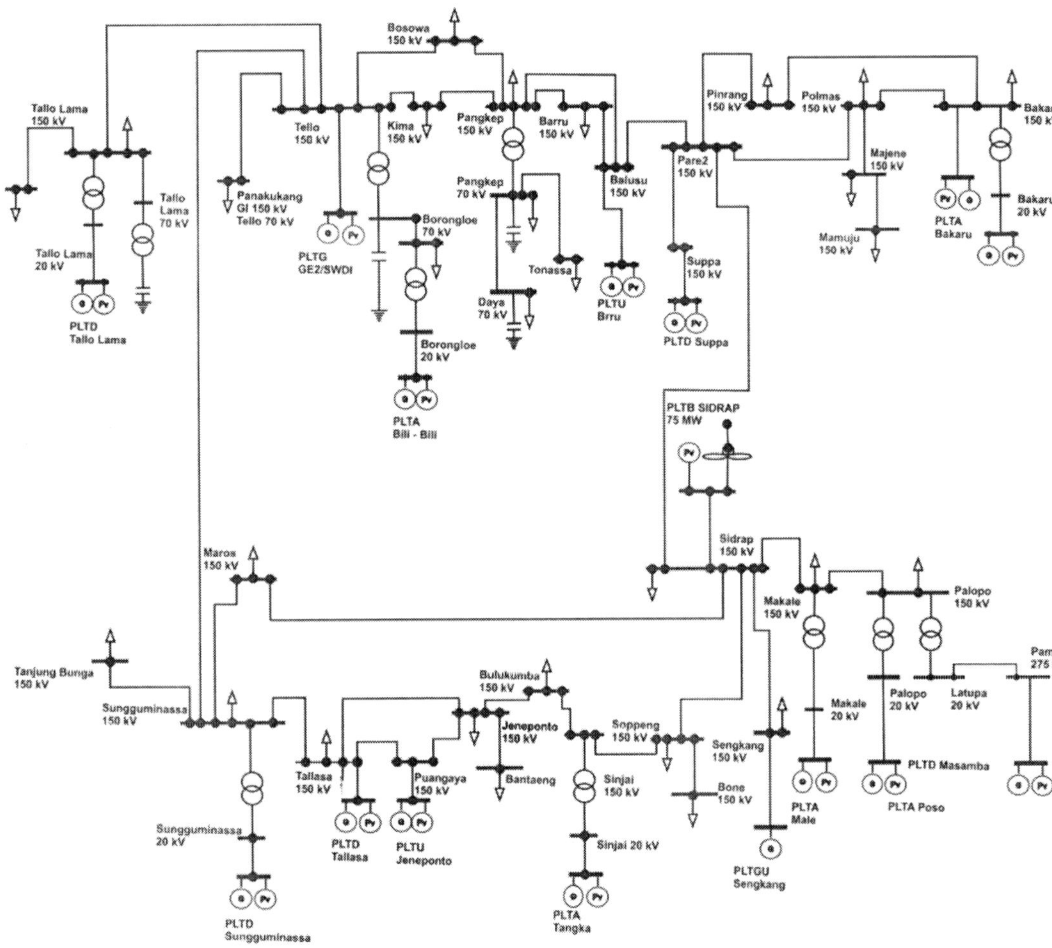

Figure 2. Single Line System 150 KV Sulselbar

For more details about the steps of the research carried out, it can be seen in the following research flowchart.

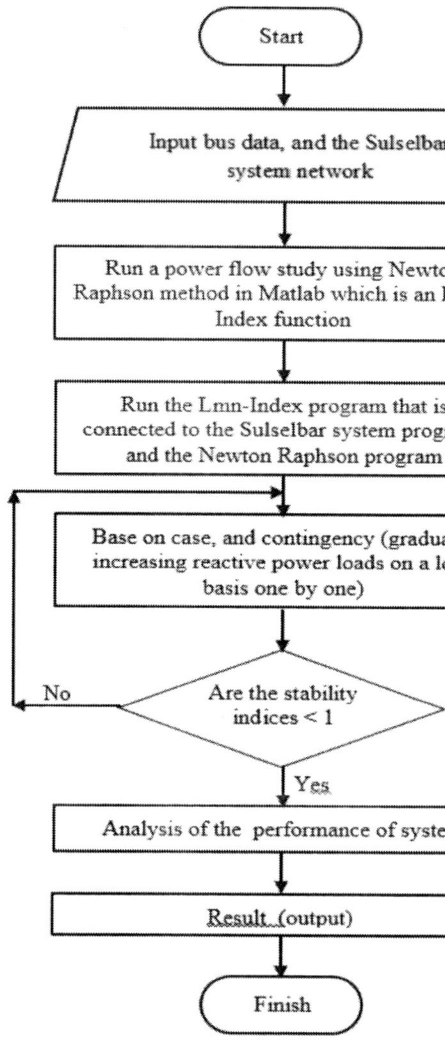

Figure 3. Flowchart of Research

4. Results and Discussion

The results obtained from the South Sulawesi 150 KV transmission system *Lmn*-Index method were analyzed and compared before and after penetration with the Sidrap and Jeneponto Wind Power Plant (PLTB). Table 2 shows the network stability index and Figure 3 shows a graph of the Lmn-index system of 150 KV, 44 buses, 14 generators, 15 generators, and 16 generators.

Table 2. Result of Analysis of Transmission System 150 KV Selselbar

Line No.	From bus	To Bus	Lmn Indeks			
			Sulselbar	Sulselbar + Sidrap	Sulselbar + Jeneponto	Sulselbar + Sid + JP
1	1	2	0.00561	0.00558	0.00558	0.00555
2	2	3	0.00339	0.00338	0.00340	0.00338
3	3	4	0.00552	0.00550	0.00553	0.00550
4	1	5	0.00229	0.00229	0.00229	0.00228
5	2	6	0.00138	0.00137	0.00138	0.00137
6	5	6	0.00647	0.00642	0.00643	0.00637
7	6	7	0.00349	0.00346	0.00349	0.00347
8	6	8	0.00362	0.00360	0.00362	0.00360
9	6	9	0.00203	0.00201	0.00203	0.00201

10	9	10	0.00470	0.00469	0.00469	0.00467
11	10	11	0.00148	0.00148	0.00148	0.00147
12	11	12	0.00143	0.00143	0.00143	0.00143
13	12	13	0.00993	0.00991	0.00991	0.00988
14	12	20	0.00590	0.00588	0.00588	0.00585
15	11	14	0.00174	0.00173	0.00173	0.00173
16	11	15	0.02208	0.02202	0.02202	0.02195
17	15	16	0.02912	0.02909	0.02909	0.02905
18	14	16	0.02223	0.02218	0.02217	0.02211
19	16	24	0.00212	0.00212	0.00212	0.00212
20	16	17	0.00332	0.00332	0.00332	0.00332
21	16	18	0.01885	0.01889	0.01889	0.01894
22	16	22	0.00254	0.00254	0.00254	0.00254
23	16	27	0.00320	0.00320	0.00320	0.00320
24	22	23	0.00906	0.00906	0.00906	0.00906
25	18	19	0.00752	0.00751	0.00751	0.00750
26	18	20	0.00615	0.00615	0.00615	0.00614
27	20	21	0.00664	0.00663	0.00663	0.00661
28	18	21	0.00757	0.00757	0.00757	0.00756
29	24	25	0.00871	0.00871	0.00871	0.00871
30	25	26	0.00739	0.00739	0.00739	0.00739
31	27	28	0.00495	0.00495	0.00495	0.00495
32	27	29	0.02046	0.02043	0.02091	0.02092
33	27	30	0.00670	0.00671	0.00671	0.00671
34	29	31	0.00820	0.00820	0.00820	0.00820
35	29	32	0.00688	0.00688	0.00689	0.00689
36	31	32	0.01075	0.01075	0.01077	0.01077
37	32	33	0.00612	0.00612	0.00589	0.00589
38	33	34	0.00199	0.00199	0.00198	0.00198
39	34	35	0.00149	0.00149	0.00149	0.00149
40	33	35	0.01836	0.01836	0.01819	0.01819
41	35	36	0.00448	0.00448	0.00453	0.00453
42	8	36	0.00397	0.00385	0.00397	0.00384
43	36	37	0.00340	0.00337	0.00342	0.00338
44	8	37	0.00564	0.00548	0.00563	0.00544
45	8	38	0.00372	0.00358	0.00373	0.00358
46	8	30	0.00289	0.00280	0.00289	0.00279
47	38	39	0.02920	0.02920	0.02920	0.02920
48	39	40	0.00827	0.00827	0.00827	0.00827
49	40	42	0.02727	0.02727	0.02727	0.02727
50	41	42	0.00059	0.00059	0.00059	0.00059
51	42	43	0.00320	0.00320	0.00320	0.00320
52	43	44	0.00630	0.00630	0.00630	0.00630

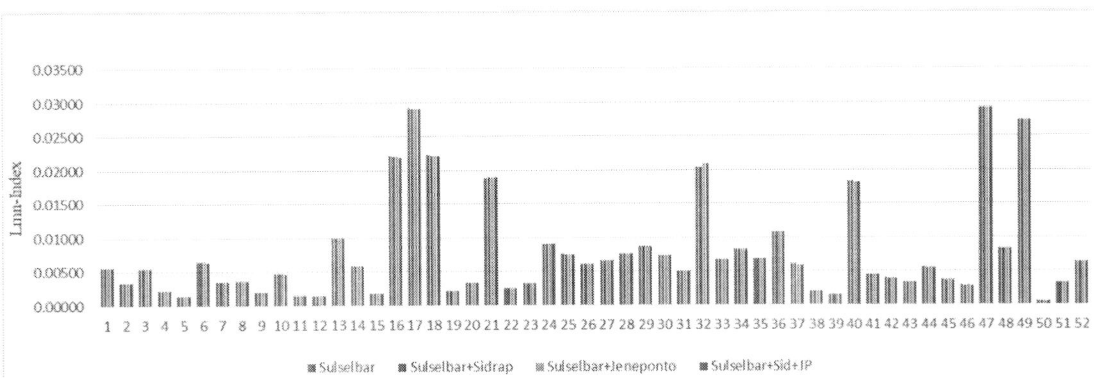

Figure 4. Grafik Lmn-Index of Transmission System 150 KV Sulselbar

Table 2 and Figure 3 show that Sulselbar's 150 KV transmission system is stable because there is no stability index for each network close to one (1). All networks, 52 of which are in the stability index range of less than one. Before penetration with PLTB, the highest Lmn index was network from bus 38 to bus 39, which was 0.0292. Second is bus 15 to bus 16 which is 0.0291, and third is bus 40 to 42 that is 0.0273. After penetration with PLTB

Sidrap and Jeneponto, the number of generators will be 16 generators. The highest Lmn index results are from buses 38 to 39 namely 0.0292, then bus 15 to bus 16 ie 0.0290 are ranked second, and buses 40 to 42 are equal to 0.0272 or third. This condition needs to be known by the operator so that the stability of the system can be maintained.

5. Conclusions

In the research that has been done it can be concluded that based on the results of the analysis using the *Lmn*-index, the Sulselbar 150 KV transmission system is stable. The results of the *Lmn* index analysis can be used as one of the variables to identify system stability.

References

[1] C. W. Taylor, "Power System Voltage Stability," 1994.

[2] I. C. Gunadin, "Transient Stability Improvement Using Allocation Power Generation Methods Based o n Moment Inertia," no. ICELTICs, pp. 147–152, 2017.

[3] F. Larki, M. Joorabian, H. Meshgin Kelk, and M. Pishvaei, "Voltage Stability Evaluation of The Khouzestan Power System in Iran Using CPF Method and Modal Analysis," in *2010 Asia-Pacific Power and Energy Engineering Conference*, 2010, pp. 1–5.

[4] P. S. Kundur, "Power System Stability and Control," vol. 46, no. 0, 2011.

[5] P. Kessel and H. Glavitsch, "Estimating the Voltage Stability of a Power System," *IEEE Trans. Power Deliv.*, vol. 1, no. 3, pp. 346–354, 1986.

[6] K.R.Vadivelu and Dr.G.V.Marutheswar, "Fast Voltage Stabilty Index Basedoptimal Reactive Power Planning Usingdifferential Evolution," *Electr. Electron. Eng.*, vol. Vol 3, No, 2014.

[7] M. Moghavvemi and F. M. Omar, "Technique for contingency monitoring and voltage collapse prediction," *IEE Proc. - Gener. Transm. Distrib.*, vol. 145, no. 6, p. 634, 1998.

[8] I. Andi Muhammad Ilyas, Ansar Suyuti, Indar Chaerah Gunadin, Agus Siswanto, "Optimal Power Flow The Sulselrabar 150 KV System Before and After The Penetration of Wind Power Plants Considering Power Loss and Generation Costs," 2019.

IOP Conf. Series: Materials Science and Engineering **875** (2020) 012046 doi:10.1088/1757-899X/875/1/012046

Performance of Routing Protocol OLSR and BATMAN in Multi-hop and Mesh Ad Hoc Network on Raspberry Pi

Wardi[1]*, Dewiani[1], M Baharuddin[1], S Panggalo[1], and M F B Gufran[1]

[1]Electrical Engineering Department, Faculty of Engineering Hasanuddin University, Gowa, Sulawesi Selatan, Indonesia 92171

*Email: wardi@unhas.ac.id

Abstract. Ad hoc network is a network where each node can act as a terminal and also as a router. Ad hoc technology does not require infrastructure that controls communication systems centrally. Therefore, each node on the ad hoc network can manage itself. This study aims to design nodes that can be used for communication on wireless ad hoc networks then compare the networks performance on OLSR and BATMAN routing protocols. There are four performance parameters used, namely Throughput, Round Trip Delay (RTD), Packet Delivery Ratio (PDR), and Convergence Time. The topology used includes P2P communication and multi-hop communication. Network performance testing is done both for P2P and multi-hop communication on MESH networks. Convergence Time Testing is done only for multi-hop communication to find out the time needed by the protocol to repair the path when a path fail occurs. The test results show that OLSR gives a better average value of throughput compared to BATMAN. Whereas RTD shows that the BATMAN routing protocol is better compared to OLSR. Both protocols, OLSR and BATMAN, successfully received all data packets sent by the source node on P2P communications. Failure to send data packets occurs in multi-hop communication as indicated by a PDR value of 98% for the OLSR protocol and 95% for the BATMAN protocol.

1. Introduction

Ad hoc network is a wireless network technology without infrastructure consisting of a group of nodes that are interconnected. Nodes in the network not only working as terminals but also function can as routers. The ad hoc network technology is very easy to install. Therefore, it is suitable for use in emergencies or only for temporary networks such as in disaster areas or used for interconnection at meetings [1].

Ad hoc nodes can be static or dynamic, move randomly, and can change topology quickly. Therefore, changes in ad hoc network topology are difficult to predict. Some routing protocols are made specifically for ad hoc networks to cope with topological changes or to find the best path. Ad hoc network protocols are divided into 3 classifications: proactive, reactive, and hybrid protocols [2, 3].

Proactive routing protocol updates its routing table by spreading information about the existence of each node to all nodes in the network periodically. Examples of proactive routing protocols are OLSR, DSDV, and BATMAN. While reactive routing protocols do not broadcast information about the presence of nodes in the network periodically. The protocol creates a route to the destination node at the time of communication (on demand operation). AODV, MAODV, and DSR are examples of types of reactive protocols. Hybrid routing protocol is a combination of the characteristics of the proactive and reactive protocols. Examples of hybrid protocols are ZRP and TORA [4].

Routing is one of the most important things of an ad hoc wireless network. Several studies on the performance of routing protocols on wireless ad hoc networks have been conducted, but only a few of these studies have been carried out in real tests. Testing with real tests gives real environmental conditions [5-7]. This research designs ad hoc nodes that can be used for wireless ad hoc network communication and compares the network performance on OLSR and BATMAN routing protocols. There are four performance parameters used, namely Throughput, Round Trip Delay (RTD), Packet Delivery Ratio (PDR), and Convergence Time. The network performance testing is done both for point to point (P2P) communication and multi-hop on MESH networks. Convergence Time Testing is carried out to determine the speed of the network system in repairing the path when a path failure.

2. Literature Review

2.1. Wireless Mesh Networks (WMN)

Wireless mesh network is one of wireless telecommunications network technology. The networks consist of a series of interconnected nodes where each node has two or more communication lines. The node in WMN topology can function as a mesh host or mesh router. The node as a mesh router functions to forward data transmission packets for other nodes that cannot communicate directly with the destination node. WMN has the ability to manage and configure its own network (self-configure/self-organize). Therefore, WMN can create and maintain its connectivity in the event of a trouble at another node. The ability of WMN makes the nodes in the network have a high level of robustness and reliability because they can always be connected to each other even though there are nodes that have been damaged [8, 9].

2.2. Raspberry Pi

Raspberry Pi is a single board small computer that operates on a Linux-based operating system or a Windows IoT-based operating system. The Raspberry Pi device was developed since 2006 by volunteers and technology academics in the UK who are members of the non-profit Raspberry Pi Foundation. The development of Raspberry Pi is very fast and already has several versions. Raspberry Pi 3 which can be seen in Figure 1 used in this study is the third generation that has the following specifications [10]:

- SoC: Broadcom BCM2837
- CPU: 1.2 GHz 64-bit Quad-Core ARMv8
- GPU: Broadcom VideoCore IV
- RAM: 1 GB LPDDR2 900 MHz
- Networking: 10/100 Ethernet, 2.4 GHz 802.11n Wireless LAN
- Bluetooth: Bluetooth 4.1 Classic, Bluetooth Low Energy (BLE)
- Storage: MicroSD
- GPIO: 40-pin header
- Ports: HDMI, 3.5mm analogue audio-video jack, 4x USB 2.0, Ethernet, Display Serial Interface (DSI)

2.3. Routing Protocol

Routing protocol is a protocol used to determine the communication path from the sender to the destination. There are 3 types of routing protocols on ad hoc network technology, namely proactive, reactive, and hybrid. This study will compare the performance of two proactive routing protocols, namely OLSR (Optimized Link State Routing) and BATMAN (Better Approach to Mobile Ad-hoc Network).

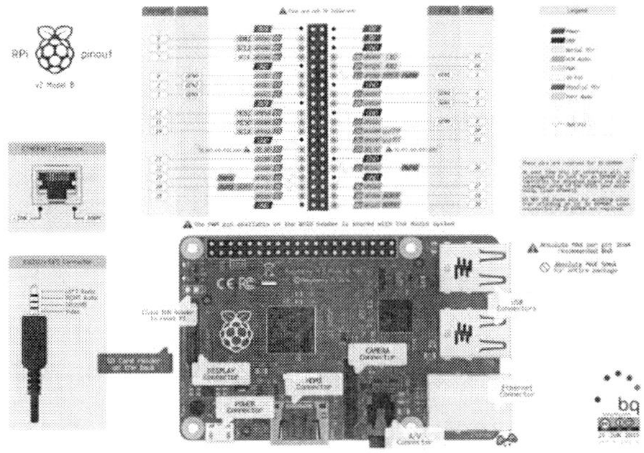

Figure 1. Pin-out diagram of raspberry pi 3

2.3.1. OLSR. The most popular link state protocol is OLSR which is standardized by the Internet Engineering Task Force (IETF) in document RFC3626. OLSR protocol is a proactive protocol that creates a route for sending data packets based on the routing table in each node. Routing tables are made based on topology information from each node at the time of the broadcasting of Topology Control (TC) packets. The TC packet is activated after each node has a list of identities of its neighboring nodes. These nodes are neighbors if and only if they can communicate directly with each other [11, 12].

OLSR selects MPR (Multi Point Relay) nodes which are used to forward messages from the source node. Only the MPR nodes can rebroadcast the messages. Therefore, not all nodes in the OLSR network can spread information sent by sending nodes. The MPR nodes are chosen so that each node in the network can reach all the nodes in the network. The MPR technique can reduce the number of TC packages.

To overcome the shortcomings of the shortest-path algorithm in the link state algorithm, OLSR has been equipped with LQ Extention. LQ Extention is the shortest path algorithm with an average packet error rate as a measurement parameter. This parameter is called ETX (Expected Transmission Count) which is defined as ETX (i) = 1 / (NI(i) × LQI(i)). NI (i) is the packet arrival rate seen by a node on the i-th link. LQI (i) is the estimated arrival rate of packets seen by a neighbor node using the i-th link. When the network has a low packet error rate, ETX is higher. LQ extension can increase packet delivery ratios due to hysteresis-based techniques.

2.3.2. BATMAN. The BATMAN routing protocol does not have a broadcasting topology message. Each node runs the following algorithm [13].

A. Sending advertisement messages regularly called OriGinator message (OGM). OGM contains the sender's IP address, IP address of the forwarding node, TTL value and sequence number (SQ).
B. Check the best neighbouring node for each destination node in the network through the ranking procedure.
C. Rebroadcasting of the OGM received through the best neighbouring nodes.

If SQ and OGM received from a particular node are in a certain area, communication is considered two-way. For example, in the time interval T, node A sends the message Tr, r is the level of the OGM message. The nodes around node A will rebroadcast the OGM node A messages and the other OGM nodes. If node A receives an OGM message from neighbouring nodes, node A checks the OGM message whether the SQ value of the most recently received message from that node is less or equal to the T_r message. If they are the same, then they have two-way communication. Otherwise, if they are not the same, then the nodes are simplex communication.

3. Research Method

Some hardware and software used in this study are presented in Table 1.

Table 1. Device list

Devices	Number	Types
Raspberry Pi 3 model B	4 pcs	Hardware
Memory micro SD 8GB	4 pcs	Hardware
USB wireless network adapter TP-Link WN722N	8 pcs	Hardware
Computer Windows	1 pc	Hardware
Raspbian 8 (Jessie) lite version	-	Software
Command Prompt	-	Software
HSMM-Pi	-	Software
Mozilla Firefox	-	Software
Hostapd	-	Software
Dnsmasq	-	Software

3.1. System Design

An ad hoc wireless network system that was built using the BATMAN or OLSR routing protocols. Each system is composed of four wireless nodes that are interconnected in an ad hoc network. System performance testing is done using 2 topologies in three scenarios. The first topology, is a MESH topology which uses 4 nodes arranged in the range of each node to be connected directly (point to point). The topology can be seen in Figure 2. The second topology, the four test nodes are arranged so that multi-hop communication occurs. The second topology is illustrated in Figure 3. The test is carried out using three scenarios below:

A. The first scenario using the first topology is intended to determine the throughput for each routing protocol in point to point communication.
B. The second scenario uses the second topology, building multi-hop communication in order to determine the throughput of the routing protocol.
C. The third scenario uses the second topology to determine the self-healing ability of each protocol.

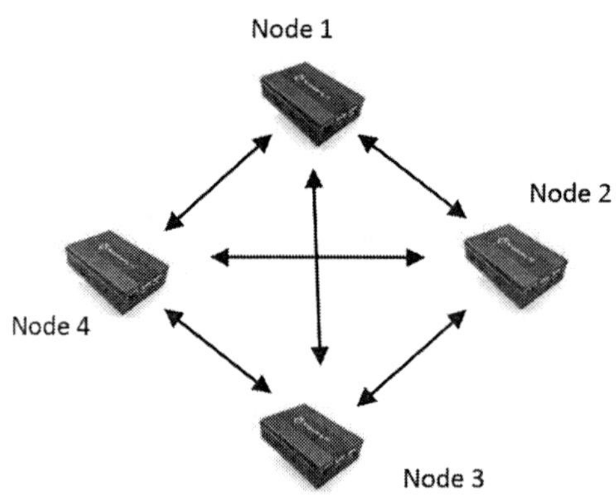

Figure 2. The MESH network topology

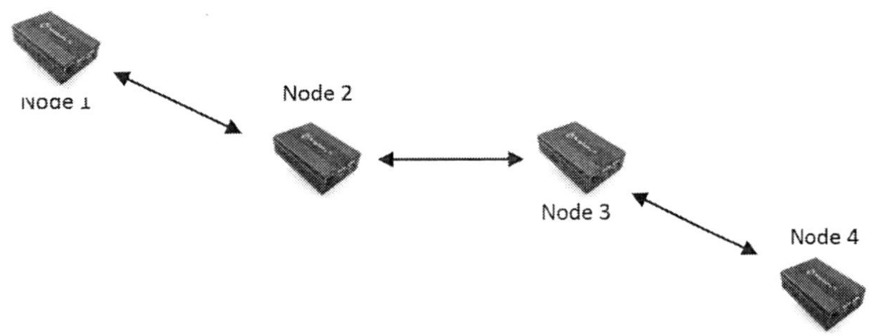

Figure 3. The Multi-hop network topology

3.2. Node Configuration

The configuration process is done by headless process, namely with the SSH command at the command prompt or through the PuTTY application. Setting up the nodes in the BATMAN routing protocol is done starting from setting up the wireless node to testing the delivery of data packets. Following are the settings for node 1 configuration with the wlan0 interface which acts as an ad hoc interface and wlan1 as the hostapd (access point) interface.

Wireless node settings and access point services

```
sudo nano /etc/network/interfaces
  auto wlan0
  iface inet wlan0 static
          address 172.168.137.21
          netmask 255.255.255.0
  auto wlan1
  iface inet wlan1 static
          address 152.168.1.1
        netmask 255.255.255.0
```

Hostapd configuration for access point:

```
sudo apt-get update
sudo apt-get install hostapd
sudo apt-get purge dns-root-data
sudo nano /etc/hostapd/hostapd.conf
interface=wlan1
driver=nl80211
ssid=BATMAN HOST 1
hw_mode=g
channel=1
wmm_enabled=0
macaddr_acl=0
auth_algs=1
ignore_broadcast_ssid=0
wpa=2
wpa_passphrase=12345678
wpa_key_mgmt=WPA-PSK
wpa_pairwise=TKIP
rsn_pairwise=CCMP
```

DHCP configuration in hostapd

```
sudo apt-get install dnsmasq
sudo nano /etc/dnsmasq.conf
interface=wlan1
bogus-priv
domain-needed
dhcp-range=152.168.1.10,152.168.1.50,255.255.255.0,1h
sudo nano /etc/sysctl.conf
```

Ad hoc Networks configuration for wireless node

```
sudo apt-get install batctl
sudo nano /root/mesh.sh
#!/bin/bash
sudo modprobe batman-adv
sudo ip link set wlan0 down
sudo ifconfig wlan0 mtu 1500
sudo iwconfig wlan0 ap any
sudo iwconfig wlan0 mode ad-hoc
sudo iwconfig wlan0 essid BATMAN
sudo iwconfig wlan0 channel 8
sleep 1s
sudo ip link set wlan0 up
sleep 1s
sudo batctl if add wlan0
sleep 1s
sudo ifconfig bat0 up
sleep 1s
sudo ifconfig bat0 132.168.168.1
sudo route add default gw 132.168.168.1
sudo route add -net 152.168.2.0 netmask 255.255.255.0 gw 132.168.168.2
sudo route add -net 152.168.3.0 netmask 255.255.255.0 gw 132.168.168.3
```

Configuration for mesh.sh

```
sudo chmod 700 /root/mesh.s/h
sudo crontab -e
@reboot /root/mesh.sh
```

All nodes perform the configuration steps with different IP address and SSID access point on each node.

3.3. System Testing

The performance of each routing protocol was evaluated using the Raspberry Pi 3 model B which is composed of four terminals. Each terminal uses the IEEE 802.11bgn protocol on the 2.4 GHz spectrum. Node 1 is specified as the origin terminal in all scenarios. The first and second scenarios perform bandwidth measurements using iperf which generates traffic in the form of TCP packets in a 10 second period. While iperf is running, the origin node also sends ICMP packets to the destination node using the fping command. This provides averaged round trip delay (RTD) and a packet delivery ratio (PDR).

The third scenario is done by turning off the node that is being passed by data from sender to receiver. Then observe the time needed (convergence time) to find a new path. The fping command is performed to send ICMP packets from the origin node (node 1) to the destination node (node 4) via one of the paths whether node 2 or node 3. Convergence time is calculated from the time from the loss of the main path until a new path appears as a maintenance for the damaged path . The convergence time value is obtained by calculating the difference in the ICMP sequence number between the last packet received through the old path and the first packet received through the new path then divided by the fping rate.

4. Result and Discussion

4.1. Throughput testing

Figure 4 shows the value of throughput for each protocol in each scenario. It appears that OLSR provides a higher average throughput when compared to BATMAN. The throughput has decreased significantly as the increasing the number of hops passed.

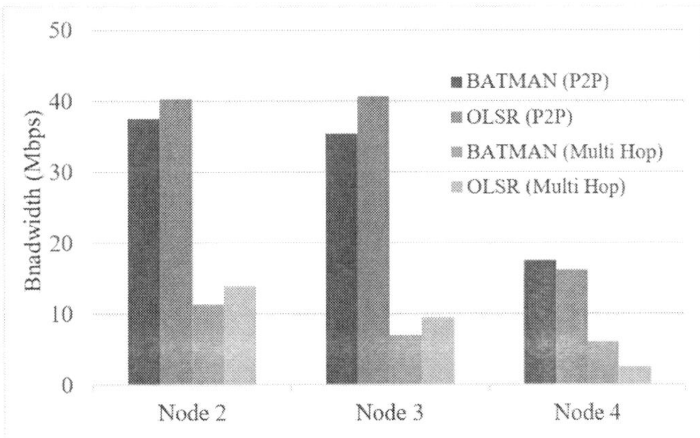

Figure 4. Bandwidth

4.2. Round Trip Delay (RTD) and Packet Delivery Ratio (PDR) Testing

The observation of round trip delay of the two scenarios for both the protocols can be seen in Figure 5. BATMAN compared to OLSR shows a better average time for both scenarios. Figure 6 shows the PDR of the two protocols does not show any packet drop in mesh network communication. The failure of sending data packets occurs in multi-hop communication. OLSR and BATMAN routing protocols experienced packet drops of 2% and 5%, respectively.

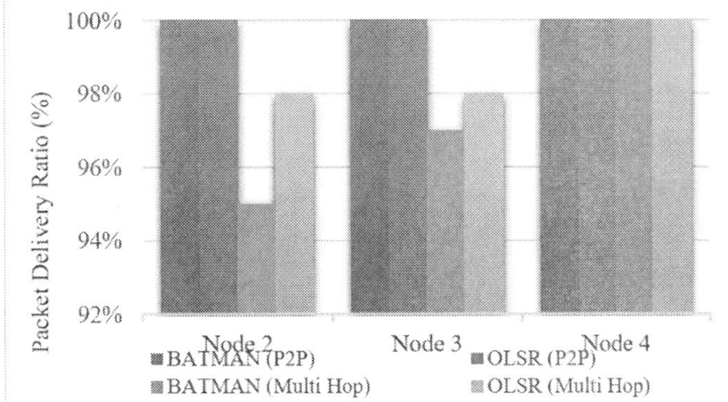

Figure 5. Packet delivery ratio (PDR)

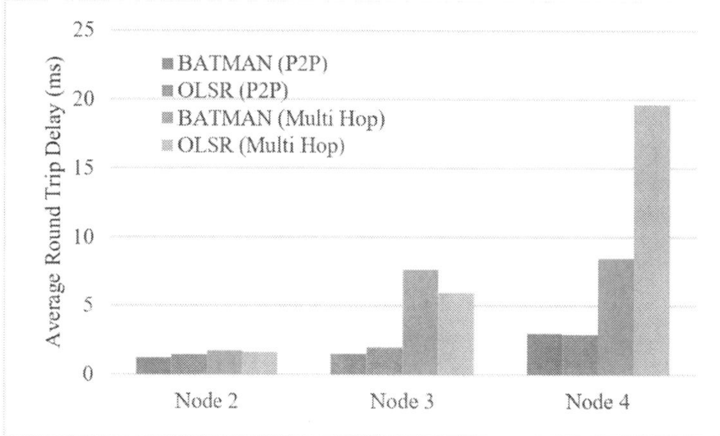

Figure 6. Round trip delay (RTD)

4.3. Convergence Time Testing Results

The results of convergence time testing are shown in Table 2. The BATMAN protocol shows a faster convergence time when compared to OLSR. The average convergence time on the BATMAN protocol is 12.75 seconds while for the OLSR protocol is 18.5 seconds. This shows that the BATMAN routing protocol is faster in finding new pathways when connection link failure occurs.

Table 2. Convergence time

Protocol	Convergence Time (second)		
	Minimum	Maximum	Average
OLSR	7	17	12.75
BATMAN	13	32	18.5

5. Conclusion

The OLSR routing protocol provides a better average value of throughput compared to the BATMAN protocol. However, the BATMAN routing protocol shows better RTD compared to the OLSR routing protocol. A 100% PDR for the two protocols is shown for MESH network communication. They indicate that all data sent arrives at the destination. The failure to send data packets occurs in multi-hop communication, where the PDR for the OLSR protocol is 98% while for the BATMAN protocol is 95%

Acknowledgement

We would like to acknowledge to Faculty of Engineering Hasanuddin Universitas for facilities and financial support for this research under grant of Laboratory-based Education (LBE) 2019.

References

[1] Naeem R, Aftab MU, Qasim MA, Omair A and Irfan M 2016 *Communications and Network* vol.8 (3) pp. 131-6.
[2] Jhum S, Binod KP, Bibudhendu P 2017 *Proc. Int. Conf. on Inventive Communication and Computational Technologies (ICICCT)* Coimbatore, India pp. 506-9.
[3] Xuan L, Zhuo L, Peng Y and Yongqiang D 2017 *Ad Hoc Networks* vol 58, April 2017, pp. 255-68.
[4] Rajeshwar S, Tarun S and Aditi K 2016 *International Journal of Computer Applications* vol 133(1) pp. 33-8.

[5] Dimitris NK 2017 *Journal of Multimedia Data Engineering and Management (IJMDEM)* vol. 8(1) pp. 42-71.
[6] Rath M, Pattanayak BK 2017 *Proc. Int. Conf. on Information and Communication Technology for Intelligent Systems (ICTIS)* pp. 579-86.
[7] Roy A and Deb T 2018 *Proc. Int. Conf. on Computing and Communication Systems* Singapore vol 24 pp. 33-48.
[8] Axel N, Ester L and Leandro N 2015 *Computer Networks* vol 93(2) pp. 308-23.
[9] Andrés VR, Luis J and Cruz L 2015 *Ad Hoc Networks* vol 24(B) pp. 34-54.
[10] Raspberry Pi 3 Specifications [Internet] [cited 20th May 2019]. Available from://www.raspberrypi.org/magpi/raspberry-pi-3-specsbenchmark.
[11] Khalil A, Fahs W, Chbib R and Raad A 2018 *Proc. Int. Arab Conference on Information Technology (ACIT)* Werdanye Lebanon pp. 1-6.
[12] Wardi, Hirata K, Higami Y and Kobayashi S 2011 *Proc. Int. Conf. on Multimedia Technology*, Hangzhou pp. 3214-17.
[13] Sliwa B, Falten S and Wietfeld C 2019 *Proc. Int. Conf on IEEE 89th Vehicular Technology (VTC2019-Spring)* Kuala Lumpur Malaysia pp. 1-7.

Design and Implementation of REST API for Academic Information System

A A Prayogi[1]*, M Niswar[1], Indrabayu[1], and M Rijal[1]

[1]Department of Informatics Engineering, Engineering Faculty, Hasanuddin University, Makassar, Indonesia

*Email: aisprayogi@unhas.ac.id

Abstract. With the increased number of information systems used in an organization, there is also an increased importance of data exchange between these systems. This research deals with prototype development and performance analysis of Rest API for academic information system. Rest API was developed using two different server technologies, NodeJS and PHP. The prototype was developed on top of a database server using one sample table that represents employee in a higher education institution. For each of the Rest API developed, there were 2 types of endpoint created. Experiment was set with one database containing a single table and utilized Apache Jmeter to simulate up to 1000 concurrent requests. The results of the experiment show NodeJS implementation of REST API consistently has better performance compared to PHP based REST API implementation. NodeJS implementation reached 100% throughput for up to 1000 concurrent requests, while PHP reached 48.70% throughput when serving the same number of concurrent requests.

1. Introduction

The use of information system especially web-based information system in the academic field especially in a higher education institution has become the norm. Information system holds important role in the various business processes in a higher education institution, ranging from new student registration process, course registration at the beginning of an academic period, end of academic period evaluation process to graduation and alumnae information management process [1].

To accommodate and ensure the flow of these business process runs smoothly, there are also multiple and diverse information systems set up. Some more common information systems used by a higher education institution including Learning Management System (LMS), Content Management System (CMS), Asset Management System, Student Registration System and Course Management System. In most institution, these different type of information systems are developed either in-house (by information-technology work unit), outsourced to a third-party vendor or customized from open-source community-based project. The maintenance of an application based on proprietary solution comes from the vendor itself. The process to develop new feature or to fix certain bugs need a lengthy administrative process that lags behind the user time schedule. Customized open-sourced based solution also faces the same problem, in which each bugs or new features must be registered first before it can be processed. These circumstances then lead to new features or bug fixing fall short of user expectation and needs [2].

Within the institution there is a need to have a common and uniform data communication between each information systems. The purpose of this uniform data communication is to prevent data duplication and data mismatch stored within each information systems. With the uniform data

IOP Conf. Series: Materials Science and Engineering **875** (2020) 012047 doi:10.1088/1757-899X/875/1/012047

communication format, new applications can be built easier and faster by using data stored in the previously established systems. The number of smartphone usage among internet users has been increased steadily every year. Smartphone users in Indonesia has reached 83.5 million users in 2018. Although most information system is easily accessed from mobile web browser, using mobile native apps has certain advantages over mobile website [3]. Mobile native apps need to communicate with the existing information system server via Application Programming Interface (API).

There are many types of architecture can be chosen to build data communication between information systems or applications. One of the options is using REST (Representational State Transfer Protocol) API architecture. With REST architecture, web service can be built using simple methods and underlying HTTP architecture. A prototype of REST API architecture for academic information system built with two types of framework technologies. The prototypes then analyzed for performances with three parameters (response time, throughput and packet loss) under multiple concurrent connections.

2. REST API

REST (Representational state transfer) is an architectural API (Application Programming Interface) which provides client-server communications for Web Applications over HTTP protocol, making it easily implemented since it is not bound to any transfer protocol. The three main design principles of REST are addressability, uniform interface, and statelessness [4]. REST addresses acceptability by defining endpoints in a directory structure [5] via different URI for extracting the data. The API works on the principle of CRUD (Create, Read, Update, Delete), which correspond to the most popular functions [4] INSERT, SELECT, UPDATE, and DELETE, in persistent data-storages such as SQL.

Calling a RESTful service over the HTTP protocol can be done in multiple programming languages. In jQuery, a framework of Javascript, a REST server can be called from an Ajax query. A common response of data is in the form of JSON (Javascript Object Notation) data. Typical JSON response can be seen in Figure 1. Data can be returned as well in the form of XML (eXtensible Markup Language) [4].

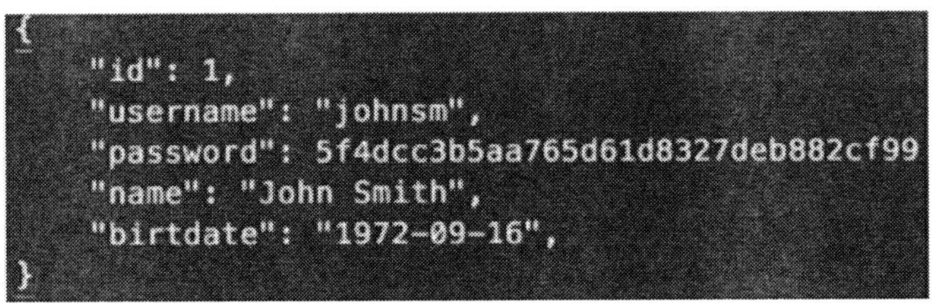

Figure 1. Examples of JSON response

3. Server Side Technologies

Developing web services with REST API architecture currently can be done with various server technologies. It is decided two server technologies most used, which are NodeJS and PHP.

3.1. NodeJS

Node JS is a web server technology based on JavaScript. It used V8 Javascript Engine with integrated module add-ons. NodeJS used event-driven architecture to handle request from clients. With the event-driven architecture, all requests from clients will be put to event queue. NodeJS single-threaded event loop then pick one of the requests to check if the request require blocking I/O operation or complex computation tasks. If it is, the request will be moved to different thread which taken from NodeJS Internal Pool Thread [6]. After thread finished processing the request either blocking I/O operation or complex computation tasks, it will send the response to back to event queue to be processed by the NodeJS single-threaded event loop later. This process is shown Figure 2.

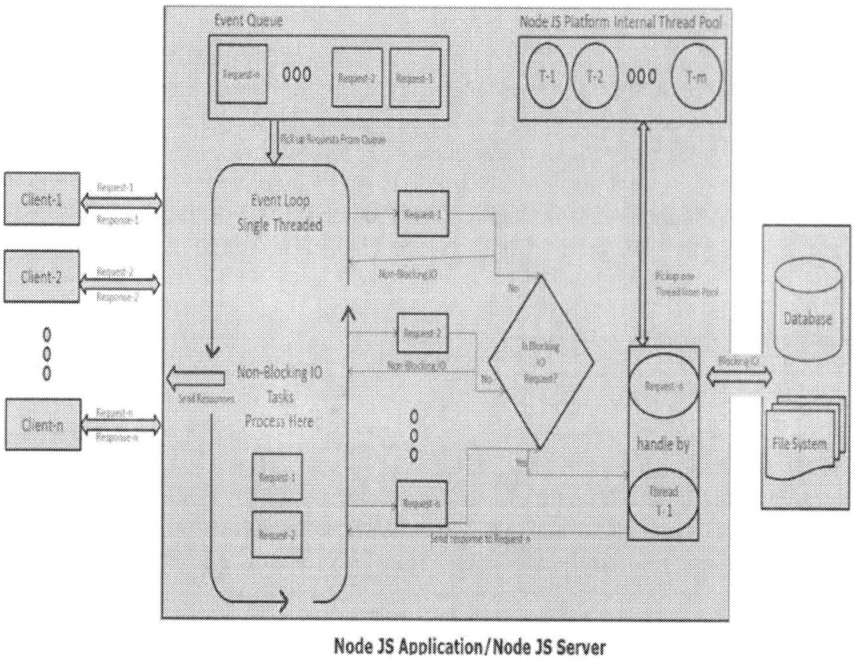

Figure 2. NodeJS architecture [6]

3.2. PHP

PHP also know as PHP:Hypertext Preprocessor is a script language developed to add dynamism to the static HTML page. It was first developed in 1994 by Rasmus Lerdorf. PHP script is processed by a certain process called PHP interpreter in a web server to generate HTML code, or binary image data. PHP evolved from procedural paradigm programming to object-oriented type programming in its latest iteration (PHP 7).

4. Experiment

The experiment was designed to show how the two server technologies used to implement REST API performed under multiple concurrent connections. There are 2 types of performances measurement used in this experiment including request time and throughput. Request time measures elapsed time between the time client send the request until the client received responses from server. Throughput is total number of processed requests in certain amount of fixed time. Packet loss is measured as percentage of packet lost with respects to packets sent.

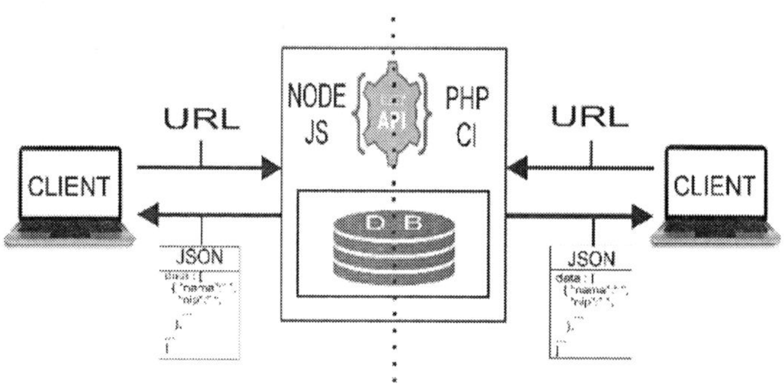

Figure 3. Prototype architecture

IOP Conf. Series: Materials Science and Engineering **875** (2020) 012047 doi:10.1088/1757-899X/875/1/012047

The prototype built as shown in figure 3. Clients send requests by accessing certain URLs provided by the server. Server was setup with two server technologies, NodeJS and PHP with Apache as web server. Every time request received either by NodeJS or PHP, the request processed based on the parameter and embedded data by the respected web server technology then return response in JSON (Javascript Object Notation) format.

A database was setup within the server to serve as data repository for both NodeJS and PHP REST API implementation. The database contains one table, called *profilpegawai* table. The table contains 5 fields and 3920 rows of data, copied from the current employee information management system implanted in Hasanuddin University. The detail structure of the table can be seen in Figure 4.

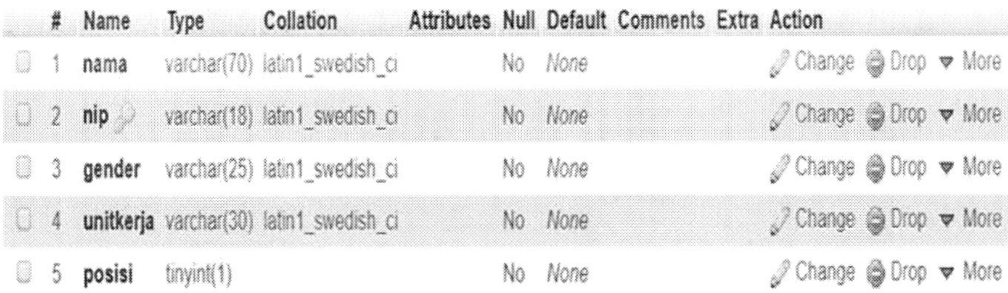

#	Name	Type	Collation	Attributes	Null	Default	Comments	Extra	Action
1	nama	varchar(70)	latin1_swedish_ci		No	None			Change Drop More
2	nip	varchar(18)	latin1_swedish_ci		No	None			Change Drop More
3	gender	varchar(25)	latin1_swedish_ci		No	None			Change Drop More
4	unitkerja	varchar(30)	latin1_swedish_ci		No	None			Change Drop More
5	posisi	tinyint(1)			No	None			Change Drop More

Figure 4. Table *profilpegawai* structure details

For this experiment, two types of REST API endpoints was created as shown in table 1. The first endpoint is used when a client wants to request all of rows, in this case employee data, from *profilpegawai* table. The second endpoint is used for a client to request certain employee data which matched the parameter NIP. Both endpoint responses are returned in JSON format.

Table 1. REST API endpoints

Parameter	URLs	
	NodeJS	PHP
GET ALL	http://localhost:3000/profilpegawai/	http://localhost:8080/rest-api/rest-ci/api/profilpegawai
GET PARAMS	http://localhost:3000/profilpegawai/:nip	http://localhost:8080/rest-api/rest-ci/api/profilpegawai?nip=NIP

After setting up database and API endpoints, next step was to prepare simulation for testing scenario. Apache JMeter was used to simulate requests from clients. To measure response time, for each NodeJS and PHP implementation, 10 requests was made to each endpoint. Each request repeated in 30 iterations and then the average of response time was calculated. Throughput was measured by simulating requests from 1, 100, 200, 300, 400, 500, 600, 700, 800, 900 and 1000 requests for each endpoint in each NodeJS and PHP implementation.

5. Result and Discussion

Response time for NodeJS REST API implementation shows significantly better result than those of PHP implementation. When testing GET ALL endpoint, average response time for NodeJS was ranging from 138ms to 156ms, while PHP implementation ranging from 244ms to 266ms. Complete results of response time measurement for GET ALL endpoint are shown in table 2. Similar result was also shown for the GET PARAMETER endpoint, in which NodeJS shows better results than its PHP counterpart.

Table 2. Response time results for GET ALL endpoint

Request	PHP			NODEJS		
	max (ms)	min (ms)	average (ms)	max (ms)	min (ms)	average (ms)
1	894	207	266	188	113	138
2	446	206	253	371	107	146
3	369	208	244	503	113	156
4	795	211	257	212	116	141
5	410	207	248	159	110	136

During throughput measurement, NodeJS implementation perform better compared to PHP implementation. When using GET ALL endpoint, NodeJS implementation consistently able to handle all requests from 1 to 1000 requests concurrently. PHP implementation on the other hand shows decreasing in throughput performance when the number of concurrent requests reached 300 requests as shown in table 3. When handling 1000 requests, PHP implementation only reached 48.70% throughput. NodeJS also show better performance in throughput value when testing with GET PARAMETER endpoint.

Table 3. Throughput measurement results for GET ALL endpoint

# of concurrent requests	THROUGHPUT	
	PHP	NodeJS
1	100%	100%
100	100%	100%
200	100%	100%
300	90%	100%
400	93,25%	100%
500	90,20%	100%
600	85%	100%
700	71,43%	100%
800	57,63%	100%
900	50,11%	100%
1000	48,70%	100%

The experiment result shown NodeJS implementation have better performance when asked to handle multiple concurrent requests. This type of situation was common real-world use case, in which a web service should handle multiple request in almost the same time. NodeJS architecture with non-blocking I/O single thread was crucial in handling simultaneous request while making sure all the requests is served in the least amount of time.

6. Conclusion

In our prototype of REST API implementation for academic information system with two server technologies, we explore the performance with two parameters, response time and throughput. Our prototype with NodeJS and PHP with two endpoints. NodeJS implementation tend to show better performance for both measurements in both endpoints. Further experiments need to be conducted to find out how the NodeJS and PHP implementation of REST API perform under more complex database setup and with more complex endpoints including I/O extensive operations.

Acknowledgement
This work has been supported by Faculty of Engineering, Universitas Hasanuddin under 2019 Labo-Based Education (LBE) grant.

References

[1] Aswati S, Mulyani N, Siagian Y and Zikra Syah A 2015 Peranan Sistem Informasi dalam Perguruan Tinggi *Jurnal Teknologi dan Sistem Informasi* **1** 79-86.
[2] Barata, R., Silva, S., Martinho, D., Cruz, L., & Guerra e Silva, L 2014 Open APIs in Information Systems for Higher Education. *Umea*.
[3] Turner-McGrievy GM, Hales SB, Schoffman DE 2017 et al Choosing between responsive-design websites versus mobile apps for your mobile behavioral intervention: presenting four case studies. *Transl Behav Med.* **7(2)** 224–232.
[4] Belqasmi, F., Singh, J., Melhem, S.Y.B., Glitho, R.H., 2012 SOAP-Based vs. RESTful Web Services: A Case Study for Multimedia Conferencing *IEEE Internet Comput.* **16** 54–63.
[5] Choi M. 2012 A Performance Analysis of RESTful Open API Information System *Future Generation Information Technology, Lecture Notes in Computer Science* Springer Berlin Heidelberg 59–64.
[6] Posa R. 2015 Node JS Architecture – Single Threaded Event Loop Retrieved from https://www.journaldev.com/7462/node-js-architecture-single-threaded-event-loop.

IOP Conf. Series: Materials Science and Engineering **875** (2020) 012048 doi:10.1088/1757-899X/875/1/012048

Stability Analysis and Fault Changes on Wind Turbine Effect in Multi Machine Power System

B A Ashad[1]*, A Siswanto[2], I C Gunadin[1], and Yusran[1]

[1]Electrical Engineering, Engineering Faculty, Hasanuddin University, Makassar, Indonesia
[2] Electrical Engineering Study Program, University of 17 Agustus 1945, Cirebon, Indonesia

*Email: bayuadrianashad@gmail.com

Abstract. This study presents stability analysis using a new approach to model the disturbances that occur in multi-machine systems connected with wind turbines. Consideration of stability and operability of conventional electric power systems planning and developing facing problems with the integration of installations in wind power plants (WPP) is needed because the current system is connected with renewable energy. This study highlighting the effects of WPP connected to a multi-engine 9 bus system from the effect of changes in disturbances occurring on several buses when operating. Disruption in multi-machine systems connected with wind turbines can affect system stability. The noise changes from bus to bus in the voltage drop effect on the system bus. The effect of wind turbine when a fault has a small moment of inertia can make the system unstable on the bus. The results of simulations carried out using ETAP to observe system stability have been carried out. The simulation results show that a system disruption that occurs in a multi engine system connected to the wind turbine does not result in a decrease in the bus near the fault then the stability decreases.

1. Introduction

Transmission models experience interference from various types from a balanced or unbalanced short circuit connected to generators and other equipment [1]. Disruption in multi machine systems is a problem of stability which leads to fluctuations in frequency stability and voltage on the power system connected to the wind turbine impact on multi machine systems [2]. However, when wind turbines emit power with insufficient capacity, when a fault occurs it can cause fluctuations in voltage and large frequencies. Many studies that discuss the interference with wind turbine connected systems are presented on the line [3, 4]. However, the problem is faced. Wind turbine technology that moves by wind speed is very dependent on natural conditions. In the supply and demand power system is demanded to always be balanced, because this affects the stability of the system. The wind turbine efficiency and DG impact on the load that has been studied by [5-7], and on [8] DG effect and placement are very influential on the system so it needs to be analyzed more deeply. In the power system there are two conditions, namely normal and abnormal. Normal conditions according to conditions within the prescribed limits. An abnormal state is when experiencing a change in load or interference. Disturbances in transmission can result in a decrease in voltage and frequency and oscillation in transient stability [9] In the study analyzing the disturbances that occur in several buses from 9 interconnected system buses in the wind turbine connected to a multi machine system. There is a disturbance point on some buses which results in changes in the condition of the bus system.

IOP Conf. Series: Materials Science and Engineering **875** (2020) 012048 doi:10.1088/1757-899X/875/1/012048

2. Power System Model

The power system model used in this study is shown in Figure 1, interconnected on 9 buses. the system uses 3 generators each with a capacity of 8.5 MW, a voltage of 6.6 kV with a capacity of 10 MVA, cosphi 85%, a primary voltage step-up transformer 6.6 kV secondary voltage 20 kV 5 MVA capacity has been winded by wind turbines on the bus the type of aluminium is divided and distributed. The length of channel 1-2 is 1 km, while the conveyor 3-6 measures 0.5 km. There are 4 loads on buses 5.6 and 8 large capacity of 7 MVA. The wind turbine has been towed by a generator in the system. Wind turbine with a capacity of 20 MW and a grid frequency is 50 Hz. Two thermal power plants (SG1 and SG2), conventional power plants are hydropower (SG3), and one Turbine wind is on bus 5.

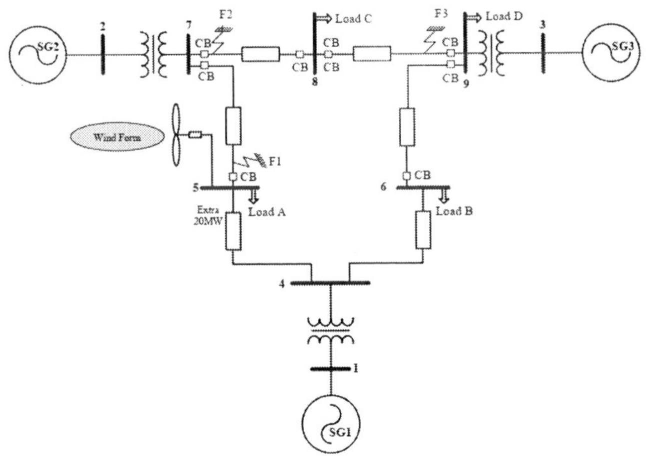

Figure 1. 9-bus test system with a wind farm [10]

3. Wind Turbine and Characteristic Model

The representation of the wind turbine model is shown in Figure 2 from ref [10]. On fig 3, shows from the MPPT Output without measuring wind speed, where it is noted that the maximum power captured can be expressed as a rotation speed cube (Pmppt= ω r3) [11] Characteristic model of wind turbine and MPPT of difference wind speed.

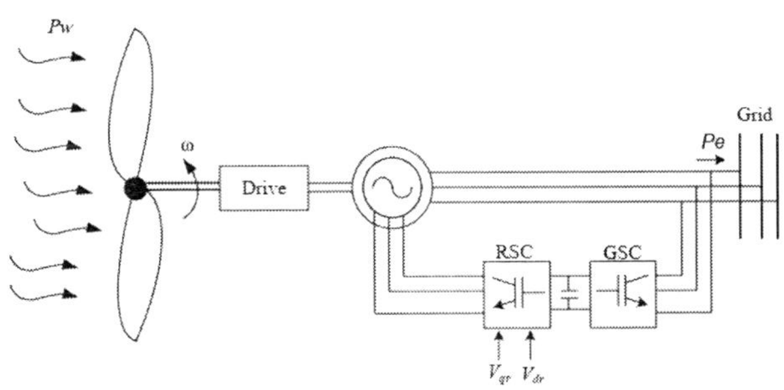

Figure 2. Wind turbine

Wind turbine output is obtained from wind speed. Significantly, the variation of wind turbine speed from the situation and conditions according to natural conditions (local area). Fluctuations in varying wind speeds affect the output of wind turbines.

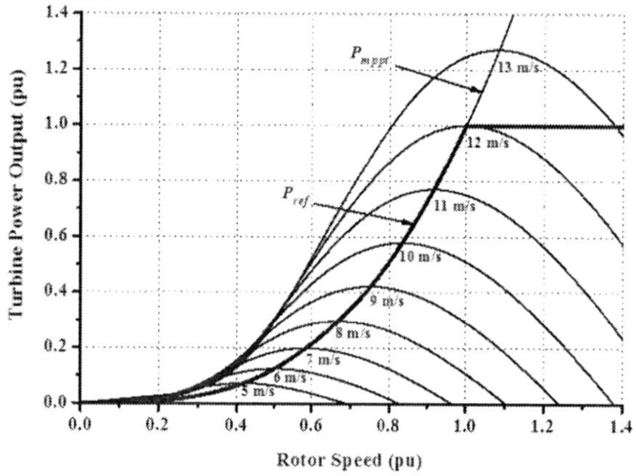

Figure 3. Characteristics of wind turbine and MPPT Curve

4. Simulation Results

In this section, shows the results of the simulation system of 9 buses connected with wind turbine and disturbances occur on buses 3, 4, and 7. Comparative analysis on the system is carried out in two scenarios. The first scenario of interference is done before the wind turbine is pulled. The scenario of both wind turbines is lost. Then an analysis of the stability and stress is carried out on each bus. In this system the voltage is 6.6 kV, the base capacity is 10MVA. 85% Cosphy. The basic capacity of bus bars 1,2 and 3 has a voltage of 6.6 kV. At 4–9 bus bar has a voltage of 20 kV. The primary voltage step up transformer is 6.6 kV. Secondary voltage is 20 kV, with a capacity of 10 MVA. Delivery of long conveyor 1, 2 aluminum is 1 km. while deliveries 3-6 have a length of 0.5 km. there are 4 loads divided on buses 5.6, 8 and 9, the size of each bus is 7 MVA. 20 MW wind turbine capacity. This is done to determine the stability and change in disturbed voltage in the system connected with wind turbine. This analysis uses ETAP, using the data in Table 1. The simulation has been performed on the personal computer Intel (R) core (TM) i5-3337U CPU @ 1.80 GHz installed memory (RAM) 4.00 GB. System type 64-bit.

4.1. Scenario #1. Before the Wind Turbine Enters

In Figure 4 shows the system before entering the wind turbine in scenario # 1 by presenting a 9 bus system, three generators on buses 1, 2 and 3, loads on buses 5, 6, 8 and 9. Short circuit on bus 3 is a bus generator so that changes in load flow have an impact on generators 1 and 2. Voltage surges in the bus type in Table 1.

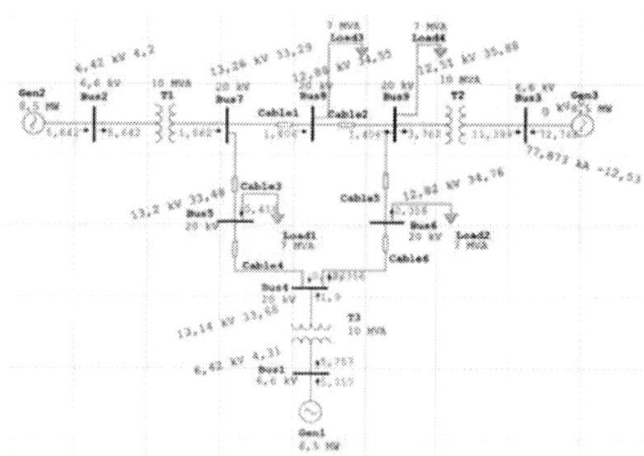

Figure 4. Fault in bus 3

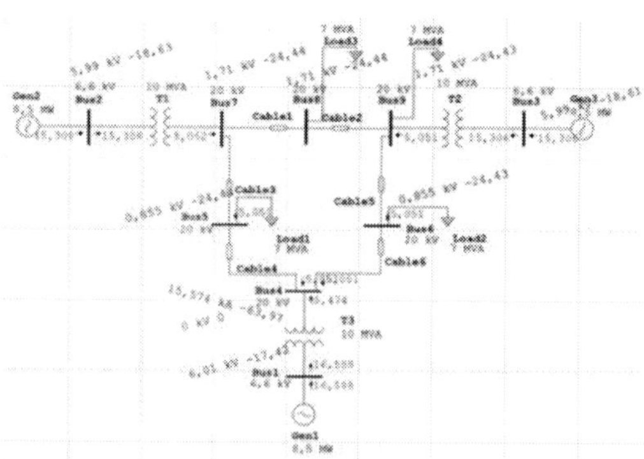

Figure 5. Fault in bus 4

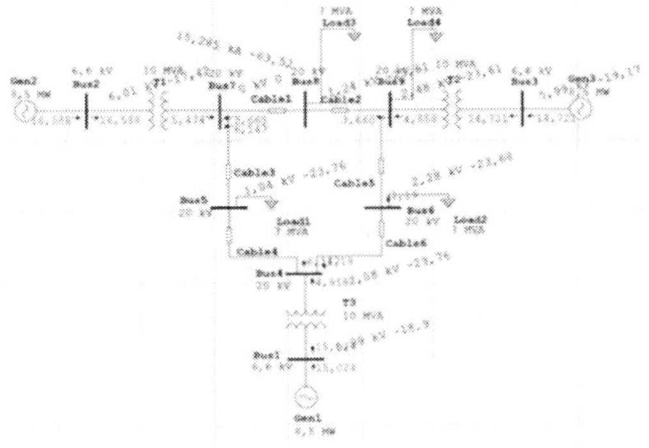

Figure 6. Fault in bus 7

Table 1. Table voltage on each bus when the fault is before the wind goes in

No. Bus	Bus voltage	kV, bus fault 3	kV, bus fault 4	kV, bus fault 7
1	6.6 kV	6.42 kV	6.01 kV	5.99 kV
2	6.6 kV	6.42 kV	5,99 kV	6.01 kV
3	6.6 kV	0 KA	5.99 kV	5.99 kV
4	20 kV	13.44 kV	0 kV	2.08 kV
5	20 kV	13.2 kV	0,855 kV	1.04 kV
6	20 kV	12.82 kV	0.855 kV	2.28 kV
7	20 kV	13.26 kV	1,71 kV	0 kV
8	20 kV	12.88 kV	1.71 kV	1.24 kV
9	20 kV	12.51 kV	1,71 kV	2.48 kV

4.2. Scenario #2. Wind Turbine Enters

The accuracy of the system model in this section has been validated using the system before the wind turbine enters the system. in the following figure shows the system model considered in this analysis. A wind turbine with a data capacity of 30 MW has been connected to the grid system in transformers and transmission tracks.

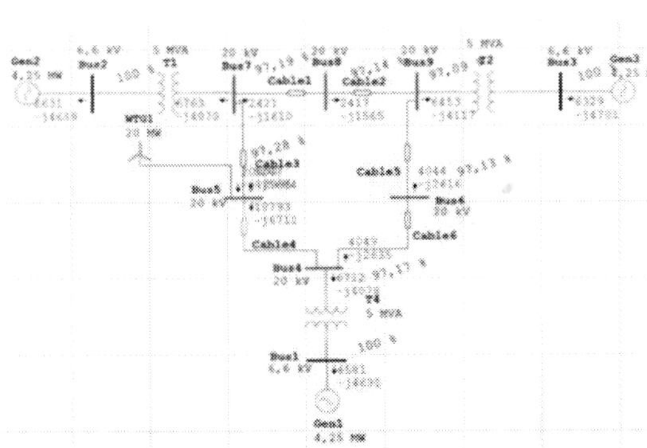

Figure 7. Fault in bus 3

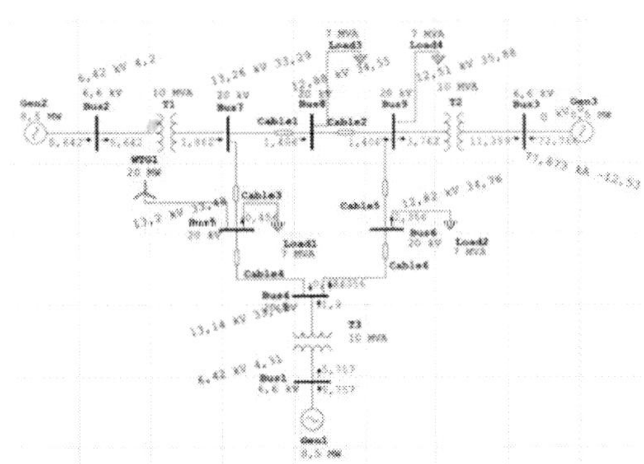

Figure 8. Fault in bus 3

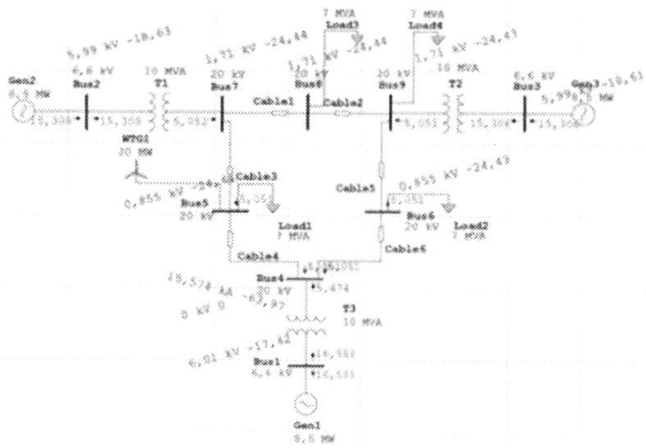

Figure 9. Fault in bus 4 with wind

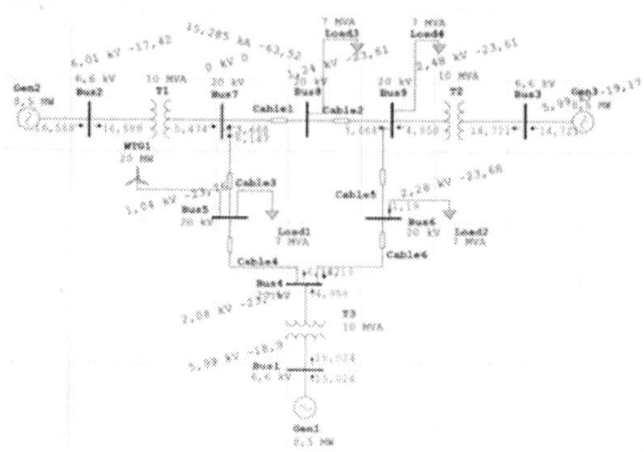

Figure 10. HS bus 7 with wind turbine

Table 2. Results The voltage table on each bus is faulted after wind is entered

No. Bus	Bus voltage	kV, bus fault 3	kV, bus fault 4	kV, bus fault 7
1	6.6 kV	6.42 kV	6.01 kV	5.99 kV
2	6.6 kV	6.42 kV	5,99 kV	6.01 kV
3	6.6 kV	0 KA	5.99 kV	5.99 kV
4	20 kV	13.14 kV	0 kV	2.08 kV
5	20 kV	13.2 kV	0,855 kV	1.04 kV
6	20 kV	12.82 kV	0.855 kV	2.28 kV
7	20 kV	13.26 kV	1,71 kV	0 kV
8	20 kV	12.88 kV	1.71 kV	1.24 kV
9	20 kV	12.51 kV	1,71 kV	2.48 kV

5. Conclusion

Effect of the wind turbine applied to the 9 bus system has been used measured using SVSI. The stability analysis is carried out on the simulation which has been carried out with the scheme occurring interference from several buses. System interruptions do not interfere with the stability of the interconnected system. the simulation results show that the response when the connection occurs occurs on bus 3 with a short circuit voltage of 5.99 kV, and bus 8 1.24 kV.

References

[1] Abhyankar, S. and Tushar. *Modeling of transmission line faults for transient stability analysis.* in *2017 North American Power Symposium (NAPS)*. 2017.

[2] Siswanto, A., et al., *Stability improvement of wind turbine penetrated using power system stabilizer (PSS) on South Sulawesi transmission system.* AIP Conference Proceedings, 2018. 1941(1): p. 020036.

[3] Tajdinian, M., A.R. Seifi, and M. Allahbakhshi, *Transient Stability of Power Grids Comprising Wind Turbines: New Formulation, Implementation, and Application in Real-Time Assessment.* IEEE Systems Journal, 2018: p. 1-12.

[4] Wang, L., et al., *Dual-Enhanced Sparse Decomposition for Wind Turbine Gearbox Fault Diagnosis.* IEEE Transactions on Instrumentation and Measurement, 2018: p. 1-12.

[5] Sainz, L., et al., *Effect of wind turbine converter control on wind power plant harmonic response and resonances.* IET Electric Power Applications, 2017. 11(2): p. 157-168.

[6] Yuli Asmi Rahman, A.S., and irwan Mahmudi, *Stability Issues in Presence Variable Distributed Generation Into Radial Distribution Network* International Conference on Industrial Electrical and Electronics (ICIEE) 2018.
[7] Yuli Asmi Rahman, e.a., *Distributed Generation's Integration Planning Involving Growht Load Model By Means of Genetic Algorithm.* Archivesof electrical Engineering 2018. 67 I(3): p. 667-682.
[8] Rahman, Y.A., et al., *Evaluating the effect placement capacitor and distributed photovoltaic generation for power system losses minimization in radial distribution system.* AIP Conference Proceedings, 2018. 1941(1): p. 020027.
[9] Gunadin, I.C., Z. Muslimin, and A. Siswanto. *Transient stability improvement using allocation power generation methods based on moment inertia.* in *2017 International Conference on Electrical Engineering and Informatics (ICELTICs)*. 2017.
[10] Bevrani, H., *Robust Power System Frequency Control.* Springer Cham Heidelberg New York Dordrecht London, 2014. Second Edition.
[11] Rosyadi, M., et al. *Fuzzy-PI controller design for PM wind generator to improve Fault Ride Through of wind farm.* in *2012 International Conference on Renewable Energy Research and Applications (ICRERA)*. 2012.

IOP Conf. Series: Materials Science and Engineering **875** (2020) 012049 doi:10.1088/1757-899X/875/1/012049

Determination of Geochemical Correlation of Cu, Pb, Zn in Stream Sediment Samples in Barru Area, South Sulawesi

A Tonggiroh[1]*, Kaharuddin[1], U R Irfan[1], and A Akbar[1]

[1]Department of Geology Engineering, Engineering Faculty, Hasanuddin University, Makassar, Indonesia

Email: *atonggiroh@gmail.com

Abstract. The Barru area is known as the rock basement location in Sulawesi Island. The disclosure of ultramafic rock, metamorphic rock, and various sedimentary rock formations in the field and outcrops of unmapping intrusive rocks on a regional scale are certainly very interesting to be used as a media experiment in statistical methods in exploration of mineral resources. This is quite reasonable because among these rock formations are revealed intrusion of dacite and diorite rocks which are altered and deposited as sulfides. The research gradually combines the pearson hybrid correlation geochemical stream sediment data and intrusive rocks on the relationship between SiO_2, TiO_2, Al_2O_3, Fe_2O_3, MnO, MgO, CaO, Na_2O, K_2O, P_2O_5, and Cr_2O_3 variables which are illustrated by multidimensional spatial correspondence analysis and anomalous sources based on Cr, Cu, Fe, Mn, Ni, Zn, As, Mo, Co, Sb, Cd, and Pb variables. The results showed that the application of the hybrid statistical analysis method was effective enough to identify elemental variables in identifying sulfide deposits and the characteristics of anomalies were quite different from diorite and dacite due to magmatic hydrothermal and tectonic activity.

1. Introduction

Stream sediment is a simple method of taking samples of various small sizes that form naturally [1]. Initially, this method was very popular in the mineral exploration in reconnaissance stage in the tropics and sub-tropics. is currently used in environmental research mainly in lake and river sedimentary materials.

Regionally, study area is occupied by complex metamorphic rocks and ultramafic rocks which form the basement rocks of Sulawesi Island [2]. Various sedimentary rock formations, intrusion of dacite, diorite and there are still some unmapping mineralized intrusive rocks. Reflections on the complex differences in the rocks cause various types of river genetically such as consequent, insequent and subsequent rivers, which are generally divided into two directions, namely flowing to north and south through various rock formations. Flow effects affect in elemental transportation [3, 4] and it is mainly associated with the formation of river deposit.

Statistics and stream sediment are a solid combination method following the elemental mobility properties that can be used as univariate or multivariate analysis, that is aiming to evaluate the geochemical variables [5], one of which is pearson correlation. The research applies multivariate pearson correlation to evaluate the strong variables between existing elements, ultimately facilitating the analysis of spatial mapping factors between elements.

The research location was chosen which was easily accessed from the highway, to investigate the dispersion characteristics of various elements and indications of the type of metal sulfide deposits. The research also evaluated the effectiveness of random sampling techniques based on river flow through various types of rock formations.

2. Regional Geology

The stratigraphy of the study area is included in the Geological Map of the Pangkajene and Western Watampone Sheets [6], namely (Figure 1):

s: Metamorf Complex; schist, gneiss, glaucophane, garnet, epidote, muscovite, chlorite, eclogite, garnet schist, amphibole schist, chlorite schist, muscovite schist, muscovite-tremoilite-actinolite schist. It generally spreads to the Northeast, rising up by fault towards the Southwest.

Ub: Ultramafic Rocks; peridotite, serpentinite, contains chromite lenses and laterite.

m: Melange Complex; graywacke, breccia, conglomerate, sandstone, gray shale, red shale, red radiolarian chert, slate, schist, ultramafic rock, basalt, diorite, and mudstone.

Kb: Balangbaru Formation; flysch-sediment types; sandstone siltstone-intercalated, claystone and shale; intercalated conglomerates, tuffs and lava; the sandstones are composed of graywacke, arkose, tuff and carbonate rock, conglomerates with basalt, andesite, diorite, shale, splattered tuff, schist, quartz, and cemented by sandstone. This formation is overlapped by Mallawa Formation and prophilitic volcanic rocks which are composite and uncomformable with the Bantimala tectonic complex.

Tem: Mallawa Formation; sandstones, conglomerates, siltstone, claystone, marl with coal lenses or beds and claystone intercalation; the sandstones are mostly quartz sandstones, and there are also arkose, graywacke, and tuffaceous rocks. It is comformable with the Temt limestone and that uncomformable with Kb sedimentary rocks and Tpv volcanic rocks.

Temt: Tonasa Formation; massive coral limestone, bioclastic limestone and calcarenite, bituminous limestone. It is unconformable with Mallawa Formation and Camba Formation, intruded by sill, dyke, and igneous rock stocks which are basal, trachyte and diorite.

Tmcv: Member of Volcanic Rocks Camba Formation; volcanic rocks are intercalated by marine sedimentary rocks; volcanic breccia, lava, volcanic conglomerates, fine grained to lapilli tuff; intercalated tuffaceous sandstones, carbonate sandstones and mudstones. It is intruded by dykes, sills, and stock of basalt and diorite arrangements, andesitic-basaltic tuffs, andesites, andesitic trachyte, and leusite basalt.

d: diorite – granodiorite; diorite intrusion, granodiorite, mainly in the form of stock, and partly in the form of dykes, mostly in porphyry textures, light gray to gray colour. Diorit, that is expose to the east of Barru, intrudes the sandstone of the Balangbaru Formation. Ultramafic rock, porphyry granodiorite, with many of biotite-amphibole phenocrysts, and it intrude Tonasa Formation's limestone and Camba Formation.

3. Methods

Random sampling of sediments and water acidity is carried out on rivers that pass through rock formations, namely Watu River, Palakka River, Pijae River, Barru River. Water acidity is measured in the field and rock samples are taken as comparative correlation controllers.

Data distribution is considered to represent the distribution of rock formations which is divided into four groups: (1) SS02, Balangbaru Formation, Tonasa Formation, Tonasa Formation, Metamorphic Complex (2) SS01, SS03 and SS04: Tonasa Formation (3) SS05, SS06 and SS07: Tonasa Formation, Camba Formation, Ultramafic Complex (4) SS08, SS09: Tonasa Formation, Camba Formation, diorite intrusion. Data on SS02, SS08, SS09 are considered to be single correlated statistics as independent factors.

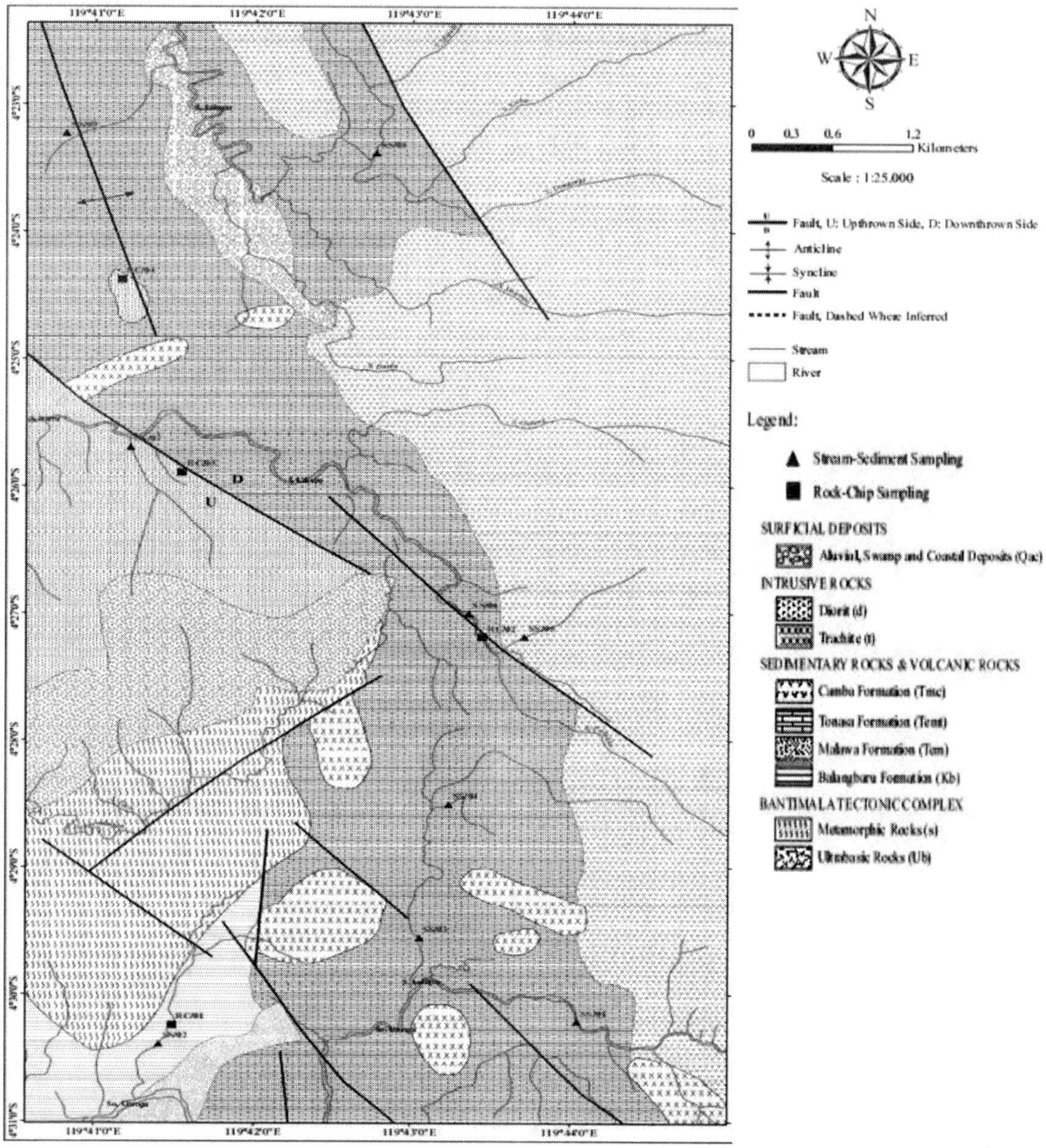

Figure 1. Regional Geology and sampling point

Taking of fresh dacite (RC01) and (RC04), diorite alteration (RC02), and dacite alteration (RC03) samples are used as a major linearity element, then composite river sediment samples at each creek on different trajectories are randomly collected from the bottom in 20 to 30 cm depth of active flow channel by using a shovel.

The samples are homogenized and that is washed by the wood pan. The panning or washing process is carried out under the surface of the water repeatedly to separate heavy particles and clays which produce heavy mineral residues. Then, the concentrate was dried in air before being submitted for XRF chemical analysis and ICP-MS analysis in Intertek Jakarta for find the value of Al_2O_3, Fe_2O_3, MnO, SiO_2, CaO, K_2O,Na_2O, TiO_2, Cr_2O_3,MgO,P_2O_5, S dan Cu, Pb,Zn,As, Fe, and Cr.

Laboratory data is converted into a standard format in the Excel database, statistics on the IBM SPSS 25 and geochemical information which is obtained by maps created using ArcGIS 10.3 software.

4. Result and Discussion

4.1. Rocks Geochemistry

Statistical description of the chemical samples of rock RC01, RC02, RC03, RC04 show the maximum and minimum range of major elements as follows: SiO_2 (29.36% -83.78%), TiO_2 (0.07% -0.23%), Al_2O_3 (2.97% -6.96%), F_2O_3 (1.88% -8.11%), MnO (0.02% -0.23%), MgO (0.32% -1.32%), CaO (0.02% -29.33%), Na_2O (0.04% -5.37%), K_2O (0.09% -2.02%), P_2O_5 (0.02% -0.10%), Cr_2O_3 (0.0% - 1.0%).

The value of SiO_2 is high enough which strengthens the interpretation that river sediment material originates from alteration of dacite and diorite rocks, although contamination can be occured by quartz sand sedimentation of the Tonasa Formation.

Scatterplot analysis of SiO_2 variables in rock samples that have showed a strong correlation can be found in altered-sulfide mineralized dacite and diorite (Figure 2).

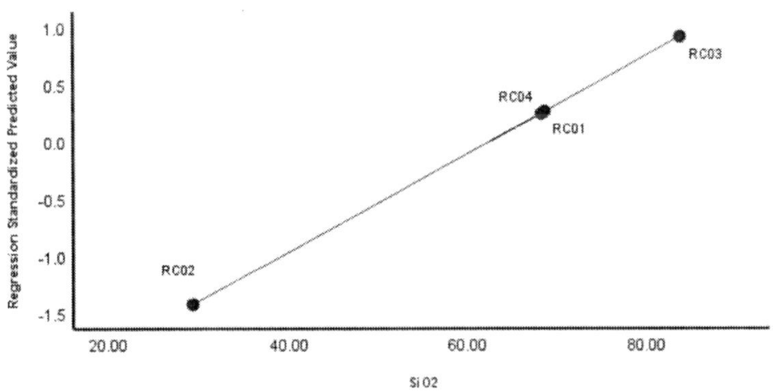

Figure 2. Scatterplot of rock point

4.2. Element Correlation

River sediment material is sourced from rock masses undergoing weathering and transportation processes which subsequently serve as guidelines in identifying source and deposit correlations. Stream sediments generally reflect the mineralogical composition and the presence of outcrops in the drainagebasin, upstream of the sampling sites [1]. Descriptions on stream sediment samples on major elements can be seen in Table 1.

Table 1. Major elements of stream sediment samples

ELEMENT(%)	SS01	SS02	SS03	SS05	SS06	SS07	DET LIMIT
SiO2	39.40	82.68	41.45	45.82	46.75	49.18	0.01
TiO2	1.78	1.70	1.66	1.22	1.09	1.29	0.01
Al2O3	5.00	4.54	4.41	6.09	8.46	8.72	0.01
Fe2O3	20.58	7.06	18.60	14.40	13.11	14.28	0.01
MnO	0.25	0.08	0.24	0.22	0.20	0.22	0.01
MgO	10.65	0.70	11.98	10.88	8.71	7.70	0.01
CaO	18.58	0.70	19.63	17.49	14.62	12.53	0.01
Na2O	0.45	0.43	0.38	0.72	0.94	1.04	0.01
K2O	0.55	0.62	0.27	1.20	2.05	2.06	0.01
P2O5	0.34	0.04	0.31	0.33	0.36	0.36	0.01
Cr2O3	0.07	0.28	0.01	0.08	0.05	0.05	0.01
S	0.02	0.01	0.01	0.02	0.03	0.02	0.01
LOI	2.20	1.00	1.10	1.90	3.10	2.20	0.01

Pearson correlation uses 0.5 as a measure of the major element variables in SS01 / SS02 / SS03 data group and SS05/06/07 data group. The results are known that are three groups and strongly correlated each others: (1) (Al_2O_3, K_2O) = (0.99) = (Al_2O_3, K_2O) = 0.99; (Al_2O_3, Na_2O) = 0.82 < (Al_2O_3, Na_2O) = 0.98; (Al_2O_3, P_2O_5) = 0.74 < (Al_2O_3, P_2O_5) = 0.97; (Cr_2O_3, Fe_2O_3) = 0.81 > (Cr_2O_3, Fe_2O_3 =0.70; (Cr_2O_3, MnO) = 0.93 > (Al_2O_3,MnO) = 0.70, (2) (Cr_2O_3,TiO_2) = 0.99; (Cr_2O_3S) = 0.99; (Fe_2O_3, MnO) = 0.97 (Cr_2O_3,S) = 0.90; (SiO_2,CaO) = 0.80 (SiO_2, MgO) = 0.99; (Na_2O, Al_2O_3) = 0.82; (Na_2O, P_2O_5) = 0.99, (3) (Al_2O_3, SiO_2) = 0.80; (Cr_2O_3, MgO) = 0.90; (Fe_2O_3, TiO_2) = 090; (K_2O, SiO_2) = 0.72; (K_2O, S) = 0.84; (P_2O_5, Al_2O_3) = 0.97; (K_2O, TiO_2) = 0.97, (MgO, CaO) = 0.99; (MgO, Cr_2O_3) = 0.88.

Figure 3, multidimensional and correlation spatial maps show that eucludian distance CaO, Na_2O, Mgo have similar distances and close to each other compared to Al_2O_3, then P_2O_5 and SiO_2 respectively in different quadrants. The position of SiO_2 and P_2O_5 in different quadrants and Cr_2O_3, Fe_2O_3 and MnO are in the same quadrant even though Cr_2O_3 is closer to Fe_2O_3 than MnO.

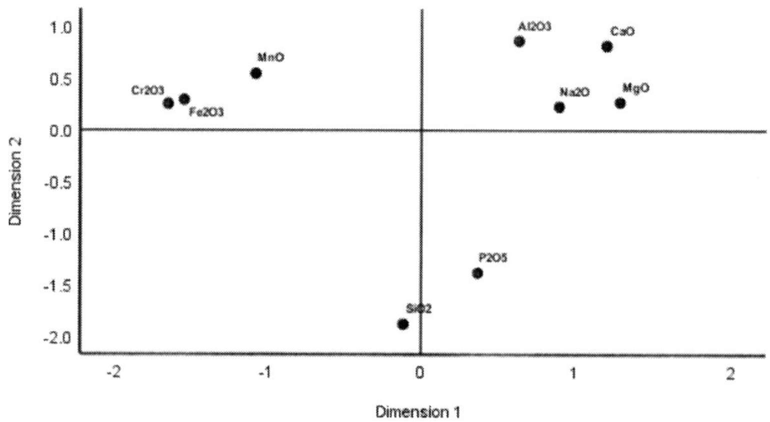

Figure 3. Spatial map of major elements

Statistic analysis in SS01 and SS08/09 as follows : Fe (5.07 ppm) < Fe (mean 22.5 ppm), Cr (1740 ppm) < Cr (mean 1201 ppm), Ni (27 ppm) < Ni (mean 89.5 ppm), Co (12 ppm) < Co (mean 96.5 ppm), Mn (631 ppm) < Mn (mean 6820 ppm). This difference is influenced by weathering of ultramafic rocks which produce ferromagnesian material in river deposits.

The present of mineralized dacite intrusion which is in contact with ultramafic and mineralized diorite in contact with limestone is characterized by an elemental contents: Cu (12 ppm) < Cu (mean 56.5 ppm), Pb (9.0 ppm) < Pb (mean 31.5 ppm), Zn (65 ppm) < Zn (mean 256 ppm), Mo (0.6 m) < Mo (mean 1.3 ppm), As (4.0 ppm) < As (mean 21 ppm) and Sb (0.4 ppm) < Sb (mean 1.35 ppm).

By using a 0.5 correlation in the sample group, we obtain six classifications of elemental properties, as follows:

(1) Equal strong correlation : Fe, Cr,Co,Ni,As, Cu,Pb,Mo,Sb,Cd,Zn,Mo,Mn
(2) Strong correlation : Zn,Pb,As
(3) Inverse Correlation : Fe, Cr,Ni,As,Zn,Pb, Co,Cu, Mn
(4) Without correlation : (Cr,Co), (As,Sb), (Cd,Pb), (Fe,Mn), (Fe,Mo), (Fe,Sb), (Fe,Mo), (Fe,As), (Fe,Cd), (Cu,Fe), (Cu,Mn), (Cr,Sb), (Cr,Fe), (Cr,Mn), (Mn,Sb), (Ni,As), (Ni,Sb), (Zn,Fe), (Zn,Sb), (Mo, Mn), (Mo,Fe), (Sb,Cr), (Sb,Ni), (Sb,Zn), (Cu,As), (Cu,Sb), (Cu,Cr), (Ni,Co), (Zn,Co), (As,Cu), (As,Cd), (As,Co), (Cd,Zn), (Cd,Mo), (Cd,Pb), (Co,As), (Mo,Cd), (Mo,As),(Sb,Cu), (Sb,Cd), (Sb,Cu) (Table 2).

Table 2. Pearson correlation trace elements

	Cr	Cu	Fe	Mn	Ni	Zn	As	Mo	Co	Sb	Cd	Pb
Cr			X	X	XX++	X	XX**		++	XX		
Co	++	XX**					**				**	**
Ni	XX						XX			XX		
Cu	**			X			++	XX	XX	++	X++	XX+
Zn	**		XX	XX				XX++	++	XX		XX
Fe				XX++		XX	X	XX**		XX	X	XX
Mn			XX			XX		**		XX		XX
As	XX**	++			XX				++	XX	++	++
Mo		XX++	XX	XX							++	XX
Sb	XX	XX			XX	XX	XX++			++	++	X++
Cd		XX				++		XX++	XX			
Pb		++					++					

Abbreviaton: Cr = chromium; Cu = cuprum; Fe = ferrum; Mn = mangan; Ni = nickel; Zn = zink; As = arsen; Mo = Molybdenum; Co = cobalt; Sb = stibium; Cd = cadmium; Pb = plumbum.

Figure 4, spatial map SS010304,050607 shows that Cu and Cr are located far apart and that are in first quadrant and have similarities to As and Fe in third quadrant. Elements of Cu, Pb, Zn and As form strong correlation groups in different quadrants (Figure 4).

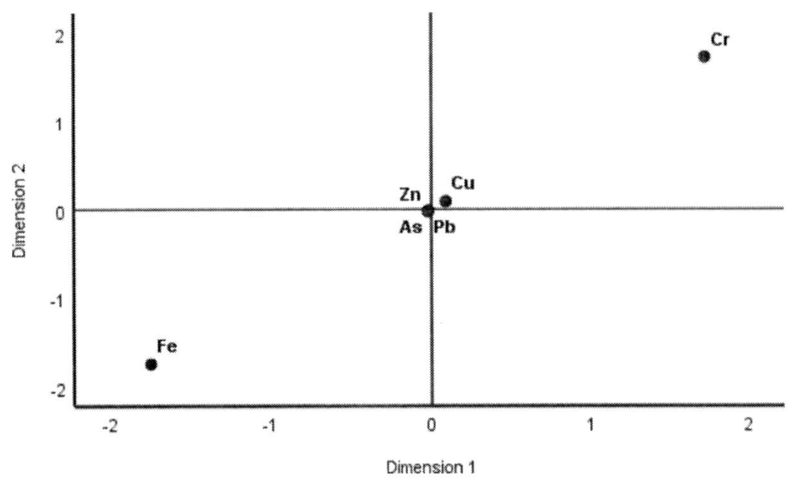

Figure 4. Cu, Zn, Pb, As, Cr and Fe Spatial Map

5. Conclusion

- The Position of Fe and Cr in different quadrant give an indication that the mineralization source are different where Fe is stronger formed as an oxide mineral than Cr forms deposits. Strong estimated that Cr, which forms deposits, is derived from ultramafic rocks so that Cu and Cr are different types of deposits.
- Position of Zn, As and Pb in the spatial image are indications of mineralization in the same rock source compared. Cu, which is close proximity to these three elements, forms different types of deposits even though there are also deposits of Zn, As, and Pb mineralization.

IOP Conf. Series: Materials Science and Engineering **875** (2020) 012049 doi:10.1088/1757-899X/875/1/012049

References

[1] Rose A W, Hawkes H E and Webb J S 1979 Geochemistry in Mineral Exploration. Academic Press, London 657.

[2] Maulana A, Ellis D J, Christy A G 2010 Petrology, geochemistry and tectonic evolution of the South Sulawesi basement rocks, Indonesia, Proced. IPA, Thirty-Fourth Annual Convention & Exhibition.

[3] Davis J C 2002 Statistics and Data Analysis in Geology. 3rd Edition, John Wiley & Sons, Inc. NY.

[4] Weber K, Davis 1990 Multivariate statistical analysis of stream-sediment geochemistry in the Grazer Paleozoikum, Austria, Mineralium Deposita 25 213-220.

[5] Grunsky E C, Drew L J, and Sutphin D M 2009 Process recognition in multi-element soil and stream-sediment geochemical data. Applied Geochemistry 24 1602–1616.

[6] Sukamto R 1982 Geological Map of Pangkajene dan Western Watampone Sheet, Sulawesi, Centre of Geological Research and Development, Directorate of General Mining, Energy and Mining Department, Bandung, Indonesia.

Sensitivity Analysis of Net Present Value due to Optimal Pit Limit in PT Ceria Nugraha Indotama, Kolaka Regency, Southeast Sulawesi Province

A V Anas[1]*, R Amalia[1], N F Qaidahiyani[1], Djamaluddin[1], and S R D Herin[1]

[1]Department of Mining Engineering, Engineering Faculty, Hasanuddin University, Makassar, Indonesia

*Email: virtanti@gmail.com

Abstract. PT Ceria Nugraha Indotama is planning to open a new mining site in Lapaopao Block area. In order to determine the most profitable pit design on this new area, pit limit optimization is required. Pit limit is optimized by considering economic parameters, geotechnical parameters, and production target. The optimal pit limit generates the highest Net Present Value (NPV) which is affected by economic parameters. The objective of this research is to analyse the NPV sensitivity of pit due to the changes of economic parameters namely mining cost, processing cost, and selling cost by using sensitivity analysis. Data used in this research were drill hole, ore and waste density, slope geometry, production target. The planned production target, ore price, recovery factor, dilution factor, mining cost, processing cost, and selling cost. The analysis results showed when mining costs increased by 20%, NPV decreased from $2,407,032 to $1,683,430 and when mining costs decreased by 20%, NPV increased from $2,407,032 to $3,130,635. When processing cost increased by 20%, NPV decreased from $2,407,032 to $2,133,523 and when processing cost decreased by 20%, NPV increased from $2,407,032 to $2,680,542. When selling cost increased by 20%, NPV decreased from $2,407,032 to $1,995,755 and when selling cost decreased by 20%, NPV increased from $2,407,032 to $2,818,310. It indicated that the economic parameters significantly affect the NPV and most sensitive to the mining cost parameter changes.

1. Introduction

PT Ceria Nugraha Indotama operates a laterite nickel mine on an area of 6,785 hectares called Lapaopao Block in Wolo District, Kolaka Regency, Southeast Sulawesi Province. One of the area in Lapaopao Block will be used as a new mining site. In order to obtain the best possible economic benefit from the resources available at this project, pit limit optimization is required. Pit limit optimization is conducted to determine the best pit design that obtain the largest profit [1]. The size and shape of pit depends on economic factors and design or production constraints. Understandably, pit optimization is the maximization of Net Present Value (NPV) for a given ore deposit subject to numerous of mining and economic constraints [2]. Pit limit is optimized by considering some parameters such as economic parameters, geotechnical parameters, and production target. The optimal pit limit generate the highest Net Present Value (NPV) which is highly affected by economic parameter such as mining cost, processing cost, and selling cost.

The purpose of this study is to find out how the changes of economic parameters affect the Net Present Value NPV, therefore sensitivity analysis is performed. Sensitivity analysis is necessary to evaluate probable effects of project parameters on economic outputs that can be used for recognizing the most critical factors and in this way, prediction of extreme optimistic and pessimistic conditions would be possible. In this process, range of variation of variables is determined [3]. Sensitivity analysis is performed to determine the impact of changes in the value of mining costs, processing costs, and selling costs on the Net Present Value (NPV). The sensitivity analysis of Net Present Value (NPV) based on economic parameters could be a key consideration for companies to anticipate future events due to changes in economic parameters.

2. Research Method

Sensitivity analysis is an analytical method for risky projects using an estimation of Net Present Value (NPV) for each situation where the intended parameters taken as the values of optimistic, pessimistic, or most likely [4]. Sensitivity analysis allows to evaluate the project performance at different values of parameters and to determine the most critical variables that have the greatest affect on the feasibility of the project [5]. In this research, sensitivity analysis was conducted to determine the sensitivity of Net Present Value (NPV) due to changes of economic parameters.

Data in this research consisted of:

- Drill hole data: collars, surveys, assays, and geology.
- Density of ore and waste: 1.6 tonnes/m3.
- Slope geometry: single slope of 60o, overall slope of 40o, bench high of 5 meters, and bench width of 2 meters.
- Production target: 5,000 tonnes per day.
- Ore price: $10/tonnes.
- Recovery factor: 95%.
- Dilution factor: 5%.
- Capital cost $174.503.
- Mining cost: $0.51/tonne.
- Mining CAF of 2.77.
- Processing cost: $1.35/tonne.
- Selling cost: $2.02/tonne.

Data were processed using Surpac 6.5.1 to reconstruct block model and to design pit limit, Whittle 4.5.1. for pit optimization, and Microsoft Excel for sensitivity analysis with following steps:

- Created a block model by adding density and drillhole data.
- Optimized pit limit by adding the block model, overall slope, production target, recovery factor, dilution factor, and economic parameter where 13 pit shell generated through this process.
- Determined optimal pit shell. The fifth pit shell selected as the optimal pit shell due to the number of revenue factor = 1.
- Designed pit limit based on the pit shell.
- Calculated the ore reserves using pit limit design and topographic map of the research area.
- Composed a cash flow statement based on amount of reserves and economic parameters.
- Calculated the Net Present Value (NPV) based on the cash flow statement.
- Analyzed the sensitivity of Net Present Value (NPV) due to the changes of mining cost, processing cost, and selling cost.

3. Results

3.1 Bock Model Reconstruction and Nickel Resources Estimation

The reconstruction of nickel resources block model is performed to get a visualisation of ore deposit geometry as shown in Figure 1. Resources estimation is performed to determine volume, tonnage, and grade of the resources. In this research, nickel resources estimation using a cut-off grade of 1.4% with Inverse Distance Weighting (IDW) method (Table 1).

Figure 1. Block model reconstruction of nickel resources

Table 1. Nickel resources estimation

Ni *Range* (%)	*Volume* (m³)	*Tonnage* (ton)	Ni *Average* (%)
0.0 – 1.4	17,207	27,531	1.40
1.4 – 1.5	310,605	496,969	1.45
1.5 – 1.7	286,875	459,000	1.58
1.7 – 2.0	74,277	118,844	1.80
>2.0	55,508	88,813	2.40
Total	744,473	1,191,157	1.61

3.2. Pit Limit Optimization

Pit limit optimization is conducted to produce a pit shell with the highest Net Present Value (NPV) by used data were block model, overall slope 40o, mining cost $0.51/tonne, mining CAF 2.77, processing cost $1.35/tonne, selling cost $3.02/tonne, ore price $10/tonne, capital cost $174.503, mining recovery fraction 0.95, mining dilution fraction 0.05, annual discounting 10%, annual production target 7.500.000 tonne/year (360 work days), and revenue factor range 0,3 to 2 using 0,1 steps. The results of pit limit optimization are shown in Table 2.

Table 2. Results of pit limit optimization

Pit Shell	Revenue Factor	NPV ($)	Ore (Tonnes)	Waste (Tonnes)	Life Years
1	0.6	10,599	51,331	51,414	0.0137
2	0.7	2,014,394	721,275	1,418,600	0.28532
3	0.8	2,683,426	1,050,450	2,612,779	0.48843
4	0.9	2,733,559	1,094,259	2,842,322	0.52488
5	1	2,738,475	1,118,136	3,010,422	0.55047
6	1.1	2,734,936	1,122,990	3,053,320	0.55684
7	1.2	2,727,692	1,127,474	3,100,587	0.56374
8	1.3	2,725,062	1,128,556	3,113,693	0.56563
9	1.4	2,715,221	1,131,374	3,153,533	0.57132
10	1.5	2,704,607	1,133,848	3,192,372	0.57683
11	1.6	2,702,005	1,134,343	3,201,065	0.57805
12	1.7-1.8	2,690,215	1,136,299	3,238,240	0.58327
13	1.9-2	2,688,439	1,136,516	3,243,273	0.58397

IOP Conf. Series: Materials Science and Engineering **875** (2020) 012050 doi:10.1088/1757-899X/875/1/012050

From Table 2, optimal pit shell with revenue factor (RF) 1 is Pit Shell 5 with NPV of $ 2,738,475. Pit shell 5 is shown in Figure 2.

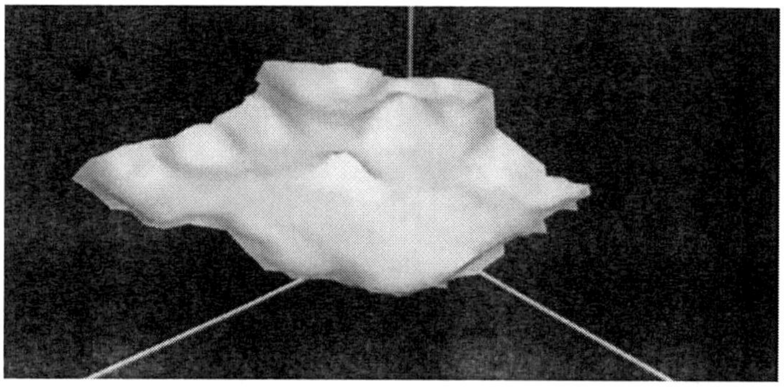

Figure 2. Pit shell 5

3.3. Sensitivity Analysis of Net Present Value (NPV) of Pit Shell 5
Sensitivity analysis is used to determine how different values of independent variable impact a particular dependent variable. The dependent variable in this research is Net Present Value (NPV), while the dependent variables are economic parameters namely mining costs, processing costs, and selling costs.

3.3.1. Sensitivity Analysis of Net Present Value (NPV) due to Mining Cost. In this sensitivity analysis, NPV is determined from different values of mining cost changes as shown in Table 3 and Figure 3.

Table 3. The sensitivity of NPV due to mining cost

Changes of Mining Cost (%)	NPV ($)
-20	3,330,093
-15	3,182,189
-10	3,034,284
-5	2,886,380
0	2,738,475
5	2,590,570
10	2,442,666
15	2,294,761
20	2,146,856

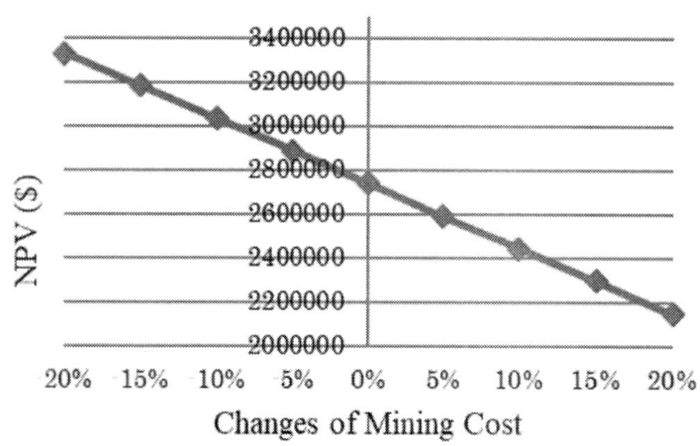

Figure 3. Graph of sensitivity of NPV due to mining cost

Table 3 and Figure 3 shows that the rate of mining cost is inversely to the NPV. The NPV decrease from \$2,738,475 to \$2,146,856 when mining cost increase by 20% and increase from \$2,738,475 to \$3,330,093 when mining cost decrease by 20%.

3.3.2. Sensitivity Analysis of Net Present Value (NPV) due to Processing Cost. In this sensitivity analysis, the sensitivity of NPV is determined from the processing cost changes. The results of this analysis are shown in Table 4 and Figure 4.

Table 4. The sensitivity of NPV due to processing cost

Changes of Processing Cost (%)	NPV (\$)
-20	3,024,941
-15	2,953,324
-10	2,881,708
-5	2,810,091
0	2,738,475
5	2,666,858
10	2,595,242
15	2,523,626
20	2,452,009

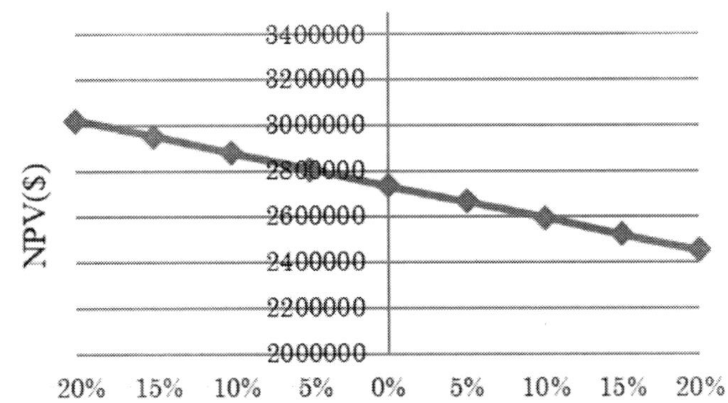

Figure 4. Graph of sensitivity of NPV due to processing cost

Table 3 shows that the NPV decrease from \$2,738,475 to \$2,452,009 when processing cost increase by 20% meanwhile the NPV increase from \$2,738,475 to \$ 3,024,941 when the processing cost decrease by 20%. Figure 4 shows that the rate of processing cost changes is inversely proportional to the NPV.

3.3.3. Sensitivity Analysis of Net Present Value (NPV) due to Selling Cost. In this sensitivity analysis, the NPV is determined from changes in selling cost as shown in Table 5 and Figure 5.

Table 5. The sensitivity of NPV due to processing cost

Changes of Selling Cost (%)	NPV ($)
-20	3,110,518
-15	3,017,507
-10	2,924,496
-5	2,831,486
-	2,738,475
5	2,645,464
10	2,552,454
15	2,459,443
20	2,366,432

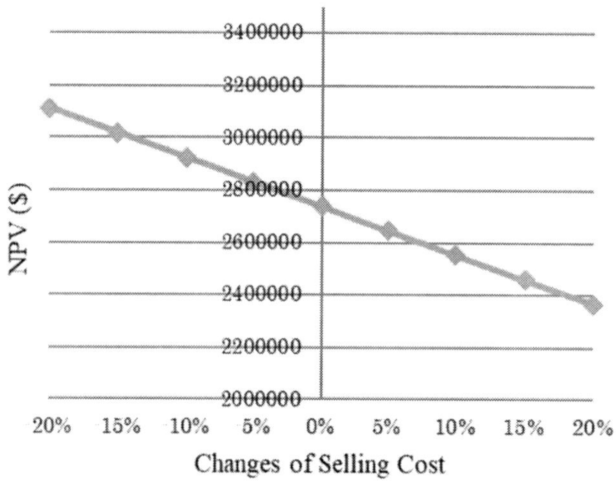

Figure 5. Graph of NPV sensitivity due to selling cost

Figure 5 shows that the rate of changes in selling cost is inversely to the rate of NPV. Table 5 shows that the NPV value decrease from $2,738,475 to $2,366,432 when selling cost increase by 20%, and the NPV increase from $ 2,738,475 to $3,110,518 when selling costs decrease by 20%. Sensitivity analysis of Net Present Value (NPV) of Pit Shell 5 based on mining cost, processing cost, and selling cost is shown in Figure 6.

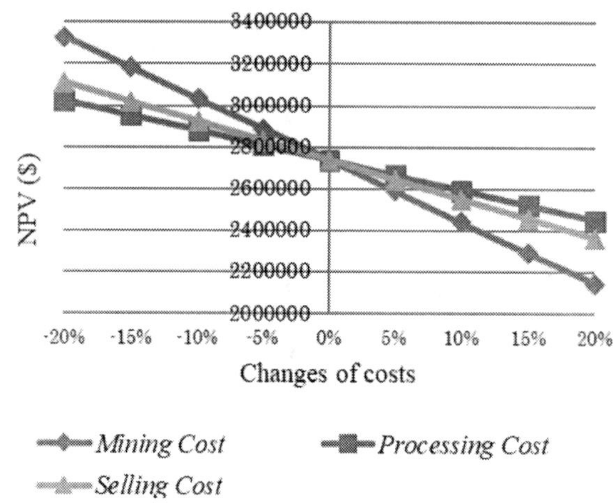

Figure 6. Graph of NPV Sensitivity of Pit Shell 5

Figure 6 shows that the Net Present Value (NPV) is most sensitive to mining cost changes than to selling cost changes and processing cost changes.

3.4. Design of Pit

The pit is designed based on pit shell 5 which requires several important parameters namely height of bench 5 m, width of bench 2 m, and single slope 60°. Design of the pit is shown in Figure 7.

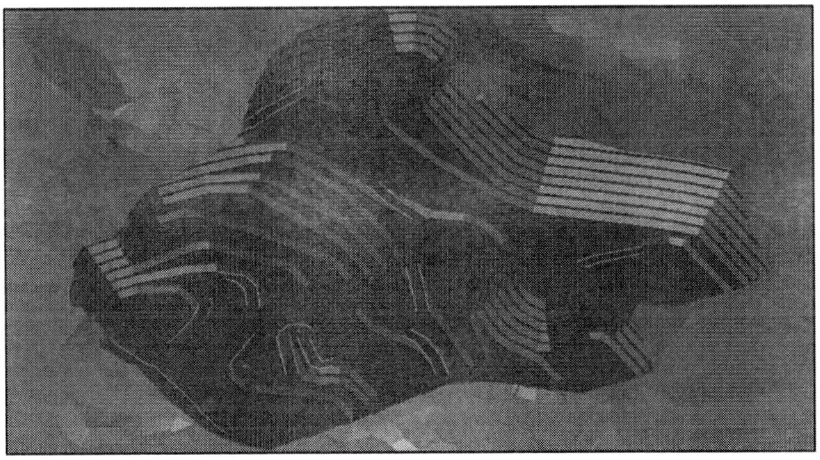

Figure 7. Design of Pit

3.5. Reserves Estimation

Reserves estimate from material between topography and pit based on the block model. Total reserves estimation is shown in Table 6.

Table 6. Total reserves estimation

Waste/Ore	Ni Range (%)	Volume (m^3)	Tonnage (tonnes)	Ni Average (%)
Waste	0.0 - 0.5	478,242	765,188	0.00
	0.5 - 0.9	762,168	1,219,469	0.81
	0.9 - 1.4	1,005,723	1,609,156	1.09
Sub-Total		2,246,133	3,593,813	0.76
Ore	1.4 - 1.5	287,715	460,344	1.45
	1.5 - 1.7	278,262	445,219	1.58
	1.7 -> 2.0	73,340	117,344	1.80
	> 2.0	55,410	88,656	2.40
Sub-Total		694,727	1,111,563	1.62
Total		2,940,860	4,705,376	0.96

3.6. Net Present Value of Pit Design

The Net Present Value (NPV) from pit design is calculate using cash flow analysis with discount rate 9.73% as shown in Table 7. Based on the cash flow in Table 7, the result of Net Present Value (NPV) from pit design is $2,407,032.

Table 7. Cash flow analysis

Year	0	1
Revenue ($)		10,559,849
Operational Cost ($)		(7,727,130)
Capital Cost ($)	(174,503)	
Cash Flow ($)	(174,503)	2,832,719
Discount Rate (%)		9.73
Discounted Cash Flow ($)	(174,503)	2,581,535
Cumulative Discounted Cash Flow (NPV) ($)		2,407,032

3.7. Sensitivity Analysis of Net Present Value of Pit Design
The sensitivity of Net Present Value (NPV) of pit design as shown in Table 7 is analyzed by using three parameters namely mining cost, processing cost, and selling cost.

3.7.1. Sensitivity Analysis of Net Present Value (NPV) due to Mining Cost. In this sensitivity analysis, NPV is determined from changes in mining cost. The results are shown in Table 8 and Figure 8.

Table 8. The sensitivity of NPV due to mining cost

Changes of Mining Cost (%)	NPV ($)
-20	3,130,635
-15	2,949,734
-10	2,768,834
-5	2,587,933
0	2,407,032
5	2,226,132
10	2,045,231
15	1,864,330
20	1,683,430

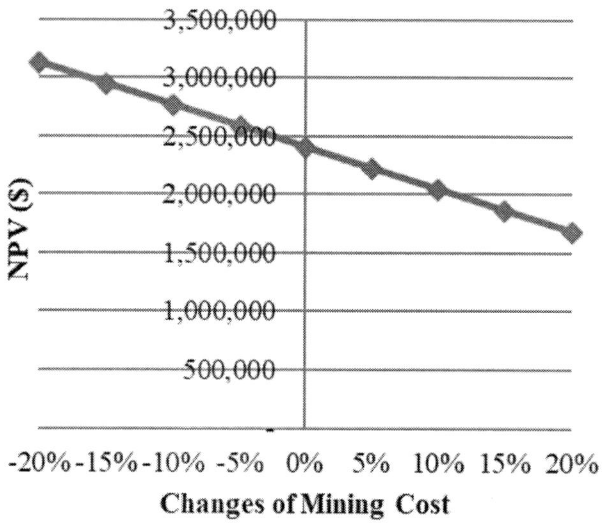

Figure 8. Graph of NPV sensitivity due to mining cost

Figure 8 shows that the mining cost changes is inversely to the rate of NPV. Table 8 shows that the NPV decrease from $2,407,032 to $1,683,430 when mining costs increase by 20% and NPV increase from $2,407,032to $3,130,635 when mining costs decrease by 20%.

3.7.2. Sensitivity Analysis of Net Present Value (NPV) due to Processing Cost. In this analysis, the NPV is determined from the changes of processing cost. The results of this analysis are shown in Table 9 and Figure 9.

Table 9. The sensitivity of NPV due to processing cost

Changes of Processing Cost (%)	NPV ($)
-20	2,680,542
-15	2,612,165
-10	2,543,787
-5	2,475,410
0	2,407,032
5	2,338,655
10	2,270,278
15	2,201,900
20	2,133,523

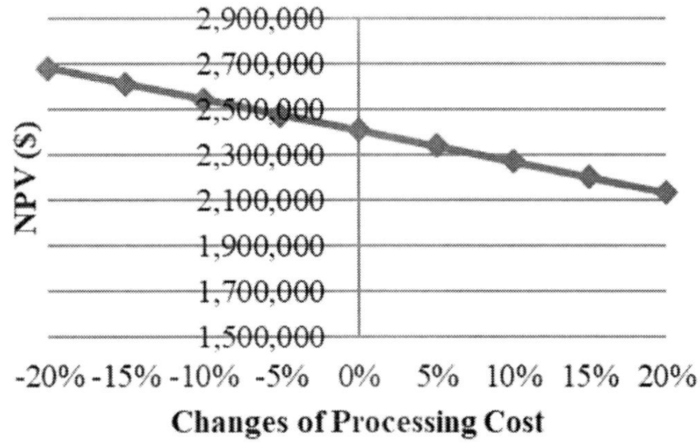

Figure 9. Graph of NPV sensitivity due to processing cost

Table 9 shows that the NPV decrease from $2,407,032 to $2,133,523 when processing cost increase by 20%, and NPV increase from $2,407,032 to $2,680,542 when the processing cost decrease by 20%.

3.7.3. Sensitivity Analysis of Net Present Value (NPV) due to Selling Cost. The sensitivity of NPV is determined from the selling cost changes which is shown in Table 10 and Figure 10.

Table 10. The sensitivity of NPV due to selling cost

Changes of Selling Cost (%)	NPV
-20	2,818,310
-15	2,715,490
-10	2,612,671
-5	2,509,852
0	2,407,032
5	2,304,213
10	2,201,394
15	2,098,574
20	1,995,755

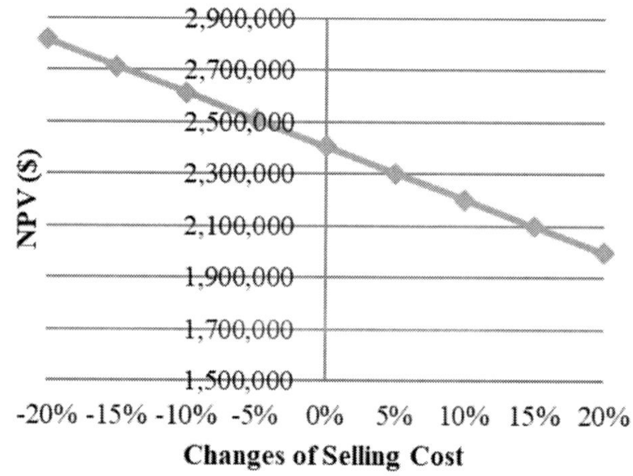

Figure 10. Graph of NPV sensitivity due to selling cost

Figure 10 shows that the rate of changes in selling cost is inversely to the rate of NPV changes. Table 10 shows that the NPV value decrease from $2,407,032 to $1,995,755 when selling costs increase by 20%, and NPV increase from $2,407,032 to $2,818,310 when selling cost decrease by 20%.

Sensitivity analysis of Net Present Value (NPV) of pit design based on mining cost, processing cost, and selling cost are shown in Figure 11.

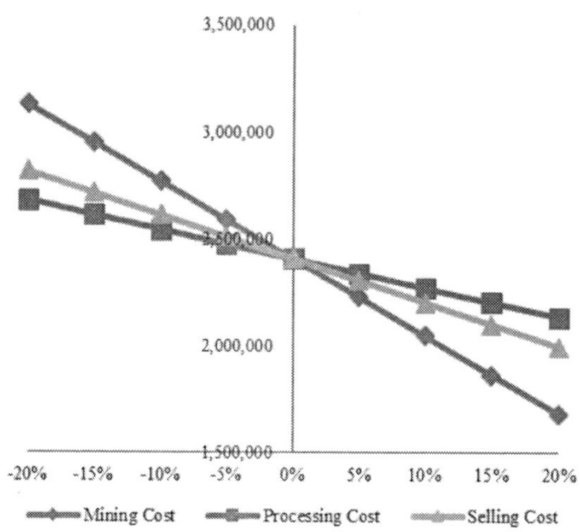

Figure 11. Graph of NPV sensitivity of pit design

Figure 11 shows that the Net Present Value (NPV) of pit design is most sensitive to mining cost changes than to selling cost changes and processing cost changes.

4. Conclusion

It is recognized that economic parameters namely mining cost, processing cost, and selling cost affect the value of Net Present Value (NPV). The results of sensitivity analysis show that when mining cost increase by 20%, NPV decrease from $2,407,032 to $1,683,430 and when mining cost decrease by 20%, NPV increase from $2,407,032 to $3,130,635. When processing cost increase by 20%, NPV decrease from $2,407,032 to $2,133,523 and when processing cost decrease by 20%, NPV increase from $2,407,032 to $2,680,542. When selling costs increase by 20%, NPV decrease from $2,407,032 to $1,995,755 and when selling cost decrease by 20%, NPV increase from $2,407,032 to $2,818,310. The Net Present Value (NPV) of pit design is most sensitive to mining cost changes than to selling cost changes and processing cost changes.

Acknowledgment

The authors would like to thank profusely to PT Ceria Nugraha Indotama which has provided all necessary data and convey special gratitude to Engineering Faculty of Hasanuddin University for LBE research fund.

5. References

[1] Hustrulid, W., Kuchta, M., Martin, R. 2013. Open Pit Mine Planning and Design Volume 1-Fundamentals, 3rd Edition. Boca Raton: CRC Press.

[2] Cardu, M., Ciccu, R., Lovera, E., Michelottl, E. 2006. Mine Planning and Equipment Selection. The Fifth International Symposium on Mine Planning and Equipment Selection, pp. 1234-1239. Torino.

[3] Oraee, K., Sayadi, A. R., Tavassoli, S. M. M. 2011. Economic Evaluation and Sensitivity-Risk Analysis of Zarshuran Gold Mine Project. Denver: SME Annual Meeting.

[4] Dayananda, D., I, Richard., Harrison, S., Herbohn, J., Rowland, P. 2002. Capital Budgeting (Financial Appraisal of Investment Projects). Press syndicate of the University of Cambridge

[5] Zenad, Y. S. 2015. Sensitivity Analysis to Know the Project's Ability to Continue. European Journal of Accounting, Auditing and Finance Research, Vol.3 No.11, pp.60-66. Baghdad.

Flood Risk Mapping of Jakarta Using Genetic Algorithm Rule-Set Production (GARP) and Quick Unbiased Efficient Statistical Tree (QUEST) Methods

R R Yusya[1]*, M R Septyandy[1], and T L Indra[2]

[1]Program Study Geology, Faculty of Mathematics and Natural Sciences, Universitas Indonesia
[2]Department of Geography, Faculty of Mathematics and Natural Sciences, Universitas Indonesia

*Email: tito.latif@sci.ui.ac.id

Abstract. Jakarta as the country's capital, has experienced a series of floods that crippled cities in 2002, 2007 and 2013 and extreme rainfall has always been the main cause of major flood with casualties and property. Given the increasing impact of flooding in Jakarta, methods for assessing current and future flood risks is needed. A model is needed that can predict and determine which areas have the potential to be affected by future floods. Several models have been carried out for prediction and assessment of flood risk in Jakarta and several other major cities, namely GIS based on SVM models, hierarchical process analysis models (AHP), machine learning with ensemble models, and multivariate discriminant, classification, and regression trees. However, from several methods there are still some weaknesses, one of which is the time to get a predictive model that is quite long, which is 3-4 hours depending on the amount of data. To overcome this problem, genetic algorithm rule-set production (GARP) and quick unbiased efficient statistical tree (QUEST) models can create a prediction model with a shorter time with the same results. Several factors that influence flooding are used as input model data: precipitation, slope, distance to river, distance to channel, depth to groundwater, land use, elevation, and flood data for 2002 - 2019. The Area under receiver-operator characteristic curve (AUC-ROC) and root mean square error (RMSE) were used as evaluations for model performance. The results showed that the GARP model (AUC-ROC = 94%, RMSE = 0.2) had higher performance accuracy than the QUEST model (AUC-ROC = 84%, RMSE = 0.4). The results also indicated that the distance to channel, land use, and elevation as important parameters in determining flood hazards.

1. Introduction

Flood is one of the natural disasters of hydrometeorology that often occurs in Indonesia [1]. Flood disasters are caused by natural factors such as geographical and topographic conditions of the region, rainfall intensity, and the shape of the river channel [1]. Social factors are also related to flooding, such as population density and land use. The area of Indonesia that often experiences floods is the Special Capital Province of Jakarta (DKI Jakarta). Jakarta experienced several major flood disasters in 2002, 2007 and 2013 [2]. Floods that occur in Jakarta are caused by overflowing rivers that occur when rainfall is very high in the rainy season. The river overflowed due to it unable to accommodate the intensity of the rain that came and the silting of the river [3]. Impacts that occur due to floods are the paralysis of life activities in Jakarta and the impact of material and non-material losses such as the emergence of disease / health outbreaks, damage to buildings, dwellings, public facilities, and infrastructure, and can cripple the economic activities of the local community. Therefore, proper risk management needs to be done so that the disaster can be reduced.

Modeling flood risk areas is a form of effort in predicting the occurrence of floods in the future. Flood risk modeling has been carried out with several GIS-based models such as the SVM (Support Vector Machine) model [4], the hierarchical analysis process model (AHP) [5], statistical models and machine learning [6], and multivariate discriminant, classification, and regression trees [7]. However, from some of these models there are some shortages, one of which is the time to get a fairly long prediction model that is around 3-4 hours depending on the amount of data used. To overcome these problems, the GARP and QUEST methods are able to create a prediction model with a shorter time with the same results [8].

Modeling using the GARP and QUEST methods has been used in predicting flood disasters in the City of Sarri, Iran [8]. Based on this research the writer wants to do research in the Jakarta area. The purpose of this study is to evaluate the factors that influence the occurrence of floods in Jakarta which will be included in the depth of the flood vulnerability map. The factors included in this model consist of seven components, namely: rainfall, land use, area height, slope, distance from large rivers, distance to river, and depth of ground water level.

2. Study Area and Input Data

2.1. Study Area

The Special Capital Region of Jakarta (5° 19 '12 "- 6° 23" 54 "S, 106° 22" 42 "- 106° 58" 18 "E) is the country's capital and largest city in Indonesia. Jakarta is located on the northwest coast of Java. Jakarta has a land area of about 661.52 km², with a population of 10,374,235 people (2017). The Jakarta area is the most populous metropolitan area in Southeast Asia and the second most populous city in the world. Jakarta was located in the lowlands at an average altitude of 8 m above sea level. The southern area of Jakarta is a mountainous area with high rainfall intensity. Jakarta is crossed by 13 large rivers which all lead to Jakarta Bay, namely: Ciliwung River, Pesanggrahan River, Angke River, Baru Barat River, Baru Timur River, Grogol River, Mookervart River, Krukut River, Cipinang River, Sunter River, Buaran River, Jati Kramat River, and Cakung River. In addition, there are two Flood Channels, namely the West Flood Canal and the East Flood Canal and two large drainages in the west and east of Jakarta, namely Cengkareng Drain and Cakung Drain. Ciliwung River is the largest river in Jakarta, located right in the middle of the city of Jakarta. Almost all of Jakarta's land areas are residential buildings or office buildings. Jakarta has hot, dry or tropical temperatures. Jakarta experiences the peak of the rainy season in January and February with an average rainfall of 350 millimeters with an average temperature of 27 °C. Rainfall between January and early February is very high, that's when Jakarta is flooded every year, and the peak of the dry season in August with an average rainfall of 60 millimeters. September and early October are very hot days in Jakarta, temperatures can reach 40 °C [9].

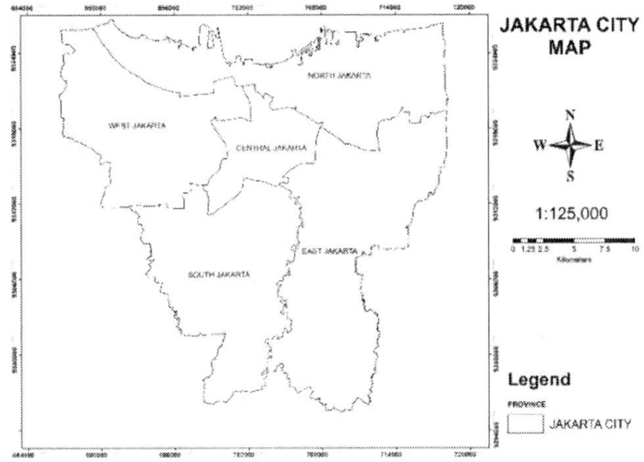

Figure 1. Administrative map of Jakarta city

IOP Conf. Series: Materials Science and Engineering **875** (2020) 012051 doi:10.1088/1757-899X/875/1/012051

2.2. Geological Study Area

In the geological map of Jakarta and the Thousand Islands [10], Jakarta is composed of the youngest to the oldest lithology: alluvium, coastal dike sediments, and alluvial fans. Alluvium in the form of clay to lump; alluvium fan in the form of fine tuff and sandy tuff intermittent tuff conglomerate; coastal embankment deposition of fine to coarse sand with mollusk shells (Figure 2).

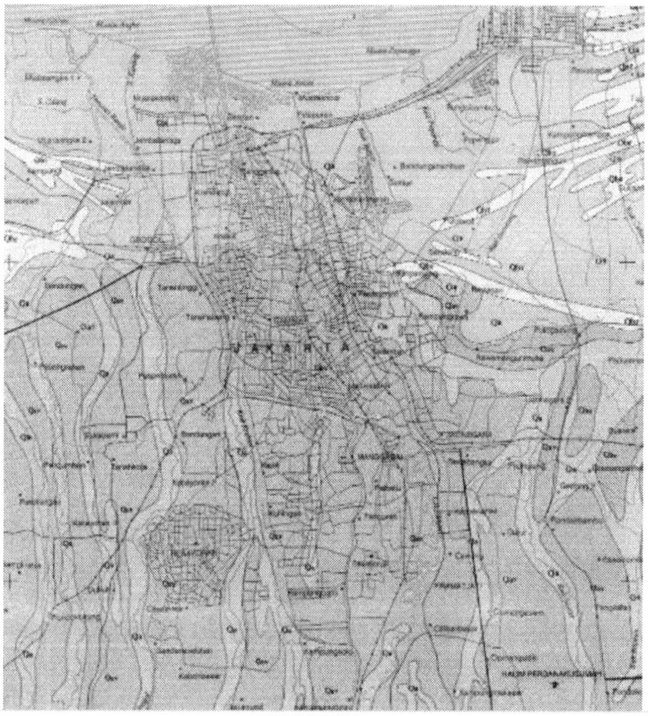

EXPLANATION

Qa	Alluvial	clay, silt, sand, gravel, pebble, and boulder
Qbr	Beach Ridge Deposit	coarse sand, well sorted, with mollusc shells
Qav	Alluvial Fan	bedded fine tuff, sandy tuff, interbedded with conglomeratic tuff

Figure 2. Regional geological map of Jakarta [10]

2.3. Data

The data used in this study amounted to 8 data consisting of 7 hydrological data which are factors that cause flooding, namely: rainfall data, land use data, area height data, slope percentage data, large river distance data, small river distance data, and groundwater level depth data [8] which are input data. As well as flood distribution data in Jakarta as external data. The input data used is shown in Table 1.

Table 1. Input data

No.	Data	Source
1	Rainfall	BMKG
2	Land Use	USGS Landsat Imagery
3	Elevation	DEM SRTM BIG
4	Slope	DEM SRTM BIG
5	Depth to Groundwater	BPBD DKI Jakarta
6	Distance to River	BPBD DKI Jakarta
7	Distance to Channel	BPBD DKI Jakarta

2.3.1. Rainfall. The data used are rainfall data obtained from the recording of 5 observation station points belonging to the Meteorology, Climatology and Geophysics Agency (BMKG). The 5 station points are in the region:

- Kedoya Selatan, West Jakarta
- Pakubuwono, South Jakarta
- Pintu Air Pulo Gadung, East Jakarta
- Pasar Minggu, South Jakarta
- Teluk Gong, North Jakarta

Five data points are processed using ArcGIS software to get the value of interpolation, so that the entire Jakarta area has a different rainfall value. The interpolation results using the kriging method in ArcGIS produce a map of the rainfall distribution in Jakarta which is shown in Figure 3. The map shows that rainfall with high intensity is in the southeast of Jakarta while low intensity rain is in the northwest of Jakarta.

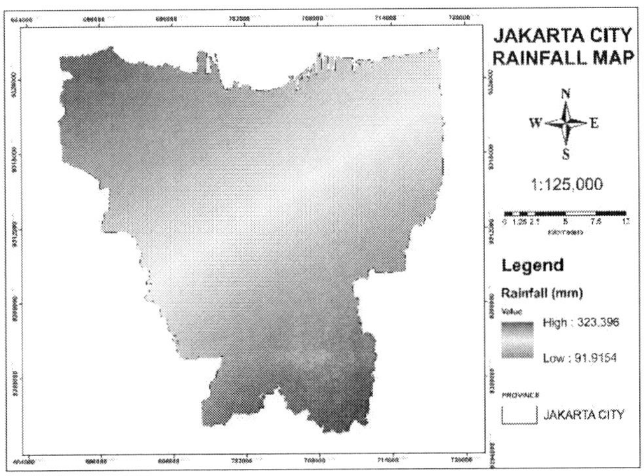

Figure 3. Rainfall distribution map of Jakarta

2.3.2. Land Use. Data used for land use were taken from USGS Landsat imagery which were then classified into four classes: built up land (settlements, buildings), water bodies (rivers, lakes and swamps), rice fields and fields, and vegetation cover. It can be seen in Figure 4 that land cover in the form of built up land is very dominant in the Jakarta area. The land cover of water bodies is mostly located in the northern part of Jakarta. Vegetation land cover and rice fields and fields only appear to be minimal.

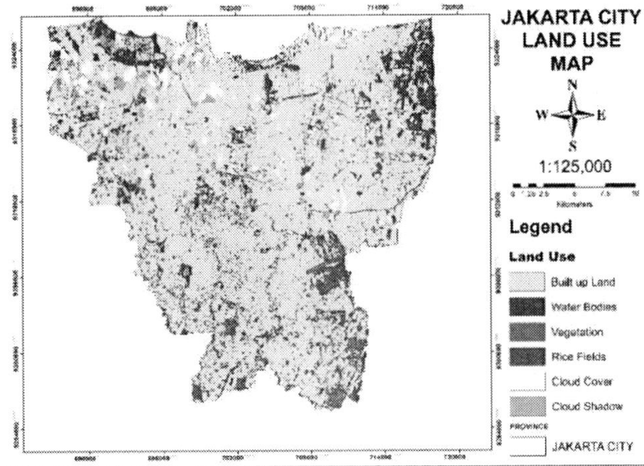

Figure 4. Land use map of Jakarta

2.3.3. Elevation. Elevation data in the Jakarta area is obtained using digital elevation model data released by the Geospatial Information Agency (BIG). The DEM data is then processed using the ArcGIS application to classify the height data in the Jakarta area. Then obtained four height classes seen in Table 2.

Table 2. Elevation classification

Class	Elevation
1	<5 m
2	5-25 m
3	25-50 m
4	>50 m

Elevation <5 m in the northern part of Jakarta including North Jakarta, part of West Jakarta and part of Central Jakarta. Elevation of 5-25 m is in the central part of Jakarta, covering parts of West Jakarta, Central Jakarta, South Jakarta and East Jakarta. Elevation of 25-50 m is in the southernmost part of Jakarta in the South Jakarta and East Jakarta regions. Elevation > 50 m is in the southern part of Jakarta bordering the Bogor area (Figure 5).

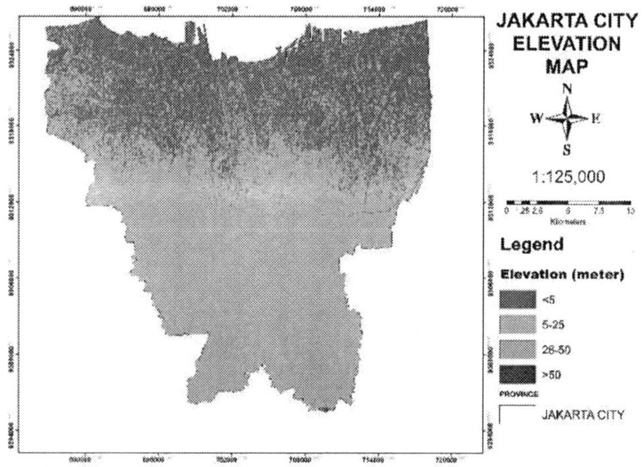

Figure 5. Elevation map of Jakarta

2.3.4. Slope. The slope data in the Jakarta area is obtained using digital elevation model data issued by the Geospatial Information Agency (BIG). The DEM data is then processed using the ArcGIS application to classify the height data in the Jakarta area. Then obtained five slope classes with percentage seen in Table 3.

Table 3. Slope classification

Class	Percentage of Slope
1	0 – 2 %
2	3 - 7 %
3	8 – 13 %
4	14 – 20 %
5	>20 %

In contrast to the elevation data, the percentage of slope in the Jakarta area is spread randomly with the dominant percentage at 3-7%, while the slope of> 20% is also dominant around the watershed in Jakarta (Figure 6).

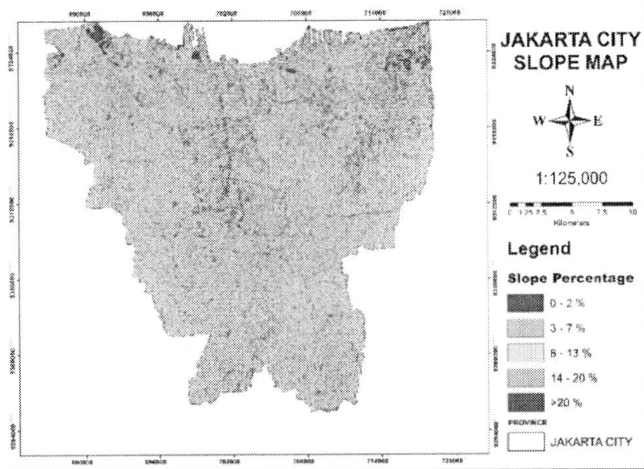

Figure 6. Slope map of Jakarta

2.3.5. Depth to Groundwater. The depth of groundwater level data was obtained from BPBD DKI Jakarta. The data are three data points:

- Cilincing District, North Jakarta
- Pademangan District, North Jakarta
- Cakung District, East Jakarta

Just like rainfall data, groundwater depth data are interpolated to get data distribution that interprets groundwater depths in all points of DKI Jakarta province. Interpolation using the kriging method. The results can be seen in Figure 7 which shows that in the western part of Jakarta the depth of the ground water level is getting deeper, while the eastern part of Jakarta is deeper and shallower.

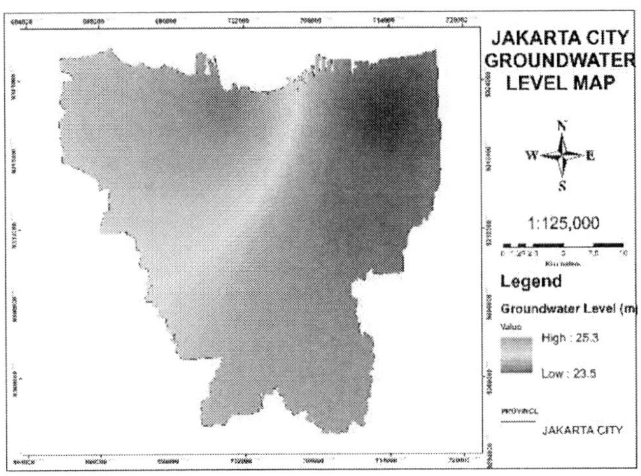

Figure 7. Groundwater level map of Jakarta

2.3.6. Distance to River. Major rivers in the Jakarta area are 13 main rivers that flow in Jakarta, namely: Ciliwung River, Pesanggrahan River, Angke River, Baru Barat River, Baru Timur River, Grogol River, Mookervart River, Krukut River, Cipinang River, Sunter River, Buaran River, Jati Kramat River, and Cakung River. Including two flood canals, namely: West Flood Canal and East Flood Canal, and two large drainages in the west and east of Jakarta, namely: Cengkareng Drain and Cakung Drain.

To find out the distance from each river using the method in ArcGIS software, namely: Multiple Ring Buffer. Before using this method, first classifying river distance class which is divided into three namely near, intermediate and far as shown in table 4.

Table 4. Distance to river classification

Class	Distance
Near	< 500 m
Intermediate	500 – 2500 m
Far	2500 - 7000 m

The results of this method are seen in Figure 8. From that map, we can be seen that most of the areas in Jakarta are located at a medium distance from a large nearby river.

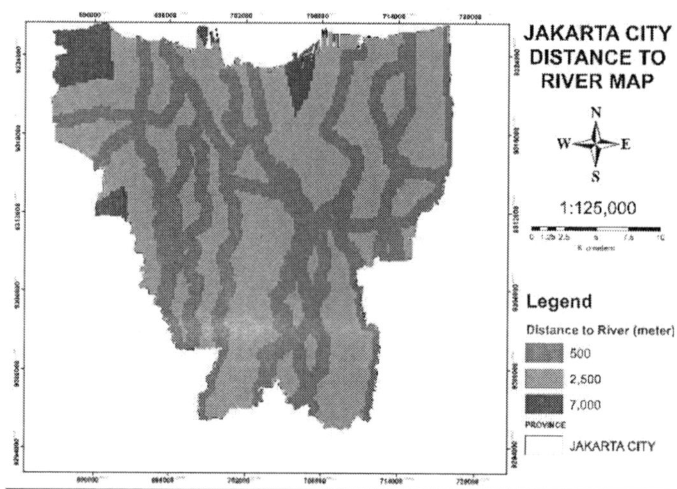

Figure 8. Distance to river map of Jakarta

2.3.7. Distance to Channel. The channel examined in this study is a tributary of the 13 major rivers in the Jakarta area. Class classification on distance data is the same as the distance on a large river but different on the size of the distance in each class is seen in Table 5

Table 5. Distance to channel classification

Class	Distance
Near	< 500 m
Intermediate	500 – 2500 m
Far	2500 – 15000 m

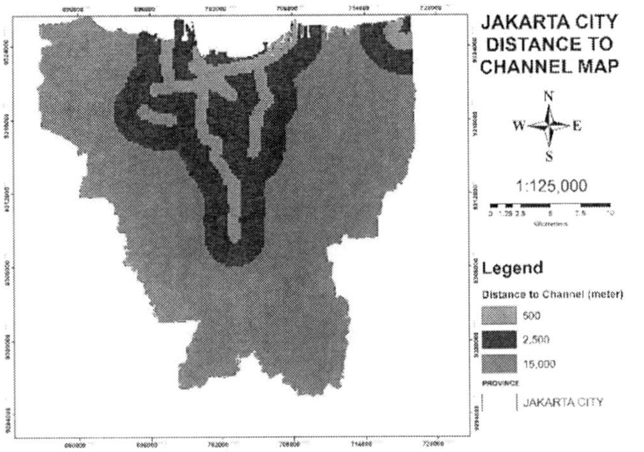

Figure 9. Distance to channel map of Jakarta

2.3.8. Flood Distribution. The data of flood point distribution was obtained from the Regional Disaster Management Agency (BPBD) of DKI Jakarta Province in 2015. It was seen that the distribution of flood points was most numerous in the northern area of Jakarta and along the main river flow (Figure 10).

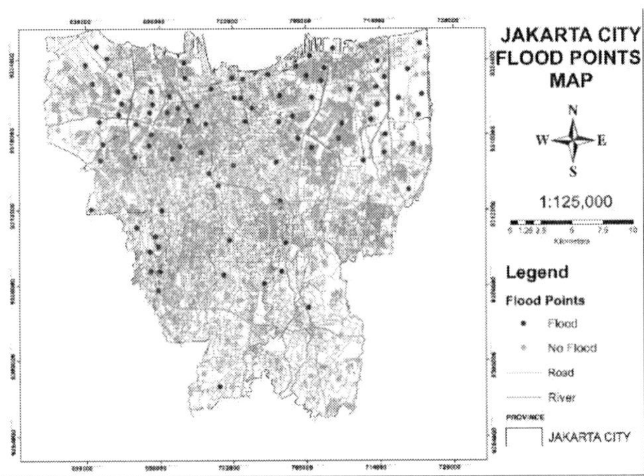

Figure 10. Flood distribution map of Jakarta in 2015

3. Methods

3.1. Genetic Algorithm for Rule-Set Prediction (GARP)

Genetic Algorithm Rule-Set Production (GARP) is a machine learning algorithm that has demonstrated excellent predictive abilities in areas such as ecological modeling [11]. The GARP algorithm [8] produces predictions of flood inundation for urban areas through an iterative process to improve the stability of the model output. Running multiple processes to obtain different model outputs and using the best subset method is important in choosing the best output with optimal parameters. The set of models that achieve harmony between the limits of error of omission (sensitivity) and commission (specificity) are defined by the user [8]. GARP output is a collection of grids from the study area, which can be used in a GIS environment to identify flood-prone areas [8]. In this case, the GARP model is run using software such as Arc GIS. In addition, the importance of flood-causing factors such as rainfall, slope, distance to river, distance to channel, depth to ground water, land use, and elevation for urban flood hazard was analyzed using the GARP model.

3.2. Quick, Unbiased, and Efficient Statistical Tree (QUEST)

QUEST [12] is a popular data processing model that produces subsets of data that are as homogenic as possible with respect to response variables. QUEST is a tree structured classification algorithm that produces a growing binary split decision tree [8]. It uses a sequential tree planting method, which uses a linear discrimination analysis method in splitting tree nodes. This has many advantages over recursive tree construction methods such as classification and regression trees (CART) [8]. The QUEST algorithm was chosen as the second model to predict flood inundation. In addition, the QUEST algorithm uses imputation instead of substitute separations to handle missing values. QUEST has negligible bias because it uses variable selection techniques that are unbiased in modelling. Therefore, QUEST can easily handle categorical and sustainable factors [8].

4. Result and Discussion

The results of the GARP model (AUC-ROC = 94%, RMSE = 0.2) show that this model uses a set of rules that have been restricted so that weights can be determined from each input data parameter used. From this model it is known that the main factors causing flooding from the weighting value of each parameter. The parameters of distance to rivers have the highest weighting of 30%, followed by height parameters of 20%, land use by 15%, distance to channels with weights of 15%, rainfall has a weight of 10%, ground water depth of 8%, and slope of 2%.

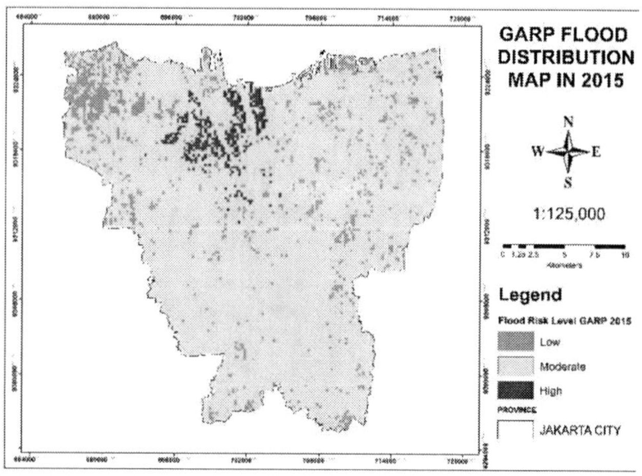

Figure 11. 2015 flood distribution map results of the GARP Model

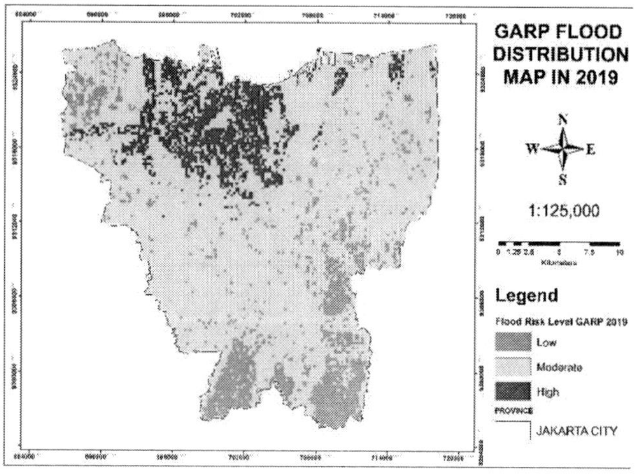

Figure 12. 2019 flood prediction map results of the GARP Model

The results of the QUEST model (AUC-ROC = 84%, RMSE = 0.4) show that this model gets worse grades than the GARP model. This is because there is a need for data clustering that can produce bias values against the input data parameters used. The resulting weight values for each parameter are: height of 25%, land use by 20%, distance to rivers by 20%, distance to channels by 15%, rainfall by 10%, depth of ground water by 9%, and slope of 1%.

The weighting value of each of these parameters results in a disaster risk map that is classified into three classes, namely: small disaster risk, medium disaster risk, and large disaster risk. The map results in Figure 11 are the results of the GARP model for the 2015 flood distribution data and Figure 12 results from the GARP model for the 2019 flood prediction.

The results of the QUEST model can be seen in Figure 13 is the result of the QUEST model for 2015 flood distribution data and Figure 14 of the QUEST model results for the 2019 flood prediction.

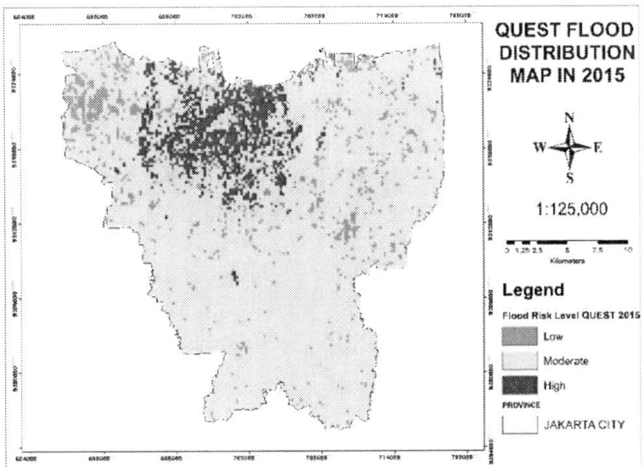

Figure 13. 2015 flood distribution map results of the QUEST Model

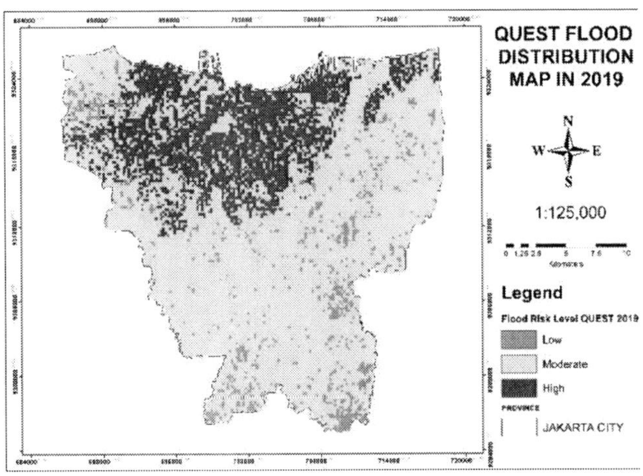

Figure 14. 2019 flood prediction map results of the QUEST Model

The results of modelling using the GARP and QUEST models show that there is a similarity between the results of the model for the 2015 flood data distribution with the 2015 flood data distribution owned by DKI Jakarta BPBD. This can be seen from the distribution of flood data, which is mostly located in the northern area of Jakarta, very similar to the map of the model results which shows that the greatest risk level is in the northern area of Jakarta.

The results of the model for the predictions of floods in 2019 show that the area that experiences the greatest risk level becomes wider than the 2015 data. This is expected due to higher rainfall data and has a broad rainfall distribution. In the southern part of Jakarta, the point of flood risk is in the smallest risk class. In the central part of Jakarta, most are in the middle risk flood class.

5. Conclusion

Province of Special Capital Region of Jakarta is an area that is at risk for flood disasters. then effective flood management is needed in predicting floods that will occur in the future. In this research, flood modelling was made using two models, namely the GARP model and the QUEST model. The GARP model and the QUEST model are used for research that produces flood risk maps in the Jakarta city area. The results of this model can be applied to flood zoning in urban areas. The results show the distance to the river is the main factor causing floods in the city of Jakarta. From these results it can be used using the GARP model and the QUEST model can be relied on with little data and is also fast in predicting flooding.

Acknowledgement

The Authors would like to thank Program Study Geology and Department of Geography Universitas Indonesia for supporting this research. This research was funded by Universitas Indonesia International Indexed Publication (PIT-9) grant with contract number NKB-0035/UN2.R3.1/HKP.05.00/2019.

References

[1] Jati, R. dan Amri, M.R., 2016. Indonesian Disaster Risk. Jakarta: Badan Nasional Penanggulangan Bencana. [in Bahasa]

[2] Liu, J., Doan, C.D., Liong, S.Y., Sanders, R., Dao, A.T. dan Fewtrell, T., 2015. Regional frequency analysis of extreme rainfall events in Jakarta. Natural Hazards, 75(2), pp.1075-1104.

[3] Putuhena, W.M. and Ginting, S., 2013. Development on Jakarta Flood Modelling System. Jurnal Teknik Hidraulik, 4:1, pp. 63-78. ResearchGate.net.

[4] Tehrany, M.S., Pradhan, B., Mansor, S. dan Ahmad, N., 2015. Flood susceptibility assessment using GIS-based support vector machine model with different kernel types. Catena, 125, pp.91-101.

[5] Kazakis, N., Kougias, I. dan Patsialis, T., 2015. Assessment of flood hazard areas at a regional scale using an index-based approach and Analytical Hierarchy Process: Application in Rhodope–Evros region, Greece. Science of the Total Environment, 538, pp.555-563.

[6] Shafizadeh-Moghadam, H., Valavi, R., Shahabi, H., Chapi, K. dan Shirzadi, A., 2018. Novel forecasting approaches using combination of machine learning and statistical models for flood susceptibility mapping. Journal of environmental management, 217, pp.1-11.

[7] Choubin, B., Moradi, E., Golshan, M., Adamowski, J., Sajedi-Hosseini, F. dan Mosavi, A., 2019. An Ensemble prediction of flood susceptibility using multivariate discriminant analysis, classification and regression trees, and support vector machines. Science of the Total Environment, 651, pp.2087-2096.

[8] Darabi, H., Choubin, B., Rahmati, O., Haghighi, A.T., Pradhan, B. dan Kløve, B., 2019. Urban flood risk mapping using the GARP and QUEST models: A comparative study of machine learning techniques. Journal of Hydrology, 569, pp.142-154.

[9] Turner, Peter, 1997. Java (1st edition). Melbourne: Lonely Planet Publications. pp. 37

[10] Turkandi, T., 1992. Geological Map of Jakarta and the Thousand Islands, Java 1: 100,000 scale. Bandung: Dit. P3G, Dit. Geologi dan Sumberdaya Mineral. [in Bahasa]

[11] Stockwell, D., 1999. The GARP modelling system: problems and solutions to automated spatial prediction. International Journal of Geographical Information Science, 13:2, pp.143-158.

[12] Loh, W.Y. and Shih, Y.S., 1997. Split Selection Methods for Classification Trees. Statistica Sinica, 7, pp.815-840.

Analysis of Soil Erosion on Mine Area

M Ramli[1]*, Purwanto[1], M Thamrin[1], Maemuna[1], and M Asrafil[1]

[1]Mining Engineering Department, Engineering Faculty, Hasanuddin University, Makassar, Indonesia

*Email: ramli@unhas.ac.id

Abstract. Mining activities disturbing large areas of land may increase erosion rate up to several hundred times greater than from undisturbed areas. The erosion process occurs in stripping overburden, excavation of rocks and minerals, dumping in stock file and waste dump, and mine reclamation. Since the eroded material/sediment becomes a big problem to the environment and for mining operations, estimation of soil erosion need to be carried out to create a good mine planning. For the early stage, a study of soil erosion was performed to classify erosion hazard level and to estimate soil erosion size. This study uses the Revised Unified Soil Loss Equation (RUSLE) method supported with Geographic Information System application for study area in the Block D1-D3 of Petea Mine Area. The study result showed that the research area can be classify into four erosion hazard level namely; low, medium, heavy, and very heavy with estimated material losses is 116,146.43 tons/year.

1. Introduction

Mining activities may produce environmental impacts such as erosion, sinkholes, loss of biodiversity or the contamination of soil, groundwater, and surface water. This paper discusses the erosion in mining area. The erosion of exposed hillsides, mine dumps, tailing dams, and resultant siltation of drainages, creeks and rivers may significantly impact the surrounding areas. The soil erosion depends on: climatic conditions, soil erodibility, overland slope and slope length, ground cover, soil conservation control practices, and catchment drainage characteristics. Mining activities tend to change radically several of these factors and severe sediment production could occur in the following locations: topsoil stockpiles, spoil piles, waste dumps, bare topsoil areas, steep out slopes, ramps, and haul roads [1]. Most methods for determining soil erosion and sediment transport have been developed based on studies with agricultural lands. Although relatively little work has been done with surface mined lands, they will in general have similar sedimentation characteristics to agricultural areas.

The most widely accepted and utilized soil erosion estimation is Unified Soil Loss Equation (USLE) developed by ARS scientists W. Wischmeier and D. Smith. The Universal Soil Loss Equation [2]:

$$A = R \cdot K \cdot LS \cdot C \cdot P \quad (1)$$

where A is the soil loss per unit area, R is a rainfall factor usually expressed as the product of rainfall energy times the maximum 30 minutes intensity for a given rainstorm, K is soil erodibility, LS is a dimensionless length slope factor, C is a dimensionless soil cover factor and P is a dimensionless conservation practice factor. Equation (1) should only be used to determine gross erosion from an area. To determine the sediment yield at a downstream point the gross erosion needs to be multiplied by a sediment delivery ratio term (sediment yield/gross erosion). Numerous methodologies have been developed to predict sediment delivery ratios. A method known as the Modified Universal Soil Loss Equation (MUSLE) overcomes the problem of determining a delivery ratio and has seen widespread application. The method is defined by the following equation:

$$Y = 11.8(Q \cdot q) - 0.56\,KLSCP \tag{2}$$

where Y is a single storm sediment yield (ton), Q is the storm runoff volume (m^3), q is the peak storm discharge (m^3/sec), and K, LS, C and P are the standard USLE terms used in equation (1).

However, the USLE was modified by incorporating the features that make its application suitable to mining lands, construction sites and other land applications and the same is popularly known as the revised universal soil loss equation (RUSLE). The RUSLE is an exceptionally well-validated empirical soil erosion prediction model which estimates average annual soil loss and sediment yield resulting from inter-rill and rill erosion. It computes the average annual soil erosion rate caused by rainfall and its associated overland flow. Galetovic et al maintained that the RUSLE included several modifications that have specific importance to mined lands, construction sites, and reclaimed land applications. For example, 'K-factor' was modified to account for the variability of soil erodibility during the year. Now, both K and C factors take into account the multivariate influence of rock fragment covers within soil profiles and fragments resting upon hill slope surfaces. Also, the equations used to estimate the LS factor were reconstituted to improve their accuracy and extended to include steeper hill slope gradients than the equations contained in the USLE [3].

The main objective of this paper is to quantify the soil erosion from the mining area by using the RUSLE method and to determine the erosion hazard level. The RUSLE method is used to estimate erosion rate, and all parameter is represented with map using Geographic Information System (GIS) Application software. The overlay is then used to determine erosion hazard level. The study area is in the Block D1-D3 of Petea area of PT Vale Indonesia Tbk, covering an area of 1,071.83 hectares with elevation about 1,075 meters above sea level.

2. Research Methodology

The research was conducted by collecting data from PT. Vale Indonesia Tbk, Environmental Office of South Sulawesi, and relevant published data on website. The collected data from PT. Vale Indonesia Tbk is topographic data/map and recorded rainfall data of the Block D1-D3 of Petea Mine Area (Hydrological Department), and pit design of the Block D1-D3 of Petea Mine Area (Department of Long Term Planning). The collected data from South Sulawesi Environmental Office is land use map, and the data from website are Geological Map of Bungku Quadrangle Sulawesi with scale 1: 250.000 developed by the Geological Resource Center, and Map of World Soil Type.

The collected data were used to generate five parameters of the RUSLE equation, which include; rainfall-runoff erosivity factor (R), soil erodibility factor (K), slope length and steepness factor (LS), land cover (C) and land support practice (P). The rainfall-runoff erosivity factor (R) was generated predicted rainfall using Gumbel Method and was defined unity for the study area. The soil erodibility factor (K) were estimated by using composition of clay, silt, and sand, and was also assumed unity. Fo the study area. The slope length (L) and steepness factor (S), the land cover and management factor (C), and the land support practices factor (P) were drawn in the GIS environmental software. Furthermore, by using overlay maps system for the RUSLE parameter, the erosion hazard level was defined, and the erosion rate for each erosion hazard area was counted.

3. Soil Erosion from Mining Area

3.1. Geological Setting

The Block D1-D3 Petea mining area is in the mountains with an altitude of 150 – 1,075 meters above sea level. This area is dominated with a steep slope in the center to the north area, and moderate slope in the south. A complex tectonic shown in this region is indicated by number of geological structures; folds, joint, and fault. The main fault is Matano Fault in northwest-southeast direction. The Matano

Fault is a sinistral fault that is considered to be connected with the Sorong Fault. The Block D1-D3 of Petea area is composed by the Ultramafic Complex (Ku).

The Ultramafic Complex (Ku) consists of harzburgite, lherzolite, wehrlite, websterite, serpentinite, dunite, diabase and gabro [4]. Harzburgite is green to black, massive, locally refined minerals, composed of fine to coarse minerals, consisting of olivine, pyroxene and serpentine. The latest mineral is alteration product from pyroxene and olivine. Lherzolite is blackish green, massive and medium to coarse grained. Wehrlite is black, massive, and fine-grained to coarse graied. The rock is composed of olivine, sometimes pyroxene. These minerals are converted into serpentine, talc and idingsit. Websterite is solid black and massive composed of olivine and pyroxene minerals, finely grained to moderate. Serpentine replaced olivine and pyroxene fills the modified crystal fracture. Serpentinite is gray to blackish green, massive composed of antigoritic minerals, clay and fine grained magnetite. These rocks are generally often found sturdy structures and there is a cesarean mirror (solid, fine-grained to moderate, composed of plagioclase minerals, orthoclase, pyroxene and ore minerals, these rocks are strongly altered in several places. Fine to moderate-grained dunite, greenish, grayish-green to blackish, composed of olivine minerals, pyroxene, plagioclase and ore minerals, olivine and pyroxene minerals are transformed into serpentine minerals, talc and chlorite, these rocks are found to be strongly altered, showing the structure of the nest of the ore. Black-spotted gabro, medium to coarse, massive, this rock-forming mineral consists of plagioclase and olivine. These rocks are found in the form of cracks that break through ultramafic rocks.

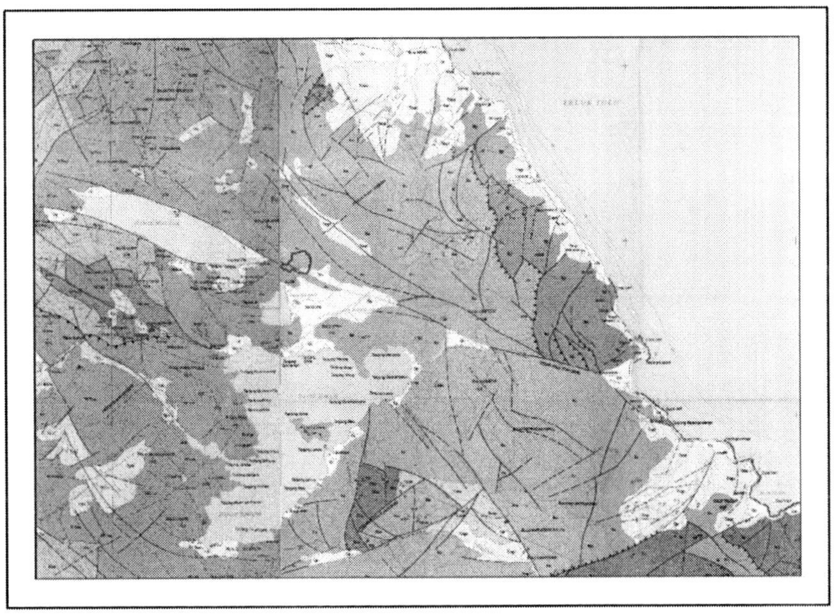

Figure 1. Geological map of study area and its surrounding (Geological Research and Development Center, Bandung, 1993)

3.2. Parameter of the RUSLE Model

The RUSLE model is an empirical model that is the most widely used for erosion modelling in mining [3]. The RUSLE equation computes the average annual erosion or average soil loss (A). The A is usually expressed in ton/ha/year. Several factors related to the A need to be generated are; the R is the rainfall-runoff erosivity factor (J/ha); the K is the soil erodibility factor (t/J); the LS is the hill slope length and steepness factor (dimensionless); the C is the cropping management factor (dimensionless); and the P is the supporting conservation practice factor (dimensionless). To estimate the soil loss in this mine area, the RUSLE model was used in a GIS environment.

The rainfall-runoff erosivity factor (R) is computed using Equation (3) proposed by Singh et al. [3];

$$R = 79 + 0.363\ (X)\ \text{(J/h)} \tag{3}$$

where X is the average annual rainfall (mm). Estimation of the R factor uses a daily rainfall data for five years (2012-2017) were collected from the study area. The rainfall data was analyzed using Gumbel Method to estimate rainfall for a desired 5 years return period. The estimated rainfall-runoff factor for the Block D1-D3 Petea mine area is 100.2 J/h. Since the rainfall intensity recorded on this area only at one gauge station, the rainfall/runoff erosivity factor are similar for the study area.

The soil erodibility factor (K) calculation was initiated Wischmeier et al., (1971) by representing susceptibility of soil to erosion and the rate of runoff, taking into account the % portion of silt, sand or organic matter, soil texture, soil structure and permeability as soil erodibility factor [2]. A simple method for indirectly estimating soil erodibility were applied by Bouyoucos equation based on soil physical properties (texture and organic matter content) as input data [5];

$$K = \frac{SAN + SIL}{CLA} \times \frac{1}{100} \tag{4}$$

where: SAN, SIL and CLA are percent sand, silt and clay, respectively. The erodibility factor is calculated based on filter analysis of 27 soil samples taken from the mine site (Table 1) for the mine area and based on soil type data for areas outside the mine. Soil laterite is characterized with highly compacted and cemented soil can easily be cut into brick-shaped blocks. This material is quick stable or difficult to destroy by raindrops. Calculation the erodibility factor using equation 4 show a various value from 0.008 – 0.081 and average is 0.012 for the mining area and 0.032 for the area outside the mine. This erodibility value is also assumed to be similar within the study area.

The LS factor considers for two parameter slope length and slope steepness. The LS factor describes the combined effects of slope length (i.e., flow length) and slope gradient (i.e., grade or relief). Slope length is defined as the distance from the point of origin of overland flow to the point where the slope decreases sufficiently for deposition to occur or to the point where runoff enters a defined channel (wet or dry). The slope steepness is the segment or site slope. Slope length and slope steepness strongly influence the transport of soil particles once the soil particles are dislodged by raindrop impact or runoff. Soil loss increases more rapidly due to slope steepness than due to slope length.

Estimation of LS factor of Block D1-D3 of Petea Mine Area was performed by driven a slope map using topographical map. The topographical map is produced by Department of Geotechnics, Hydrology, Reclamation and Environment. The map represents surface relief with contour interval of 5 meters. Determination of the length and slope map is obtained based on the results of the raster calculator processing in Arcmap 10.3 Software, the value entered is the value of accumulated flow grid, grid slope percentage, and grid size.

The slope is one of the factors in determining the LS value. The results of the processed data on the processed raster calculator can be seen in Figure 2. Based on the equation entered into the raster calculator in Arcmap 10.3 Software, the LS value varies from 1 m to 16 m in length. Figure 1 shows that a high LS value is at a high level of steep steepness marked with a reddish orange color and a low LS value at a gentle steepness level marked with grays.

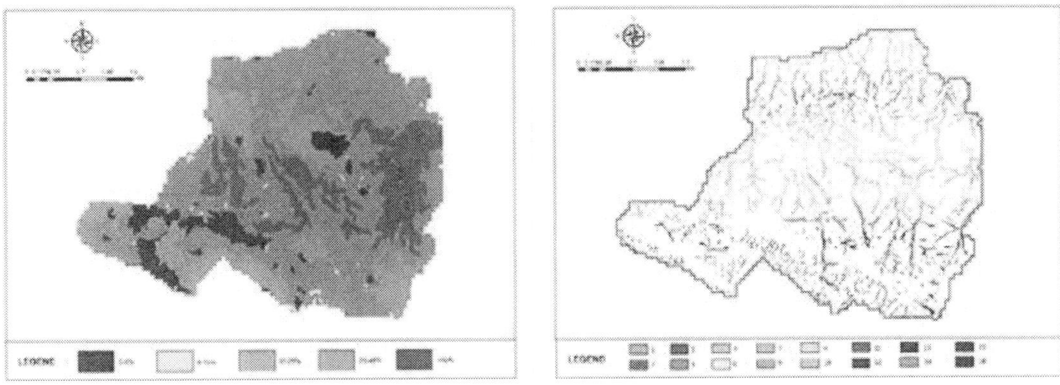

Figure 2. Map of estimated LS factor; Slope Steepness (left), and Slope Length (right)

Table 1. Slope steepness classification of the Blok D1-D3 of Petea Mine Area

No	Slope (%)	Class	Area (Ha)	Area (%)
1	0 - 8	Little or none	76.78	7.16
2	8 - 15	Gentle	13.83	1.29
3	15 - 25	Moderate	219.48	20.48
4	25 - 45	Steep	605.75	56.52
5	> 45	Extremely steep	102.54	14.55
		Total	1071.83	100.00

The C factor reflects the effect of cropping and management practices on soil erosion rates. Since the C factor represents the effect of plants, soil cover, soil biomass and soil disturbing activities to erosion, each land cover type can correspond to an estimated C value. Figure 3 shows the C-factor values for the different land use categories. The study area is classified into four types of land cover namely open land, shrubs, forests, and mining area. The C value of each land cover can be seen in Table 2.

Table 2. Cover factor value of the Blok D1-D3 Petea Mine Area

No	Land use	C value	Area (Ha)	Area (%)
1	Open (soil) area	0.600	670.50	62.56
2	Shrubs	0.001	45.46	4.24
3	Forest	0.010	132.42	12.35
4	Mining	0.034	223.45	20.85
	Total		1071.83	100

Based on the Table 2, it is shown that land use in the Block D1-D3 of Petea Mine Area is dominated with open (soil) area covering 683.21 ha with a value of C (0.6), mining land with an area of 223.06 ha and a value of C of (0.034), forest Land with an area of 135.37 ha and a C value of (0.01), and shrubs with an area of 53.15 ha and a C value of (0.001). Distribution of land cover in the Catchment Area can be seen in Figure 3.

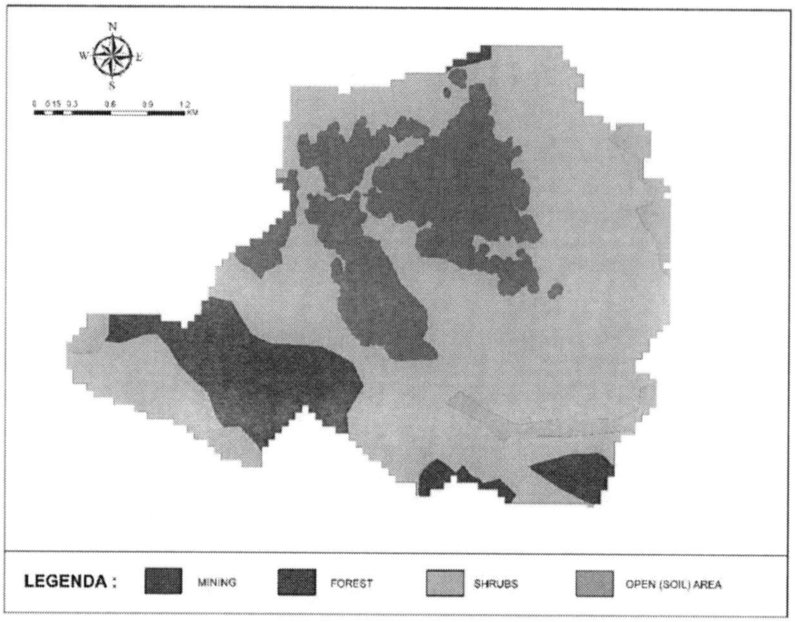

Figure 3. Land use map

The P factor reflects the impact of support practices on the average annual erosion rate. As with the other factors, the P-factor differentiates between cropland and rangeland or permanent pasture. Since the Block D1-D3 of Petea Mine Area on going progress for mining activities, there is no soil treatment and conservation carried out, specifically for the D1 and D2 Block areas. Reclamation activity will be performed while the Block D3 is mined. Therefore, there is no processing activity and soil conservation for handling the deteriorated soil in this area. Therefore, estimation the P factor value is based on topographic values and slope values [6]. Table 3 shows the estimation of P factor value based slope and topographical condition. This shows that the handling of land conservation in the steep sloping area is greater than the area with the gentle slope.

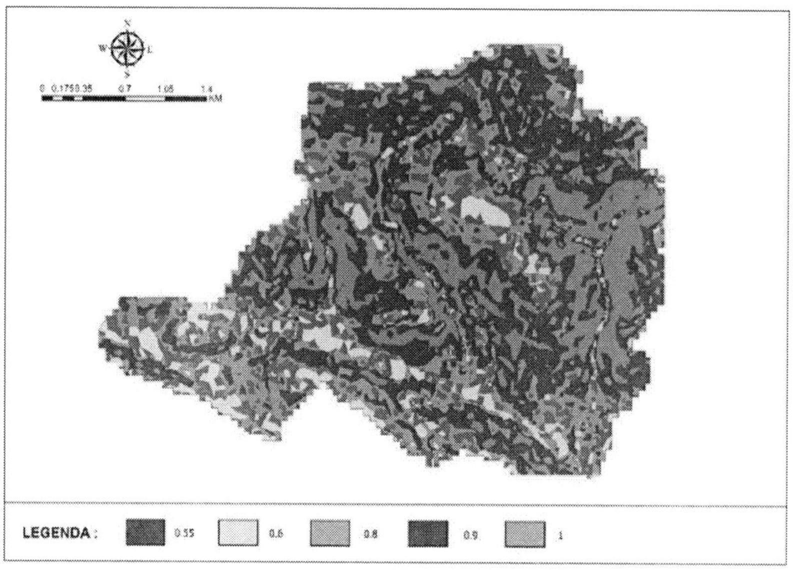

Figure 4. Map of land conservation

Table 3. Land conservation factor based on slope steepness

Slope (%)	Elevation	*Strip Cropping*	Area (ha)	Area (%)
0.0 – 7.0	0.55	0.27	406.50	37.93
7.0 – 11.3	0.60	0.30	125.20	11.68
11.3 – 17.6	0.80	0.40	265.00	24.72
17.6 – 26.8	0.90	0.45	97.78	9.12
< 26.8	1.00	0.50	177.35	16.55
Total			1071.83	100

3.3. Mine Erosion Hazard Level

The major causes of erosion problems at a surface mining operation are the extent of the disturbed areas, poor drainage plans and the lack of a sediment control strategy which is integrated with the mining operations. In developing a sediment control plan the following basic approach should be adopted: minimize the area which is disturbed at any one time; develop a drainage control system for the mine lease area; integrate drainage, erosion and sediment control into each stage of the mining operation; develop a mining and rehabilitation plan prior to initiating mining activities; construct drainage and erosion controls in advance of mining activities ; divert storm runoff away from areas with high erosion potential; incorporate measures to reduce the flow velocity of storm runoff; limit the handling of spoil and topsoil materials; rehabilitate areas as soon as possible; maintain drainage and erosion control measures [1].

Estimation of erosion rate in this research was calculated based on map of mine erosion hazard level with support by GIS software. The erosion hazard level map was generated by overlaying maps developed with parameter of the RUSLE equation (R, K, LS, C, P). The overlay parameters produced four hazard levels, namely; low level, medium level, high level, and very high hazard level. The estimation of average erosion rate for each level from low to very high hazard are 100.42 ton/ha/year, 113.58 ton/ha/year, 158.85 ton/ha/year, and 201.93 ton/ha/year, respectively. Furthermore, the widest area distribution is the criteria of low hazard level, then highest hazard level, medium hazard level, and high hazard level with an area of 949.35 ha, 75.03 ha, 41.43 ha, and 6.02 ha, respectively.

Table 4. Detail information for each erosion hazard level

No.	Hazard Level	Erosion Rate (ton/ha/year)	Area (ha)	Material Losses (ton/year)
1	Low	100.42	949.35	95,333.73
2	Medium	113.58	41.43	4,705.62
3	High	158.85	6.02	956.28
4	Very High	201.93	75.03	15,150.81
	Total erosion		1071.83	116,146.43

Based on data on the Table 4, in the total erosion from the Petea Watershed is 116,146.43 tons/year. This material loss volume was mainly caused by erosion from low hazard level area with area 949.35 ha to produce 95,333.73 ton/year. This due to wide of area is 88.57 % of the research area. The smallest material loss was caused by erosion from the high hazard level with area of 6.02 ha. Two other hazard level produced material loss volume of 4,705.62 tons/year for medium category, and 15,150.81 tons/year for the very high category. The analysis result show that the material loss volume depends on the wide area of each hazard level category. Therefore, the erosion hazard level represents the rate erosion but is not to indicate the volume of material loses or transported.

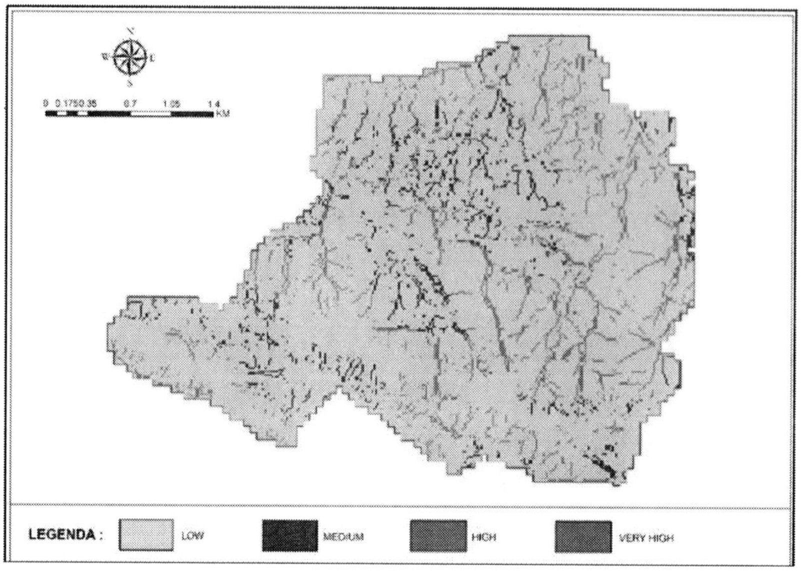

Figure 5. Map of erosion hazard level

4. Conclusions

The study of soil erosion on mining area illustrates that without careful management and the development of a sediment control scheme, the erosion from a surface mine can be severe. By using overlay maps of the RUSLE equation parameter on geographic information system, it was generated the erosion hazard level in the Mine Block D1-D3 area. The erosion hazard map categorizes hazard level into four categories, namely: low, moderate, severe, and very heavy erosion hazard levels. In the mine area, it estimated that erosion rates various from 100.42 – 201.93 ton/ha/year, with total material losses is 116,146.43 tons/year.

Acknowledgements

We gratefully thank the Faculty of Engineering -Hasanuddin University for financial support to this research through Laboratory-Based Education scheme.

References

[1] Ward A., Smith A., and Caldwell J., 1984, Surface erosion and sediment control at opencast mines in southern Africa Challenges in African Hydrology and Water Resources, *Proceedings of the Harare Symposium*, IAHS Publ. no. 144.

[2] Sigalos G., Loukaidi V., Dasaklis S., and Alexouli-Livaditi A., 2010, Assessment of the quantity of the material transported downstread of Spercios River, Central Greec, *Bulletin of the Geological Society of Greece*.

[3] Yellishettya M., Muddb G.M., and Shuklab R., 2013, Prediction of soil erosion from waste dumps of opencast mines and evaluation of their impacts on the environment, *International Journal of Mining, Reclamation and Environment*, Vol. 27, No. 2, pp. 88–102.

[4] Simandjuntak T.O., Rusmana E., Supandjono J.B., and Koswara., 1993, *Geological map of Bungku Quadrangle Sulawesi*, Geological Research and Development Center, Bandung.

[5] Anache J.A.A., Bacchi C.G.V., Panachuki E., Sobrinho T.A., 2015, Assessment of methods for predicting soil erodibility in soil loss modeling , Geociências, v. 34, n. 1, p.32-40.

[6] Arsyad S., 2006, Soil and Water Conservation, IPB Press, Bogor. [in Bahasa]

IOP Conf. Series: Materials Science and Engineering **875** (2020) 012053 doi:10.1088/1757-899X/875/1/012053

Extraction of Nickel and Cobalt from Sulawesi Limonite Ore in Nitric Acid Solution at Atmospheric Pressure

Sufriadin*, S Widodo, I Nur, A Ilyas, M Y Ashari

Department of Mining Engineering, Universitas Hasanuddin
Jl. Poros Malino, km. 6 Gowa, Sulawesi Selatan, 92171, Indonesia

*Email: sufriadin.as@gmail.com

Abstract. Characterization and leaching study of a limonitic laterite ore from Sulawesi has been undertaken in order to find out mineralogical nature and Ni and Co concentration along with dissolution behavior during leaching process in nitric acid. Extraction rates of Ni and Co in different acid concentration and reaction time were also assessed. Mineralogical identification was conducted using optical microscopy and X-ray diffraction (XRD) techniques; whereas Ni and Co content was determined by means of atomic absorption spectrometry (AAS). Results of analyses show that minerals found in lateritic nickel ore studied were dominated by goethite with less quartz, talc, and trace spinel. Goethite was found to be high dissolution rate in nitric acid whereas quartz, talc and spinel were not dissolved during the leaching process. The highest extraction rates of Ni and Co with different acid concentrations at 1 hour leaching time were 13.22% and 24.17% respectively; whereas the maximum Ni and Co extraction rates under 2 molar nitric acid with different leaching times were obtained at 90 minutes with recovery of 24.17% and 56.53% respectively.

1. Introduction

Lateritic nickel ore is formed by chemical weathering of ultramafic rocks situated in the tropical regions. Ore types are classified on the basis of their mineralogy namely: hydrous silicate, clay silicate and oxide ores [1-3]. Global nickel reserves in 2018 reached up to 89 million tonnes, of which about 21 million tonnes or 23.59% originated from nickel laterite deposits in Indonesia [4]. Nickel laterite deposits in Indonesia are distributed in Eastern Sulawesi, Halmahera and West Papua.

At present, nickel laterite ores in Indonesia are processed by using conventional smelting to produce nickel matte and ferro nickel. However, smelting prosess requires ore with higher grade (>1.5 wt%) of Ni to be economic viable. Therefore, lower grade ores, generally limonite, are not profitable to be processed and they are discarded to the waste dump. Hydrometallurgical route is known to be an alternative option in processing the lower grade Ni-ore so that resources of deposit will be expected to increase. Some advantages associate with using leaching method such as operating at lower temperature and some other valuable metals can be recovered [5]. Some leaching studies on lateritic nickel ores for both saprolite and limonite from Indonesia have been carried out using different leaching agents such as sulfuric acid [6], hydrochloric acid [7], nitric acid [8] and citric acid [9].

The objective of this paper is to find out mineralogical and chemical nature of a limonite ore sample, to analyze dissolution behavior of this Ni-laterite ore and to estimate recovery of Ni and Co during leaching process using nitric acid at atmospheric pressure condition.

2. Sample and Methods

Nickel laterite ore sample used in this study was collected from limonite zone of an abandoned mine in the Latowu village, Batuputih subdistrict of North Kolaka Regency, Southeast Sulawesi Province (Fig.1). The mineralogy of the sample including texture, abundance and types of minerals was initially characterized by means of polarizing microscope (Nikon Eclipse LV100POL) and X-ray diffractometry (XRD; Shimadzu Maxima X-7000); whereas chemical compositions of the sample mainly Ni and Co were analyzed using atomic absorbtion spectrometry (AAS).

Leaching experiment was performed by using magnetic stirer (LSH) with the temperature of 85°C, grain size of 100#, liquid to solid ratio of 10:1 and stirring rate 250 rpm. Leaching process was carried out in nitric acid (HNO_3) with different concentrations of 0.5; 1.0; 1.5; 2.0; 2.5; and 3.0 M and leaching time of 30; 60; 90; 120; 150 and 80 minutes. After leaching, pregnant solutions and solid residues were separated by membrane filter with diameter 0.45 micron. Pregnant leach solutions (PLS) were further analyzed by AAS whereas solid residues were characterized by XRD.

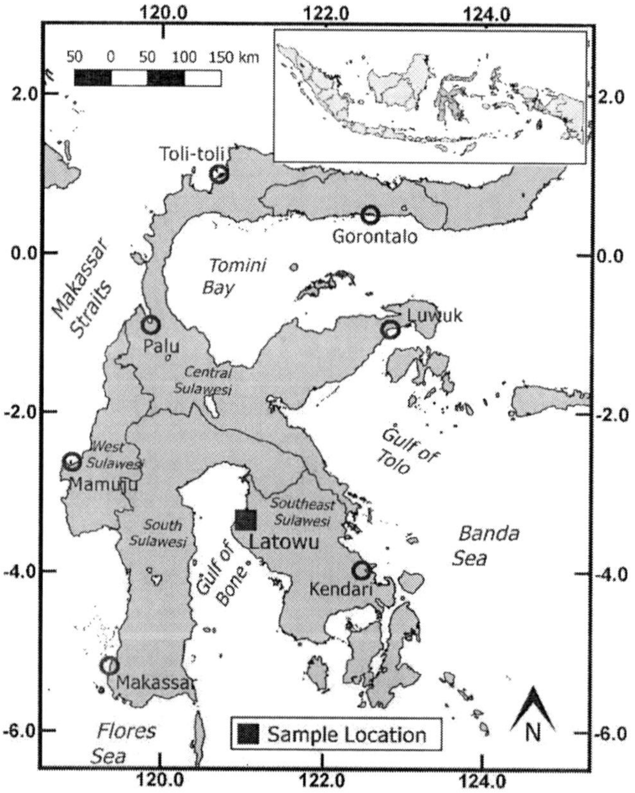

Figure 1 Map showing sample location

3. Results and Discussion

3.1 Ore Characterization

Photomicrographs of a limonite sample are depicted in Figure 2, while the XRD pattern is shown in Figure 3. Results of microscopic observation exhibits that sample predominantly contains goethite [α-FeOOH]. The occurrence of goethite was assigned to the irregular and acicular texture. It shows reddish brown and locally yellowish in color. Goethite was also detected with the presence of basal spacing with d_{hkl}~4.18Å; 2.69Å and 2.45Å.

Other minerals identified in association with goethite were quartz [SiO_2], talc [$Mg_3Si_4O_{10}(OH)_2$] and spinel [$MgAlCrSi_2O_4$]. Microscopic appearance of quartz was characterized by darker in color with relatively rounded in texture. The presence of basal reflections with d_{hkl} values of 4.25Å; 3.34Å and 1.82Å indicate positive identification of quartz. Talc was detected in sample based on XRD data with reflections intensity occurring at d_{hkl} values of 9.40 Å; 4.70 Å and 3.12 Å. Talc is a phyllosilicate mineral having 2:1 structure. It might be formed by hydrothermal or weathering process of mafic minerals mainly pyroxene. Spinel was also identified in sample in trace amount. It is shown by bright in color with unhedral crystal.

Result of chemical analysis of sample by means of AAS shows that sample contains 1.23% Ni. This value indicates low grade so that the ore is not amenable to be processed by means of conventional smelting due to the high cost in operation. Majority of mining companies in Indonesia exploit the Ni laterite ore with minimum of 1.50% Ni. Meanwhile, sample contains about 1.23% Co, indicating relatively high grade. The mean grade of Co in limonite ore is generally about 0.10%.

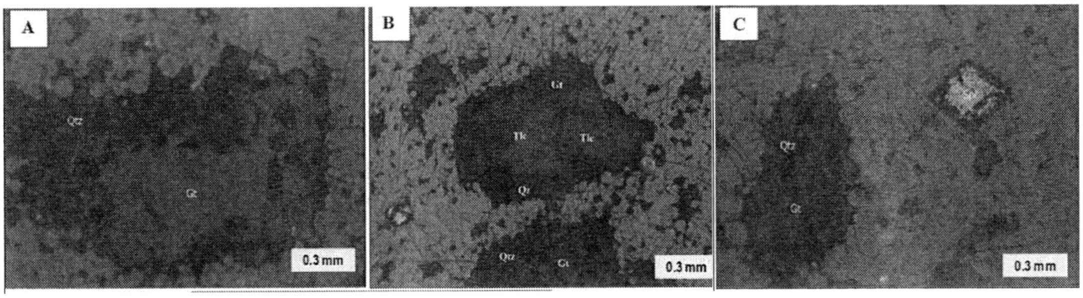

Figure 2. Photomicrographs of limomite ore sample showing the irregular texture of goethite (Gt) in association with quartz (Qtz), (A). Talc also present with goethite replacing the rim (B). Euhedral spinel, possibly chromite still preserves the crystal form, however the color start to change on the margin of spinel crystal into goethite (C)

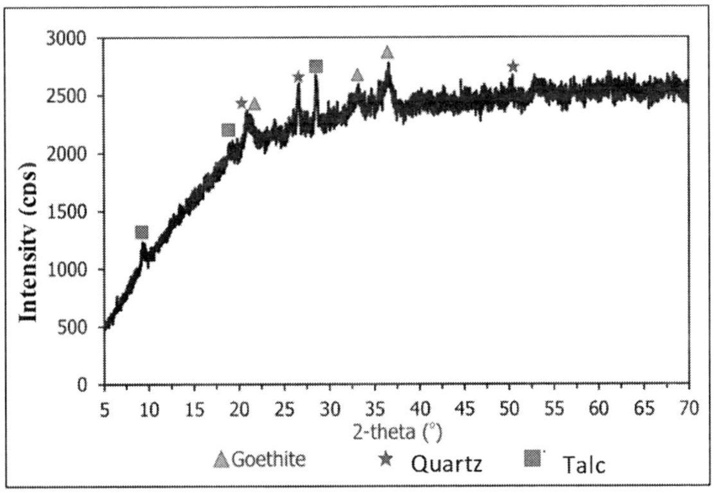

Figure 3. X-ray powder diffraction pattern of limonite ore sample used in this study

3.2 Dissolution Behavior of Limonite Ore

The XRD patterns of some solid residues after leaching of limonite ore in nitric acid solution are provided in Figure 4. It was shown that reflection intensity with d_{hkl}~4.16Å decreases after leaching process. It indicated that goethite as Ni and Co bearing mineral in limonite ore partially dissolve during reaction with nitric acid which called incongruent dissolution. In contrast, intensity of peaks with d_{hkl}~9.30Å and ~3.12Å strongly increase as compared with the peaks at raw ore sample. This indicate that talc did not dissolved. Similarly, quartz with basal spacing occurs at d_{hkl} 3.34Å exhibits the increasing intensity in solid residue.

Dissolution rate of minerals in acid solution is affected by some factors such as crystalinity, impurities, grain size and porosity [10]. Goethite is characterized by poorly crystalline minerals as the result of chemical weathering of ultramafic rocks. Goethite structure is formed by hexagonal closed-packing arrangement of O^{2-} and OH^- anions, stacked with Fe^{3+} occupying half of octahedral site [11]. Isomorphic subtitution of divalent cations (Ni, Co and Zn) in goethite is not only affecting the physical its properties such as color, unit cell dimension and magnetic susceptibility but also they can modify the solubility and dissolution kinetic. Crystal shape, crystal size, and crystallinity may also influence the dissolution rate of goethite.

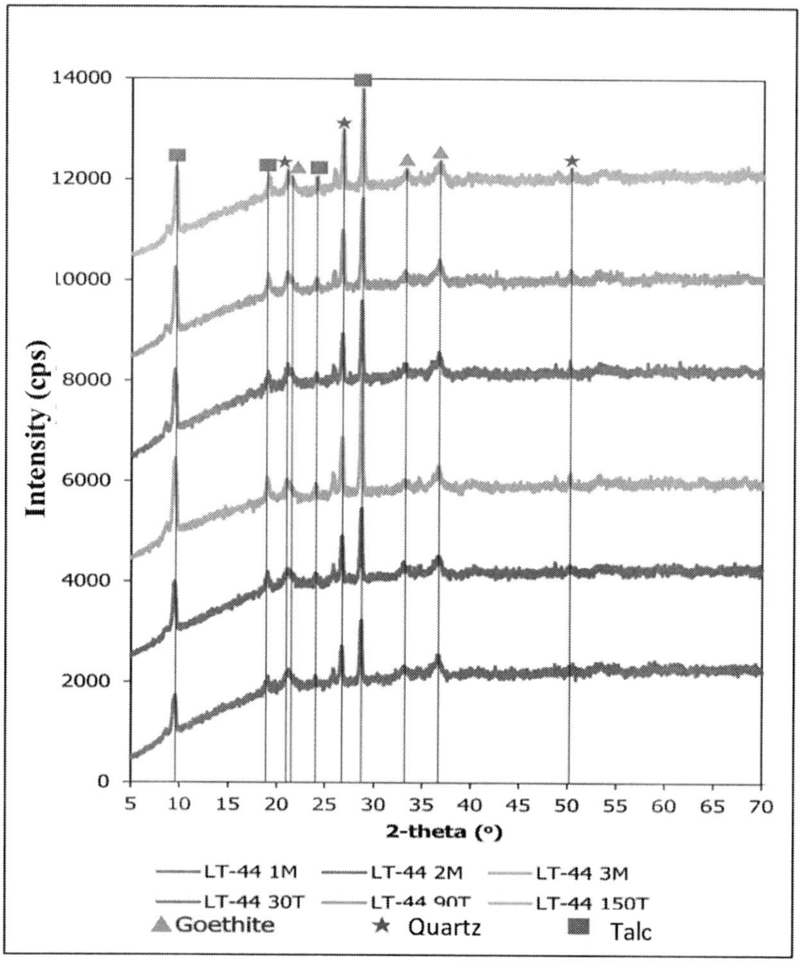

Figure 4. X-ray powder patterns of solid residues after leaching at different concentrations and reaction times

It is suggested that Ni and Co is generally present in goethite in three modes: (i) associated with poorly crystalline or amorphous goethite, (ii) adsorbed onto goethite surface, and (iii) substituted for Fe in goethite structure. Nickel and cobalt atom that located in the single coordination hydroxyl groups would be easier to liberate by acid than in double or triple coordination hydroxyl groups [12].

In the case of talc, quartz and spinel, these minerals do not not expected to be the host of Ni and Co. Hence, they are not necessary to dissolve during leaching process.

3.3 Extraction of Ni and Co from Limonite Ore

The influence of nitric acid concentration on extraction rate of Ni and Co was investigated at temperature 80°C and leaching time of 1 hours. It is shown that with an increase of HNO_3 concentration, followed by the increase of extraction rate of Ni and Co (Figure 5). However, leaching rate of Co was more twice than Ni. Cobalt likely present in association with manganese wad or asbolane. During leaching process, Co and Mn were released from host minerals. Eventually, Co extraction from laterite ore is easier than Ni due to the high concentration of Fe and relaively easy dissolution of manganese in acid [13].

The effect of leaching time on the extraction rate of Ni and Co was also investigated. As shown in Figure 6, it was indicated that the highest extraction rate for both Ni and Co took place at 90 minutes reaction in 2 M nitric acid. Maximum 24.17% of Ni extraction rate were reached at these conditions. This value is far lower than extraction rate of Co which attain up to 56.53%. However, the increase of leaching time was not accompanied by the increase in extraction rate, even it shows strongly decrease in leaching rate either Ni or Co at 120 minute reaction (Figure 6). Extraction rates of Ni and Co reveal slowly depletion after 2 hours reaction.

Results of leaching experiment exhibit that extraction rates of Ni and Co were still low. This might be due to incomplete dissolution of goethite as Ni and Co carrier. It is therefore suggested that dissolved Ni and Co in pregnant leach solution (PLS) most likely derived from adsorbed metals on goethite surface. Another possibility of low extraction rates of Ni and Co in this experiment is that formation of fine grained silica during laterization process or formation of amorphous silica during leaching process might readsorb Ni and Co from PLS. Cobalt extraction was much better than nickel recovery in this experiment.

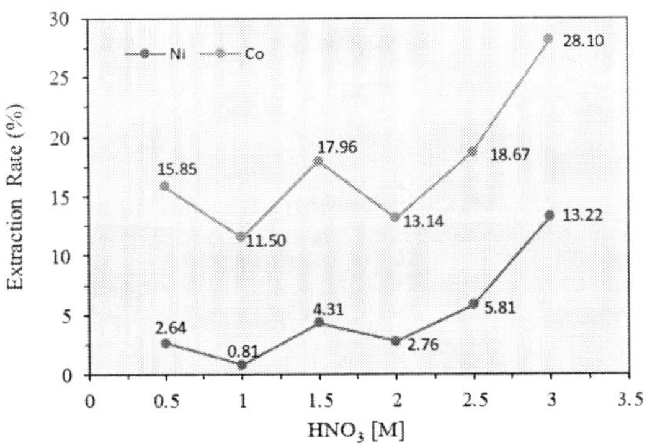

Figure 5. Graph displaying the influence of nitric acid concentration on the extraction rates of Ni and Co from limonite ore

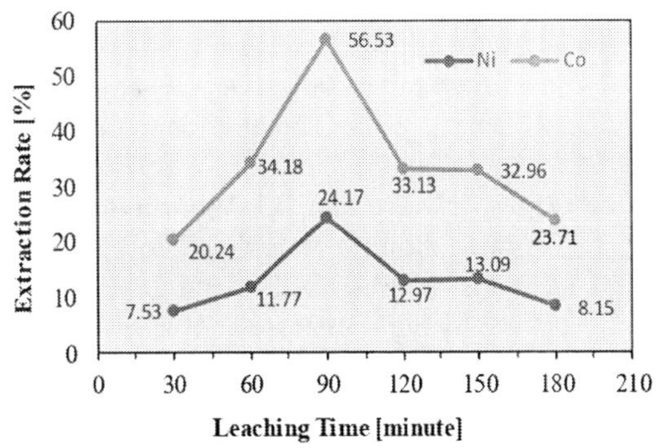

Figure 6. Grapah showing the effect of leaching time on the extraction rates of Ni and Co

4. Conclusions

Based on analysis and discussion in previous section, some conclusions can be drawn as the following:

1. Studied limonite ore contains mainly goethite [α-FeOOH] with subordinate quartz, talc and trace spinel.
2. Goethite as Ni and Co bearing phase shows partial dissolution in nitric acid leading to release less nickel and cobalt from goethite crystal, in contrast, quartz, talc and spinel did not indicate dissolution.
3. The highest extraction rate of Ni was 24.17 % more than twice lower than Co extraction of 56.53% which were achieved with using 2 M HNO_3 at 90 minutes reaction. However, Co recovery was much higher than Ni.

Acknowledgement

The authors would like thanks to Dean of Engineering Faculty, Hasanuddin University for providing financial support under the scheme of LBE Research Grant 2019.

References

[1] N.W. Brand., C.R.M. Butt., and M. Elias., 1998, Nickel Laterite : classification and feature. *AGSO Journal of Australian Gology & Geophysics*, 17 (4), 81- 88.

[2] P.H. Freyssinet., C.R.M. Butt., R.C. Morris., and P. Piantone., 2005, Ore-forming processes related to lateritic weathering. *Economic Geology 100th Anniversary Volume*, pp. 681 – 722.

[3] S.A. Gleeson., C.R.M. Butt., and M. Elias., 2003, Nickel laterite: a review, *SEG Newsletter*, 54, pp. 9-16.

[4] M.E. McRae., 2019, *Mineral Commodity Summaries*: Nickel, US Geological Survey.

[5] K. Liu., Q.Y. Chen., H.P. Hu., and Z. Yin., 2010, Characterization and leaching behavior of lizardite in Yunjiang laterite ore, *Applied Clay Science*, pp. 311 – 16.
[6] Sufriadin, A. Idrus, S. Pramumijoyo, I.W. Warmada, and A. Imai, 2011. Study on mineralogy and chemistry of the saprolitic nickel ores from Soroako, Sulawesi, Indonesia: Implication for the lateritic ore processing. *Journal of Southeast Asian Applied Geology*, Vol.3. No. 1, (23 -33).
[7] B. Wang., Q. Guo., G. Wei., P. Zhang., J. Qu., and T. Qi., 2012, Characterization and atmoepheric hydrochloric acid leaching of a limonitic laterite from Indeonesia, *Hydrometallurgy*, 129-130, pp. 7-13.
[8] B. Ma., C. Wang., W. Yang., B. Yang., and Y. Zhang., 2013. Selective pressure leaching of Fe (II)-rich limonitic laterite ores from Indonesia using nitric acid, *Mineral Engineering*, 45, 151-158.
[9] W. Astuti., T. Hirajima., K. Sasaki., and N. Okibe., 2016, Comparison of effectiveness of citric acid and other acids in leaching of low-grade Indonesian saprolitic ores, *Minerals Engineering*, 85, 1-16.
[10] W. Chesword., 2008, *Encyclopedia of soil science*, Springer-Verlag, Berlin, pp. 363 – 369.
[11] B. Terry., 1983, The acid decomposition of silicate minerals Part I. Reactivities and modes of dissolution of silicates, *Hydrometallurgy*, 10, 135 – 150.
[12] R.G. McDonal and B.I. Whittington., 2008, Atmospheric acid leaching of nickel laterites review Part I. Sulphuric acid technologies, *Hydrometallurgy* 91, 35-55.
[13] X.Guo., W. Shi., D. Li., and Q. Tian., 2011, Leaching behavior of metals from limonitic laterite ore by high pressure acid leaching, *Trans. Nonferrous Met. Soc. China*, 21, 191-195.

IOP Conf. Series: Materials Science and Engineering **875** (2020) 012054 doi:10.1088/1757-899X/875/1/012054

Sediment Profiles and Distribution of Losari beach Ujungpandang District Makassar City

H Sirajuddin[1]*, Kaharuddin[1], and Sultan[1]

[1]Geology Engineering Department, Engineering Faculty, Hasanuddin University, Makassar, Indonesia

*Email: haerany_sirajuddin@yahoo.com

Abstract. Research on Losari beach sediment profiles and distribution aims to analyze subsurface conditions and composition of coastal sediment materials based on hydrodynamic processes consisting of currents, waves and tides, analysis of sediment textures and subsurface profiles. The research method used in the form of primary data collection consisting of hydrodynamic and coastal characteristics data, and sediment data collection using SES 2000 and grab sampler. Bathymetry measurements show depths ranging from 0.5 to 13 m. Current direction from west to east and southwest to north at speeds of 0.05 - 0.25 m / sec, wave direction from west to east with speeds between 1.50 - 6.50 m / sec and height of 0.30 - 1.00 m. The distribution of the type of sedimentary material consists of sand, silty sand, silty clay and clay. This research is expected to be useful in adding evidence and information regarding the profile and distribution of coastal sediments, both to the local government and the surrounding community, so that it can become a reference for local government and as a material for consideration in efforts to deal with coastal damage and serve as a basis for decision making and policies to preserve nature and develop tourism.

1. Introduction

As a maritime country supported by data from the Ministry of Maritime Affairs and Fisheries (2011), that Indonesia has around 17,504 islands and about 70% of its territory is a sea with a shoreline of 104,000 km or is fourth in the world after the United States, Canada and Russia. Objective conditions with large coastal areas and have considerable potential such as aquaculture, industry, beach tourism and others. Changing coastal environment, working processes and shoreline change history are very important variables to be used in explaining the level of coastal vulnerability [1]. Although the index obtained cannot be compared with specific physical changes, the action to determine it is the first step to examine the factors that contribute to changes in the coast. Several factors that influence the development of the coast, especially the West coast of South Sulawesi, can be reviewed based on marine disasters combined with approaches to landforms that contain information in the form of coastal morphology, coastal material, geomorphic processes, geological and chronological structures.

The marine disaster is based on the factors that cause it, namely morphodynamics, hydrodynamics, ecodynamics and geodynamics [2]. Research on the profile and distribution of the coast is the first step in management and utilization efforts in developing the coastal environment, then the data and information obtained can be used to make a map of the coastal environment zone that is useful in composing steps to manage the coastal area [3]. Losari Beach has a coastline of ± 1 km that stretches from south to north in an area that has been widely used as a tourist site, and at several locations reclamation has been carried out for the utilization of infrastructure and several other coastal structures so that it has an impact on the coastal environment both in profile and distribution of sediment. The condition and physiography of the island as such occur by the hydrodynamic activity of coastal waters towards the island, especially the west and east seasons. This tends to occur on all islands in the area of Sangkarang, Makassar Strait [4].

Seeing this phenomenon, it is necessary to make a study to explain and analyze the profile and distribution of coastal sediments as one of the first steps towards coastal protection efforts so that this area is protected from damage, considering that this area has a very strategic value both in economic and social terms. Considering the potential hazards and damage that can be caused by the presence of coastal processes in the study area, especially if there are neighboring buildings or coastal protection structures, this research is expected to be useful in adding evidence and information regarding the profile and distribution of coastal sediments and their relation to the environment beach, both to the local government and the surrounding community.

Through the information obtained is expected to be a reference for local government and as a material consideration in efforts to deal with coastal damage and serve as a basis for decision making and policies to preserve nature and tourism development [5]. Through the results of this study, it is hoped that the handling steps can be carried out early based on an analysis of the coastal environment. The more important thing is to be able to increase knowledge about the profile and distribution of coastal sediments along with their causal factors.

2. Overview of Losari Beach

Losari Beach according to [6], composed of volcanic rocks. Generally composed of alluvial coastal deposits that are prone to abrasion, so that beach and building damage and roads occur quickly. The beach topography is relatively gentle with a slope of ± 25o, but at the sea level about 0.5-1 km the coastline is relatively steep.

Losari Beach is one of the tourist sites and centers of community activities in the city of Makassar, which is visited daily by the surrounding community and from outside the city (Figure 1). Besides that, as one of the objects of research by lecturers and students of the Department of Geology, Faculty of Engineering, Hasanuddin University. It is clear that changes in the coast and its buildings due to reclamation that has been going on for decades, so that the impact on the coastal environment, including the underwater profile and distribution of sediment. It can be estimated, how much is the cost spent by the government in the reclamation effort. The understanding of beaches is essential to coastal management [7].

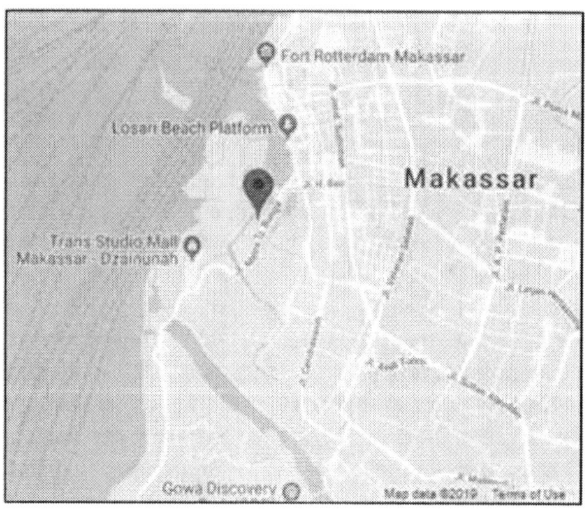

Figure 1. Location map of Losari Beach

Reclamation can have both positive and negative impacts on people and coastal and marine ecosystems. These impacts can be short-term and long-term depending on the type of impact and condition of the ecosystem and the community in the reclamation site and its surroundings. The positive impacts of reclamation activities include an increase in the quality and economic value of the coastal area, reducing land that is considered less productive, additional areas, protection from erosion, improving the condition of aquatic habitats, improving the hydraulic regime of the coastal area, employment and others. While the negative impacts of reclamation activities on the environment include physical impacts such as hydro-oceanographic changes, coastal erosion, sedimentation, increased turbidity, sea pollution, changes in groundwater regimes, increased flooding potential and inundation in coastal areas; biological impacts such as disruption of ecosystems (mangrove ecosystems, coral reefs, seagrass beds, estuaries and others) and degradation of biodiversity.

Losari Beach is located west of Makassar on Jalan Penghibur. The coastal area or coastal plain of Makassar City was formed by sediment transport of the Tallo river 6.5 km from the northern city center and the Jeneberang River with a situation of 4.5 km south of the city. The sediment is dominated by clay and silt, a small fraction of fine sand. The discharge flow of the Jeneberang River in the rainy season can reach 2,800 m3 / s, far greater than the Tallo River discharge.

In identifying coastal problems and predicting erosion of coastal sedimentation in South Sulawesi [8], the Kraus equation is used with parameters consisting of significant wave heights (H) and wave periods (T), in addition to diameter parameters grain (D50) and sediment fall velocity (Ws). Predict shoreline changes using Google Earth.

The shape of the beach profile is strongly influenced by wave attack, sedimentary properties such as mass density and resistance to erosion, particle size and shape, wave and current conditions and bathymetry. The processes of coastal dynamics of the result of interaction between different hydrodynamic process in the spring and summer east west along the year, will have an impact on the vulnerability of abrasion [9]. In general, in the west season (October to April), the energy of waves and currents much stronger than east monsoon (June to August).

3. Results and Discussion

3.1. Oceanography and Sedimentation

Based on the current conditions, Losari beach is no longer original because it has been reclaimed by the city government and several other civilian buildings. However, it still can be explained the appearance or general form which is part of the coast of Makassar, Maros, Gowa and Takalar. Losari Beach is formed by the composition of river sediment and a small amount of marine material from coral sand in a Jeneberang delta system which is classified as a primary beach [10], a beach whose constituent material is sourced from inland activities. The field observation and laboratory analysis has been done to complete this study. The purpose field observation is to see the beach appearance both the composition and topography [11]. Laboratory analysis is done by texture observation. All of sand samples were collected at 20 sandy beaches.

3.1.1. Oceanography. Before being reclaimed, the Losari beach still revealed a form of sediment in the form of spit, lagonal and several sand piles (bars) and the Laelae Island (Figures 2a and 2b). The basic morphology of the Losari coast is relatively gentle, about ± 2 km west to the Laelae island, and outward with a greater depth of 5 m, it is relatively steep.

Between Laelae Island and Losari beach there is a small basin with a depth of 3 - 3.5 m, the bottom of the beach is covered by loose sediment in the form of sand, silt sand and silt clay in the form of mud bonds. Losari Beach receives sediments throughout the year from the Jeneberang and Tallo rivers which form the physiography of bird foot deltas.

Human activities and coastal reclamation (Figure 3), including the presence of the Bili-bili dam, cause the balance of the beach environment to be disturbed and totally changed from its original form, so that no body forms of sand deposits have even been lost.

Figure 2. The appearance of Losari Beach satellite imagery before (a. 2008)) and after (b. 2019) reclamation

Figure 3. Appearance of several buildings on the reclaimed land to the west of Losari beach

Measurement of the physical properties of seawater is one of the methods in this study, besides taking sediment material samples using grab sampler [12]. The physical condition of the measured Losari beach consists of temperatures ranging from 22 - 31°C during the day, salinity around 20 - 29 ‰ (dry season), pH 7.5 - 8.0. Sea currents in the morning from north to south, southwest to north (N45 - 60°E) with varying speeds of 0.05 - 0.25 m/s. Wave direction from west to east with speeds between 1.50-6.50 m/sec and height of 0.30-1.00 m (Table 1, 2 and Figure 4, 5).

3.1.2. Sedimentation. The main source of sediment in the Losari coastal area comes from the Jeneberang river sediment material spill following the Tallo river. These sedimentary waters arrive in coastal waters and are then spread by currents and waves around the coast, both on the mainland and on the Laelae island. This sediment is slightly mixed with coral sand, especially on the Laelae island and outward in the west. The sediment distribution map shows that there are 4 (four) types of sediment that cover this area, consisting of:

- Sand, covering the shallow terrain and edge of the Laelae island and spread out relatively long along the shoreline
- Silty sand, spread in the middle of shoreline
- Silty clay, covering the middle and the edge of the beach which is closed lagonal area
- Clayey silt, occupying the outer and southern parts of the coastal area that extends westward to the depth region

Table 1. Measurement results of Losari beach sediment type and current

St.	Depth (m)	Current		Sediment (types)
		Direction (N°E)	V (m/sec)	
1	0.8	45	0.4	Sand
2	1.25	60	0.4	silty sand
3	0.6	90	0.4	Sand
4	0.7	65	0.35	Sand
5	2.5	65	0.35	silty sand
6	3.5	60	0.35	silty clay
7	2.5	-	-	clayey silt
8	3	120	0.25	silty sand
9	5	115	0.25	clayey silt
10	11	110	0.2	silty clay
11	13	155	0.2	silty clay
12	5.1	160	0.15	silty clay
13	2.2	135	0.15	clayey silt
14	0.5	137	0.1	silty sand
15	0.7	95	0.1	sand
16	0.6	-	-	clayey silt
17	0.6	-	-	clayey silt
18	0.8	145	0.07	sand
19	2	125	0.05	silty clay
20	8.5	125	0.1	silty clay

Table 2. The results of the calculation of texture analysis on sand sediments

Stasion	Sands (%)	Silt (%)	Clay (%)	Aperture (μm)							Total (gr)
				2000	1000	500	250	125	62.5	2	
ST-01	47.59	31.05	20.63	1.19	3.87	14.18	18.75	9.60	31.05	20.63	99.27

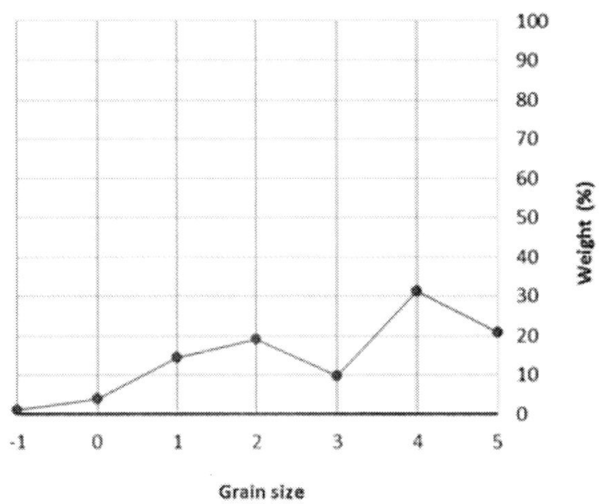

Figure 4. Sand frequency curve

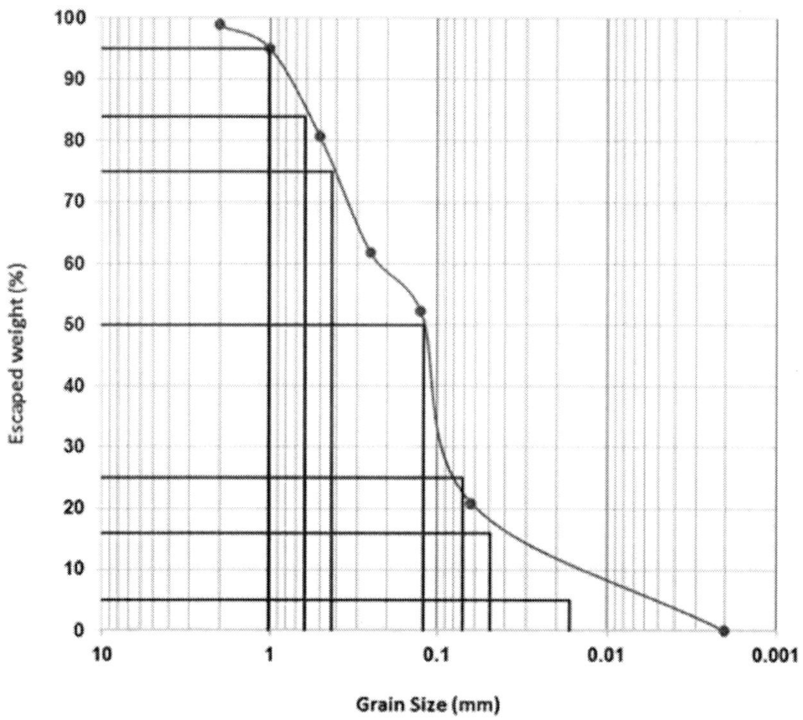

Figure 5. Sand semilogarithmic curves

Losari Beach is the western part of Makassar city that is in contact with sea water, including Laelae Island, Kayangan Island and Samalona Island as part of the Sangkarang island group or also called the Spermonde Archipelago [13], but what is discussed in this study only reaches Laelae Island because it is representative for the observation location. The development of the coast is influenced by the factor of the amount of sediment supply from the Tallo river in the north and Jeneberang river in the south, as well as the influence of coastal hydrodynamics, namely wave activity, currents and tides. Jeneberang and Tallo rivers spill sediment material every year on the coast, especially during the western monsoon, when it is a rainy and flood season.

The sediment will be redistributed by hydrodynamic agents along the coast in the form of spits, bars and tidal flats. The activity of westerly winds tends to be trending from the southwest to the northeast during the rainy season, giving rise to the direction of the waves in the direction of the wind, causing the growth of sedimentary bodies to the north formed by longshore current, which moves from north to south at a speed 1 m/s in the morning (5am - 10am), and from south to north at a speed of 0.5 m/sec in the afternoon (12:30 - 15:00).

3.2. Conditions of the Beach Profile

In the Losari beach, Laelae Island is located about 1 km away, formed by coral reefs covered by a mixture of river sand and coral sand (Figure 6). The bottom of the seabed coastline based on the profile on the sediment distribution map and data collection underwater profiles using SES 2000 (Figure 7a and b), shows the relative morphology of the sloping especially near the shoreline, being outward with a depth of about 5 m indicates a relatively steep relief changes. This relatively steep topography is a trace or former erosion of currents and waves on the base sediment deposits, which during the flood season is covered by sediment. Interpreted that in the west or the rainy season there is an expansion of sediment exposure which is gently sloping by sediment piles from the mainland, then spread by currents and waves along the coast.

Figure 6. The appearance of the Laelae island (↙) on the west of Losari beach

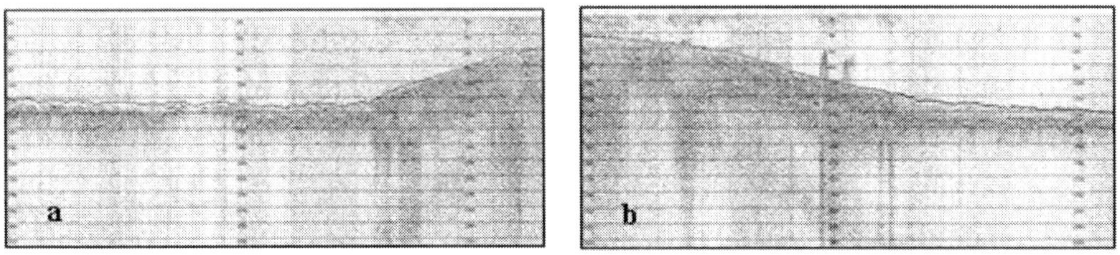

Figure 7. Sea bottom profile near the Losari shoreline (a) and at a depth of about 5 m (b)

While in the east monsoon (dry season), the supply of river sediments decreases, so that in coastal areas subsurface erosion occurs causing steep terrain to advance near the coastline. On the other words, in the shoreline zone, sedimentation occurs which forms a spit, bar and beach. The bottom surface of the beach is covered by sand sediment, silt sand, silt silt and silt clay in the form of active sediment which is still loose so that at any time it can be moved (retransportation) by currents and waves.

3.3. Sediment Distribution

Sediment material shed by the Tallo and Jeneberang rivers in their estuaries will be retransported and spread by currents and waves. The shape and extent of its distribution depends on the amount of energy between the river and the sea that works and interacts with each other, and the amount of sediment supply carried by the two rivers.

Based on the results of oceanographic measurements (Figure 8) show river energy is greater than sea energy at the time of flooding. River current velocity is 1.28 m/sec while sea currents are around 0.05-0.25 m/ sec and wave velocity is 0.50-6.50 m/sec. Marine energy is significant enough to offset the influence of river energy, but due to the large supply of sediment during the flood season, sedimentary deposits develop into the sea forming a bird foot delta. Current and wave conditions in the Losari beach at the time of measurement on August 17, 2019 at 11:00 to 13.30, indicate in the northern part of the current direction from west to northeast with a speed of about 0.35-0.4 m/sec. While in the middle and south direction of the current from west to east and turning southeast with a speed of about 0.05-0.25 m/sec.

The difference in direction and velocity of the current is caused by the influence of the island of Laelae, where refraction of currents and waves occurs which causes the diverting of the direction of the current and increasing its speed in the north under narrow waters.

This large hydrodynamic energy in the north causes intensive washing/sorting of sediments and causes sand material to be left dominant in the north, while silt and clay materials are transported and deposited in the central, western and southern of coastal. Based on the analysis of sediment grain size as outlined in the sediment distribution map, there are 4 (four) types of sediment that cover the bottom of Losari beach, namely:

- Sand, covering Laelae island and Losari beach with relatively large current and wave energy and good sorting
- Silty sand, occupying the northern and middle part with medium sorting
- Clayey silt, spread in the west, east and center in the condition of energy-weak with moderate sorting
- Silty clay, covering the west and south with moderate relative energy and medium sorting

The appearance of bottom sediment distribution in the coastal waters on the incision/profile of the sediment distribution, generally shows in areas that are relatively >5m deep and the energy is weak - moderate, covered by fine sediments namely clay and silt. While in the shallow area, where relatively large energy is deposited, coarse sediments are being sand, silty sand and silty clay (Figure 9).

The presence of the island of Laelae as part of the Losari coast causes disrupted water conditions and distribution of coastal bottom sediments, deviating from the distribution of coastal system sediments and river mouths based on the shape of the beach bottom profile. Ideally, from river mouths to sea with increasing depth will show coarse to fine sediments, which is different from the distribution of sediments in the waters of the Losari beach due to its beach profile due to the Laelae island.

3.4. Coastal Environmental Conditions

The natural condition of Losari Beach has now been disrupted even some have been damaged due to beach reclamation and other physical development. Before the reclamation, around the 1990s, the natural shape of the Losari beach was still visible with the growth of spit, barbed sand, tidal flat exposure and visible coastal terrace clear along the coast and at the mouth of the river.

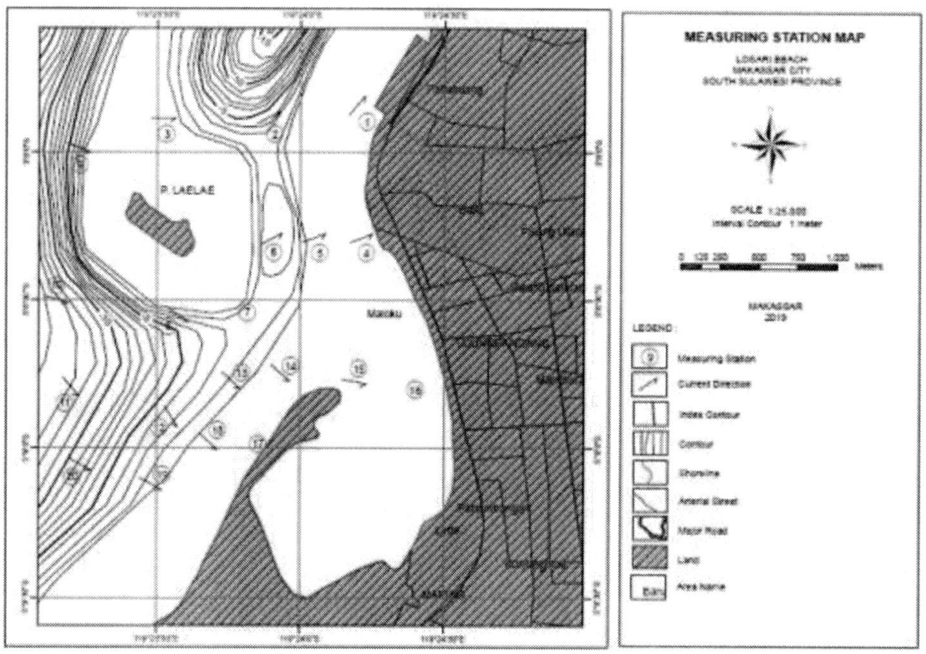

Figure 8. Measuring station map

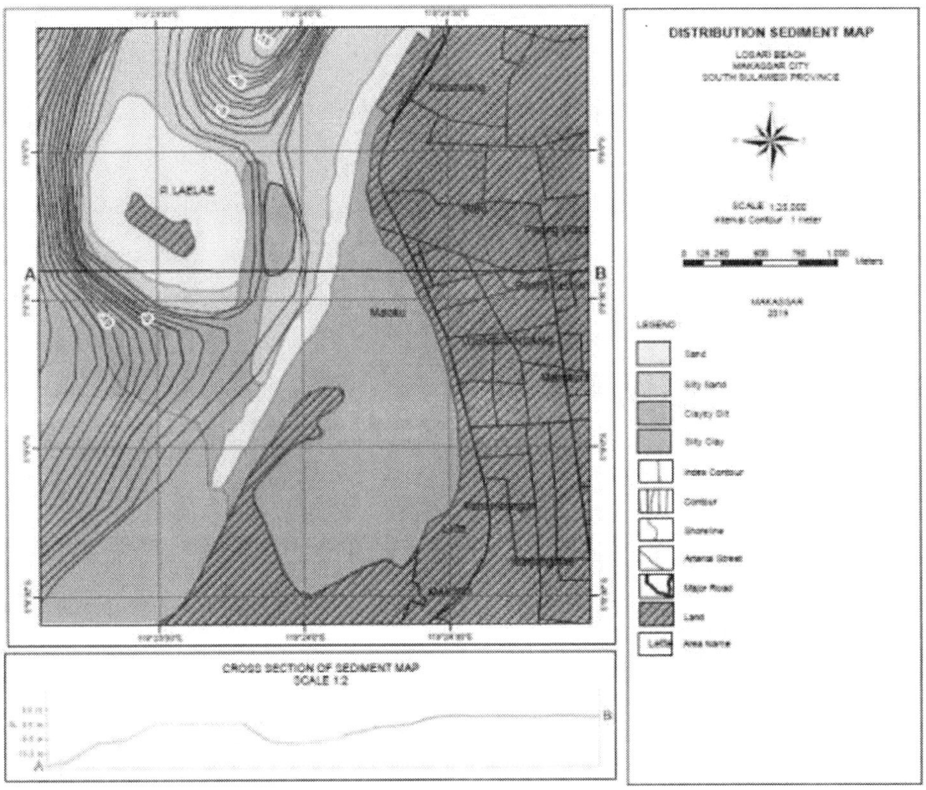

Figure 9. Distribution sediment map

After reclamation and other physical development, the physiographic forms of the beach have been damaged or even gone, leaving only the exposure covered by material deposits from the mainland [14]. This can increase the turbidity of coastal waters and threaten the disruption of the coastal ecosystem. The accumulation of coastal from soil material, is easily eroded and transported by currents and waves (Figure 10 and 11), especially during the western season where the wave current will work proximaously. Heap material which is overturned and transported by currents and waves to the north will influence or accelerate the silting and turbidity of the Soekarno Hatta harbor.

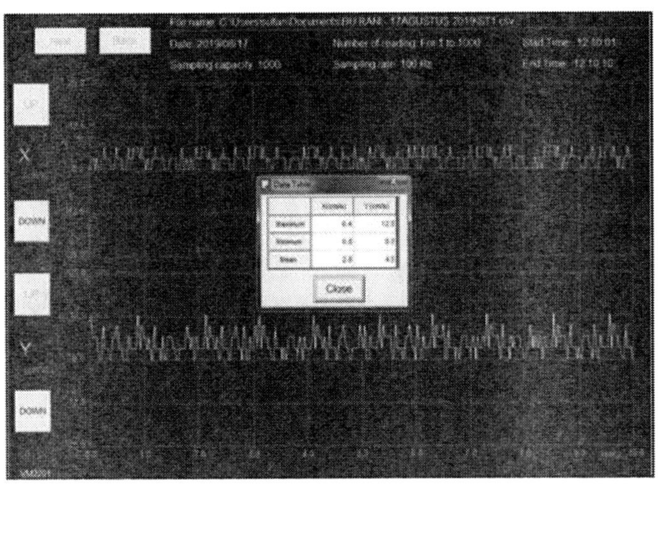

Figure 10. Graph of the results of measurements of currents and waves at station 1

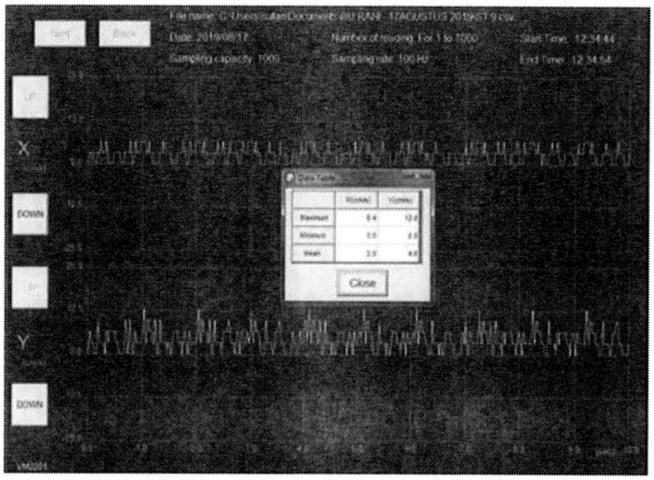

Figure 11. Graph of the results of measurements of currents and waves at station 1

4. Conclusion

1) The physical condition of the measured Losari beach consists of temperatures ranging from 22 - 31°C during the day, salinity around 20 - 29 ‰ (dry season), pH 7.5 - 8.0. Sea currents in the morning from north to south and southwest to north (N45 - 60oE) with varying speeds of 0.05 - 0.25 m/s. Wave direction from west to east with speeds between 1.50-6.50 m/sec and height of 0.30-1.00 m
2) The appearance of bottom sediment distribution generally shows in areas that are relatively >5m deep and the energy is weak - moderate, covered by fine sediments namely clay and silt. The presence of the island of Laelae as part of the Losari coast causes disrupted water conditions and distribution of coastal bottom sediments
3) The natural condition of Losari Beach has now been disrupted even some have been damaged due to beach reclamation and other physical development. The natural shape of the Losari beach was not as the growth of spit, barbed sand, tidal flat exposure and visible coastal terrace clear along the coast and at the mouth of the river.

Acknowledgments
The author are grateful to my institutes through LBE grants which has provided motivation and support in the preparation of this journal. We would also like to thank Kaharuddin, Anggit Kurnia, Muhammad Kinandjar, Adhayani Hajarinanda, Sumiati, Fadel Muhammad, Dian Pratiwi and others student also assistance at Department of Geology for help in the field and data processing. This study is contributed to the research in Geology Engineering Hasanuddin University especially for Marine Geology and for scientific progress.

References

[1] Pendleton, E.A., Thieler, E.R., Williams, S.J., 2010, Importance of Coastal Changes Variables in Determaning Vulnerability to Sea and Lake Level Change, Journal of Coastal Research, Vol 26 No 1, p. 176 – 183.

[2] Wahyudi, Haryanto,T., Suntoyo., 2009, Coastal Vulnerability Analysis in the Coastal Areas of the North Coast of East Java, SENTA Journal 2009, FTK-ITS Surabaya.

[3] Koddeng, B, 2011, Zonation of the Makassar Coastal Area Based on Disaster Mitigation (Case Study of the Barombong Beach-Celebes Convention Center), Prosiding Hasil Penelitian Fakultas Teknik, Vol. 5, Grup Teknik Arsitektur.

[4] Kaharuddin, Husain, R., Busthan, 2018, The Mitigation of Coastal Abrasion On Islands, Special Reference to The Kodingareng Keke Island Makassar City, Indonesia, International Journal of Engineering & Technology IJET-IJENS Vol:18 No:01.

[5] Kaharuddin, Mawaleda, M., Husain, R., Busthan, 2016, The Vulnerability of Coastal Abrasion in the Islands Area Case Studeis in Kodingareng Keke Island Makassar City, International OPEN ACCESS Journal of Modern Engineering Research (IJMER), Vol. 6 Iss. 12 Dec. 2016 (1).

[6] Sukamto. R., 1982, The Geology of Ujungpandang, Benteng and Sinjai Quadrangle, Research Centre and Development of Geology, Bandung.

[7] Abanades,J., Greaves, D., Iglesias, G., 2015, Wave Farm Impact, on Beach Modal State, Marine Geology 361 (2015) 126-135.

[8] Umar, H., Rahmana, S., Baeda, A.Y., Klara, S., 2015, Identification of Coastal Problem and Prediction of Coastal Erosion Sedimentation in South Sulawesi, Procedia Engineering 8th International Conference on Asian and Pacific Coasts (APAC 2015) 116, 125 – 133, 1877-7058 Published by Elsevier Ltd, www.sciencedirect.com.

[9] Kaharuddin, Imran, A.M., Abdullah, C.I., Jaya, A., 2019, Olistostrome and its Implication to Geological disaster on Coastal Area with Special Reference to the Bantimala Tectonic Complex, Pangkep Regency South Sulawesi Province, OP Conf. Series: Earth and Environmental Science 235 (2019) 012043 IOP Publishing.

[10] Shepard, 1973, F.P., Submarine Geology, Harper and Row Publisher, New York, Everston, San Fransisco, London.

[11] Sirajuddin, H., Suriamihardja, D.A., Imran, A.M., Thaha, A., 2015, Influence of Sediment Texture on Beach Morphodynamic State, International Journal of Development Research Vol. 05, Issue 08, pp. 5249 – 5254.

[12] Sirajuddin, H., Suriamihardja, D.A., Imran, A.M., Thaha, A., 2015, Coastal Vulnerability Zonation of Pinrang District and Parepare City Based on Morphodynamic Review, International Journal of Inventive Engineering and Sciences Volume-3 Issue-12, pp. 23-26, December 2015.

[13] Imran, A.M., Kaharuddin, M.S., Suriamihardja, D.A., Sirajuddin, H., 2013, Geology of Spermonde Platform, Proceedings of the 7th International Conference on Asian and Pacific Coasts (APAC 2013), Bali, Indonesia, September 24 – 26, 2013.

[14] Saengsupavanich, C., 2012, Assessing and Mitigating Impacts of Shore Revetment on Neighboring Coastline, International Conference on Environment Science and Engieering IPCBEE vol.3 (2), Singapoore.

IOP Conf. Series: Materials Science and Engineering **875** (2020) 012055 doi:10.1088/1757-899X/875/1/012055

Ultrahigh-Pressure Metamorphic Rocks in the Bantimala Mélange Complex and Its Implication to Cretaceous Tectonics in the South Arm of Sulawesi

M Mawaleda[1]*, H Umar[1], J R Husain[1], A Maulana[1], A Jaya[1], U R Irvan[1], and D D Permana[1]

[1]Department of Geology Engineering, Faculty of Engineering, Hasanuddin University Jl. Poros Malino Km.6 Bontomaranu, Gowa, Indonesia

*Email: musri.mawaleda@unhas.ac.id

Abstract. Bantimala Mélange Complex is about 75 Km to the North of Makassar, the capital city of South Sulawesi Province. The presence of High-pressure Metamorphic Rocks, and Ultrahigh Pressure in the Bantimala Mélange Complex, has been known and studied by a number of geologists. Eclogite and Blue Schist are Ultrahigh, and High-pressure metamorphic rocks in the Bantimala Mélange Complex which have important significance in explaining the Cretaceous tectonics in this area. Furthermore, Bantimala Eclogite is the only Ultrahigh-pressure metamorphic rock known to be exposed in Sulawesi. The results of thin section analysis of Eclogite show the presence of Omphacite Minerals, and Coesite as an index mineral for Ultrahigh pressure metamorphic rocks. While in Blue Schist with Glaucophane Minerals are found which are partially replaced by Chlorite which indicates the presence of Poly-metamorphism as a result of retrogradation from Ultrahigh pressure to High pressure and to Medium pressure metamorphic rocks. Radiometric age dating using the Ar^{40}/Ar^{39} method shows an age spectrum of 100 Ma -114 Ma (Early Cretaceous). The 114 Ma, age is interpreted as the age of Eclogite's Ultrahigh pressure metamorphic rocks formation with respect to the peak of subduction tectonics. Whereas the age of 100 Ma before now is interpreted as the age of exhumation or collision as the end of the Cretaceous tectonic subduction in the South Arm of Sulawesi or in the East or Southeast of Sundaland.

1. Introduction

Sulawesi Island has a unique shape and has or there are at least nine metamorphic complexes. Is the mainland or lithosphere with the most metamorphic rock complexes in the world [1]. Bantimala Mélange Complex is one of them, also known as the accretionary complex, is about 75 Km to the Northeast of Makassar, the capital city of South Sulawesi Province (Figure 1). It stretches relatively North to South along approximately 55 Km with a width of about 15 Km [2].

The Bantimala Mélange Complex consists of two blocks namely the Bantimala Block in the south and the Barru Block in the North [3]. High and Ultrahigh-pressure metamorphic rock outcrops associated with Mélange, well exposed at a number of locations in this complex. So far as much research has been done both in relation to the presence of mélange, and related to tectonics event and metamorphic rocks which are very varied. A number of geologists who have conducted research in this area, including [2, 3, 5-13].

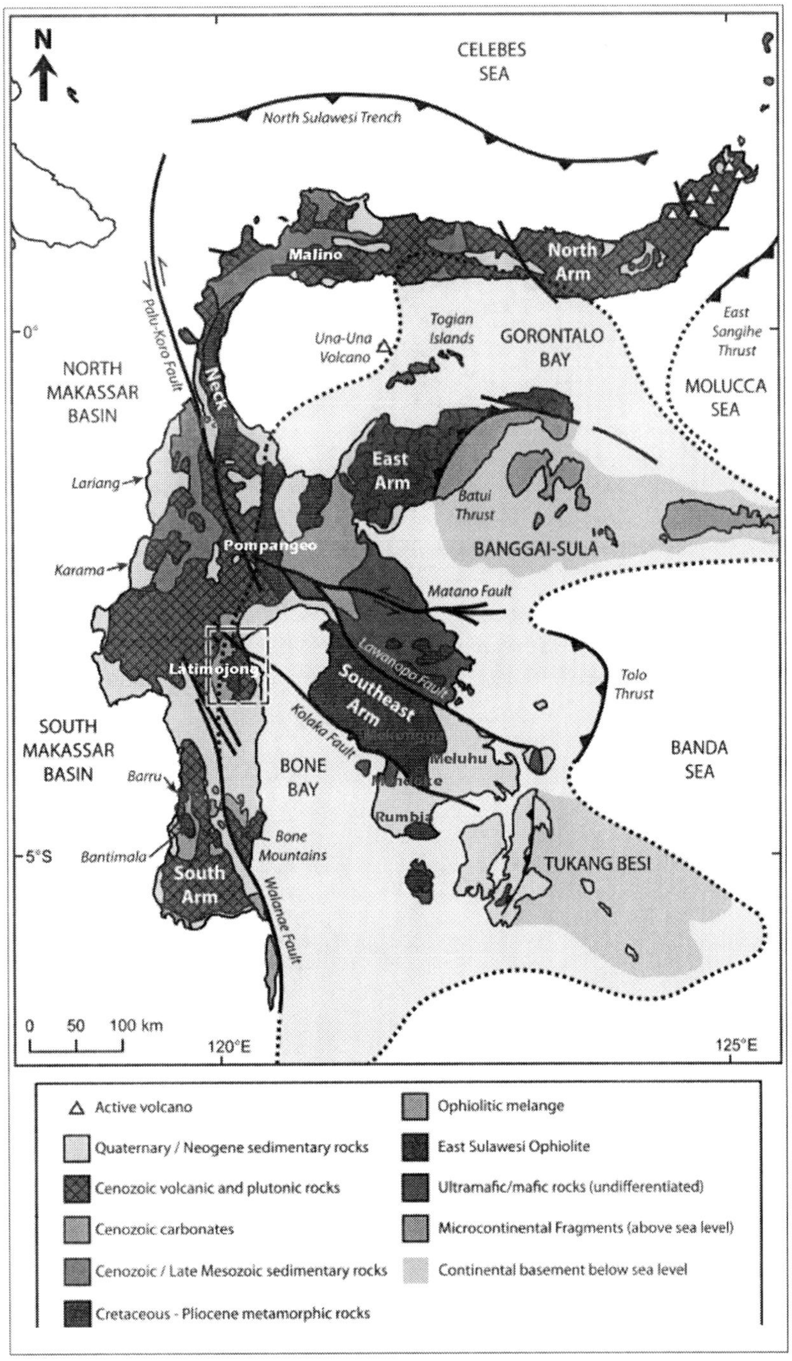

Figure 1. Geological map of Sulawesi Island (Modified from [4])

2. Research Location

The research location is about 75 Km to the north of Makassar City, the Capital City of South Sulawesi Province (Figure 2). This area can be reached by vehicles for approximately 1.5 hours from Makassar City or one hour from Hasanuddin International Airport in Maros, South Sulawesi.

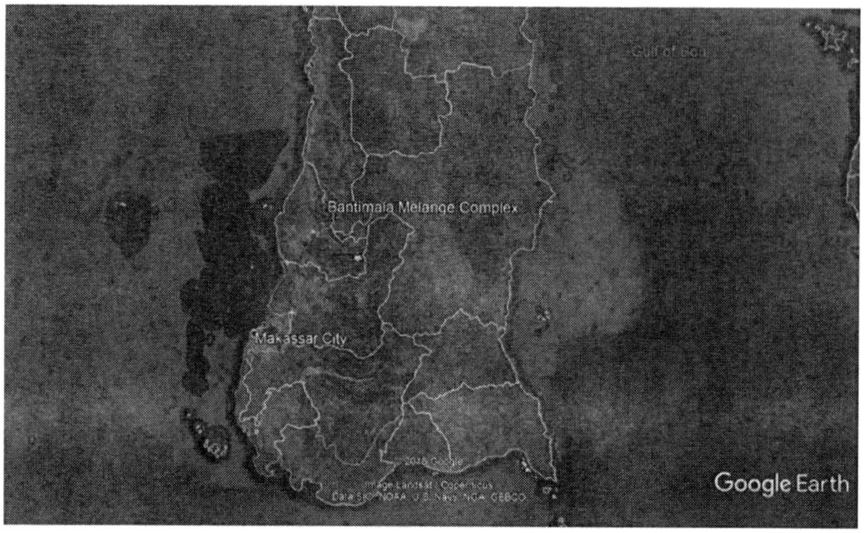

Figure 2. Research location (Bantimala Mélange Complex). The red square is Bantimala Mélange Complex.

3. The aims and objectives of the study

The purpose and objective of this research are to know and understand the Cretaceous tectonics in the South Arm of Sulawesi and/or Southeast Sundaland based on metamorphic rock data.

4. Geology of the Bantimala Mélange Complex

In the Sulawesi Island, there are at least nine segments of the accretionary complex [1], one of which is the Bantimala Complex [5-8]. Other accretion complexes namely the Pompangeo Complex in Central Sulawesi [9, 10], Latimojong Complex in the Northern part of the South Arm [2, 4, 11]. Mekongga, Rumbia, Mendoke and Meluhu Complex in the Southeast Arm of Sulawesi [1, 12, 13], Malino Complex in the Western part of the North Arm of Sulawesi [14], and the Palu Complex in Western part of Central Sulawesi [15].

It is noted that between the Bantimala Mélange Complex in South Sulawesi and the Meratus Complex in South Kalimantan it is assumed to be the Cretaceous accretion complex that merges before the Makassar Strait opening in Paleogen [2, 6, 16]. This assumption explains that both complexes are part of the same accretion complex in Southeast Sundaland. This can be seen from the appearance of spatial distribution, and between the two complexes there is the Paternoster Platform [2]. The Paternoster platform was reported as a continental fragment that moved to North after detachment from Gondwanaland during the opening of Meso-Tethys [2, 17-19]. Other researchers explain that Bantimala mélange complex is the same accretion pathway as the Karang Sambung accretion complex in Central Java [2].

4.1. Lithology Unit of Bantimala Complex

The Bantimala Complex is composed by Mesozoic basement which is overlain unconformably of Late Cretaceous sedimentary rock formations to Cenozoic sedimentary and volcanic rocks [2]. Mesozoic rocks as a basement exposed in the South Arm of Sulawesi especially in the Bantimala Complex consist of ultramafic, metamorphic rocks, and Mélange [2, 3, 6, 8]. At the top of the basement, the Cretaceous sedimentary rocks are overlain the same direction as the intersection between re-crystallized sandstone and shale. This unit is known as the Balangbaru Formation with a thickness at least 2500 m [2, 8].

4.1.1. Mesozoic Metamorphic Basement Rocks

The Mesozoic metamorphic rock unit in the Bantimala Mélange Complex consists of eclogite, glaucophane schist, garnet schist, amphibolite schist, muscovite schist, quartz-feldspar gneiss [2, 3, 8]. This unit exhibits medium to ultra-high pressure and medium-grade metamorphisms [8, 18, 19, 20]. Experienced metamorphism under pressure conditions 16- 17 Kbar and 24-27 K-bar, and temperatures 580-620°C and 580-650°C [20].

4.1.2 Mélange

Mélange outcrops consist of deformed sandstones, shales, conglomerates, diorites and schist breccias. The lithology unit is classified as mélange [2, 3, 8]. The stratigraphic relationship between its components is not clear, between unit contact cannot be defined. Foliation tends to tilt to the northeast and southwest. The thickness of this unit is estimated to be not less than 1,750 m [2, 8]. The Breccia-schist overlies with metamorphic schist at the top the schist-breccias develops into sandstones which are crushed by radiolarian chert. Radiolarian chert is conformably overlain by shales alternating with sandstones with conglomerate intercalation. Chert and limestone which are components in mélange were deposited in the Cretaceous Period [2, 21, 22].

4.1.3 The Late Cretaceous Sediment

Sedimentary rock units which are in the Late Cretaceous, are shown by flysch-type deep-sea sediments and turbidite sediments consisting of solid and shale sandstone intercalations. This lithology unit is known as the Balangbaru Formation. Deposited unconformably above the Bantimala Mélange Complex in the Late Cretaceous [2]. Paleo-current studies show that the Balangbaru Formation originated from the west highlands to the northwest [10]. In the upper reaches of the Bontoriu River, this unit serves an imbrication between the blueschist and greenschist and is considered a crystalline basement which proposed as a Jurassic Paremba Sandstone [8]. The Balangabaru Formation intercalated with the Marada Formation which was deposited in the shallow marine during the Late Cretaceous Period [2]. Volcanic activity at that time was responsible for the presence of volcaniclastics intercalations between the Marada Formation and the flysch sediments of Balangbaru Formation [2, 8, 10].

4.1.4 Tertiary Sediment and Volcanic

Tertiary sediment deposits on the top of the Pre-Tertiary rocks are represented by sedimentary and volcanogenic rocks, which are separated into two groups representing different events. The first is the sedimentation and volcanic activity during Paleogene to the Lower Miocene when compression and orogens occur in the South Arm of Sulawesi. This event was shown by Langi volcanic with Calc-alkaline affinity and some shallow marine sediment sequences in the paralic environment namely the Malawa, Tonasa and Salo Kalupang Formations [2, 7, 8, 23, 24]. During the Oligocene - Early Miocene period, important tectonic events occurred which were followed by the emplacement of the Kalamiseng Formation as the Back- Arc Basalt associated with oceanic crust obduction in the South Arm of Sulawesi at the end of subduction in this region [2, 24]. The second is sedimentation and volcanic activity after the Early Miocene when extensional regimes dominated in this region. Miocene sedimentation in the South Arm of Sulawesi, dominated by volcanic activity. The Camba Formation, which occurred during this area, consisted of marine sediments intercalation with volcanic rocks. K-Ar dating of wall rocks alkaline basal, dacite, andesite and basaltic rocks from the Camba Formation obtained age 17.7 Ma; 8.93 Ma; 9.29 Ma and 6.2 Ma [24, 25]. It was explained that the volcanic rocks older than the Middle Miocene were of an orogenic type, while the younger ones of the Middle Miocene were said to be unrelated to subduction. It seems that the magmatic arc formed from partial melting after collision due to thickening crust [25].

In the Middle Miocene to the Pliocene in this region, it shows the spread of the Walanae Formation in the north and east and laterally in the form of thicker limestone and is known as the Tacipi Formation. During the Miocene, volcanic and plutonic rocks predominantly formed in the Southern Arm of Sulawesi. In this case, such as Baturape and Pare-Pare volcanic rocks [2]. This is interpreted as a magmatic belt produced by the development of volcanic arcs associated with subduction [6, 8]. While Yuwono, et.al [25] interpret the magmatic arc as in this context of plate dimensions during Tertiary subduction, which ended before the main volcanic period. Other researchers suspect that magmatic arcs result from melting of the lithosphere due to collision [26].

4.2. Metamorphic Rocks of Bantimala Mélange Complex

As explained earlier that high pressure metamorphic rocks (glaucophane schist) and ultrahigh-pressure (eclogite) in the Bantimala Mélange Complex are found in the Koraja-Cempaga River, Bontoriu, Tinjau Bali, Elle and Pateteyang. Consisting of amphibole eclogite, amphibole eclogite epidote, epidote eclogites; zoisite blueschist, tourmaline blueschist, lawsonite blueschist, epidote blueschists; Phyllitic blueschist, greenschist, phyllitic greenschist, piemontite phyllite and graphite phyllite [2, 8].

Eclogite is commonly found as a boulder, and some show it has been retrogradation to glaucophane schist and chlorite schist. Some of the eclogite is the result of a prograde and finally after the peak of metamorphism, followed by retrogradation [2]. This is reflected in the replacement as omphacite minerals by glaucophane and lawsonite. Furthermore, by chlorite and epidote. Eclogite as lenses of dimensions from a few centimeters to several meters. Boulder eclogite is found along the Pateteang River and Elle River, with a diameter of up to 3 meters (Figure 3).

4.3 Methodology

This research includes survey and rock sampling, petrographic analysis and radiometric age dating. Survey and sampling of high and ultrahigh-pressure metamorphic rocks (glaucophane schist and eclogite) were collected from the Cempaga River, Pateteang River and Pangkajene River where various eclogite and glaucophane schists. Subsequently, the samples collected were analyzed by petrography using a polarizing microscope at the Department of Geology, Faculty of Engineering, Hasanuddin University. Whereas radiometric age dating used $^{40}Ar/^{39}Ar$ in collaboration with the *Research School of Earth Sciences, The Australian National University, Canberra, Australia.*

Dating $^{40}Ar/^{39}Ar$ is very good for the study of the evolution of metamorphic rocks because this method can detect the possibility of contamination. In particular, $^{40}K^*$ isotopes can also form $^{40}Ca^*$ isotopes and $^{40}Ar^*$ isotopes. Therefore, although the K-Ar dating method has so far contributed to radiometric age data that are known to have a half-life of 1,250 years, a number of weaknesses and weaknesses are now being discovered. However, the $^{40}Ar/^{39}Ar$ method is based on the ^{40}K isotope or in other words, K-Ar is the basis for developing the $^{40}Ar/^{39}Ar$ dating method.

4.3.1 Field data

The geological survey and rock sampling were carried out in the Bantimala Complex namely in the Pangkajene River, Pateteang River and Cempaga River. Field data show that high to ultrahigh-pressure metamorphic rocks were found in bolder form along the Pangkajene River. Large bolder of more than 3 meters is found in Cempaga and Pateteang River. Eclogite is found in various types including garnet eclogite, lawsonite-glaucophane eclogite, glaucophane eclogite, tourmaline-glaucophane eclogite. In some spots showing omphacite minerals replaced by glaucophane to chlorite. Also found outcrops with an abundance of garnets with relatively fine crystal sizes forming an orientation measuring 0.5-2 mm. While garnet crystals up to 2 cm in size, show relatively euhedral crystalline shapes and spread in rocks as porpyroblasts that are present between omphacite, tourmaline and glaucophane crystals. Eclogite, which is present with partial replacement of omphacite minerals by glaucophane and/or chlorite, exhibits polymetamorphism symptoms. This condition is found and is well exposed near the branches of the Cempaga River and Pateteang River. Visually eclogite outcrops and glaucophane schist can be seen (Figure 4). It appears that eclogite cannot be mapped on a regional scale. Eclogite exists as a subordinate of metamorphic rocks. However, the existence of eclogite provides important significance in the history and tectonic evolution of the Bantimala Mélange Complex and/or in the South Arm of Sulawesi.

The distribution of ultrahigh-pressure metamorphic rocks in the Bantimala Mélange Complex, appears to be located or subordinated to metamorphic rocks of high-pressure (glaucophane schist) and medium pressure (chlorite schist) and mélange. The geology of the Bantimala Mélange Complex is shown in the geological map of the western part of the South Arm of Sulawesi (Figure 5).

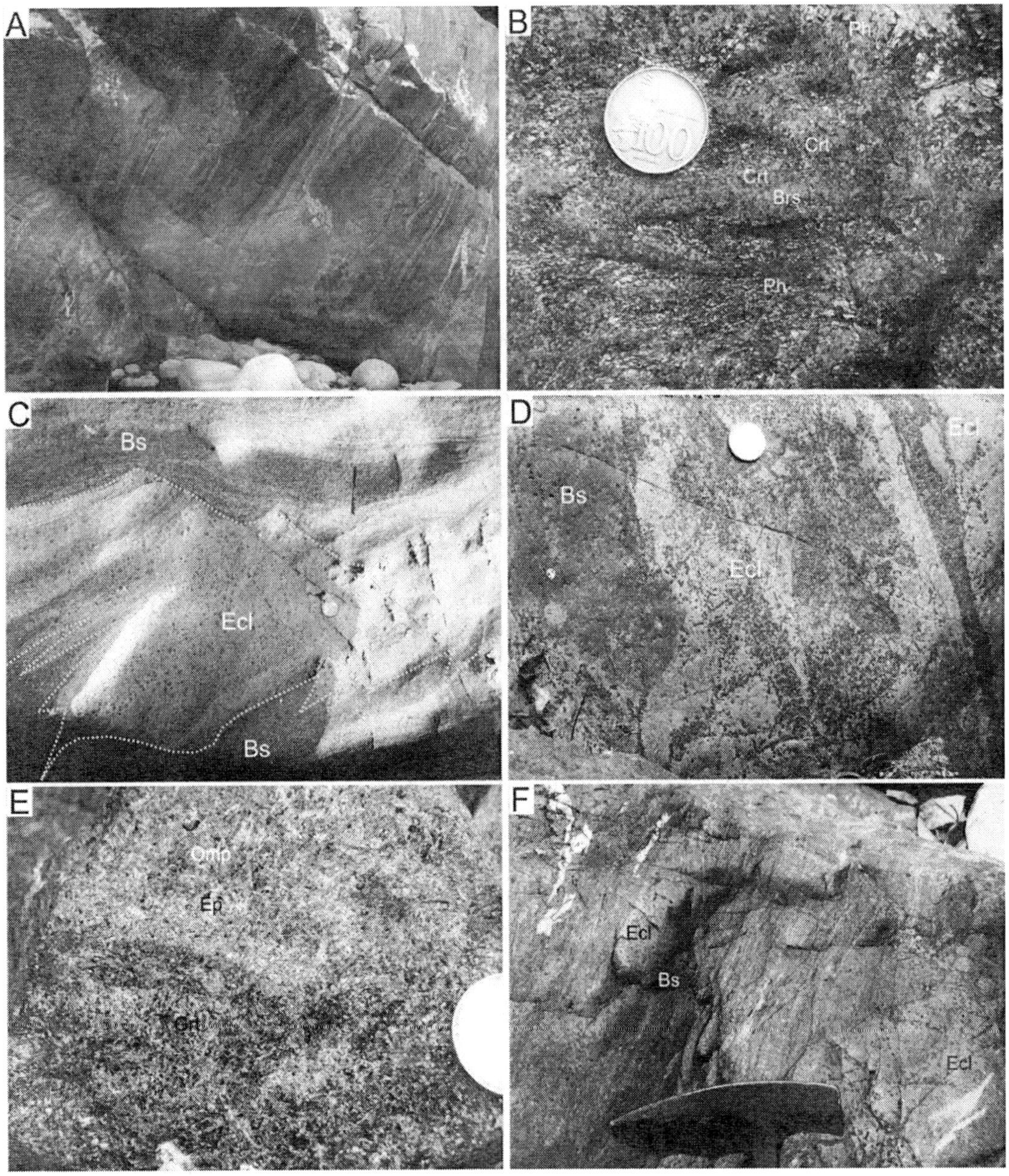

Figure 3. Outcrop of eclogite at Koraja River (all except E) and at Tinjau Bali River (E). This eclogite outcrops are common along Station BTM 134 until BTM 142 at Koraja River. Eclogite lenses are commonly found inside blueschist (C and F) while in some part remnant of blueschist is still preserved during prograde toward eclogite facies (D). Figure E is epidote eclogite at Tinjau Bali River (After [2])

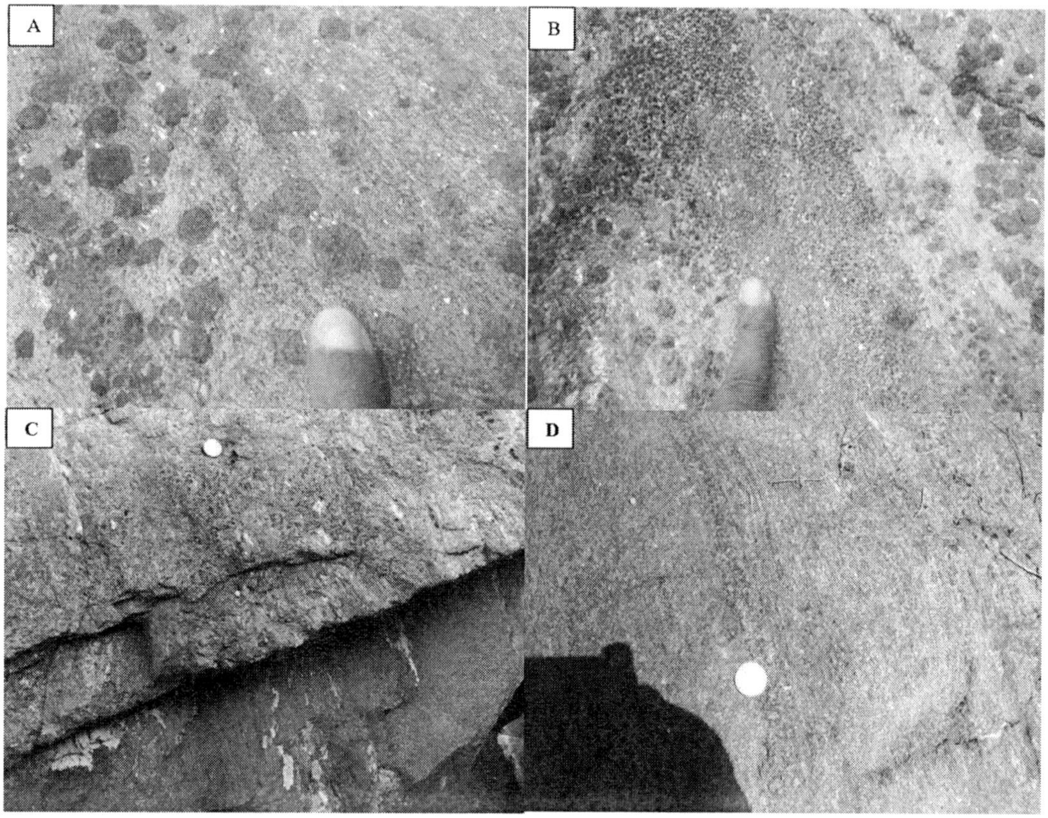

Figure 4. Eclogite outcrop at Cempaga River; (A) Garnet-tourmalin Eclogite with big garnet porphyroblasts, and some of garnet wrap by chloriote; (B) Garnet-Eclogite with small garnet forming layers; (C) Glaucophane schist with small garnet, and some showing glaucophane replaced omphacite; (D) Glaucophane schist showing glaucophane and chlorite replaced ophacite

4.3.2 Petrography Analysis of High and Ultrahigh-Pressure Metamorphic Rocks

A total of 5 eclogite and 5 glaucophane schist samples were prepared using a thin section (0.01 -0.03) mm. Then observed and analyzed using a polarizing microscope. Petrographic observations and results show various types of eclogite namely: Garnet Eclogite, Garnet-tourmaline Eclogite, Garnet-glaucophane Eclogite, Epidote-glaucophane Eclogite and Tourmaline-hornblende Eclogite. Some exhibit mortar textures which are interpreted regarding the event of enlargement and deformation after the metamorphism phase. Five variants of eclogite can be grouped based on the specific abundant minerals found. Eclogite with abundant epidote; epidote-amphibole; amphibolite; quartz (coesite) and lawsonite. Common mineral assemblages of eclogites in the Bantimala Complex are garnet + omphacite + phengite + rutile + glaucophane + small amounts of quartz with several epidemics ± tremolite-actinolite ± titanite ± zoisite ± clinozoisite ± lawsonite and opaque minerals. Eclogite variations seem to reflect that most eclogites in the Bantimala Complex experience polymetamorphism. Also enhanced is the local presence of high temperature metamorphic rocks (Amphibolite and Aphibolite-Chlorite schist). It can be interpreted that the polymetamorphism occurs with regard to the decrease in pressure and temperature accompanying the end of the subduction event at the Bantimala Complex followed by collision and exhumation events. This can be proven by the presence of medium to low pressure minerals replacing ultrahigh-pressure and high-pressure minerals. Likewise, high temperature minerals by medium temperature minerals. Visually shown in (Figures 6 and 7).

IOP Conf. Series: Materials Science and Engineering **875** (2020) 012055 doi:10.1088/1757-899X/875/1/012055

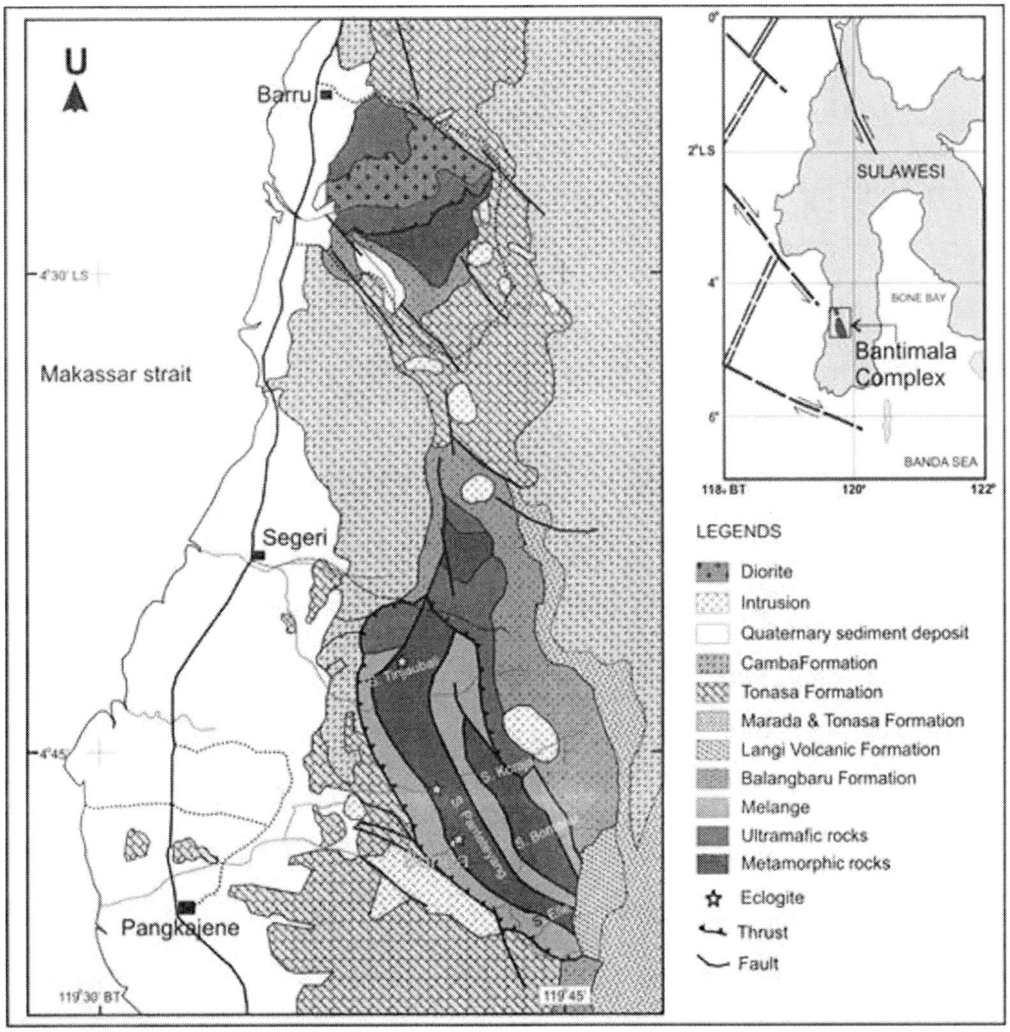

Figure 5. Geological map of Western part of South Sulawesi Indonesia shows metamorphic and ultramafic units are surrounded by Cretaceous sediment and Tertiary [2]

Glaucophane are common mineral in eclogite and blueschist of the Bantimala Melange Complex, either present as inclusion or matrix. Its lavender blue to blue color is typical for glaucophane mineral. Besides present in matrix of eclogite, some glaucophane also present as an inclusion in garnet.

4.4 Geochronology and Cretaceous Tectonics

Radiometric age dating results for metamorphic rocks of high pressure and ultrahigh-pressure of the Bantimala Mélange Complex from various methods can be described as follows:

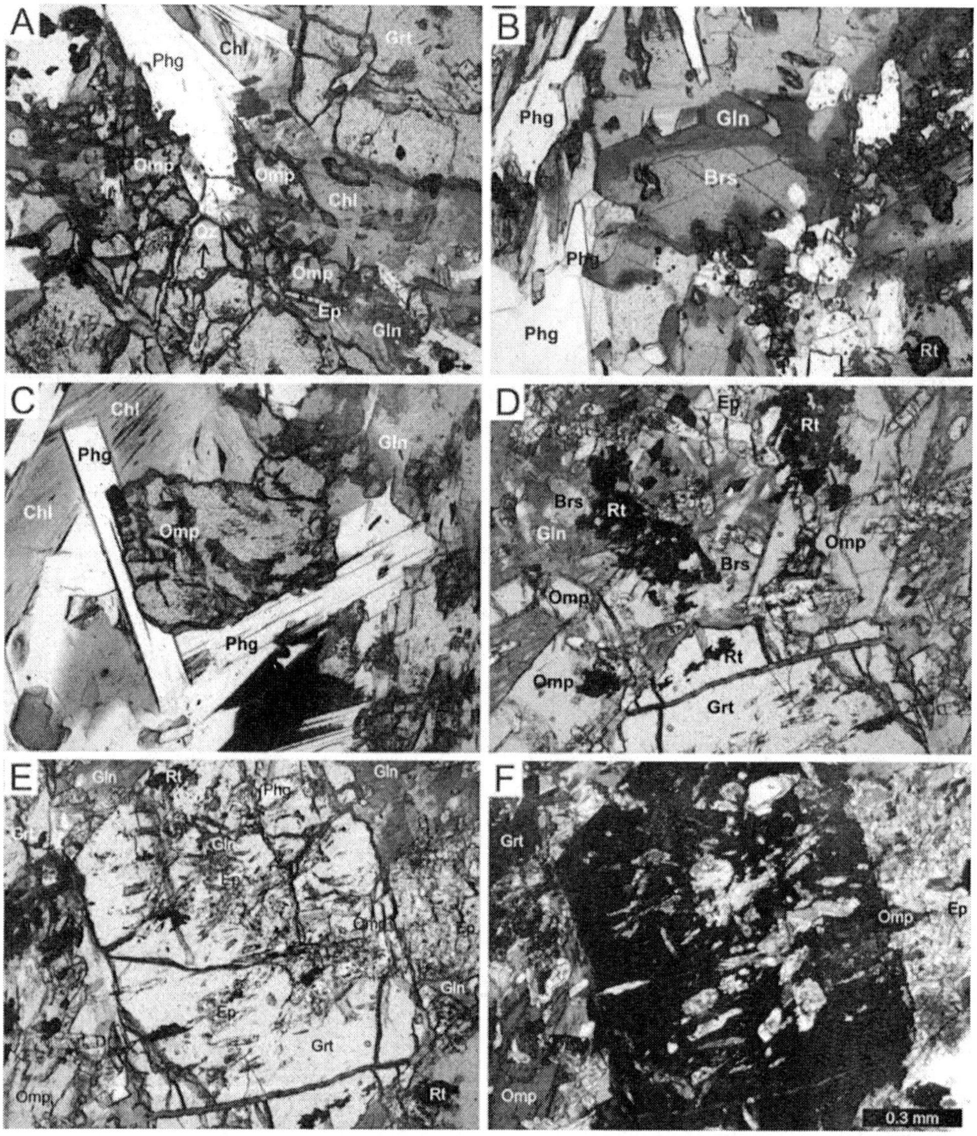

Figure 6. Microphotograph of eclogites taken from the Bantimala Melange Complex. A, B and C are taken from Epidote Amphibole Eclogite of BTM 134A while D and E are taken from Amphibole Eclogite of BTM 140A. F is cross polar of Figure E. Assemblage of garnet-omphacite- phengite-rutile-glaucophane designate this rock. Inclusions of Ph, Crt, Rt, Ep, Qz are common. Chorite present as replacement of amphibole and garnet (After [2])

4.4.1 K-Ar Dating

K-Ar age dating of phengite minerals from blueschist and eclogite [21, 28], indicates Valanginian age through the Cretaceous Aptian Age. Whereas K-Ar on the muscovite mineral from blueschist yielded 111 Ma [27]. The most recent K-Ar age data taken from pengite minerals of garnet-glaucophane schist from Barru Blok provide age spectrum 132 ± 7 Ma, 124 ± 6 Ma, 115 ± 6 Ma, and 114 ± 6 Ma and 106 ± 5 Ma [28]. Whereas the eclogite collected from Bantimala produced K-Ar age of 137 ± 3 Ma [18].

4.4.2 U-Pb Zircon SHRIMP

Radiometric age dating of U-Pb from zircon minerals by the *SHRIMP* method gives results of 819.4 ± 9.3 Ma, 758.8 ± 8.7 Ma, and 296 ± 2 Ma [2]. In detail the *SHRIMP U-Pb Zircon* age spectrum can be shown in Figure 8.

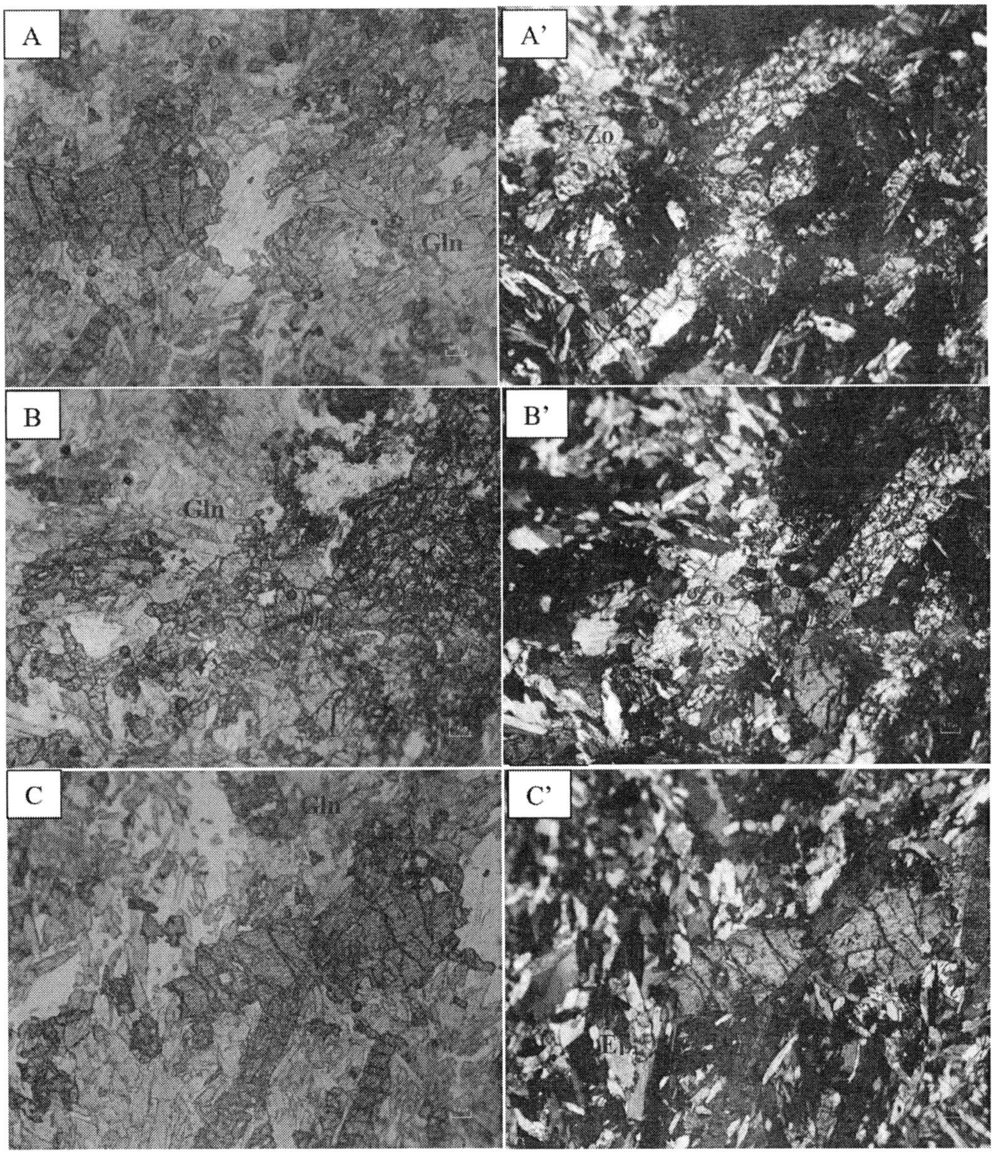

Figure 7. Microphotograph of thin section of Eclogite from Cempaga river. Paralel nicol (PPL) in left, and Cross nicol (XPL) in right. It appears that the mineral Omphacite (Omp) was partially replaced by Glaucophane (Gln) and the zoisite and epidote (Ep). The conditions reflect polymetamorphism from ultrahigh-pressure to high-pressure - medium temperature.

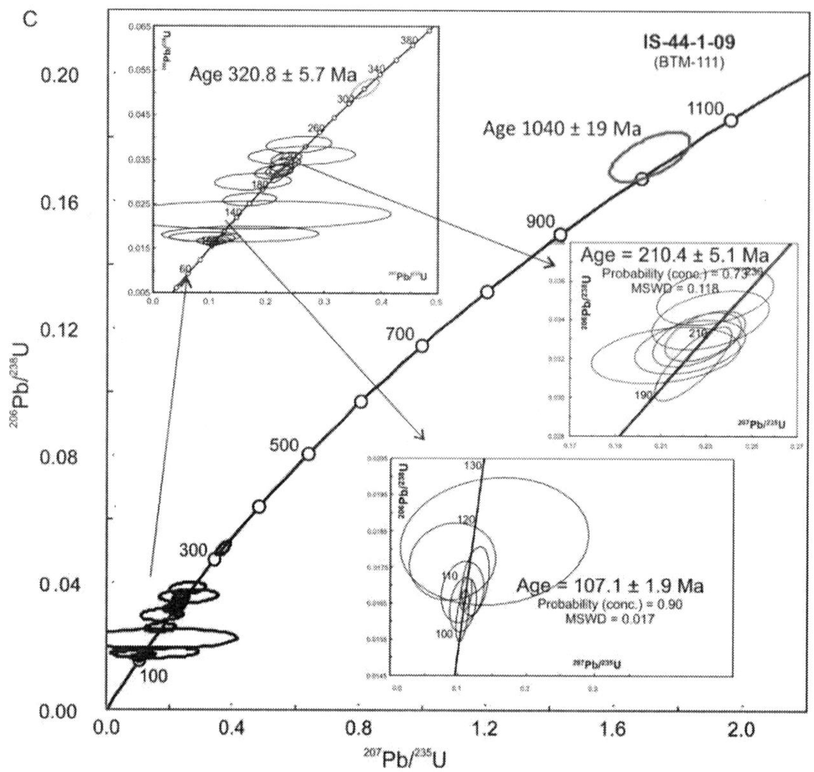

Figure 8. Concordia diagram with U-Pb SHRIMP analysis of zircon from the Bantimala Melange Complex. Eclogite is MORB geochemistry (above left). Diorite block in mélange (above right). Below is detrital zircon of the Balangbaru sandstone. Some zooms are made to display the detail concordant ages (After [2]).

4.4.3 $^{40}Ar/^{39}Ar$ Geochronology

$^{40}Ar/^{39}Ar$ age dating of white mica (phengite) from samples of phenocrysts with a crystal size of 150 >250 μm. Dating results gives an age spectrum of 114 Ma - 100 Ma. The age of 114 Ma can be interpreted as the age of eclogite formation which is the peak of metamorphism or as the final or maximum subduction. Whereas the age of 100 Ma can be interpreted as the age of exhumation or collision as the end of subduction tectonics in the Bantimala Complex.

5. Conclusions and Discussion

Based on the results of radiometric age dating and petrographic analysis of high and ultrahigh-pressure metamorphic rocks, that in the Cretaceous Period in the Bantimala Complex has experienced subduction and collision tectonics followed by exhumation. This event, allegedly occurred in the South Arm of Sulawesi, or in the Southeastern part of Sundaland, which caused Sundaland to increase. This tectonic phase occurred in the Cretaceous Period (137-100 Ma). Collision occurred at around 100 Ma which is interpreted as the age of exhumation or collision regarding the end of subduction tectonics in the South Arm of Sulawesi for the tectonics of Cretaceous Period. The results of the *U-Pb Zircon SHRIMP* 819.4 ± 9.3 Ma, 758.8 ± 8.7 Ma originating from zircon detrital in the Balangbaru Formation, were interpreted as zircon detrital originating from the Gondwana Continental and experiencing transportation and deposition with the Balangbaru Formation during the Cretaceous Period. Whereas zircon ages 320.2 ± 5.7 Ma and 210.4 ± 5.1 are interpreted as zircon magmatism before or during the initial phase of subduction tectonics, regarding the disappearance or consumption of the Tethys Ocean under the Eurasian Continent.

References

[1] Musri, 2015. Metamorphic rocks evolution of Rumbia Complex, Southeast Arm of Sulawesi and its relation to gold mineralization, ITB Dissertation (Unpublished).

[2] Soesilo, J., 2012. New cretaceous tectonic setting of southeast Sundaland based on metamorphic evolution, A PhD Dissertation of ITB (Unpublished).

[3] Maulana, A., Christy, A., Ellis, D., Imai, A., Watanabe, K., 2013a. Geochemistry of eclogite- and blueschist-facies rocks from the Bantimala Complex, South Sulawesi, Indonesia: protolith origin and tectonic setting. The Island Arc 22 (4), 427–452.

[4] Lloyd T. W., Hall, R., Amstrong, R. A., Barber, A.J., Fadel, M. B., Baxter, A., Wakita, K., Manning, C., Soesilo, J., 2017. The geological history of the Latimojong region of western Sulawesi, Indonesia, Journal of Asian Earth Sciences 138 (2017) 72–91.

[5] Katili, J.A., 1978. Past and Present Geotectonic Position of Sulawesi, Indonesia, *ectonophysics*, **45**, pp. 289-332.

[6] Hamilton W., 1979. Tectonics of the Indonesian Region. *U.S. Geological Survey Professional Paper* **1078**, pp. 345.

[7] Van Leeuwen, T.M., 1981. The geology of south-west Sulawesi with special reference to the Biru area. In: In: Barber, A.J., Wiryosujono, S. (Eds.), Geology and Tectonics of Eastern Indonesia Special Publications, vol. 2. Geological Research Development Centre, Bandung, Indonesia, pp. 277–304.

[8] Sukamto, R., 1982. *The Geology of the Pangkajene and western part of Watampone, Sulawesi*, Geological map, 1:250.000 with explanatory note, Geological Survey of Indonesia in cooperation with the USGS.

[9] Parkinson, C.D., 1996. *The petrology, structure and geologic history of the metamorphic rocks of Central Sulawesi, Indonesia*, PhD thesis, University of London, London, UK.

[10] Parkinson. C.D., 1998. An Outline of the petrology, structure and age of the Pompangeo Schist Complex of Central Sulawesi, Indonesia, *The Island Arc* 7, 231-245.

[11] Soesilo, J., 1998. Metamorphism in the Latimojong Complex, South Sulawesi and its tectonic significances, *unpublished Magister thesis, Institute of Technology Bandung*, (In Indonesian, with English abstract), pp. 85.

[12] De Roever, W.P., 1956. Some additional data on the crystalline schists of the Rumbia and Mendoke mountains, Southeast Celebes, *Verh. K. ned.geol.-mijnb. Genoot, XVI*: 385-393.

[13] Helmers, H., Sopaheluwakan, J., Nila, E.S. and Tjokrosapoetro, 1989. Blueschist evolution in Southeast Sulawesi, Indonesia, *Netherlands Journal of Sea Research,* 24, (213), 373-381.

[14] Van Leeuwen, T., Allen, C.M., Kadarusman, A., Elburg, M., Palin, J.M., Muhardjo , Suwijanto, 2007. Petrologic, isotopic, and radiometric age constraints on the origin and tectonic history of the Malino Metamorphic Complex, NW Sulawesi, Indonesia.

[15] Van Leeuwen, T., Allen, C.M., Elburg, M., Massome, H-J., Palin, J.M., Hennig, J., 2016. The Palu Metamorphic Complex, NW Sulawesi, Indonesia: Origin and evolution of a young metamorphic terrane with links to Gondwana and Sundaland.

[16] Wakita, K., 2000. Cretaceous accretionary-collision complexes in central Indonesia, *Journal of Asian Earth Sciences,* **18**, pp. 739-749.

[17] Metcalfe, I., (1996); Pre-Cretaceous evolution of SE Asian terranes, In Hall, R and Blundell, D.J., (eds). *Tectonic Evolution of Southeast Asia, 97-122.,*

[18] Parkinson, C.D., Miyazaki K., Wakita K., Barber A J. and Carswell D A., 1998. An Overview and Tectonic Syntesis of the PreTertiary Very High-Pressure Metamorphic and Associated Rocks of Java, Sulawesi and Kalimantan, Indonesia, *Island Arc*, **7**, pp. 184-200.

[19] Wakita K, Miyazaki K, Zulkarnain I, Sopaheluwakan J, and Sanyoto P., 1998. Tectonic implications of new age data for the Meratus Complex of South Kalimantan, Indonesia, *The Island Arc*, 7, pp. 202-222.

[20] Miyazaki, K., Zulkarnain, I., Sopaheluwakan, J.and Wakita, K., 1996. Pressure-temperature conditions and retrograde paths of eclogites, garnet glaucophane rock and schists from South Sulawesi, Indonesia. *Journal of Metamorphic Geology* **14**, pp. 549-563.

[21] Wakita, K. Munasri, Sopaheluwakan, J., Zulkarnain, I and Miyazaki, K., 1994a. Early Cretaceous tectonic events implied in the time lag between the age of radiolarian chert and its metamorphic basement in Bantimala area, South Sulawesi, Indonesia. *Island Arc* **3**, pp. 90-102.

[22] Hasan, K., 1990. The Upper Cretaceous flysch succession of the Balangbaru Formatio Southwest Sulawesi, *unpublished PhD thesis of RHBNC University of London*, pp. 312.

[23] Yuwono, Y.S., Bellon, H., Soeria-Atmadja, R and Maury, R.C., 1986. Neogene and Pleistocene volcanism in South Sulawesi. *Proc. of 14th IAGI Ann conv.*, Jakarta 1985, pp. 169-179.

[24] Polve, M., Maury, R.C., Bellon, H., Rangin, R., Priadi, B., Yuwono, S., Joron, J.L. and Soeria Atmadja, R., 1997. Magmatic evolution of Sulawesi (Indonesia): Constrains on the Cenozoic geodynamic history of the Sundaland active margin, *Tectonophysics*, **272**, pp. 69 – 92.

[25] Yuwono, Y.S., Priyomarsono, S., Maury,R.C., Rampnoux, J.P., Soeria-Atmadja, R., Bellon, H. and Chotin, P.,(1988): Petrology of The Cretaceous Magmatic Rocks From Meratus Range, Southeast Kalimantan, *Journal of Southeast Asian Earth Sciences*, Vol **2**, No. 1, pp. 15-22.

[26] Bergman, S.C., Coffield, D.Q., Talbot, J.P. and Garrard, R.J., 1996. Tertiary tectonic and magmatic evolution of western Sulawesi and the Makassar Strait, Indonesia: Evidence for a Miocene Continent Collision. In Hall, R. and Blundell,D.J. (eds.) *Tectonic Evolution of SE Asia*, Geological Society, London, Special publications, **106**, pp. 391-430.

[27] Sukamto, Rab, (1978); The structure of Sulawesi in the light of plate tectonics, Pros. Reg. Conf. Geol. Min. Res. SE Asia, pp. 121-141.

[28] Wakita, K., Sopaheluwakan, J., Miyazaki K., Zulkarnain, I and Munasri, 1996. Tectonic Evolution of The Bantimala Complex, South Sulawesi, Indonesia. In Hall, R & Blundell, D. (eds), 1996.

IOP Conf. Series: Materials Science and Engineering **875** (2020) 012056 doi:10.1088/1757-899X/875/1/012056

Identification of Landslide Disaster Potential based on Weathering Grade of Rock in Parepare City South Sulawesi, Indonesia

Busthan[1]*, Safruddim[1], and H Pachri[1]

[1]Department of Geological Engineering, Engineering Faculty, Hasanuddin University, Makassar, Indonesia

*Email: bazikin@yahoo.com

Abstract, The purpose of this research is to identify the potential of landslide disaster based on weathering grade of rocks from selected areas. There are three aspects of field observation method used in this research, consist of slope geometry, lithology, and weathering grade of rocks aspects. The result shows that the rocks on research area have been slightly to completely weathered, some areas have became residual soil and slope angles of the research area varieties from steep to very steep. With those condition of weathering and slope angle, the research area will be slightly to very susceptible to landslide. The landslide models or types that will occur are debris slide and debris fall.

1. Introduction

Increase of population in a city or regency results in the expansion of land for housing and other infrastructures. Parepare City is an agrarian city that becomes city of service with high increase in population. For example, in 2017, population in Parepare increased to 142,097 people from 132,699 people in 2015. Increase of population affected the increase of residential land and other land needs such as road infrastructure, etc. [1]. Utilization of land from forest land into residential land can no longer be limited to land plains, mountainous land to coastal areas. Excessive use of the area has a negative impact, which results in disaster. One of the disasters that will occur if land extensification is carried out for infrastructure is the occurrence of landslides.

Parepare City is occupied by three rock formations namely: Parepare Volcanic Rocks, Camba Formation, and Alluvial Deposit [2]. Parepare volcanic rocks consist of breccia and tuff. Alluvial deposit consists of deposits of sand to clay-sized materials. Some of the rocks has been weathered into residual soil and alluvial.

Based on information above, it is very important to identify areas with landslide potential in Parepare and determine model or type of landslide that will occur. Based on background of the research, the problem equation of the research are: (1) how the weathering grade of the rocks affect the landslide disaster and, (2) what model or type of landslide will occur.

Therefore, the purposes of this research are to identify the effect of weathering grade on the susceptibility of landslide and to analyze and determine model or type of the landslide that will occur.

2. Literature Review

2.1. Geological Setting

Regionally, the research area belongs to geological map of Pangkajene and Western Part of Watampone in 1:250,000 [2]. The following is the explanation of the regional geology of the research area, consists of geomorphological conditions, stratigraphy, and geological structures. In Pangkajene and West Part of Watampone sheet, there are two rows of mountain that extend in north northwest and separated by Walanae River valley. Western part of the mountain occupies nearly half of the area, widen in southern part (50 km) and narrow in northern part (22 km). The highest peak of the area is 1694 meters, whereas average elevation of the area is 1500 meters. Forming rocks are mostly volcanic rocks. Northern part of this mountain has karst topography where the surface is mostly conical. The northern boundary of the area is Bone plain, which occupies nearly a third of northern area. Walanae Valley which separate the mountains is 35 kilometers wide, but only 10 kilometers in southern part.

Regional stratigraphy of research area is based on previous researchers, according to [2], regional stratigraphy of research area belongs to geological map of Pangkajene and Western Part of Watampone. Mostly mountains, both western and eastern part have volcanic rocks. Most of western part of the mountain formed from Camba Formation, which is approximately 5000 meters thick, uncomformably laying above Tonasa Formation, meanwhile Walanae Formation is interfingering with upper part of Camba Formation.

Alluvium and coastal deposit (Qac): Clay, silt, sand, and gravel along the big river, and along the coastal line. At some areas, coastal deposits contains clam shell remains and coral limestone.

Parepare Volcanic Rocks (Tppv): Volcanic breccia consists of trachyte and andesite fragments; pumice, tuffaceous sandstone, conglomerate, and tuffaceous breccia; intruded by trachyte and andesite stocks. The age of this unit is Pliocene, based on radiometry dating on trachyte and tuff in Parepare, which results 4.25 and 4.95 Ma [2].

Camba Volcanic Rocks (Tmcv): Volcanic rocks alternating with marine sedimentary rocks; volcanic breccia, lava, and fine-grained to lapili tuff; lime sandstone, limestone, and marl. The rocks consist of andesite and basalt; intruded by basalt-composed sill and stock; light gray, dark gray, and brown in color. The rocks mostly deposited in neritic marine environment as Camba Formation volcanic facies, some volcanic breccias contain limestone.

Regional structure in research area is based on previous researcher, according to [2], belongs to geological map of Pangkajene and Western Part of Watampone. Developing structures are affected by Walanae regional fault in northwest – southeast direction, actively, resulting in deformation such as fold, joint, and fault. North northwest-directed major fault formed in Middle Miocene, and grew until Post-Pliocene. Fold directed in almost parallel with major fault estimated to be formed because of horizontal pressure in east – west direction before Upper Pliocene. The faulting relatively smaller in western part of west mountain directing in northwest – southeast and southwest; and have fault contact with surrounding rock units.

2.2. Grade of Rock Weathering

According to [3], stated that weathering is a changing and breaking down process of rocks and soil on the or near the surface of earth by physical, chemical, and biological process into clay, iron oxide, and other weathering products.

There are 3 types of weathering, those are physical (mechanical) weathering, chemical weathering, and weathering caused by organism activity. According to [4], weathering grade is divided into 6 grades, from fresh rocks (unweathered) to residual soil. Weathering grade will be explained in Table 1 as follows:

Table 1. Terminology of weathering grade of rocks [4]

Grade	Term	Description
I	Fresh	No visible sign of rock material weathering; perhaps slight discoloration on major discontinuity surfaces
II	Slightly weathered	Discoloration indicates weathering of rock material and discontinuity surfaces. All the rock material may be discolored by weathering and may be somewhat weaker externally than in its fresh condition.
III	Moderately weathered	Less than half of the rock material is decomposed and/or disintegrated to a soil. Fresh or discolored rock is present either as a continuous frame work or as core stones.
IV	Highly weathered	More than half of the rock material is decomposed and/or disintegrated to a soil. Fresh or discolored rock is present either as a discontinuous frame work or as core stones
V	Completely weathered	All rock material is decomposed and/or disintegrated to soil. The original mass structure is still largely intact
VI	Residual soil	All rock material is converted to soil. The mass structure and material fabric are destroyed. There is large change in volume, but the soil has not been significantly transported.

According to [5], one of the causing factor of landslide is weathering because the higher the weathering grade, the more the rock forming minerals disintegrate and if that happens in a slope area, that area will be susceptible to landslide.

According to [6], landslide can be caused by interaction of some conditions like morphology, geology, hydrogeology, and land use. Those conditions affect each other, making a slope has a tendency to move. Furthermore [6] stated that such a slope condition is called susceptible to move. So, the definition of susceptible is when a slope has a potential or tendency to move, but has not moved yet.

Movement on a slope can happen when there is a movement trigger. Movement triggers are natural or unnatural processes that can change condition of the slope from susceptible to start moving. Movement triggers are rain, vibration, human activity on slope such as slope excavation and cutting, blasting, overburden, or water entering into slope through leaks on channel or ponds.

The causing factor of landslide [6] can be seen in Figure 1.

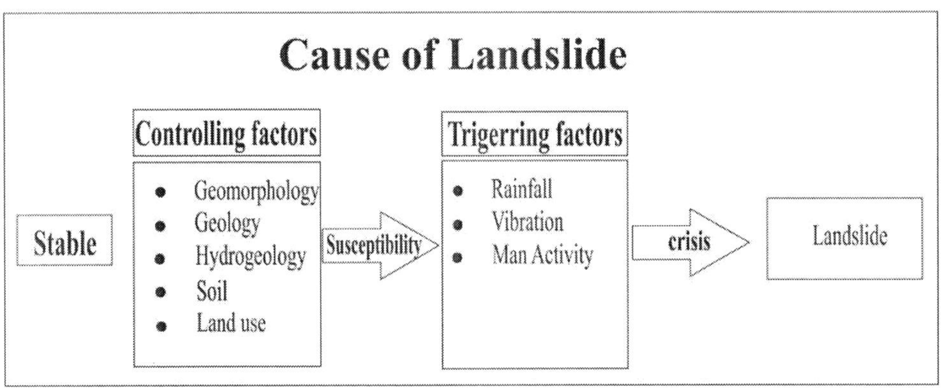

Figure 1. Causing factors of landslide (controlling factors and triggering factors) [6]

In Figure 1 above, we can see that there are 4 stage of the process of landslide, those are:

- Stable stage
- Susceptible stage

- Critical stage
- Landslide stage

Furthermore, [6] explained that Figure 1 also shows that the cause of landslide can be divided into direct cause (controlling factors) and indirect cause (triggering factors). Landslide classification [7] is based on movement mechanism and moving materials. Classified the landslides into 6 types, those are fall, topples, slides, lateral spread, flow, and complex/compound [7].

According to [6], landslide susceptibility is a phenomenon that makes slope become potential or susceptible to move although it is stable at the moment (hasn't moving or land sliding yet). A slope susceptible to landslide will move when there is an interference triggering the movement.

3. Research Methods

3.1. Research Area

The research area is located in Parepare City, about 150 km from Makassar City to the north.

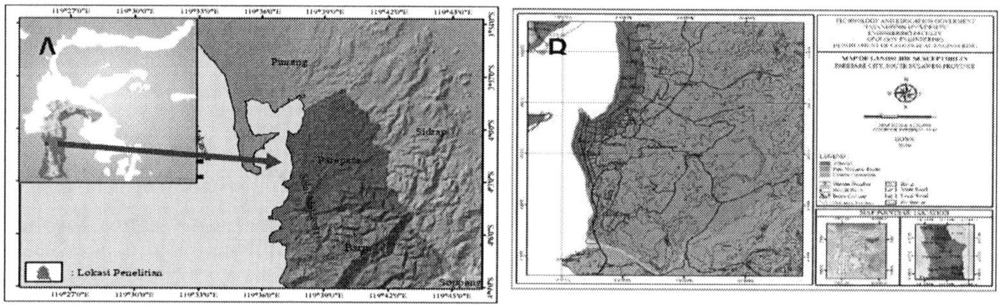

Figure 2. (a) Administrative map of research area, (b) Geological map of research area

3.2. Retrieving Field Data

This research is explorative, where all retrieved data are field data. Field data consist of:

1. Geomorphological Data
   Geomorphological data contains slope geometry, such as slope height, slope length, slope width, and slope angle
2. Lithological Data
   Lithological data contains: the position of the outcrops, rocks' type, and the name of the rocks
3. Weathering Grade Condition

Next observation is observing weathering, which is observing weathering such as fresh rocks, slightly weathered, moderately weathered, highly weathered, and completely weathered.

4. Result and Discussion

Result of observation and measurement of the slope geometry in research area is shown in Table 2 and Figure 2.

Table 2. Slope geometry data of research area

Location	Position		Slope Condition				Explanation
	x	y	Slope Height (Meter)	Slope Width (Meter)	Slope Direction (N….°E)	Slope Dip (….°)	
1	119°37'50.43"	04°01'39.17"	10	70	225	67	Jalan Patung Kuda
2	119°37'49.92"	04°01'57.25"	13	35	240	56	Geddongnge Bagian Timur

Location	Position		Slope Condition				Explanation
	x	y	Slope Height (Meter)	Slope Width (Meter)	Slope Direction (N….°E)	Slope Dip (….°)	
3	119°37'41.76"	04°01'48.17"	22	100	60	34	Geddongnge Bagian Barat
4	119°37'26.10"	04°02'57.04"	13	20	305	75	Toraya Kel. Lumpue Bacukiki Barat
5	119°38'24.46"	03°59'58.98"	9	40	85	87	Jalan Lauleng (MAN 1)
6	119°38'30.45"	03°59'07.67"	15	45	180	48-60	Jalan Lauleng IAIN Parepare

The result of weathering grade observation of each observed location is shown in Table 3 below.

Table 3. Weathering grade of research area

Location of Observation	Grade of Weathering				
	Fresh	Slightly Weathered	Moderately Weathered	Highly Weathered	Completely Weathered
Location-1			√	√	√
Location-2			√	√	√
Location-3		√	√	√	√
Location.4		√		√	√
Location-5		√	√		
Location-6		√	√	√	√

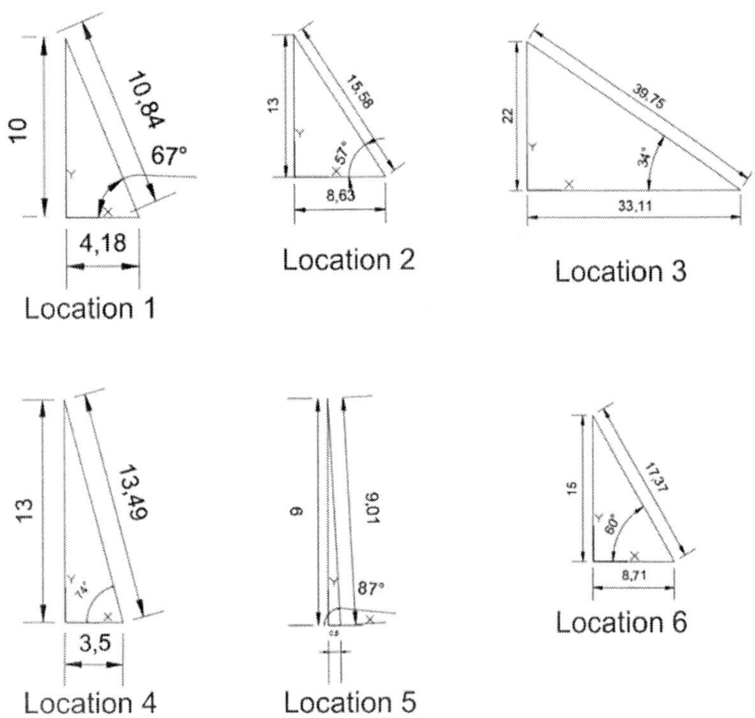

Figure 3. Slope geometry of research area

Below is the explanation of each observed locations and potential of occurrence of landslide.

4.1. Location-1
Location-1 is located in Patug Kuda Street with slope geometry consists of: (1) slope height: 10 meters; (2) slope width: 70 meters; (3) slope length: 10.84 meters; (4) slope direction: N225°E; and (5) slope angle: 67 degrees. Slope forming rocks are volcanic rocks, consist of alternating tuff, lapilli tuff, volcanic breccia, and agglomerate. Those rocks are moderately weathered and locally have completely weathered (Table 3), locally occurred residual soil were also found in this area.

Based on the data above, location-1 is moderately susceptible to landslide because of steep slope (67 degrees), but the weathering is not evenly distributed to completely weathered, and the soil is thin. If landslide is about to happen, the model/type of landslide that will happen is debris slide.

4.2. Location-2
Location-2 is located in Northern Part of Geddonge with slope geometry consists of: (1) slope height: 13 meters; (2) slope width: 35 meters; (3) slope length: 15.58 meters; (4) slope direction: N240°E; and (5) slope angle: 56 degrees. Slope forming rocks are volcanic rock, consists of bedded coarse tuff, moderately to completely weathered and locally has become residual soil.

Based on the data above, location-2 is very susceptible to landslide because of steep slope angle (56 degrees) and evenly distributed weathering into completely weathered. Moreover, there are buildings like offices and residents' houses above the slope which became burden to the slope. If landslide is about to happen, the model/type of landslide that will happen is debris slide.

4.3. Location-3
Location-3 is located in Western Part of Gaddonge with slope geometry consists of: (1) slope height: 22 meters; (2) slope width: 100 meters; (3) slope length: 39.75 meters; (4) slope direction: N60°E; and (5) slope angle: 34 degrees. Slope forming rocks are volcanic rocks, consists of bedded coarse tuff, slightly to completely weathered (Table 3).

Based on the data above, location-3 is moderately susceptible to landslide because of moderately steep slope (34 degrees) and unevenly distributed completely weathered rocks and also thin soil. If landslide is about to happen, the model/type of landslide that will happen is debris slide.

4.4. Location-4
Location-4 is located in Tonraya Western Part of Bacukiki with slope geometry consists of: (1) slope height: 13 meters; (2) slope width: 20 meters; (3) slope direction: N305°; and (4) slope angle: 74 degrees. Slope forming rocks are volcanic rocks consists of lapilli tuff and agglomerate, moderately to completely weathered (Table 3).

Based on the data above, location-4 is very susceptible to landslide because of steep slope (74 degrees) and distributed weathering into moderately to completely weathered. If landslide is about to happen, the model/type of landslide that will happen is debris fall.

4.5. Location-5
Location-5 is located in Lauleng Street with slope geometry consists of: (1) slope height: 9 meters; (2) slope width: 40 meters; (3) slope length: 9.01 meters; (4) slope direction: N85°E; and (5) slope angle: 87 degrees. Slope forming are volcanic rocks, consists of coarse tuff, fine tuff, and locally lapilli tuff, slightly to moderately weathered (Table 3).

Based on the data above, location-5 is moderately susceptible to landslide because of very steep slope (87 degrees) and slightly to completely weathered rocks, but will be very susceptible if the weathering goes on until the rocks are completely weathered. If landslide is about to happen, the model/type of landslide that will happen is debris fall.

4.6. Location-6

Location-6 is located in Lauleng Street (IAIN Parepare) with slope geometry consists of: (1) Slope height: 15 meters; (2) slope width: 45 meters; (3) slope length: 17.37 meters; (4) slope direction: N180°E; and (5) slope angle: 48 to 60 degrees. Slope forming rocks are volcanic rocks, consists of coarse tuff and fine tuff, slightly to completely weathered (Table 3).

Based on the data above, location-6 is very vulnerable to landslide because of very steep slope (48 to 60 degrees) and moderately to completely weathered. If landslide is about to happen, the model/type of landslide that will happen is debris slide.

5. Conclusions

- Parepare City official or instances has to pay attention to all six of research location, because weathering will go on over time and will result in weaker slope strength into critical slope
- If landslide happens, the model/type of landslide that will happen is debris fall in location 1, 2, and 6 and debris slide in location 4 and 5
- To prevent the slope from becoming more critical, mitigation has to be done in all six locations, structurally and non-structurally.

Acknowledgement

The authors are grateful to Dean of Engineering Faculty Hasanuddin University for permission and expense to carry out this research. Thanks are directed to Mr. Burhanuddin, Regional Disaster Management Institution for being companion during research. Thanks are also due to Gunawan, a student of Geological Engineering Hasanuddin University for his assistance during the research.

References

[1] Central Bureau of Statistics Parepare City, 2018, Parepare City in Numbers Data.
[2] Sukamto, R.& Supriatna, S.,1982, *Geology of Pangkajene and Watampone Western Part*, Center of Research and Development of Geology, Directorate of General Mining, Energy and Mining Department, Bandung, Indonesia. (In Bahasa)
[3] Undul O., 2012, Weathering of Ultramafic Rocks, Istanbul University, Geological Engineering, ETH, Zurich.
[4] Wyllie D.C. and Mah Ch. W., 2004., Rock Slope Engineering, 4th Edition, Spon Press, Taylor and Francis Group, London.
[5] Busthan, 2015, Analysis of Landslide Susceptibility of Rupture Surface Based On Weathering Stage of Volcanic Rock, Dissertation, Post Graduate Program, Hasanuddin University, Makassar. (In Bahasa)
[6] Karnawati, D. 2005., Natural Disaster of Earth Mass Movement In Indonesia and Prevention Effort, Department Of Geological Engineering, Engineering Faculty, Gadjamada University, Yogyakarta. (In Bahasa)
[7] Varnes, D.J., 1978. *Slope Movement Types and Processes*, In Schuster, R.L. Krizek RJ., Landslide Analysis and Control, Transportation Research Board, Special Report 176, National Academic of Science USA.

IOP Conf. Series: Materials Science and Engineering **875** (2020) 012057 doi:10.1088/1757-899X/875/1/012057

The Roughness Characteristic of AA6061-F, AA6061-O and AA6061-T6 after Machining Process

H Arsyad[1]*, L H Arma[1], and Januari[1]

[1]Mechanical Engineering Department, Universitas Hasanuddin

*Email: arsyadhairul@yahoo.com

Abstract. Generally, the aims of the machining process are to produce a smooth surface according to a specific standard. To get a good surface quality can be implied by improving the machinability of the raw material. In addition, the selection of machining methods and machining parameters is also an important factor. This study aims to obtain a relationship between heat treatment and surface quality of AA6061 alloy after machining. Heat treatment is expected to improve the machinability of the AA6061 alloy to produce a surface with low roughness. The study was conducted using two heat treatment methods, namely AA6061-T6 and AA6061-O and comparing it with AA6061-F. After the heat treatment process, the AA6061 alloy was machined using a lathe with 1500 rpm spindle speed, 0.05 mm/rev of feed rate and 2 mm depth of cut. The condition of the turning process was varied with and without lubrication. The effect of heat treatment on the surface quality after machining is in terms of surface roughness, microstructure, hardness value, and chip shape. From the results, it was found that Al6061-O has the best surface quality with lowest hardness and lowest surface roughness.

1. Introduction

Product quality is the goal of each manufacturing process. The machining process is one of the commonly used process in manufacturing processes. Machining process works by removing part of the work pieces by cutting using the appropriate tool and doing it on the machining equipment's. Machining quality can be seen from the accuracy of the product dimensions, the level of surface roughness and conformity with the desired shape. The stages of surface finishing in each machining process are very important to note.

The quality of surface finishing is influenced by many factors including machining parameters such as cutting speed, feed rate, and cutting depth. The choice of tools and use of cooling fluid is also one of the main considerations. However, another factor that is not less important is the machinability of the raw materials. Improving the machinability is often an option if there is no other material can be choose. The machining process plays a large portion in the manufacture of metal products. Machining is generally carried out after the main manufacturing process. The aim is to provide a good surface quality and dimensional accuracy. The importance of the machining process in making metal products makes machining topics are attractive to explore. This is to get the most appropriate combination of the many machining parameters involved to produce high quality products.

Aluminum is a metal that has the best combination of properties for applications. Its strength can be improved, the metal is lightweight, has excellent corrosion resistance, adequate thermal and electrical conductivity. This is the main reason why an aluminum is so widely used in every field and application. Many components are made from aluminum as machined parts. Aluminum has a weakness in strength but is overcome by the ability of aluminum to obtain microstructural modifications. The strength of aluminum alloy can be increased by heat treatment, work hardening and the addition of alloying element.

The historical treatment of aluminum as a raw material certainly affects the behaviour during the machining process. The problem is to find the most appropriate manufacturing route for aluminum alloy with typical micro structure condition in manufacturing process to meet the desired product requirements.

Improving the machinability is one of the way to get best surface quality in machining. Machinability is related to tool life, surface roughness, cutting force, machining power and type of chips obtained. There are so many studies on aluminum alloys that are related to heat treatment and machining processes. Good quality of machined surface will increase fatigue strength of material, corrosion resistance and lower friction [1]. Demir et al has investigated about the effect of heat treatment (SHS) and SHTA on machinability of material AA6061. They found that differences in aging time and cutting speed significantly affected the value of surface roughness. However, the SHTA treatment did not significantly affect the cutting force except for SHT which resulted low hardness [2]. However, F. Frost et al found that reducing the processing time during the heat treatment stages had little effect on the mechanical properties that obtained [3].

Similar research was also conducted by Kaya et al studied effect of aging on machinability of AA7075 materials with variations in aging time (1, 6, 12, and 24 hours at a temperature of 180°C) and machining conditions. Changes in tool temperature were examined to see the effect on chips resulted. The results of the study showed that the condition of aging 6 hours had the best machinability with the lowest surface roughness value [4].

Another study was carried out by Rashad et al with five variations of heat treatment namely natural aging, artificial aging, annealed and artificial aging with superimposed free forging and as received. The effect of heat treatment on the machinability of alloy 7116 was focused on the value of surface roughness. By varying the cutting parameters such as cutting conditions, cutting speed, feeding, and cutting depth it was found that artificial aging conditions with superimposed free forging have the best machinability [5].

Imhade et al conducted study of the relationship between machining parameters using the end milling process on AA6061 alloys. Sequentially it was found that the most influential on surface roughness was spindle rotation speed, feed rate and depth of cut [6]. The combination of spindle rotation variations and feed rate also produces an effective influence on the level of surface roughness. Similar research is also carried out by [7], the presence of burrs defects and surface quality during the milling process in AA6061-T6 alloys can be reduced by controlling the feed per tooth whereas the cutting speed gives minimal effect. Another study was conducted by Adnan, he examined various cutting techniques in pure Aluminum alloys and 6061 then proposed cutting with water jet cutting techniques as the best method, the reason was that this technique did not produce significant changes in microstructure and there was no significant decrease in hardness after the cutting process [8]. However, it is still necessary to conduct an in-depth study of the relationship between heat treatment and machining conditions in the AA6061 alloy on machining quality, hardness and the shape of the chips produced

2. Materials and Method

AA6061 cylinder was used as specimen which is dimension 150 mm length and 32 mm diameter. Specimens were divided in three categories, namely AA6061-F, AA6061-O and AA6061-T6. Figure 1 shows the heat treatment mode for AA 6061-O and AA6061-T6. The first specimen, AA 6061-O was heated to 440°C and then hold for 2 hours, after that the sample was cooling down to the room temperature with slow cooling. The second specimen, AA6061-T6 was heated to 550°C and hold it for 2 hours and then rapid cooling to 29 °C then artificial aging for 8 hours at 176 °C. The third specimen of AA6061 was as fabricated (AA6061-F). After heat treatment process, the specimen was machined in turning lathe using GEDEE WEILER LZ-330 G with HSS as a tool steel, see figure 2. There were two type of turning condition without cooling media and with cooling media.

The hardness test was conducted after heat treatment process using hardness leeb portable instrument before machining process. The machining variable was set up in 1500 rpm spindle speed, 0.05 mm/rev feed rate and 2 mm of depth of cut for 80 mm length. For microstructural observation, the specimen was etched using *Keller reagent* consist of 25 ml aquades, 5 ml HF, 7.5 ml HCL and12.5 HNO_3 for 30 second. The microstructure was evaluated using Optical Microscope. The surface roughness was measure by using *surface roughness tester* Mitutoyo SJ-310.

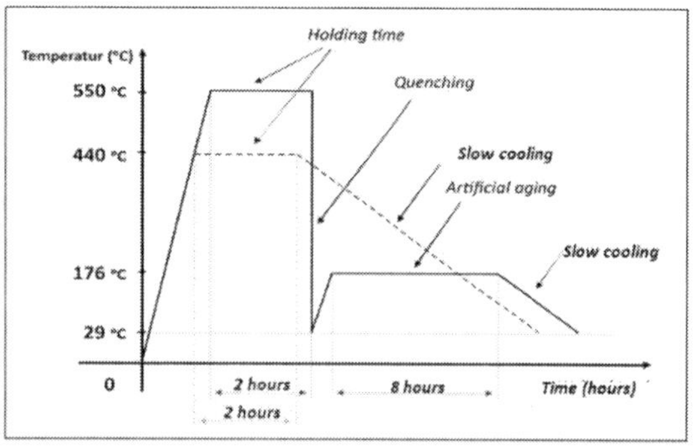

Figure 1. The schema of heat treatment process for AA6061-O (dashed line) and AA6061-T6 (bold line)

Figure 2. Machine lathe, Gedee Weiler LZ-330 G

3. Result and Discussion

Figure 3 shows the hardness value of each specimen. AA6061-T6 has the highest hardness (412.6 HL) and the lowest hardness value is AA6061-O (337 HL). AA6061-T6 has increases in hardness due to microstructural change during heat treatment process. The changes on microstructure of AA6061 during aging is the causes of increases hardness mainly due to the presence of precipitate state. Similar result was obtained by Ozturk et al, where variations in strength of AA6061 occurred with different levels of aging process. This variation in mechanical strength is caused by structural evolution during the aging process [9]. In the other hand for AA6061-O has solution heat treatment where the hardness was decreased compare to AA6061-F. The decreases of AA6061-O hardness because of softening in microstructure during solution treatment, the same result with SHT treatment conduct by Demir et al. [2].

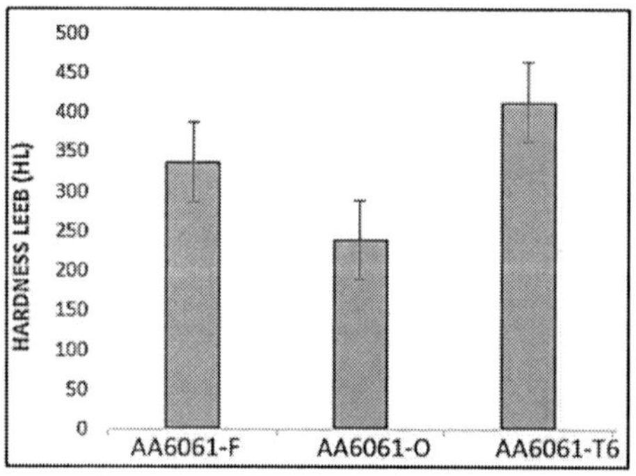

Figure 3. Hardness value of AA6061-F, AA6061-O and AA6061-T6

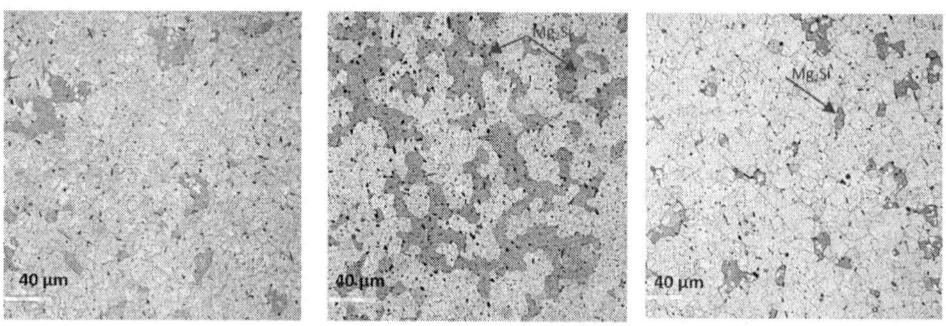

Figure 4. Microstructure of AA6061-F, AA6061-O and AA6061-T6

Figure 4 shows the different of microstructure of each specimen AA6061-F. AA6061-O and AA6061-T6. Figure 4a show typical microstructure of AA6061 consist of α-Al matrix with unclear of grain boundary. Figure 4b show large area of Mg_2Si surrounding by α-Al matrix and then 4c shows that large of Mg_2Si precipitate dissoluble in aluminum matrix. Three different type of microstructure causing different value of hardness, which is AA6061-T6 give the highest hardness.

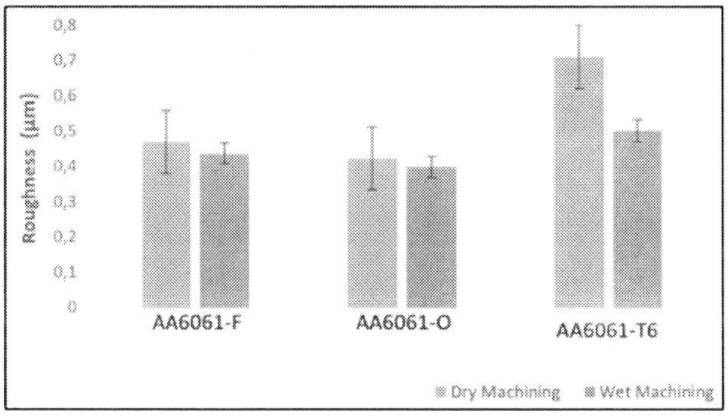

Figure 5. Roughness value of AA6061-F, AA6061-O and AA6061-T6 after turning process.

Figure 5 shows the roughness value for each type of specimen after turning process using lathe with 1500 rpm spindle speed and 0.05 mm/rev of feed rate. For each type of aluminum specimen, the roughness measurement was repeated five times, and it was found that the lower roughness was 0.396 µm for AA6061-O with wet machining and in the other hand the highest roughness was 0.710 µm for AA6061-T6 with no lubrication (dry machining). AA6061-T6 show a significant roughness decreases by using wet machining condition, from figure 5 also show that in overall, wet machining effectively stabilize the value of roughness for each specimen. The uses of lubricant will be reduced energy consumption during cutting process and as result the chip temperature decreases and also better improve in a controlled surface [10]. Moreover, according to Sreejith, that machining with minimal lubrication conditions can actually improve the characteristics of the machine's ability so that the use of flooded can be reduced [11]

However, from roughness measurement it seems that AA6061-O has the best quality of surface indicated by low surface roughness which also has lower hardness. The different result was obtained by Sevim et al, where the quality of surface roughness was improved by increasing the specimen hardness due to presence of precipitate. The presence of soluble precipitate in AA6061-T6 play important role to increase hardness [12], however this precipitate also probably produce distortion and deflection during shear deformation in cutting process and produce bad quality of surface roughness. Walid et al mentioned that if the interaction between surface hardness and cutting speed produce built-up edge chip condition then the surface quality became undesired [13].

The other reason can be explained by residual stress, AA6061 specimen was relatively free of residual stress due to softening, Assini et al stated that the residual stress will give unpredictable machining behaviour, especially for large work piece [14]. As mentioned that work piece hardness has significant effect to the surface finish quality [1], but the effect of the hardness on the surface quality could be play different mechanism and different result. As reported by Kaya et al that workpiece with high hardness tend to produce BUE chip condition and it is insufficient to meet a better quality of surface [4]. Khamer et al has different explanation, he stated that material with high resist of plasticity will be difficult to facilitate material flow during cutting or machining process and then surface finish will improved.

Figure 6 shows the chips produced for each type of specimen by turning process. The result shows that the chips have different type such as continued spiral chip, continued chip, discontinued spiral chip and discontinued chip. For dry machining condition, AA6061-F and AA6061-O has continued chips, otherwise AA6061-T6 has discontinued chips. However, for wet machining each type of specimen produce discontinued chips. As mentioned above that for AA6061-F and AA6061-O both has continued chips but AA6061-F produce continued spiral chips for dry machining condition. AA6061-O with the lowest hardness produce relatively straight continued chip but AA6061-T6 with the highest hardness produce discontinued chip. It is implied that the hardness of specimen also influence the form and type of chips produced. Straight continued chips indicated good surface finish and steady cutting force. From figure 6 also shows that for AA6061-F and AA6061-O, the lubrication had significant effect to the chips form. In dry machining, the chips tend to be continuous form but in wet machining the chips tend to be discontinued.

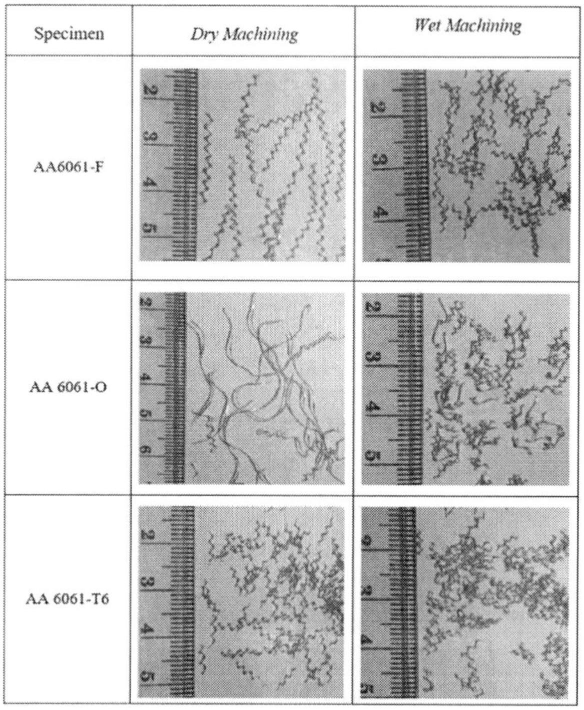

Figure 6. The chips produced for each type of specimen by turning process with and without cooling media

According to Kaya et al, that chips temperature during cutting for AA7075 is in the range 52°C-92°C [4]. So that, the energy consumption due to cutting process is transferred as heat on the tool tip and chips. Wet machining condition influence cutting energy consumption and gives effect to the chip temperature. The cutting energy consumption has decreased by using cooling media and ultimately improve machined surfaces [15]. For discontinued chips is resulted from fluctuating cutting force, and may also cause vibration and chatter so that surface roughness increases. Chip grain size is also a function of cutting conditions and gives an effect on chip temperature which affects the microstructure of the chip during the machining process [15].

4. Conclusion

This study presents the experimental results of the roughness characteristic for AA6061-F, AA6061-O and AA6061-T6 after turning process. Experimental results show that AA6061-O has the lowest value of roughness and AA6061 has the highest value of roughness. Based on results obtained in hardness test, AA6061-O also has the lowest hardness due to softening process during heat treatment. It is indicated that AA6061-O has good machinability compare to AA6061-T6 and AA6061-F. The presence of precipitate on AA6061-T6 may cause distortion and vibration as the reason of high roughness. The effect of lubrication significantly decrease the roughness value of AA-6061-T6, but not for AA6061-F and AA6061-O. For Specimen AA6061-O, the form of chips is continued and relatively straight, however for AA6061-T6 produce discontinued spiral chips type.

Acknowledgment
The authors gratefully acknowledge to the Engineering Faculty of Universitas Hasanuddin for funding this research by LBE fund 2019.

Reference

[1] T. Özel., T. K. Hsu., and E. Zeren., 2005, Effects of cutting edge geometry, workpiece hardness, feed rate and cutting speed on surface roughness and forces in finish turning of hardened AISI H13 steel, *Int. J. Adv. Manuf. Technology*. 25 pp. 262-269.

[2] H. Demir., and S. Gündüz., 2009, The effects of aging on machinability of 6061 aluminium alloy Materials & Design Volume 30, Issue 5, pp. 1480-1483.

[3] F. Frost., and V. Karri., 1999, Effect of Heat Treatment on Mechanical Properties and Surface Roughness of Aluminum Alloy 601, *Journal of Testing and Evaluation* 27, No. 4: pp. 273-281.

[4] H. Kaya., M. Uçar., A. Cengiz., D. Özyürek., A. Çalişkan., and R. E. Ergün., 2012, The effect of aging on the machinability of AA7075 aluminium alloy, *Scientific Research and Essays* Vol. 7(27), pp. 2424-2430.

[5] R. M. Rashad., and T. M. El-Hossainy., 2006. Machinability of 7116 Structural Aluminum Alloy, *Materials and Manufacturing Processes,* Volume 21, pp. 23-27.

[6] O. Imhade., O. Ugochukwu., and O. Chinenye., 2013, Cutting Parameters Effects on Surface Roughness During End Milling of Aluminium 6061 Alloy Under Dry Machining Operation, *International Journal of Science and Research (IJSR).*

[7] S. A. Niknam., and V. Songmene., 2013, Simultaneous optimization of burrs size and surface finish when milling 6061-T6 aluminium alloy, *International Journal of Precision Engineering and Manufacturing,* Volume 14, Issue 8, pp. 1311–1320.

[8] A. Akkurt., 2015, The effect of cutting process on surface microstructure and hardness of pure and Al 6061 aluminium alloy, *International Journal of Engineering Science and Technology,* Vol. 18, Issue 3, pp. 303-308.

[9] F. Ozturk., A.Sisman., and T. K.Picu., 2010, Influence of Aging Treatment on Mechanical Properties of 6061 Aluminum Alloy, *Materials and Design,* Volume 31 Issue 2, pp. 972-975.

[10] J. Kouam, V. Songmene, M. Balazinski and P. Hendrick, 2012, Dry, Semi-Dry and Wet Machining of 6061-T6 Aluminium Alloy, *Aluminium Alloys - New Trends in Fabrication and Applications.*

[11] P. S.Sreejith., 2008, Machining of 6061 aluminium alloy with MQL, dry and flooded lubricant conditions, *Materials Letters,* Volume 62, Issue 2, , pp. 276-278.

[12] I. Sevim., S. Sahin., H. Cug., E Cevik., F. Hayat., and M. Karali., 2014, Effect of Aging Treatment on Surface Roughness, Mechanical Properties, and Fracture Behavior of 6XXX and 7XXX Aluminum Alloys, *Strength of Materials,* Vol. 46, Issue 2, pp 190–197.

[13] M. W. Azizi., S. Belhadi., M.A. Yallese., T. Mabrouki., and J.F. Rigal., 2012, Surface roughness and cutting forces modeling for optimization of machining condition in finish hard turning of AISI 52100 steel, *Journal of Mechanical Science and Technology* pp. 4105~4114.

[14] S. Hassini., B. Blaysat., and E. Duc., 2017, Methodology for aluminium part machining quality improvement considering mechanical properties and process conditions, *CIRP Journal of Manufacturing Science and Technology,* Volume 18, pp. 18-38.

[15] J. Kouam., V. Songmene., M. Balazinski., and P. Hendrick., 2012, Dry, Semi-Dry and Wet Machining of 6061-T6 Aluminium Alloy, *Aluminium Alloys - New Trends in Fabrication and Applications.*

Cognitive Work Analysis to Comprehend Operations in Self Check-in Counter

Lukmandono[1*], H C Suroso[1], and A H Nasution[1]

[1]Adhi Tama Institute of Technology, Surabaya, Indonesia

*Email: lukmandono@itats.ac.id

Abstract. Self service check-in is a new service that has been given by the airlines for passenger. This service is able to minimize the use of manpower in the airport and reduce the time consuming for passenger to do the requirement before flying. Cognitive Work Analysis (CWA) is expected to extract the phases and explore the parts that implied to the system. CWA is a method to analyse an industrial complex system which is useful for industrial design. CWA divided into five stages which outlined different function in each stages those are work domain analysis, control task analysis, strategies analysis, social organization and cooperation analysis, and worker competencies analysis. This research used brainstorming as qualitative method and got several results such as there are five categories that implied in the WDA, five categories involved on CTA, seven categories on SOCA, and nine categories implied on WCA.

1. Introduction

As technology become very sophisticated and human required to do their task as quick as possible, the system in public places should be improved to reduce waste time and increase the effectivity. Effectivity means reducing the variability of the process, simplify the action, and reduce several waste [1, 2]. Travelling by the airplane must be given enough spare time. Passengers should pass the x-ray inspection, baggage check, baggage storage, and check in. All the steps to enter the airplane is really takes time. The average time for doing all the stages is approximately 30 minutes when there will no long queuing. Moreover, check in process is a crucial part since it is the first operational step for passengers during their travel [3]. To reduce this critical situation, this paper will outline the developing of self-check in system to reduce the check in time by using cognitive work analysis. Check-in at the airport usually being long queuing because the limited of counter and customer services. By Cognitive Work Analysis, I will create self-check in system which able to use in a check-in machine which provide by the ailines company and any devices such as smartphone, tablet, and laptop a few hours before their flight will be due.

Cognitive work analysis (CWA) is a design, analyse, and evaluation from a complex sociotechnical. The aim of CWA is to improve the system design by exploring parts that implied with the function of it [4]. Sociotechnical is a theory for developing work system, moreover it has strong connection between human and machine [5, 6]. The output of CWA is to breakdown the support design rather than doing the design and the benefit of it in this case is the ability of CWA to generate some function to support developing the self-check in machine. To create a system which used by a human there should be many consideration such as, easy to use, safety, and efficient. The designer always needs developer team who has important role to succeed the system, and the developer team will need related stakeholder to collect some data and tools. All of these schemes will be connected together and will give impact to each other's.

Cognitive Work Analysis is a multi-phases process, it has five main stages which will be concern to their respective duties, those are work domain analysis, activity analysis, strategies analysis, social organisation and cooperation analysis, and worker competencies analysis. It would be hard to run the system when the designer does not have the complete supporting system to make it happen. However, some people might not accept the change of the system, it has high risk to refuse the improvement of the system. Due to this possibility of risk, the system should be planned as well as possible and also should involve the users as the main object of the cases.

2. Theory

2.1. Cognitive Work Analysis

The purpose of CWA is to analyse the parts of industrial complex system and divide them into five stages with specific level on each stages. Those stages are work domain control (WDA), control task, strategy, social organization and cooperation (SOCA), and Worker's competencies. Each level has the different methodology to interpret and collect the data with quantitative method used by experts in the field [4, 7].

2.1.1. Level 1: Work Domain Analysis (WDA)

General characteristics of Work Domain Analysis will be highlighted in this part. In this case, the objective is to unfold the system of self service check-in counter and explore the task of each stakeholders [7]. WDA is developed by five level such as the top level is domain purpose, the second level is domain value, the third level is domain function, the fourth level is technical functions and contextual effects, and the fifth level is physical resources and constraints [9]. The result of this level will give some details in the self service check-in system on the basis of the constrains affecting equipment and human behavior [8].

2.1.2. Level 2: Control Task Analysis (CTA)

Control task analysis will evaluate the essence of tasks, relevant to the functional purposes determined in the level above (WDA) so that every aspects of the technological design can be identified. The specific tasks are listed to accomplish the goals within a work domain. CAT is a good approach to show the personal involves in the control task [10].

2.1.3. Level 3: Strategies Analysis

This is the third level of CWA. The strategies analysis illustrates about specific factors which able to prevent task from completion and uncalculated management risks. This stage helps to find out alternative strategies by which the control tasks can be implemented [8, 10, 11].

2.1.4. Level 4: Social Organisation and Cooperation Analysis (SOCA)

The fourth level of WDA aims to identify of how individuals interact in the constraints on the mine site, and the overall team performance. It creates the interactions in the organizations to optimize the collaboration and communication. This stage mainly outlines the relationship of operators in the organization by verbal processes [11, 8, 10].

2.1.5. Level 5: Worker Competencies Analysis

At the final stage of WDA called by worker competencies analysis which gives some details of cognitive awareness of the individuals and their behaviors when having external situations coming from upper level. A specific tools from Rasmussen called skills, rules, and knowledges (SRK) will be implemented in this stage.

SRK tools categories the human behaviors within relationship of various restrictions in a workplace and the knowledge required for each strategiec task. The final result of this stage is an extraction of cognitive processes, and it will be used for improvement in the system design.

3. Materials and Method

3.1. Study Design Overview

The researchers has done the brainstorming with people who have a great impact to the design ofself check-in counter. Obviously the person who involved in the brainstorming should haveexperience related with airport system especially airlines system booking and check-in. The result of discussion has been recorded in voice and has been noted to be a consideration when creating Cognitive Work Analysis.

3.2. Brainstorming

Brainstorming is one of the most famous tools for creative thinking [12]. Creative thinking is needed for establish the industrial design which have a great contribution in the developing concepts and specifications for users and manufacturers. Industrial designers are the people who deliver the ideas to an engineer and they are not people who handle things by themself on creating products so that in creativity has the big impact on industrial design. This is convinced also by Dorval that small group of brainstorming will contribute on a better result for many practitioners in the problem solving are [13] Brainstorming is a rules which creates to maximize the productivity within the groups and generates some ideas by minimize the production loss. The main objectives of this tools is to improve the creativity in organization [14].

3.3. Participant

The participants that involve in this brainstorming activity are staff research and development facility of the airlines, a ticketing staff airline, and a consumer service staff.

3.4. Brainstorming Procedure

- Identification person in charge of developementself service check-in counter
- Creates an ethics form procedure before collecting the data
- Ask the person in charge to approve the ethics form
- Start the brainstorming in small group

4. Result

4.1. Work Domain Analysis

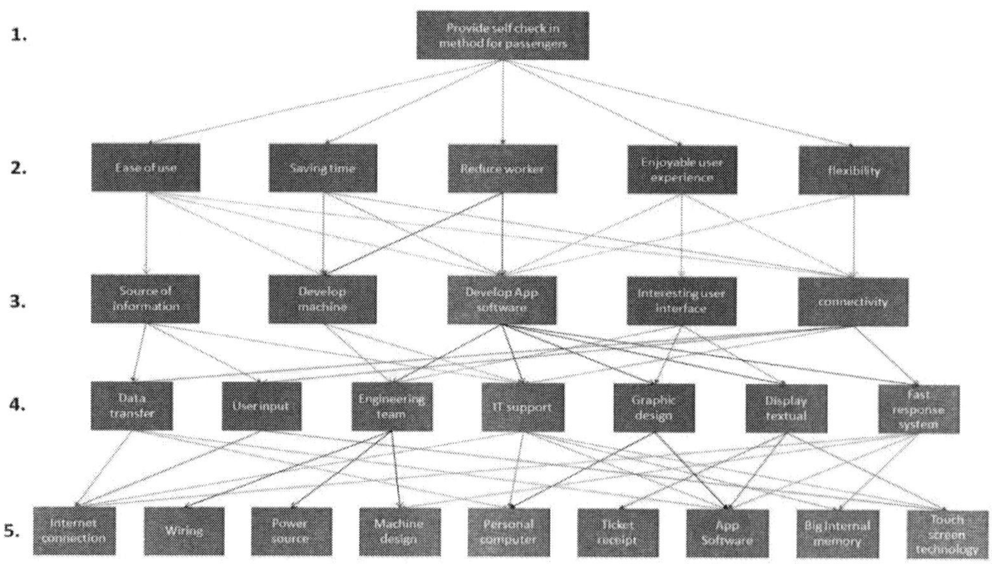

Figure 1. Work domain analysis

IOP Conf. Series: Materials Science and Engineering **875** (2020) 012058 doi:10.1088/1757-899X/875/1/012058

The first stage of Cognitive Work Analysis is Work Domain Analysis (WDA). The WDA concept was developed by Rasmussen and colleagues as a way of structuring information in terms of the means ends links that form the relationships within a system and the representations organise information in a systematic fashion that will support system design. This stage is divided into 5 level of abstractions hierarchy (AH), those are domain purpose, domain values, domain function, physical function, and physical object. The five level of AH will be linked from previous level to the next level to get the interaction between them. The aims of this stage is to understand the environment of the system that will be developed and create some constrains to be focused on the parts that might influence the system.

The first level of Work Domain Analysis (WDA) in this case is to provide self check in system for airplane passengers. The first level will indicate the final goal of the cognitive work analysis. This is the overall system that will be conducted by this case, and it called as domain purpose of work domain analysis. The second level of WDA is domain values which dispatched into five parts. This level will judge the values that will be get from the first level as a domain purpose. Domain values also can be the priority measurement of successful from the system that will be created. By this case, the domain values are the easiness of use by the passengers, the saving time of using that system, could reduce worker or employer from the airport side, enjoyable user experience, and flexibility that users will achieve. The third level is the purpose related function on the second level. This level captures general function of the system necessary in order to achieve system purposes. The example in this case is the function of machine and app software will be provided in the check-in system. The second last level is object related process which have aims to describe the object that will be needed to develop the function. Moreover, the object related process is helpful to dispatched particular object in the last level. Physical object is the lowest level of WDA. It will outline several the objects that will be needed to support another level to make the final goal happen. The object should be physically things and does not suggest to be abstract. The example of the last level are personal computer, internet connection, ticket receipt and etc.

4.2. Control Task Analysis

The second stage of CWA is control task analysis (CTA). The aims of control task analysis is to give the possibility recurring activities that might happen in the complex system. The Figure 2 shows it divides into two part, the horizontal axis as a situation and the vertical axis as a function. The function in this CTA are taken from the purpose related function in the third level of work domain analysis.

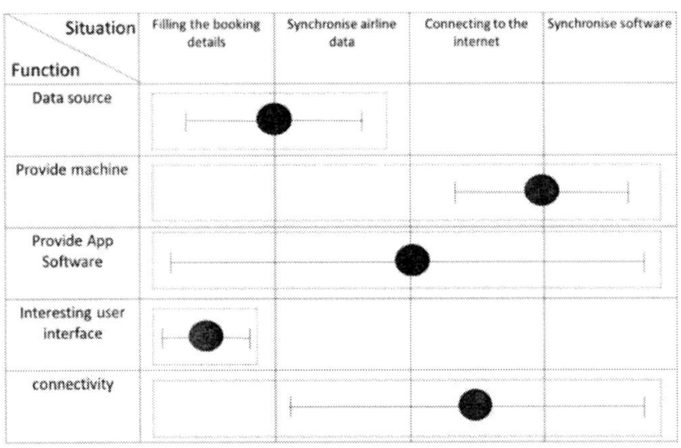

Figure 2. Control task analysis

Inside the table of control task analysis shows the relationship between situation and function. The dashes mean "can be relevant", the circle mean "usually relevant" and the blank box mean "impossible". For example in the first row, there are four situation that might happen such as filling the booking details, synchronise airline data, connecting to internet, and synchronise software. The four of this situation will be observed about their relationship with the first function, that is data source. The result is dashes and dots only located in the first and second situation. It means that the filling booking details and synchronise airline data usually relevant with data source. In the other side, the connecting to internet and synchronise software are impossible relevant with the data source.

4.3. Strategies Analysis

Strategies Analysis is the third stage of cognitive work analysis. The purpose of this stage is to know how the things can be done by several steps. There might be some different steps that possible to take in the same destination. In this case, I choose two different goals related to the check in system and connected to the network.

The purpose of strategies analysis in Figure 3 is to know the steps to check in and they got three diferent step. The first step is by using the counter check in, the second is by using the application in the devices such as smartphone, tablet, or passengers' laptop, and the last step is using the self check in machine method. The longest step is using passengers' devices, it is contain of 4 steps to get their self check in. However this step does not determine the time that will be spent.

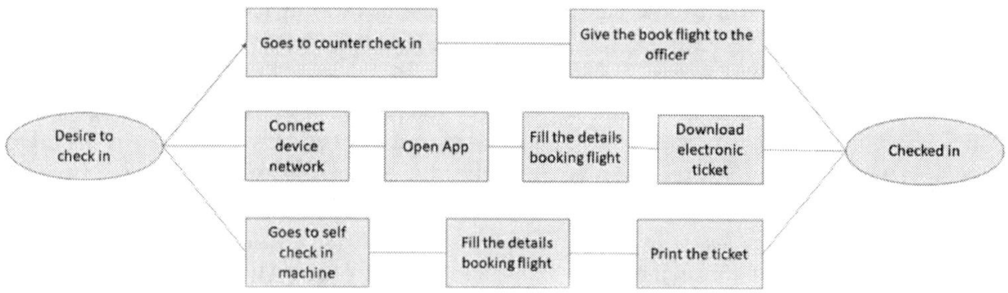

Figure 3. Strategies analysis for check in

The second strategies is how to connect the network in order to access the app software to check in as can be seen at Figure 4. There are three different step to be connected with the network. The main differences is by using the Wi-fi installation, SIM card, and local area network. The longest step is using the LAN network which contains of four step to be connected.

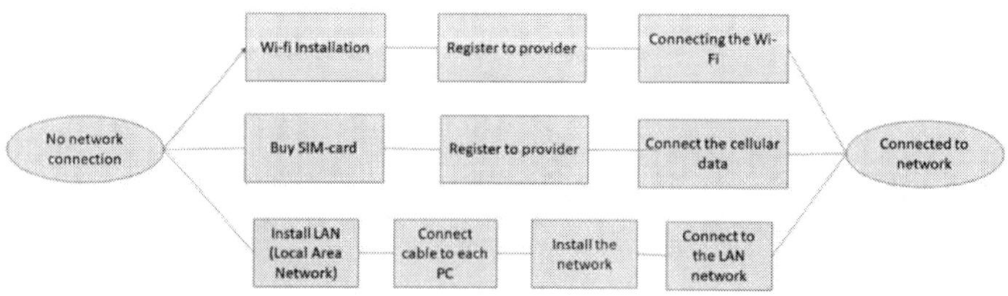

Figure 4. Strategies analysis for connected to network

4.4. Social Organisation and Cooperation Analysis

This stage will show the staff that will be needed in each situation and function of the case. Sometimes there will be more than one staff in each situation and function. It will help to cooperate each other to succeed the system. In this case there are five different staff which been interpreted in five different colour as be seen at figure 6. There are airport officer, technician, IT controller, provider inspector, and airline ticketing staff. Each staff has their own responsibility, for example the airport officer has responsibility to control the passenger who will use the self check in system, sometimes passengers do not know how to use it and need some guidance from the staff. Technician is needed to provide the machine and also provide frequently maintenance the machine. IT controller will check the general problem about the app software in the device and report some problem related connection to provider inspector. However provider inspector should checked the connectivity frequently to prevent technical problem while passengers are using the system. The last one is airline ticketing staff who has responsibility to each data that will be used in the system. They should be cooperate each other and know the part of their responsibility to make the system running appropriately.

The diagram that used here is same with stages 2, the difference is only the interpretation of the colour in each box. If we have a look at the fourth row, the relationship between situation "filling the booking details" and the function "interesting user interface" created the colour red and yellow. This means airport officer and IT controller will be responsible in this part.

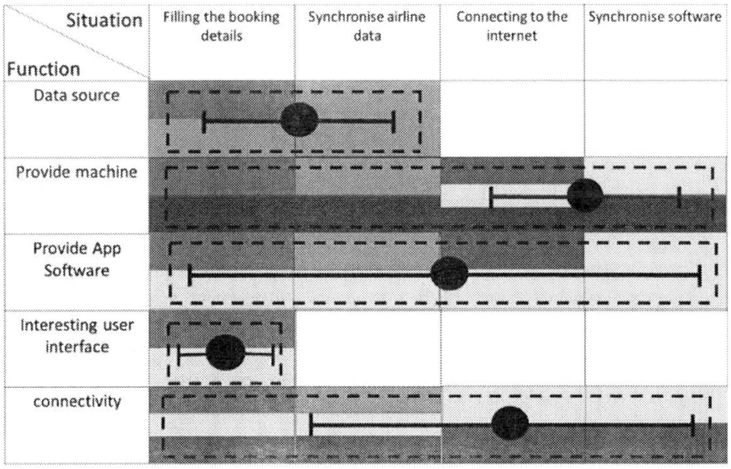

Figure 5. Social organisation and cooperation analysis (SOCA)

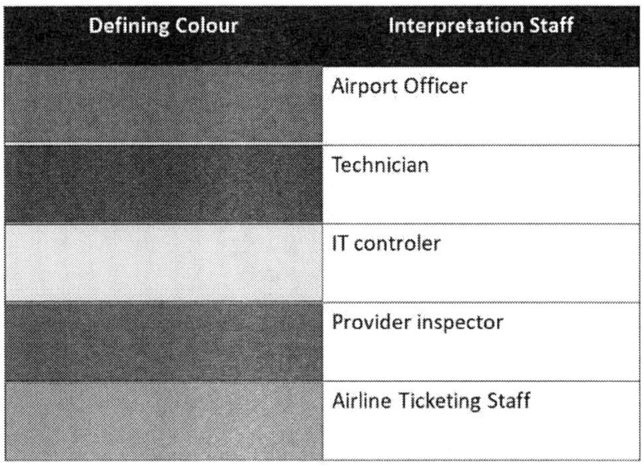

Figure 6. Interpretation staff of SOCA

4.5 Worker Competencies Analysis

The last stage of cognitive work analysis will outline the competencies for each worker that required by the system. In this phase, the purpose related function in the first stage will be used for the main consideration to create competencies of worker. The competencies divided into three part such as skill, rule, and knowledge. Skill is automatic performances to deal with their job and problems. Skill has strong relationship with experience, the more they solve the problem the more skill that will they gain.

Rule is procedures and guidance to do their job. The rule is made to guide the worker achieve the goals that has been define before the system running. Therefore knowledge is an understanding of situations or facts which have learned before.

Table 1. Worker competencies analysis

Object Related Process	Skill	Rule	Knowledge
Data Source	Check the upcoming data of booking frequently	Update and synchronise every new booking flight ticket and remove the cancellation ticket, also re-create the new booking details for change time flight	Understand about creating the booking details and know how to change booking details into ticket receipt, so people could use their ticket to enter the gate
Provide Machine	Assembly parts become a machine	Observe the drawing parts of machine, then create the parts correspond with their dimensions, therefore assembly it correctly	Understand about dimensions of machine in some projections and could install each part according to the draft so the machine runs properly
Provide App	Make the programming language become a software design	Design the app based on the purpose, collecting the data that will be used in the app, and developing the software by the programming language	Understand data processing system which transferred from data sources for communicating the passengers with their flight
Interesting user interface	Designing the display of software	Observe users interaction with software interface, interpreting users desire to use the software, and create the comfortable design on it	Understand the necessary content on display and could simplify the content to make it feasible for users without ignoring any important details
Connectivity	Connected the device to the network	Count on the numbers of devices and the numbers of possible passengers that might use the device, choose the broadband capacity and install the network into devices.	Understand the network capacity during pick hour and understand how to increase the speed to prevent long queuing

5. Conclusion

The final result of this research is a model of cognitive work domain which has been outline in the result. The work domain analysis produced categories ease of use, saving time, reduce worker, enjoyable user experience, and flexibility. Control task analysis found out source of information, develop machine, develop app software, interesting user interface, and connectivity. SOCA produced several categories such as data transfer, user input, engineering team, IT support, graphic design, display textual, and fast response system. The last stage WCA which used SRK by Rasmussen taxonomy found out internet connection, wiring, power source, machine design, personal computer, ticket receipt, app software, big intenal memory, and touch screen technology.

Acknowledgments

We would like to thank Directorate General of Learning and Student Affair, Ministry of Research Technology and Higher Education for giving us the financial support in this publication as implementation of grant program in the postgraduate improvement quality by the contract number (No.036/B4/PPK-SPK/VII/2019).

IOP Conf. Series: Materials Science and Engineering **875** (2020) 012058 doi:10.1088/1757-899X/875/1/012058

References

[1] L. Di Pietro, R. Guglielmetti, F. Musella, P. Vicard, and M. F. Renzi, "Monitoring an airport check-in process by using Bayesian networks," Transp. Res. Part A Policy Pract., vol. Volume 106, p. Pages 235–247, Dec. 2017.

[2] L. Lukmandono, R. Soesilo, M. Basuki, and M. J. Hidayat, "Minimize the Time for Mold Replacement with Lean Manufacture & Single Minute Exchange of Dies Approach (SMED)," *Simp. Nas. RAPI XVII*, vol. 1, pp. 89–96, 2018. [In Bahasa]

[3] A. Correia and S. C. Wirasinghe, "Evaluating Level of Service at Airport Passenger Terminals: Review of Research Approaches," Transp. Res. Rec., vol. 1888, no. 1, pp. 1–6, Jan. 2004.

[4] K. J. Vicente, "Cognitive Work Analysis Toward Safe, Productive, and Healthy Computer-Based Work [Book Review]," IEEE Trans. Prof. Commun., vol. 46, no. 1, pp. 63–65, 2003.

[5] D. P. Jenkins, N. A. Stanton, G. H. Walker, P. M. Salmon, and M. S. Young, "Applying cognitive work analysis to the design of rapidly reconfigurable interfaces in complex networks," Theor. Issues Ergon. Sci., vol. 9, no. 4, pp. 273–295, Jul. 2008.

[6] H. C. Suroso and C. E. Revadi, "The Main Factors that Affect Pilot Attention and Decision Making During Landing Operation Leading to Runway Incursion," vol. 171, no. Icoemis, pp. 95–102, 2019.

[7] G. Jenkins, D., Stanton, N., Walker, Cognitive Work Analysis: Coping with Complexity. London: CRC Press, 2009.

[8] S. Demir, E. Abou-Jaoude, and M. Kumral, "Cognitive work analysis to comprehend operations and organizations in the mining industry," Int. J. Min. Sci. Technol., vol. 27, no. 4, pp. 605–609, Jul. 2017.

[9] P. Sanderson and C. Burns, "Rasmussen and the boundaries of empirical evaluation," Appl. Ergon., vol. 59, pp. 649–656, 2017.

[10] M. Ashoori, C. M. Burns, B. d'Entremont, and K. Momtahan, "Using team cognitive work analysis to reveal healthcare team interactions in a birthing unit," Ergonomics, vol. 57, no. 7, pp. 973–986, 2014.

[11] M. A. Regan, G. Lintern, R. Hutchinson, and C. Turetschek, "Use of Cognitive Work Analysis for exploration of safety management in the operation of motorcycles and scooters," Accid.Anal.Prev., vol. 74, pp. 279–289, 2015.

[12] S. G. Isaksen, S. Gryskiewicz, S. J. Parnes, and M. I. Stein, "A Review of Brainstorming Research: Six Critical Issues for Inquiry The author would like to thank Stanley," 1998.

[13] K. B. Dorval, "Strengthening the 'Heartbeat' of creative problem solving- Strategies for facilitating small groups," in European Conference on Creativity and Innovation, 1999.

[14] A. S. Zainol, W. Z. M. Yusof, K. A. Mastor, Z. M. Sanusi, and N. M. Ramli, "Using Group Brainstorming in Industrial Design Context: Factors Inhibit and Exhibit," Procedia - Soc. Behav. Sci., vol. 49, pp. 106–119, 2012.

IOP Conf. Series: Materials Science and Engineering **875** (2020) 012059 doi:10.1088/1757-899X/875/1/012059

Characteristic of Variable Compression Ratio Diesel Engine Operating on Diesel and Biodiesel B20

N Amaliyah[1]*, B Mire[1], A E E Putra[1], M Syam[1], and H Zakariah[1]

[1]Department of Mechanical Engineering, Faculty of Engineering, Hasanuddin University, Gowa, Indonesia

*Email: novriany@unhas.ac.id

Abstract: The limitation of natural resources especially crude oil employed in automotive engine has made an attention to switch over to biodiesel. Biodiesel is known as carbon neutral and produce no carbon dioxide leads to decrease of pollution and potentially improve the engine performance. In this study, an experimental investigation on variable compression ratio diesel engine is observed based on the engine performance using the VCR Engine test setup 1 cylinder, 4 strokes with EGR (computerized). The effect of compression ratio change on the torque, specific fuel consumption (SFC), indicated power, and brake thermal efficiency has been studied using mineral diesel and biodiesel B20. When increasing the compression ratio from 13 to 18, it was found that torque was significantly increased when using biodiesel B20. As the torque increase resulted in higher indicated power. The decreasing of SFC were about 44% and 11% for mineral diesel and biodiesel B20 respectively. Specific fuel consumption was optimum at 0.21 g/kW.h at a compression ratio of 16 using mineral diesel.

1. Introduction

In recent years, the limitation of natural resources such as crude oil, coal and so forth that are employed in power plants, boilers, and some automotive engine has been growing fast. These condition forced to the transition towards alternative fuels. Also, there is a huge amount of pollution in the atmosphere due to the burning of those natural resources. The combustion of fossil fuel in Internal Combustion (IC) engine is one of the major sources that contribute to air pollution.

As a clean fuel produced from domestic and renewable resources, the utilization of biodiesel is gaining significant attention nowadays. Biodiesel is derived from vegetable oils or animal fats and can be blended in any proportion with diesel. It is also more environmentally friendly and non-toxic compared to ordinary diesel and can be used in diesel engines with minimal modification [1, 2]. Biodiesel can be produced from many feedstocks such as Jojoba oil [3], Jatropha oil [4], Mustard oil [5], waste frying oil [2], and palm oil [6] to name a few. The yield of biodiesel oil not only depend on the feedstock but also the catalyst and process used.

The Indonesian biofuels mandate is one of the most aggressive in the world especially biodiesel by setting an ambitious blending target. Out of 64 countries having biofuels targets and/or mandates, fewer than 10 impose biodiesel mandates. From those that do, except Costa Rica that has 20% biodiesel mix mandate, none of the countries mandate more than 10% mix. Based on the Ministry of Energy and Mineral Resources regulation 12/2015, percent of biofuel blending required is expected to increase from 15% in 2015 to 30% in 2025 [7].

Recently, the researchers have heightened that the utilization of biodiesel has a significant effect on the performance and emission characteristic of a diesel engine. In this study, the performance characteristic of the variable compression ratio diesel engine was investigated using mineral diesel and biodiesel B20 produced by PERTAMINA.

2. Literature Review

Performance of engine in terms of power output, mechanical efficiency, exhaust temperature, specific fuel consumption and thermal efficiency of biodiesel and its blend have been studied by many researchers. The effect of change in compression ratio (CR), exhaust gas recirculation (EGR) and EGR temperature on the performance and emission characteristic has been studied using diesel-biogas dual fuel. Higher compression ratio was found to improve energy and exergy efficiency and on the emission side showed a significant decrease in HC, CO, and NOx emission [8].

An investigation of the effect of compression ratio on a single-cylinder, four-stroke and direct injection of the diesel engine was conducted using biodiesel produced from mineral diesel blend with a waste fried oil methyl ester. A different blends of B0, B50, and B70 was observed at 14.5, 16.5, and 17.5 of compression ratio. It was found that higher compression ratios improve engine efficiencies such as specific fuel consumption, brake power, and brake thermal efficiency [9]. A performance test in a single cylinder four strokes variable compression ratio was conducted at 14:1, 16:1, and 18:1 using mustard oil methyl ester. It was observed that specific fuel consumption slightly increase and the brake thermal efficiency was maximum at full load when using B20 blended [5].

Experimental investigation on the effect of different blends of Jojoba Methyl Ester (JME) in diesel engine performance has been observed at different compression ratio from 18 to 23. The result showed, that compared with mineral diesel, JME at higher CR indicated greater improvement for biodiesel combustion, however, the emission produced when using JME was higher than mineral diesel [10]. The performance and emission of the diesel engine were studied using diesel and Sal seed oil methyl ester (SME). When the volume fraction of SME increase, brake thermal efficiency was found to decrease. However, in full load, brake specific energy consumption increased with the increase in the blend [11].

3. Research Method

The test was conducted using a VCR Engine test setup 1 cylinder, 4 strokes with EGR (computerized). Diesel engine connected to eddy current type dynamometer for loading. The compression ratio can be changed without stopping the engine and without altering the combustion chamber geometry by specially designed tilting cylinder block arrangement. The experiment was conducted at a constant engine speed of 1450 rpm and load 9 kg. Compression ratio was set at 13, 14, 15, 16, 17, and 18. Setup is provided with necessary instruments for combustion pressure and crank-angle measurements. The setup enables the study of VCR engine performance. Labview based Engine Performance Analysis software package "Enginesoft" is provided for online performance evaluation. The major specification is given in Table 1. The schematic diagram and engine picture are shown in Figures 1a and 1b.

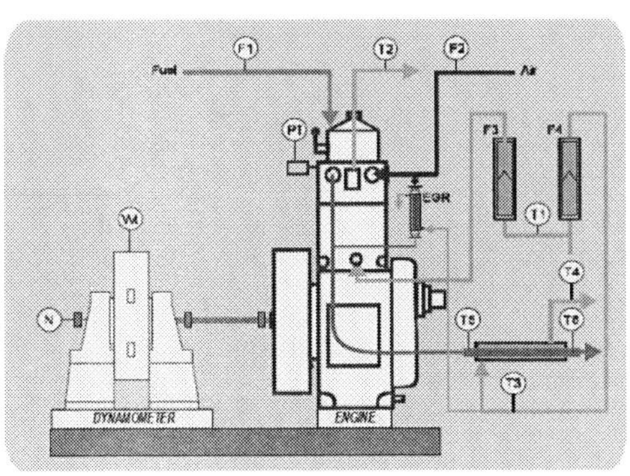

Figure 1a. Schematic diagram of engine test up

Figure 1b. Picture of engine test up

Table 1. Engine specification

Parameters	Technical Specification
Product	VCR Engine test up 1 cylinder, 4 strokes, Diesel with EGR (computerized)
Engine	Make Kirloskar, Type 1 cylinder, 4 stroke diesel, water-cooled, power 3.5 kW at 1500 rpm, stroke 110 mm, bore 87.5 mm, 661 cc, CR 17.5. Modified to VCR engine CR range 12 to 18
Fuel Tank	Capacity 15 lit with glass fuel metering column
EGR	Water cooled, ss 304, range 0-15%
Data Acquisition Device	NI USB-6210, 16 bit, 250kS/s
Temperature Sensor	Type RTD, PT100 and thermocouple type K
Software	"Enginesoft" engine performance analysis software
Overall Dimension	W 2000 x D 2500 x H 1500 mm

4. Result and Discussion

4.1. Engine Torque

The torque values of mineral diesel and biodiesel B20 as a function of compression ratio are shown in fig.2. It is revealed that the torque increased as the compression ratio increases for biodiesel B20 and slightly stable at mineral diesel. The increasing compression ratio resulted in increasing temperature and pressure in the cylinder causing torque to increase. Compared to the mineral diesel, the torque of biodiesel B20 was lower and in a range of 16.20 to 16.43 while mineral diesel was in a range of 16.39 to 16.42 as the compression ratio increased from 13 to 18. A similar result of the torque was observed in a range of 16-20 when B20 was used at a compression ratio of 14 to 18 using supercharged VCR diesel engine[12]. Due to the low volatility and higher viscosity of biodiesel B20, the increasing compression ratio had more effect on biodiesel B20 than mineral diesel.

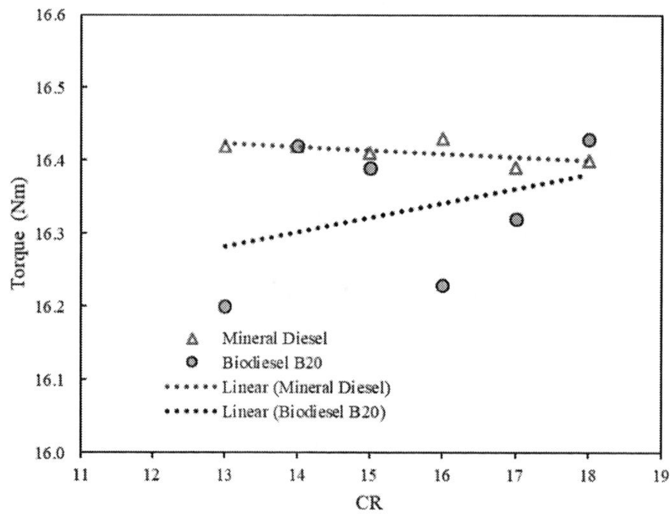

Figure 2. Variation of compression ratio on torque

4.2. Specific Fuel Consumption

The variation of specific fuel consumption on the increasing compression ratio is shown in fig. 3. The SFC is tended to decrease as the increasing of compression ratio. When using mineral diesel, the SFC is found to be lower than biodiesel B20 also the effect of increasing compression ratio had more benefit. It is assumed that in biodiesel B20 the cetane number and viscosity is lower than mineral diesel. As the compression ratio increase from 13 to 18, the decreasing of SFC were about 44% and 11% for mineral diesel and biodiesel B20 respectively.

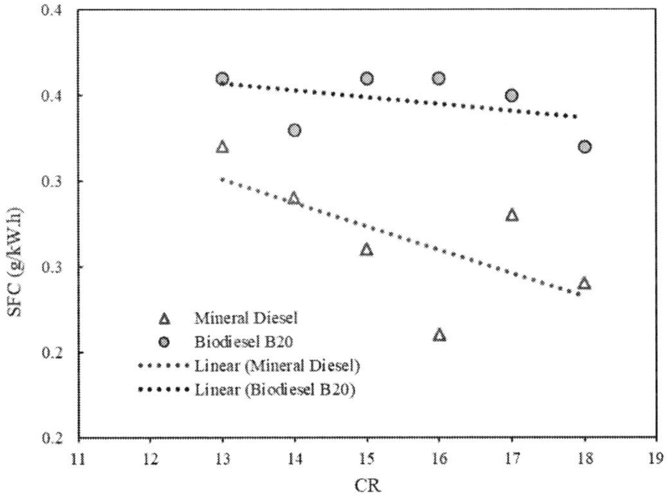

Figure 3. Variation of compression ratio on specific fuel consumption (SFC)

4.3. Indicated Power

Indicated power was defined as the power production in the cylinder. Indicated power was found to decrease by about 16% and 2.9% in mineral diesel and biodiesel B20 respectively as the increasing of compression ratio from 13 to 18 as shown in fig. 4. However, indicated power in mineral diesel is observed higher than biodiesel B20. The torque increase as the compression ratio increase due to the higher pressure inside the combustion chamber and so higher indicated power will result [13].

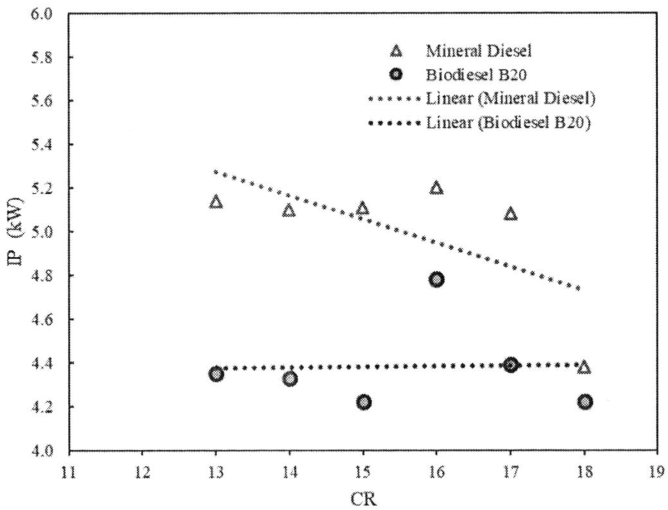

Figure 4. Variation of compression ratio on indicated power (IP)

4.4. Brake Thermal Efficiency (BTheff)

The brake thermal efficiency (BTheff) as a function of compression ratio using mineral diesel and biodieselB20 is shown in fig. 5. As the compression ratio increased from 13 to 18, brake thermal efficiency increased by 17% for mineral diesel and 12% for biodiesel B20. This could be attributed to the lower volatility of biodiesel B20 compared to mineral diesel. Diesel engine performance improvement at higher compression ratio because of oxygen content that resulted in the complete combustion. In accordance with the present result, the previous study on engine performance using pure diesel and biodiesel from waste cooking oil of B10,B20, B30, and B50 showed that brake thermal efficiency of all biodiesel blend was lower than pure diesel [13].

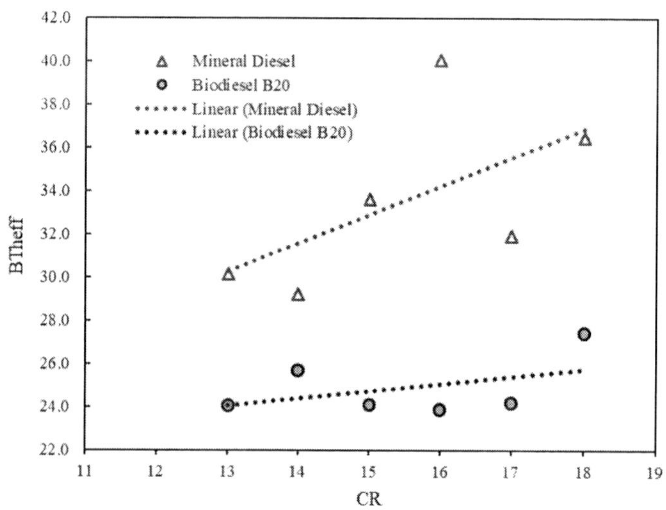

Figure 5. Variation of compression ratio on brake thermal efficiency (BTheff)

5. Conclusion

The effect of compression ratio on torque, specific fuel consumption, indicated power, and brake thermal efficiency of diesel engine fueled by biodiesel B20 have been investigated and compared with mineral diesel. Increasing compression ratio from 13 to 18 indicated a significant effect of increasing torque on biodiesel rather than mineral diesel resulted in higher indicated power. Brake thermal efficiency was higher when using mineral diesel and specific fuel consumption was optimum also when using mineral diesel. A further investigation will be conducted to study the combustion and emission of VCR diesel engine.

References

[1] V. Bansal and S. K. Bhargava, "Ionic Liquids as Designer Solvents for the Synthesis of Metal Nanoparticles," 2009.

[2] J. M. Fonseca, J. G. Teleken, V. D. C. Almeida, and C. Silva, "Biodiesel from waste frying oils : Methods of production and puri fi cation," vol. 184, no. December 2018, pp. 205–218, 2019.

[3] A. M. Ashraful *et al.*, "Production and comparison of fuel properties, engine performance, and emission characteristics of biodiesel from various non-edible vegetable oils : A review," *ENERGY Convers. Manag.*, vol. 80, pp. 202–228, 2014.

[4] F. Guo, Z. Fang, X. Tian, Y. Long, and L. Jiang, "Bioresource Technology "' One-step production of biodiesel from Jatropha oil with high-acid value in ionic liquids "' [Bioresour . Technol . 102 (11) (2011)] q," *Bioresour. Technol.*, vol. 140, pp. 447–450, 2013.

[5] K. Basavaraju and G. Jamunarani, "PerformanceAndEmissionCharacteristicsOfAVariableCompressionRatio Diesel Engine Using Methyl Esters Of Mustard Biodiesel Blends," *Int. J. Eng. Res. Appl.*, vol. 4, no. 11, pp. 20–28, 2014.

[6] M. K. Lam, N. A. Jamalluddin, and K. T. Lee, *Production of Biodiesel Using Palm Oil*, 2nd ed. Elsevier Inc., 2019.

[7] A. Karina, C. Malins, and S. Searle, "Biofuels Policy In Indonesia : Overview And Status Report," *Int. Counc. Clean Transp.*, no. August, 2016.

[8] S. Verma, L. M. Das, S. C. Kaushik, and S. S. Bhatti, *The effects of compression ratio and EGR on the performance and emission characteristics of diesel-biogas dual fuel engine*. Elsevier Ltd, 2019.

[9] J. Hirkude and A. S. Padalkar, "Experimental investigation of the effect of compression ratio on performance and emissions of CI engine operated with waste fried oil methyl ester blend," *Fuel Process. Technol.*, vol. 128, pp. 367–375, 2014.

[10] M. Hawi, A. Elwardany, S. Ookawara, and M. Ahmed, "Effect of compression ratio on performance, combustion and emissions characteristics of compression ignition engine fuelled with jojoba methyl ester," *Renew. Energy*, 2019.

[11] H. S. Pali, N. Kumar, and Y. Alhassan, "Performance and emission characteristics of an agricultural diesel engine fueled with blends of Sal methyl esters and diesel," *Energy Convers. Manag.*, vol. 90, pp. 146–153, 2015.

[12] M. El-adawy, M. El-, and Y. A. Eldrainy, "Performance characteristics of a supercharged variable compression ratio diesel engine fueled by biodiesel blends," *Alexandria Eng. J.*, 2018.

[13] M. E. L. Kassaby and M. A. Nemit, "Studying the effect of compression ratio on an engine fueled with waste oil produced biodiesel / diesel fuel," *Alexandria Eng. J.*, vol. 52, no. 1, pp. 1–11, 2013.

IOP Conf. Series: Materials Science and Engineering **875** (2020) 012060 doi:10.1088/1757-899X/875/1/012060

Energy Absorption Analysis of aluminum Filled Foam Tube Under Axial Load using Finite Element Method with Cross Section Variations.

I Renreng[1], F Djamaluddin[1]*, and F Furqani[1]

[1]Department of Mechanical Engineering, Faculty of Engineering, Hasanuddin University, Gowa, Indonesia

*Email: fauzanman_77@yahoo.com

Abstract. This study aims to determine the effect of the use of aluminum-filled foam tubes on crash boxes and the effect of cross-sectional variations in the use of aluminum-filled foam tubes on crash boxes. There are several parameters to determine crashworthiness performances such as energy absorption and deformation. Test specimens used are single walled (SW), single walled foam filled (SWFF), and double walled foam filled (DWFF), where DWFF is varied into 4 different wall thicknesses. All models have validated with experiment data from relevant reference and it evident that there are good agreement between simulation and experiment results. Tubes are impacted by rigid wall as impactor with velocity 0.4 m/s. From the simulation results, it was concluded that the aluminum crushbox construction added with aluminum foam is able to absorb more energy, where the recommended thickness is neither too thin nor too thick. Tube of DWFF 3 has energy absorption of 11.911 Joule, however the highest of deformation value is 178.66 mm for SWFF.

1. Introduction

The design of a new lateral safety system with better crashworthiness is the most relevant thing to reduce the risk of death and injury in traffic accidents and improve vehicle performance in energy absorption. Crashworthiness is the ability of the vehicle to absorb impact energy and protect vehicle passengers when an accident occurs. The car is designed with several protective systems that aim to increase the vehicle's suitability [1].

Crash Box is one of the passive safety system technology developments that has been widely studied. The function of this crash box is as a device in kinetic energy absorption when a car experiences a collision or accident, both collisions from the front or collisions from behind. The design of the device component crash box serves to reduce the occurrence of forces distributed to the entire body of the vehicle during a collision or collision. The use of a crash box is placed between the buffer and frame of the vehicle [2].

Various studies had been conducted with the purpose to develop devices that reduce the risk of accidents in collisions, one of which is testing on a crash box device where collisions are given at a predetermined rate of change in plastic deformation. It could be concluded that this incident shows the energy absorption in the crash box. However, the phenomenon obtained shows that the faster the plastic deformation changes in the crash box, the remaining speed due to collisions is still a lot that can endanger the mainframe of the vehicle, so this phenomenon considered to be still low for the driver's safety level [3].

The structure of thin-walled test specimens is known as a promising and efficient way of absorbing energy from an impact force on structures in the frame of the concept car, train structure and helicopter frame [4]. There are two main considerations that need to consider in designing the structure of the specimen to present the function as a device that can absorb energy, namely; how to achieve the right ductility to reduce the force of a collision, and how to reduce the extreme force at the first interaction on a collision. The first case relates to the collision resistance capacity of the structure, while the second case relates to the level of safety for passengers.

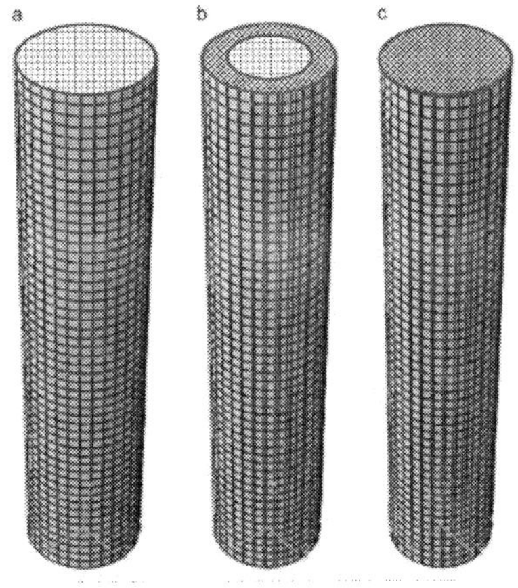

Figure 1. Test specimen model is tube-shaped

In general, the simplest way to increase the resistance to collisions with axial forces is to adjust the wall thickness of the structural frame, although this is not the best way because the mass of the structure will automatically become heavier too. One effective solution to reduce the extreme force of impetus in the first interaction on a collision, usually how to overcome it is to implement a thin wall structure. With the discovery of new material, as an alternative, which becomes a modification of parts of the structure by adding material that has a light density can be a solution in increasing resistance to collisions that occur with axial forces and on the other hand this material is lightweight so that it can streamline weight and overcome force extreme push on the first interaction on a collision. An alternative way is to fill polyurethane foam to strengthen thin wall components which are recommended by Thornton [5].

The use of aluminum foam material as one of the materials in a crash box device is one solution to prevent traffic accident victims from the side of the vehicle. Aluminum foam is lightweight, strong, and anti-rust. In recent decades, metal foam such as aluminum foam had been developed as a very light engineering material. This material has unique mechanical properties, which can withstand large deformation stresses while this material has only a small constant tension. One application of this material is a structure that absorbs energy. Filling the structure with aluminum metal foam can improve the energy absorption properties and stabilize the curve of the structure. Great efficiency in weight had been obtained by combining the structure with a thin wall and filled with metal foam as a core, as studied by Santosa et al. [6]. In another study, it was revealed about the analysis of structures that been filled with metal foam and given axial force interaction

Table 1. Specimen wall thickness

SW & SWFF		DWFF			
thickness outer h_0 (mm)	length outer l_0 (mm)	thickness outer h_0 (mm)	length outer l_0 (mm)	thickness outer h_1 (mm)	length outer l_1 (mm)
3	210	3	210	3	210

The average ability to withstand the force of collisions resulting from structures that been filled with metal foam is higher than the average ability to withstand the force of collisions produced from each of which is only metal foam or thin-walled structures, which is confirmed by Kavi et al.[7].

In maximizing the potential of implementing metal foam as a lightweight fill material at the core of the structure, an alternative concept that can be done is to duplicate the structure with walls that filled with metal foam, where the metal foam is placed between the structure with the inner wall and the structure with the inner wall externally, this solution has been shown to improve efficiency in the weight of suitable structures according to Santosa and Wierzbicki [8], but so far not much information is available regarding the nature of these double-walled structural parts given collisions with general axial force directions. One of the publications regarding the nature of the dual structure is Santosa and Wierzbicki using a non-linear dynamic finite element analysis approach. In its publication, the level of resistance to impulses in collisions from structures with double foam-filled structural members of the wall is shown to be significantly higher, compared to structures that are only covered by a foam-filled wall which gets the same loading.

A study on experiments with quasi-static loading of structures with double walls filled with metal foam with different materials, dimensions and structural shapes, was investigated by Seitzberger et al [9]. The results of the experiment show that the structure with double walls filled with metal foam is an efficient device for absorbing energy.

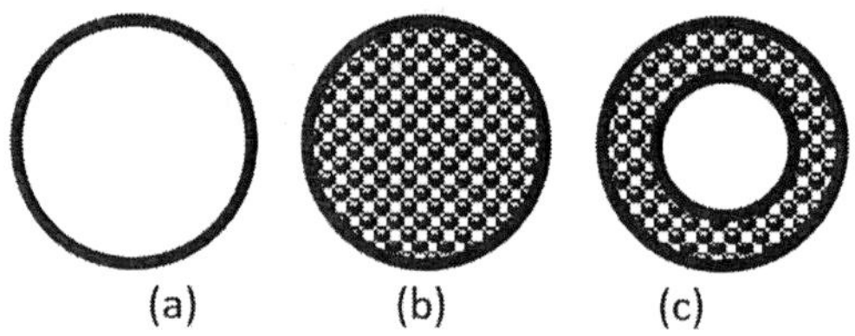

Figure 2. Geometry of specimen cross-section, (a) *single-walled* (SW), (b) *single-walled foam-filled* (SWFF), and (c) *double-walled foam-filled* (DWFF).

A study on quasi-static loading experiments with axial emphasis has also been investigated by Zhibin Li, Rong Chen, Fangyun Lu [10] wherewith two different types, namely round and box structures and with variations of structures filled with Aluminum Foam, including structures round filled with metal foam and metal foam filled with tube boxes, double walled round structure filled with metal foam and double walled box structure filled with metal foam, and metal foam filled with box structure at the corners. The results show that the dimensions of the structural part with walls on the type of structure that is double walled and filled with metal foam in between can have a very significant effect in responding to a collision.

In the research of Niknejad et al [11], published several validated theories of experiments that predict the response to collisions of rectangular structures filled with metal foam (polyurethane) during the initial fold formation under initial axial force collisions. The analysis considers the effect of the interaction between the metal foam and the inner walls of the structure. The results show that there is an influence of interactions on the nature of the collision.

Zheng et al. [12], compared the energy absorption properties of a single structure filled with metal foam with quality density and looked at the effects of density distribution for specific energy absorption capacities. The results show that 12% is better in the specific energy absorption capacity by a single structure filled with metal foam with quality density compared to the use of materials with normal density where the same loading is given.

Table 2. Detail of geometry specimen

	Code	thickness h_0 (mm)	Diameter outer d_0 (mm)	Diameter inner d_1 (mm)
			Single-walled	
	SW	(as in Table 1)	110	-
	SWFF		110	-
			Double-walled	
	DWFF 1	(as in Table 1)	110	82
	DWFF 2		110	60
	DWFF 3		110	40
	DWFF 4		110	20

2. Geometrical Specimen

In this study, there are three configuration models based on the previous research references, namely single-walled (SW), single-walled foam-filled (SWFF), and double-walled foam-filled (DWFF). The results of the previous studies regarding SW, SWFF, and DWFF were conducted by aerospace et al. [13] who used square-shaped specimens. What makes it difficult in this study is to use a tubular specimen. The geometry of the test tube according to Figure 1. The length of all test specimens is 210 mm. The tube wall of the test specimen used is aluminum. Where in this test the condition of the tube wall was filled with metal foam material. The metal foam material used is aluminum foam. In this test compare the quality of the specimen in absorbing energy when given the impact force from the axial direction with several conditions, namely specimens without foam and specimens using foam which with 3 different thickness variations.

Single-walled (SW), single-walled foam-filled (SWFF), and double-walled foam-filled (DWFF) geometries are detailed in accordance with Figure 2. All dimensions of specimen wall thickness are the same, namely 3 mm because they use a type of material that is the same, as can be seen in Table 1. In this study, the specimens were analyzed in simulation software to determine the effect of aluminum tubes without being filled with metal foam on the ability of the specimens to absorb energy from the impact force applied. Specimen size specifications in this first test are in accordance with Table 2.

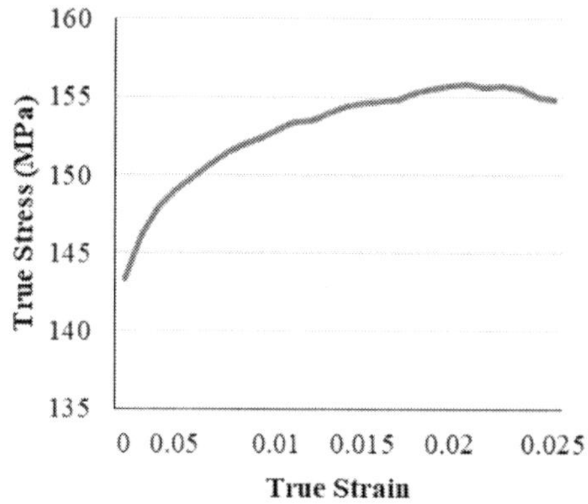

Figure 3. Value of aluminum plasticity.

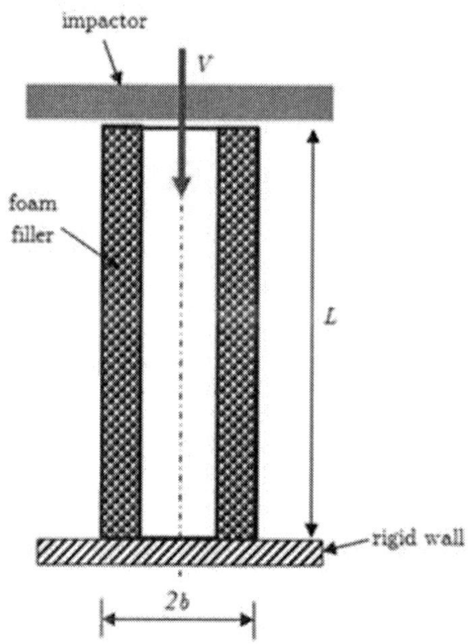

Figure 4. Test specimen model and cross-sectional area (front and rear).

In this study, there are two categories of testing, the first of which is to compare the effect of the use of metal foam and the number of walls in a specimen tube on the ability to absorb energy as shown in Table 1. The second category is to compare the effect of metal foam thickness on the ability to absorb energy as shown in Table 2.

3. Mechanical Properties

In this study, the material used is aluminum and metal foam made from aluminum. The type of aluminum foam used as a test specimen is a material with a density of ~ 480 kg / m³. The other parameters used relating to aluminum foam material are crushable foam with a compression yield stress ratio of ~ 1732 and a modulus of elasticity with a value of ~ 625 MPa, where the equation used is:

$$E = \frac{\sigma}{\varepsilon} \qquad (1)$$

The second material is aluminum, where the type of aluminum used as a test specimen is a material with a density value of ~ 270 kg/m^3. Other parameters used relating to aluminum material are aluminum elasticity value ~ 68200 MPa and plastic values according to Figure 3.

4. Finite element model

The calculation of the finite element method in this study uses simulations to get the right results. The simulation run on Abaqus CAE, which can operate analytical simulations with explicit non-linear dynamic types. The finite element testing model is tubular as in Figure 1.

In this study, the model was made in Abaqus CAE. All tests use two of additional plates that touch directly on the front and back of the specimen according to Figure 4, where the front plate moves and gives a boost to the specimen.

The process of making parts carried out in this study in Abaqus CAE has different amounts in all specimens tested according to the number of parts in single-walled (SW), single-walled foam-filled (SWFF), and double-walled foam-filled (DWFF).

The meshing process carried out in this study in Abaqus CAE uses sizing controls ~ 4 and the largest deviation used is 0.1.

In the interaction process, there are two interactions included in the simulation. The first interaction is the surface to surface, where this interaction is arranged on the surface of parts that touch each other directly. The second interaction is the self-interaction, where this interaction arranged in one whole part because it is only intended for the results of interactions that occur in a part.

Speed determination is also carried out in carrying out simulations on Abaqus CAE with a speed value of ~ 0.4 m /s.

Table 3. Comparison between experimental test and finite element

	Impact Angle (°)	FE	Li ZB at al [13]	Error (%)
Energy Absorption (j)	0	3427.34	3524	2.74
Specific Energy Absorption (j/g)		25.93	26.7	2.88
Energy Absorption (j)	15	3197.28	3286	2.69
Specific Energy Absorption (j/g)		24.25	24.8	2.19

5. Model validation

The finite element model being compared uses references from experiments that been carried out by Li et al. [13] to ensure that our design tests are quite accurate with their experiments. The previously validated model is a double circular aluminum foam tubular with oblique loading [14-17]. The material used in the test specimen is AA6063T6 aluminum alloy [13]. Loading is given constant with a speed of 0.09 mm/s for double circular aluminum foam tubular. The geometric length of the pecimen is 90 mm, with an outer diameter of 38 mm and an inner diameter of 24 mm. The thickness of the specimen tube wall is inner 2.0 mm and outer 1.2 mm. A comparison between finite element method research [13] and our design testing with the same specifications and loading can be clearly seen in Table 3.

IOP Conf. Series: Materials Science and Engineering **875** (2020) 012060 doi:10.1088/1757-899X/875/1/012060

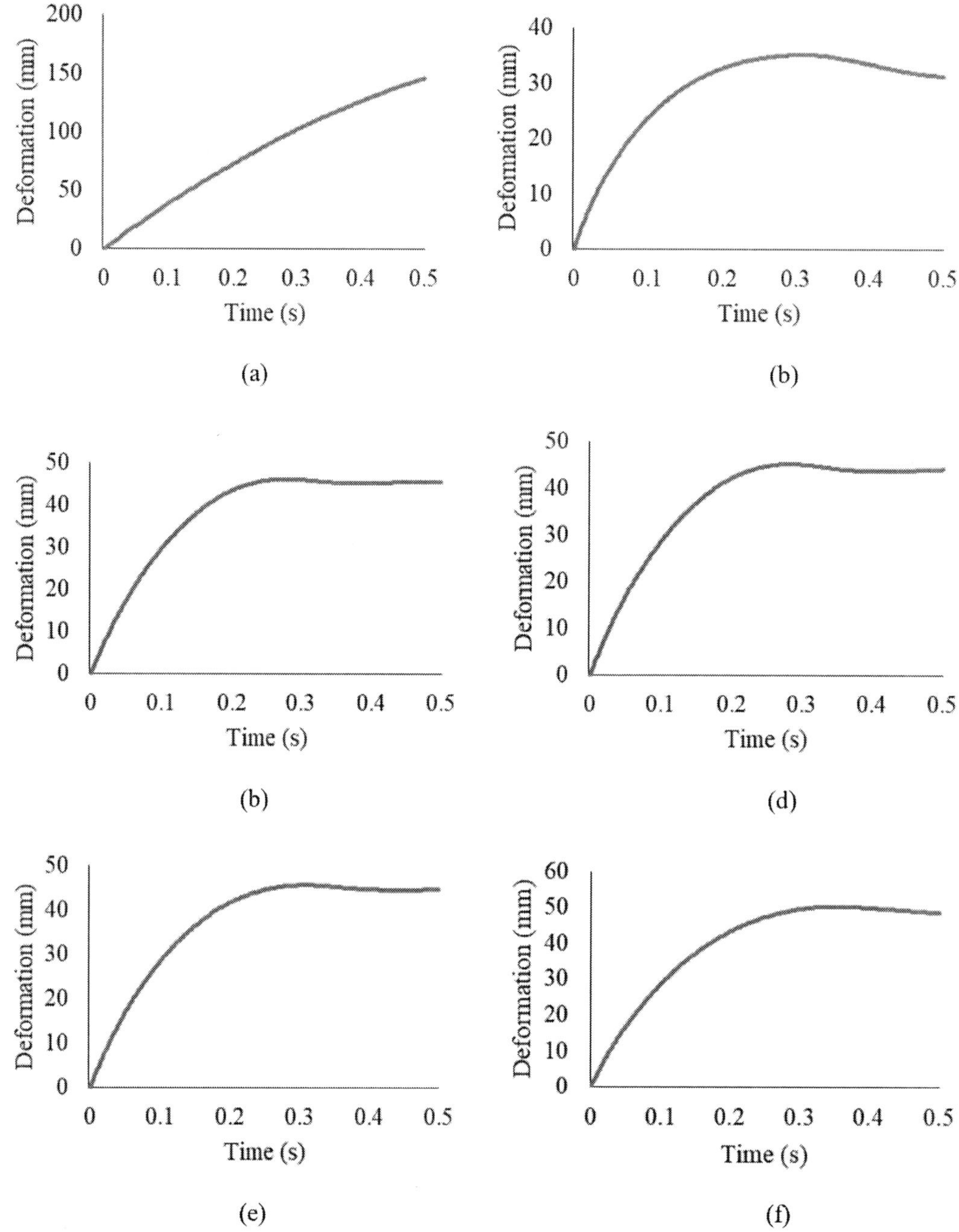

Figure 5. Comparison between deformation and time, (a) SW, (b) DWFF, (c) DWFF 1, (d) DWFF 2, (e) DWFF 3, (f) DWFF 4.

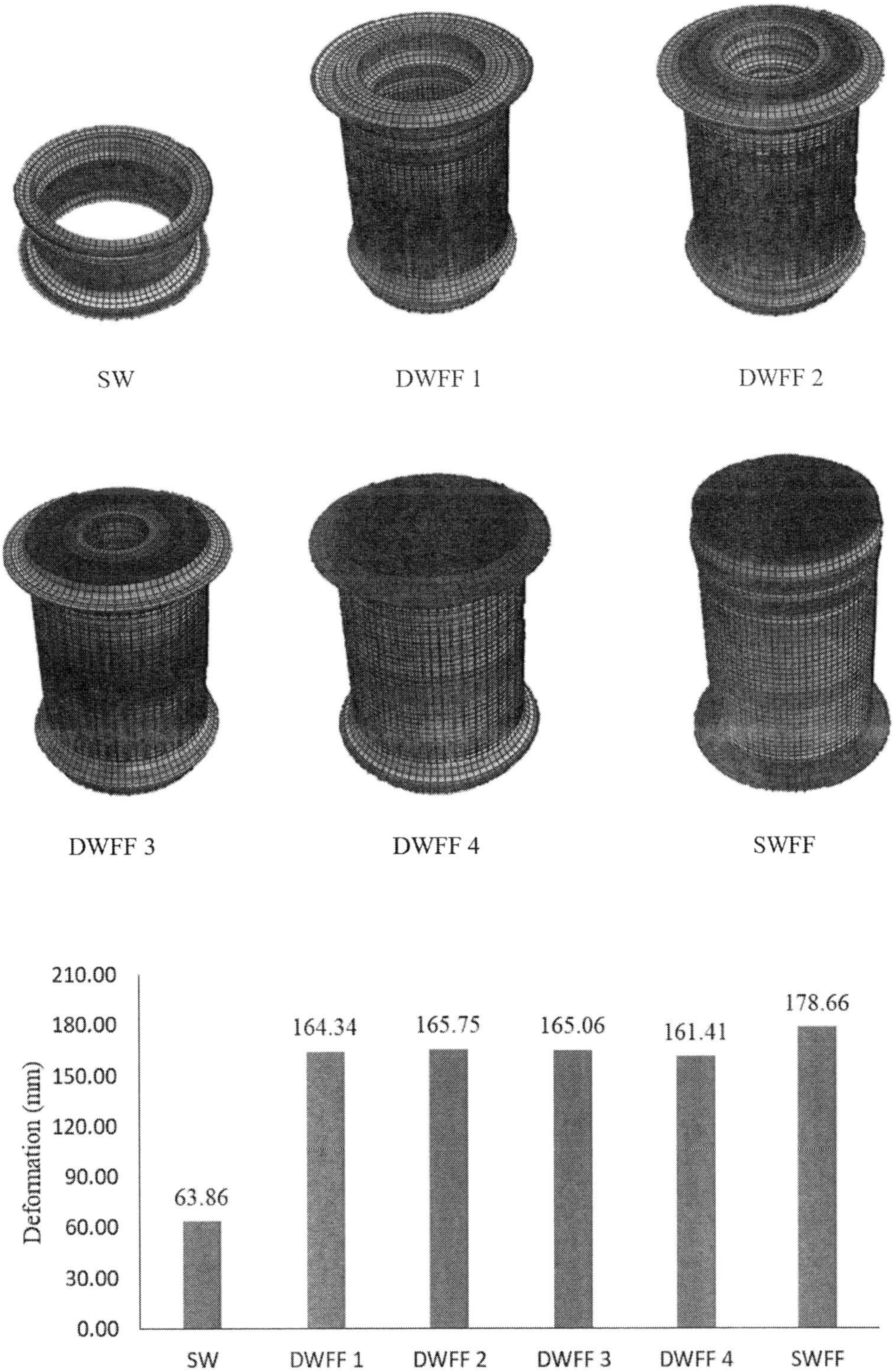

Figure 6. The final length of all specimens.

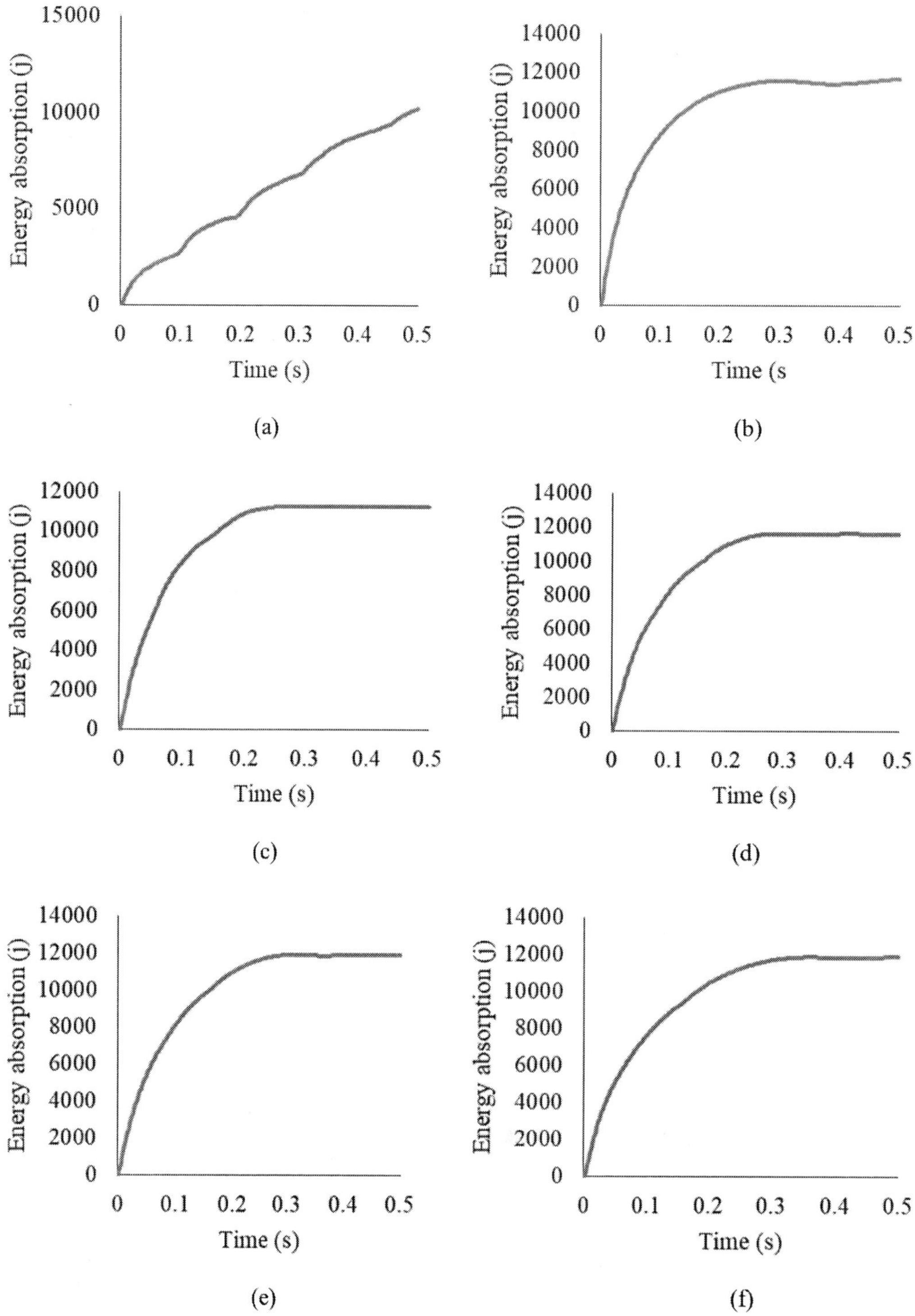

Figure 7. Comparison between energy absorption and time, (a) SW, (b) DWFF, (c) DWFF 1, (d) DWFF 2, (e) DWFF 3, (f) DWFF 4.

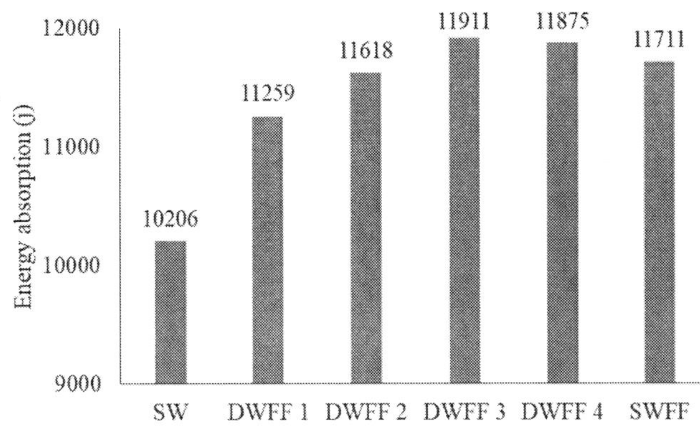

Figure 8. Highest energy absorption of all specimens.

6. Simulation Result

The data collection technique used in this study was to simulate the test material in the form of an aluminum wall filled with aluminum foam with several variations of different wall thicknesses. Specimens used in this experiment include single walled (SW), single walled foam filled (SWFF), and double walled foam filled (DWFF). DWFF specimens varied into four different wall thicknesses. The object under study is designed on Abaqus CAE software where the test material in the form of a pipe with aluminum foam is then measured its quality of resistance by simulating the test material that been made.

6.1. Deformation

One of the results of the analysis of simulation experiments on Abaqus CAE with variations in the thickness of the wall diameter of this specimen is the deformation that occurs so that the length of the specimen reduced because of the force from the axial direction. In testing single walled specimens (SW), the relationship between specimen deformation and time can be clearly seen in Figure 5(a). Figure 5(a) shows that the constant is increasing, which means that deformation is constant over time. This phenomenon occurs because of the influence of the blank wall of the specimen without the aluminum foam being filled in the middle. The final length of a single wallled (SW) after experiencing deformation is 63.8 mm. In the single wallled foam filled (SWFF) specimen, the relationship between specimen deformation and time can be clearly seen in Figure 5(b). In Figure 5(b) shows that the deformation occurs fluctuatively, meaning that the phenomenon that occurs in the test specimen is not only a reduction in length but also an increase in length. This phenomenon occurs because of the elastic nature of the specimen so that the length of the specimen has returned after shortening, before finally experiencing a reduction in length again. The final length of the single-wallled foam filled (SWFF) after being deformed is 178.8 mm. In testing the DWFF 1 specimen, the specimens used are aluminum outer diameter of 110 mm and aluminum inner diameter of 82 mm filled with aluminum foam between inner and outer.

The relationship between specimen deformation and time can be clearly seen in Figure 5(c). In Figure 5(c) shows that deformation occurs fluctuatively. The final length of DWFF 1 after being deformed is 164.33 mm. In testing DWFF 2 specimen, the specimens used are aluminum outer diameter of 110 mm and aluminum inner diameter of 60 mm filled with aluminum foam between inner and outer. The relationship between specimen deformation and time can be clearly seen in Figure 5(d). In Figure 5(d) shows that deformation occurs fluctuatively. The final length of DWFF 2 after being deformed is 164.33 mm. In testing the DWFF 3 specimen, the specimens used are aluminum outer diameter of 110 mm and aluminum inner diameter of 40 mm filled with aluminum foam between inner and outer. The relationship between specimen deformation and time can be clearly seen in Figure 5(e). In Figure 5(e) shows that deformation occurs fluctuatively. The final length of DWFF 3 after experiencing deformation is 165.06 mm.

In testing the DWFF 4 specimen, the specimens used are aluminum outer diameter of 110 mm and aluminum inner diameter of 20 mm filled with aluminum foam between the inner and outer.

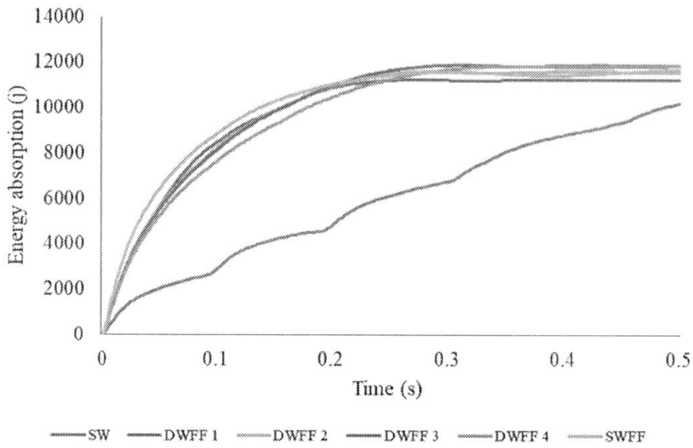

Figure 9. Comparison between all energy absorption of specimens and time.

The relationship between specimen deformation and time can be clearly seen in Figure 5(f). In Figure 5(f) shows that deformation occurs fluctuatively. The final length of DWFF 4 after experiencing deformation is 161.41 mm. Overall the final length of all specimens after the simulation can be seen based on the results of data processing in Figure 6. In Figure 6 shows that the shortest end of the specimen occurs in the type of SW specimen testing with a length of 63.8 mm and the longest end of the longest specimen occurs in the type of SWFF specimen testing with a length of 178.8 mm. Tests of DWFF specimens tend to be the same and the interval values between DWFF are not far adrift. The longer the specimen after testing shows that the less the reduction in the length that occurs due to pressure by the force from the axial direction and vice versa, the shorter the specimen after the test shows that the more reduction in the length that occurs due to the pressure by the force from the axial direction [18].

6.2. Energy Absorption

Other analysis results from the simulation experiments on Abaqus CAE with variations in wall diameter thickness of this specimen are the amount of energy that can be absorbed by the specimen when given treatment. The relationship between deformation and time is clearly shown in Figure 7. The SW specimen, from Figure 7(a) shows that energy absorption occurs fluctuatively and the maximum energy that can be absorbed by the SW specimen in this test is 10.206 joules. In the SWFF specimen, from Figure 7(b) shows that the energy absorption occurs constantly increasing and the maximum energy that can be absorbed by the SW specimens in this test is 11.711 joules. In all DWFF specimens, from Figure 7 (c), (d), (e), and (f) show that energy absorption occurs fluctuatively and maximum energy that can be absorbed by specimens DWFF 1, DWFF 2, DWFF 3, DWFF 4 in this test, each was 11.259 joules, 11.618 joules, 11.911 joules, 11.875 joules. Overall, the maximum energy that can be absorbed by the specimens in this test can be seen in Figure 8 based on the results of data processing in Figure 9. Figure 9 illustrates that all specimens have similar energy absorption characteristics, which tend to increase, except in the SW specimen, whsich shows fluctuating energy absorption characteristics and is far below other specimens [19].

7. Conclusion

The use of aluminum foam in crush box construction affects the level of energy absorption that occurs during testing, where the crush box construction that added with aluminum foam is able to absorb more energy. The thickness of aluminum foam in crush box construction also affects the level of energy absorption, where the recommended thickness is DWFF 3 which is neither too thin nor too thick. The least energy-absorbing specimen in this study was SW compared to other specimens, where the energy absorbed was 10,206 joules. The least deformation so that the final length of the specimen is the longest in this study, namely the SWFF specimen with a final length of 178.8 mm. The most deformation occurred in the SWFF specimen, so that the final length of the specimen was the shortest with a final length value of 63.8 mm.

References

[1] Zhang Xiong, Zhang Hui, Wang Zang, 2015. *Collapse of Square Tube With Variable Thickness*. (Department of Mechanics, Huazhong University of Science and Technology).

[2] I Putu Ali Putra. 2014. *Knowing of the Safety System in Cars*. Indonesia. [in Bahasa].

[3] Jatmiko, A. 2014. *The effect of variations in the geometry of the 2 segment crash box on the ability to absorb impact energy with computer simulations* (Thesis. Departement of mechanical engineering, Universitas Brawijaya).

[4] Hayashi. 2005. *Impact resistant structure for the helicopter and energy absorber used for the same*. (US Patent), US006959894B2.

[5] Thornton, P. 1980. *Energy absorption by foam–filled structures*. (SAE paper 800081).

[6] Santosa S.P, T. Wierzbicki, A.G. Hanssen, M. Langseth, 2000. *Experimental and numerical studies of foam–filled sections*. (Int. J. Impact Eng.) 24 (5) 509–534.

[7] Kavi H, A.K. Toksoy, M. Guden. 2006. *Predicting energy absorption in a foam–filled thin–walled aluminum tube based on experimentally determined strengthening coefficient*. (Mater). (4) 263–269.

[8] Santosa S.P, T. Wierzbicki. 1999. *The concept of double–walled sandwich columns for energy absorption*. (Int. J. Crashworthiness) 4 (2) 175–198.

[9] Seitzberger M, F.G. Rammerstorfer, R. Gradinger, H.P. Degischer, M. Blaimschein, C. Walch. 2000. *Experimental studies on the quasi–static axial crushing of steel columns filled with aluminium foam*. (Int. J. Solids Struct.) 37 4125–4147.

[10] Zhibin Li, Rong Chen, Fangyun Lu. 2018. *Comparative analysis of crashworthiness of empty and foam–filled thinwalled tubes*. (Thin–Walled Struct.) 124 343–349.

[11] Niknejad, M.M. Abedi, G.H. Liaghat, M. Zamani Nejad. 2015. *Absorbed energy by foam–filled quadrangle tubes during the crushing process by considering the interaction effects*. (Arch. Civil. Mech. Eng.) 15 376–391.

[12] Zheng G, S. Wu, G. Sun, G. Li, Q. Li. 2014. *Crushing analysis of foam–filled single and bitubal polygonal thin–walled tubes*. (Int. J. Mech. Sci.) 87 226–240.

[13] Li ZB, Yu J L, Guo L W. 2012. *Deformation and energy absorption of aluminium foam-filled tubes subjected to oblique loading*. (Int J Mech Sci) 48–56.

[14] Djamaluddin, F, Abdullah, S, Arrifin Ak, Nopiah Zm, 2015. *Modeling And Optimization Of Aluminum Foam Cylindrical Double Tubes Under Axial Impact,* (Journal Of Mechanical Engineering And Science) 8, 1383-1392.

[15] Djamaluddin, F, Abdullah, S, Arrifin Ak, Nopiah Zm, 2015.*Finite element analysis and crashworthiness optimization of foam-filled double circular under oblique loading, (*Latin American Journal of Solids and Structures) 13 (11), 2176-2189.

[16] Djamaluddin, F, Abdullah, S, Arrifin Ak, Nopiah Zm, 2018 *Crush analysis of the foam-filled bitubal circular tube under oblique impact,* (IOP Conference Series: Materials Science and Engineering) 308 (1), 012040

[17] Djamaluddin, F, Abdullah, S, Arrifin Ak, Nopiah Zm, *Optimisation and validation of full and half foam filled double circular tube under multiple load cases* (International Journal of Crashworthiness) 24 (4), 389-398.

[18] Yang S, Qi C. 2013. *Multi objective optimization for empty and foam-filled square columns under oblique impact loading*. (Int J Impact Eng):177–91.

[19] Fang J, Gao Y, Sun G, Zhang Y, Li Q. 2014. *Crashworthiness design of foam-filled bitubal structures with uncertainty*. (Int J Linear Mech.).

Potential of Work Related Musculoskeletal Disorder in Traditional Salt Farmers

I Bakri[1]*, R D Mudiatuti[1], Mulyadi[1], A Darmawan[1], and Akmal[1]

[1]Industrial Engineering Department, Engineering Faculty, Hasanuddin University, Makassar, Indonesia

*E-mail: ilhambakri@unhas.ac.id

Abstract. The purpose of this study was to identify work postures and determine the risk of health problems due to the body postures of traditional salt farmers. The sample used in this study was 37 traditional salt farmers from Pundata Baji, Pangkep Regency - Indonesia. The analysis techniques used were the Rapid Upper Limb Assessment (RULA) and Rapid Entire Body Assessment (REBA) methods. The results show that based on the RULA score, there were two work postures that require investigation and change as soon as possible (urgently) (Level 4). As for the REBA score, there was one posture that is included in the very high-risk category, which needs improvement right now. Subjective complaints data show that the most complaints from traditional salt farmers occur in the back area with a percentage of 72.97%. This is caused by the fact that most work postures are carried out in a bent position for a long time.

1. Background

All industries, both formal and informal, are expected to be able to apply occupational safety and health in carrying out their duties so that workers feel comfortable and safe at work, free from work-related diseases, and avoid work accidents [1]. One of the occupational diseases that can appear at any time is musculoskeletal disorders (MSDs) [2, 3]. MSDs are a disease caused by incorrect position or work attitude. MSDs are disorders of normal functioning of muscles, tendons, nerves, blood vessels, bones and ligaments, due to changes in the structure or the musculoskeletal system in a short or long time [4].

The traditional agricultural sector is one type of work that has a high risk for workers [5]. Extreme environmental conditions and the use of technology in managing land that is still quite behind compared to other work determine the level of health and safety of farmers [6]. Activities undertaken by farmers, including preparing land (hoeing), planting, and post-harvest have different levels of risk.

Pundata Baji Village is one of the areas in South Sulawesi - Indonesia where the majority of the population's livelihoods are as milk fish farmers. But when the dry season arrives and their ponds are drought, some of the fishers turn to traditional salt farmers and make their ponds a place for making salt. Salt farmers in Pundata Baji still use manual methods without machining aids in the process of production and harvest, and post it. As with other traditional farms, in each of these processes, the dominant position of traditional salt farmers is to bend. This position is not very ergonomic because it will lead to complaints, aches, or aches in the back of the body. This study aims to identify work postures and determine the risk of health problems due to the body postures of traditional salt farmers during harvest and post-harvest handling in Pundata Baji.

2. Methodology

This research was conducted on 37 traditional salt farmers in Pundata Baji. Data collection was done through the method of observation on the object under study. The observation method was carried out to obtain data on the work posture of workers by recording or photographing each work posture that occurs when the farmer did his activities. After each work posture was obtained, a portrait was then given a marker and measurements of body segmentation angles. The results of measurements of body segmentation angles were then analyzed using the Rapid Upper Limb Assessment (RULA) and Rapid Entire Body Assessment (REBA) methods to obtain the risk of MSDs in the salt harvest and post harvest processes [7]. In addition to these two methods, this study also used a questionnaire method to map the location and level of complaint of skeletal muscle disorders in traditional salt farmers. The questionnaire used was the Nordic Body Map (NBM) questionnaire [8].

3. Result

After observing the activities carried out by traditional salt farmers in Pundata Baji, especially in harvesting and post-harvest activities, three working postures were observed as presented in Table 1.

Table 1. Activities and its related work postures

No	Activities	Work Postures
1.	Salt harvesting	1
2.	Collecting salt to basket	2
3.	Moving sack of salt	3

In the Work Posture 1 activity, the farmer did the salt harvesting activity in the area of the salt plots pool with a slouchy working attitude for about one hour to collect salt to the edge of the plots with the workload on the arms holding the chips weighing about three kilograms. Every salt plot pond, there were one to three farmers who harvested with the yield of each salt plot of 150 kg. Figure 1 shows the body segment of a traditional salt farmer when harvesting salt. After being harvested by stacking, the salt is then collected and put in a basket (Figure 2).

Figure 1. Salt harvesting (Work Posture 1)

Figure 2. Collecting salt to basket (Work Posture 2)

To conduct RULA analysis on Work Posture 1, two assistive tables were used to determine the weight of the forearm, arm, and wrist (Group A) (Group 1) and the neck, back, and leg posture (Group B).

Table 2. RULA weighting Group A (forearm, arm, and wrist) for Work Posture 1

Table A		Wrist							
		1 (Neutral)		2 (0°-15°)		3 (>15°)		4	
		Bended Wrist		Bended Wrist		Bended Wrist		Bended Wrist	
Forearm	Arm	1	2	1	2	1	2	1	2
1 (20°)	1	1	2	2	2	2	3	3	3
	2	2	2	2	2	3	3	3	3
	3	2	3	2	3	3	3	4	4
2 (20°-45°)	1	2	2	2	3	3	3	4	4
	2	2	2	2	3	3	3	4	4
	3	2	3	3	3	3	4	4	5
3 (45°-90°)	1	2	3	3	3	4	4	5	5
	2	2	3	3	3	4	4	5	5
	3	2	3	3	4	4	4	5	5
4 (>90°)	1	3	4	4	4	4	4	5	5
	2	3	4	4	4	4	4	5	5
	3	3	4	4	5	5	5	6	6

Table 3. RULA weighting Group B (neck, back, and foot) for Work Posture 1

Table B	Back											
	1 (sitting)		2 (0° - 20°)		3 (20° - 60°)		4 (>60°)		5		6	
	Foot		Foot		Foot		Foot		Foot		Foot	
Neck	1	2	1	2	1	2	1	2	1	2	1	2
1 (0°-10°)	1	3	2	3	3	4	5	5	6	6	7	7
2 (10°-20°)	2	3	2	3	4	5	5	5	6	7	7	7
3 (>20°)	3	3	3	4	4	5	5	6	6	7	7	7
4 (extension)	5	5	5	6	6	7	7	7	7	7	8	8

By means of the same analysis of work risk in Work Posture 1, work risk in Work Posture 2 can be found to be at the same level as work posture 1.

Table 4 shows that Work Posture 1 for farmers who harvest and collect salt had a RULA score of 7 which means that investigation and changes to work posture are needed as soon as possible (urgently).

Table 4. RULA grand total score for Work Posture 1

Score C	Grand Total Score Score D = score from table B + Muscle Use Score + Force								
	1	2	3	4	5	6	7	8	9
1	1	2	3	3	4	5	5	5	5
2	2	2	3	4	4	5	5	5	5
3	3	3	3	4	4	5	6	6	6
4	3	3	3	4	5	6	6	6	6
5	4	4	4	5	6	7	7	7	7
6	4	4	5	6	6	7	7	7	7
7	5	5	6	6	7	7	7	7	7
8	5	5	6	7	7	7	7	7	7
9	5	5	6	7	7	7	7	7	7

Table 5. REBA weighting Group A (neck, back, and foot)

Table A		Back				
		1	2	3	4	5
	Foot					
Neck = 1	1	1	2	2	3	4
	2	2	3	4	5	6
	3	3	4	5	6	7
	4	4	5	6	7	8
	Foot					
Neck = 2	1	1	3	4	5	6
	2	2	4	5	6	7
	3	3	5	6	7	8
	4	4	6	7	8	9
	Foot					
Neck = 3	1	3	4	5	6	7
	2	3	5	6	7	8
	3	5	6	7	8	9
	4	6	7	8	9	9

Figure 3 shows the Work Posture 3 where farmers underwent activities with a bent work attitude on the edge of the salt mound pond area to lift the sacks of salt to the vehicle that will transport the sacks of salt to the storage shed. Each sack of salt weighs bout 50 kg. To analyze work risk in this work posture, two REBA assist tables (tables 5 and 6) were used to produce a RULA grand score table (table 7). Table 8 shows the occupational risk categories of the REBA method. From this table, it is revealed that work risks in Work Posture 3 are in the very high risk level category and require urgent work method changes.

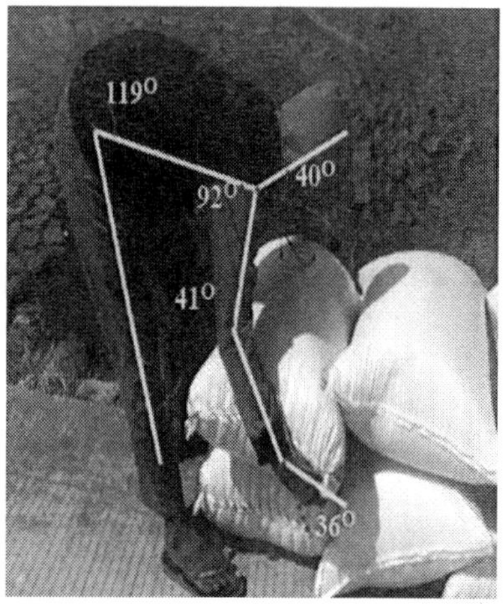

Figure 3. Moving sack of salt

Table 6. REBA weighting Group B (wrist, upper arm, arm)

Table B		Upper Arm					
		1	2	3	4	5	6
	Pergelangan						
Arm = 1	1	1	1	3	4	6	7
	2	2	2	4	5	7	8
	3	2	3	5	5	8	8
	Pergelangan						
Arm = 2	1	1	2	4	5	7	8
	2	2	3	5	6	8	9
	3	3	4	6	7	8	9

Table 7. RULA grand total score for Work Posture 3

		Skor A											
		1	2	3	4	5	6	7	8	9	10	11	12
Skor B	1	1	1	2	3	4	6	7	8	9	10	11	12
	2	1	2	3	4	4	6	7	8	9	10	11	12
	3	1	2	3	4	4	6	7	8	9	10	11	12
	4	2	3	3	4	5	7	8	9	10	11	11	12
	5	3	4	4	5	6	8	9	10	10	11	12	12
	6	3	4	5	6	7	8	9	10	10	11	12	12
	7	4	5	6	7	8	9	9	10	11	11	12	12
	8	5	6	7	8	8	9	10	10	11	12	12	12
	9	6	6	7	8	9	10	10	10	11	12	12	12
	10	7	7	8	9	9	10	11	11	12	12	12	12
	11	7	7	8	9	9	10	11	11	12	12	12	12
	12	7	8	8	9	9	10	11	11	12	12	12	12

Table 8. Work risk category of REBA's method

REBA Score	*Risk Level*	Actions
1	-	-
2-3	Low	Maybe needed
4-7	Medium	Needed
8-10	High	Needed immediately
11-15	Very High	Needed now

Next, Table 9 shows the recapitulation results of subjective complaints caught through the NBM questionnaire felt by 37 farmers.

Table 9. Subjective complain (N=37)

No	Location	Frequency	Presentation
1	Tailbone	0	0.00 %
2	Buttocks	20	54.05 %
3	Back	27	72,97 %
4	Neck	19	51,35 %
5	Head	2	5,41%
6	Shoulder	0	0.00 %
7	Hip	4	10.81%
9	Forearm	23	62.16%
10	Elbow	0	0.00%
11	Arm	10	27.03%
12	Wrist	14	37.84%
13	Palm	0	0.00 %
14	Thigh (back)	0	0.00 %
15	Knee	0	0.00 %
16	Calf	13	35.14%
17	Ankle	0	0.00 %
18	Foot	0	0.00 %
19	Chest (upper part)	0	0.00 %
20	Chest (lower part)	0	0.00 %
21	Stomach	0	0.00 %
22	Thigh (front)	0	0.00 %

Based on the complaint data obtained using the NBP questionnaire, it is seen that traditional salt farmers had the biggest complaint on their back with a percentage of 72.97%. This was caused by the fact that most work postures were carried out in a bent position for a long time so that the burden was more felt on the back. The next biggest complaint was in the upper arm as much as 62.16% due to the position in using aids and lifting position which was still wrong and carried out repeatedly, resulting in pain in the upper arm. There were also complaints as much as 54.05% in the buttocks caused by the weight of the burden lifted and the harvest process was carried out in a bent position repeatedly within 1-2 hours. Complaints on farmers' necks also had a high percentage (51.35%) caused by the position of the neck which forms a slightly raised angle in quite a long time.

Table 10 provides work posture recommendations for traditional salt farmers which can avoid working risks from work posture errors during harvest and post-harvest salt.

IOP Conf. Series: Materials Science and Engineering **875** (2020) 012061 doi:10.1088/1757-899X/875/1/012061

Table 10. Recommendations for improving work posture

No	Activity/Posture	Initial Work Posture	Recommended Work Posture	Recommended Posture Image
1		Activities were carried out with the back position slightly bent and the neck forming an angle with the head slightly raised. RULA Score : 7	The activity of collecting salt should be done in a standing body position. Back does not form a normal angle and neck position so that there is no fatigue in the back RULA Score : 4	
2		The activity of putting salt into the basket was done with the position of the legs bent and the back forming a very large angle, and the position of the arms that form a large angle. RULA Score : 7	The activity of putting salt into the basket should be done by bending the knees and squatting, keeping the back and neck straight so that the angles of the arms formed are also not too large and the body position is more stable. RULA Score : 4	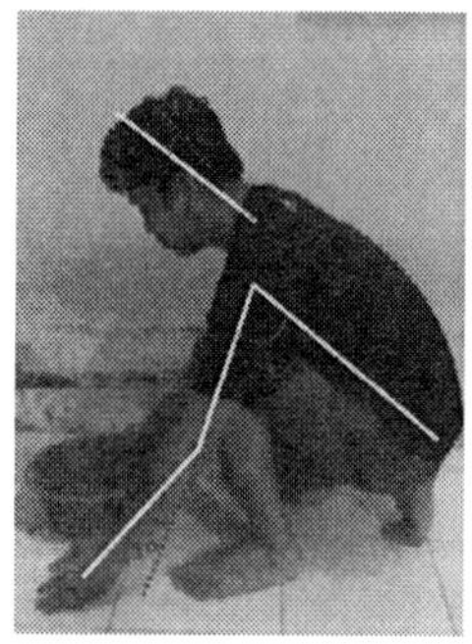
3		Activity The activity of lifting salt was carried out with the back bent, the foot slightly bent so that it supports a weight of > 10 kg and the neck forms an angle. REBA Score : 11	Activities when you want to lift d should be done in a squat position, so as to keep your back straight so that the weight of the load is spread evenly on the body. REBA Score : 3	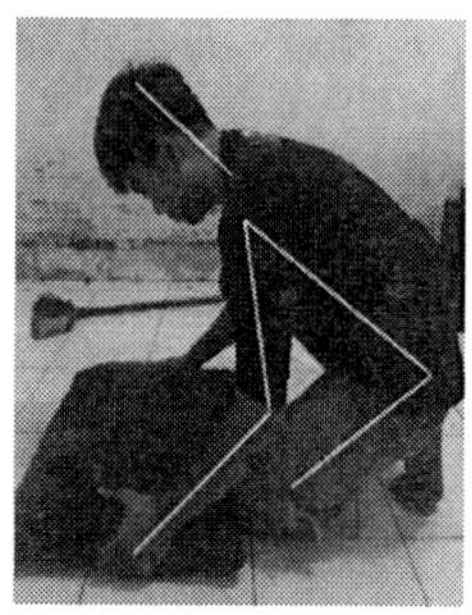

4. Conclusion

Based on observations and data processing that had been carried out on 37 salt farmers in Pundata Baji, it is concluded that there were three working postures in the traditional salt harvest process. Based on the RULA scoring, there were 2 postures (work posture collecting harvested salt and work poster putting salt into the basket) which fall in level 4 category, namely work posture that requires investigation and changes were needed as soon as possible (urgent). Meanwhile, based on REBA scoring, there was one posture that is included in the category of very high risk that requires improvement now. Based on the complaint data obtained using the NBP questionnaire, it is seen that traditional salt farmers had the biggest complaint on the back (72.97%) followed by the upper arm area (62.16%), buttocks (54.05%) and on the farmer's neck (51.35%).

Acknowledgement
This research was funded by Engineering Faculty Universitas Hasanuddin under LBE Research Fund Scheme. We also would like to thank all participants of this study in Pundata Baji Village.

References

[1] Sulzer-Azaroff B, Loafman B, Merante R J and Hlavacek A C 1990 Improving occupational safety in a large industrial plant: A systematic replication *J. Organ. Behav. Manage.* **11** 99–120.
[2] da Costa B R and Vieira E R 2010 Risk factors for work-related musculoskeletal disorders: a systematic review of recent longitudinal studies *Am. J. Ind. Med.* **53** 285–323.
[3] Denis D, St-Vincent M, Imbeau D, Jette C and Nastasia I 2008 Intervention practices in musculoskeletal disorder prevention: a critical literature review *Appl. Ergon.* **39** 1–14.
[4] Buckle P 2005 Ergonomics and musculoskeletal disorders: overview *Occup. Med. (Chic. Ill).* **55** 164–7.
[5] Jo H, Baek S, Park H, Lee S-A, Moon J, Yang J E, Kim K S, Kim J Y and Kang E K 2016 Farmers' cohort for agricultural work-related musculoskeletal disorders (farm) study: study design, methods, and baseline characteristics of enrolled subjects *J. Epidemiol.* JE20140271.
[6] De Clercq M, Vats A and Biel A 2018 Agriculture 4.0: The Future of Farming Technology *Proc. World Gov. Summit, Dubai, UAE* 11–3.
[7] Kong Y-K, Lee S, Lee K-S and Kim D-M 2018 Comparisons of ergonomic evaluation tools (ALLA, RULA, REBA and OWAS) for farm work *Int. J. Occup. Saf. Ergon.* **24** 218–23.
[8] Hermawati S, Lawson G and Sutarto A P 2014 Mapping ergonomics application to improve SMEs working condition in industrially developing countries: a critical review *Ergonomics* **57** 1771–94.

IOP Conf. Series: Materials Science and Engineering **875** (2020) 012062 doi:10.1088/1757-899X/875/1/012062

Hardness Improvement of Aluminum Reinforced with Carbon Nanotubes through Forging and Friction Surfacing Process

L H Arma[1]*, H Arsyad[1], M Syahid[1], E Iskandar[1], Nurhudaedah[1], and Asruddin[1]

Mechanical Engineering Department, Faculty of Engineering, Universitas Hasanuddin, Gowa, Indonesia

*E-mail: armalh@yahoo.com

Abstract. This research deals with the improvement of hardness property on the composite material made from aluminum matrix reinforced with carbon nanotubes (Al-CNTs). There were two experiments conducted to enhance hardness property; forging and friction surfacing process. An axial mechanical force forged the Al-CNTs composite up to 40% of height of a cylindrical aluminum. A thermo-mechanical force was also applied in producing Al-CNTs coating by the friction surfacing process using Al-CNTs composite rod. Then, the hardness property obtained from these two processes was investigated. Results of the experiments showed that the hardness of Al-CNTs after the forging process improved 40–50% due to high amount of dislocation during forging which avoids deformation. This was indicating that 1 wt% CNTs does not decrease the hardenability of aluminum alloys. Meanwhile, the hardness Al-CNTs coating was not altered too much by the process of friction surfacing. This is becaused that the homogenous dispersion of the CNTs have occurred during the stir casting process.

1. Introduction

Aluminum alloys are mostly selected in engineering applications when required to meet a material with lightweight and high strength properties. However, low hardness and high friction coefficient have limited the application of aluminum alloys in many sliding and rubbing components. The way to enhance the hardness of aluminum is by adding particulate reinforcement in the aluminum matrix as well as by surface modification.

It was reported that hardness increases with the increase of reinforcement content due to the high hardness and strength of the reinforcement phase. The addition of silicon carbide [1] and alumina [2] to aluminum alloys improved their hardness and wear resistance. Many kinds of particulate reinforcements have been successfully used to improve tribological properties of aluminum alloys such as aluminum nitride [3], granite [4], glass [5], titanium dioxide [6]. Also, several surface modification techniques such as high energy laser melt treatment [7], high energy electron beam irradiation [8], plasma spraying [9] and friction surfacing [10] have been developed to deposit metal matrix composites on aluminum alloy surfaces.

Unlike other surface modification processes aforementioned, deposition of metal matrix composite in friction surfacing is carried out at a temperature below the melting point of aluminum alloys. Therefore, the presence of undesirable and detrimental phases [11] can be avoided. In friction surfacing, an axial force pressed a high rotational speed consumable rod on the surface of a moving substrate, which generates frictional heat energy in the interface to soften the rod. Thereby, a viscoplastic shearing layer is continuously deposited between the rod and the substrate (Figure 1).

Recently, research dealing with the effect of nano reinforcement such as carbon nanotubes (CNTs) to enhance mechanical properties has been extended to aluminum and its alloys. Such aluminum matrix with CNTs reinforcement has been introduced as a composite with high mechanical properties as well as wear resistance [12].

In this present work, the hardness of aluminum-carbon nanotubes (Al-CNTs) composite after subjected to mechanical force such as forging process and thermo-mechanical force like friction surfacing process was investigated. The composite of Al-CNTs was produced using the optimum process parameter of stir casting method that has been investigated in our previous research (not published yet).

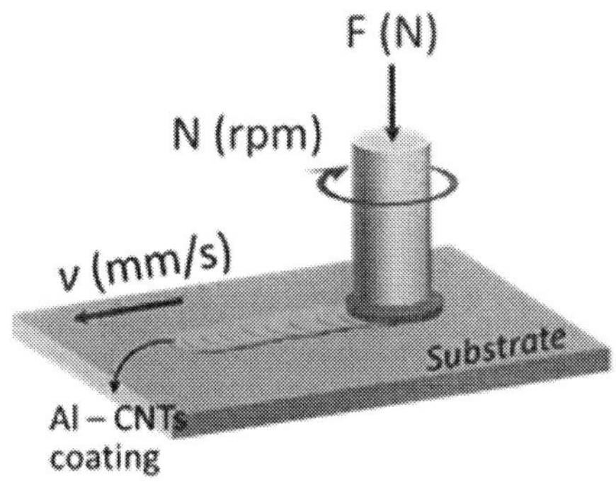

Figure 1. Friction surfacing process

2. Experimental

2.1. Casting Process

A consumable rod shown in Figure 2 made from Al-CNTs composite was produced using stir casting process (Figure 3). The matrix and reinforcement used in this study were AA6061 alloys and multiwall carbon nanotubes (MWCNT), respectively. The chemical composition of AA6061 characterized by XRF can be seen in Table 1.

Tabel 1. Chemical composition of AA6061

Element	Al	Mg	Si	Fe	Cu	Cr	Zn
wt%	Balance	0.9	0.7	0.6	0.3	0.25	0.2

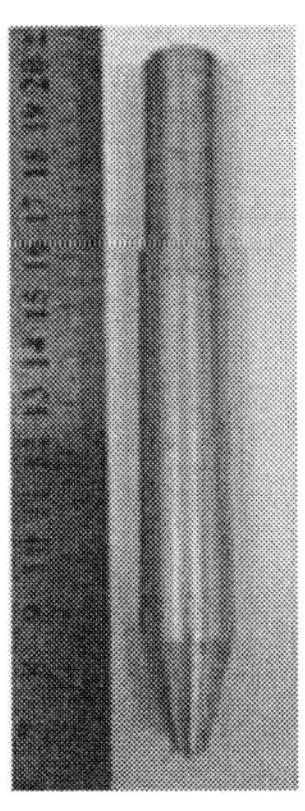

Figure 2. Consumable rod

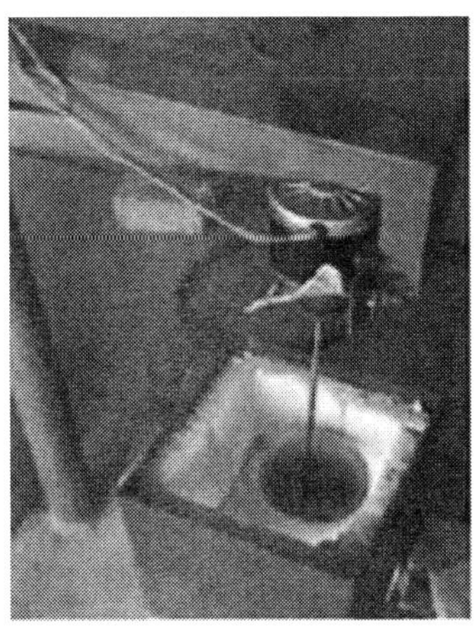

Figure 3. Stir casting apparatus

The MWCNT has an outer diameter >50 nm, length 0.5–2.0 mm, and purity > 7%. The stir casting process was done by heating AA6061 in the crucible up to 750°C. After melting, argon gas was given to clean the slag. Then the molten aluminum was stirred for 5 minutes at a speed of 500 rpm while entering 0.5 and 1.0 wt% of MWCNT powder. Then, the molten aluminum was poured into heated metal molds forming a casted aluminum rod. The pouring temperature was 750°C. Finally, the rod was processed to meet the dimension of the consumable rod using a turning machine.

2.2. Friction Surfacing Process
An axial drilling-milling machine (Figure 4) was employed to deposit the rod material to be a coating onto a substrate of steel plate. The consumable rod made from Al-CNTs composite was 15 mm in diameter. The rod was mounted in the drill chuck of the machine. The rotary speed selected was 3000 rpm. The substrate of the steel plate was mounted in a vise of the machine. The coating was deposited on it by rotating consumable rod on the moving substrate. It is possible to deposit a thin layer of coating by utilizing heat energy generated by friction between the consumable rod and the substrate.

Figure 4. Drilling-Milling machine used for friction surfacing process

2.3. Forging Process
The samples of the as-cast rod were subjected to mechanical force. An axial compression force from a universal testing machine forged the samples up to a 40% reduction of height dimension in order to investigate the hardenability of Al-CNTs composite. After forging, the hardness of the samples was tested.

2.4. Testing Process
Then, the consumable rod, the friction coatings, and the residual rod were subjected to hardness test using Rockwell hardness test with HRF scale, indentor diameter is 1.588 mm with a load of 60 kgf.

3. Results and Discussions
3.1. Effect of Forging on the Hardness of Al-CNTs Composite
The hardness value of the forged rod as well as the as-cast rod (before forging) is shown in Figure 5. It was revealed that all of the forged rod samples showed higher hardness than the as-cast rod. The hardness significantly increased to about 40–50%. This is due to the high amount of dislocation during forging which avoids deformation [13].

This result indicates that the addition of CNTs to aluminum matrix up to 1 wt% does not decrease the hardenability of Aluminum alloys. The Al-CNTs composite even can be hardened by subjecting to mechanical force.

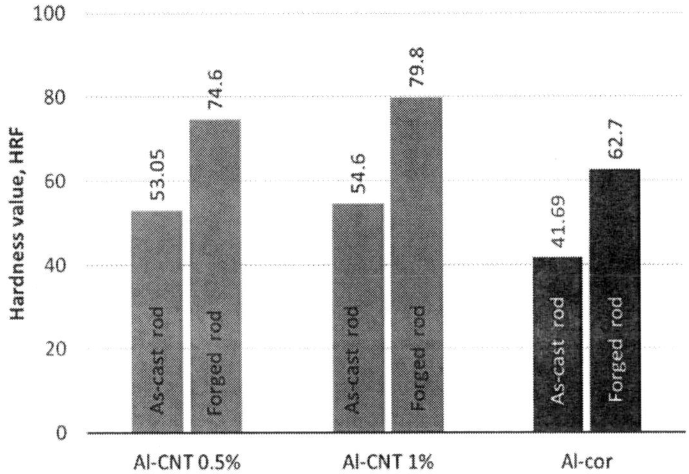

Figure 5. The hardness value of Al-CNTs composite

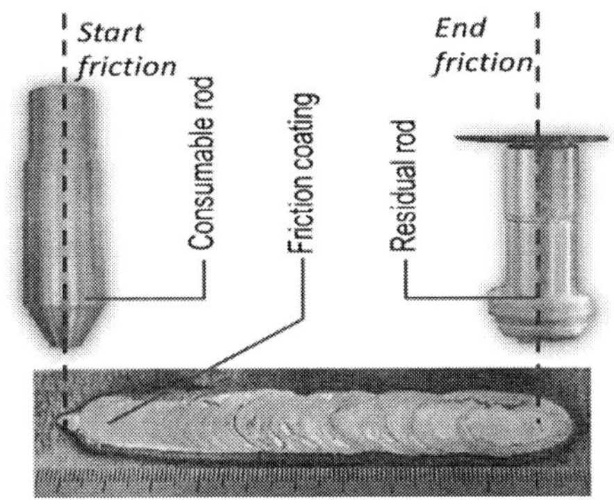

Figure 6. Friction surfacing coating

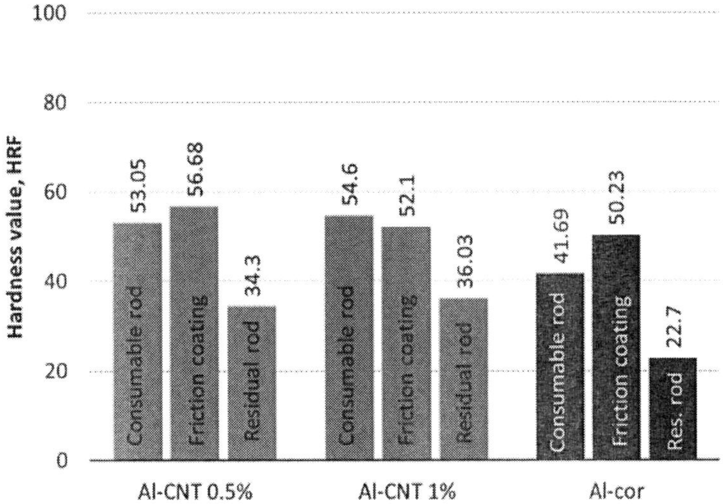

Figure 7. The hardness value of friction surfacing products

3.2. Effect of Friction Surfacing on the Hardness of Friction Coating

The hardness value of friction surfacing products: consumable rod, friction coating, and the residual rod (Figure 6) are shown in Figure 7. The friction coating deposited from the consumable rod made of Al-cor showed a significant increase in hardness value. On the other side, there were no significant changes in hardness values of friction coatings for both of the consumable rods composite. The increasing hardness of friction coating was due to homogenous fine-grained coating formed during the visco-plastic deformation [14] and may eliminate the defect or imperfection formed during solidification. Meanwhile, the Al-CNTs composites have only light respond with thermo-mechanical force. In fact, the enhancement of mechanical properties for Al-CNTs composite relies on the homogenous dispersion of the CNTs [10]. The stir casting method used to produce the Al-CNTs composite rods has already well dispersed the CNTs in the aluminum matrix. Therefore, the effect of friction surfacing on distributing CNTs homogeneously was not significant. As consequent, the hardness values did not significantly alter. In addition, the hardness of all rods (residual rod) after the friction surfacing process was decreased. Residual rods experienced conductive heating from the friction process. This may cause the strengthening precipitate dissolved to the aluminum matrix that brought about softening the rods.

4. Conclusion

Aluminum reinforced with carbon nanotubes has good hardenability through the forging process. Its hardness improved significantly. Meanwhile, its hardness was not altered too much by the process of friction surfacing.

Acknowledgment

The research team at the department of mechanical engineering Universitas Hasanuddin thanks to the JICA C-BEST for the support and sponsoring the research with the topic of the enhancement of wear properties of lightweight materials for railway application using aluminum composite coating reinforced with carbon nanotubes fabricated by friction surfacing process.

References

[1] L Huei-Long, L Wun-Hwa, and Sammy Lap-Ip Chan 1992. Abrasive wear of powder metallurgy Al alloy 6061-SiC particle composites, *Wear*, vol **159** no 2 pp 223–231.

[2] R. Sahil and R, Rakshaskar 2016. Mechanical Property Analysis of Aluminum Alumina Composite By Varing the Percentage of Alumina 1 Bartakke Nikhil N *Int. J. Mech. Prod. Eng. ISSN 2320-2092,* vol. Vol **4**, no 6 pp 42–45.

[3] A. E. Jiménez, M. D. Bermúdez, J. Cintas, and E. J. Herrera 2009. Dry wear of NiAl3-reinforced mechanically alloyed aluminium with different microstructure *Wear*, vol **266** no 1–2 pp 255–265.

[4] M Singh, B K Prasad, D P Mondal, and A K Jha 2001. Dry sliding wear behaviour of an aluminium alloy-granite particle composite *Tribol. Int.*, vol **34** no 8 pp 557–567.

[5] J Hemanth, 2002. Wear behavior of chilled (metallic and non-metallic) aluminum alloy-glass particulate composite *Mater. Des.* vol **23** no 5 pp 479–487.

[6] C S Ramesh, A R A Khan, N Ravikumar, and P Savanprabhu 2005. Prediction of wear coefficient of Al6061-TiO2 composites *Wear* vol **259** no 1–6 pp 602–608.

[7] S Nath, S Pityana and J Dutta Majumdar 2012. Laser surface alloying of aluminium with WC+Co+NiCr for improved wear resistance *Surf. Coatings Technol.* vol **206** no 15 pp 3333–3341.

[8] S H Choo, S Lee and S J Kwon 1999. Surface hardening of a gray cast iron used for a diesel engine cylinder block using high-energy electron beam irradiation *Metall. Mater. Trans. A Phys. Metall. Mater. Sci.*, vol **30** no 5 pp 1211–1221.

[9] M Gui and S B Kang 2000. 6061Al/Al-SiCp bi-layer composites produced by plasma-spraying process *Mater. Lett.* vol **46** no 5 pp 296–302.

[10] Q Liu, L Ke, F Liu, C Huang, and L Xing 2013. Microstructure and mechanical property of multi-walled carbon nanotubes reinforced aluminum matrix composites fabricated by friction stir processing *Mater. Des.* vol **45** pp 343–348.

[11] L A B Mabhali, L Pityana, and N Sacks 2010. Laser surface alloying of aluminum (AA1200) with Ni and SiC powders *Mater. Manuf. Process.* vol 25 no **12** pp 1397–1403.

[12] A M Al-Qutub, A Khalil, N Saheb, and A S Hakeem 2013. Wear and friction behavior of Al6061 alloy reinforced with carbon nanotubes *Wear* vol **297** no 1–2 pp 752–761.

[13] R Arreola-Herrera, A Cruz-Ramírez, M Á Suárez-Rosales, and R G Sánchez-Álvarado 2014. The effect of cold forming on structure and properties of 32 CDV 13 Steel by radial forging process *Mater. Res.* vol 17 no **2** pp 445–450.

[14] D Yadav and R Bauri 2012. Effect of friction stir processing on microstructure and mechanical properties of aluminium *Mater. Sci. Eng. A*, vol **539** no March pp 85–92.

Inventories Analysis of Animal Feed Raw Materials by Using the Silver Meal Method and Wagner within Algorithm (Case Study of PT. XYZ Makassar)

S Asmal[1*], I Setiawan[1], N Ikasari[1], and Y Adriani[1]

[1]Department of Industrial Engineering, Faculty of Engineering, Hasanuddin University, Makassar, Indonesia

*Email: saptaasmal@yahoo.com

Abstract. PT. XYZ Indonesia is a national company engaged in animal feed production. through the Production Planning Inventory Control (PPIC) department has implemented the Material Requirement Planning (MRP) system for inventory control but in the implementation of the company's MRP system, the PPIC department only considers the quantity of raw materials that can meet demand without regard to the issue of lot size inventory efficiency, so that the cost of raw material inventory is high. The purpose of this research is to compare the cost of procuring raw materials that have been carried out by the company with the cost of procuring raw materials using the silver meal method and Wagner Within algorithm. From the results of research and analysis conducted a comparison between the methods used by the company with the lot size silver meal method and Wagner Within algorithm in order to obtain the total inventory cost of raw materials in 2016, where yellow corn and rice bran raw materials have the same inventory costs between the methods used by the company, the silver meal method and the Wagner Within algorithm method are Rp. 1,080,000 so they do not have cost savings. As for the inventory of lime stone raw materials, it is more optimal to use the silver meal method and Wagner Within algorithm with a total inventory cost of Rp. 606,394 and can save costs by 35.08%.

1. Introduction

In a manufacturing industry, the availability of raw materials becomes very important to support production activities [1]. The process of procuring raw materials and their availability can support the smooth production process. Inventory Control section in this case plays an important role in controlling raw material inventory, therefore a good analysis is needed in determining the quantity and time of ordering raw materials. Determination of quantity and good order time will not interfere with the smooth production and can minimize the total inventory cost.

According to [2] inventory is a general term that indicates everything or organizational resources that are stored in anticipation of fulfilling demand. Without inventory, the entrepreneur will be faced with the risk that his company will not be able to meet the demands of its customers. Therefore, inventory has a meaning and a very important role in the company. Meanwhile, according to [3] states that, the final goal of inventory management is to minimize costs in changing inventory levels.

In this study will be discussed about the supply of raw materials at PT. XYZ Makassar Branch where the product produced is animal feed. This company produces many types of animal feed products ranging from feed broilers, breeders, hobbies and others. The types of feed produced are in the form of pellets (granules) and crumble (fine grains). To produce these products, PT. XYZ Makassar Branch requires a diverse supply of raw materials such as Yellow corn, Rice bran, Lime stone, and others. The marketing area of PT. XYZ Makassar Branch is Kalimantan and eastern Indonesia. With this wide marketing area, the supply of raw materials is very important especially for limited materials and can only be imported from outside the region or from abroad. The business risks are availability and fluctuation of raw material prices. So far, PT. XYZ Makassar Branch through the PPIC department implements the MRP system for inventory control which starts from forecasting demand. The forecasting process carried out by the company is the result of coordination between the marketing department who knows the market conditions with the PPIC department. Forecast by the company is not able to describe the actual condition of the product demand. Inaccurate calculation of forecasts can cause a lack of raw material stock in warehouses that will cause losses that disrupt the production process and also result in loss of opportunity to earn profits if it turns out that the demand for the actual conditions exceeds the demand predicted and also the excess raw material stock which causes high holding costs .

In implementing the company's MRP system, the PPIC department only considers the quantity of raw materials that can meet demand without regard to the issue of lot size inventory efficiency. In addition the company also has never evaluated and compared the existing inventory system with other methods.

From the description above, the researcher is interested in conducting a study of controlling the supply of animal feed raw materials in the company by using the method of comparing the cost of procuring raw materials that have been carried out by the company with the cost of procuring raw materials using the silver meal method and the Wagner Within algorithm for the company.

2. Study Literature

Material Requirement Planning (MRP)

MRP system is a system of planning and scheduling material requirements that require several stages of the process. The production plan for the final product is translated into raw material requirements by calculating the waiting time so that it can be determined when and how much raw material must be ordered for the product to be made [2]

To implement MRP, basically there are four basic prerequisites that must be met, namely:

1. Availability of master production schedule.
   The master production schedule is a detailed plan about the amount of goods to be produced at several time units in the planning horizon. The master production schedule is a cost optimization that takes into account the capacity available in the demand forecast to achieve production which will minimize the total production cost.
2. Product structure in the bill of material
   Each product unit and component must have a clear and unique identification so that it is useful during computerization. The product structure contains information about the relationships between components in assembly. This information is important in determining the gross needs and net requirements of a component. Furthermore, the product structure also contains information about all units.
3. Clarity and accuracy of inventory records.
   The MRP system is based on the accuracy of the inventory held so that the decision to make or order goods at one time can be done as well as possible. For this reason, inventory levels of components and materials must always be observed. MRP is not possible without accurate inventory.
4. Lead time.
   Waiting time is the time needed from the time the unit order is made until the time the unit is received and ready for use, both the product unit that must be made by itself or the product units ordered from

outside the company. The MRP system can be implemented well if the waiting time for ordering is known.

The Silver Meal Method

The Silver-Meal method was developed by Edward Silver and Harlan Meal based on the cost period. Silver Meal is one of the heuristic methods, which is a method with an approach that is easy to use, and from repetition of work will get good results when compared with other heuristics. The Silver Meal method is used as a request as a basis for repetition of variables in subsequent periods, then the total demand is above the planning limit.

The criterion of the Silver-Meal technique is that the lot size chosen must be able to minimize the total cost per period. Requests for successive periods are accumulated into a lot size (tentative lot size) until the total carrying cost and setup cost of the lot divided by the number of periods involved increases. This method tries to find a minimum average cost for each period for a number of planned periods [4]. General formulas that can be used are as follows:

$$K(m) = 1 / m (A + h + 2hD3 + ... + (m-1) h \quad (1)$$

Calculate K (m), m = 1,2,3, ..., m, and stop the count if K (m + 1)> K (m) Note:
Dm = Demand in the month period (D1, D2, D3, ..., Dm)
K (m) = Average inventory cost per unit time
m = Period
A = Order cost
h = Cost of saving each unit / period

This Silver Meal method is used for problems where the variation in demand from one time period to the next is quite high.

Simply stated, the steps in the calculation use the Silver Meal Method as follows:

- Step I:
  Calculating the total cost for ordering a first time period, also calculating the costs for ordering the first two time periods. If the second cost is higher than the first cost, then it is concluded that the order is made for the first time period. But, if the second fee is not greater than the first cost or the second fee is smaller than the first cost, then proceed to the calculation of the total cost for ordering the first three time periods. If this third fee is higher than the second cost, then the booking is concluded for the second time period. But if the third fee is not greater than the second fee or the third fee is smaller than the second fee, then proceed to the calculation of the total cost for ordering the first four time periods. And so on until the condition is obtained where the total cost for the next order is higher than the total cost for the previous order.
- Step II:
  If there are still time periods that have not been calculated, then step 1 is repeated for the remaining time periods.

It is generally seen that this method has longer calculations than the other lot sizing methods, but the researchers conclude that this method is closer to optimal. This is a common symptom in management science that the better a heuristic method, the longer and more expensive the calculation, but still simpler and cheaper than the method for finding the best solution.

Wagner Within (AWW) Algorithm Method

The Wagner-Within Algorithm method is an optimization based on a dynamic programming model [5]. The goal is to obtain an optimum ordering strategy for the entire schedule of net needs, by minimizing total procurement costs and savings costs. This algorithm was developed by Wagner and Within in 1958 to provide the optimum solution for the problem of deterministic order sizes at a certain time period where the needs of all periods must be met. Number of orders and time of order remains.

This method uses two ways to determine the number of orders, the first by ordering adjusted to the number of economic orders (must be close to the number of economic orders) by communicating the demand every month until approaching the number of economic orders after that do a trial and error until getting the lowest cost. The second method uses Wagner- Within algorithm calculations.

The Wagner-Within algorithm obtains a maximum number of solutions to the data that minimizes the problem of dynamic order size over limited planning. Several theories are used to simplify the calculation by following the steps in the following procedure:

- Step 1:

  Calculate the total cost of the matrix variable for all alternative horizon bookings for the time being which consists of N periods. Total variable costs include ordering and holding costs. Describe CEZ as the total variable cost in periods c through e placing an order in period c which creates needs in periods c to e:

$$ZCE = C + \text{Hp} \sum_{k=c}^{e} (Qce - Qct \; For \; 1 \leq c \leq e\text{l} \leq N \qquad (2)$$

e

Where:
C: Per-order Message Cost
h: Save Cost
P: Unit Purchase Cost

$$\text{Qce} = \sum_{k=c}^{e} R_k \qquad (3)$$

Rk: Value of Request in period k

- Step 2:

  Describe fe as the minimum cost in periods 1 through e, that the inventory level at the end of period e is zero. The algorithm starts with f0 and calculates f1, f2, fn on the order. Then fe is calculated in the order of the order upwards using the formula.

$$F_e = \text{Min} \; (Z_{ce} + f_{c-1}) \text{ for } c = 1,2,\ldots\ldots, e, \qquad (4)$$

  In other words, because in each period all alternative bookings are combined and compared in the replacement strategy fe. The best (lowest cost) combination is like the fe strategy to meet the needs for periods 1 to e. So the value of fe is the optimal cost of the order schedule.

- Step 3:

$f_e = Z_{w\,N} + f_{w-1}$ The last sequence occurred in the period w and sufficient to satisfy requests w to n.

$f_{w-1} = Zv_{w-1} + f_{v-1}$ Prior to the last order sequence occurred in period v and sufficient to meet demand in periods v to w 1.

$f_{u-1} = Z_{1u-1} + f_o$ The first order occurred in period 1 and is sufficient to meet demand in periods 1 through u 1.

3. Research Method

In planning the supply of raw materials, the company not only considers the quantity of raw materials that can meet demand, but must consider the problem of determining the size of the lot in order to produce an optimal order at a low cost. In calculating the lot size, there are various techniques which are divided into two major groups, namely the static lot sizing model and the dynamic lot sizing model. Where in static lot sizing there are several methods used, one of them is the Economic Order Quantity (EOQ) method. Whereas in dynamic lot sizing there are several methods used, including the Silver meal and AWW methods. For the level of demand with numbers up and down (random) dynamic lot sizing models are used, including the silver meal and AWW methods. The silver meal and AWW methods have not been widely used but can produce solutions that are close to optimal. After calculating the lot size of raw materials, it can be done planning raw material inventory.

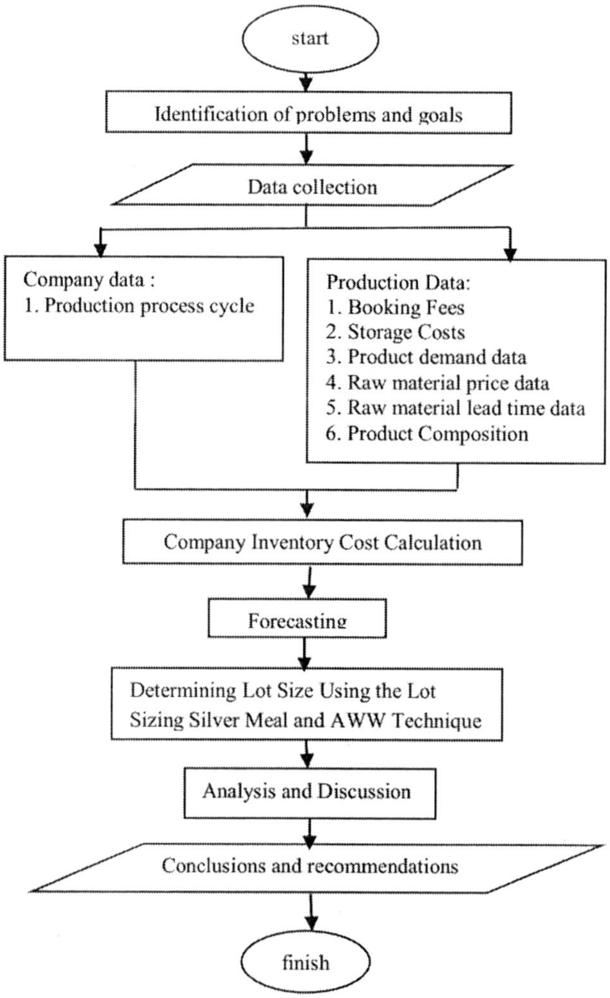

Figure 1. Schematic of a research framework

4. Result and Discussion

Table 1. Product demand data for January - December 2015

No.	month	Broiler (Ton)	Breeder (Ton)	Hobby (Ton)
1	January 2015	2,688.20	2,755.00	20.00
2	February 2015	2,351.75	2,410.00	20.00
3	March 2015	3,231.20	1,995.00	20.00
4	April 2015	3,556.20	1,742.50	20.00
5	Mei 2015	3,201.10	1,287.00	25.00
6	June 2015	3,701.90	1,851.60	25.00
7	July 2015	2,879.65	1,585.00	25.00
8	August 2015	2,895.00	1,540.00	25.00
9	September 2015	3,000.00	1,940.00	30.00
10	October 2015	3,495.00	70.00	35.00
11	November 2015	3,164.60	470.00	30.00
12	December 2015	3,077.50	705.00	35.00

The pattern of data on demand for broiler products from January 2015 to December 2015 presented in graphical form can be seen in Figure 2:

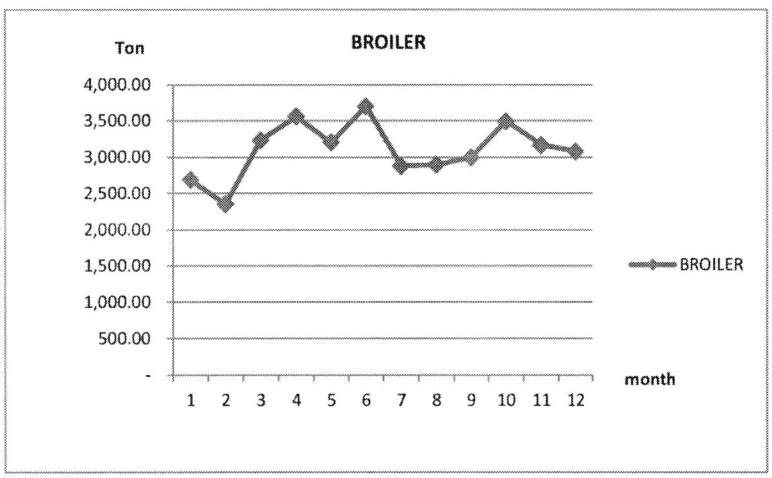

Figure 2. Pattern of Broiler feed product demand for January - December 2015

In Figure 2, it showed that the demand for Broiler feed products is random. where the shape of the data pattern cannot be known because the demand for broiler feed every month increases or decreases. The data patterns of demand for breeder feed products from January 2015 to December 2015 presented in graphical form can be seen in Figure 3:

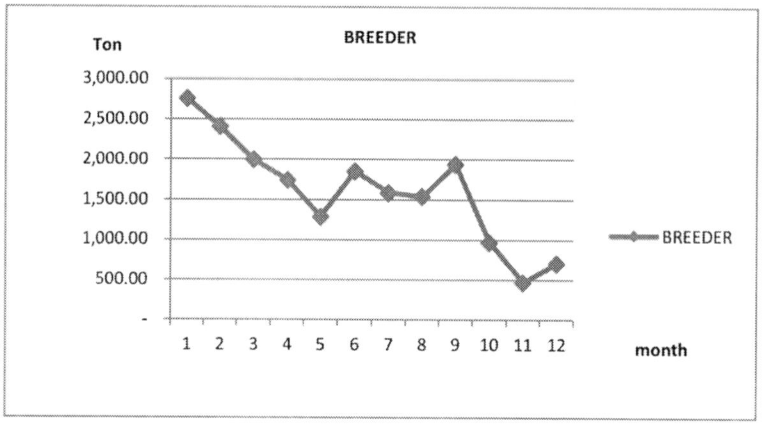

Figure 3. The pattern of demand for breeder feed products January 2015 - December 2015

In Figure 3 it showed that the demand for Breeder feed products is also random, it appears that there is a fluctuation in product demand every month. This can be seen in the month of June 2015 breeder feed demand increased and then decreased in July to August 2015 then in September 2015 demand again increased and decreased in October to November 2015 and increased again in December 2015.

While the pattern of demand data for Hobby feed products from January 2015 to December 2015 presented in graphical form can be seen in Figure 4. :

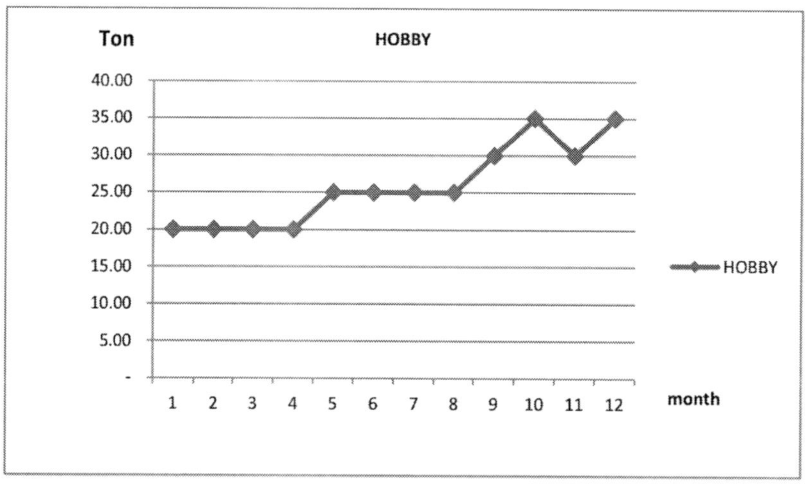

Figure 4. Patterns of demand for Hobby feed products January 2015 - December 2015

In Figure 4. The graph showed that the demand for Hobby feed products is random (random). This can be seen in May 2015 Hobby feed demand increased and demand remained / stable until August 2015 then again increased in September to October 2015 and again declined in November and increased again in December 2015.

Product composition is used to determine the amount of raw materials contained in each product. In this research, three main raw materials are discussed, namely Yellow Corn, Rice Bran, Lime Stone. The composition used can be seen in Table 2 :

Table 2. Composition of animal feed products

Raw material	Feed		
	Broiler (% use)	Breeder (% use)	Hobby (% use)
Yellow corn	55.17	63.98	47.27
Rice bran	6.86	5.00	4.96
Lime stone	3.00	4.50	5.00

source: PPIC Dept. PT. CPI Makassar

Inventory Costs applied by the Company PT. XYZ

a. Cost of ordering Yellow corn, Rice bran and Lime Stone raw materials

Cost of ordering raw materials at PT. Makassar CPI consists of telephone costs. For Yellow corn, Rice bran and Lime stone transportation costs are borne by the supplier. Data on ordering costs of Yellow corn, Rice bran, and Lime stone raw materials can be seen in Table 3:

Table 3. Cost of ordering Yellow corn, Rice bran and Lime stone raw materials During 2016 the company method

Raw material	Booking Fee / month (Rp)	Order Frequency / year	Total Booking Fee (Rp)
Yellow corn	90,000	12	1,080,000
Rice bran	90,000	12	1,080,000
Lime stone	90,000	4	360,000
Total	**270,000**	**28**	**2,520,000**

b. Cost of purchasing Yellow corn, Rice bran and Lime stone raw materials

High and low costs of purchasing raw materials is influenced by the order quantity. Where the purchase cost is multiplication of the order quantity with the price / ton of raw materials. Data on the cost of purchasing Yellow corn, Rice bran, and Lime stone raw materials can be seen in the Table 4

Table 4. Cost of purchasing Yellow corn, Rice bran and Lime stone raw materials for the company 2016 method

Raw material	Order Quantity (ton)	Price / ton (Rp)	Purchase cost
Yellow corn	33,135	4,500,000	149,109,255,000
Rice bran	3,542	2,500,000	8,855,025,000
Lime stone	1,008	500,000	504,025,000
Total	**37,685**	**7,500,000**	**158,468,305,000**

c. The cost of storing raw materials for Yellow corn, Rice bran and Lime stone.

Raw material storage costs consist of material handling / maintenance costs. Data on the cost of storing raw materials for Yellow corn, Rice bran, and Lime stone can be seen in the Table 5:

Table 5. The method of storing raw materials for Yellow corn, Rice bran, and Lime stone in 2016 is the company's method

Raw material	Total Inventory	Storage costs / ton (Rp)	Total storage costs / year (Rp)
Yellow corn	0	20,000	0
Rice bran	0	12,000	0
Lime stone	2,871	200	574,204
Total	**2,871**	**Rp 32,200**	**574,204**

d. Inventory costs for Yellow corn, Rice bran and Lime stone.

Inventory costs of Yellow corn, Rice bran, and Lime stone raw materials with the company method in 2016 are the total of the total cost of ordering with the cost of storing raw materials in 2016. Data on inventory costs of Yellow corn, Rice bran, and Lime stone can be seen in Table 6 :

Table 6. Cost of raw materials for Yellow corn, Rice bran and Lime stone in 2016 is the company's method

Raw material	Order cost/year (Rp)	Holding cost/year (Rp)	Total inventory cost / year (Rp)
Yellow corn	1,080,000	0	1,080,000
Rice bran	1,080,000	0	1,080,000
Lime stone	360,000	574,204	934,204
Total	**2,520,000**	**574,204**	**3,094,204**

Comparative Analysis of Inventory Control Methods

Table 7. Comparison of the cost of ordering raw materials

Raw material	Company Method (Rp / year)	Silver meal method (Rp / year)	*Algoritma wagner within method* (Rp/year)
Yellow corn	1,080,000	1,080,000	1,080,000
Rice bran	1,080,000	1,080,000	1,080,000
Lime stone	360,000	450,000	450,000

Comparison of raw material inventory costs

Raw material inventory costs are the sum of ordering costs and storage costs so that the amount depends on the two components. The comparison of Yellow corn, Rice bran and Lime stone raw material inventory costs between the company method, the silver meal method and the Wagner Within algorithm can be seen in Table 8:

Table 8. Comparison of raw material inventory costs

Raw material	Company Method (Rp / year)	Silver meal method (Rp / year)	*Algoritma wagner within method* (Rp/year)
Yellow corn	1,080,000	1,080,000	1,080,000
Rice bran	1,080,000	1,080,000	1,080,000
Lime stone	934,204	606,394	606,394

Table 9. Percentage of cost savings in controlling inventory of raw materials using the silver meal method, Wagner Algorithm within the company method

Raw material	Silver meal method (Rp / year)	(%)	Company Method (Rp / year)	(%)	*Algoritma wagner within method* (Rp/year)
Yellow corn	1,080,000	0	1,080,000	0	1,080,000
Rice bran	1,080,000	0	1,080,000	0	1,080,000
Lime stone	606,394	35.08	934,204	35.08	606,394

Based on Table 9 above, it can be seen that for Yellow corn and Rice bran raw materials have the same inventory costs between the methods used by the company with the silver meal method and Wagner Within algorithm method so that there is no cost difference while for Lime stone raw materials cost savings occur using the Silver meal method and Wagner Within algorithm method is 35.08% of the company's method. This proves that the Silver meal method and Wagner Within Algorithm method can be an alternative for companies in managing Lime stone raw material inventory policies whose demand fluctuates because this method can meet raw material requirements precisely and with minimum inventory costs.

5. Conclusion

Based on the results of data processing and analysis conducted, it can be concluded as follows:
The Silver meal method and Wagner Within Algorithm method, a lot size recommendation is obtained for ordering the three main raw materials. Where the total inventory of Yellow corn raw materials is 33,135.39 tons with an order frequency of 12 times a year. The total inventory of rice bran raw materials is 3,542.01 tons with a frequency of ordering 12 times a year. The total inventory of Lime stone raw materials is 2,746.31 tons with an order frequency of 5 times a year. Comparison of raw material inventory costs between the methods used by the company with the Silver meal method and Wagner Within algorithm as follows: a). Ordering costs for, Silver meal, and Wagner Algorithm within Rp 1,080,000 while for lime stone raw materials with the company method amounting to Rp 360,000 and using the Silver meal method, and Wagner within algorithm Rp. 450,000. b). The storage cost for Silver meal, and Wagner Algorithm within Rp. 0 while for lime stone raw materials, the company method is Rp. 574,204 and using the Silver meal method, and the Wagner Within algorithm. Rp. 156,394, c). Inventory costs for Silver meal, and Wagner Algorithm within Rp 1,080,000 while for lime stone raw materials the company method is Rp 934,204 and using the Silver meal method and Wagner Algorithm within Rp. 606,394. And from the calculation of inventory costs, that the exact lot size method of lime stone used is the Silver meal method and Wagner Within Algorithm method due to inventory cost savings of 35.08% compared to using the method used by the company.

Acknowledgment

Thank you for the engineering faculty of Hasanuddin University for providing funding assistance for this research activity through the Learning Based Education (LBE) program.

References

[1] Pramana, F. G. (2011). Application of ABC Analysis in Fabric Raw Material Inventory Control at PT. Batik Danar Hadi Surakarta Printing Division. Universitas Sebelas Maret. (in Bahasa)
[2] Handoko T. Hani. 1984. Basics of production management and operations First edition. Faculty of Economics, Gajah Mada University, Yogyakarta.
[3] Yamit Zulian. 2003. Production and Operations Management. Faculty of Economics, Gajah Mada University, Yogyakarta.
[4] Rangkuti, F. (2007). Inventory Management: Applications in the Business Field. Issue 2. Jakarta: PT. Raja Grafindo Persada
[5] Tersine, R. J., 1994, Principles of Inventory and Material Management, Edition, Englewood Cliffs, New Jersey: Prentice-Hall, Inc.,

Fatigue Strength Improvement of Low Carbon Steel through Carburizing Process with Coconut Shell Charcoal

M Syahid[1]*, A Hayat[1], S Arief[1], and Rudi[1]
[1]Department of Mechanical Engineering, Engineering Faculty, Hasanuddin University, Makassar Indonesia

*Email: syahid@unhas.ac.id

Abstract. The fatigue strength of low carbon steel is a limitation that is often considered in machine construction, especially in rotating component parts. Low carbon steel has special properties that are soft, good welding and good machining, as well as relatively inexpensive prices but have low fatigue strength. For this reason, in this study, a carburizing process was carried out with the media of coconut shell charcoal as a source of carbon and shell powder as a catalyst. The addition of carbon using the carburizing method is relatively easy and does not require a large cost because the materials used are abundant in nature. The carburization process was carried out at 950°C for 4 hours. Furthermore, mechanical properties were tested with hardness, tensile testing and fatigue testing. The carburization process with charcoal media is expected to increase hardness, tensile strength and fatigue strength. Fatigue testing is carried out by the reverse bending method with a load varying 70%, 60%, 50% and 45% of the tensile strength of the material. In addition, microstructure examination and fault surface observation are also carried out to analyze the phenomenon of changes in material structure due to carburization. The hardness value of ST37 steel before carburizing ie 118 HRB increased to 127 HRB after undergoing carburizing treatment. Tensile test results also experienced an increase in which the initial tensile strength value of ST37 steel by 356.66 MPa increased to 541.15 MPa after the carburizing process. Likewise in fatigue testing, the fatigue limit has increased significantly, where the ST37 steel fatigue limit before the carburizing pack is at a stress level of 161 MPa. After experiencing the carburizing pack process the ST37 steel fatigue limits are at a voltage level of 232 MPa.

1. Introduction

Steel is always a top priority in the selection of materials of construction machines because it has good mechanical properties. The hard and ductile nature of steel is a factor that is always taken into consideration in planning machine construction, especially in rotating component parts. For example, steel is often used in cutting tools, chisels, gears or construction machinery. The nature of steel is generally influenced by the content of carbon, other alloying elements and microstructure [1, 2]. The difference in carbon content in steel is one way to classify steel. Based on its carbon content, steel is divided into three types namely high carbon steel, medium carbon steel and low carbon steel. Low carbon steel has special properties that are soft, good welding and good machinery, as well as relatively inexpensive prices but has low fatigue strength [3].

Fatigue failure is a form of failure that occurs in the structure of the material due to dynamic loads that fluctuate below yield strength that occurs in a long time and repeatedly [3, 4]. One example of a form of material failure is a fracture. Generally, the fracture begins with a stress concentration on the crack on the surface of the material. Because the load is fluctuating, the cracks propagate then breakage occurs. Based on the fracture mechanism, the surface condition on the specimen is one of the factors causing the fatigue strength of the material [5-7]. The surface condition in question is the surface hardness of the material, where the higher the hardness of a surface, the possibility of cracks or defects will decrease. Surface hardness can be increased by carburizing and heat treatment [8].

In high carbon steel and medium carbon steel, the surface hardness can be increased by heat treatment. Whereas in low carbon steel, the hardness of the surface cannot be increased by heat treatment it is because of the low carbon content. The way to increase surface hardness in low carbon steel is by carburizing. Carburizing is a process of adding carbon elements to the surface of steel by diffusion to improve its physical and mechanical properties [8].

The carburizing process is relatively easy and inexpensive because the materials used in the carburizing process are available abundantly in nature. Natural materials that are often used as carburizing media are wood charcoal, coconut shell charcoal, used battery charcoal, mangrove tree charcoal and rice husk charcoal. In this study, we use coconut shell charcoal as an active carbon element and shells (calcium carbonate) as a catalyst. Based on this, the researcher wanted to analyze how "Increasing the Strength of Static Steel 37 through the carburization process with coconut shell charcoal media.

2. Research Method

The material used in this study is ST37 steel in the form of rods. The material is cut to become a tensile test specimen in accordance with ASTM E-8 standard and a fatigue test specimen according to ASTM E-466 standard. Then the specimen is carburized using coconut shell and shell media as a catalyst. The carburizing treatment is carried out by inserting the specimen into the carburizing box then inserting the coconut shell charcoal powder mixed with shell powder (10% of the total weight of the coconut shell charcoal powder) until it covers the entire surface of the specimen and then closes the carburizing box again with the lid. After that the carburizing box is put into a heating furnace at a temperature of 950^0C held for 4 hours then the specimen is quenched in seawater media at room temperature. Furnace and carburization boxes can be seen in Figure 1. Furthermore, hardness testing, tensile testing and fatigue testing are performed (Figure 2). Also examined the microstructure and fracture surface using an optical microscope

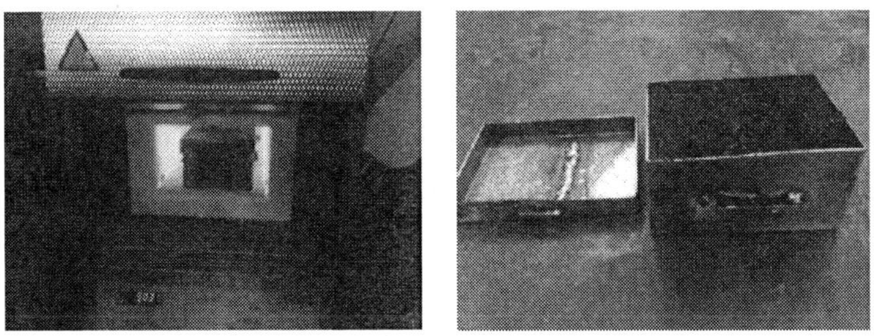

Figure 1. a) Furnace b) Pack carburizing

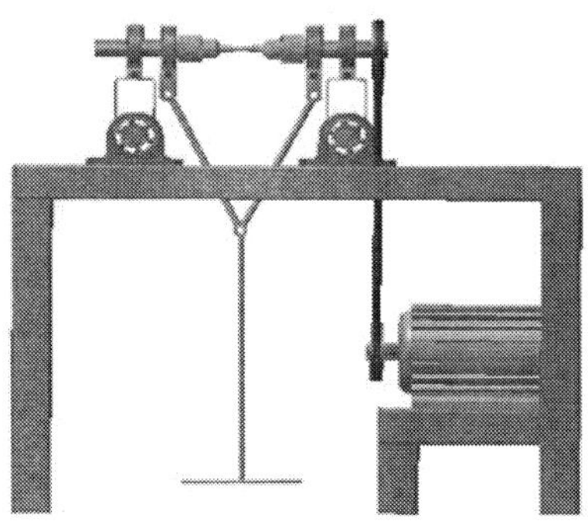

Figure 2. Three Point Bending reverse fatigue machine

3. Result and Discussion

3.1. Effect Carburizing on Hardness Value

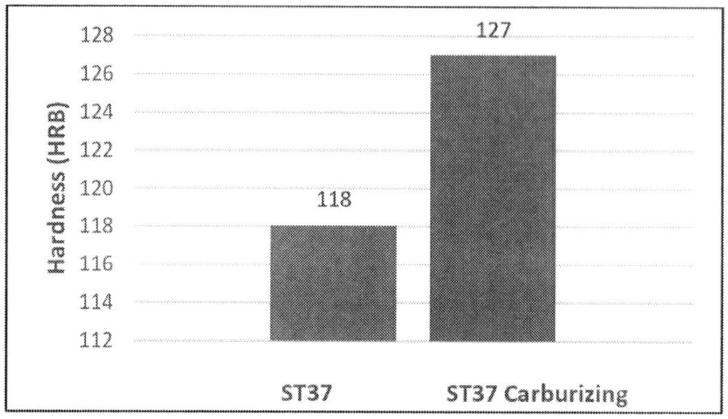

Figure 3. Effect of carburizing on hardness of low carbon steel ST37

Figure 3 shows the value of ST37 hardness before and after the carburizing process. It can be seen that the value of ST37 hardness increases after a carburizing process. This is due to the diffusion of carbon which causes the addition of carbon atoms in the material from the outer side to the inside and continues when the carburizing process is carried out with a temperature of 950°C. It is clear that the hardness of the material is very dependent on the amount of carbon which means more and more carbon content in a material, the harder the material is [2]. In addition, the cooling process also affects the hardness of the ST37 material where a rapid cooling process is carried out in order to obtain a martensitic structure. Seawater is a very fast and evenly distributed cooling medium because seawater contains salinity. The salt will settle around the specimen and tend to prevent/shorten the formation of the vapor layer. This formed precipitate causes the coolant to come into direct contact with the surface of the specimen so that the cooling rate becomes very effective and quickly absorbs heat in high-temperature regions. The effective definition here is that carbon steel is cooled at a rate of cooling that can prevent a softer structure or get a hard, even structure throughout the surface. Thus the effective cooling rate of seawater is at a slower cooling rate in the temperature range of martensitic formation in ST37 material and evenly distributed across the surface. The initial hardness of the ST37 material before the pack carburizing process was an average of 118 HRB and after experiencing the carburizing pack process using coconut shell charcoal media at a temperature of 950°C for 4 hours using seawater as a quenching medium, the hardness value increased to 127 HRB.

3.2. Microstructure Examination

Figure 4 is a photo of the ST37 microstructure before and after the carburizing process. The figure shows that there is a difference in color, which is the color of Figure 4 (b) has a darker color than the color of Figure 4 (a). This is due to the presence of carbon atoms which diffuse into iron at austenite temperature (carburizing temperature). Diffusion of carbon atoms results in the formation of cementite consisting of three (3) Fe atoms binding one (1) atom C. This bond forms new cementite, the growth of cementite mixed with ferrite into pearlite crystals. Rapid cooling of specimens results from the formation of the martensitic phase. The more carbon elements, then the martensitic structure that is formed will also be more and more [2]. This is because the atoms which are separated from the bonds do not have time to be distributed into the bonds to bind the atoms making up the metal and the atoms that are enlarged not to have time to shrink. So here there is a very fast recrystallization process. With a structure like this, it causes a bond that is not strong between atoms with other atoms so that the nature is brittle and hard because the granules are enlarged to fill the material space.

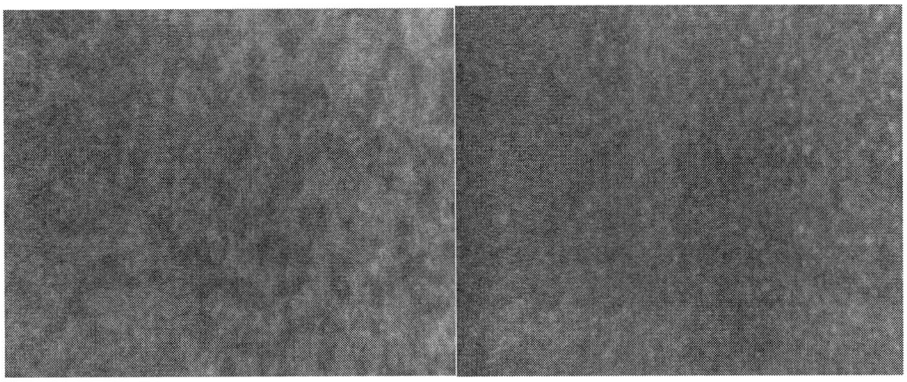

Figure 4. The photographic microstructure of low carbon steel ST37 : (a) Before carburizing; (b) After carburizing

3.3. Effect of Carburization on the Results of the Tensile Test

Figure 4 is a graph of the tensile strength values before and after the pack carburizing process. The results show that the tensile strength value of ST37 material has increased after the carburizing pack process. This is due to an increase in the value of the ST37 hardness after undergoing a carburizing process. The hardness of a material is directly proportional to the tensile strength because the understanding of violence and tensile strength are the same. Equally means material resistance to plastic deformation. It's just that violence is material resistance to plastic deformation locally (surface) while tensile strength is material resistance to plastic deformation that occurs throughout the surface of the material. So that if a part of the material is resistant to plastic deformation, then automatically all parts of the material will be resistant to plastic deformation. The harder material is the higher tensile strength. The average tensile strength value of ST37 is 356.66 N/mm^2 and after undergoing a carburizing process the tensile strength value becomes 541.15 N/mm^2. As a study conducted by Zuchry et al with the results that ST37 which undergoes a pack carburizing process at temperatures of 900°C and 950°C for 3 hours, 6 hours and 9 hours has increased the value of tensile strength. However, in general, an increase in the tensile strength of material causes strain to decrease [1].

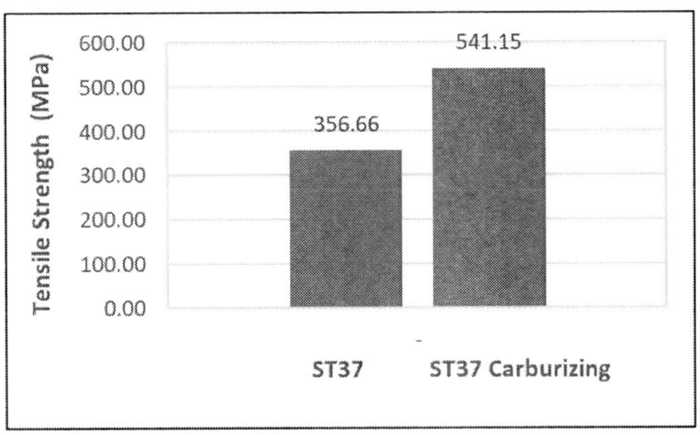

Figure 5. Effect of Carburizing towards low carbon steel ST37 tensile strength

Figure 5 is a ST37 stress-strain chart before and after the carburizing pack process. The results show that tensile stress is inversely proportional to strain. This is because the increased hardness value causes the material to become more brittle after the carburizing process due to the formed martensitic phase.

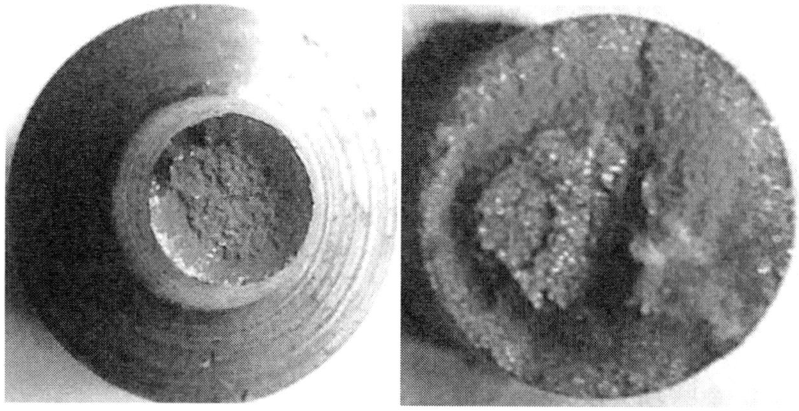

Figure 6. Macro photograph of surface fracture of tensile test specimen ST 37
(a) Without carburizing, (b) Carburizing

Figure 6 is a macro photograph of the fracture surface of the ST37 tensile test material before and after undergoing the carburizing pack process. On the fracture surface of the tensile test specimen (static load), there are two types of fractures namely brittle fracture and ductile fracture. In Figure 6 (a) it is a brittle fracture where the brittle fracture has a characteristic on the fracture that there are no cups and cones. In brittle fracture, necking does not occur so that the specimen is broken immediately if it is given a load above it. In addition, the fracture forms an angle of 900 to the normal axis of the specimen. In Figure 6 (b) it is a ductile fracture, in a ductile fracture there are cups and cones caused by shear stress. The fracture angle forms an angle of 450 to the normal axis of the specimen. Where the tensile load that works raises the maximum stress. If the stress applied to the specimen exceeds the limit, the movement of this dislocation will reach the surface. Dislocation movement until it reaches the surface is called plastic deformation. This plastic deformation which causes length increase in the specimen is fixed. If the amount of stress applied to the specimen reaches the Ultimate point, then the specimen begins to experience local wasting in the middle called the necking [9].

3.4. Effect of Carburization on Fatigue Strength

Figure 7 is a graph of stress vs. the ST37 cycle before and after undergoing the carburizing process. The graph shows that. ST37's fatigue strength increases after undergoing a carburizing pack process. ST37 material fatigue strength before the carburizing pack process is at 161 MPa (45% of the ultimate tensile strength). While the fatigue strength of the ST37 material after undergoing the carburizing process is at 232 MPa (65% of the ultimate tensile strength). This is due to differences in the level of violence in the two test samples. As research conducted by [8] with the result that fatigue strength increased after carbon steel AISI 1020 hardened by the carburizing process followed by quenching [8].

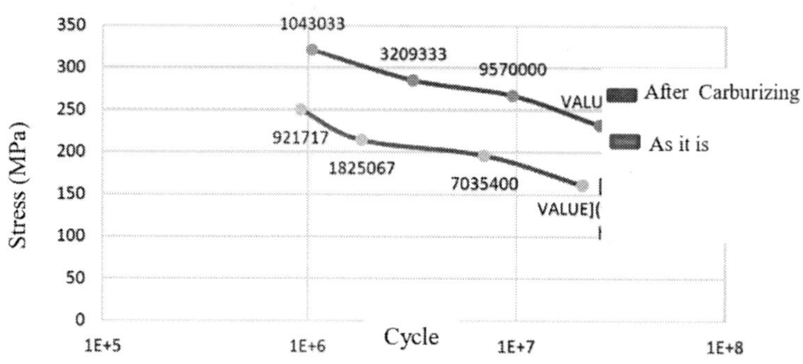

Figure 7. S-N Curve of material ST37 as it is and after carburizing

The higher tensile is fatigue strength. It is because tensile strength and fatigue strength are material resistance to deformation that occurs on the entire surface due to loading. It's just that the loading on the tensile test is a static load while the loading on the fatigue test is a dynamic load. Dynamic loading causes a failure at a voltage much lower than the voltage needed to cause failure at static loading. This is because the fatigue limit stress is half of the ultimate tensile strength value for metals containing ferrous [3].

a) **(b)**

Figure 8. Macro photograph of surface fracture specimen fatigue test of ST37,
(a) Without carburizing (b) With carburizing

Figure 8 is a macro photograph of the surface fracture of the ST37 fatigue material test before and after undergoing the carburizing pack process. In this picture, we can clearly see that the initial crack (initiation crack) occurs at the edge which then continues to propagate (propagation crack) due to continuous loading resulting in fracture (static fracture) because the material is no longer able to accept the load. The area between the crack propagation stage and fracture failure quantitatively can show the amount of stress that works. If the area of the propagation crack stage is greater than the area of the fracture failure, It indicates the working stress is low.

4. Conclusion

Based on the results of research and discussion about the effect of the pack carburizing process on the strength of fatigue of ST37, several conclusions can be drawn as follows: 1). The process of pack carburizing material ST37 at a temperature of 950°C for 4 hours, using seawater as a quenching medium can increase surface hardness by 7.6%. The initial hardness of the ST37 material was 118 HRB and increased to 127 HRB after undergoing a carburizing process. 2). The tensile strength is straight-line with hardness, the initial tensile strength of ST37 material is 356.66 N/mm^2 with a strain value of 8.33% increasing to 541.15 N/mm^2 with a strain value of 4.17%. 3). The fatigue test also shows an increase in fatigue strength where ST37's fatigue strength is at a stress of 161 MPa (45% of the value of ultimate tensile strength) increased to 232 MPa (65% of the value of ultimate tensile strength).

Acknowledgment

The authors would like to thanks to LBE Project of Engineering Faculty of Hasanuddin University as the funding sponsor for the research.

References

[1] Zuchry, M. 2011. Effect of Temperature Carburizing and Hold Time on the Tensile Strength of Carbon Steel with Variations Media Cooling.
[2] Wattimena, W.M.E.. Louhenapessy, Jandri. 2014. The Effect of Holding Time and Quenching on the Hardness of Carbon Steel ST 37 in the Carburizing Pack Process Using Nutmeg Shell (Myristica fagrans).
[3] Akuan, Abrianto. 2007. Metal Fatigue. Universitas Jenderal Achmad Yani, Bandung.
[4] Collins,J.A., 1981, Failure of Material in Mechanical Design, Analysis Predection and Prevention, John Willey & Son, Inc US.
[5] Haftirman, "Fatigue Strength of Steel in High Humidity Environment", Transaction of the Japan Society of Mechanical Engineers, Japan, 1995.
[6] Hisashi Ogawa, An Analysis on the Growth of a Surface Fatigue Crack under Rotary Bending in terms of Fracture Mechanics.
[7] Iwamoto, K. 1989., On the S-N Curve of Carbon Steel Under Rotary Bending Condition in City Water. Transaction of the Japan Society of Mechanical Engineers, Japan.
[8] Sriyanto, Nanang Budi. 2014. Effect of Carburizing Pack Process with Coconut-Barium Carbonate Shell Charcoal on AISI 1020 Steel Fatigue Failure.
[9] Fadllu, Alaya H.M, 2007, The Influence of Metric Thread Notch Depth on Low Carbon Steel Tired Strength, Universitas Negeri Semarang, Semarang.

Six Sigma Implementation in Quality Evaluation of Raw Material: A Case Study

A Darmawan[1]*, S Bahri[1], and A T B Putra[1]
[1]Industrial Engineering Department, Faculty of Engineering, Hasanuddin University, Gowa, Indonesia

*Email: darmawanarmin@gmail.com

Abstract. The study aims to evaluate the quality level of raw material in order to improve its quality. This study is constructed based on the implementation of six sigma methodology. Six Sigma is one powerful tool that used in this study to identify the problem, determine the quality level, and to develop the improvement comprehensively. Five phases of the model has been structured by stages respectively. Six sigma had been implemented in many kinds of industry, particularly in production process phase. However, there is a challenge to implement the methodology in other phase in supply chain in the context of developing country. The implementation was conducted on receiving raw material process part in the industry. The result showed some defects identified dominantly; damage seed, broken seed, insects, and moldy seed. There are 6734.13 possibilities of raw material in experiencing defect per million opportunities (DPMO). Lack of knowledge of the workers about the raw material and lack of supporting equipment are the two main causes of poor quality of receiving raw material process.

1. Introduction

Quality is one important indicator to ensure that product meet the requirements along supply chain in order to satisfy the customer. The awareness about the quality had been increased since some decades ago. Quality can determine overall success of company [1]. The quality management had experienced development from the meaning to ensure the product meet the requirement by product inspection and then go through meet the customer satisfaction. In its journey, many methodologies were also developed as tools in planning, controlling, improving and assuring the quality.

One tool that used generally and relatively powerful in order to enhance the quality of production in industry is Six Sigma. Six Sigma has been recognized as a comprehensive tool in improving the quality of process structurally. This method has been known as primary factors for the success of many industries. TQM has also been supported by six sigma in attaining continues improvement by contributing the developed systems and method. Many scholars have conducted many research regarding the research development and application of Six Sigma in many kinds of industries in developed countries [2]. Factors that impacted in application of six sigma must be concerned by developing countries. Furthermore, the study related to implementation of six sigma and its advantage in a developing country, such as India [3]. Involvement and commitment of top management, training and education, cultural change and industrial infrastructure are factors that successfully contributed in six sigma implementation in developing country (Indonesia). The implementation of six sigma affected positively to the performance of small and medium industries in Indonesia as well [4].

Planning and evaluation in order to improve the quality of production became the important part in production system. It is general that quality evaluation was conducted holistically in production floor (production systems). Moreover, in production floor, the material experienced transformation process to be an added value product that different from its origin. There is quality control along the transformation process of product in ensuring that product meet the specific requirements. Quality control existed as a guardian in keeping and evaluating the quality of product. However, generally in wider scale in supply chain management, beside of production process in manufacturer, supply of raw material also plays important role and vital in supporting production process by contributing high quality raw material. Studies in this context is still limited, however some studies have been developed related to the implementation of six sigma on raw material such as the quality of water as raw material in food production [5], and raw material in pharmaceutical industry [6].

Corn seed is one of the material that most widely used in this industry. For this reason, it is important to pay attention in supply process of material where the raw material distributed from its origin to manufacturer. Raw material referred in this study is corn seed that used to produce fodder for animal.

Figure 1. Supply chain management

As figurVed out in Figure 1, supply chain integrating activities of material from their source to the end user. In this part included purchasing raw material, manufacturing, storage, distribution, retailing, demand planning, supply planning, and supply chain management [7].

This research aimed to minimize the number of defects in the raw material during receiving process, to improve the quality of raw material processing, and to enhance customer satisfaction. Research objectives consist of three parts: mapping the level of quality by identifying, measuring performance dimension of raw material through analysis the process capability, identify the factors that caused the defects, and develop some action plan to improve the quality of raw material.

2. Methodology

Study that related to the implementation of six sigma conducted since late eighties where Motorola used this methodology in the first time in 1986. Six sigma is a tool that extremely important to improve the quality [8]. Particularly, this methodology had been implemented in many kinds of manufacture industry, and then applied in services industry as well [9-11]. Six Sigma is one comprehensive tool used to define the quality in several of ways. Motorola defined six sigma as a quality improvement method in reducing the number of defects to 3.4 parts per million opportunities.

The research was conducted in one industry that produced grain/fodder/forage in Makassar, Sulawesi Selatan. Data was collected by identifying the types of defect, and calculate the number of defects of each type of defect. Furthermore, the result of measuring process then analysed using fishbone diagram. Root caused analysis deeply by interviewing the key people that involved in receiving raw material in the industry.

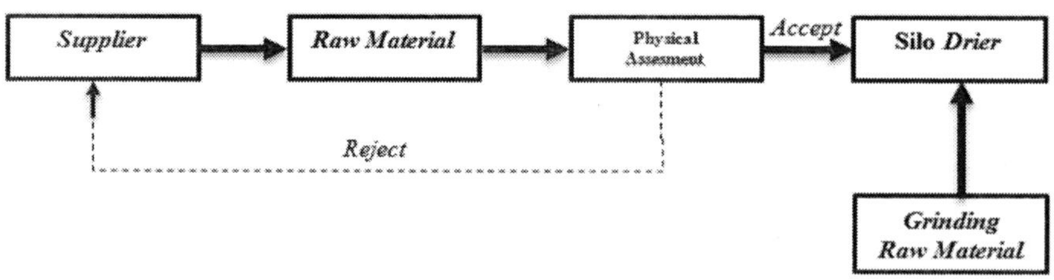

Figure 2. Flow of raw material

IOP Conf. Series: Materials Science and Engineering **875** (2020) 012065 doi:10.1088/1757-899X/875/1/012065

Raw materials are distributed by suppliers from their origin to the factory. In the process of receiving materials as shown in Figure 2, there are several stages, namely: receipt of material from suppliers, carried out tests by physical inspection of raw materials (decision to reject and accept at this phase), the material received is then taken to the silo drier for grinding process. At the physical test stage, raw materials play an important role to determine the quality of raw materials to the next process, some of the quality standards measured in this test are water content, damage/broken seed, moldy seed, other material content beside corn. Raw materials that have passed the physical test will be stored in storage to wait for requests from production department. The rejected raw material will be returned to the supplier.

Plant manager is fully responsible in the production process cycle at the manufacturer. There are two kinds of raw material that will be mixed in production process. First, the raw material in the silo containing corn will be put into a rough bin and then it will be grinding or mashed and put into a fine bin. Second, raw materials in the form of wheat grains, soybean stalks, copra and several other materials that are in the warehouse (bulk / bag storage) are put into a rough bin and then crushed or grinding, after that it goes into the fine bin. Raw materials that have been refined will be directly inserted into the refined bin. Then all the raw materials in the refined bin will be mixed and will be formed into pellets (pills, granules) and then packed. After packaging, the final products will be put in the storage for finished products.

3. Result Analysis

3.1. Define

Identifying the process, number and types of defects of raw material are the major thing that purposed in this phase [12, 13]. This stage is based on collecting data and interview conducted with the silo dryer head and observation.. There are six types of defects that appeared frequently in this industry, which are:. damage seed, insects, moldy seed, broken seed, drier damage, and foreign material.

Damage seed

This type of defect is caused by the activity of respiration or hot air caused by an artificial fainting method, its characteristics are usually brownish on the germ and soft and acidic. In the case of severely damaged seeds the corn seeds swell and peel.

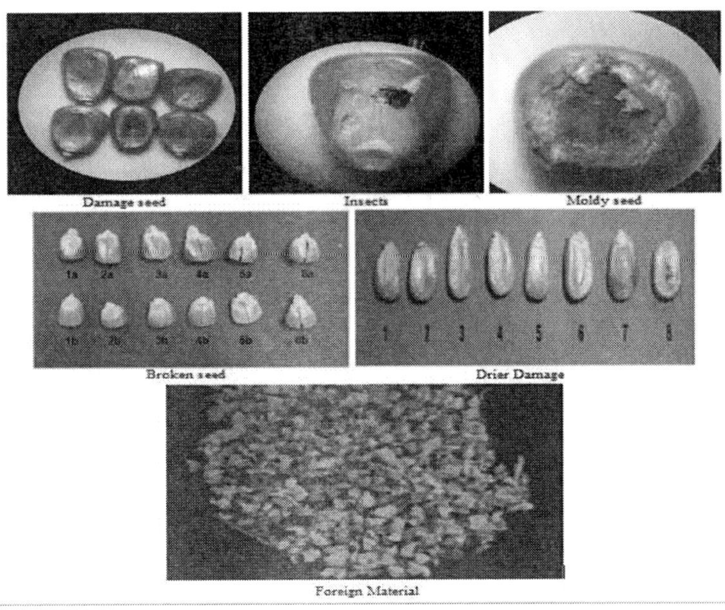

Figure 3. Defects of raw material

Insects
This type of defect is caused by an insect attack which results a hole in the seed or there is a white line of the path from the insect.

Moldy seed
This type of defect is caused by the humid and warm weather so that the fungus on the seeds will easily grow. Generally, the seeds are green or grey in colour.

Broken seed
This type of defect is caused by truck transportation, usually the characteristics of broken / crushed seeds are that there are cracked one third of the corn seeds or completely separated.

Drier damage
This type of defect is commonly due to extreme external drying methods, usually this defect arises if the corn kernels are stored too long in the silo and exposed to excessive heat.

Foreign Material
This type of damage is caused by unwanted contaminants in corn, such as corncobs, blades, gravel, grass weeds and other contaminants.

3.2. Measure
Output of this stage was resulted by measuring the quality level by involving DPMO (Defect Per Million Opportunities) as a base in identifying industry performance on quality side. In this stage data were collected from 48 sub group sample in 48 weeks. Data then plotted to pareto diagram to determine the critical to quality of defect as in Figure 4.

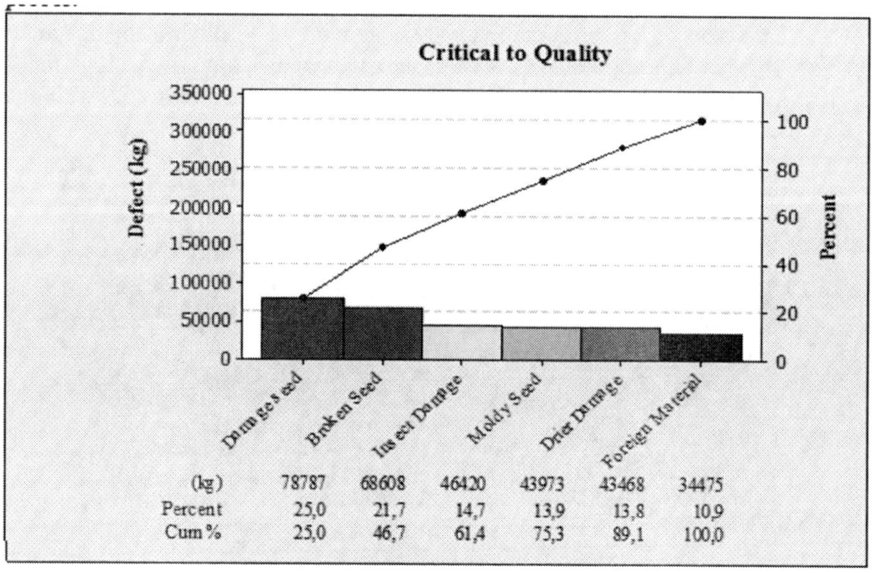

(kg)	78787	68608	46420	43973	43468	34475
Percent	25,0	21,7	14,7	13,9	13,8	10,9
Cum %	25,0	46,7	61,4	75,3	89,1	100,0

Figure 4. Critical to quality

Based on pareto diagram in Figure 4, the number of defects were dominated by four kinds of defects. Whereas the damage seed attained the most defects, and then followed by broken seed, insect damage, and moldy seeds respectively. And the rests are drier damage and foreign material.

IOP Conf. Series: Materials Science and Engineering **875** (2020) 012065 doi:10.1088/1757-899X/875/1/012065

Capability Process

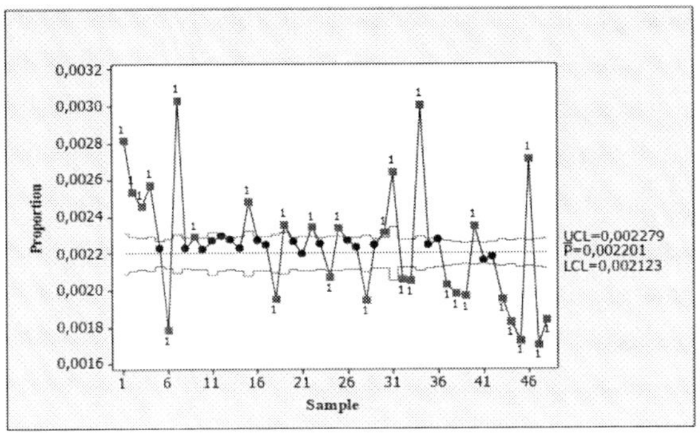

Figure 5. P chart of defect

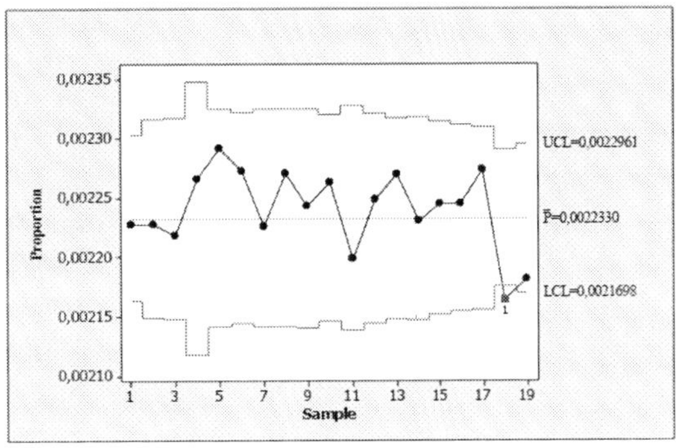

Figure 6. P chart of defect (I)

Observing P chart in Figure 5, it can be seen that the number of defect for 29 data has jumped out of UCL and LCL. On further investigation, that some of sample need to observed the cause. To ensure that to all samples is in control, the sample existed out of LCL and UCL were excluded from data provided from calculation. The data then recalculated and p chart were plotted in Figure 4.

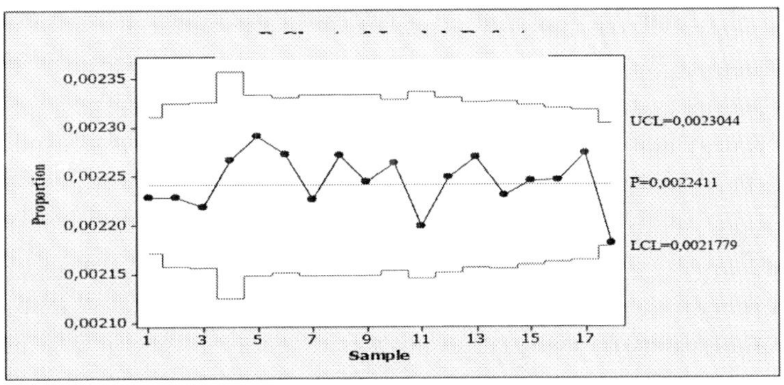

Figure 7. P chart of defect (II)

In this calculation, one data (week 41) still jumped out from LCL, simply then the data was excluded from calculation and recalculated. P chart was plotted in Figure 7. The process was within statistical control after excluding data in week 41.

In overall, DPMO and Sigma level was calculated from 48 group samples. From calculations, DPMO was valued as 6734.13. This indicates that there are 6734.13 possibilities of raw material in experiencing defect per million opportunities (DPMO).

3.3. Analyze

In analyze stage, the result of measurement was as a base of analyzing and conclusion. Fishbone diagram as one tool from seven tools of quality is relatively powerful to identify the cause in detail. The cause of defect was explored by using fishbone diagram involved five factors. By this result, the improvement was set by each factor. To further reduce the defects, some options were set to improve the process. Prior to that the cause and effect was analyzed, in particularly for data that existed jumped out from LCL and UCL. Using fishbone diagram to analyse the cause and effect by observation and interviewing the personnel and key people involved in process directly. There were five factors that caused the defect of raw material.

Some of the causes of defects in raw materials in terms of several factors, namely:

a. Humans: lack of basic knowledge of workers related to raw materials, weaknesses in the sorting process, employee indiscipline, and worker fatigue.
b. Method: Material buildup exceeds load limits, humid warehouse conditions, water content in raw materials exceeds standard levels,
c. Machine: maintenance of machines that are not optimal, especially material collector.

3.4. Improve and Control

This phase is a stage of improving the quality of raw materials by proposing some recommendations for improvement, analyzing the causes and providing solutions with the 5W + 1H kaizen approach (what, why, when, where, who, and how). Improvement carried out related to defects that occur in raw material in the silo drier. These factors consisted of human factor, machines, methods, raw materials, and the environment.

1. Providing assistance and intensive training for workers in order to enhance their knowledge and accuracy in measurement.
2. Replacing proximity sensor which is not feasible for use or damaged so that the measurement of moisture content can be accurate
3. Adding Luprosil NC machine to minimize insect nesting in seeds.
4. Procuring of automatic sensor which can detect or sort of foreign material attached to the seed
5. Increase the awareness of raw material which does not meet the qualifications. For this reason, the QC must be selective in sorting raw material.

4. Discussion

The implementation of six sigma providing methodology in gaining information in some stages structurally. This can enrich in analyzing and enhancing the understanding of process that was reflected in improvement of process. In context of raw material processing, particularly of some raw material that used dominantly in a production process, must be the main concern. Raw materials are the primary factor in assuring the quality of product and production process. Some studies have been developed related to the implementation of six sigma on raw material such as the quality of water as raw material in production of food [5], and raw material in pharmaceutical industry [6]. However, the study that related to quality improvement in the process of receiving raw material is still limited. For that reason, it is important to study widely in this area. Of course the process and procedures have to concern to satisfy the customer requirements and customer delight.

References

[1] Jenab, K., Wu, C., & Moslehpour, S. (2018). Design for Six Sigma: A review. *Management Science Letter*, 1-18.
[2] Venkatesh, N., & Sumangala, C. (2018). Succes of manufacturing industries - Role of Six Sigma. *MATEC Web of Conferences* (p. 144). RiMES 2017.
[3] Padhy, R. K., S.Sahu, & Das, R. (2011). Implementation of Six Sigma in India Industries - a delphy study. *International Journal Computer Aided Engineering and Technology*, 19-33.
[4] Zagloel, T. Y., Ardi, R., & Poncotoyo, W. (2018). Six sigma Implementation model based on critical success factor (CSFs) for Indonesia Small Medium Industries. *ICIEE 2018* (p. 218). MATEC Web of Conference.
[5] Rimantho, D., & Mariani, D. M. (2017). Penerapan Metode Six Sigma Pada Pengendalian Kualitas Air Baku pada Produksi Makanan. *Jurnal Ilmiah Teknik Industri*, 1-12.
[6] Rimantho, D., Hernadi, D., Cahyadi, B., Prasetyani, R., & Kurniawan, Y. (2017). The Application of six sigma in process cotnrol of raw water quality on pharmaceutical industry at Indonesia. *International Journal of Applied Engineering Research*, 848-860.

[7] Elbermawy, M. F., Manhawy, A. A., & Ibrahim, H. (2014). Implementation of Lean Six Sigma For Improving Supply Chain Processes in a Pharmaceutical Industry. *Internatinal Journal of Scientific & Engineering Research* , 519-529.

[8] Goh, T. N. (2010). Six triumphs and six strategies of Six Sigma. *Quality Engineering*, 299-305.

[9] Al-Qatawneh, L., Abdallah, A. A., & Zalloum, S. S. (2019). Six Sigma Application in Healthcare Logistics: A Framework and A Case Study. *Journal of Healthcare Engineering*, 2-12.

[10] Antunes, D. L., Souza, S. D., & Nunes, E. (2013). Using Project Six Sigma and Lean Concepts in Internal Logistics. *World Congress on Engineering*. London: Proceedings of the World Congress on Engineering.

[11] Erbiyik, H., & Saru, M. (2015). Six Sigma Implementation in Supply Chain: An Application for an Automotive Subsidiary Industry in Bursa in Turkey . *Word Conference on Technology, Innovation, and Entrepreneurship* (pp. 2556-2565). Istanbul: Elsevier Ltd.

[12] Fauziah, A., Harsonno, A., & Liansari, G. (2014). Usulan Perbaikan Kualitas Menggunakan Metode Six Sigma Untuk Mengurang Jumlah Cacat Produk Tahu Pada Perusahaan Pengrajin Tahu Boga Rasa. *Jurnal Online Institut Teknologi Nasional*.

[13] Purnama, Dwi Adi., Shinta, Riadho Clara., Helia, Vembri Noor. (2018). Quality Improvement on Creative Industry by using Six Sigma: a study case. *Matec Web of Conference*.

IOP Conf. Series: Materials Science and Engineering **875** (2020) 012066 doi:10.1088/1757-899X/875/1/012066

Distribution Route Optimization of a Capacitated Vehicle Routing Problem by Sweep Algorithm

R Hanafi[1]*, M Rusman[1], F Mardin[1], S M Parenreng[1], and A Azzazli[1]
[1]Industrial Engineering Department, Engineering Faculty, Hasanuddin University, Gowa, Indonesia

*Email: rhanafi@unhas.ac.id

Abstract. This paper presents the application of a sweep algorithm to optimize distribution route of a Capacitated Vehicle Routing Problem (CVRP). The objective of this research is to find a set of optimal routes that minimizes the number of vehicles required and total distance travelled for all vehicles to serve customers. The algorithm consists of two phases, a clustering phase and a route generation phase. Experimental results from its application to a real-world case of PT Eastern Pearl Flour Mills, Indonesia are presented. The obtained vehicle routes which are compared against the existing daily distribution routes showing that significant savings can be gained on the daily cost of transportation related expenses. The number of vehicle routes and vehicle used for delivery are reduced and therefore, the distance travelled and the utilization of the vehicles are also improved. The algorithm applied in this research would be suitable for solving practical vehicle routing optimization problems in other logistics companies.

1. Introduction

The vehicle routing problem (VRP) aims to determine a set of optimal vehicle routes through a number of customer locations in which each route originates and terminates at the same depot. The problem was firstly introduced by Dantzig and Ramser [1] as a generalized of travelling salesman problem (TSP) and since then several variants of VRP have been studied by researchers in different areas such as Operations Research, Computer Science, Decision Support Systems, and Artificial Intelligence. One of the basic versions of VRP is Capacitated Vehicle Routing Problem (CVRP) in which customers with known demand are visited by a homogeneous fleet of vehicles with limited capacity. The number of customers to serve in a route is considerably constrained by capacity limitation of vehicles. The simplest form of CVRP considers one depot, several customers with known demand and known locations to be served by vehicles with equal capacity. The objective is to find a solution that minimizes either the number of vehicles and/or total travelling cost (time or distance) for all vehicles to serve customers.

The object of study and observation is PT Eastern Pearl Flour Mills (EPFM). PT EPFM is the largest of Interflour's mills, located in Makassar on the west coast of the island of Sulawesi in Eastern part of Indonesia. It currently operates two flour mills, one *sea-side* and the other *city-side* mills joined by a conveyor system. The wheat flour-milling factory is one of the primary producers of flour in Indonesia with its products distributed across the nation. EPFM's flour product range supplies to all demands from 1kg consumer packs, 25kg and 50kg packs in the domestic market. The factory sells many different types of flour directly to large institutional buyers as well as small to medium sized enterprises (SMEs), while also utilizing a widespread distributor network (distributors, wholesalers, and retailers). Currently, a fleet of vehicles outsourced for distribution of a number of products to a set of customers according to their requirements. The delivery is made on daily basis directly to dealers or to end consumers. In this study, the fleet for this delivery operation is considered homogeneous with each truck has a capacity of 15 tons. There are no time windows at the customers, but it is necessary to conform with the factory working hours. This research was carried out in Makassar city and data collected only for daily distribution of the company in this region.

There have been various types of algorithms developed as an attempt to solve the vehicle routing problems. Metaheuristics such as tabu search and simulated annealing have been applied to provide the best path comply with the needs of the customers to find the optimized routes for delivering the packets to the customers [2]. An integrated modelling and optimization framework for solving complex and practical VRP had also been proposed in [3]. Variants of the sweep algorithm have been proposed in [4] to find optimal tours for vehicles with given capacity constraints to distribute goods to customers within allocated time windows. This research applies a sweep algorithm as a solution method to find the most optimal set of routes that minimizes the number of vehicles required and total distance travelled for all vehicles to serve customers.

2. Solving CVRP using Sweep Algorithm

Sweep algorithm is a two-phase constructive algorithm consisting of two stages known as cluster-first, route-second [5]. Sweep algorithm is applied for clustering the given customer nodes. In the first phase the algorithm decomposes the problem by clustering customers based on their positions in polar coordinates. Location of the warehouse (depot) is the coordinate center $(0,0)$ of the two-dimensional Cartesians plane (x, y). Polar coordinates (r, θ) of each individual node are computed with respect to the coordinate center and the nodes are ordered according to polar angle. The criteria for clustering customers are the minimal polar angle and the capacity of the vehicle assigned to the cluster. Cluster formation begins with the least polar angle nodes and continues sweeping customer nodes by increasing polar angle. Sweeping is discontinued when the included next customer node would violate maximum limit of the vehicle capacity. The demands of all the selected customers in one cluster has to be less than or equal to the capacity of the vehicle assigned to the cluster. A new cluster is created by continuing the sweeping where the last one left off. The process of adding customer nodes into cluster is repeated until all customer nodes have been included.

Table 1. The existing daily distribution route

Route	Vehicle ID	Routes	Capacity (tons)	Distance (km)
1	DD 8411 XI	0-3-4-0	5	13.8
2	DD 8454 MJ	0-20-24-0	7.5	22.1
3	DD 8719 MW	0-22-0	10	3.2
4	DD 8708 MO	0-21-0	10	16
5	DD 8756 XD	0-12-11-0	8.01	17.5
6	DD 8815 KB	0-1-7-0	15	38.5
7	DD 8901 KM	0-9-10-0	1.38	37.8
8	DD 8922 MJ	0-23-0	7.8	9.9
9	DD 8961 KF	0-5-0	12.5	8.3
10	DD 9031 CY	0-2-0	5	0.64
11	DD 9451 XU	0-19-0	15	5.8
12	DD 9633 OV	0-6-0	5	20.1
13	DD 9924 XS	0-14-18-0	6.25	30.2
14	DD 9925 XS	0-13-8-16-15-17-0	4.63	32
TOTAL				**255.84**

There are nine clusters formed based on the requirement that the total demand in one cluster must be less than the maximum truck capacity of 15 tons: Cluster 1 = (23), Cluster 2 = (12, 16, 3, 4), Cluster 3 = (5), Cluster 4 = (19), Cluster 5 = (15, 14, 22, 11, 10), Cluster 6 = (24, 6, 17, 7), Cluster 7 = (2, 13, 20, 8), Cluster 8 = (21), Cluster 9 = (1, 9, 18).

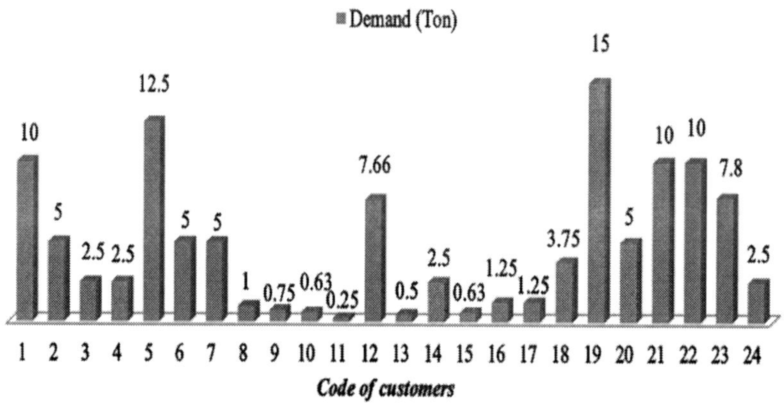

Figure 1. The daily demand for flour products of PT EPFM

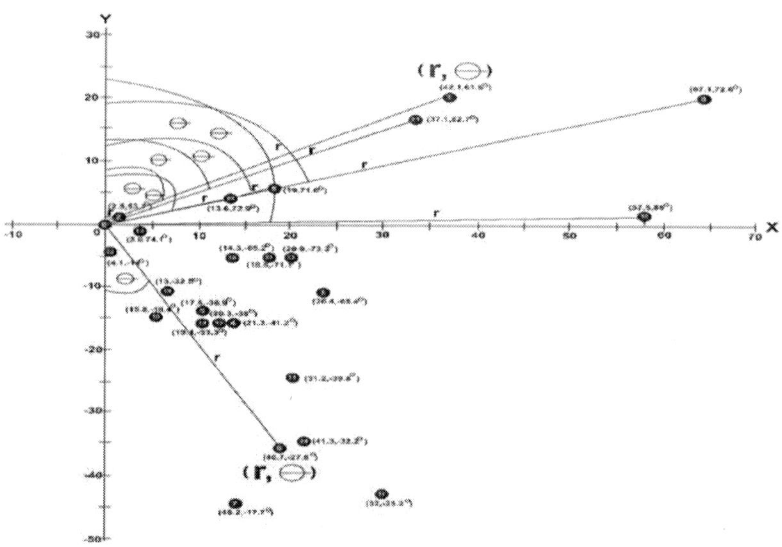

Figure 2. The depot and customers location in Polar coordinate diagram

The next stage is to apply nearest neighbour algorithm. The location of customers in the clusters formed from the previous stage will be sequenced and scheduled for vehicle tours. A tour is built based on the traveling distance from the vehicle last-visited point (node) to the next point closest in the network. The procedure begins with the base depot node and then finding the node which is the closest to the last included node in the path. This procedure is repeated until all nodes have been added to the path. The first and the last nodes in the path are then connected to make a complete tour.

3. Experimental Studies

In order to implement the sweep algorithm, locations and demands information of customers are firstly defined. The number of vehicles and the existing distribution routes taken by vehicles to serve customers are also determined as well as the distance from depot to every distribution point. Distance matrix is prepared by using data obtained from the coordinates of the customer's address on Google maps. The procedure is performed by determining customer's locations which are points in a plane geometry with distances represents the traveling costs. The distance between coordinate (x_i, y_i) and (x_j, y_j) is then calculated. The polar coordinates of each customer are determined with respect to the location of the base depot and sorted by increasing polar angle. Capacity limitation of the vehicle assigned to the cluster is considered as constraint and sweep is terminated when including the next customer node would violate maximum limit of the vehicle capacity. Through this procedure the customers have been clustered by vehicles.

Application of sweep algorithm generated nine distribution routes. Table 2 shows the nine vehicle routes that have been generated by applying the sweep algorithm. Cluster 1 covers total demand of 7.8 tons even though vehicle capacity 15 tons. Total distance travelled to serve customers in a day by nine vehicles is 142,8 km. Total demands in this case are supplied by nine routes which correspond to the number of vehicles. Among the nine routes obtained, the fourth route serves customers with truck capacity exactly the same as the customers' demand. Figure 3 shows optimized routes taken using Google maps.

Table 2. Distribution routes of vehicles generated by sweep algorithm

Routes	Sequence of visit	Vehicle Capacity (tons)	Distance Travelled (km)
1	0-23-0	7.8	9.9
2	0-16-12-3-4-0	13.91	15
3	0-5-0	12.5	8.3
4	0-19-0	15	5,8
5	0-22-10-14-15-11-0	14.01	19.1
6	0-24-6-17-7-0	13.75	25.3
7	0-2-13-20-8-0	11.5	8.9
8	0-21-0	10	16
9	0-1-18-9-0	14.5	34.5
	TOTAL		**142.8**

Computational results show that total demands can be fulfilled by nine clusters that is equal to the number of vehicles. The total distance travelled daily can be reduced from 255.84 km to 142.8 km. The number of vehicle fleets is also reduced from 14 vehicles to 9 vehicles. The obtained vehicle routes which are compared against the existing company's daily distribution routes indicating that, the daily cost of transportation related expenses may be reduced as there is a significant reduction in the total distance travelled and the number of vehicles used for the daily delivery of the company.

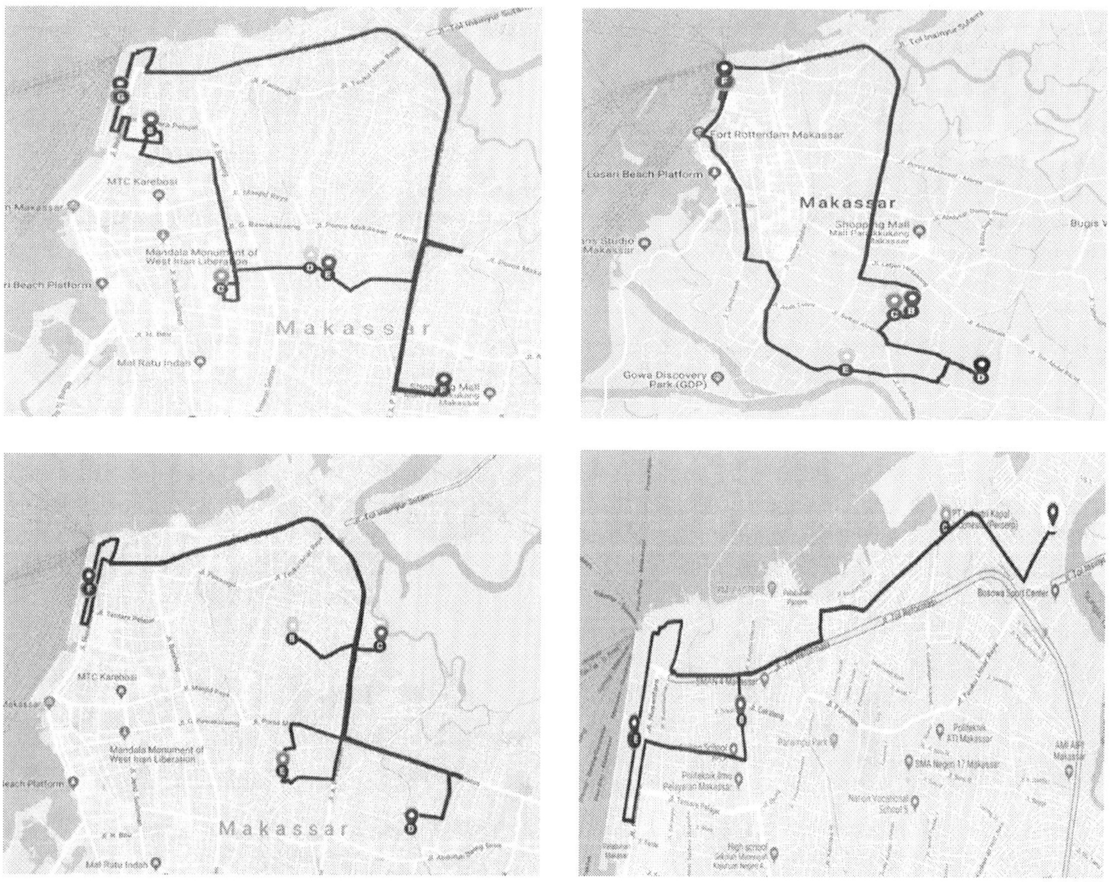

Figure 3. Optimal routes of the Capacitated Vehicle Routing Problem

4. Conclusion

This paper presents an application of sweep algorithm to solve CVRP. The objective is to find a set of optimal vehicle routes with given capacity constraints to distribute products to customers. Case study from a wheat flour industry is taken and its distribution performance is evaluated with daily delivery points of 24 customers spread across the city. The optimized vehicle routes which are compared against the existing daily distribution routes indicating that significant savings can be gained on the daily cost of transportation related expenses. The sweep algorithm applied for generating vehicle routes is capable of obtaining shorter vehicle routes than the existing routes operated by the company. The total distance travelled for the fleet of vehicles to distribute flour products to customers can be reduced significantly and therefore the number of vehicles utilized also can be minimized to five vehicles only. The sweep algorithm applied in this paper provides better solutions in terms of the traveling distance, traveling time, the number of vehicles utilized, and the daily work capacity balances under the specified constraints considered in the work. The company's daily distribution and logistics performance may be improved and the algorithm used in this study can be applied to practical vehicle routing problems arising in many other logistics and distribution companies.

Acknowledgment
This research is supported by Laboratory Based Education (LBE) Grant 2019 Engineering Faculty, Hasanuddin University, Indonesia.

IOP Conf. Series: Materials Science and Engineering **875** (2020) 012066 doi:10.1088/1757-899X/875/1/012066

References

[1] Dantzig G and Ramser J 1959 The truck dispatching problem. Manag. Sci. 6: 80–91.
[2] Kirci P 2016 An optimization algorithm for a capacitated vehicle routing problem with time windows. Indian Academy of Sciences Vol. 41 No. 5 pp. 519–529.
[3] Tonči Carić, Ante Galić, Juraj Fosin, Hrvoje Gold and Andreas Reinholz 2008 A Modelling and Optimization Framework for Real-World Vehicle Routing Problems. ISBN 978-953-7619-09-1 pp. 142.
[4] C. Hertrich, P. Hungerländer, and C. Truden 2019 Sweep Algorithms for the Capacitated Vehicle Routing Problem with Structured Time Windows.
[5] Gillett B and Miller L 1974 A heuristic algorithm for the vehicle dispatch problem. Operations Research 22, 340–349.

IOP Conf. Series: Materials Science and Engineering **875** (2020) 012067 doi:10.1088/1757-899X/875/1/012067

Improvement of Production Scheduling Performance using Simulated Annealing Algorithm

S Mangngenre[1]*, N Syamsul[1], S Asmal[1], I P Marco[1], N I Syamsul[1], and A B R Indah[1]

[1]Department of Industrial Engineering, Faculty of Engineering, Hasanuddin University, Gowa, Indonesia

*Email: saiful.ti@unhas.ac.id

Abstract. Production scheduling aims to minimize processing time, waiting time, and inventory levels, as well as efficient use of facilities, labor, and also equipment. In producing these products there is usually ineffectiveness in the scheduling process, such as sorting jobs that are not optimal so that the production requires a long time to produce products in large quantities. Therefore improving performance in order to increase productivity must be carried out continuously considering tremendously strict competition in the industry and manufacturing sector. This study is aimed to minimize the makespan value by scheduling the job sequence optimally. In this study, simulated Annealing Algorithm is used as one of the algorithms for generic optimization that is then compared to the algorithm used by companies. Based on the data processing, the maximal iteration has been reached with 2 iterations. The last iteration is selected which has the smallest makespan value, namely J2-J4-J3-J1 schedule with makespan value of 18,811.2 minutes. The makespan value obtain by simulated annealing algorithm is considered 9.03% more efficient compared to the companies method.

1. Introduction

Improving performance in order to increase productivity must be carried out continuously considering tremendously strict competition in the industry and manufacturing sector. One way to increase productivity is by scheduling production. In order to gain an optimal result, all production activities must first be planned properly. Production scheduling is strived to manage an effective job assignment at each work station in order to reduce the processing time in a production schedule so that the production results obtained can be maximized. Existence of more effective production scheduling and job processing is required.

Proper scheduling of production processes can reduce idle time in production units and minimize work in process. One measure of performance in scheduling is the completion time (makespan) to a minimum [1].

Scheduling objectives are to increase the effectiveness and efficiency of resource usage, so that total of processing time can be reduced and productivity can be increased, reducing the inventory of semi-finished goods or reducing the number of jobs lining in the queue when the existing resources are still performing other tasks, reducing some delays in work with a deadline to minimize penalty costs, and assisting in making decisions about planning plant capacity and the type of capacity required so that the addition of expensive costs can be avoided [1].

In this study, simulated Annealing Algorithm is used as one of the algorithms for generic optimization that can be applied to find approaches to the global optimum solution of a problem [2]. The advantage of simulated annealing compared to other methods is that its ability to avoid local optimal traps. The algorithm is a random search algorithm, but it does not only accept objective values that always go down.

2. Method

This research was conducted on Butsudan (a household Buddhist altar shelf) production with the following steps:

a. Secondary data collection

Secondary data used is data that has been prepared by the company. In this case, the data obtained is as follows:

- Type of machine used
- Number of machines
- Engine capacity
- Types of products
- Number of constituent parts
- Dimensions of each product

The data can be seen in Tables 1 and Table 2 which is used to obtain the standard processing time of each product.

Table 1. Data of machines

No	Machine	Code	Number of Machine (*Jm*)	Machine Capacity (*Km*)
1	Panel Saw	*M1*	1	1
2	Hot Press	*M2*	1	1
3	Yokosuri	*M3*	1	1
4	Double Saw	*M4*	1	1
5	NC Router	*M5*	2	1
6	Kazaban	*M6*	2	1
7	Panel Press	*M7*	3	5
8	Strock Belt Sander	*M8*	2	1
9	Spray Gun Painting	*M9*	4	6
10	Body Press	*M10*	1	1

Table 2. Type of products

Job	Type of Butsudan	Size	Number of constituent part (*Jb*)	Demand (*P*)
1	Dareto- L	18-18	63	27
2	Dareto- L	16-48	51	34
3	Dareto- L	18-40	65	20
4	Dareto- L	18-50	64	60

b. Primary data collection

Primary data is data directly collected which obtained data as follows:

- Production Process flow
- The standard processing time of each constituent and product called as a job.

Table 3 reveals the primary data which have been collected.

Table 3. The standard processing data

Machine	*Job* (seconds)			
	J1	*J2*	*J3*	*J4*
M1	47.02	35.97	42.90	41.57
M2	36.34	36.19	36.10	36.24
M3	84.44	77.18	79.47	80.80
M4	6.70	5.11	5.05	5.41
M5	1,842.82	1,833.97	1,836.20	1,837.89
M6	1,050.63	1,048.89	1,048.74	1,049.25
M7	1,842.82	1,833.97	1,836.20	1,837.89
M8	1,050.63	1,048.89	1,048.74	1,049.25
M9	115.87	104.81	111.75	114.23
M10	2,056.37	1,896.28	1,964.73	2,010.57
Total	**8,133.63**	**7,921.28**	**8,009.89**	**8,063.10**

c. Data Processing

Data processing is performed by using Simulated Annealing Algorithm.

2.1. Simulated Annealing Algorithm

1. The parameters used in Simulated Annealing are as follows [3]:
   a. System state
   System state is defined as a possible solution. For instance, given flow shop scheduling with 4 jobs, so one of the possible scheduling sequence is *J4, J1, J3, J2*.
   b. Energy
   Energy is defined as how much the minimum objective functions of a combination of system states. In terms of function minimization, suppose that the current solution is x and the function value is $f(x)$, similar to the energy status of a thermodynamic system, energy for flow shop scheduling problems, energy is defined as the value of the makespan of each scheduled solution.
   c. Temperature
   Temperature is a control value that makes the condition move or not. Atoms move freely at high temperatures and become increasingly restricted when the temperature drops. If the temperature is slowly and regularly lowered, the atoms produce crystals with a proper arrangement with minimal energy. Conversely, if the temperature is lowered quickly it will produce polycrystalline (imperfect crystals) with energy that is not minimal. In terms of the flow shop scheduling problem, temperature is defined as a control parameter.
   d. Cooling rate
   In terms of the flow shop scheduling, the cooling rate function is defined as to how quickly the final solution is achieved. The cooling rate is used in the temperature reduction process. Temperature reduction is shown by Equation (1).

$$T_b = a \times T_0 \qquad (1)$$

where T_0 = initial temperatur
T_b = New Temperature
a = temperature reduction factor (a<1)

2. Simulated Annealing Algorithm mechanism in flow shop scheduling is defined as follows[3]:
   a. Generating the initial solution
   The initial solution in scheduling is to generate a random schedule. At this stage, the objective value is determined. The result obtained is defined as the best solution for the initial stage.

b. Searching for new solutions
The next stage is to look for new solutions from the best solutions produced in the previous stager which is called the insertion method. New solutions are formed by randomly selecting one part of the structure to be moved to another part.
c. Checking the control parameter
Determination of control parameters reduction is performed to decide if the control parameters require to be lowered. If the control parameter is lowered, then proceed to step d. Otherwise, step b and c are repeated.
d. Determination of the best new solution
At this stage, several new solutions are found that are accepted. One of them is next chosen to be the best solution. This best solution will be the initial solution if the control parameter is changed.
e. Searching for evaluation of new solution
Evaluation of new solutions is aimed to evaluate whether the new solutions are accepted. The evaluation criteria for the solution are shown by Equation (2).

$$\Delta E = E(X_{i+1}) - E(X_i) \quad (2)$$

where ΔE = the objective value change
$E(X_{i+1})$ = Objective value of the new solution
$E(X_i)$ = Objective value of the initial solution

If the objective value of the new solution is less than the objective value of the initial solution ($\Delta E \leq 0$) then the new solution is accepted. Otherwise, then new solutions might still be chosen with probability [3].

$$P(\Delta E) = e^{-\Delta E/T} > r \quad (3)$$

where T = control parameter
r = random value between 0 and 1
f. Control parameter reduction
To decrease the control parameter is used equation (1)
g. Determination of the maximal iteration
Determination of the maximum iteration is to decide whether the iteration is complete or not. If the maximum iteration is complete, it means the best solution is the optimal solution. If the iteration is not optimal then repeat step **a** to step **g**.

3. Result and Discussion

Calculating the processing time of each product by considering the number of product demands, number of constituent parts of each product, number of machines, machine capacity and the standard processing time of each product with the following equation:

$$TM = \frac{\text{P} \times \text{W}b \times \text{Jb}}{\text{Jm} \times \text{Km}} \quad (4)$$

TM_n = processing time of product n on machine m
P = number of demand of product n
Wb = Standard processing time of product n
Jb = number of constituent parts of product n
Jm = number of machine m
Km = machine capacity
n = type of jobs; 1, 2, 3, 4
m = type of machines; 1, 2, 3……10

The values of *P, Wb, Jb, Jm* and *Km* for the calculation of product processing time 1 on machine 1 can be seen in tables 1 and 2, while the calculation is as follows;

$$TM = \frac{P\ x\ Wb\ x\ Jb}{Jm\ x\ Km}$$

$$TM = \frac{27\ x\ 63\ x\ 47.02}{1\ x\ 1}$$

TM = 79,981.06 seconds or equal to 1,332.92 minutes

The following **Table 4** reveals the calculation results of processing time for the whole products on each machine.

Table 4. Processing time of job *n* on machine *m*

Machine	***J1*(Minute)**	***J2* (Minute)**	***J3* (Minute)**	***J4* (Minute)**
TM1	1,332.92	1,039.41	929.45	2,904.51
TM2	206.03	209.18	156.41	463.86
TM3	2,394.01	2,230.63	1,721.85	5,171.03
TM4	189.85	147.60	109.47	346.36
TM5	414.63	519.63	306.03	918.94
TM6	472.78	594.37	349.58	1,049.25
TM7	328.48	302.91	242.12	731.08
TM8	976.93	972.92	728.81	2,164.31
TM9	161.74	156.13	119.12	355.39
TM10	925.36	1,165.19	685.38	2,056.27

3.1. *Simulated Annealing Algorithm*

Scheduling completion steps using the Simulated Annealing algorithm are briefly described as below:

a. Generating an initial solution

In this stage, determination of the initial solution is performed by calculating the smallest makespan value using 3 methods [4]:

- FCFS (*First Come First Serve*),
- SPT (*Shortest Processing Time*),
- LPT (*Longest Processing Time*)

Based on the calculation result obtained by the methods, it is found out *LPT* method produce the smallest makespan value which is 18,936.94 minutes with job sequences of *J4-J1-J2-J3*. The value then serves as the initial solution.

b. Generating new solution

Generation of a new solution is performing by using the insertion method. In this method, the number of exchanges (replication) is determined. In this study, replication is carried out 5 times based on several references followed. The replication value is a number indicating the number of loops which must be done before decreasing the control parameters. Job sequences obtained from the replication results are:

Table 5. Replication result of Insertion method

Job sequence				*Makespan (minutes)*
4	3	1	2	18,936.9
1	4	2	3	19,344.5
2	4	1	3	18,811.2
3	4	2	1	19,181.0
2	3	1	4	20,387.7

Among 5 job sequences shown on Table 5, it is found out that the new best solution is the job sequence of *J2-J4-J1-J* which can be performed in 18,811.2 minutes. The makespan value serves as the new solution which is then compared to the initial solution consuming time of 18,936.94 minutes gained by Equation 2.

$$\Delta E = E(X_{i+1}) - E(X_i)$$
$$= -125.7$$
where $E(X_{i+1}) = 18{,}811.2$ and $E(X_i) = 18{,}936.9$

As the criteria if $\Delta E \leq 0$ then the new solution is accepted. Because of $\Delta E = -125.5$ means that the new solution is accepted.

c. Control parameter value reduction
Reduction of control parameters is carried out to obtain the possibility of makespan values smaller than the new solution. The initial predetermined is 100 with a reduction factor of 0.45. A reduction factor is a number used to decrease parameters in a gradual and controlled manner, with the following results:

Table 6. Parameter value reduction

Number of iteration	Control Parameter	Job seq.				*Makespan(minute)*
Iteration 1	100	4	3	1	2	18,936.9
		1	4	2	3	19,344.5
		2	4	1	3	18,811.2
		3	4	2	1	19,181.0
		2	3	1	4	20,387.7
chosen solution		**2**	**4**	**1**	**3**	**18,811.2**
Iteration 2	45	2	1	3	4	20,387.7
		1	3	4	2	19,612.6
		3	2	4	1	19,055.2
		4	2	3	1	18,936.9
		2	4	3	1	18,811.2
Chosen solution		**2**	**4**	**3**	**1**	**18,811.2**

d. Evaluation of the new solution
In the state of 100 control parameter value, the new solution selected is the schedule sequence with a makespan of 18,811.2 minutes. This solution is compared to the initial solution namely makespan of 18,936.94 minutes. The new solution produced is better than the initial conditions so that the initial solution is accepted to be the new solution of job schedule *J2-J4-J1-J3* with a makespan value of 18,811.2 minutes.

e. Determination of the maximal iteration
The maximal iteration has been reached, namely 2 iterations so that the iteration has been fulfilled. From the overall sequences obtained, the last iteration is selected which has the smallest makespan value, namely *J2-J4-J3-J1* schedule with 18,811.2 minutes makespan value.

Table 7. Makespan value using simulated annealing algorithm

Job		Machine (minute)									
		M1	*M2*	*M3*	*M4*	*M5*	*M6*	*M7*	*M8*	*M9*	*M10*
J2	*Start*	0	1,039.41	1,248.59	3,479.22	3,626.82	4,146.44	4,740.81	5,043.73	6,016.65	6,172.78
	End	1,039.41	1,248.59	3,479.22	3,626.82	4,146.44	4,740.81	5,043.73	6,016.65	6,172.78	7,337.97
J4	*Start*	1,039.41	3,943.92	4,407.78	9,578.82	9,925.18	10,844.12	11,893.37	1,.624.46	14,788.76	15,144.16
	End	3,943.92	4,407.78	9,578.82	9,925.18	10,844.12	11,893.37	12,624.46	14,788.76	15,144.16	17,200.42
J3	*Start*	3,943.92	4,873.38	9,578.82	11,300.66	11,410.13	11,716.17	12,624.46	14,788.76	15,517.58	17,200.42
	End	4,873.38	5,029.79	11,300.66	11,410.13	11,716.17	12,065.75	12,866.57	15,517.58	15,636.69	17,885.80
J1	*Start*	4,873.38	6,206.30	11,300.66	13,694.67	13,884.52	14,299.16	14,771.94	15,517.58	16,494.51	17,885.80
	End	6,206.30	6,412.33	13,694.67	13,884.52	14,299.16	14,771.94	15,100.42	16,494.51	16,656.25	**18,811.16**

Figure 1 reveals Gantt chart of Scheduling by Simulated Annealing algorithm.

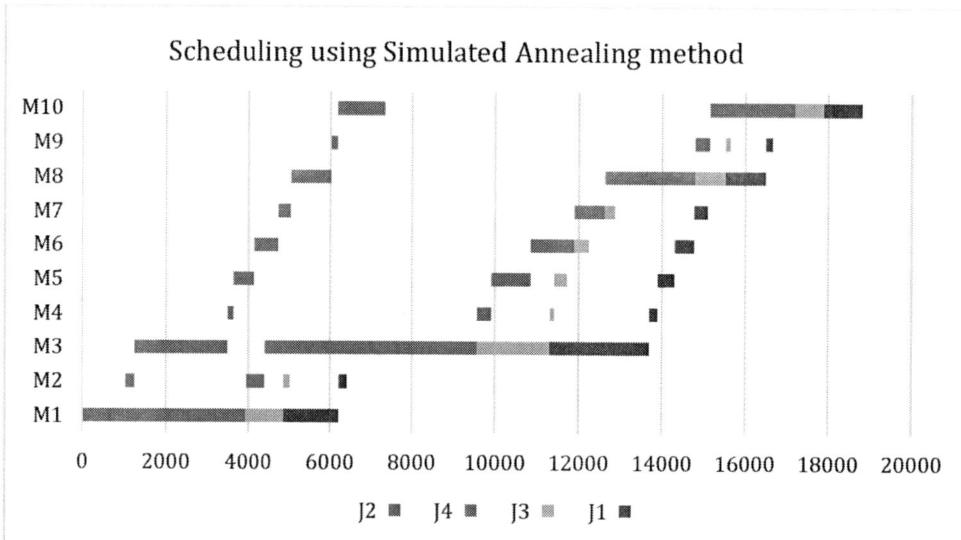

Figure 1. Gantt Chart using Simulated Annealing Algorithm

The simulated annealing algorithm is used to find approaches to the global optimum solution of production scheduling. In this case, simulated annealing algorithm provides a solution in terms of makespan.

4. Conclusion

It can be concluded that the first based on data processing performed by using the Simulated Annealing Algorithm method (4 jobs and 10 machines), the optimal makespan obtained is 18,811.2 minutes. The second comparison analysis of the makespan value of the Simulated Annealing Algorithm method to the company method is as follows, which is included with the Simulated Annealing Algorithm method obtained 18,811.2 minutes of makespan value which is 9.03% more efficient makespan compared to the Gupta algorithm which is used by companies, and the optimal job sequence of the Simulated Annealing Algorithm method is *J2-J4-J3-J1*. Therefore, Simulated Annealing algorithm is powerful and could be recommended to be used in companies to increase the productivity.

Acknowledgements
This work was supported by Engineering Faculty of Hasanuddin University (LBE 2019 grant scheme – UNHAS).

References

[1] Chen, X., Zhang, J., Production control and supplier selection under demand disruptions (2010), Journal of Industrial Engineering and Management. doi:10.3926/jiem.2010.v3n3.p421-446.
[2] Baskan, O., Optimization Algorithm, Method and Application (2016).
[3] Laarhoven P, J., *Simulated Annealing:* Theory and Applications 1st Edition (Netherlands : Springer)
[4] Yingni, Zhai, Chanjun Liu, Wei Chu, Ruifeing Guo, Cunliang Liu., A Decomposition Heuristic based on multi-bottleneck machines for large-scale job shop scheduling problems (2014), Journal of Industrial Engineering and Management. Vol.7, No. 5.

IOP Conf. Series: Materials Science and Engineering **875** (2020) 012068 doi:10.1088/1757-899X/875/1/012068

The Performance of Clay Furnace with Variation in the Diameters of the Briquette Burning Chamber

Z Djafar[1]*, N Amaliyah[1], S Suluh[2], M Isra[1] and W H Piarah[1]

[1] Mechanical Engineering Department, Engineering Faculty, Hasanuddin University, Makassar, Indonesia
[2] Mechanical Engineering Department, Universitas Kristen Indonesia Toraja, Indonesia

*Email: zuryatidjafar@unhas.ac.id

Abstract. Biomass fuel is an alternative energy source used in the form of briquettes using clay stoves. Utilization of clay as a briquette stove often results are not satisfactory as is the use of other fuel stoves, this is because the amount of air in the furnace burning chamber has not been determined and focused properly, so the purpose of this study is to determine the thermal efficiency of the furnace at variations in the diameter of the aluminum sleeve in the furnace burning chamber. The method used is to test 5 variations in diameter of the furnace burning chamber (cylinder). The results showed the best thermal efficiency was obtained at 52.87% at 180 mm slab diameter, boiling time at 9 minutes, with a heat loss of only 2.02 kW.

1. Introduction

As population growth is so rapid and followed by the depletion of the amount of fossil fuel energy reserves, it is necessary to broaden the investigation of the use of non-fossil alternative energy sources. One of them is the use of biomass energy. Biomass energy is very interesting to use due to its abundant availability. The biomass used in this research is plantation waste, which is coconut shell waste that is used as fuel in the form of briquettes. Coconut shell waste has become material in several research objects, such as Phonphuak and Thiansem [1] in their research, making the waste as an additive for brick making which aims to increase water absorption before burning. Jamilatun [2] has examined the heat content of coconut shell briquettes and obtained a maximum value of 5779.11 kcal/gram. Budi [3] in the same study noted value of 76.52% coconut shell carbon content, and these results are a reference that coconut shell has the potential to become fuel. Higher heating values were obtained in research conducted by Sallolo and Pineng [4] at a heating value of 5839.33 cal/gram with a thermal efficiency of 64.15%.

The problem that often arises in the utilization of biomass briquettes is not from the type of briquette material but in the furnace burning chamber which is not effective in transferring heat to the cooking device. This is largely influenced by the amount of air in the fuel chamber which is often uneven in the arrangement of the briquettes, so this paper is focused on modifying the burning chamber by adding sleeves so that the heat transfer process can be better, so as to provide a better furnace performance. The use of aluminum sleeve is used because the thermal efficiency is quite good at 70.73% [5] and 54.12% [4]. In this study, the diameter of the sleeve will be varied to get the best diameter in producing the best efficiency.

Cookstoves made of clay or also called 'brazier' traditionally use firewood as fuel, which naturally has a lot of exhaust emissions. Therefore, the Technology Assessment and Application Board (BPPT) has already socialized the use of low-emissions non-fossil fuels, especially carbon emissions and particulates. Haryanto and Triyono [6] have made experimental comparisons of several types of household stoves such as coal, kerosene, LPG stoves, thick pot furnaces and brick/biomass stoves in terms of CO, NO_2, SO_2 and particulate gas emissions. Kerosene stoves have the highest CO emissions, while thick pot furnaces have the lowest SO_2 emissions. In this study, the Proximation test and the heating value of the briquettes were also carried out.

2. Materials and Methods

In this study, the clay stove (brazier) taken is the result of Takalar Pottery Center handicraft as shown in Figure 1 below. The brazier dimensions; 300 mm high, 220 mm outside diameter, 200 mm inner diameter, 10 mm distance from the base of the stove.

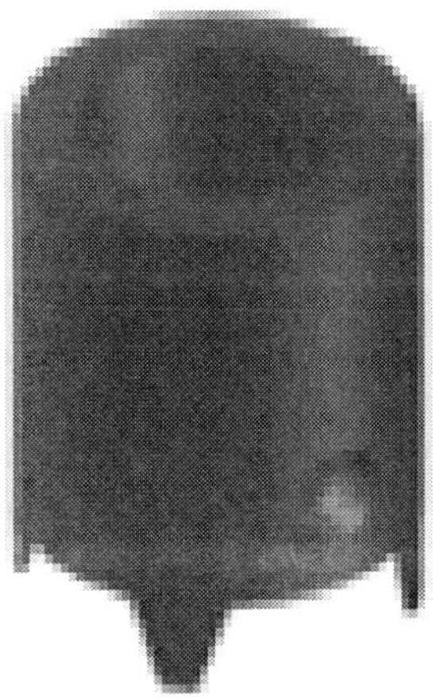

Figure 1. Clay furnace (Brazier)

The research method used is an experimental method by utilizing coconut shell charcoal briquette waste as the fuel of the stove that is tested. The briquette used was in the form of a wasp nest because it has a larger flame surface area [7]. Tests of estimation, heating value, burning process, thermal efficiency and burning efficiency of briquettes were carried out on several variations of the sleeve diameter placed in the furnace burning chamber. Sleeves that made of Aluminum plate material, have air holes with a diameter of 10 mm with a distance of 20 mm between holes along the diameter of the upper cylinder (see Figure 2).

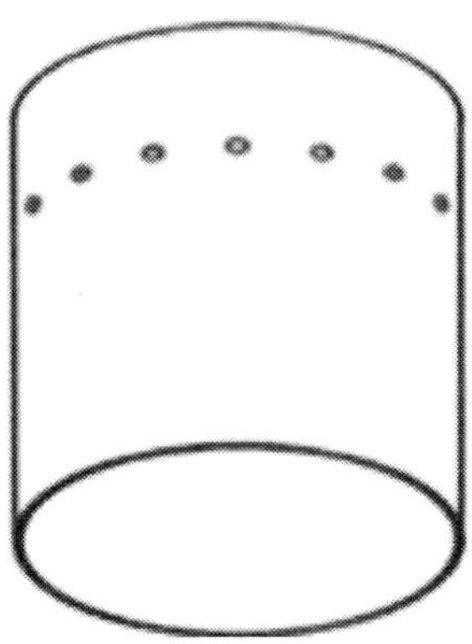

Figure 2. Furnace burning wall cylinders

With the addition of the sleeves, it is expected to improve the burning process in the burning chamber and also reduce heat loss to the radial wall. The diameter of the cylinder varied from 140mm, 150mm, 160mm, 170mm and 180mm.

3. Result and Discussions

This research begins with the manufacture of briquettes, estimation testing, and Calorific Value and then proceed with the Burning Test (performance) with a variation of 5 sleeve diameter sizes. The test parameters are the duration (time) of burning, fire temperature, fuel-burning chamber temperature, boiling time, fuel consumption and efficiency. The fire temperature as a function of burning time can be seen in Figure 3 below.

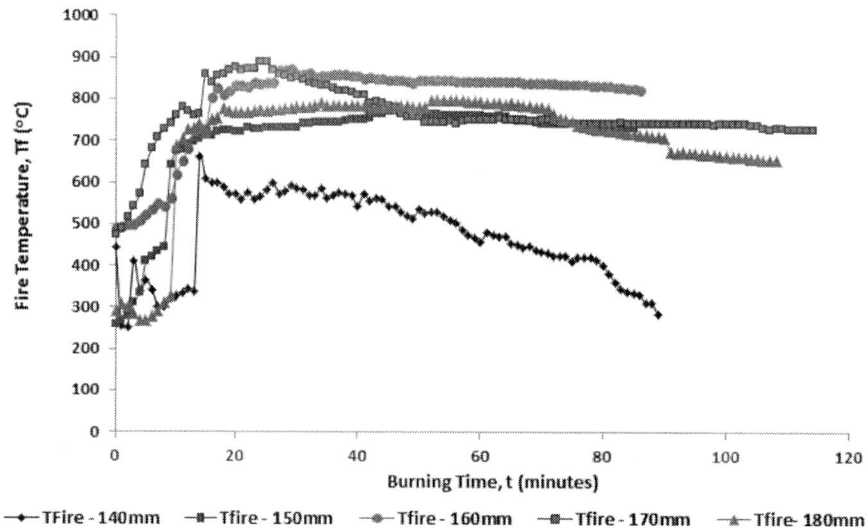

Figure 3. The fire temperature to the burning time

Figure 3 shows that the maximum temperature occurs at 160 mm sleeve diameter of 854°C. Meanwhile, for the diameter of 170 mm sleeves, although it has a relatively lower temperature (790°C), it has a longer burning time (114 minutes) and is able to boil as much as 30 liters of water.

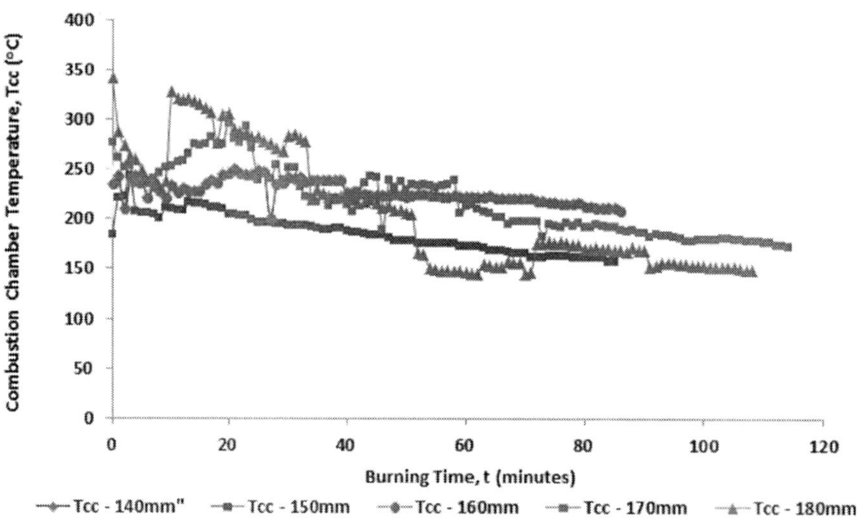

Figure 4. The burning chamber temperature to the burning time

IOP Conf. Series: Materials Science and Engineering **875** (2020) 012068 doi:10.1088/1757-899X/875/1/012068

Figure 4 shows the phenomenon of burning chamber temperature which is identical to the temperature in Figure 3. The highest temperature occurs at 160 mm sleeve diameter and the longest combustion at 170 mm diameter. It can be said, the larger the diameter is, the longer the heat in the combustion chamber can be maintained.

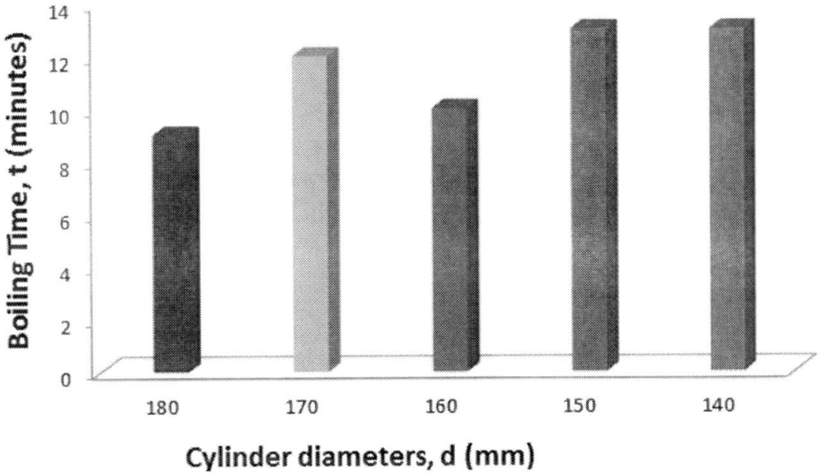

Figure 5. Boiling time to the changes in the diameter of the furnace cylinder

In Figure 5 we can see the relationship between boiling time and sleeve diameter variation. The results show that 180 mm sleeve has the best boiling time. It can be said, the wider the surface area of the sleeve diameter is, the faster the heat transfer process of the briquettes to the cooking utensil. Because the supply of air entering the cylinder burning chamber is more and more, the burning rate is also faster.

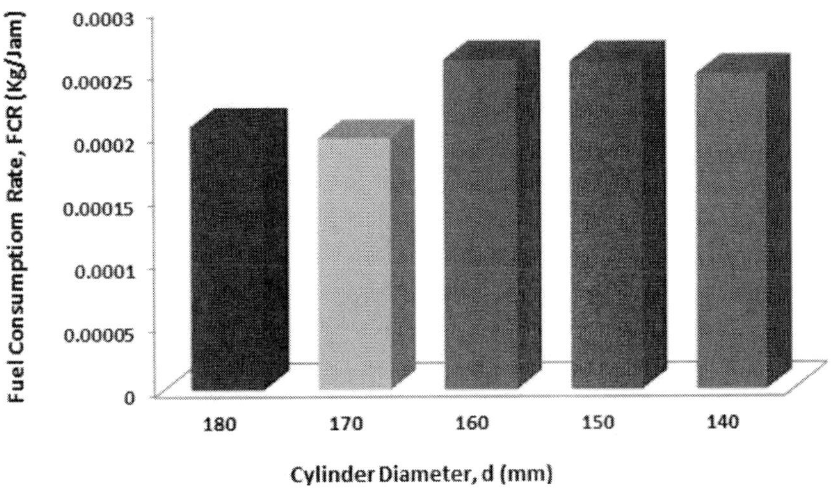

Figure 6. The fuel combustions rate to the cylinder diameter

The best effect of fuel consumption (at least) in Figure 6 is obtained in the 170 mm cylinder of 0.0002 kg/hour and the greatest consumption in the cylinder diameter of 160 mm and 150 mm cylinder of 0.000026 kg/hour. The more surface area of the cylinder diameter, the better fuel consumption. In general, if the rate of air entering the cylinder gets smaller, the fuel burning process will be faster and run out.

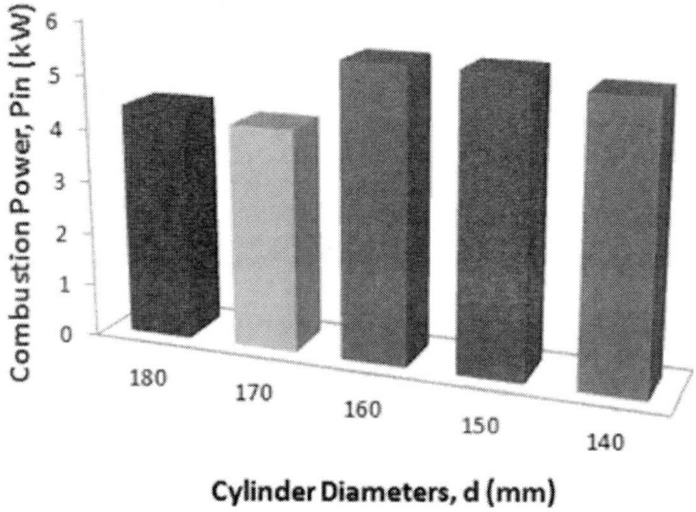

Figure 7. Furnace burning power on 5 various cylinder diameters

Figure 7 shows the effect of burning on variations in cylinder diameter, which shows that the best burning power is obtained in a 160 mm cylinder of 5.50 kW, the lowest in a 170 mm cylinder. The effect of fuel-burning power is not too significant on the area or diameter of the cylinder but depends on the factor of the incoming air supply affecting the burning process.

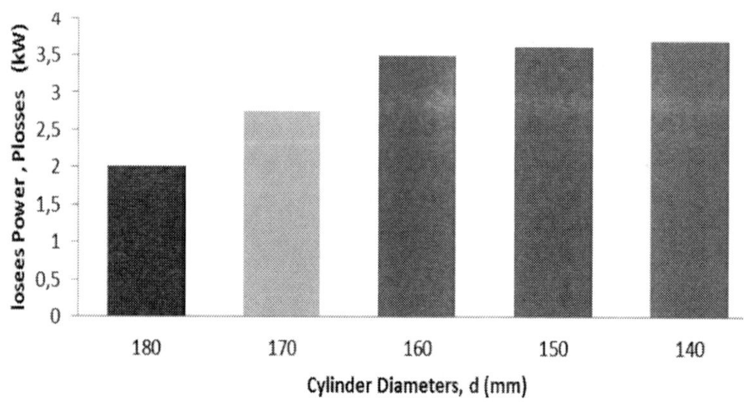

Figure 8. The heat losses to the cylinder diameter

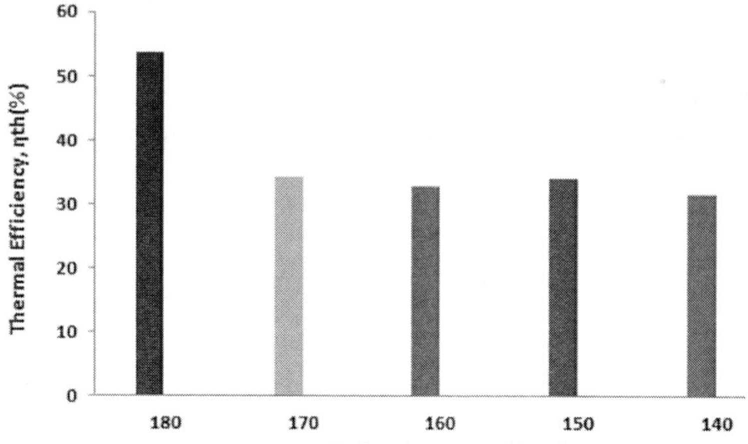

Figure 9. The thermal efficiency to the cylinder diameter

In Figures 8 and 9, we can see the effect of diameter variations on the loss and thermal efficiency. The best thermal efficiency is obtained at a cylinder diameter of 180 mm which is 53.87% with heat losses of only 2.02 kW and the worst thermal efficiency of a cylinder diameter of 140 mm which is 31.7% with heat losses reaching 3.69 kW. The more surface area of the cylinder diameter, the better the thermal efficiency, inversely proportional to the power losses obtained, and vice versa. The larger the diameter of the cylinder, the more supply of air that affects burning and the faster the rate of burning, so that it can reduce the heat loss.

4. Conclusion

From the results of calculations and analysis, it can be concluded that the best burning temperature is at 160 mm sleeve diameter and the best burning time occurs at 170 mm sleeve diameter. Meanwhile, the best thermal efficiency was 53.87% on a 180 mm cylinder diameter with the best heat losses of 2.02 kW and had the fastest boiling time in 9 minutes.

References

[1] N. Phonphuak, S. Thiansem, 2011. *Effects of charcoal on physical and mechanical properties of fired test briquettes*, ScienceAsia, 37, pp 120–124.

[2] S. Jamilatun S, 2011. *Quality properties of burning coconut shell, teak sawdust briquettes, husk briquettes and coal briquettes*. Proceedings of the National Chemical Engineering seminar Universitas Ahmad Dahlan Yogyakarta. [In Bahasa]

[3] B. Esmar, 2011. *Overview of the Process of Forming and Using Coconut Shell Charcoal as Fuel* Journal of Science Research, Vol. 4, No. 3(B). [In Bahasa]

[4] S. Suluh and P. Martina, 2017. *Analysis of Coconut Shell Briquettes as an Alternative Energy Source Reviewed from the Adhesive Material,* Proceedings of the Politeknik Ujung Pandang (Ujung Pandang Polytechnic) Research Seminar, Makassar Sulawesi Selatan. [In Bahasa]

[5] S. Suluh, 2014. *Study of Performance Improvement of Various Stoves with Waste Biomass Briquettes Fuel*, 1th The International Symposium On Smart Material and Mechatronics, Universitas Hasanuddin, pp 39-45, (2014).

[6] A. Haryanto, S. Triyono, 2012. *Household Stove Emissions Study*, *AGRITECH*, Vol. 32, No. 4, NOVEMBER, pp. 425-431. [In Bahasa]

[7] Musabbikhah, H. Saptoadi, Subarmono and M. A. Wibisono, 2015 *Optimization of Biomass Briquette Making Process Using Taguchi Method To Meet the Needs of Alternative Fuels that are Environmentally Friendly*, Journal of Human and Environment, Vol. 22, No. 1, pp. 121-128. [in Bahasa]

The Effect of Tool Geometry of Hole Towards Tool Life in Hard Material Holes

A Y Aminy*[1], S Arief[1], and Sulaiman[1]

[1] Department of Mechanical Engineering, Engineering Faculty, Hasanuddin University, Makassar, Indonesia

*E-mail: ahmadyusrana@yahoo.co.id

Abstract. This study studied the characteristics of machining hard material milling by using carbide Endmill tool with variation machining and carried out without using cooling water. The method used in this study was experimental by providing machining variables (rotation = 191 rpm, 398 rpm, 764 rpm and 1,082 rpm; feeding speed (Vf) = 0.1; 0.18 and 0.26 mm / min). After machining, a tool wear is measured using a digital microscope with magnification 1,000 times. The results of Endmill tool wear measurements show that maximum wear at Vf = 0.1 mm / min = 0.42 mm; Vf = 0.18 mm / min = 0.428 mm and Vf = 0.26 mm / min = 0.4325 mm / min, the greater the feeding speed, the fewer the number of holes produced. While the effect on roughness is that the speed of feed is more influential on roughness compared to changes in workpiece rotation and the rake angle. The roughness data obtained ranged from 0.495 μm to 2.095 μm, this means that the surface roughness value is still in the standard area (Ni ÷ N9 = 0.025 ÷ 6.3 μm) or it can be said that the selected machining variable can be used as the machining variable perforating hard material

1. Introduction

The latest development of machining currently consists of two types, namely: high speed machining and hard machining. The machining of hard material milling has many problems, including tool that are used too quickly to wear out so that production costs rise. High speed machining is triggered by increasing demand to increase productivity with low production costs because this type of machining can increase high cuts, the volume of material release from parent material will increase so that significant savings in machining time will be obtained. In addition, high speed machining is capable of producing smoother products and more precise sizes. Whereas hard machining is carried out on workpieces that have a hardness greater than 40 HRC. The hard milling process can be performed on various types of metals such as alloy steel (alloy steel), steel for bearings (steel), hot and cold steel working tools, high speed steel, die steel, and hardened cast steel [1]. To support the latest machining, the cutting tool used must be reliable and have good material properties, wear resistance and high temperature. One of the development of cutting tools is the use of hard material (hard metal).

Hard material cutting tool is a cutting tool that can minimize lag time because this type of tool is easily replaced when worn. The use of hard metal cutting tool has resulted in maximum productivity that will increase tool change time, precision products, reduce production time and production costs so that product selling prices will be affordable and will ultimately strengthen the company's ability to compete with competitors. The hard metal cutting tool consists of two main parts, the tool handle and the insert tool holder. Tool handles are rectangular or cylindrical rods where one end is clamped to the machine part (lathe or milling) while the other end is used to peel the material (insert tool) which can be replaced.

Studies on the use of tools in milling machineries have been carried out by several researchers, such as those investigated [1] using hard materials AISI 440 C martensitic stainless steel and SCM 440 alloy steel, variations in cutting rates, feeds, and medium cutting depth. The results of his research concluded that the cutting speed and the combination of feeds in the cutting process will increase the working stress of the cutting tool and consequently the cutting tool will blunt rapidly. With blunt cutting tools, product quality and productivity decline. Ghani [1], using a carbide tool coated with Titanium alloy, using various cutting rates, feeding rates and cutting depth levels. And concluded that at the level of simple cutting tool wear is determined by the friction between the tool and the workpiece.

Some researchers have conducted research on the use of carving inserts in the machining process, especially for hard machining. Various methods have been used in the machining process to determine cutting parameters, cutting tool types, measured quantities and analytical methods. All methods used lead to the optimization of machining, which is to get a low cutting force and a small surface roughness. In addition, the development of metal cutting processes with machine tools is also directed at tool endurance, low production costs and product accuracy. One solution is to use a carbide insert tool.

Venkatesan [2] conducted a study of the effect of cutting parameters of Ni-Cr, Inconel 625 alloy, using a tool coated carbide inserts (AlTiN PVD) and using a dry machining process. In this study measurements of cutting force and surface roughness were measured. This is intended to analyze the effectiveness of the temporary machine to search for cutting optimization and to analyze the effect of cutting parameters. Signal-to-noise (S/N) comparison analysis, variance analysis (ANOVA), and regression analysis are performed. This analysis was carried out to obtain lower cutting forces and good surface quality in dry cutting conditions.

Wet machining is machining which in the process is carried out with a cooler. The main function of the cooler is to reduce the cutting temperature by reducing the frictional force and as a heat carrier medium from the cutting area, it also functions as a furious carrier. When the cutting temperature decreases, the tool life will increase. This will be different in the dry machining process where in the dry machining process there is found a large friction force so that the cutting temperature becomes very high. This high cutting temperature will reach the melting point of the tool material and can change the micro structure of the tool tip. Because the tool tip is soft and subjected to pressure as a result of cutting parameters which causes the tool's strength to decrease. This will affect machining products.

Based on the description above, a more in-depth study of the use of inserts is required in the dry machining process of hard materials under conditions of cutting parameters (feeding, cutting depth, cutting speed). In this study, it is expected to show a picture of the quality of milling machining products and their relationship with the surface roughness of dry machining products on hard materials.

2. Machinery Theory

Cutting material with machine tools is an industrial activity that aims to form machine components or other equipment. To facilitate human work and to produce products quickly, effectively and economically, is the basis for the development of chisels and machine tools.

The turning process is the process of changing the shape of the product using a cutting tool and a milling machine. In the process of changing shape, all energy is converted into heat, where heat is caused by friction between growls and chisels and between chisels and workpieces. High pressure due to cutting forces and high temperatures results in a great material release process. This is the impact of the friction force on the tool's friction surface with the workpiece. At very high temperatures can

IOP Conf. Series: Materials Science and Engineering **875** (2020) 012069 doi:10.1088/1757-899X/875/1/012069

damage the tool and will affect the quality of production, also to some extent can affect the work of milling machines.

The quality of machining products is also influenced by the cutting parameters and cutting temperature (*Tp*). Cutting parameters, namely: cutting speed (*Vc*), cutting depth (*a*) and feeding (f). In the machining process it is important to know the values: the cutting force (*F*), the cutting temperature (*Tp*), the rate of furious formation (*Z*) and the cutting time (*t*) [3, 4].

The cutting speed is very influential on the quality of the machining product where the determination of the speed value is very important to know and the cutting speed is directly proportional to the rotation of the workpiece. The relationship between cutting speed and rotation is shown by the following equation [5]:

$$V_c = \frac{\pi \cdot d \cdot n}{1.000} \text{ [mm/min]} \tag{1}$$

The following equation is used to determine the characteristics of machining, namely:

- Feeding speed,

$$V_f = f \cdot n \text{ [mm/min]} \tag{2}$$

- Cutting time,

$$T_c = \frac{l_t}{v_f} \text{ [mm/min]} \tag{3}$$

- Chips formation speed

$$Z = A \cdot V_c \text{ [mm}^3\text{/min]} \tag{4}$$

3. Research Methods

In this study the workpiece material ST 60 is used, the dimensions are 100 x 50 x 10 mm, with a tensile strength of 194.56 kg/mm^2 while the hole means using End Mill Two Flute Carbide. The machining process used is milling and without cooling water. Engine variables selected: rotation = 191 rpm, 398 rpm, 764 rpm, 1,082 rpm; feeding speed = 0.1 mm/min, 0.18 mm/min, 0.26 mm/min and tool geometry changes are made at the rake angle of 2° and 3°.

Figure 1 shows the End mill two flute tool used. Cutter Diameter (d) = 8 mm and Shank diameter (D) = 8 mm. Figure 2 shows the End Mill tool geometry, while Figure 3 shows the Digital Microscope which is used to view the tool type.

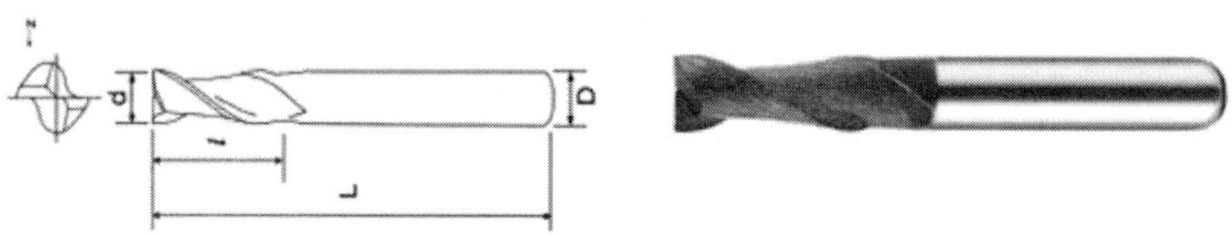

Figure 1. Two flute end mill

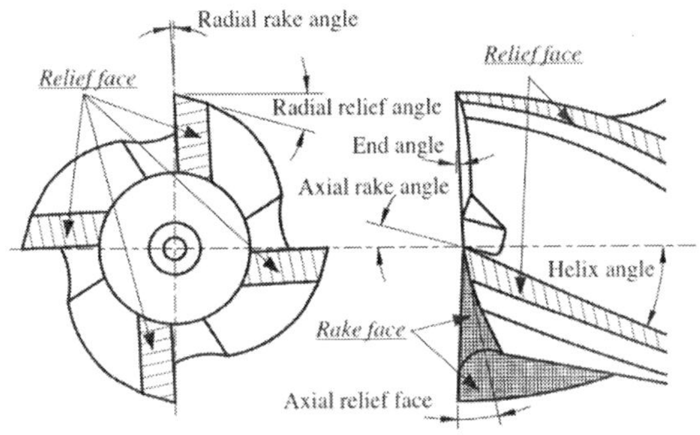

Figure 2. Endmill cutting tool geometry

Figure 3. Digital microscope

The machining process is carried out with 1 set of treatments (cutting angle, rotation, feeding and rake angle), then the tool wear and product surface roughness is measured. Furthermore, the measurement results are illustrated in graphical relation to the machining variables.

The machining of perforation using a milling machine on ST 60 material is carried out to a depth of 10 mm with variations in cutting angle, rotation and feeding speed. The perforation is done first by adjusting the rotation and feed speed then changing the rake angle of rotation and the feeding speed does not change. After that, the rotation changes are carried out as much as 4 times the variation. After completing the punching process, measurements of tool wear and surface roughness are carried out every 1 data set. In this case roughness is done 12 times (data set = 12). While tool wear measurements are carried out every 3 (three) minutes of use until wear reaches 0.38 mm. Surface roughness measurements are carried out using Mitutoyo Surftest Surface Tester 301, while cutting tool wear measurements use a digital microscope.

4. Result

Roughness measurement results are made in the form of a graph of the relationship between rotation and surface roughness at variations in feed speed and cutting angle. The results are shown in Figure 4. While the tool wear results are made in the form of graphs of the relationship of wear to the time of hollowing and are shown in Figure 5.

Figure 4 shows that the higher the round perforation, the roughness value decreases and the variable that is very influential is into the cut. When compared with the cut angle (rake angle), the roughness value at 2° rake angle is lower than the roughness value at 3° rake angle, but the decrease in the roughness value at 3° rake angle is smaller than the roughness value at 2° rake angle. At the 2° rake angle the smallest roughness value was obtained 0.58 µm and the largest roughness value was 2.08 µm. Whereas at the rake angle of 3° the smallest roughness value was obtained 0.495 µm and the largest roughness value was 2.095 µm.

While the tool wear measurement results are made in the form of a graph of the rotation relationship with tool wear on variations in feed speed and cutting angle. The results are shown in Figure 5.

In Figure 5, it appears that the greater the speed of feed, the greater the level of tool wear, as well as the shorter use time of the tool. The greater the cut angle (rake angle), the greater the tool wear while the tool use time has the same value.

At rake angle of 3° and perforation time of 36 minutes there was a decrease in wear especially at the feeding speed Vf = 0.1 mm/min and Vf = 0.18 mm/min. Whereas at 2° rake angle, this did not happen. Perforated tool wear increases with increasing time for drilling.

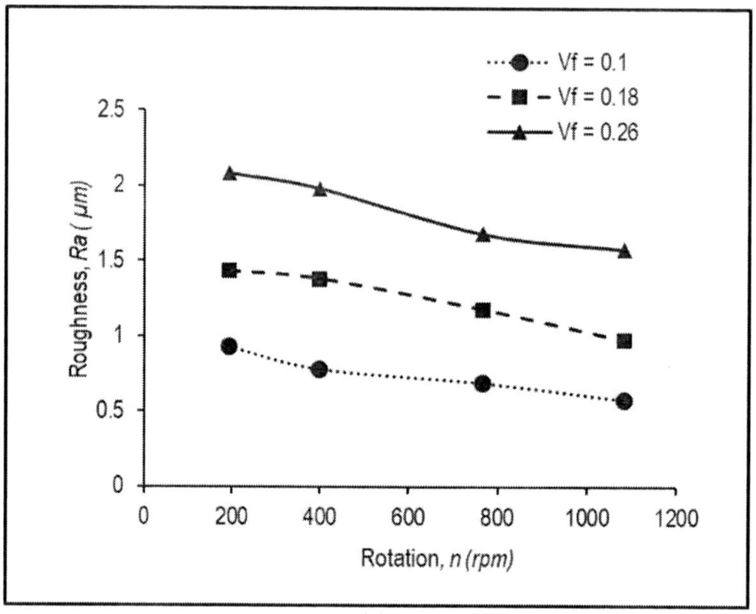

(a)

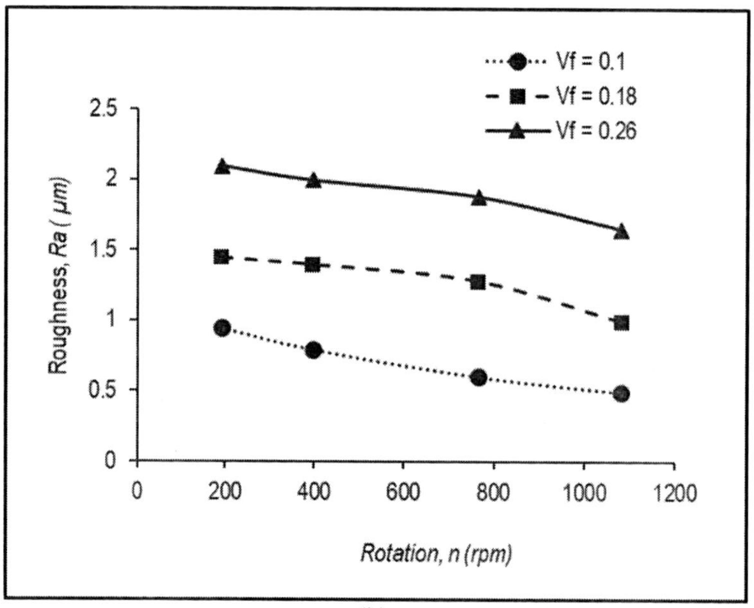

(b)

Figure 4. Graph of the relationship of rotation with surface roughness at various variations in cutting depth and rotation, (a) Cutting angle 2° and (b) cutting angle 3°

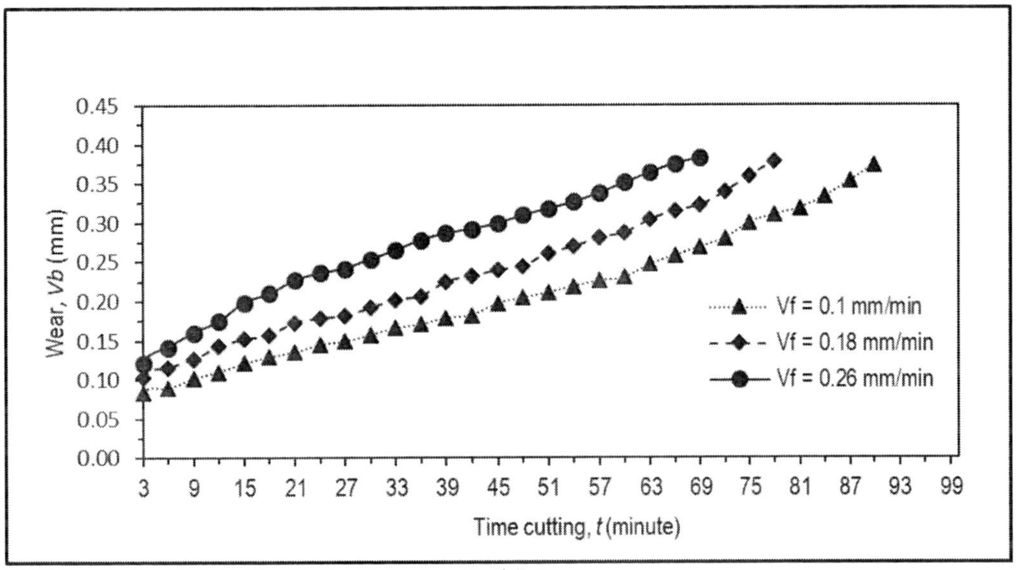

(a)

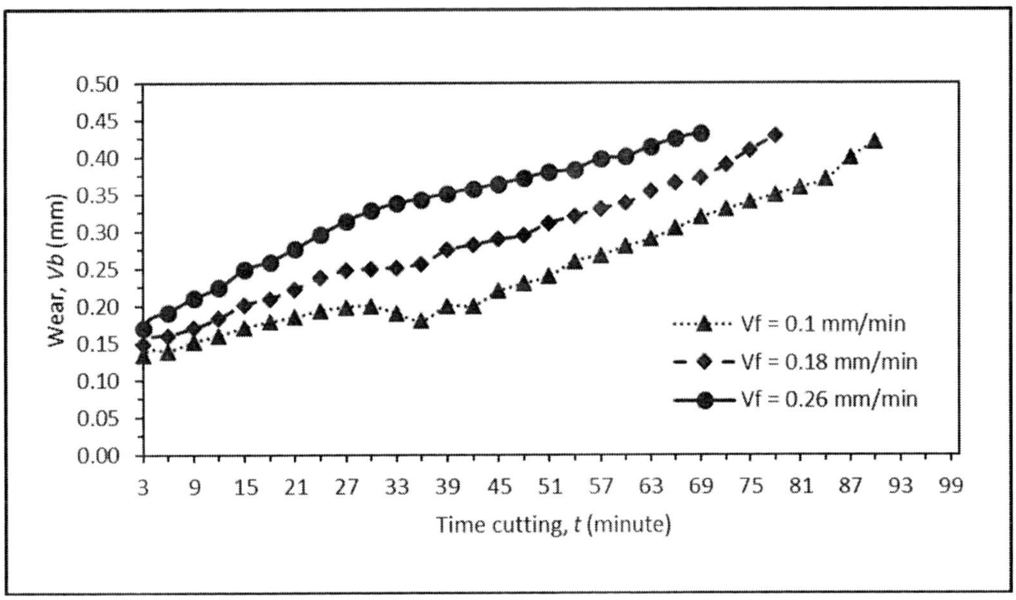

(b)

Figure 5. Graph of the relationship of rotation with tool wear on various variations of feed and rotation, (a) Cutting angle 2° and (b) cutting angle 3°

5. Conclusion

The results showed that the higher the turning hole of the hole, the smaller the surface roughness, while at a large rake angle, the surface roughness was also greater. The variable which is very influential on roughness is the feeding speed (*Vf*). The greater the value of Vf, the greater the resulting roughness perforation. The smallest surface roughness values were obtained at rake angle of 3°, feeding speed of 0.1 mm/min and rotation of the chisel hole 1082 rpm. While the greatest surface roughness value was also obtained at rake angle of 3°, feeding speed of 0.25 mm/min and turning hole of turning tool 191 rpm.

IOP Conf. Series: Materials Science and Engineering **875** (2020) 012069 doi:10.1088/1757-899X/875/1/012069

Surface roughness values obtained in this study were between 0.495 μm to 2.095 μm. This shows that the surface roughness value is still in the standard standard area (Ni ÷ N9 = 0.025 ÷ 6.3 μm) or it can be said that the chosen research variable can be used as a variable of hard material milling machining

References

[1] Abhang L.B. & Hameedullah M., 2010. *"Chip-Tool Interface Temperature Prediction Model for Turning Process"*. International Journal of Engineering Science and Teknology. Vol. 2(4), 2010, pp.382-393.

[2] Ben Salem S., Bayraktar E., Boujelbene M., Katundi D., 2012. "*Effect of Cutting Parameters on Chip Formation in Orthogonal Cutting*". Journal of Achievements in Material and Manufacturing Engineering, AMME. Vol. 50, 2012.

[3] Che Haron C.H., A. Ginting, Arshad, 2006. *"Performance of Alloyed Uncoated and CVD-Coated Carbide H. Tools in Dry Milling of Titanium Alloy Ti-6242S"*. Journal of Materials Processing Technology (2006), Elsevier.

[4] Thamizhmanii S., Hasan S., 2006. "*Analyses of Roughness, Forces and Wear in Turning Gray Cast Iron*", Journal of Achievements in Material and Manufacturing Engineering, AMME. Vol. 17, 2006.

[5] Boothroyd, G., *"Fundamentals of Metal Machining and Machine Tools"*, International Student ed., McGraw-Hill, New York, 1981.

IOP Conf. Series: Materials Science and Engineering **875** (2020) 012070 doi:10.1088/1757-899X/875/1/012070

Structural Modeling Using 2D Shell Element to Predict the Initial Plastic Condition of Tubular Frame Structures

Asykur[1]*, Y S Hadiwidodo[1], and Handayanu[1]

[1]Department of Ocean Engineering, Faculty of Marine Technology, Institut Teknologi Sepuluh Nopember (ITS), Surabaya, Indonesia

*Email: mas.asykur@gmail.com

Abstract. A Fixed Jacket Platform is usually designed using tubular structures in which every steel pipe has a thickness and diameter. Generally, for the structural analysis of this structure, the tubular members are represented using 1D beam elements. But stress distribution, plastic strain and also *ovalization* around the tubular member cannot be determined by a beam element. In this research, all tubular structures were modeled using 2D shell elements to identify how the stress distribution around a tubular member is. And the non-linear analysis was carried out to identify the initial plastic condition. This study was focusing on the splash zone where this area was assumed to be exposed to the extreme wave in 8 directions according to the API RP2A-WSD code, in addition to the structure weight and deck load. Due to each direction of the wave loads, the stress distribution and initial plastic condition are studied. The modeling and non-linear static analysis in this study were done using *Altair Hyperworks* software.

1. Introduction

The tubular frame is one type of structure commonly used in many constructions such as vehicle frames, cranes, and towers including the offshore platform structure. In this research, the strength of a fixed jacket platform was studied. The fixed jacket platform is a huge steel-framed structure used for the exploration and extraction of oil and gas from the earth's crust in shallow water depth that typically consists of tubular members of various diameters and wall thicknesses.

The offshore structures standards i.e. API (American Petroleum Institute) RP2A LRFD and RP2A WSD that normally used for designing Offshore Jacket Platforms provides detailed procedures to assess the performance of the jacket platform [1]. In API RP2A LRFD (Load and Resistance Factor Design) there are many parameters that mandatory to be used for planning, designing and constructing fixed offshore platforms including assessment of loads and specific events that could occur for the selected platform over its intended service live and service functions, while API RP2A WSD (Working Stress Design) provides detailed procedures for the assessment of existing platform structures.

Base on that standard, assessment of jacket platforms has rarely been studied. Abdel Raheem [2] evaluated nonlinear response of fixed jacket offshore platform under structural and wave loads and used SAP2000 software for the analysis. Alessi et al. [3] proposed a new simplified tool used to study the structural analysis of the jacket structure and the retrofitted wind tower configuration using MATLAB and Strand7. Fayazi and Aghakouchak [4] detailed reliability-based Assessment of existing fixed offshore platforms using The Software for Analysis of Ultimate Strength for Framed Offshore

Structures (USFOS). Zadeh et al. [5] analyzed linear response of a fixed offshore platform under the wave forces using SAP2000 software. Sandhya [6] described multiple loads on offshore jacket structure analysis using SACS software. Alhasan and Ibrahim [7] studied the specific parameters for the elements of an existing fixed platform using SACS software. Taheri and Jahangir [8] described the nonlinear static pushover analysis of an existing fixed platform using SACS software. Hadiwidodo and Prastianto [9] studied the response of a fixed offshore platform subjected to extreme wind and wave loading (lateral loads) under corrosion conditions using SACS software. Ali Sari et al. [10] illustrated structural corrosion modeling of fixed offshore platform for life Extension Study using Abacus software.

Except for the last mentioned, all of the above studies are using 1D beam element for simulating the tubular structure, which cross-section characteristics of the beam such as cross-section area, inertia, and torsional constant are the most important parameter. For tubular structures with small diameters, modeling as 1d beam elements may be sufficient, but for tubular structures with relatively large diameters such as fixed jacket platforms, it is not enough because stress concentrations can start from one side of the tubular pipe surface, and also corrosion usually occurs on a part of the tubular pipe surface that cannot be represented by 1D beam element accurately.

2. Literature review

According to Cook [11], there are the advantage of using 2D shell element.

2.1. 2D Plate and Shell Element

Plates may be considered similar to beams, however:

- Plates can bend in two directions
- Plates are flat with a thickness (can't have an interesting cross-section)

Shell elements are different from plate elements in that:

- They carry membrane and bending forces.
- They can be curved.

The most simple shell element combines a bending element with a membrane element.

- E.g., combines a plate element and a plane stress element.
- These elements are flat, therefore it is important that elements are not all co-planar where they meet at a node.

To do a proper FE analysis, the analyst must understand how the structure is likely to behave and how elements are able to behave. In some cases it is more appropriate to use shell elements rather than beam elements. A curved I-beam reacts to moments as shown, therefore shell elements would more accurate than beam elements. Pipe bends react to moments as shown. Use shell elements or specialized beam elements with correction factors.

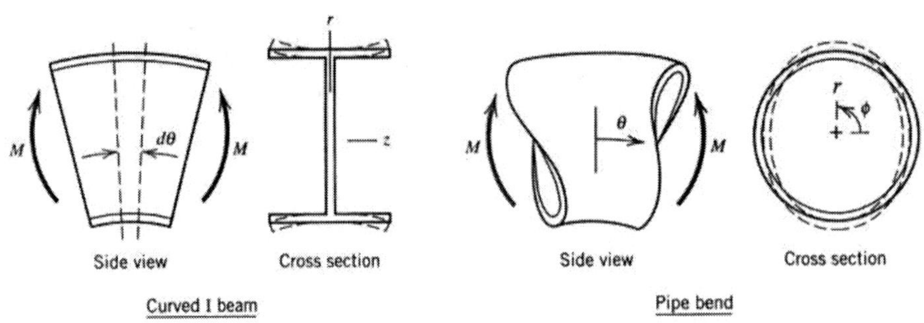

Figure 1. Beam moment

According to above theory, to get more accurate result, the tubular structure must be modeled using 2D Shell element than beam element.

IOP Conf. Series: Materials Science and Engineering **875** (2020) 012070 doi:10.1088/1757-899X/875/1/012070

In this study we were analyzed the initial plastic condition of a fixed jacket platform structure, and non-linear analysis was carried out to identify the initial plastic conditions. This study focused on the splash zone area (the area immediately above and below the mean water level) where this area is assumed to be exposed to the extreme wave in 8 directions according to the API RP2A Working Stress Design (WSD) code, in addition to the structure weight and deck load.

3. Analysis Approach

3.1. Analysis Software

The program used in this research is *Atair Hyperworks* which is based on finite element method (FEM) modeling, the phase of modeling and plastic strength analysis using non-linear static analysis. This software has the ability to offer a wide array of analysis and design to its user, it has many tool for modeling a structure using 1D, 2D, and 3D element, checking element quality, and its can be modified easily. This software is also able to solve non-linear problem and report the structural responses caused by wind, sea waves, and other related loads. One of the important features of this module is the file size and required system are relatively smaller than other module in same case, so this can analyze a complex structure and its loading condition such a giant vessel or other structure by the combination of beam element and shell element.

3.2. Analysis Units

English Engineering units was used in this study. Inch, pound force, Psi are used for length, force, and stress/pressure consecutively.

3.3. Platform Structural Modeling

The platform structure in this study consisted of four leg with 34-in diameter and had five horizontal framing at EL. (-) 180'-0", EL. (-) 125'-0", EL. (-) 75'-0", EL. (-) 30'-0", and EL. (+) 12'-0". The jacket work point elevation was located at EL. (+) 17'-0". All structures were modeled using shell element, except deck beam face was modeled using beam element. Depend on pipe diameter, mesh size were about 5 in - 10 in, and mesh size at splash zone (the part near the water surface) was about 1 in, with fixed boundary condition at the bottom of jacket. The structure section drawing was shown in Figure 5, and structural modeling using *Altair Hyperworks* was shown in Figure 2.

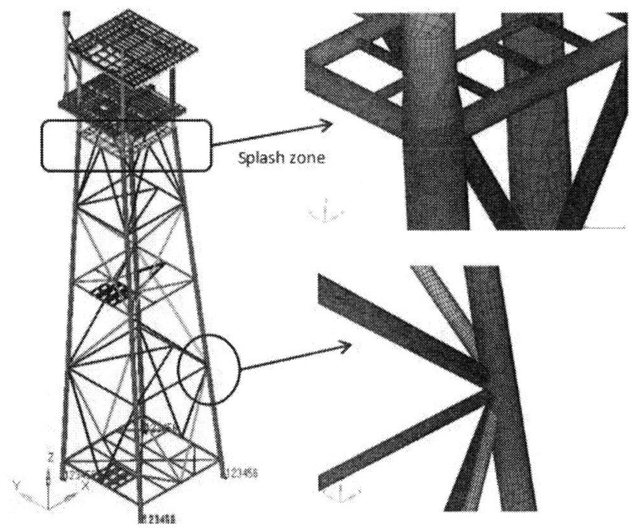

Figure 2. Fixed jacket platform modeling

All steel plate, bars and tubular with $\varnothing < 18$ in was using ASTM A36 steel grade, and API 5L Grade B for tubular with $\varnothing > 18$.

Table 1. Properties of mechanical steel used in the platform

Property Type	ASTM A36	API 5L Grade B
Specific Weight	0.284 lb/in^3	0.284 lb/in^3
Elastic Modulus	2.9 x 10^7 psi	2.9 x 10^7 psi
Shear modulus	1.16 x 10^7 psi	1.16 x 10^7 psi
Poisson ratio	0.26	0.26
Yield Strength	3.63 x 10^4 psi	3.5 x 10^4 psi
Ultimate Strength	5.8 x 10^4 psi	6.0 x 10^4 psi

3.4. Loading Condition

Gravity load, on deck load, and extreme wave load in 8 direction were applied at this analysis. Gravity load was calculated automatically by inputting specific gravity into the model. On deck load was modeled as distributed force, and wave load was modeled as pressure and current flow directions of 20°, 65°, 110°, 155°, 200°, 245°, 290°, and 335°. Non-linear static analysis was applied for evaluating initial plastic condition.

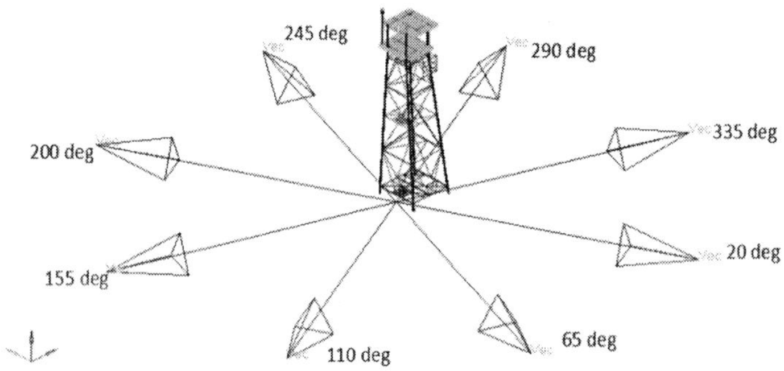

Figure 3. Wave load direction

4. Analysis Result

These analysis results focused on the stress distribution and initial plastic condition at the splash zone due to extreme wave condition.

4.1. Stress Distribution

The equivalent stress was evaluated in this study. Figures 4-11 present the equivalent stress distribution on tubular structure related to wave direction. The stress value are in *Psi* and the part in red means the stress value in that parts are exceed yield stress of material. As indicated in the figure, the most of stress concentration are in the tubular joint between working deck members and diagonal branch members of the platform.

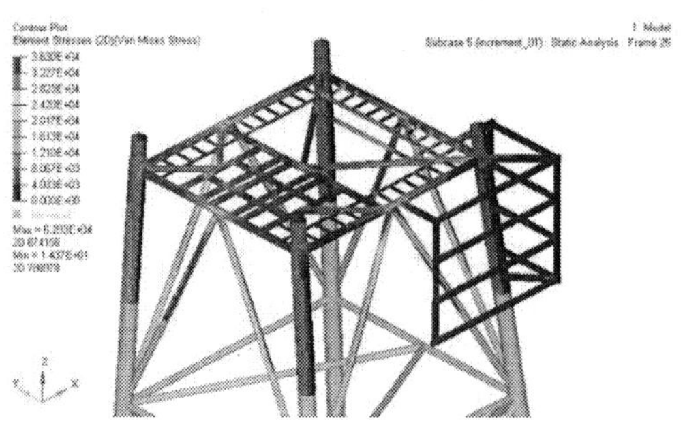

Figure 4. Stress distribution (wave load 20°)

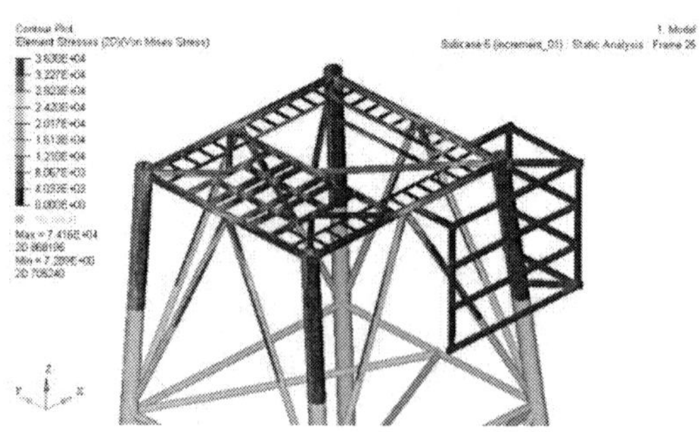

Figure 5. Stress distribution (wave load 65°)

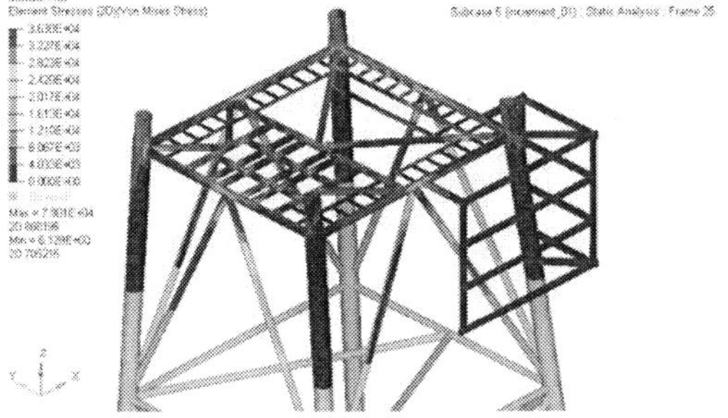

Figure 6. Stress distribution (wave load 110°)

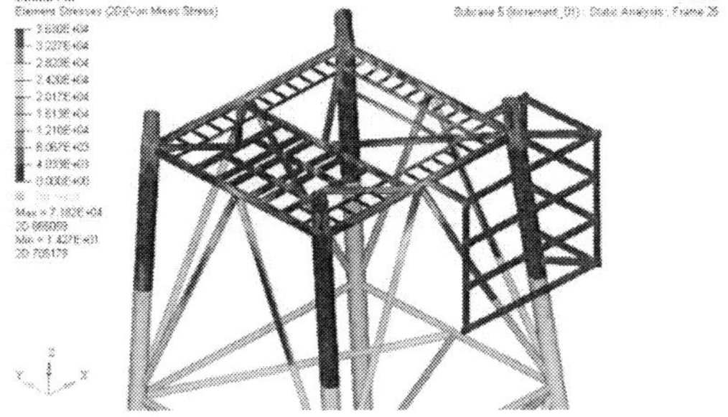

Figure 7. Stress distribution (wave load 155°)

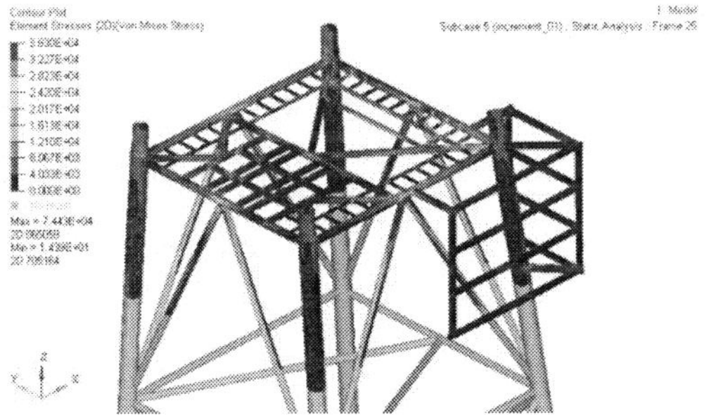

Figure 8. Stress distribution (wave load 200°)

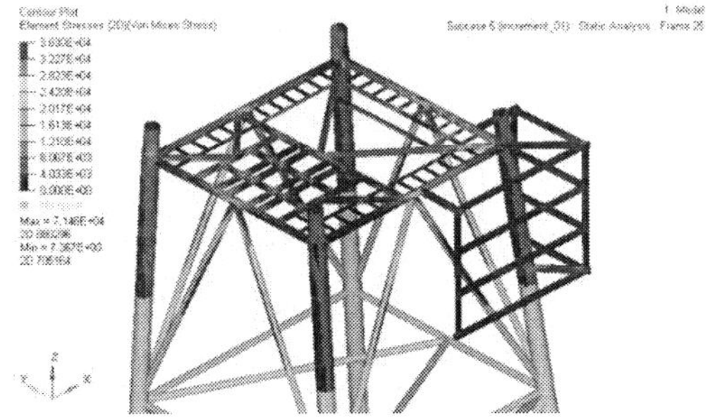

Figure 9. Stress distribution (wave load 245°)

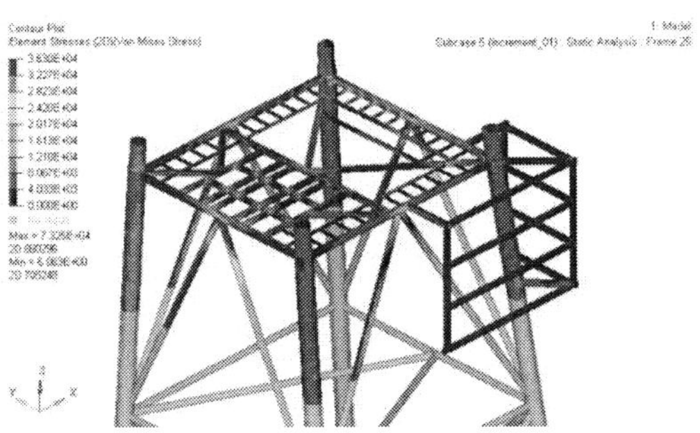

Figure 10. Stress distribution (wave load 290°)

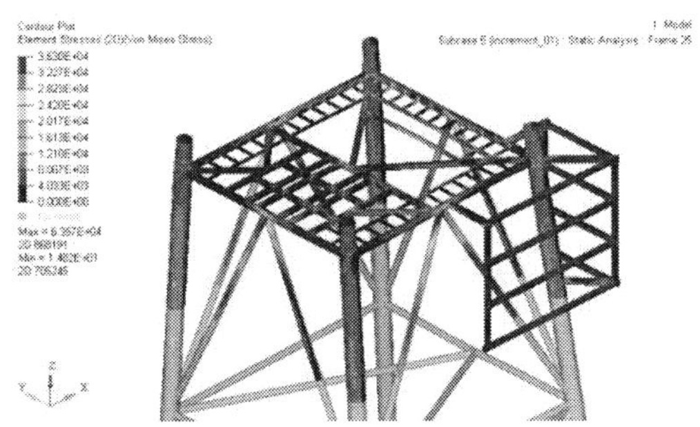

Figure 11. Stress distribution (wave load 335°)

4.2. Plastic Condition

Based on above stress results, the plastic condition most likely occurred in the tubular joint between working deck members and diagonal branch members of the platform. Four locations as shown in Figure 16 were considered to be evaluated. The plastic strain result was summarized in Tables 2 - 4 and the maximum plastic strain in each location are shown in Figures 13 - 16.

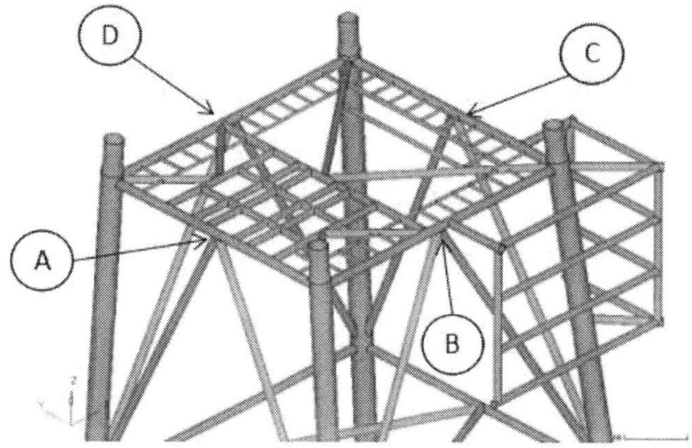

Figure 12. Plastic evaluation location

4.2.1. Summary of Plastic Strain Result. According to the result of this analysis, the most initial plastic condition has occurred in the above-mentioned location in any wave direction. It is indicated by stress value in that location exceeded the ultimate strength of the material. And by using non-linear plastic analysis in software *Hyperworks*, the plastic condition is indicated by plastic strain value. If the value is between 0 to 0.001, it means the strain is elasto-plastic, but if it exceeds 0.001, it means that locations are in plastic condition. Plastic strain results at 4 locations in 8 loading directions are shown in Table 2.

Table 2. Summary of plastic strain result

load direction	location	stress (psi)	plastic strain	deformation
20 deg	A	62930	0.0012	plastic
	B	56910	0.0010	plastic
	C	58068	0.0010	plastic
	D	54250	0.0008	elasto-plastic
65 deg	A	74163	0.0017	plastic
	B	55917	0.0009	elasto-plastic
	C	66912	0.0014	plastic
	D	55657	0.0007	elasto-plastic
110 deg	A	79009	0.0019	plastic
	B	62028	0.0012	plastic
	C	68008	0.0014	plastic
	D	55206	0.0009	elasto-plastic
155 deg	A	60054	0.0011	plastic
	B	71819	0.0016	plastic
	C	53087	0.0008	elasto-plastic
	D	68554	0.0015	plastic
200 deg	A	56848	0.0010	plastic
	B	74430	0.0018	plastic
	C	55636	0.0009	elasto-plastic
	D	70297	0.0016	plastic

load direction	location	stress (psi)	plastic strain	deformation
245 deg	A	67012	0.0014	plastic
	B	56241	0.0010	plastic
	C	71458	0.0016	plastic
	D	56357	0.0010	plastic
290 deg	A	72965	0.0017	plastic
	B	50359	0.0007	elasto-plastic
	C	73253	0.0017	plastic
	D	53419	0.0008	elasto-plastic
235 deg	A	63517	0.0013	plastic
	B	51993	0.0008	elasto-plastic
	C	59526	0.0011	plastic
	D	52068	0.0008	elasto-plastic

From the above data, the loading directions were sorted based on the plastic strain value in Table 3, and the maximum plastic strain value in each location are shown in Table 4.

Table 3. Maximum plastic strain by loading direction

No.	plastic strain	load direction
1	0.00193	110 deg
2	0.00176	200 deg
3	0.00173	65 deg
4	0.00167	290 deg
5	0.00164	155 deg
6	0.00162	245 deg
7	0.00126	335 deg
8	0.00123	20 deg

Table 4. Maximum plastic strain by location

No.	plastic strain	location	Load direction
1	0.00193	A	110 deg
2	0.00176	B	200 deg
3	0.00167	C	290 deg
4	0.00157	D	200 deg

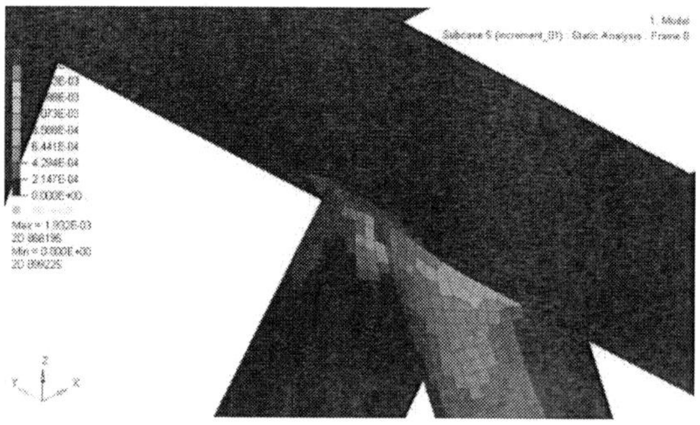

Figure 13. Plastic condition at location A

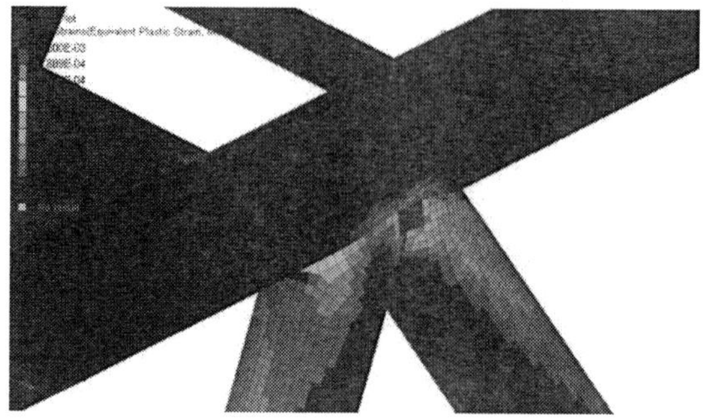

Figure 14. Plastic condition at location B

Figure 15. Plastic condition at location C

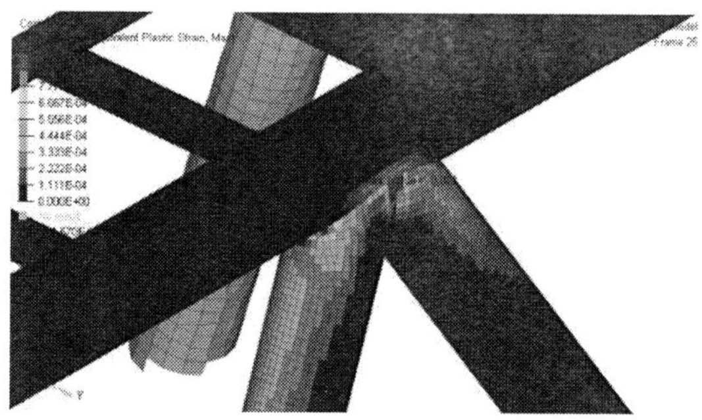

Figure 16. Plastic condition at location D

At the above figures, the plastic condition based on the above figures, the initial plastic condition has occurred in the intersection between working deck members and diagonal branch members of the platform.

4.2.2. Overall Plastic Analysis Result. In addition, the plastic analysis result of the whole platform structure were resumed in Table 5 and Figure 17. The plastic strain was constant at a certain value depending on the loading conditions.

Table 5. Maximum plastic strain

Iteration	Wave direction							
	20 deg	65 deg	110 deg	155 deg	200 deg	245 deg	290 deg	335 deg
0	0	0	0	0	0	0	0	0
1	4.48E-03	2.01E-03	2.22E-03	4.97E-03	5.05E-03	3.28E-03	2.66E-03	4.11E-03
2	8.00E-03	2.85E-03	3.81E-03	9.19E-03	9.18E-03	3.66E-03	2.82E-03	7.39E-03
3	8.06E-03	2.89E-03	3.86E-03	9.27E-03	9.25E-03	3.77E-03	2.82E-03	7.45E-03
4	8.06E-03	2.89E-03	3.86E-03	9.27E-03	9.25E-03	3.77E-03	2.82E-03	7.45E-03

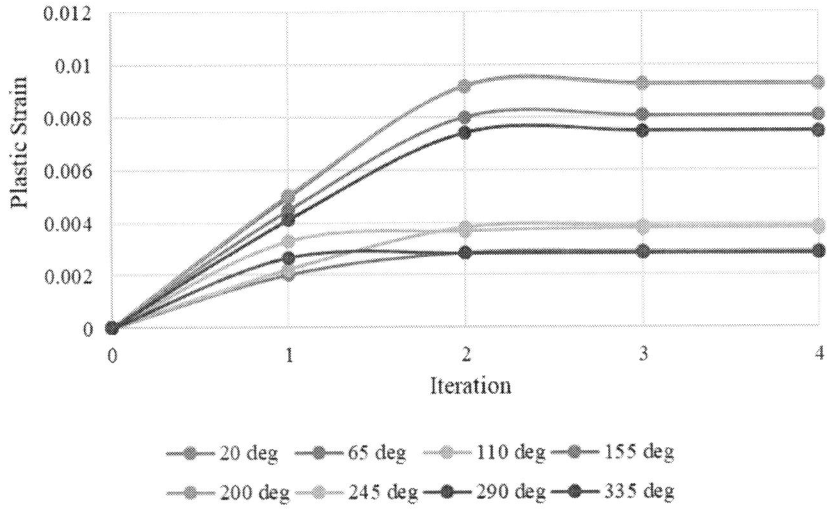

Figure 17. Plastic strain curve

4.3. Ovalization of Tubular Frame

The difference between the beam frame and tubular frame is which an extreme load can cause the ovalization of tubular frame. Ovalization is the deviation of the tubular shape from a circle to an elliptic cross-section or the process of changing to an oval shape that can be evaluated by 2D shell elements modeling as shown in Figure 18.

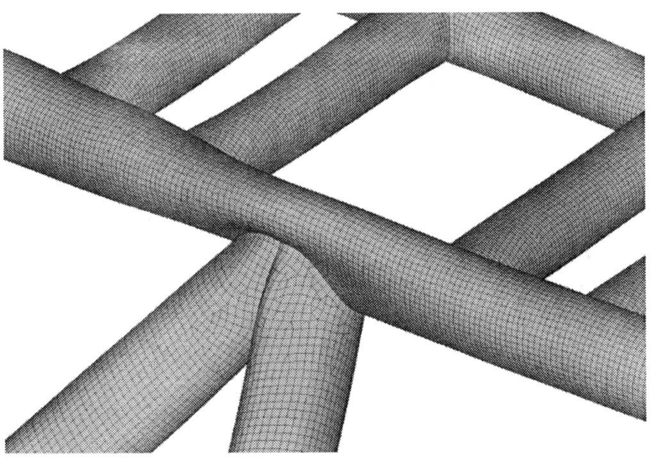

Figure 18. Process of ovalization

The maximum ovalization of each evaluation locations were summarized in Table 6.

Table 6. Ovalization of tubular frame

Location	Diameter of horizontal frame (in)	Ovalization (in)
A	16	0.02
B	16	0.65
C	16	0.06
D	16	0.06

5. Conclusion

A methodology for structural modeling in strength assessment of tubular structure using 2D shell elements was studied. The approach was applied in case studies in which extreme wave loads in 8 directions were assumed applied on a four legs fixed jacket platform. The analysis results indicate that the critical point of this jacket platform structure is the tubular joint between working deck members and diagonal branch members of the platform. In this analysis, the ovalization value of the tubular frame is quite small compared with its diameter, but when an extreme concentration load is working on a part of the tubular frame, the ovalization can be more significant.

On a big tubular structure modeling such as fixed jacket platform, using 2D shell element is more suitable than using 1d beam element to get more accurate and detail result. Also, due to some tubular member must be watertight in which hydro-static pressure work on it, it can be represented only by 2D element or 3d element. Furthermore, Strength assessment of existing structure under corrosion can be done more accurately by modeling corrosion's shape and thickness using 2D shell element based on an inspection data. But, the effectiveness of this analysis depends on the software performance and hardware capacity related to the modeling and analysis time.

Acknowledgments

The authors highly appreciate to PT. Terafulk Ship Design, Surabaya, Indonesia who have allowed using software *Altair Hyperworks* for this study.

References

[1] American Petroleum Institute 2000 *Recommended Practice for Planning Designing and Constructing Fixed Offshore Platforms - Working Stress Design* 21st edition (Washington, District of Columbia USA API Publishing Services) chapter 17.
[2] Abdel Raheem S E 2013 Nonlinear Response of Fixed Jacket Offshore Platform Under Structural and Wave Loads *Coupled Systems Mechanics* Vol 2 No. 1 pp 111-126.
[3] Lorenzo Alessi José A F O Correia and Nicholas Fantuzzi 2019 Initial Design Phase and Tender Designs of a Jacket Structure Converted into a Retrofitted Offshore Wind Turbine *Energies 2019 12,* 659.
[4] Alireza Fayazi and Aliakbar Aghakouchak 2015 Reliability Based Assessment of Existing Fixed Offshore Platforms Located in the Persian Gulf *International Journal of Maritime Technology* vol 4 pp 37-50.
[5] Zadeh S M G Payambarnia P Baghdar R S Alikhani E T 2015 Finite Element Analysis of Offshore Jacket Affected by Marine Forces *Bulletin of Environment, Pharmacology and Life Sciences* Vol 4 [Spl issue 1] pp 152-159.
[6] Gande Sandhya 2018 Analysis of Offshore Jacket Structure *International Research Journal of Engineering and Technology* vol 5 pp 722-729.
[7] Mahar Jasem Alhasan and Adnan Mohamed Ibrahim 2018 *American Journal of Innovative Research and Applied Sciences* ISSN 2429-5396 pp 91-96
[8] Abdolrahim Taheri and Mohammad Jahangir 2017 *International Journal of Coastal & Offshore Engineering* vol 1 No 3 pp 57-64.
[9] Hadiwidodo Y H and Prastianto R W 2018 Structural Corrosion Modeling for Reserve Strength Assessment of Fixed Jacket Platform *International Journal of Civil Engineering and Technology* vol 9 pp 1006-1012.
[10] Ali Sari, Ekkirala R, Nguyen H, Arablouei A and Azimov U 2016 Structural Corrosion Modeling for Strength Assessment of Fixed Offshore Platforms for Life Extension Study *Offshore Technology Conference* OTC-26521-MS.
[11] Cook R D Malkus D and Plesha M E 1996 *Concepts and Applications of Finite Element Analysis* 4th Edition (New York: Wiley–Interscience).

IOP Conf. Series: Materials Science and Engineering **875** (2020) 012071 doi:10.1088/1757-899X/875/1/012071

Plate Deflection Analysis with Different Mesh

M Z M Alie[1]*, Juswan[1], D Paroka[1], and M R S Rakib[1]

[1]Department of Ocean Engineering, Engineering Faculty, Hasanuddin University, Makassar, Indonesia

*E-mail: zubair.m@eng.unhas.ac.id

Abstract. Ocean structures like a jacket, jack-up, and other floating structures consist of many structural elements to support and strengthen from internal and external loads. The structural element, like a beam, plate, and the stiffened plate, has a significant influence on the ultimate strength of the global structure. So that all these elements components should be evaluated and analyzed for the local and global of the structural strength. The objective of the present study is to analyze the plate deflection due to different mesh dimension. The plate is modeled by shell elements with different mesh dimensions; those are 8 x 8 and 16 x 16. The clamped edges as boundary conditions are applied to all sides of the plated. The applied load is located at the center of the plate. The Finite Element Method is used to analyze the plate deflection with different mesh dimensions under axial load acts at the center of the plate. The result obtained by FE Method is therefore compared to the analytical solution. It is found that the plate deflection obtained by the FE method is in good agreement with the analytical solution, and the behavior of the plate is also presented in this study.

1. Introduction

Generally, the ocean structure is very complex and consists of many subdivision components such as beam, plate, stiffened plate, and so on. All these components must be analyzed and evaluated due to the applied load on the structure. The loads could be internal and external loads, for instance, the structure itself, wind, waves, currents, etc. The ocean environment is severe, complex, and continuously varying, especially waves that always acting periodically in the horizontal direction. On the other hand, the load of the structural component is also applied in the vertical direction. Therefore, due to applied load acting on the structure and the components itself, the behavior in terms of deflection and deformation must be taken into consideration.

The analysis of plate behavior such as deflection or deformation on the structural engineering like a ship and other ocean structures has been conducted by some papers. Vaz [1] presented an experimental campaign and a numerical finite element model to obtain the ultimate strength of tubular structures with circular perforated damage subjected to axial compression. Tekgoz [2] analyzed the effect of structural damage and associated neutral axis translation and rotation of the residual load-carrying capacity of a container ship hull subjected to asymmetrical bending loading. The influence of nonlinear finite element method models on the ultimate bending moment for hull girder was studied by Xu [3]. The residual strength of an Aframax-class double hull oil tanker damaged in the collision had been assessed by Parunov [4]. The ultimate strength of ship hull girder strength was conducted by Muis Alie [5] considering the critical element at the deck part under sagging condition. Muis Alie [6] analyzed the effect of symmetrical and unsymmetrical configuration shapes on buckling and fatigue strength analysis of the fixed offshore platform. Two models of the fixed offshore structure were taken to be analyzed with the same dimension but different configuration shapes. The numerical calculation was performed to analyze the buckling and fatigue strength of both structures. Muis Alie [7] analyzed the residual strength of asymmetrically damaged ship hull girder using the beam finite element method. The Bulk Carrier ship model was taken to be analyzed.

In addition, a computational model for analysis of local buckling and postbuckling of stiffened panels were derived by Byklum [8]. An approximate analytical method to determine the large-deflection behavior of rectangular simply supported thin plates under transverse loading was described by Bakker [9]. The comparison of the ultimate strength of T and Y stiffeners subjected to lateral load was conducted by Badran [10]. Tanaka [11] performed a series of collapse analyses by applying nonlinear FEM on stiffened panels subjected to longitudinal thrust. Lee [12] applied the nonlinear finite element analysis to examine the ultimate strength characteristics of steel brackets, and to develop a simple design formula to predict the ultimate strength of a steel bracket. The ultimate strength behavior of steel concrete steel sandwich plate under concentrated load was studied by Yan [13]. The development of an assessing formula for ultimate strength of hull plate with pitting corrosion damage under combined loading was done by Zhang [14]. The ultimate strength assessment method of corroded plates based on corroded volume loss had been used proposed by the theoretical derivation and finite element analysis by Zhang [15].

The objective of the present study is to analyze the plate deflection due to different mesh dimension. The plate is modeled by shell elements with different mesh dimensions; those are 8 x 8 and 16 x 16. The clamped edges as boundary conditions are applied to all sides of the plated. The applied load is located at the center of the plate. The Finite Element Method is used to analyze the plate deflection with different mesh dimensions under axial load acts at the center of the plate. The result obtained by FE Method is therefore compared to the analytical solution. It is found that the plate deflection obtained by the FE method is in good agreement with the analytical solution, and the behavior of the plate is also presented in this study.

2. Fundamental Study

The deflection can be analyzed by using an example like a plate with the thickness as a parameter. The thickness is constant to all edges of the plate. Assuming that y-axes are the long axes and z-axes in the positive direction to downward, as shown in Figure 1. If the breadth of the plate is l, so that the small element of the plate may be assumed as a beam with the rectangular cross-section having a length (l) and depth (h), respectively.

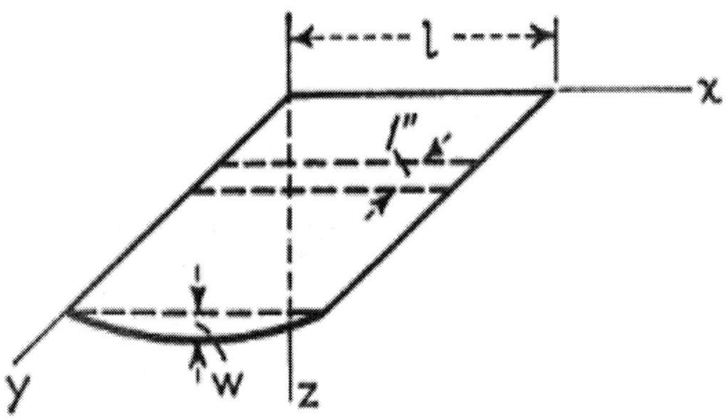

Figure 1. Deflection on plate

To calculate the stress-strain on a plate, it is assumed that the beam theory can be used and the cross-section remained plane so that it can be deflected during bending. Then, these cross-sections rotate only to its neutral plane. If the normal forces do not apply to the edge of the cross-section, then the beam surface is concise with the center surface. The curvature of the beam deflection can be taken as $-\frac{\partial^2 w}{\partial x^2}$ where w is deflection on beam in the z-direction, and this is small than beam length (l). The unit strain ε_x of the layer in the z-direction of the center surface, as shown in Figure 1 is $-\frac{z\,\partial^2 w}{\partial x^2}$. By applying Hooke's law, the strain ε_x and ε_y defined in the normal stresses σ_x and σ_y acting on the element as shown in Figure 1 can be determined by the following equation,

$$\begin{cases} \varepsilon_x = \dfrac{\sigma_x}{E} - \dfrac{\upsilon \sigma_y}{E} = 0 \\ \varepsilon_y = \dfrac{\sigma_y}{E} - \dfrac{\upsilon \sigma_x}{E} = 0 \end{cases} \tag{1}$$

The lateral strain in y-direction must be zero to keep continuity on the plate during bending, so that $\sigma_y = \upsilon \sigma_x$. Substitute this value into equation (1) yields,

$$\begin{cases} \varepsilon_x = \dfrac{\left(1-\upsilon^2\right)\sigma_x}{E} \\ \sigma_x = \dfrac{E\,\varepsilon_x}{1-\upsilon^2} = \dfrac{Ez}{1-\upsilon^2}\dfrac{d^2 w}{dx^2} \end{cases} \tag{2}$$

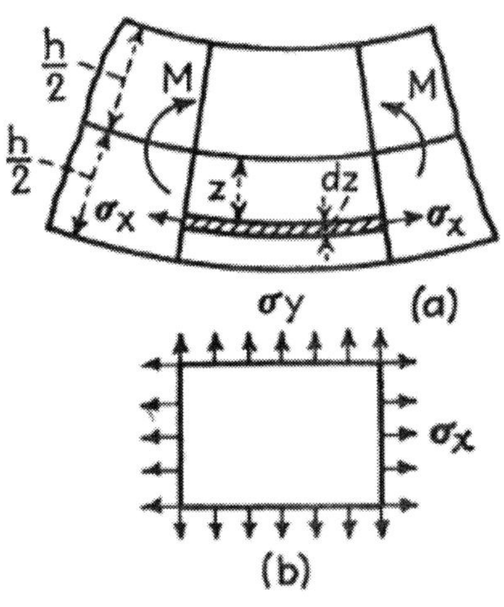

Figure 2. Cross-section of the plate with bending

After obtaining the bending stress σ_x, by integrating the equation the bending moment can be obtained on the element layer,

$$M = \int_{-\frac{h}{2}}^{\frac{h}{2}} \sigma_x \, z\, dz = \int_{-\frac{h}{2}}^{\frac{h}{2}} \frac{E z^2}{1-\upsilon^2}\frac{d^2 w}{dx^2} dz = -\frac{E h^3}{12\left(1-\upsilon^2\right)}\frac{d^2 w}{dx^2} \tag{3}$$

where,

$$\frac{E h^3}{12\left(1-\upsilon^2\right)} = D \tag{4}$$

Therefore, the equation of deflection curve on the element can be determined as follow,

$$D\frac{d^2w}{dx^2} = -M \tag{5}$$

Where D changed to EI on beam called the flexural rigidity of plate.

For the general equation of plate, the notations are expressed in Figure 3 as follow,

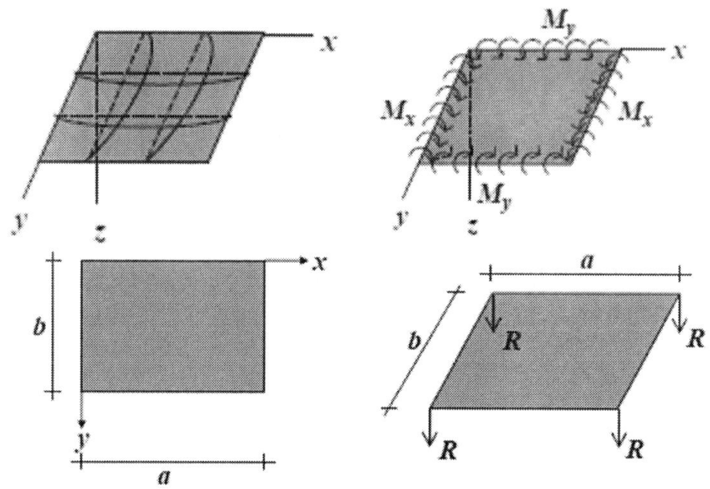

Figure 3. Notations on plate

The general equation of the plate is written by

$$\frac{\partial^4 w}{\partial x^4} + \frac{\partial^4 w}{\partial y^4} + 2\frac{\partial^4 w}{\partial x^2 \partial y^2} = \frac{q}{2} \tag{6}$$

where,

w = deflection
x, y = distance
q = load

$$q = q_0 \sin\frac{\pi x}{a} \sin\frac{\pi y}{b} \tag{7}$$

q_0 = load intensity at the plate center
D = bending stiffness

By giving the load, the equation of the plate becomes,

$$\frac{\partial^4 w}{\partial x^4} + \frac{\partial^4 w}{\partial y^4} + 2\frac{\partial^4 w}{\partial x^2 \partial y^2} = \frac{q_0}{D} \sin\frac{\pi x}{a} \sin\frac{\pi y}{b} \tag{8}$$

The boundary conditions are given by,

$$\begin{cases} x = 0\, and\, x = a, w = 0\, M_x = 0 \\ y = 0\, and\, y = b, w = 0\, M_y = 0 \end{cases}$$

Deflection of plate required the boundary conditions is determined by the equation as,

$$w = c \sin\frac{\pi x}{a} \sin\frac{\pi y}{b} \tag{9}$$

The constant c must be calculated by considering the boundary conditions, yields,

$$c = \frac{q_0}{D\pi^4} \frac{1}{\left(\frac{1}{a^2} + \frac{1}{b^2}\right)^2} \tag{10}$$

Finally, the deflection equation becomes,

$$w = \frac{q_0}{D\pi^4} \frac{1}{\left(\frac{1}{a^2} + \frac{1}{b^2}\right)^2} \sin\frac{\pi x}{a} \sin\frac{\pi y}{b} \tag{11}$$

3. Results and Discussions

In the present study, plate deflection is analyzed using the Finite Element Method, and the result is compared to the analytical solution. The plate is modeled by two kinds of different mesh dimensions, namely 8 x 8 and 16 x 16. Figures 4 and 5 show the plate deflection obtained by FE Method.

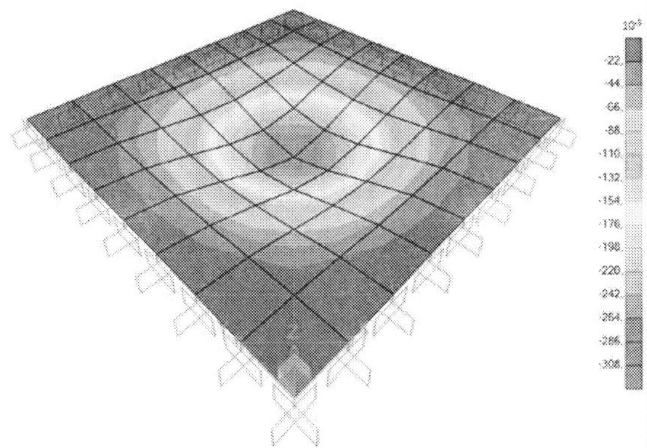

Figure 4. Deflection with mesh 8 x 8

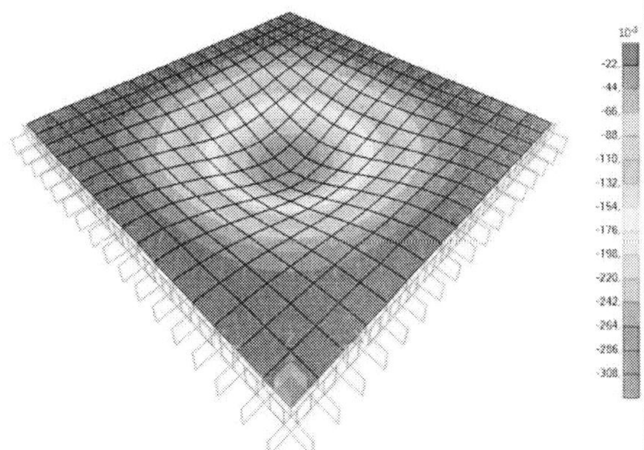

Figure 5. Deflection with mesh 16 x 16

The comparison of the plate deflections between FEM and analytical solution with different mesh dimensions are summarized in Tables 1 and 2 as follow,

Table 1. Comparison of deflection with meshing 8 x 8

h	Deflection		Error Percentage
	Analytical Solution	FEM	
1	0.000306	0.000322	5.31%
0.8	0.000725	0.000763	5.28%
0.5	0.002446	0.002575	5.27%
0.3	0.019569	0.020600	5.27%
0.1	0.305760	0.321881	5.27%
Mean			5.28%

Table 2. Comparison of deflection with meshing 16 x 16

h	Deflection		Error Percentage
	Analytical Solution	FEM	
1	0.000306	0.000311	1.71%
0.8	0.000725	0.000738	1.83%
0.5	0.002446	0.002490	1.80%
0.3	0.019569	0.019920	1.80%
0.1	0.305760	0.311245	1.79%
Mean			1.78%

4. Conclusions

The plate deflection analysis has been conducted using the Finite Element Method considering different mesh dimensions. The following conclusion can be drawn; the result of plate deflection obtained by the finite element method is almost identic with the analytical solution, the plate model using mesh 16 x 16 is more accurate than mesh 8 x 8 to obtain the deflection value.

References

[1] Vaz M A, Cyrino J C R, Hernández I D, Zegarra V D, Martinez J L and Liang D A 2018 Experimental and numerical analyses of the ultimate compressive strength of perforated offshore tubular members *Mar. Struct.* **58** 1–17.

[2] Tekgoz M, Garbatov Y and Guedes Soares C 2018 Strength assessment of an intact and damaged container ship subjected to asymmetrical bending loadings *Mar. Struct.* **58** 172–98.

[3] Xu M C, Song Z J, Pan J and Soares C G 2017 Ultimate strength assessment of continuous stiffened panels under combined longitudinal compressive load and lateral pressure *Ocean Eng.* **139** 39–53.

[4] Parunov J, Rudan S and Bužančić Primorac B 2017 Residual ultimate strength assessment of double hull oil tanker after collision *Eng. Struct.* **148** 704–17.

[5] Muis Alie M Z 2018 Simplified approach on the ultimate hull girder strength of asymmetrically damaged ships *Int. J. Offshore Polar Eng.* **28** 200–5.

[6] Muis Alie M Z 2016 The Effect of Symmetrical and Asymmetrical Configuration Shapes on Buckling and Fatigue Strength Analysis of Fixed Offshore Platforms *Int. J. Technol.* **7** 1107

[7] Muis Alie M Z 2016 Residual Strength Analysisof Asymmetrically Damaged Ship Hull GirderUsing Beam Finite Element Method *Makara J. Technol.* **20** 7.

[8] Byklum E and Amdahl J 2002 A simplified method for elastic large deflection analysis of plates and stiffened panels due to local buckling **40** 925–53.

[9] Bakker M C M Ã, Rosmanit M and Hofmeyer H 2008 Thin-Walled Structures Approximate large-deflection analysis of simply supported rectangular plates under transverse loading

using plate post-buckling solutions **46** 1224–35.

[10] Badran S F, Saddek A B and Leheta H W 2013 Ultimate strength of y and T stiffeners subjected to lateral loads with three different levels of initial imperfection *Ocean Eng.* **61** 12–25.

[11] Tanaka S, Yanagihara D, Yasuoka A, Harada M, Okazawa S, Fujikubo M and Yao T 2014 Evaluation of ultimate strength of stiffened panels under longitudinal thrust *Mar. Struct.* **36** 21–50.

[12] Lee S E, Thayamballi A K and Paik J K 2015 Ultimate strength of steel brackets in ship structures *Ocean Eng.* **101** 182–200.

[13] Yan J B, Wang J Y, Liew J Y R, Qian X and Zong L 2016 Ultimate strength behaviour of steel-concrete-steel sandwich plate under concentrated loads *Ocean Eng.* **118** 41–57.

[14] Zhang Y, Huang Y, Zhang Q and Liu G 2016 Ultimate strength of hull structural plate with pitting corrosion damnification under combined loading *Ocean Eng.* **116** 273–85.

[15] Zhang Y, Huang Y and Meng F 2017 Ultimate strength of hull structural stiffened plate with pitting corrosion damage under unaxial compression *Mar. Struct.* **56** 117–36.

IOP Conf. Series: Materials Science and Engineering **875** (2020) 012072 doi:10.1088/1757-899X/875/1/012072

NC3 CPP Jacket Launching Analysis by Using Numerical Simulation (Case Study of NC3 Gas Field, SK316 Block Bintulu Sarawak Malaysia)

M Saeful M H[1]*, Juswan[1], and D Paroka[1]

[1]Department of Ocean Engineering, Engineering Faculty, Hasanuddin University, Makassar, Indonesia

*Email: saefulmh@gmail.com

Abstract. In most relatively shallow waters, fixed platforms are the optimum economical choice. One of the important aspects of offshore fixed platform design is their installation procedure. Among the several conditions during installation, launching is probably the most critical. This operation has to be modelled and examined carefully in order to insure safe separation of jacket from barge. Sensitivity studies are performed to clarify the influences of the initial condition of launch barge to the jacket launch process. Launch analysis is carried out by using numerical simulation methods that supported by MOSES software. Jacket structure and launch barge are modelled on SACS and MOSES respectively. The jacket structure is laid on top of barge with the position of the top of jacket in the stern launch barge. Simulation results on the base case with a barge initial trim is 3.5° and a draft at midship is 6.1 m show that the jacket launch process can take place safely. Sensitivity analysis shows that the greater the trim angle and the initial draft of the barge the shorter the sliding speed and the reaction of the rocker arm decreases. While changes in the LCG jacket and the additional weight of the jacket have a significant effect on the bottom clearance and jacket free floating conditions after being separated from the launch barge.

1. Introduction

Jacket platform is one of the most widely used platforms in the offshore engineering field. The size of jackets becomes larger and larger with the development of oil and gas exploration. The jackets are transported to the operation water with barges after they have been fabricated in the shipyard, then the jackets slip into the water. This slipping process is called launch and it is the most hazardous stage during the installation of a jacket. Jacket is firstly fixed to the barge at the beginning of a launch operation. Secondly, the ballast of the barge is adjusted to ensure the barge with a predetermined trim. Then, the jacket is released and slides into the sea driven by its gravity or an external force depending on the specific case. Finally, the jacket leaves the barge and reaches a balance position after certain duration. The duration of jacket launch only lasts from 1 to 2 minutes. Moreover, the state of the jacket is out of control after the jacket is released. Therefore, evaluations on every aspect of this process must be made carefully before the launch at sea.

Vasicek and Lu [1] presented a numerical model of jacket launching. They used an iterative finite difference scheme to solve the governing equations of motions. Sphaier et al. [2] describe the theoretical backgrounds of a launch modelling software. In their work, the system of governing differential-algebraic equations is solved (algebraically) at the beginning of time step with accelerations and reaction forces as unknowns. These values are assumed constant through the time step and displacements at the end of time step are calculated. Nelson et al. [3] presents a numerical modelling of a similar case; lifeboat launching. Similar to Sphaier et al. [2], accelerations and reaction forces are calculated at the beginning of each time step. Reactions are considered constant through the step and the resulting system of differential equations is solved using the Runge-Kutta method. Jo et al. [4] found that with large trim and draft, structural loads of jacket tend to decrease. And trim contributing more than draft on structural load. He also did a sensitive analysis on trim, draft of barge and centre of gravity etc. Jo et al. [5] built numerical models to analysis various conditions for the launch procedure and discussed and investigated the launching criteria.

In this article, we conducted a numerical simulation method that supported by MOSES software and performed a sensitivity studies to clarify the influences of the initial condition of launch barge to the jacket launch process.

2. General Equation

The Newton-Euler equations of motion state that rate of change of system momentum equals the forces acting on system. We consider jacket and barge as two separate systems, with the common reaction forces acting on both. In practice jacket is placed on barge deck such that their CG's lay in a vertical plane above each other, eliminating any yawing moments. Therefore, launch is essentially a two-dimensional problem. In this regard and by assuming jacket and barge as rigid bodies, position of each body can be represented by three components, namely X, Z and θ coordinates of their CG. Equations of motion for these six degrees of freedom are as follows [6]:

$$\left.\begin{aligned} m_j\ddot{X}_j &= P_c^X + P_w^X + P_{F,j}^X \\ m_j\ddot{Z}_j &= P_c^Z + P_w^Z + P_{F,j}^Z - W_j \\ I_j\ddot{\theta}_j &= M_c^Y + M_{F,j}^Y + \left(P_c^X + P_w^X\right)\times\left(Z_c - Z_j\right) - \left(P_c^Z + P_w^Z\right)\times\left(X_c - X_j\right) \end{aligned}\right\} \quad (1)$$

$$\left.\begin{aligned} m_b\ddot{X}_b &= P_c^X + P_w^X + P_{F,b}^X \\ m_b\ddot{Z}_b &= P_c^Z + P_w^Z + P_{F,b}^Z - W_b \\ I_b\ddot{\theta}_b &= M_c^Y + M_{F,j}^Y + \left(P_c^X + P_w^X\right)\times\left(Z_c - Z_b\right) - \left(P_c^Z + P_w^Z\right)\times\left(X_c - X_b\right) \end{aligned}\right\} \quad (2)$$

where *m* is mass, *I* is moment of inertia about CG of body. While *X, Z, θ* is component of position vector, *P* is forces, *M* is moment, *W* is weight. Subscript *j, b*, denote jacket and barge respectively and subscript *c, w, F* denote contact, winch and fluid (hydrodynamic and hydrostatic) forces respectively. Double dot denotes second derivative with respect to time. The hydrodynamic force acting on the jacket barge during the launch process can be expressed with respect to drag force and inertial force as follows [7]:

$$F_h = F_d + F_a \quad (3)$$

$$F_d = \frac{1}{2} C_d \rho A_w V_n \left|V_n\right| \quad (4)$$

$$F_a = -M' A_n \frac{dM'}{dt} V_n \quad (5)$$

where F_d is drag force, F_a is added mass force, C_d is drag coefficient, ρ is water density, A_w is the submerged area, V_n is the normal velocity component, M' is added inertia mass and A_n is normal acceleration.

3. Jacket Launching System

The jacket was successfully launched and installed in 145.27 m deep waters with the assistance of the launching barge. To describe the motion of the jacket launching system, two coordinate system are introduced [8]: the global coordinate system ($o_g - x_g y_g z_g$) and body fixed coordinate system. As shown in Figure 1, the former system is fixed with respect to the earth, and its x-y-plane coincide with the water surface. The jacket fixed coordinate system ($o_j - x_j y_j z_j$) moves with the jacket and its x-y-plane coincides with the jacket waterline. In addition, the barge fixed coordinate system ($o_b - x_b y_b z_b$) moves with the barge and its x-y-plane coincides with the barge keel. The origin of each coordinate system is located in the central plane for simplicity.

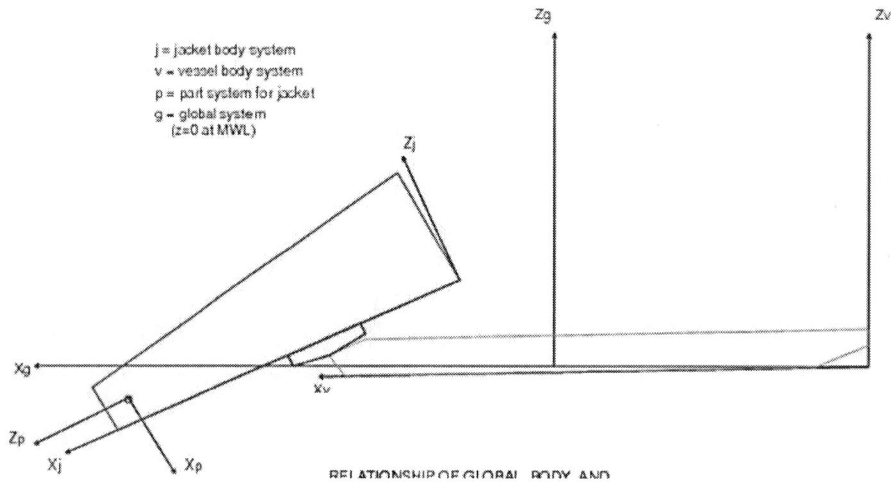

Figure 1. Launch coordinate system

3.1. *Launching Barge and Jacket*

The jacket is designed for the area near the NC3 and NC8 gas field in Sarawak Malaysia Sea. The barge was built by Heerema Marine Contractors. Detailed information for the jacket and barge is given in the Tables 1 and 2.

Table 1. Main dimension of the jacket

Jacket Details	
Number of Legs	8
Number of Horizontal Levels	6
Water Depth at MSL	104.270 m
Overall Height	115.120 m
Mudline Horizontal to Top Horizontal	103.820 m
Top Horizontal Above MSL	6 m
Length x Breadth at Jacket Top	65 m x 24 m
Length x Breadth at Jacket Bottom	65 m x 48 m

Table 2. Main dimension of the barge

Description	Data	
Principal Dimension		
Length Over All	165	M
Breadth	42	M
Depth	10.7	M
Lightship Parameter		
Lightship Weight	11267	MT
LCG	3.65	m Aft of Midship
TCG	0.00	m from centerline
VCG	8.62	m from baseline
Deck Loading	20	MT/m^2
Maximum Allowable Rocker arm Reaction (each)	7500	MT
Rocker arm Height	2	m
Rocker arm Width	2	m
Rocker arm Length	10	m
Skid Beam Height	2.05	m
Skid Beam Width	2	m
Skid Beam Length	145	m
Skid Beam Spread	13 m (min) to 35 m (max)	

3.2. Launching Condition

To initiate the launching process, the barge was ballasted to the launching condition with the mid-ship draft of 6.1 m and trim of 3.5° for base condition. The initial conditions include the barge trim and draft, the longitudinal position of the jacket COG (LCG) and the jacket weight. To quantify the effect of these parameters, the critical launching process, including the jacket pitch motion, the launch velocity, the bottom clearance of jacket and the maximum rocker arm force under the specified condition, are investigated. The non-dimensional values describing every type of launching response under various conditions depend on the ratio of every value to its maximum value. The sensitivity analyses are conducted using the numerical method and the relevant initial condition are listed in the Table 3 as follows:

Table 3. Launch initial condition

Items	Cases	Trim (deg)	Draft (m)	LCG deviation (m)	Weight % (ton)
Base Condition	Case 0	3.5	1.0	0.0	0.0
Initial trim condition	Case A-1	2.0	1.0	0.0	0.0
	Case A-2	3.0	1.0	0.0	0.0
	Case A-3	4.0	1.0	0.0	0.0
	Case A-4	5.0	1.0	0.0	0.0
	Case A-5	6.0	1.0	0.0	0.0
	Case A-6	7.0	1.0	0.0	0.0
	Case A-7	8.0	1.0	0.0	0.0
Initial draft condition	Case B-1	3.5	1.5	0.0	0.0
	Case B-2	3.5	2.0	0.0	0.0
	Case B-3	3.5	2.5	0.0	0.0
	Case B-4	3.5	3.0	0.0	0.0
	Case B-5	3.5	3.5	0.0	0.0

Items	Cases	Trim (deg)	Draft (m)	LCG deviation (m)	Weight % (ton)
LCG Condition	Case C-1	3.5	1.0	+1.75	0.0
	Case C-2	3.5	1.0	+0.75	0.0
	Case C-3	3.5	1.0	0.0	0.0
	Case C-4	3.5	1.0	-0.75	0.0
	Case C-5	3.5	1.0	-1.75	0.0
Jacket Weight	Case D-1	3.5	1.0	0.0	+5%
	Case D-2	3.5	1.0	0.0	-5%

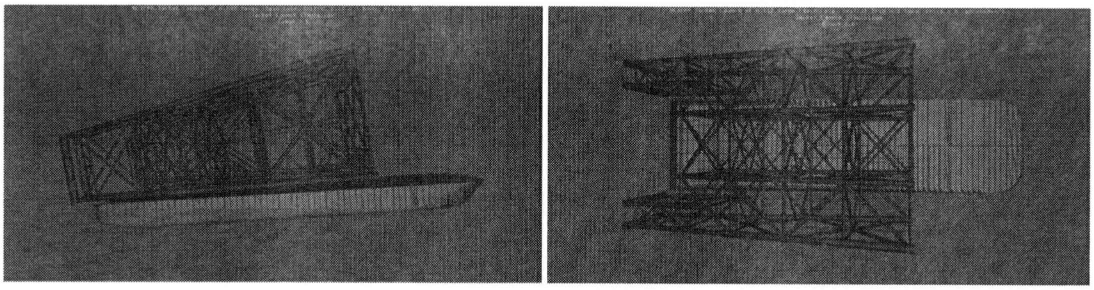

Figure 2. Initial launch condition

3.3. Launching Stages
The launching operation can be divided into four stages [9]:

- Ballasting stage: the barge is ballasted to achieve the desired trim and draft.
- Sliding stage: the jacket slides on the barge due to it self weight without tipping of the rocker arms.
- Tipping stage: the jacket slides with tipping of the rocker arms until the rocker arms rotate up to the maximum allowable angle and the jacket is launched into the sea.
- Self-righting stage: the jacket once freed from the barge, oscillates a few times and comes to rest.

4. Sensitivity Analysis
Sensitivity analysis are carried out to quantify the effect of the initial launch conditions such as the trim and draft barge, the longitudinal position of the jacket COG, and the jacket weight to the critical launching process, including the jacket pitch motion, the launch velocity, the bottom clearance of jacket and the maximum rocker arm force [9, 10].

4.1. Barge Trim and Draft
For a successful launching operation, the pitch motions should be investigated to insure the stability of the jacket [8]. Figure 3 shows the time series of the pitch motions that obtained from the numerical simulation. The tipping time occurs when the relative angle of the jacket begins to increase significantly.

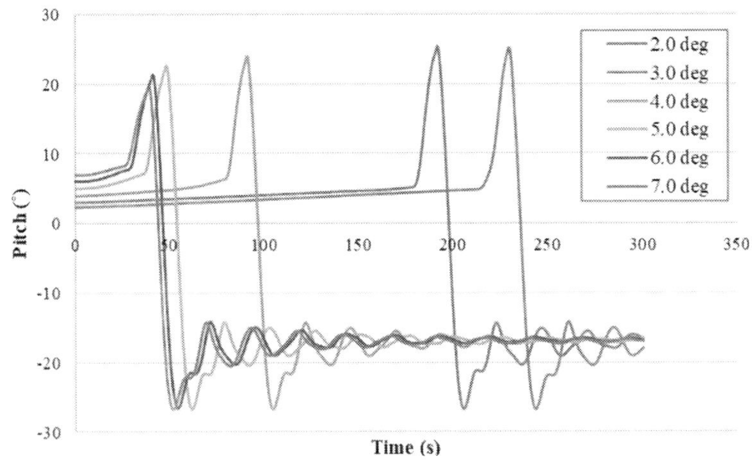

Figure 3. Comparison of the jacket pitch motion to time

Jacket begins to separates from the launch barge when the relative angle of the jacket is decrease significantly. As the initial trim angle increases, the time taken for the jacket to slide and then separate from the barge is also getting shorter. For an initial jacket angle of 7.0° it only takes 40 s from the jacket begin to slide until the jacket separated from the barge. Whereas the initial trim angle of 2.0° takes 180 s from the jacket begin to slide until the jacket separated from the barge. Figure 4 shows a graph of the comparison of pitch jacket motion to the position of the longitudinal centre of gravity jacket. The jacket's maximum trim angle decreases as the initial trim angle of the barge increases. This is due to the large trim angle conditions, some of the jacket structure has been submerged in water and has buoyancy so that the gravity of the structure that acting is corrected by the buoyancy force at the initial launch conditions.

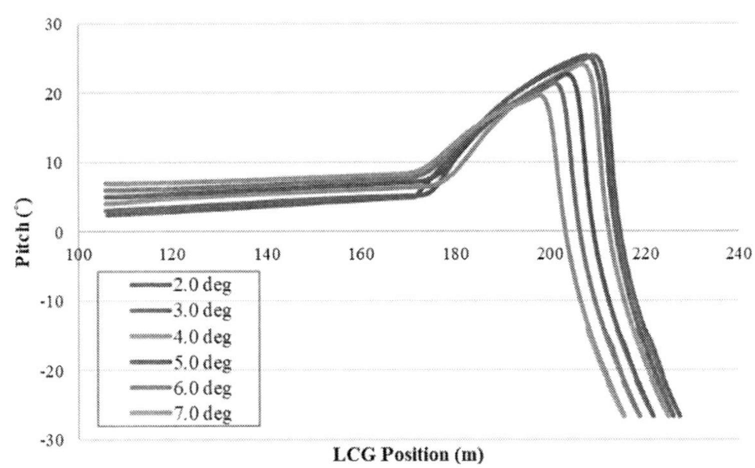

Figure 4. Comparison of the jacket pitch motion to LCG position

Variation of the critical launching responses against the barge trim are presented in Figure 5. Through a comparison of these eight trim conditions, one can conclude that the initial trim has a slight effect on the maximum rocker force. Taking the maximum rocker arm force under the trim of 2.0° as the reference, the corresponding values in case A-2, case 0, A-3, A-4, A-5, A-6 and A-7 are reduced by 4.0%, 11%, 21%, 33%, 42%, 48% and 52% respectively. Clearly, a large trim lead to a reduction in the maximum rocker arm force. Similar trends are evident in the free sliding, tipping time, and separate time. A large trim corresponds with more rapid launching.

By contrast, the trim exerts no noticeable effect on the bottom clearance. A possible contribution to this interesting phenomenon may be the jacket gravity. As the initial trim angle increases, the friction on the jacket decreases, which result in a higher sliding velocity. Moreover, the effect of increasing displacement and velocity during the tipping stage, thereby increasing the hydrodynamic force on the jacket, directly reduces the maximum rocker arm force.

The same analysis is also carried out on the maximum trim barge due to trim variations. The maximum trim barge increases with increasing initial trim angle. This is caused by the shift of the jacket COG above the barge. Movement of the jacket over the barge will cause a moment so that the location of the barge COG will shift into another point which causes the barge will have a trim by stern [10].

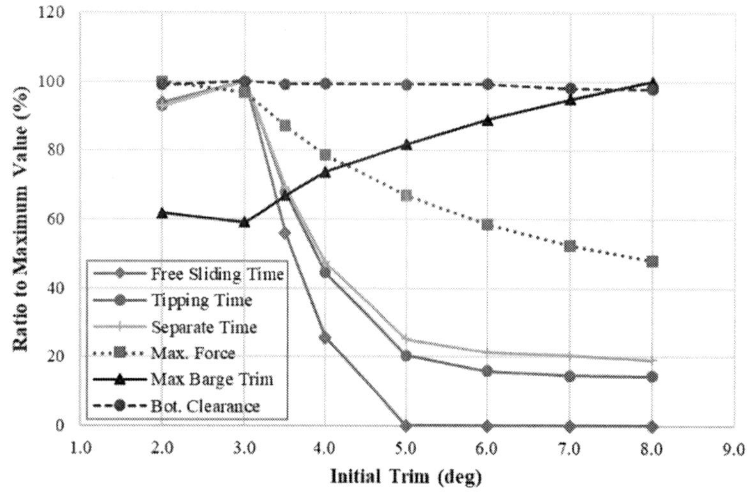

Figure 5. The effect of initial trim variation to launch responses

Figure 6 shows the influences of the barge draft on the jacket motion and maximum rocker arm force. A slight correlation is observed between the draft and the maximum rocker arm force. More specifically compared with case 0 the maximum rocker loads in case B-1, B-2, B-3, B-4 and B-5 are reduced by 4.0%, 8.0%, 11%, 14% and 17% respectively. This indicates that a large draft can also reduce the maximum rocker arm force. The opposite relationship is observed between the tipping time and the draft. In addition, there is no explicit correlation between the bottom clearance and the draft. One possible reason for this phenomenon may be that the buoyancy increases due to the larger displacement, which results in a longer tipping time. Correspondingly, the maximum rocker arm force decreases slightly with a larger draft due to the compensation for the jacket weight by the buoyancy.

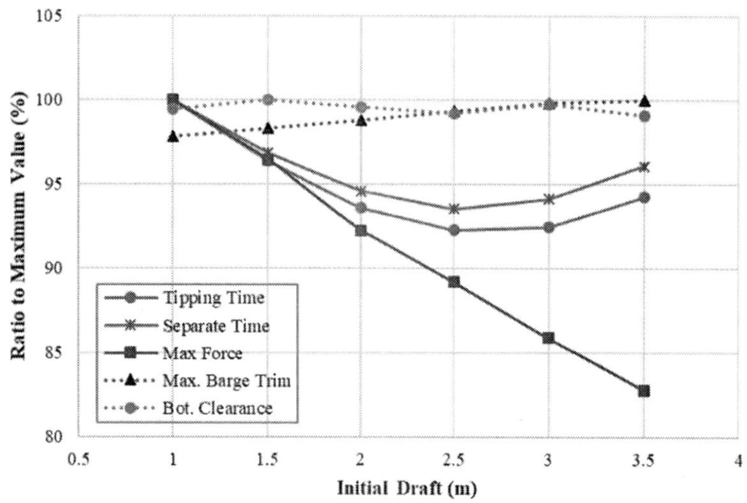

Figure 6. The effect of initial draft variation to launch responses

4.2. Longitudinal Position of the Jacket COG

Figure 7 shows the influences of the jacket LCG on the motion responses and rocker arm force. A positive LCG deviation means that it moves closer to the barge stern, whereas a negative value indicates a farther distance from the barge stern. As shown in Figure 7, the bottom clearance is reduced by 15 m as the deviation of the jacket LCG changes from +1.75 to -1.75. In addition, a longer duration will be generated when the jacket COG moves farther from the barge stern. From the structural load perspective, the maximum value of the rocker arm force is 3038 MT. The maximum rocker arm force from LCG position of -1.75 to +0.75 are increase by 3.1% to the maximum value.

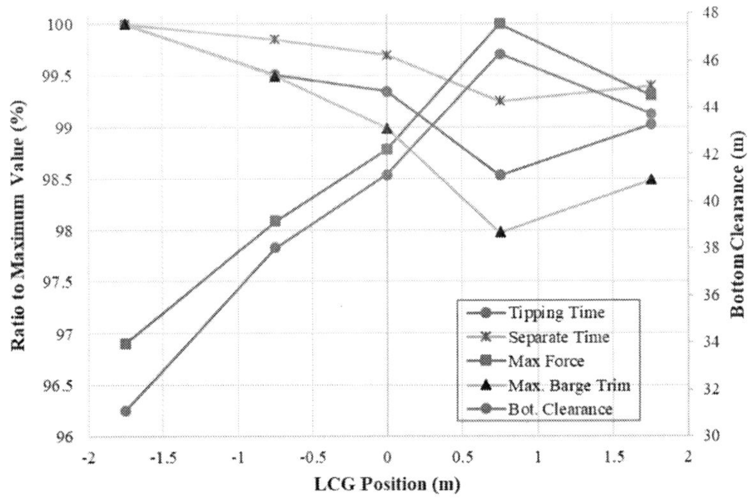

Figure 7. The effect of initial LCG variation to launch response

4.3. Jacket Weight

Analysis of adding a jacket weight to the launching process aims to determine whether the final condition of the jacket after the launch process is complete still meets the safety requirements or not. Table 4 is a comparison data of the launching responses on cases of structural weight changes.

Table 4. The launching responses on cases of structural weight changes

Items	Units	Base Case	Case D-1 (+5%)	Case D-2 (-5%)	Allowable Value	Remarks
Jacket Weight	MT	7620	8001	7239	-	
Max. Rocker arm	MT	3001	3162	2833	7500	OK
Bottom Clearance	m	41	31.02	43.13	10.427	OK
Floating Weight	MT	7620	8001	7239	-	
Total Buoyancy	MT	8983.48	8983.48	8983.48	-	
Reserve Buoyancy	%	17.89	12.28	24.10	10	OK

The maximum reaction that occurs in the rocker arm with the addition of 5% by weight of the jacket structure is 3162 tons. This value still meets the maximum reaction limit of the H-254 rocker arm barge which is 7500 tons. Likewise, the minimum reserve buoyancy and bottom clearance required by Nobel Denton. Nobel Denton requires that the reserve buoyancy jacket in the intact condition must be greater than 10% and the minimum bottom clearance is 10% of the water depth to prevent damage to the jacket structure due to collision with mudline [9]. Based on the simulation data in Table 4 shows that the reserve buoyancy jacket with the addition of 5% structure weight is 12.28% and the bottom clearance is 31.02 m. This value still meets the minimum requirements so the jacket is safe to launch.

5. Conclusion

The greater the trim angle and the initial draft of the barge the shorter the launch speed and the reaction of the rocker arm will be smaller and the bottom clearance of the jacket decreases if the LCG jacket moves towards the bottom of the jacket (the LCG jacket deviation is negative). Bottom clearance for various changes in LCG jacket still meets the minimum required value.

References

[1] Vasicek, Daniel and Lu, Cheng-Heng (1979). "Launch and floatation analysis of offshore structures part 2 – barge and jacket interaction on launch analysis." Petroleum Engineer International, Vol. 51, No. 6, PP. 10-16.

[2] Sphaier, S. H., Vasconcellos, J. M., Esperanca, P. T. T. and Ferreira, M. D. A. (1985). "The study of jacket installation using INPLA system." Proc. Of the 5th International Symposium on Offshore Engineering, Federal University of Rio de Janeiro, Brazil, PP. 541-573.

[3] Nelson, J. K., Fallon, D. J. and Hirsch, T. J. (1991). "Mathematical modeling of free-fall lifeboat launch behavior." Proc. of International Conference of Offshore Mechanics and Arctic Engineering, PP. 695-702.

[4] Jo, C. H.,Kim, K. S.,Kim, J. H., etc. Criterion of offshore jacket launching analysis[A]. In 2001.

[5] Jo, C. H.,Kim, K. S.,Lee, S. H. Parametric study on offshore jacket launching[J]. Ocean Engineering, 2002, 29 (15): 1959-1979.

[6] Nourpanah, N., Pirooz, M. D. 2008. Numerical Modeling of Launching Offshore Jackets from Transportation Barge & the Significance of Water Entry Forces on Horizontal Jacket Members. Journal of Faculty of Engineering. Vol. 42, No. 6. PP. 809-821.

[7] R., H. M.,D., P. M.,R., B. M. A physical and numerical modeling for launching of jackets (case study on Balal PLQ Platform)[J]. Journal of Offshore Mechanics and Arctic Engineering, 2008, 130 (3).

[8] Ultramarine Inc. 2001. Reference Manual for Moses.

[9] NOBLE DENTON NO:0028/NDI. 2005. Guidelines for The Transportation and Installation of Steel Jackets.

[10] Hu, Z., Li, X., Li, J., Yang, J.M. 2016. *Comparative Study on a Jacket Launching Operation in South China Sea*. Ocean Engineering 111 (2016) 335–347.

Analysis of the Ship and Cargo Service Facilities at the Local Port of Sorong

S Dewa[1]*, M Idrus[1], A S Chairunnisa[1], W Djafar[1], A H Djalante[1], and A Karyadi[1]

[1]Department of Naval Architecture, Engineering Faculty, Hasanuddin University, Makassar, Indonesia

*Email: syarifuddindewa@gmail.com

Abstract. Sorong local port is one of the ports in Sorong City that provides services for pioneer ships, machine-sail ships, and fast passenger ship for tourists. Currently, mooring locations for the pioneer ships and machine-sail ships are nearby the mooring location of fast passenger ships. This condition affects the safety and amenities of passengers, mainly when the loading and unloading of cargoes and passengers take place at the same time. Besides, their ave not been proper supporting facilities such as passenger terminal, warehouse, and parking area. Therefore, this research aims to analyze the performance of existing facilities and the required capacity of supporting facilities for service improvement purposes. Some methods are being applied to assess the performance of berth (Berth Occupancy Ratio or BOR), passenger terminal, and warehouse (Shed Occupancy Ratio or SOR). The result shows that the ship calls grow by 13.33 % annually, where the highest number of visiting ships is the fast passenger ship (47.62%). Current BOR of Sorong local port can reach up to 77.88 %, which is higher than the standard BOR 70 %, while SOR of the warehouse is 20.98 %, which indicates that the utility of the warehouse is low. In the meantime, the required area for the passenger terminal is 617 m^2.

I. INTRODUCTION

The local port of Sorong is in the Sorong District, Sororng City, West Papua Province. Astronomically, the port located at 0° 53’22.92” S and 131° 16.5’33” E. The local Sorong seaport is a port that serves pioneering transport, machine-sail ships (KLM), passenger ships, and fast tourist vessels.

The pioneer ships that visit the local port of Sorong include the sea toll ships, namely MV. Sabuk Nusantara 60 and MV. Sabuk Nusantara 42, which are the ro-ro passenger ships. Fast passenger ships include MV. Marina Express, MV. Express Bahari and others serving the Raja Ampat Islands include Waisai, Misool, Muturi, Sausapor, and Raja Ampat.

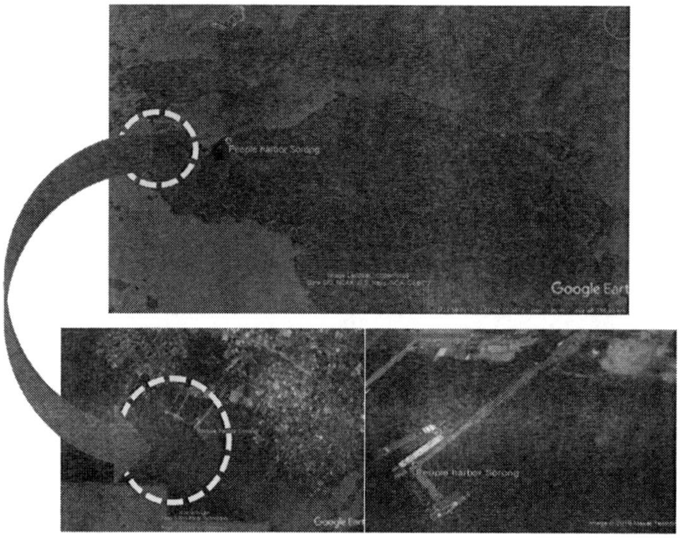

Figure 1. The Local Port of Sorong

The facilities in the local port of Sorong include a causeway, trestle, berth, back up area, and warehouse. The passenger terminal facility, which is needed by tourists who will visit Raja Ampat and local passengers, has not been available. The number of ship visits at the local port of Sorong in 2017 was 1,699 units. The highest number of visits by the fast passenger vessels of 918 units, with an average growth of visits 13.33% in the last five years.

The high number of ship visits at the local port of Sorong requires adequate mooring facilities. Besides, several problems found at the Sorong Local Port, including:

- Current ship mooring conditions that often occur where there are 2 to 4 ships in the queue.
- The location of the pioneer ship mooring is in the trestle.
- Mooring locations for the pioneer ships and KLM are nearby the mooring location of fast passenger ships. This condition affects the safety and amenities of passengers, mainly when the loading and unloading of cargoes and passengers take place at the same time.
- The fast passenger and tourist ships must moor at the end of the berth

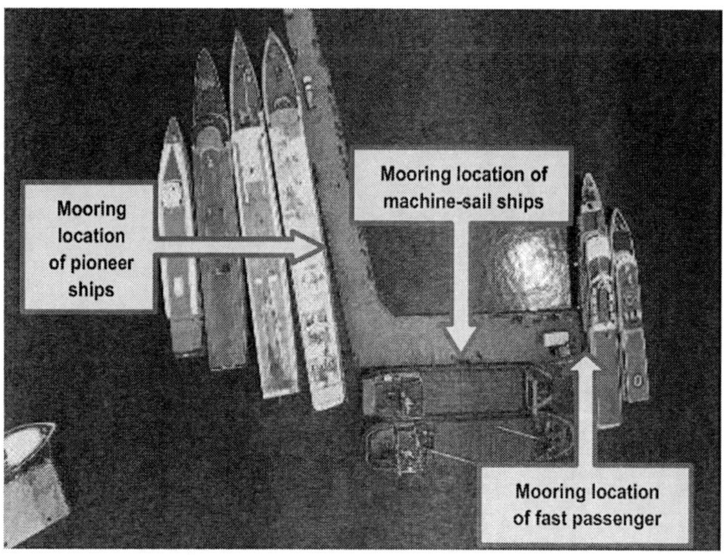

Figure 2. Ship mooring condition at the Local Port of Sorong

Based on some of the problems, the objectives of this research are:

a) Identifying the availability of facilities in the Local Port of Sorong
b) Analyze the needs of land and water facilities at the Local Port of Sorong

2. Literature Review

2.1. General Review of Port

According to Government Regulation No 61 the Year 2009, a port is a place consisting of land and/or waters with specific limits as a place for government activities and business activities used as a place for ships to berth, embarking and disembarking passengers, and / or loading and unloading of goods, as the terminals and berths which are equipped with shipping safety and security facilities and supporting facilities as well as the intra-modes and inter-modes of transportation [1].

Port is an area that has the infrastructure (facilities and infrastructure) to support operational activities. The infrastructure is a facility that must be available at a port to support port operations or business. Port infrastructure or facilities consist of basic facilities and supporting facilities [2].

Comprehensively, the role of a port is not merely from its existence and future development. However, it is closely related to aspects of planning and management in supporting regional development, between regions/islands/ports to facilitate the interaction between development resources, such as population, natural resources (sectoral), investment, technology, and other development resources. Besides, as an infrastructure, a port becomes the transportation node and the place to switch various modes of transportation [3]. The critical role carried by the port in international, national, and local shipping, according to its function, requires operational strengthening presented in Figure 3.

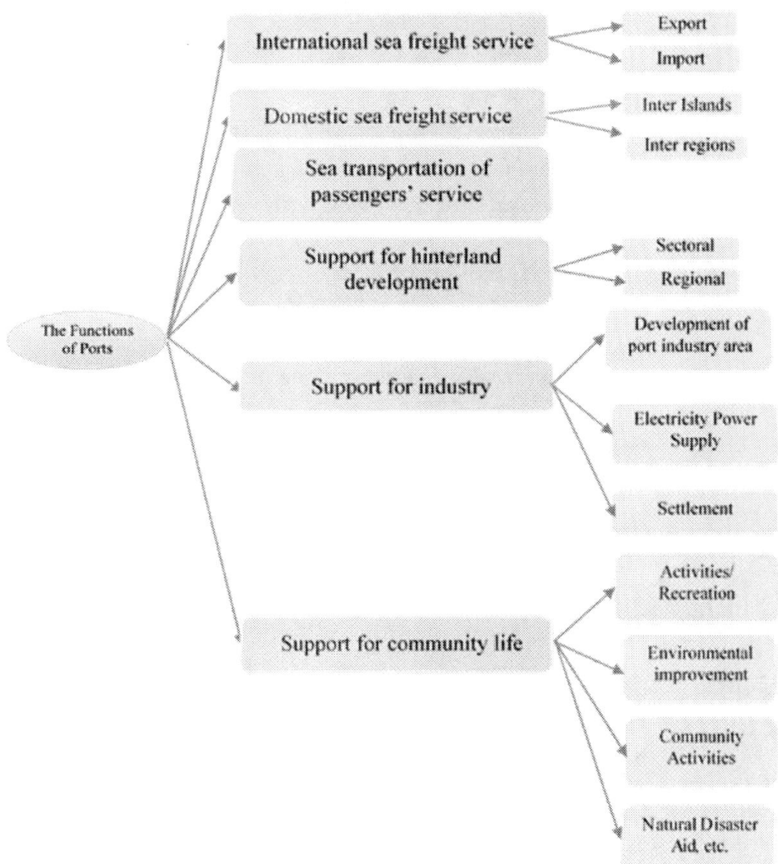

Figure 3. The functions of ports

2.2. Port Facilities
For safety and seamless activities, the port requires two types of facilities, which are the main facilities and supporting facilities.

a) The main facilities are mainly used to ensure the smooth operation of port activities. The main facilities include berth, line I of warehouse, line I of yard, passenger terminal, container terminal, ro-ro terminal, waste storage and treatment facilities, bunker, fire engine facility, warehouse and hazardous and toxic materials/goods facilities, maintenance and repair facilities, as well as navigation equipment and navigation aids.
b) The supporting facilities are complementary facilities to support activities in the port. The supporting facilities include office areas, post and telecommunication facilities, tourism and hospitality facilities, clean water installations, electricity, and telecommunications, road, and rail network, sewage water, drainage, and rubbish networks, port development areas, waiting for areas for vehicles, trade areas, industrial areas and other public facilities.

2.3. Port Operational Service Performance
Based on the Decree of Directorate General of Marine Transportation Number UM.002 / 38/18/DJPL-11 that concerns the standard of port operational service performance, operational service performance is the measurable work results at ports in carrying out ship and goods services, facility utilities and equipment in specific periods and units [4]. The service performance indicators related to port services which consist of:

- Waiting time/*WT*; the time from mooring request after the ship arrived at the anchor location until the ship moves towards the mooring.
- Approach Time/*AT*; the used time for a ship to move from anchor location to tie the rope at the mooring or vice versa.
- Effective Time/*ET*; the actual time for loading and unloading of a ship at the berth.
- Berth Time/*BT*; ready for operation time to serve ship at the berth.
- Receiving/Delivery of container; the speed of delivery/receiving service at the container terminal, which begins from the time when the conveyance moves into and out of the port.
- Berth Occupancy Ratio/*BOR*; the ratio of occupied berth time by vessel to the available operation time of berth within a specific period.
- Shed Occupancy ratio/*SOR*; the ratio of occupied room to the available room in ton day or m^3 day.
- Yard Occupancy Ratio/YOR; the ratio of occupied yard space to the available yard space in ton day or m^3 day.
- The readiness of equipment; the ratio of ready equipment to the available equipment in a specific period.

3. Research Methods
3.1. Berth
The BOR calculated using the following equation, which depends on the type of pier [5].

$$BOR = \frac{\text{X}\left(\text{LOA} + \text{distance between vessels}\right)\times\text{berth time}}{\text{available operation time of berth} \times \text{length of berth}} \times 100\% \tag{1}$$

where

LOA	: the length overall of the ship (m)
Distance between vessels	: allowed space between vessels at berth, 10 m for small vessels, and 20 m for big vessels.
The length of berth	: length of berth area that vessel can use for mooring

3.2. Passenger Terminal

The passenger terminal of the port is a building area, including the passenger ship berth in which the passengers embark and disembark and other passenger activities. The passenger ships from different regions in Indonesia can visit the terminal.

The passenger terminal is for people who will travel using passenger ships. The terminal also functions as the facility to check baggage of passengers, so that security at the port can be effectively carried out.

The passenger terminal area can be calculated by the following equation:

$$A_T = \beta \cdot \psi \frac{\mathrm{P}}{\varphi \cdot \mathrm{Call}} \tag{2}$$

where

A_T = a required field of the passenger terminal (m^2)
β = coefficient of passenger peak season (1,25)
ψ = required area per passenger (m^2/person)
ϕ = a coefficient of supporting facility (200)
call = the number of visiting ships

3.3. Port Warehouse

The warehouse is behind the berth, where the goods from inland transportation are stored before they loaded onto the ship or vice versa. Warehouses in the port can be classified based on their functions and uses, which are:

- The line I of the warehouse; customs area or transit shed.
- Line II of the warehouse; the warehouse is located behind line I. The goods in this warehouse waiting to be moved out from the port or typical long-stored cargoes.
- 'Verlengstuk' warehouse; the warehouse is functioned temporarily as warehouse line II.
- The entrepot warehouse; the warehouse is located outside the port, which is functioned as warehouse line I.

Ocean warehouse (or custom warehouse, the first line of warehouse transit shed) is on the edge of the harbor water area. This warehouse stores the goods that have just unloaded from the ship and which loaded onto the ship so that the cargoes protected from the rain and the sun.

Ocean warehouse stores the goods temporally before transported to the next destination. A time period of stored cargoes in this warehouse is a maximum 15 days for inland distribution and 30 days for goods which distributed to other seaports (with other ship). Some considered variables include:

A : warehouse area (m^2)
T : throughput per year (cargo passing each year)
TrT : transit time/dwelling time (transit time, day)
Sf : storage factor (the volume required to stow a given weight of commodities or goods)
Sth : stacking height (maximum height of cargoes can be safely stacked, m)
BS : broken stowage of cargo (The amount of space that among the stacked cargo and the space for forklift and handling activities, %)
365 : the number of days in a year [5].

4. Analysis

4.1. Brief Description of the Local Port of Sorong

The existing condition of the local port of Sorong, as mention earlier, only KLM moored at the pier. While the pioneer ships moor on the trestle, which is a bridge that connects the land with the pier. And for fast passenger and tourist ships moored at the end of the pier.

Figure 4. Activities in the Local Port of Sorong

The facilities in the local port of Sorong include causeway, trestle, pier, warehouse, and backup area, as presented in Table 1.

Table 1. Existing facilities in the Local Port of Sorong

No	Facility	Dimension	Description
1	Causeway	106 x 6 m^2	
2	Trestle	77 x 5 m^2	
3	Pier	50 x 10 m^2	Pile, Beam and Concrete Plates
4	Back-Up Area	1,8 hectare	
5	Warehouse	400 m^2	

Source: KSOP Class 1 Sorong

4.2. Port Operational

Ships visiting the local port of Sorong are pioneer ships, RORO ships, KLM, fast passenger ships, and tourist ships. However, since December 2017, the Roro ship no longer has visited the port as the RORO ship has diverted to the Arar Port. The number of call ships, embarking and disembarking passengers, as well as loading and unloading cargoes in 2013-2017 can be seen in Figure 5.

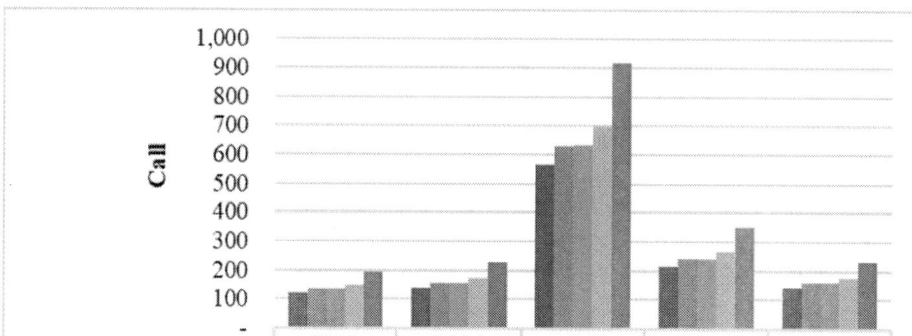

	2013	2014	2015	2016	2017
Pioneer Ship	122	141	568	217	144
RoRo	135	157	629	241	159
Fast Passenger Ship	136	158	633	243	160
Tourist Ship	149	173	695	266	176
KLM	197	229	918	352	233

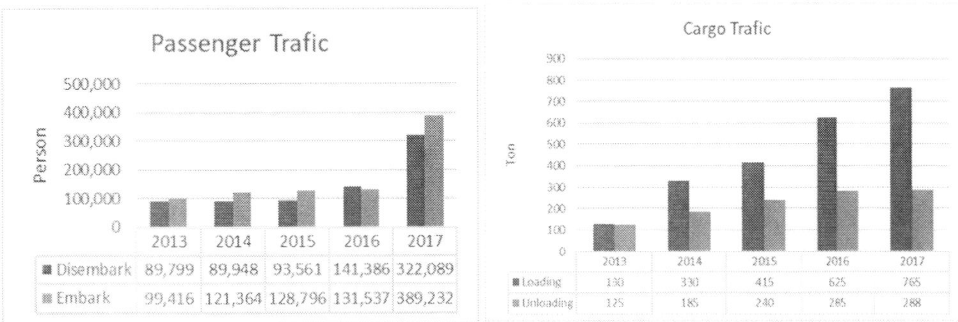

Figure 5. The Chart of call ships, embarking and disembarking passengers, as well as loading and unloading cargoes in the Local Port of Sorong

Figure 5 shows that the growth of the visiting ship is 13.33 %. The highest number of visits is gained by the fast passenger ship which is 47.62 % of the total of call ships.

4.3. Port Operational Performance

Based on the historical call ship data, the utility of berth in 2019 is higher than the BOR standard (70%) [6], which is 72.92%. Regarding the highest call ship frequency of the fast passenger ships with LOA 34.14 m and considering the length of the pier for mooring, which is 100 m, and average berthing time 18 hours, the utility of berth provided in Table 2.

Table 2. Berth performance of the Local Port of Sorong

Year	Call Ship (unit)	BT (hour)	LOA + space (meter)	Berth Length	BOR (%)
2019	1809	18	44.14	100	72.92
2022	1971	18	44.14	100	79.46
2027	2173	18	44.14	100	87.61
2037	2458	18	44.14	100	99.08

4.4. The Required Facilities

To improve the current mooring condition of ships, in the development plan, the berth for different berthing purposes will be carried out, such as the fast passenger ships and the tourist ships, while the KLM with the pioneer ships.

The required berth for the fast passenger ships calculated by considering forecasting results of call fast passenger ships, the highest number of visiting fast passenger ships, which is MV. Marina Perkasa with LOA 34.14 meters and average berthing time 12.09 hours. Regarding BOR standard 70%, the required berth length shown in Table 3.

Table 3. The required berth length for fast passenger and tourist ship

Year	Required Berth of Fast Passenger Ship (meter)	The Required Berth of Tourist Ship (meter)	Total Required Berth (meter)
2019	1809	18	44.14
2022	1971	18	44.14
2027	2173	18	44.14
2037	2458	18	44.14

The required berth for the tourist ships calculated by considering the forecasting result of call tourist ships, the highest frequency number of visiting tourist ships, which is MV. Blue Manta with LOA 42.53 meters, 595 GT and average berthing time 23.44 hours, the required berth length provided in Table 3.

For the pioneer ships and KLM, in the development plan, the existing berth is still used. The berth length for pioneer ships and KLM presented in Tables 4 and 5.

Table 4. The required berth length for pioneer ship

Year	Call Ship (unit)	BT (hour)	LOA + space (meter)	Standard BOR (%)	BOR (%)
2019	185	18	77.42	70	41.94
2022	201	18	44.14	70	45.70
2027	222	18	44.14	70	50.39
2037	251	18	44.14	70	56.99

Table 5. The required berth length for KLM

Year	Call Ship (unit)	BT (hour)	LOA + space (meter)	BOR Standard (%)	BOR (%)
2019	218	28	34.36	70	34.25
2022	238	28	34.36	70	37.32
2027	262	28	34.36	70	41.35
2037	297	28	34.36	70	46.53

Analysis of the required passenger terminal facility is for the fast passenger and tourist ship's purpose. The analysis considering the coefficient of passenger peak season, which is 1.25 with the required area for each passenger, is 1.5 m^2. The required area of the passenger terminal provided in Tabel 6.

Table 6. The required passenger terminal facility

Year	Passenger/Year (person)	Call Ship	Passengers/call ship	The Required Area of Passenger Terminal (m^2)
2019	518,924	1278	406	1269.17
2022	648,814	1392	466	1456.26
2027	836,407	1535	545	1702.68
2037	1,151,836	1736	663	2073.61

The utility of warehouse considering the loading/unloading of cargoes and several assumptions such as the stored cargoes in the warehouse is 10% of the total loading/unloading cargoes, storage factor 1 m^3/ton, stack height 3 m, the average time of stored cargoes in warehouse 7 days, and available area of warehouse 400 m^2. The shed occupancy ratio presented in Table 7.

Table 7. Warehouse facility performance

Year	Loading/unloading Cargoes Ton/m^3	10% Loading/unloading Cargoes Ton/m^3	SOR (%)
2019	1160	116	31
2022	1334	133	36
2027	1532	153	41
2037	1782	178	47

Table 7 shows that SOR is still adequate for extended periods; however, it needs to relocate to the reclamation area. Parking area planning for four-wheeled and two-wheeled vehicles and cargo trucks. The analysis of the required parking area, it considers the number of passengers per call ship, the car carrying capacity, and the required space for each type of vehicle. They need a domain for passenger vehicles is 12.5 m^2 per vehicle, for a two-wheeled vehicle requires 1.5 m^2· and a truck requires 25 m^2.

From the analysis results, the required parking area is 2,440 m^2 for a short period, 2,486 for the medium period, and 3,006 for the long periods.

5. Conclusions

The stage of terminal development in the local port of Sorong which based on its required facility analysis consists of the short-term development plan; pier construction for fast passenger and tourist vessels along the 80 meters that can be utilized on the right and left the side of the pier for mooring, causeway construction 108.54 meters, trestle construction 98.4 meters, passenger terminal and office construction 1,500 m^2. The medium-term development plan; development of warehouse facility 400 m^2, expansion of parking area 3,000 m^2, construction of passenger terminal. The long-term development plan; development of passenger terminal, extension of length fast passenger and tourist ship berth 20 meters.

Acknowledgement

The researcher presents his sincere appreciation goes to LBE research fund.

References

[1] Peraturan Pemerintah Republik Indonesia No. 61 Tahun 2009 Tentang Kepelabuhanan. Jakarta.
[2] Nasution. 1996. *Manajemen Transportasi*. Penerbit Ghalia Indonesia.
[3] Pelabuhan Indonesia. 2000. Referensi Kepelabuhanan Seri 10 Terminologi Kepelabuhanan dan Pelayaran. Pelabuhan Indonesia.
[4] Direktorat Jenderal Perhubungan laut. 2011. Surat Keputusan Direktorat Jenderal Perhubungan Laut No. UM.002/38/18/DJPL-11 tanggal 15 Desember 2011 tentang Standar Kinerja Pelayanan Operasional Pelabuhan. Jakarta.
[5] Triatmodjo, B. 2009. *Perencanaan Pelabuhan*. Beta Offset: Yogyakarta.
[6] Suparsa, I Gusti Putu. 2009. *Optimasi Kinerja Pelabuhan Penyeberangan Ketapang–Gilimanuk*. Jurnal. Teknik Sipil. Universitas Udayana. Denpasar.

IOP Conf. Series: Materials Science and Engineering **875** (2020) 012074 doi:10.1088/1757-899X/875/1/012074

Strength Behavior Analysis of Insert Plate for Placing Corner Fitting due to Container Load

A M Nugraha[1]*, Hamzah[1], G Sitepu[1], A Ardianti[1], Ifhanuddin[1], and S Baso[1]
[1]Department of Naval Architect, Engineering Faculty, Hasanuddin University, Makassar, Indonesia

*Email: andi.mursid08@gmail.com

Abstract. Currently, the ship container generation has allowed for the carriage of containers on deck up to four tiers high. The container loads in tiers high affect on deck structure. Consequently, an insert plate is used and located on the top of the ship deck structure for placing the container corner fitting. Sometimes container fitting is placed between longitudinal deck beams under without supporting construction such as pillar or carling. Therefore, this matter should be analyzed to obtain the strength behavior of the insert plate. This study describes the strength analysis using the finite element method (FEM) through ANSYS software. Based on the simulation results, the maximum Von Mises stress of the insert plate was obtained exceeds 181 N/mm^2 and 235 N/mm^2 in the initial condition, whereas the insert plate could support only 15% loads per corner fitting (three tiers high). Then, the thickness addition of the insert plate until 24 mm reduced the stress, however, the reduction was a small magnitude. By adding two carling construction is similar to the size of a transverse deck beam, under the insert plate, the stress was obtained 138.54 N/mm^2.

1. Introduction

Local stress on an element of container ship construction is a very interesting topic to be studied, mainly the main deck. Some studies related to stresses on container deck member structure have been carried out because the stresses at deck level are typically large such as the deck longitudinal [1]. In a previous study, Rosmani et al. [2] studied about the response of the container deck structure due to laying all container on deck. This study found out that ship construction experiences the greatest stress on the deck precisely on the deck beam. Another element of ship structure components that must be considered is the deck plate. The deck plate is one of the structural components of the ship that first receives a loading response before being forwarded to the deck beam. For certain types of ships, the deck plate must withstand considerable loads. As happened in container ships whose cargo is on the deck. Containers as a unit of load on a ship have treads at each angle. Container tread is a component that directly supports containers. Containable containers at each corner make the container load concentrated at each corner of the base.

Then, some container ships have decks with different plate thicknesses. Thickness differences occur in each area at the bottom of the container shoes. Thickening of the deck plate is not caused by a doubling plate but a whole plate without doubling or called an insert plate. Insert plate has the size or dimensions specified following the plan. In this research, insert plate thickness is estimated to have a major effect on the strength of the insert plate under the container tread.

2. Literatur Review

2.1 Container Ship

Container ships are freighter designed for the transportation of goods inside the container. The engine room and accommodation are usually in the rear to maximize container storage, see Figure 1 [3].

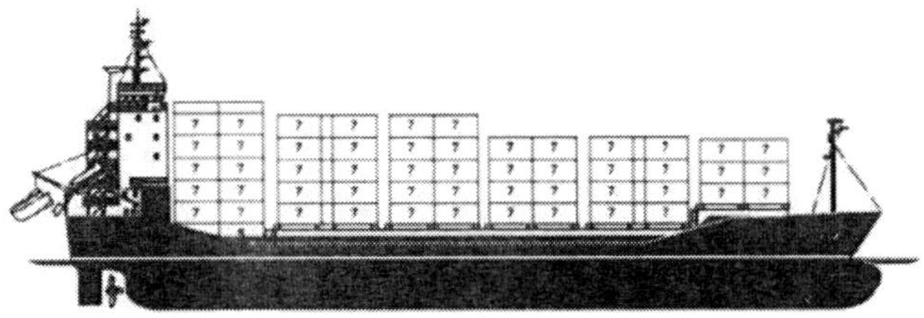

Figure 1. Container ship

Container ships are designed to meet the needs, including the desired speed, and the number of containers arranged according to the size, type and weight of the calculated cargo. The maximum number of containers is calculated according to the capacity available on the ship. Supporting features or equipment is provided to increase port productivity by enabling rapid loading and unloading. If cranes are not available at some of the ports to be visited, the ship must be equipped with equipment capable of being used for loading and unloading. Meanwhile, container loading and unloading equipment are very important, where competition requires ships to be reliable, operated efficiently and constructed and operated economically [4].

2.2 Insert Plate

Insert plates on the deck of the container ship are distributed according to the planned placement of the container holder. Planning for ship container placement is arranged horizontally and vertically with the number of piles from front to back is not the same. For the part in front of the ship's midship, there are two piles of containers and for the parts behind the midship, there are three piles of containers, see figure 2. The under deck construction component for each Insert plate has a different position and type. Some insert plates are placed with a container holder just above the deck girder, web beam, or bulkhead with their respective position differences. There is also the placement of the Insert plate with a container holder that is not right above the web beam or girders. Placement of the Insert plate and container ship deck construction can be seen in Figures 2 and 3.

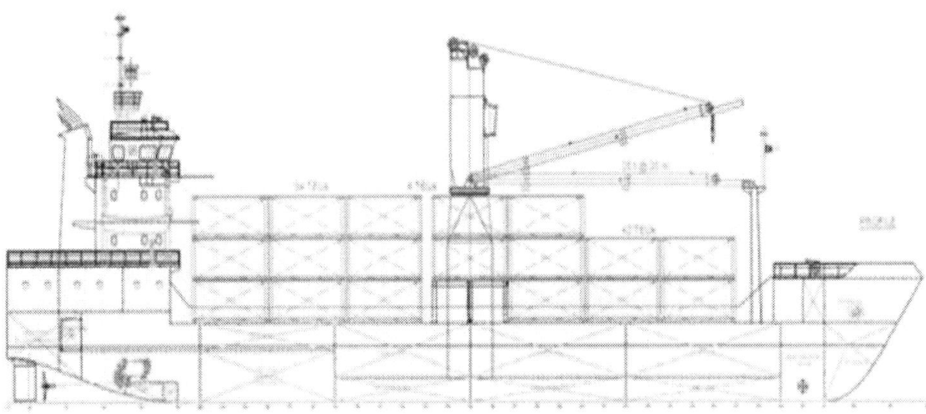

Figure 2. Container formation

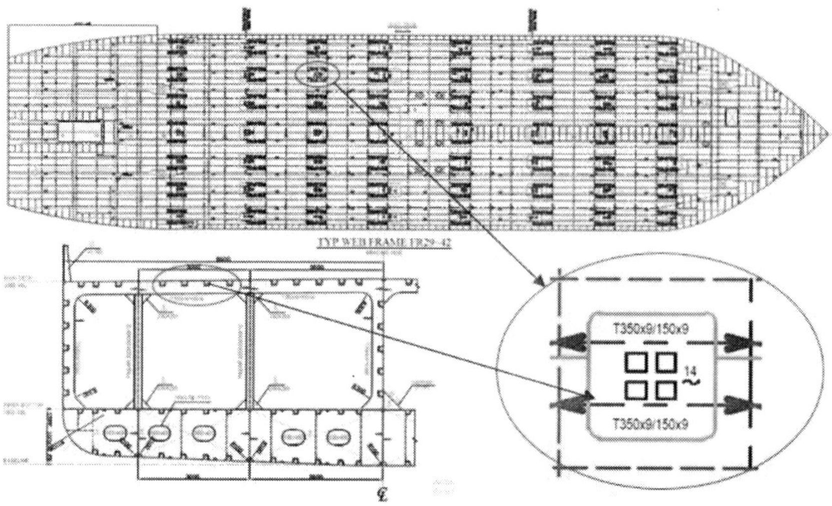

Figure 3. Construction profile and midship section

2.3 Container Load

Containers are reusable boxes made of steel or aluminum. Containers come in various types and sizes. Details can be obtained from the website of one of the manufacturers. Nominal dimensions have lengths of 20, 40 and 45 feet, the width of 8 feet and a height of 8.5 or 9.5 feet. The inner volume and weight of items that can be carried vary according to the material. For general purpose steel containers, it has a size of 20 ft with an internal capacity of around 33 m^3, an empty weight of around 2.3 tons and a maximum load of around 21.7 tons. The 40 ft container has an internal capacity of around 68 m^3, an empty weight of around 3.8 tons and a maximum load of around 26.7 tons. Aluminum containers have the same volume but are lighter in weight and capable of carrying larger loads than steel containers. Containers can be used for transportation of general goods and liquids [5]. Container dimesion are in accordance with ISO (International Organization for Standardization) standart, namely containers 10 ft, 20 ft, 30 ft and 40 ft, see Figure 4.

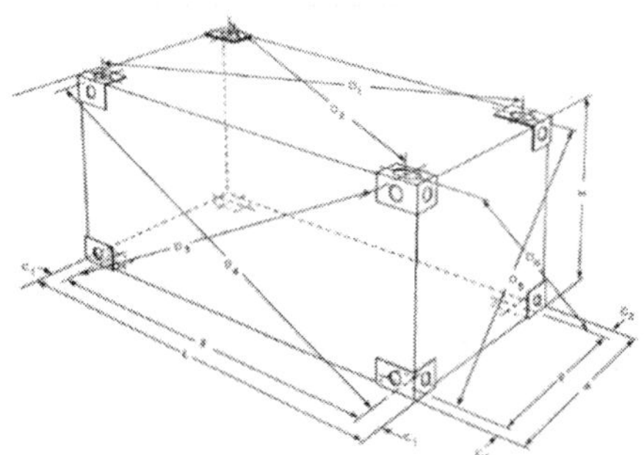

Figure 4. ISO standard of container dimension [6]

2.4 Stress

Stress is defined as the amount of force acting per unit area. There are two types of stress: normal stress and shear stress. Normal stress (σ) is a measure of normal force or axial force per unit area [7]. Mathematically the definition can be written as:

$$\sigma = \frac{F}{A} \tag{1}$$

defining:
σ = Stress (N/mm²)
F = Force (N)
A = Sectional Area (mm²)

The normal stress equation for the three-dimensional plane is as follows :

$$\sigma_x = \frac{E}{(1+v)(1-2v)}[\varepsilon_x(1-v) + v(\varepsilon_y + \varepsilon_z)] \tag{2}$$
$$\sigma_y = \frac{E}{(1+v)(1-2v)}[\varepsilon_y(1-v) + v(\varepsilon_x + \varepsilon_z)] \tag{3}$$
$$\sigma_z = \frac{E}{(1+v)(1-2v)}[\varepsilon_x(1-v) + v(\varepsilon_x + \varepsilon_y)] \tag{4}$$

Finite element software analysis usually has the advantage of being able to produce Von Mises stress values or equivalent stresses, that is the type of stress that results in failure of the material structure formulated by its inventor named Von Mises. To determine the Von Mises stress, the main stress acting on the structure with equation (3–4) is calculated first, after the main stresses are found, the Von Mises stress can be obtained by the equation:

$$\sigma = \left\{\frac{[\sigma_1-\sigma_2]^2+[\sigma_2-\sigma_3]^2+[\sigma_3-\sigma_1]^2}{2}\right\}^{1/2} \tag{5}$$

2.5 Permissible Stress

Permit stress is the stress that results in a construction experiencing a large deflection where the deflection is the limit of construction is still safe in overcoming the burden that occurs or who works on it. If the permissible stress from the construction concerned is less than the maximum stress that occurs, then the construction is not safe.

The normal strength of the hull steel structure is the hull steel structure having a minimum yield point of REH 235 N/mm² and tensile strength of Rm 400 – 520 N/mm² [8]. Tension in the area of ship structure and substructure for containers as well as the cell guide system and lashing device on the hatch cover of the cargo deck not to exceed the following values:

$$\sigma_b = \frac{ReH}{1.5} \ [N/mm^2] \tag{6}$$

$$\sigma = \frac{ReH}{2,3} \ [N/mm^2] \tag{7}$$

$$\sigma_v = \sqrt{\sigma_b^2 + 3\tau^2} \leq \frac{ReH}{1.3} \ [N/mm^2] \tag{8}$$

defining,
σ_b = bending stress [N/mm²]
σ = shear stress [N/mm²]
σ_v = equivalent stress [N/mm²]
ReH = yelling minimum limit of material

2.6 Finite Element Method
Finite element analysis is a numerical technique. In this method, all the complexity of the problem, such as various forms, boundary conditions and loads are maintained but the solution obtained is an estimate. Because of its diversity and flexibility as an analytical tool, it receives a lot of attention in the engineering field. Rapid improvements in computer hardware technology and reduced computer costs have pushed this method because computers are the basic requirement for implementing this method. Using this package, one can analyze several complex structures [9].

The new method based on virtual displacement is the last solution in the world. Applied mathematical problems and solutions are still being developed along with the development of computer science [4]. Types of structural problems include [10]:

a. Stress analysis, including Truss and frame analysis
b. Buckling
c. Vibration analysis.

In problems involving complex geometry, such as the problem of imposing complex structures, it is generally difficult to solve through mathematical analysis. This is because mathematical analysis requires quantities or prices that must be known at each point in the structure under study. Completion of the analysis of a differential equation of complex geometry, complex loading, is not easy to obtain, the formulation of the finite element method can be used to overcome this problem [10].

3. Research Methodology

3.1 Ship Container Data
The ship has a frame distance of 600 mm and using a longitudinal construction system. Container loads are placed on the main deck of the ship. The ship does not have a container load hole in the hull and the ship is not equipped with a torsion box. The data needed for this research is the main dimensions of the ship, as follow:

Ship Type	: Container Ship
LOA	: 74.05 m
LBP	: 69.20 m
B	: 17.20 m
H	: 4.90 m
T	: 3.50 m

Plate material on ships using grade A normal strength steel base on BKI Class.

3.2 Structure Modeling
Data in structural modeling is obtained through profile construction data and midship construction, as shown in Figure 5. Construction components that are modeled with the Insert plate include deck plates, longitudinal deck beams, and Intercostal beams. The dimensions of the structural components being modeled such as Insert plate, deck plate, Intercostal deck beam, and longitudinal deck beam are as follows:

a. Intercostal beam with profile size T 350 mm × 9 mm / 150 mm × 9 mm.
b. Longitudinal deck beam with profile size L 120 mm × 120 mm × 11 mm.
c. Deck plate with a thickness of 10 mm.
d. Insert plate with a thickness of 14 mm and the area of A is 1.44 m^2.

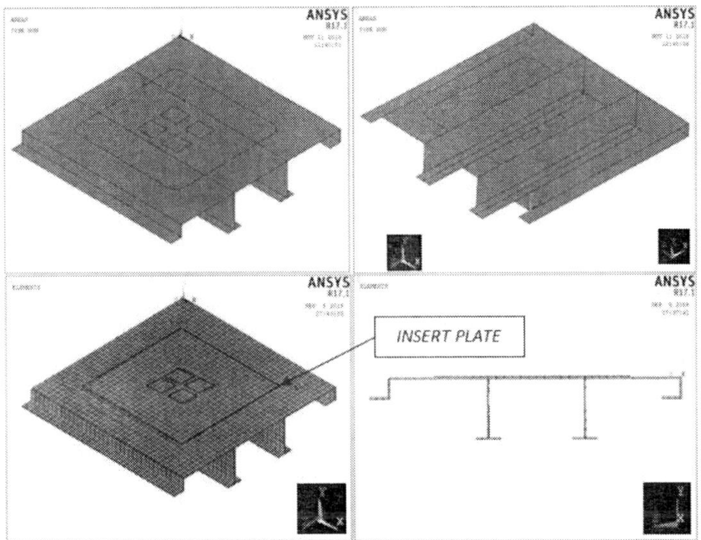

Figure 5. Insert plate model on the container deck

3.3 Loading Variation

The load is given to the model in the form of a uniform load (pressure). The evenly distributed load is obtained from the division between point P load and the area of loading. P point load is obtained from the calculation of deck load with a container load, while the loading area is the container tread area at each corner (container corner fitting). Container loads on deck are determined based on the following equation:

$$P = P_E (1 + a_v) \quad [kN]$$

where,

P_E = maximum container load x G m/s^2
G = gravitational acceleration (9.81 m/s^2)
P_E = 24 ton x 9.81 m/s^2
= 235.44 kN
G = gravitational acceleration (9.81 m/s^2)
v_o = 12 knot
F = 0.11 [12 /69.20$^{0.5}$]
= 0.159
m = 1.0 for $0.2 < x/L \leq 0.7$
a_v = 0.159 x 1,0
= 0.159
P = 235.44 kN (1 + 0.159)
= 272.8 kN

The container load is then divided by four according to the number of corners (corner fittings) at the bottom of the container, which the calculation is as follows :

P	= 272.8 kN	(load of one container)
¼ P	= 272.8 kN x ¼	(for each *corner fitting*)
	= 68.2 kN	

There are three piles of containers above the Insert plate. Thus, the total load for one container site (corner fitting) is the accumulated load of the three piles. The total load for one tread container (corner fitting) as follows:

¼ P total = ¼ P x 3 = 68.2 kN x 3
= 204.6 kN ~ 204600 N

The total container load for one container site (corner fitting) is then converted to a pressure load according to the container tread area (the dimensions of the container site can be seen in Figure 6. Pressure load calculation for one tread container (corner fitting) as follows:

pressure q = ¼ Ptotal / Load Area

The loading area is the container tread area (corner fitting). From Figure 6, the container tread size (corner fitting) is 178 mm x 162 mm so that the loading area (A) and pressure load (q) can be calculated as follows:

Loading Area (A) = 178 mm × 162 mm
= 28836 mm^2.
Pressure Load (q) = ¼ Ptotal / A
= 204600 N/28836 mm^2 = 7.0953 N/mm^2

From the calculation above the pressure load applied to each container site (corner fitting) for three stacks is 7.0953 N/mm^2. Meanwhile, the value of the pressure load based on the percentage load of 3 piles of containers can be seen in Table 1.

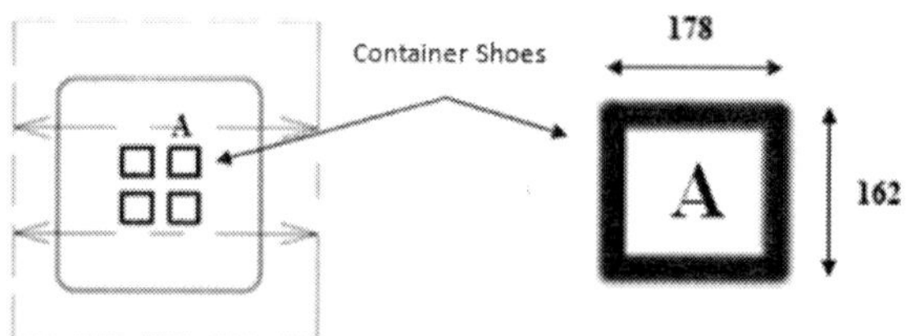

Figure 6. Loading area

Table 1. Pressure load on container shoes base on percentation of container weigth

Percentage Container Load	Container Weigth (w) [ton]	PE = (w x 9.81 m/s^2) [kN]	P = {P_E (1 + a_v)} [kN]	¼ P = (¼ P x 1000) [N]	Pressure Load [N/mm^2]
100 %	72	706.32	818.62	204600	7.0953
80 %	57.6	565.06	654.90	163700	5.6769
60 %	43.2	423.79	491.17	122700	4.2551
40 %	28.8	282.53	327.45	81800	2.8367
20 %	14.4	141.26	163.72	40900	1.4184
15 %	10.8	105.95	122.79	30600	1.0611
14 %	10.1	32.96	38.20	9500	0.3294
13 %	9.4	30.61	35.47	8800	0.3051
12 %	8.6	84.76	98.23	24500	0.8496
11 %	7.9	77.70	90.05	22500	0.7802
10 %	7.2	70.63	81.86	20400	0.7074

3.4 Boundary Condition

Restraints or supports are given based on the identification of the meeting between the construction components that are modeled with the construction components that border with it. Translational or rotational restraints are given in accordance with the shape of the modeled construction cut. The translation is divided into three namely x, y, and z translations. The rotation has 3 degrees of freedom, namely rotation x, y, and z. Each degree of freedom is used according to the shape of the construction cut being modeled.

All deck construction components bordering the transverse bulkhead are fix supported (see Figure 7), i.e., translation x, y, and z are not free or equal to 0 so that translational motion does not occur. Meanwhile, x, y, and z rotations are 0 (not free), so the model cannot rotate. Then at the end of the intercostal beam, which borders frame 30, a pinch is also provided. For longitudinal deck beams and deck plates bordering frame 30 (see Figure 7) a support is placed where translation x = 0, translation y = 0, and rotation z = 0. Then for deck plates lengthwise (see Figure 7) given a support where translation y = 0, translation z = 0, and rotation x = 0. A value of 0 (zero) indicates that part of the structure is not given the freedom to make movements according to the shape of the movement (translation or rotation) and in accordance with the direction of movement (x, y, or z).

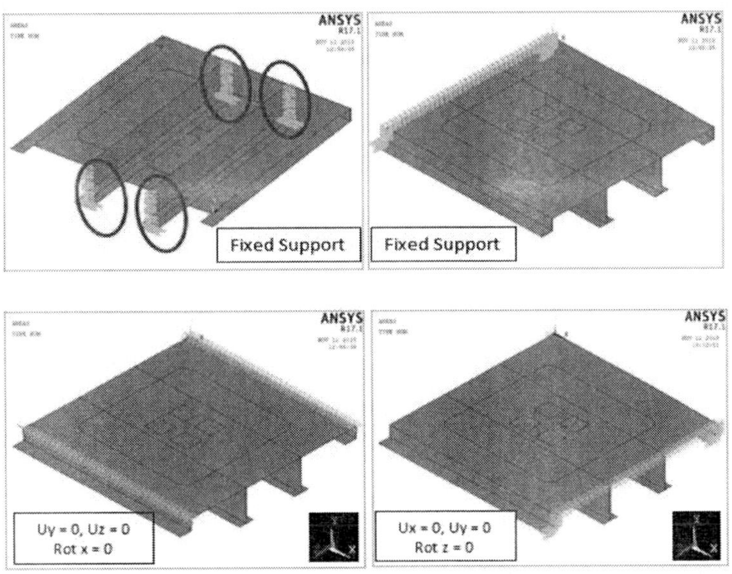

Figure 7. Restraint on insert plate model

4. Numerical Results

4.1 Initial Condition

Structural analysis in the initial conditions is analysis without adding other construction components such as carlings and so on. At this stage, the analysis is carried out based on the displacement value and the stress value that occurs on the Insert plate.

4.1.1 Displacement

Some parts of the structure being modeled will experience displacement (Displacement) when a load is working on it. Displacement is a displacement or deflection that occurs in several parts of the structure. Caused by the forces acting on the structure. The maximum displacement of the modeled structure occurs around the encumbered area. To prove that the maximum displacement occurs around the loading area can be seen in Figure 8.

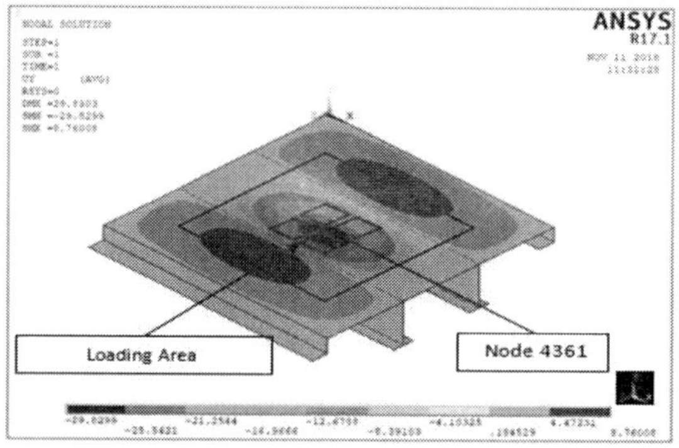

Figure 8. Displacement maximum on y-direction

Based on figure 8, the Displacement of the maximum y-direction occurs at node 4361. Node 4361 is in the meeting area of the four container sites. The maximum displacement value of y-direction for the initial conditions of the Insert plate is -29.83 mm.

4.1.2 Stress

Figure 9 shows the stress gradation that occurs in the initial conditions of the structure. When the foundation container under the container corner fittings is evenly loaded with maximum load, the maximum Von Mises stress occurs in the area of the plate is loaded. This can be seen from the gradation of red, shown in Figure 9. The red gradation in the Von Mises voltage contour represents the maximum stress experienced by the structure. The maximum stress value that occurs is 1306.82 N/mm^2 at node 5639. This value exceeds the limit of the REH material, yelling, which is 235 N/mm^2.

Then a load reduction simulation is performed to determine the load value by obtaining stress less than the yelling material limit and the BKI permit stress.

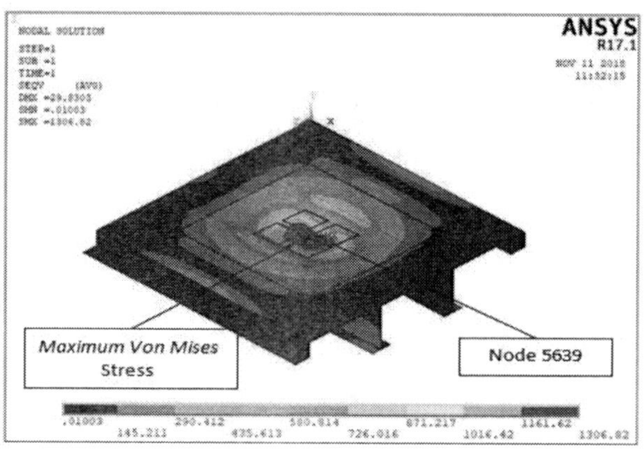

Figure 9. Stress gradation on initial condition

The stress values obtained through the results of the simulation of load reduction can be seen in Table 2.

Table 2. Von Mises stress maximum base on container load percentage

Container Load Percentage	**Weigth of 3 Container (ton)**	**Node**	**σ vmax (N/mm^2)**
100 %	72	5639	1306
80 %	57.6	5639	1045
60 %	43.2	5639	783
40 %	28.8	5639	522
20 %	14.4	5639	261
15 %	10.8	5639	195
14 %	10.1	5639	182
13 %	9.4	5639	169
12 %	8.6	5639	156
11 %	7.9	5639	143
10 %	7.2	5639	130

Based on Table 2, yelling material limits can be met when a given load <15% of the maximum load of the container (load and construction). Then the BKI permissible stress of 181 N/mm^2 can be met when the weight of the container and its load is less than <14% of the maximum load of the container (load and construction). To meet the BKI permit stress with a load of 100%, one way to reduce the stress experienced by the structure is to increase the thickness of the Insert plate or add carlings under the deck plate.

4.2 First Solution: Added Thickness of Insert plate

The addition of a thick Insert plate is done to increase the strength of the deck structure so that the stresses experienced can be reduced, and the stress values obtained below the BKI permissible stress. To find out the amount of the addition of insert plate thickness with a stress value less than the BKI permissible stress, a simulation of adding thickness with a difference of 2 mm is carried out. The stress values obtained from the simulation results can be seen in Table 3.

Table 3. Maximum Von Mises stress causes an added thickness of the *insert plate.*

Thickness (mm)	**Node**	**σ v max (N/mm^2)**
14	5639	1306.70
16	5639	1035.50
18	5639	844.68
20	5639	705.22
22	5639	599.89
24	5639	518.04

Based on Table 3, the value of the stress that meets the BKI permissible stress is at the thickness of the 46 mm insert plate. While the initial thickness of the insert plate in Table 3 is 14 mm. Thus, the addition of an insert plate thickness from the initial thickness is 32 mm.

For linear analysis, the pattern of stress changes due to the addition of insert plate thickness in accordance with the given thickness variation can only be reviewed if the structural stress is less than the yelling material (REH) limit of 235 N/mm^2. The load value for the stress is less than the yelling material limit of 1.0611 N/mm^2. The stress value for each variation of thickness addition can be seen in the following Table 4.

Table 4. Maximum Von Mises stress on every added thickness of the insert plate

Thickness (mm)	Node	σv max (N/mm^2)
14	5639	195.43
16	5639	154.90
18	5639	126.39
20	5639	105.56
22	5639	89.84
24	5639	77.61

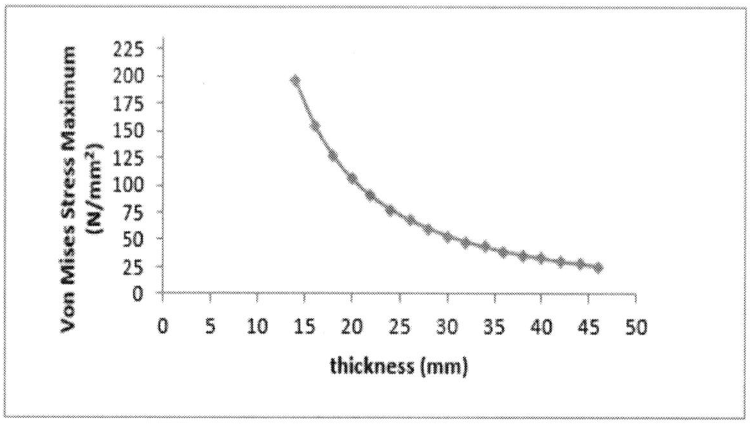

Figure 10. Von Mises Stress maximum base on the addition of insert plate thickness

The graph of the influence of the addition of carlings accompanied by the addition of thick insert plates to changes in stress that occurs in the structure can be seen in Fig. 10. Based on the curve in figure 10 shows that with the addition of thickness to the Insert plate, the stress can be reduced. The greater the thickness value is given, the smaller the stress obtained.

4.3 Second Solution: Added Carlings Construction

At the bottom of the deck plate, which is loaded with containers, structural components such as carlings are added to add strength to the deck vessels, see Figure 11. Carlings are blocks that are placed transversely below the deck of a ship. Carlings connects one deck beam to another deck beam. In this case, carlings are placed under the Insert plate.

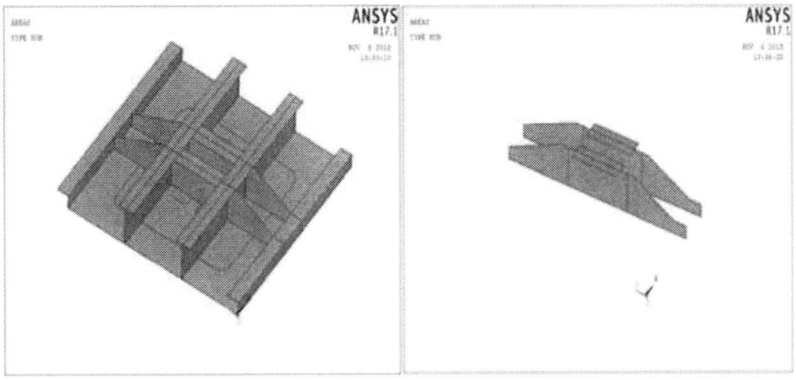

Figure 11. Addition carlings construction under the insert plate

Carlings connect between the two intercostal beams under the Insert plate. The dimensions of the carlings are determined following the dimensions of the intercostal beam with a profile size of 350 mm × 150 mm × 9 mm. The number of carlings installed is 2 (two) carlings transversely.

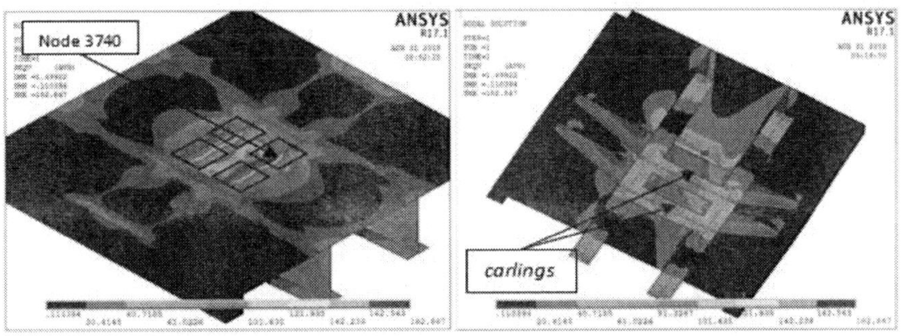

Figure 12. Stress gradation with addition carlings

Based on Figure 12, the maximum Von Mises stress experienced by the structure after the addition of carlings under the deck of the container holder occurs at node 3470. At the node 3470 is the meeting between the Insert plate and the carlings below. The maximum stress value that occurs at node 3470 is 185.24 N/mm^2. The addition of carlings causes the stress experienced by the structure to decrease to close to the BKI permissible stress. BKI permit stress is 181 N/mm^2 has a difference of 4.24 N/mm^2 from the stress experienced by the structure with the addition of carlings. To fulfill the permitted stress of BKI, which is 181 N/mm^2, the addition of thickness insert plates or longitudinal carlings are added.

Addition of Insert Plate Thickness

Based on the first simulation at point 4.1 and the second simulation at point 4.2, it can be seen that the addition of thick Insert plate and the addition of carlings causes the structural stress to decrease. Both simulations at points 4.1 and 4.2 can be used as a basis for the implementation of the next simulation by adding carlings and adding thick Insert plates. From the test results in this simulation, it is known that the stress experienced by the structure decreases after the addition of carlings accompanied by the addition of thick Insert plates. The stress values obtained based on the addition of carlings and variations in the thickness of the Insert plate can be seen in Table 5.

Table 5. Maximum Von Mises cause added carlings and thickness of the insert plate

Thickness (mm)	Node	σv max (N/mm^2)
14	3470	185.24
16	3469	138.54
18	3566	125.22
20	3566	122.90
22	3566	120.50

The graph of the influence of the addition of carlings accompanied by the addition of thick insert plates to changes in stress that occurs in the structure as follows.

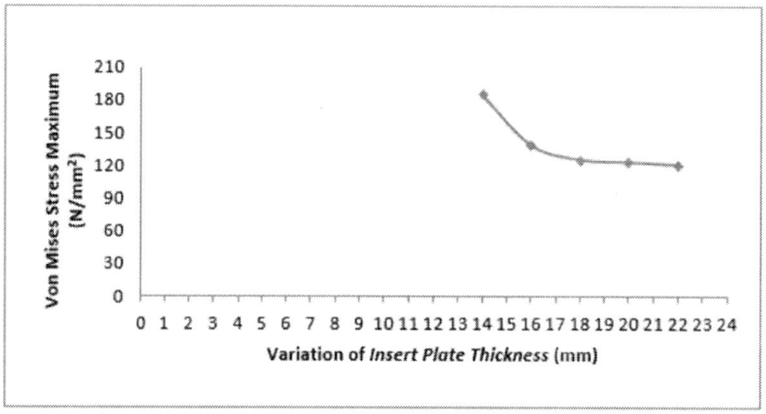

Figure 13. Von Mises stress maximum base on addition carlings and insert plate thickness

5. Conclusion

This research conducted using ANSYS 17.1 software and analysis by using the Finite Element Method. Based on the result of this research can be concluded as below:

1). The maximum Von Mises stress obtained in the initial condition by maximum loading exceeds 181 N/mm^2 (BKI permit stress) and 235 N/mm^2 (REH material yelling limit). The yelling material limit can be met when the load is given less than 15% of loading for three container layers (load and construction). Meanwhile, the BKI permit stress can be fulfilled when the burden is less than 14% of loading for three container layers (load and construction).

2). The addition of insert plate thickness affects the stress reduction experienced by the structure. The greater the thickness that is added, the stresses obtained and the percentage of stress reduction is smaller. 3). The BKI permit stress (181 N/mm^2) can be fulfilled when adding 2 Carlings transversely under the Insert Plate with the addition of 2 mm thick Insert Plate or with the addition of 2 Carlings longitudinally. The dimensions of Carlings follow the dimensions of intercostal beams with profile sizes of 350 mm × 150 mm × 9 mm. The stress obtained after the addition of 2 transverse Carlings accompanied by the addition of Insert Plate thickness is 138.54 N/mm^2. Meanwhile, the stress obtained after the addition of transverse and longitudinal Carlings is carried out at 141.38 N/mm^2.

References

[1] Zhiyuan Li.2013. Fatigue assessment of container ships – a contribution to direct calculation procedures. Chalmers University of Technology in Gothenburg, Sweden.

[2] Munandar, R., Ardianti, A., Sitepu, G., Nugraha, A., Hamzah, H., & Hayatuddin, M. Study on Transverse Strength of the Deck-Container Ships due to Laying All Containers on Deck. *EPI International Journal of Engineering*, *2*(2), 178-184.

[3] Shama, Mohamed. 2013. Buckling of Ship Structures. Springer. Heidelberg.

[4] Lamb, Thomas. 2004. Ship Design and Construction. SNAME. New Jersey.

[5] Molland, Anthony F. 2008. The Maritime Engineering Reference Books – A guide to Ship Design, Construction and Operation. Butterworth-heinemann. Oxford.

[6] Taggart, Robert. 1980. Ship Design and Construction. SNAME. New Jersey.

[7] Patnaik, Surya N. and Dale A Hopkins. 2004. Strength of Materials. Elsevier. Amsterdam

[8] Biro Klasifikasi Indonesia. 2014. Rules For The Classification and Construction Seagoing Ship, Volume II. BKI. Jakarta.

[9] Bhavikatti, S.S. 2005. Finite Element Analysis. New age International Publisher. New Delhi.

[10] Susatio, Y. 2004. Metode Elemen Hingga. ITS. Surabaya.

IOP Conf. Series: Materials Science and Engineering **875** (2020) 012075 doi:10.1088/1757-899X/875/1/012075

Study of Longshore Sediment Transport on Erosion and Sedimentation Beaches in South Sulawesi

H Umar[1]*, A Y Baeda[1], F Husain[1], and Taufiqurrahman[1]

[1]Department of Ocean Engineering, Engineering Faculty, Hasanuddin University, Makassar, Indonesia

*Email: hasdinar.umar@gmail.com

Abstract. Accumulation of sediment on the beaches absorbs and reflects energy which mainly comes from waves. If all wave energy is absorbed, the beaches is in a balanced condition. Conversely, if the beaches is in an unbalanced condition, erosion and sedimentation will occur which in turn will cause damage to the coastal line. To overcome the problem of erosion and sedimentation that occurs usually the first step that must be done is to look for the causes of erosion and sedimentation, so that by knowing the cause, then you can determine how to overcome them. To find out what approaches can be used to overcome erosion and sedimentation problems, a survey and study of erosion and sedimentation that occur on the beaches must first be conducted. This research was conducted by first analyzing the characteristics of coastal sediment grains then analyzing the predicted erosion and sedimentation on the beaches and calculating the magnitude of longshore sediment transport using empirical equations. Result of the research shows that Padongko Beach Mangempang Village Barru Regency has sedimentation with total longshore sediment transport was 201.71 m^3/years, and Desa Aeng Batu-Batu Beach Galesong Utara Takalar Regency has sedimentation with longshore sediment transport was 169.79 m^3/years

1. Introduction

The dynamic process will always adjust the shape of the profile so that it can destroy the incoming wave energy. This shape adjustment is the natural dynamic response of the beaches to the sea. Under normal wave conditions that occur for a longer time, wave energy will be easily destroyed by natural coastal defense mechanisms. However, during a storm where waves that have great energy will result in the beach's natural defenses not being able to withstand waves, so the beaches can be eroded. After the big waves subside, the beaches will return to its original shape by normal waves, but there are times when the eroded beach does not return to its original shape because the beach-forming material is carried by the current to another place and does not return to its original location. The material carried by the current will settle in quieter areas, such as in river mouths, bays, ports, etc, resulting in sedimentation in the area. In addition, the use of coastal land has also become one of the causes of several problems in coastal areas, such as: erosion, sedimentation, diverting and silting of river mouths, pollution and sea water intrusion [1]. Several studies related to coastal erosion and sedimentation have been carried out by several previous researchers. [2], The sedimentation and erosion of the Coast of the District of Teluk Segara Bengkulu City caused by a crib building originating from the old port of Bengkulu which was not used anymore, the crib was blocking the sediment rate along the coast eastward [2]. The prediction of erosion and sedimentation of beaches in

Mattirotasi Pare-Pare Beach in South Sulawesi, show that most of the Mattirotation Beach area has sedimentation, although some areas also experience erosion [3]. Longshore sediment transport at Tanjung Bayang beach was 341,37.10^3 m^3/year eastward (due to wave from the west) without the installlment of permeable groin, and by using 40% of permeability of groin density, the longshore sedimentation transport at Tanjung Bayang beach can be reduced up to 170,68.10^3 m^3/year [4]. This is caused by the presence of islands in front of the beach and beach reclamation work. South Sulawesi Province which has a capital city of Makassar with a total area of 45,574.48 km^2, with a coastline length of 1,973.7 km, consists of 23 regencies/cities spread over diverse topography and there are small islands with potential for natural resources both on land and in the very big sea. Some beaches in South Sulawesi experience erosion and sedimentation problems, but only 2 locations will be investigated, namely Padongko Beach Mangempang Village Barru Regency and Desa Aeng Batu-Batu Beach, Galesong Utara, Takalar Regency. These locations are representative of coastal areas that experience erosion and sedimentation problems.

2. Erosion and Sedimentation

Coastal erosion is one of the problems in the coastal area that has to get attention because coastal erosion can cause huge losses due to damage to residential areas and facilities in the area. To overcome coastal erosion the first cause must be sought first, so the solution can be found. One solution to erosion is to build a coastal protection structure, the building is used to protect the coast from waves and currents.

Sedimentation in its definition includes a variety of definitions that almost everyone has a different view in providing an understanding or definition of sedimentation, such as [1], states that Sedimentation is the entry of sediment loads into a particular aquatic environment, through water media and deposited in the media the environment, whereas sediment is the main material forming coastal morphology (topography and bathymetry). While [5] defines sedimentation as the process of forming sediments or sedimentary rocks caused by the deposition of the forming material or its origin in a place called the depositional environment in the form of a river, estuaries, lakes, deltas, estuaries, shallow seas to deep seas. As according to [6] defines marine sediments as the accumulation of minerals and rock fragments that are mixed with broken shells and bones of marine organisms and some other particles are formed through chemical processes that occur in the ocean.

3. Sediment Transport

Coastal sediment transport is the movement of sediments in coastal areas caused by waves and currents that they generate. Sediment transports studied in this case are those that occur in the area between the breaking waves and the coastal line. Coastal sediments can be classified as onshore-offshore transport and longshore transport. Transportation to and from the coast has an average direction perpendicular to the coastline, while transportation along the coast has an average direction along the coast. Coastal sediment transport is calculated using the longshore sediment transport empirical equation. The longshore sediment transport equation as a function of the velocity of the shoreline currents that will be used to calculate the amount of longshore sediment transport is derived from the CERC equation [7],

$$Q_l = KP_l^n \tag{1}$$

$$P_l = \frac{\rho \cdot g}{8} H_b{}^2 C_b \sin\alpha_b \cos\alpha_b \tag{2}$$

$$Q_l = 1290 P_l \text{ (m}^3\text{/year)} \tag{3}$$

where Q_l is longshore sediment transport (m^3/day), K is the empirical coefficient, P_l is longshore flux wave energy (Nm/s/m), ρ is the density of water/seawater (kg/m^3), ρs is the density of sedimentary mass (kg/m^3), n is the porosity of the sediment (n = 0.4), H_b is the height of the breaking wave (m), αb is the angle of the breaking wave (o), Cb is breaking wave velocity (m/s).

4. Prediction of Erosion and Sedimentation

Measuring height and wave period data, and particle size of sediment particles are used to predict whether the beach has erosion or sedimentation. Determination of erosion or sedimentation was done using the criteria of [8], namely:

a) Erosion if,

$$\frac{H}{W,T} \geq 3.2 \tag{4}$$

b) Sedimentation if,

$$\frac{H}{W,T} < 3.2 \tag{5}$$

where H is the significant wave height, Ws is the sedimentation velocity of sediment particles, T is the wave period.

5. Experimental Setup

This research was field research. Field survey measuring instruments that was sediment traps, and laboratory measuring instruments that was 1 set of sediment filters and digital scales. The field survey was conducted at Padongko Beach Mangempang Village Barru Regency and Desa Aeng Batu Batu Beach, Galesong Utara, Takalar Beach. The analysis conducted in this study was the analysis of the distribution of sediment grains in areas of erosion and sedimentation. The research data was used to determine the predicted erosion and sedimentation events and the amount of longshore sediment transport in the area.

6. Experimental Result

The occurrence of waves on the beaches was one of the factors that determine the erosion or sedimentation of the beaches. Large waves with a small period as was usually the case during storms will cause erosion on the beaches. So that the wave parameters that influence the determination of erosion and sedimentation events at the beaches were significant wave height (H) and wave period (T). In addition to the wave parameters, the characteristics of coastal sediment grains such as the diameter of the sediment grains (D50) and sediment fall velocity (W_s) are also parameters that influence in determining erosion and sedimentation at the beaches. Prediction of coastal erosion and sedimentation used in this study was a prediction using the characteristics of [8]. In the event of large waves (maximum) the beaches will tend to experience erosion and calm wave conditions (normal) beaches will tend to experience sedimentation. The results of erosion and sedimentation predictions from 2 (two) study sites were shown in the following table.

Table 1. Erosion prediction and sedimentation of Padongko Beach Mangempang Village Barru Regency

Point	Grain size, D50 (mm)	Fall Velocity (W_s)	Wave height significant (H_s), m	Wave period (T), second	$H/W_s.T$	Prediction
PP1A	0.19	0.117	1.3126	3.48	3.23	Erosion
PP1B	0.13	0.097	1.3126	3.48	3.91	Erosion
PP1C	0.18	0.114	1.3126	3.48	3.32	Erosion
PP2A	0.17	0.111	1.3126	3.48	3.42	Erosion

Point	Grain size, D50 (mm)	Fall Velocity (W_s)	Wave height significant (H_s), m	Wave period (T), second	$H/W_s.T$	Prediction
PP2B	0.18	0.114	1.3126	3.48	3.32	Erosion
PP2C	0.19	0.117	1.3126	3.48	3.23	Sedimentation
PP3A	0.20	0.120	1.3126	3.48	3.15	Sedimentation
PP3B	0.21	0.123	1.3126	3.48	3.10	Sedimentation
PP3C	0.20	0.120	1.3126	3.48	3.12	Sedimentation
PP4A	0.22	0.126	1.3126	3.48	3.00	Sedimentation
PP4B	0.18	0.114	1.3126	3.48	3.32	Erosion
PP4C	0.29	0.144	1.3126	3.48	2.61	Sedimentation
PP5A	0.19	0.117	1.3126	3.48	3.23	Erosion
PP5B	0.20	0.120	1.3126	3.48	3.15	Sedimentation
PP5C	0.50	0.189	1.3126	3.48	2.23	Sedimentation

Table 1, shows that in the condition of normal waves (H_s) the coast is relatively sedimentation and longshore sediment transport of Padongko Beach Mangempang Village Barru Regency were shown in the following table.

Table 2. Longshore sediment transport of Padongko Beach Mangempang Village Barru Regency

Point	Grain size, D50 (mm)	H_o	H_b	α_b (rad)	Q_l (m^3/years)
PP 1A	0.19	1.5	0.5	0.327	14.07
PP 1B	0.13	1.5	0.5	0.327	16.35
PP 1C	0.18	1.5	0.5	0.327	14.43
PP 2A	0.17	1.5	0.5	0.327	14.80
PP 2B	0.18	1.5	0.5	0.327	14.43
PP 2C	0.19	1.5	0.5	0.327	14.07
PP 3A	0.20	1.5	0.5	0.327	13.72
PP 3B	0.21	1.5	0.5	0.327	13.39
PP 3C	0.20	1.5	0.5	0.327	13.72
PP 4A	0.22	1.5	0.5	0.327	13.06
PP 4B	0.18	1.5	0.5	0.327	14.43
PP 4C	0.29	1.5	0.5	0.327	10.96
PP 5A	0.19	1.5	0.5	0.327	14.07
PP 5B	0.20	1.5	0.5	0.327	13.73
PP 5C	0.50	1.5	0.5	0.327	6.48
Total longshore sediment transport to Southeast					201.71

Total longshore sediment transport at Padongko Beach Mangempang Village Barru Regency was 201.71 m^3/years to the southeast. The prediction of erosion and sedimentation at Pantai Desa Aeng Batu-Batu Galesong Utara Takalar Regency were shown in Table 3 below.

Table 3. Erosion prediction and sedimentation of Desa Aeng Batu-Batu Beach, Galesong Utara Takalar Regency

Point	Grain size, D50 (mm)	Fall Velocity (W_s)	Wave height significant (H_s), m	Wave period (T), second	$H/W_s.T$	Prediction
PP1A	0.19	0.117	1.28	3.43	2.96	Sedimentation
PP1B	0.13	0.097	1.28	3.43	2.17	Sedimentation
PP1C	0.18	0.114	1.28	3.43	3.10	Sedimentation
PP2A	0.17	0.111	1.28	3.43	2.58	Sedimentation
PP2B	0.18	0.114	1.28	3.43	1.96	Sedimentation
PP2C	0.19	0.117	1.28	3.43	3.37	Erosion
PP3A	0.20	0.120	1.28	3.43	2.89	Sedimentation
PP3B	0.21	0.123	1.28	3.43	2.96	Sedimentation
PP3C	0.20	0.120	1.28	3.43	2.50	Sedimentation
PP4A	0.22	0.126	1.28	3.43	2.22	Sedimentation
PP4B	0.18	0.114	1.28	3.43	2.17	Sedimentation
PP4C	0.29	0.144	1.28	3.43	3.10	Sedimentation
PP5A	0.19	0.117	1.28	3.43	3.27	Erosion
PP5B	0.20	0.120	1.28	3.43	2.50	Sedimentation
PP5C	0.50	0.189	1.28	3.43	3.18	Sedimentation

Table 3, shows that in the condition of normal waves (H_s) the coast was relatively sedimentation and longshore sediment transport of Desa Aeng Batu-Batu Beach, Galesong Utara Takalar Regency were shown in the following table.

Table 4. Longshore sediment transport of Desa Aeng Batu-Batu Beach, Galesong Utara Takalar Regency

Point	Grain size, D50 (mm)	H_s	H_b	α_b (rad)	Q_l (m^3/years)
PP 1A	0.19	1.78	0.5	0.322	12.85
PP 1B	0.13	1.78	0.5	0.322	7.99
PP 1C	0.18	1.78	0.5	0.322	13.51
PP 2A	0.17	1.78	0.5	0.322	10.80
PP 2B	0.18	1.78	0.5	0.322	6.22
PP 2C	0.19	1.78	0.5	0.322	14.60
PP 3A	0.20	1.78	0.5	0.322	12.533
PP 3B	0.21	1.78	0.5	0.322	12.85
PP 3C	0.20	1.78	0.5	0.322	10.26
PP 4A	0.22	1.78	0.5	0.322	8.40
PP 4B	0.18	1.78	0.5	0.322	7.99
PP 4C	0.29	1.78	0.5	0.322	13.51
PP 5A	0.19	1.78	0.5	0.322	14.20
PP 5B	0.20	1.78	0.5	0.322	10.26
PP 5C	0.50	1.78	0.5	0.322	13.85
Total longshore sediment transport to Southeast					169.79

Total longshore sediment transport at Desa Aeng Batu-Batu Beach, Galesong Utara Kabupaten Takalar was 169.79 m^3/years to the southeast.

7. Conclusion

Based on prediction results, it shows that Padongko Beach Mangempang Village Barru Regency has sediemntation with total lonsghore sediment transport was 201.71 m^3/years to the Southeast. The prediction results shows that Desa Aeng Batu Batu Beach, Galesong Utara, Takalar Regency has sedimentation too and total longshore sediment transport was 169.79 m^3/years to the Southeast.

Acknowledgments

The authors would like to thank to Engineering Faculty of Universitas Hasanuddin for the grant of LBE 2019 and Coastal Engineering and Environmental Laboratory, Department of Ocean Engineering, Faculty of Engineering, Universitas Hasanuddin.

References

[1] Triatmodjo B, 1999, Coastal Engineering, Beta Offset, Yogyakarta.
[2] Kesumajaya. 2005. Sedimentation and Erosion at Segara Bay District Beach, Bengkulu, Magister Thesis ITB, Bandung.
[3] Hasdinar, 2014, Identification of Coastal Problems and Erosion Predictions and Coastal Sedimentation in South Sulawesi, Final Report of the Excellence Research for Ocean Engineering Study Program, Faculty of Engineering, Universitas Hasanuddin.
[4] Hasdinar, 2018, In Use of Permeable Groin for Reducing Longshore Sediment Transport at Tanjung Bayang Beach of South Sulawesi, EPI International Journal of Engineering Volume 1, Number 2, August 2018, pp. 70-73.
[5] Pettijhon, F.J. 1975. Sedimentary Rocks Second Edition. Harper and Row Publisher.New York. pp. 626.
[6] Gross, M.G. 1990. Oceanography : A View of Earth. Prentice Hall, Inc. Englewood Cliff. New Jersey.
[7] CERC, 1984, Shore Protection Manual, US Army Coastal Engineering Research Center. Washington (SPM, 1984).
[8] Kraus, N.C., Larson, M., and Kreibel, D.L. 1991. Evaluation of beach erosion and accretion predictors. Proc. Coastal Sediments '91, ASCE, 572-587.

IOP Conf. Series: Materials Science and Engineering **875** (2020) 012076 doi:10.1088/1757-899X/875/1/012076

Investigating the Performance Characteristics of a Semi-Planing Ship Hull at High Speed

S Baso[1]*, L Bochary[1], Rosmani[1], M Hasbullah[1], A D E Anggriani[1], and A Ardianti[1]

[1]Department of Naval Architecture, Engineering Faculty, Hasanuddin University, Makassar, Indonesia

*Email: s.baso@eng.unhas.ac.id

Abstract. Nowadays, the monohull and catamaran design as the high-speed vessel (HSV) become the convenient ships in sailing for the use of civilians, cars, buses, large trucks, and freight. However, some accidents by this kind ship due to its performances at high speed still occurred and required serious attention. The resistance and stability of a semi-planing hull must be reasonably considered because sometimes these are contradictory in making decisions for proper design. Therefore, this study has investigated the performance characteristics of a semi-planing ship hull at high speed. The semi-planing hull was modeled into three trim by stern conditions. The trim by stern of the semi-planing hull increases air cavity length and effects on resistance reduction. The decreased resistance is significant in FnV > 1.5 and the resistance reduction was averaged 34%. The powering requirement of the semi-planing hull would reduce due to trim by stern to then shifts on higher FnV. On the other hand, the stability range of semi-planing hull for all trim conditions is reduced due to the increase of trim by stern. Also, the maximum righting arm (GZ) shifts due to trim by stern as well where the increase of maximum GZ was averaged 3.72%. The GM decreases in increasing the angle of stern trim averaged 9.12%.

1. Introduction

Nowadays, convenient sailings using high-speed vessels are still popular for the use of civilians, cars, buses, large trucks, and freight. As known, in the 1990s, monohull and catamaran designs as a high-speed vessel (HSV) become widely used. However, some accidents due to performance still occurred and required serious attention.

The HSVs, hull shape underwater surface, are classified into semi-planing hull and planing hull. The performance and hydrodynamic characteristics of the HSVs have been interesting to be studied. Some studies have attempted various approaches to investigate the performance of the HSV. The modeling of vertical plane motion of an air cavity ship was modeled [1], and then artificial cavitation for reducing ship drag was applied [2]. However, these researches were continued to investigate more [3-6], then the steady hydrodynamic of semi-planing air cavity hulls was modeled based on linearized potential-flow theory [7]. However, the result had not validated yet with experimental results. These all research has shown that interaction between planing hull-water-air cavity contributes complex hydrodynamics matter. Moreover, the surface wave contours associated with the forebody wake of stepped planing hulls was investigated where the longitudinal surface wake profiles aft of prismatic hulls was defined and empirical equations are that quantitatively was developed as well [8].

This study has introduced to the concept of stepped planing hull that could contribute hydrodynamic advantages and then it was highlighted that the stepped planing hull was firstly studied and introduced. Correspondingly, the stepped planing hull was continuously studied [9-11]. Nevertheless, both the planing hull without stepped and the stepped planing hull are still used for the designs of monohull, catamaran, and trimaran.

On the other hand, the performances of the HSVs have been studied as well. The resistance variations related to different hull forms, slenderness ratio and other hull characteristics along with suggested design criteria for forecasting the threshold of dynamic transverse stability [12]. However, the semi-planing hull is sometimes sailing at high speed until 30 knots. The characteristic of resistance and hull form of high-speed trimaran planing hull were investigated [13], where the two auxiliary side hulls increased aerodynamic lift at high-speed motion, which improves the hydrodynamic performance. The trimaran planing hull also has excellent longitudinal stability and low wave-making action. The calm-water resistance of hard chine hulls in the pre-planning regime was predicted by using mathematical model [14], and the model could be used in the concept design phase. The hydrodynamic of the stepped planing hull was investigated by using both experimental and numerical analyses [15]. This study focused on the vertical flow phenomena featured by planing hull.

On the other hand, the related studies focused on the stability of the ship at high speed had been discussed [16-18]. Correspondingly, the mechanism of a transverse stability loss of planing craft was investigated [19]. The towing tank test was used for investigation and then estimation method of stability criteria was proposed. Therefore, the stability of a ship at high speed becomes important concern to study more detail.

Based on the discussion above, there are two understanding points of hydrodynamics view that are resistance and stability of a semi-planing hull at high speed. Those points are important matters in ship design for semi-planing hull. However, the proper resistance and good stability at high speed are sometimes contradictory in decision making for ship design. The speed of semi-planing hull increases in decreasing resistance due to increasing the angle of stern trim, however, this condition decreases absolutely stability. Regardless, the studies of the performance and hydrodynamic characteristics of semi-planing hull at high speed are still devoted widely to obtain some interpretations and knowledge for proper ship design. Therefore, this present study has investigated resistance and stability of a semi-planing hull at high speed.

2. Methods

Here, the semi-planing hull which was investigated has the main dimensions as provided in Table 1 and the body lines plan shown in Figure 1. The semi-planing hull probably shifts to high speed and then trim by stern is resulted. Therefore, the semi-planing hull was modeled and set into three trim by stern conditions, i.e. 1 degree, 2 degrees, and 3 degrees as shown in Figure 2. Those trim conditions were kept during computation process. Firstly, the semi-planing was modeled by using Maxsurf Modeller application [20] as shown in Figure 2. The investigation of resistance and stability of the semi-planing hull was conducted by using Maxsurf Resistance and Maxsurf Stability application [20].

The resistance was computed by using the Wyman method. The speed of the semi-planing hull was set until higher volume Froude number (FnV) around 4.0 with efficiency 60%. Ship water depth (T) is similar to the displacement given by the sum of Lwt and Dwt in even keel conditions.

Table 1. The main dimensions of the semi-planing hull

Length Over All (LOA)	19.06 m
Length Between Perpendicular (LBP)	17.72 m
Length Water Line (LWL)	17.77 m
Breadth (B)	4.51 m
Height (H)	1.80 m
Draft (T)	0.45 m

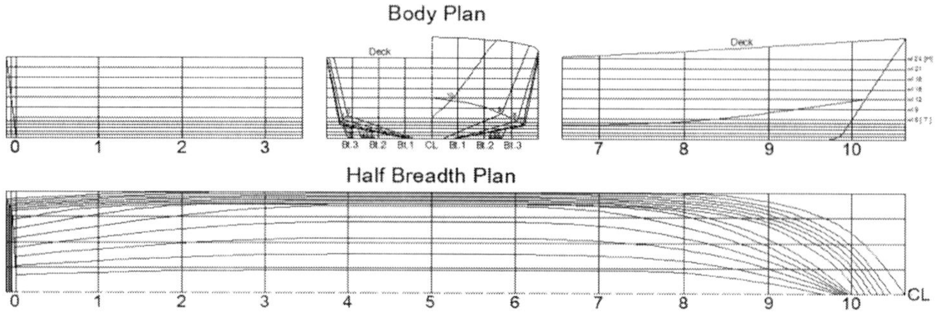

Figure 1. The main dimensions of the semi-planing hull

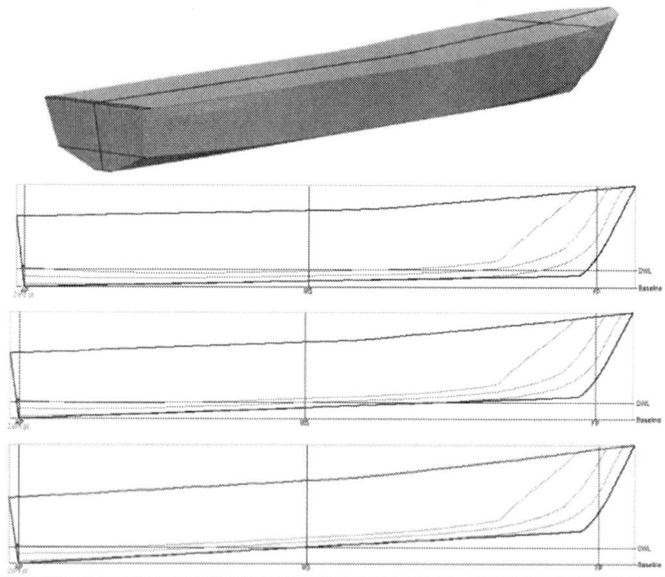

Figure 2. The model of semi-planing hull and their conditions in trim by the stern 1 degree, 2 degrees, and 3 degrees

For stability investigation, the center gravity point of the semi-planing hull was considered by calculation of weight moment. The deadweight ton (Dwt) of the semi-planing hull included passengers, luggage, fuel oil, fuel diesel oil, freshwater. Then, the lightweight ton (Lwt) included overall hull, outboard engine, outfitting component, accommodation equipment, navigation equipment, deck machinery, and lifesaving appliance and equipment. The weight distributions of Dwt and Lwt components were accorded to general arrangement. Then, the weight distributions were input in load case of Maxsurf Stability. The stability parameters were then corrected by using the criteria of IMO–HSC 2000 provided by Maxsurf Stability application.

3. Results and Discussion

The semi-planing hull was modeled successfully using Maxsurf Modeller. Then, the model was computed using Maxsurf Resistance and hereafter, it was computed using Maxsurf Stability. The followings are the discussion of the obtained results of the resistance and stability of the semi-planing hull.

3.1. Resistance Analysis

The several parameters of hydrostatic for each trim by the stern condition are shown in Table 2. The water displaced of the semi-planing hull decreased in increasing the angle of trim by stern due to shifting FnV where immersed depth was similar each trim condition. Water length also decreased in increasing the angle of trim by stern, therefore, the length air cavities of trim by stern 1 degree, 2 degrees, and 3 degrees resulted approximately 0.31 m, 4.75 m, and 9,04 m respectively.

Hereafter, the resistance of the semi-planing in increasing FnV for each trim by stern is shown in Figure 3. The resistance seems an increase in increasing FnV, however, the resistance increases significantly in FnV more than 1.5 for each stern trim condition. However, the effect of trim by stern could be reduced resistances and then results different resistance in increasing the angle of stern trim.

Table 2. The hydrostatic parameters of the semi-planing hull in each trim by the stern condition

Parameter	Trim by stern (degree)			
	Even keel	1	2	3
Displacement (ton)	20.19	11.78	6.271	3.979
Volume displaced (m^3)	19,696	11,495	6,118	3,882
Draft Amidships (m)	0.45	0.45	0.45	0.45
Immersed depth (m)	0.45	0.45	0.45	0.45
WL Length (m)	17.7	17.392	12.953	8.661
Wetted Area (m^2)	68,232	57,706	37,794	24,376
Waterplane Area (m^2)	62,898	55,231	36,243	23,052
Prismatic coefficient (Cp)	0.835	0.661	0.522	0.95
Block coefficient (Cb)	0.619	0.372	0.286	0.297

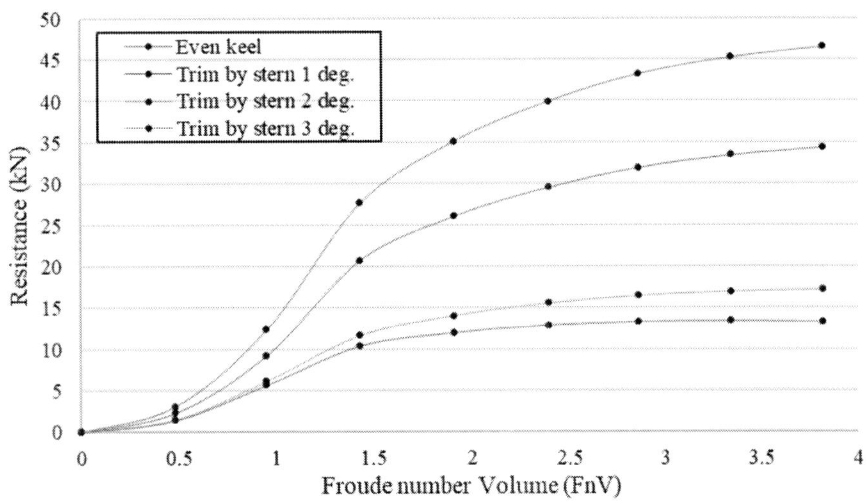

Figure 3. The resistance of the semi-planing hull in increasing Froude number volume (FnV); even keel, 1 degree, 2 degrees, and 3 degrees

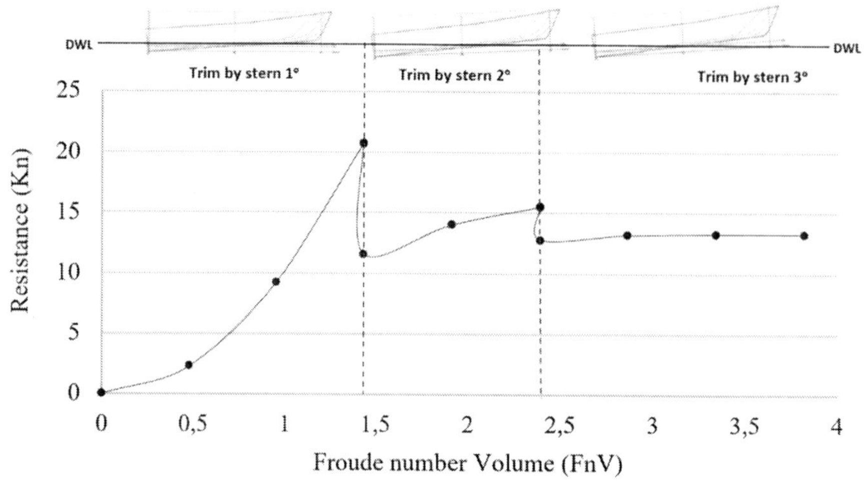

Figure 4. The resistance of the semi-planing hull in each trim by stern and fluid flow regime based on Froude number volume

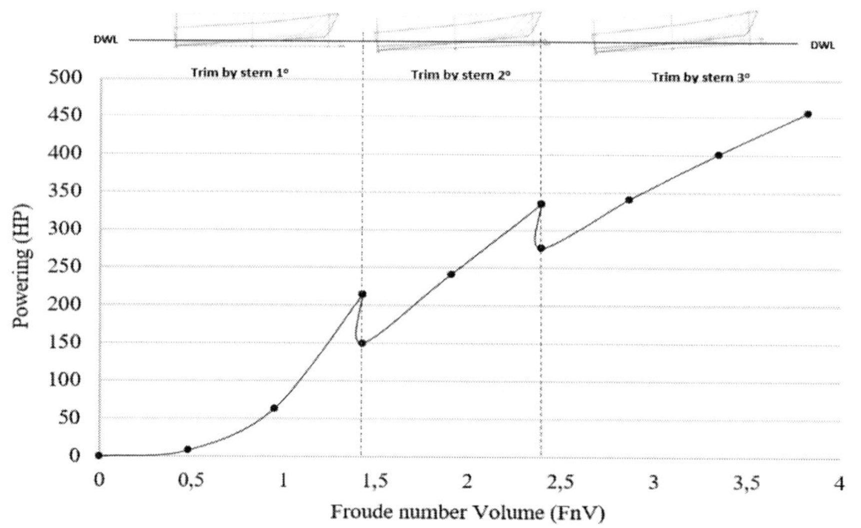

Figure 5. The powering of the semi-planing hull in each trim by stern and flow velocity regime based on Froude number volume

Figure 4 shows the resistance and shifting speed (FnV) of the semi-planing hull in each trim by stern. Correspondingly, Figure 4 describes the effect of trim by stern on resistance and speed contribution through the water. The resistance of the semi-planing hull by stern trim 1 degree reached approximately 13 kN in FnV around 1.15 and it reached 20 kN in FnV around 1.43. However, the stern trim 2 degrees reached 11.6 kN in FnV around 1.43 and the resistance 15 kN reached FnV 2.39 but it was only 12.5 kN by stern trim 3 degrees. Moreover, the resistance approximately 12.5 kN contributed FnV 1.2 by stern trim 1 degree, FnV 1.7 by stern trim 2 degrees, and FnV $\geq$ 2.39 by stern trim 3 degrees. This means the increased stern trim angle or the increased air cavity length effects on resistance reduction or increase of FnV.

On the other hand, the resistance and FnV as discussed previously effects on resulted powering and then it could be used could decide properly powering taking into account economical operation and performance safety. Figure 5 shows the powering of the semi-planing hull in each trim by stern and flow velocity regime based on Froude number volume. The powering semi-planing hull by stern trim 1 degree is required approximately 214 Horse Power (HP) when FnV reaches 1.43, however, it is only required approximately 150 HP by stern degree 2. In addition, the reached FnV 2.39 is required powering approximately 334 HP by stern trim 2 degrees but it is only required approximately 276 HP by stern trim 3 degrees. Similarly, the required power 214 HP reaches FnV 1.43 by stern trim 1 degree and shifts FnV around 1.78 by stern trim 2 degrees.

3.2. Static Stability Analysis

The stability of the semi-planing hull was analyzed successfully using Maxsurf Stability in full loading condition. Figure 6 shows examples of the statical stability curves in still water that were produced for even keel condition and stern trim 1 degree by measuring the righting arm or arm lever (GZ) for each heel angle.

The stability range for all trim conditions seems the same from 0 degrees to 82 degrees, In fact, it has a different because the vanishing stability angles that were obtained are small different magnitude. The righting arm (GZ) of the semi-planing hull instability range each stern trim condition is provided in Table 3. The maximum righting arm (GZ) of even keel condition was given 0.663 meter which was in 40 heeling degrees but it moved to 35 degrees for all stern trim conditions that were 0.656 m, 0.640 m, and 608 m respectively. Those seem little increase averaged 3.72%.

The heel angel larger after the maximum point is strong decrease of righting arm to zero. Also, those seem the stability range reduction where the trim by stern 3 degrees has the lowest stability range. Then, the stable upside-down seems in overall trim conditions and the highest negative stability is in stern trim 3 degrees as shown in Table 3.

As discussed the righting arm in range stability, the intact stability of the semi-planing hull in each stern criteria hull was corrected to fulfill stability criteria as shown in Table 4. The intact stability criteria were given by the code of High-Speed Craft Monohull 2000 which is provided into the Masxurf application. Overall criteria requirements were accepted for all trim conditions. The metacentric height (GM) is the most vital parameter for intact stability where the GM decreases in increasing the angle of stern trim of semi-planing hull averaged 9.12%. Then, overall stern trim conditions describe positive GM or GM> 0.

The GM of the semi-planing hull is affected by body form or shape underwater surface.

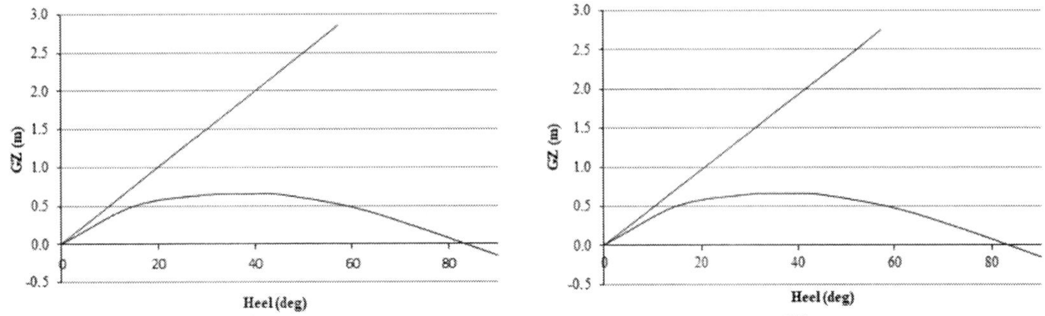

Figure 6. Example static stability curves; a). even keel condition, b). trim by stern 1 degree

Table 3. Righting arm of the semi-planing hull in stability range each stern trim condition

Heel degree / Conditon	Righting arm (GZ) in meter								
	0	15	30	35	40	45	60	75	90
Even Keel	0.000	0.504	0.649	0.661	0.663	0.653	0.486	0.189	-0.155
1 Degree	0.000	0.501	0.646	0.656	0.655	0.639	0.471	0.181	-0.155
2 Degree	0.000	0.486	0.628	0.640	0.638	0.617	0.455	0.179	-0.143
3 Degree	0.000	0.454	0.596	0.608	0.603	0.581	0.428	0.171	-0.131

Table 4. The stability criteria requirements of the semi-planing hull in each stern trim condition

Code	Criteria	Value	Trim by stern (degree)			
			Even Keel	1	2	3
HSC 2000 Annex 8 Monohull Intact	1.1 Weather criterion from IMO A.749(18) Angle of steady heel shall not be greater than (<=)	16	2	2.1	2.3	2.7
	Angle of steady heel/Margin line immersion angle shall be less than (<)	80	5.21	5.71	6.9	9.35
	1.2 Area 0 to 30 or GZmax	3.15	12.42	12.36	11.99	11.27
	1.3 Area 30 to 40	1.71	6.58	6.53	6.36	6.03
	1.4 Max GZ at 30 or greater	0.2	0.663	0.656	0.64	0.60
	1.5 Angle of maximum GZ	15	40	35	35	35
	1.6 Initial GMt	0.15	2.86	2.75	2.48	2.14

The above discussion highlighted that resistance and stability are important considerations for the design of semi-planing hull. The body form of semi-planing hull underwater surface affects its performances, therefore, a good understanding of the performance characteristics would be produced proper design of semi-planing hull.

4. Conclusions

The performance characteristics of the semi-planing hull at high speed, resistance dan stability, were successfully investigated using Maxurf Resistance and Stability application. The investigations of performance characteristics are obtained as follows: 1). The resistance of the semi-planing hull increases in increasing FnV for each stern trim. In high speed, the different increase of resistance is significant in FnV > 1.5. The trim by the stern of the semi-planing hull increases air cavity length, therefore, it affects resistance reduction when shifting FnV. 2). The powering requirement of the semi-planing hull would reduce due to the increased trim by stern. For instance, the required power 214 HP reaches FnV 1.43 by stern trim 1 degree and then shifts to FnV around 1.78 by stern trim 2 degrees. 3). The stability range of the semi-planing hull for all trim conditions is reduced due to the increased trim by stern. The vanishing stability angle in each stern trim condition has a small different magnitude. The maximum righting arm or arm lever (GZ) shifts due to trim by stern where the maximum GZ of even keel condition is in 40 heeling degrees, other conditions move to 35 degrees. The increase in maximum GZ is averaged 3.72%. 4). The intact stability of the semi-planing was corrected to fulfill stability criteria given by the code of High-Speed Craft Monohull 2000. The GM decreases in increasing the angle of stern trim averaged 9.12%. 5). The resistance and stability are important considerations for the design of the semi-planing hull. The performance characteristics of the semi-planing hull could be improved depending on body form underwater surface. Therefore, a good understanding of the performance characteristics would be produced proper design of semi-planing hull.

Acknowledgments
This research was supported partially by the Labo Based Education (LBE) 2019 grant, Faculty of Engineering, Hasanuddin University. We would like to thank our colleagues from the Naval Architecture Department, who provided insight and expertise that greatly assistance. Also, we would like to extend our appreciation to students their kind help during the research work.

References

[1] Matveev K I 1999 Modeling of vertical plane motion of an air cavity ship *Proceedings of the 5th International Conference on Fast Sea Transportation*, Seattle, WA, pp.463-470.
[2] Matveev K I 2005 Application of artificial cavitation for reducing ship drag *Oceanic Eng. Int.*, 9(1), pp.35-41.
[3] Matveev K I 2007 Three-dimensional wave patterns in long air cavities on a horizontal plane *Ocean Eng.*, 34(13), pp.1882-1891.
[4] Matveev K I, Burnett T and Ockfen A 2009 Study of air-ventilated cavity under model hull on water surface *Ocean Eng.*, 36(12-13), pp.930-940.
[5] Matveev K I and Ockfen A 2009 Modeling of hard-chine hulls in transitional and early planing regimes by hydrodynamic point sources *Int. Shipbuil. Prog.*, 56(1-2), pp.1-13.
[6] Matveev K I and Miller M J 2011 Air cavity with variable length under model hull *J. of Eng. for the Mar. Enviro.*, 225(2), pp.161-169.
[7] Matveev K I 2015 Hydrodynamic modelling of semi-planing hulls with air cavities *Int. J. Nav. Ocean Eng.* 7, pp.500-508.
[8] Savitsky D and Morabito M 2010 Surface wave contours associated with the forebody wake of stepped planing hulls, *Mar. Technol.*, 47, pp.1–16.
[9] Makasyeyev M V 2009 Numerical modeling of cavity flow on bottom of a stepped planing hull *Proc. of the 7th Int. Symp. on Cav.*, CAV2009, Michigan, USA, pp.
[10] Matveev K I 2012 Two-dimensional modeling of stepped planing hulls with open and pressurized air cavities *Int. J. of Nav. Arch. and Ocean Eng.*, 4, pp.162-171.
[11] Lotfi P, Ashrafizaadeh M, Esfahan R K 2015 Numerical investigation of a stepped planing hull in calm water *Ocean Eng.*, 94, pp.103–110.
[12] Blount D L and McGrath J A 2009 Resistance characteristics of semi-displacement mega yacht hull forms *Intl. J. Small Craft Tech.*, Vol. 151, Part B2.
[13] Ma W, Sun H, Zou J and Yang H 2013 Test research on the resistance performance of high-speed trimaran planing hull *Polish Mar. Res.* 4(80), Vol. 20, pp.45-51.
[14] Radojcic D, Zgradic A, Kalajdzic M and Simic A 2014 Resistance prediction for hard chine hulls in the pre-planing regime *Polish Mar. Res.* 2(82), Vol. 21, pp.9-26.
[15] Marco A D, Mancini S, Miranda S, Scognamiglio and Vitieloo 2017 Experimental and numerical hydrodynamic analysis of a stepped planing hull *Appl. Ocean Eng.* 64, pp.135-154.
[16] Smiley R F 1952 A theoretical and experimental investigation of the effects of yaw on pressures, forces, and moments during seaplane landing and planing *NACA Technical Note* 2817.
[17] Payne P R 1994 Recent developments in added-mass planing theory *Ocean Engi.*, Vol.21, No.3, pp.257-309.
[18] Lewandowski E M 1996 Prediction of the dynamic roll stability of hard-chine planing craft *J. of Ship Res.*, Vol.40, No.2, pp.144-148.
[19] Katayama T, Fujimoto M and Ikeda Y 2007 A study on transverse stability loss of planing craft at super high forward speed *Intl. Shipbuild. Prog.*, Vol. 54, No. 4, pp. 365-377.
[20] Maxsurf Enterprise 2014, Bentley Software, http:www.bentley.com/myselected/.

An Improvement for the Design of the Small Fiberglass Seaweed Farmers Boat to Minimize the Sinkable Risk

W Mustafa[1]*, S Asri[1], M R Firmansyah[1], F Fachruddin[1], and G Apelaby[1]

[1]Department of Naval Architecture Engineering Faculty, Hasanuddin University, Makassar, Indonesia

*E-mail: wahyuddin.mustafa@unhas.ac.id

Abstract. The boat plays significant role in supporting the household business scale for the seaweed farmers. Their current boat are made from fiberglass reinforced plastic (FRP) which is lighter then the wooden boat. However, based on the field survey in the seaweed farmers boats in the area of Pattontongan village, Jeneponto regency, their boat is susceptible for sinking if any leak is occur on the boat body since they are not designed with the reserved buoyancy system. Geometry and the reserved buoyancy of their boat become the serious issues. Hence, an improved design for their boat is need to be developed with the criteria of the boat capacity and safety. This paper discusses two of the improved design of the boats which are ND_1 (design 1) and ND_2 (design 2). The result shows an improved design with the additional of the boat reserved buoyancy while considering the boat capacity. The parameter of the boat design is the relationship between the boat reserved buoyancy volume and the boat weight which is called anti sinking ratio. The anti sinking ratio for ND1 with the boat length 6 meters, width 0.85 meter, and height 0.5 meter is ≥ 1. The ratio is obtained in the condition of the boat reserved buoyancy of $0.066m^3$ and the boat carrying capacity of $0.969m^3$. Meanwhile, the anti sinking ratio for ND2 with the boat length of 4.2 meters, width 1 meter and height 0.5 meter is ≥ 1 in the condition of the boat reserved buoyancy of $0.055m^3$ and the boat carrying capacity of $0.936m^3$.

1. Introduction

Small boats are plays significant role in a seaweed cultivation process. The use of the boats are mainly for spreading the seaweed seed and during harvesting time. The safety of the boats are important since its frequently being used during the cultivation process. The material for the boats are from wooden and FRP. The latter is being mostly used compared to the wooden boat since it is lighter and simpler in maintenance.

Geometry of both the wooden and FRP boats is comparatively the same. The shape of the boats are log like. The proportion of the boat length and width is big while the proportion of the boat width and height is small. Hence, the boat stability is not good. Outriggers on the left and right of the boat are required to maintain the boat stability. The distance of the outriggers from the main body of the boat are 1 meter each. The current seaweed farmer's wooden boat and FRP boat in the area of Pattontongan, Jeneponto regency can be seen in Figure 1.

Based on the visual observation and interviews, the current seaweed farmers boats are susceptible for sinking if any leak is occur on the boat body since they are not designed with the reserved bouyancy system. An experiment of sinking one boat have been conducted in order to prove the boat susceptibility (Figure 2).

Besides the lacking of the boat reserved buoyancy, the geometry of the current boat especially on the boat stem shape is raise another problem as well. The angle of the stem from the base line is around 45° and it's restricted the movement of the farmers. The stem angle have been complained by some seaweed farmers since the success during the seaweed seed spreading and seaweed harvesting are depending on this stem angle.

Figure 1. The wooden and FRP boat which have been used for seaweed cultivation in the area of Pattontongan, Jeneponto regency

Figure 2. Sinking the boat test to see the boat susceptibility in the area of Pattontongan, Jeneponto regency

Since the geometry and the buoyancy of the current boat are being the main problems of the seaweed farmers, their boat design must be improved by considering their safety and working characteristics.

2. Review of the boat

Defining geometry of dimension and shape of any physical object is the first step before any production or manufacturing of the object begin. Ship and boat are one of the physical object to produce by mankind. The successful planning and production of the ship and boat is depending on the prior description of their geometry [1, 2]. However, the ship design is not only about geometry. The ability of the ship to operate safely in their environment is depending on the many ship characteristics and criteria such as stability, buoyancy, resistance, maneuvering and the ship structures itself. These ship characteristics cannot be seen directly merely from its geometry but instead it needs to be analyzed.

Currently, some research in the naval architectures are focusing on the development of practical engineering method for predicting the ship characteristics. Each analysis method is based on the ship geometrical foundation as each method requires ship geometrical representation as their input. The analysis can be conducted if the respective ship geometry have been determined [2-6]. Ship geometry can be described as ship particulars and it refers to the ship description of dimension, volume and other ship capacity such as Length Overall (LOA), Waterline Length (LWL), Length Between Perpendiculars (LBP or LPP), Beam (B), Draft (D), Displacement, Tonnage, Form coefficients [1, 2, 4-6].

Three of the ship analysis which is based on the ship geometry can be conducted on the ship weight analysis, ship hydrostatic analysis and ship hydrodynamic analysis. The ship weight analysis is one of the important component of the ship design which connect ship weight and capacity. The analysis requires calculation on the length, area, volume, center of gravity for surface and space as well as requires knowledge on the weight units such as weight/length, weight/area and weight/volume of the construction material [2, 4].

The hydrostatic analysis is the most common ship geometrical evaluation being used. It evaluates forces and moments as the result of the ship static balance as well as static fluid pressure variable which working on the ship surface and on the free surface area in the ship tank [2, 4, 5, 6]. While hydrodynamic analysis predicting forces, movement and structural load as the result of the ship movement on the water, the water flow around the ship including the effect of the sea wave. Though the hydrodynamic analysis is complex but simplicity and prediction of the fluid flow as well as the ship geometry can be applied [2, 4, 7].

3. Research Method

Methodology for solving the fiberglass boat geometry and reserve buoyancy problems can be divided into several stages which are measuring and collecting the "base/sample boats" dimension, determining the "base boat" particulars and then determining the improved boat design particular. Based on the latter information, the correlation between the boat capsizing ratio and the boat carrying capacity is determined.

4. Result and Discussion

The improved boat design was developed based on the three boat samples which being used by the seaweed farmers ("which become the base boats"). Based on the current boat direct measurement, weigth and hydrostatic analysis, the characteristic geometry or the "base boats" particular are determined. The base boat lines plan and the boat particulars can be seen in Figure 3 and Table 1.

Table 1. The main particulars of the three base boats (BB)

Designation	Symbol	Unit	BB_1	BB_2	BB_3
Length Overall	*LOA*	m	6.00	6.50	7.30
Length Between Perpendiculars	*LBP*	m	5.66	6.13	6.54
Breadth /Beam	*B*	m	0.70	0.65	0.65
Depth	*D*	m	0.60	0.60	0.64
Draft	*T*	m	0.41	0.50	0.50
Block coefficient	C_B		0.52	0.56	0.52
Midship section coefficient	C_M		0.81	0.86	0.80
Water plane coefficient	C_{WP}		0.71	0.75	0.76
Prismatic coefficient	C_P		0.64	0.66	0.65
Volume of displacement	∇	m^3	0.73	0.86	0.97
Displacement	Δ	kg	747.80	881.30	998.10
L/B			8.09	9.43	10.06
B/T			1.71	1.30	1.30
L/D			9.43	10.22	10.22

Considering the base boat particulars as well as the farmer's needs, the improved boat designs are developed based on the displacement volume constant value of 0.73 m^3. The value is selected in order to keep the boat weight of 80 kgs or less so that it can be carried by at least two seaweed farmers.
The result of the boat weight and hydrostatic analysis in order to obtain the boat particulars as well as the boat lines plan can be seen in Figure 4 and Table 2.

IOP Conf. Series: Materials Science and Engineering **875** (2020) 012077 doi:10.1088/1757-899X/875/1/012077

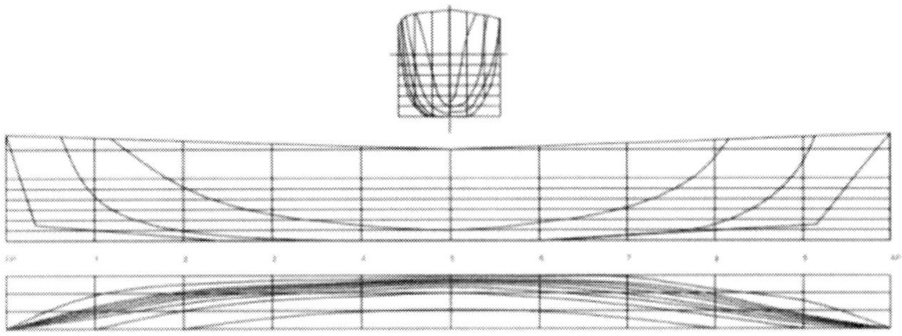

Base Boat 1 (BB_1)

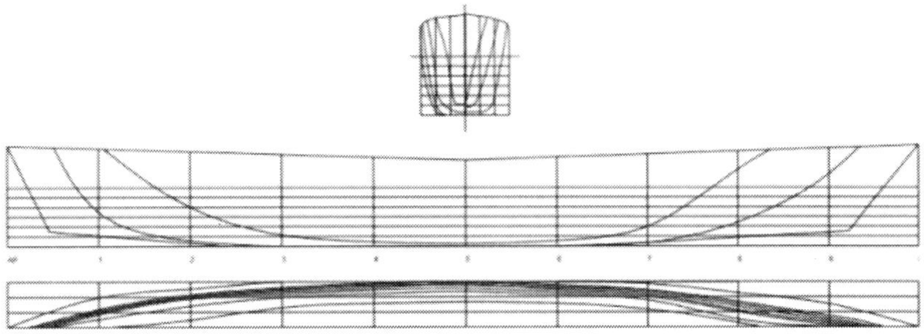

Base Boat 2 (BB_2)

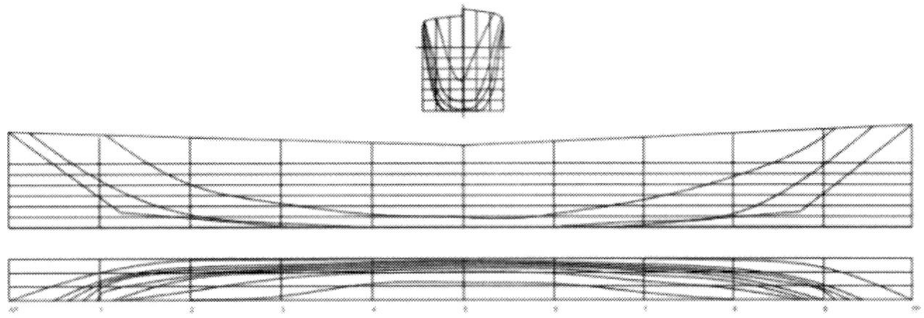

Base Boat 3 (BB_3)

Figure 3. The lines plans of the three base boats.

Table 2. The main particulars of the two improved boat design

Designation	Symbol	Unit	Design 1 (ND_1)	Design 2 (ND_2)
Length Overall	*LOA*	m	6.00	4.20
Length Between Perpendiculars	*LBP*	m	5.60	4.00
Breadth /Beam	*B*	m	0.85	1.00
Depth	*D*	m	0.50	0.50
Draft	*T*	m	0.35	0.35
Block coefficient	C_B		049	0.52
Midship section coefficient	C_M		0.80	0.88
Water plane coefficient	C_{WP}		0.66	0.65
Prismatic coefficient	C_P		0.62	0.59
Volume of displacement	∇	m^3	0.73	0.73

Designation	Symbol	Unit	Design 1 (ND_1)	Design 2 (ND_2)
Displacement	Δ	kg	748.10	748.5
L/B			7.22	4.07
B/T			2.24	2.84
L/D			12.00	8.40

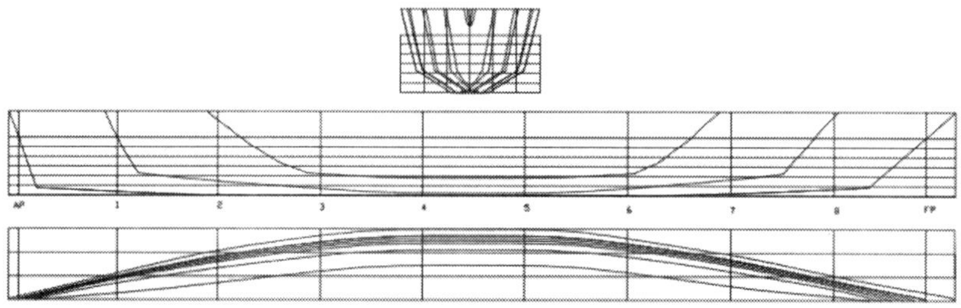

Design 1 (ND_1)

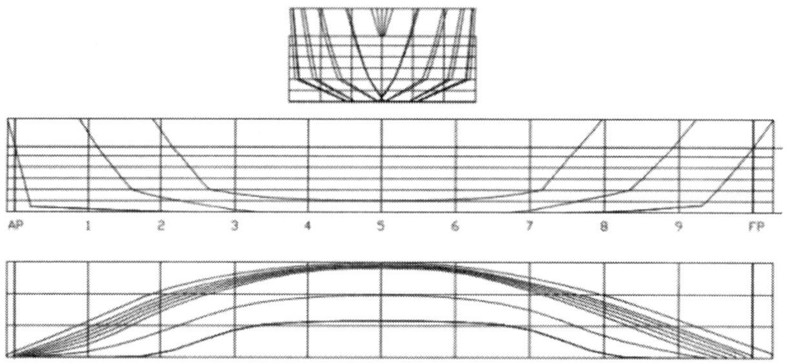

Design 2 (ND_2)

Figure 4. The lines plans of the two improved boat designs

The basic difference of the improved boat design to compare with the base boats is in the midship section shape and in the angle of the boat stem. In the improved boat designs, the chine shape of the boat have been applied to get the wider boat base. The purpose of the application is to ease the farmers to dock the boat on the beach. The angle of the boat stem is to be made steep as well in order to minimize the accident risk to the farmers while working.

The other problem to be solved in the improved boat design is to minimize the risk of the boat to be sinking in case of any leak in the boat. Hence, the improved design must be added with the boat reserved buoyancy while keep considering the boat capacity. The improved boat design must have the anti sinking ratio more than 1. The anti sinking ratio is the relationship between the boat reserved buoyancy volume and the boat hull weight. Figure 5 shows the arrangement of the empty tanks as the source of the boat reserved buoyancy.

Based on the general arrangement (GA) drawing as can be seen in Figure 5, the value of the boat reserved buoyancy and the boat weight are being simulated and the result of the analysis can be seen in Table 3 and Figure 6.

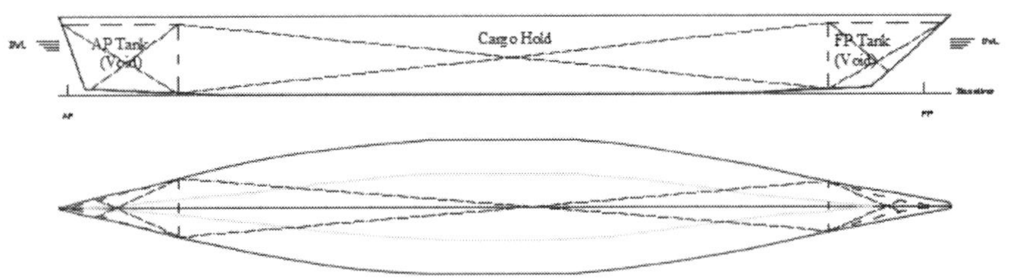

General arrangement of design 1 (ND_1)

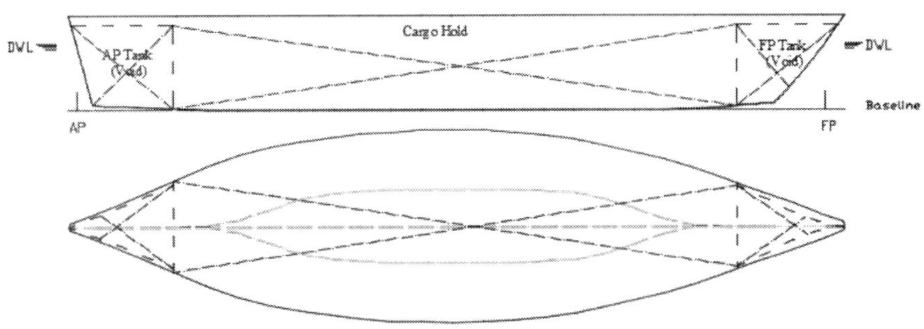

General arrangement of design 2 (ND_2)

Figure 5. The general arrangement of the both improved boat design

Table 3. Anti sinking ratio of the improved boat design

No	Length of tank (m)		Reserved Buoyancy Volume (m3)			Boat Hull Weight (ton)	Anti Sinking Ratio	Cargo Hold Volume (m3)
	AP	FP	AP	FP	Total			
1	*2*	*3*	*4*	*5*	*6=4+5*	*7*	*8=6/7*	*9*
Improved Design 1 (ND_1)								
1	0.65	0.65	0.031	0.017	0.048	0.065	0.742	0.987
2	0.70	0.70	0.036	0.021	0.057	0.065	0.881	0.978
3	0.75	0.75	0.041	0.025	0.066	0.065	1.021	0.969
4	0.80	0.80	0.047	0.030	0.077	0.065	1.191	0.958
5	0.85	0.85	0.053	0.035	0.088	0.065	1.361	0.947
6	0.90	0.90	0.060	0.040	0.100	0.065	1.546	0.935
7	0.95	0.95	0.067	0.046	0.113	0.065	1.748	0.922
8	1.00	1.00	0.075	0.052	0.127	0.065	1.964	0.908
9	1.05	1.05	0.082	0.058	0.140	0.065	2.165	0.895
Improved Design 2 (ND_2)								
1	0.45	0.45	0.019	0.015	0.034	0.052	0.65	0.957
2	0.50	0.50	0.024	0.020	0.044	0.052	0.84	0.947
3	0.55	0.55	0.030	0.025	0.055	0.052	1.05	0.936
4	0.60	0.60	0.037	0.031	0.068	0.052	1.30	0.923
5	0.65	0.65	0.044	0.038	0.082	0.052	1.57	0.909
6	0.70	0.70	0.053	0.045	0.098	0.052	1.87	0.893
7	0.75	0.75	0.062	0.053	0.115	0.052	2.20	0.876
8	0.80	0.80	0.072	0.062	0.134	0.052	2.56	0.857
9	0.85	0.85	0.083	0.072	0.155	0.052	2.96	0.836

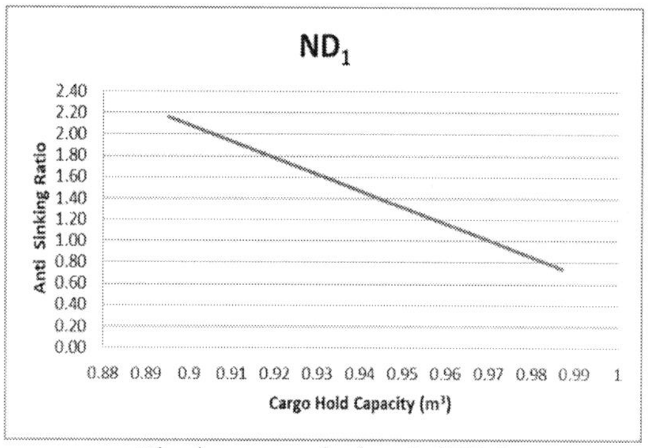

The improve design 1 (ND_1)

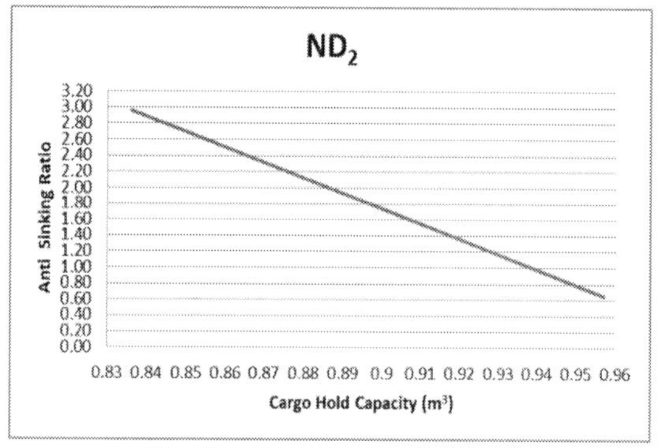

The improve design 2 (ND_2)

Figure 6. The relationship between anti sinking ratio with the boat carrying capacity of both of the improved boat design.

Ratio number of ≥ 1 on Figure 6 means that in case of any leak on the boat and the seawater entering the boat up to the boat deck, the boat cannot be sinking into the seabed but instead the boat is still floating even in the upside down position. The value of the boat reserved buoyancy is influencing the boat carrying capacity (cargo hold capacity). If the value of the boat reserved buoyancy increase, then the boat carrying capacity will be decrease.

Based on the both graphs in Figure 6, the improved design 2 (ND_2) seem to be more realistic to be applied as the length of the boat is smaller than ND_1 with almost the same capacity.

5. Conclusion

The result shows an improved design with the additional of the boat reserved buoyancy while considering the boat capacity. The parameter of the boat design is the relationship between the boat reserved buoyancy volume and the boat weight which is called anti sinking ratio. The anti sinking ratio for ND_1 with the boat length 6 meters, width 0.85 meter, and height 0.5 meter is ≥ 1. The ratio is obtained in the condition of the boat reserved buoyancy of 0.066 m^3 and the boat carrying capacity of 0.969 m^3. Meanwhile, the anti sinking ratio for ND_2 with the boat length of 4.2 meters, width 1 meter and height 0.5 meter is ≥ 1 in the condition of the boat reserved buoyancy of 0.055 m^3 and the boat carrying capacity of 0.936 m^3.

Acknowledgments

The authors would like to thanks the Faculty of Engineering, Universitas Hasanuddin for its research grant under scheme of Labo Base Education (LBE) research 2019 so as this research work can be done. The appreciation is given as well to the members of the seaweed farmers group in Pattontongan City for their help and cooperation during data collection and interviewing process in location.

References

[1] Hamlin N A 1988 Ship Geometri *Principles of Naval Architecture Second Revision Volume I Stability and Strength* ed Lewis E V (Jersey City, NJ: SNAME) chapter 1 pp 15–28.
[2] Letcher Jr J S 2009 The Geometric of Ship *Principles of Naval Architecture Series* ed Paulling J R (Jersey City, NJ: SNAME).
[3] Francescutto A and Papanikolaou A D 2011 Buoyancy, Stability, and Subdivision: From Archimedes to SOLAS 2009 and the Way Ahead *Proceedings of the Institution of Mechanical Engineers, Part M: Journal of Engineering for the Maritime Environment, 225*(1), 17 32.
[4] Parsons M 2009 Parametric Design *Ship Design and Contruction Vol. 1-2* ed Lamb T (Jersey City, NJ: SNAME) chapter 11 pp 1-46.
[5] Papanikolaou A 2014 Ship Design *Methodologies of Preliminary Ship Design* (London;Springer).

[6] Asri S, Firmansyah M R, Lage F F, Wahyuddin Mustafa, Yusuf Z, Paroka D, Muil Ali M Z 2017, Designing and constructing a FRP boat for seaweed farmer in Jeneponto regency, *Proceeding of the 4th International Marine and Fisheries Symposium*, pp. 155-164 .

[7] Keuning J A and van Walree 2006, The Comparison of the Hydrodynamic Behaviour of Three Fast Patrol Boats with Special Hull Geometries [online]. In*: HIPER 06: 5th International Conference on High-performance Marine Vehicles* pp 137-152.

IOP Conf. Series: Materials Science and Engineering **875** (2020) 012078 doi:10.1088/1757-899X/875/1/012078

Risk Assessment for Fishing Boats Operating in the Makassar Strait

M R Alwi[1]*, A H Muhammad[1], Rahimuddin[1], and H Hasan[1]

[1]Department of Marine Engineering, Engineering Faculty, Hasanuddin University, Makassar, Indonesia

*Email: mrusydi.alra@gmail.com

Abstract. The main cause of marine accidents that lead to loss of human life is purely human error. Specifically, in fishery activities, as much as 80 percent of marine accident factors are caused by human negligence. The purpose of this study is to inventory and describe all hazards and their possible consequences in every aspect and stage, to inventory and identify aspects related to fisherman safety management, both institutionally, existing regulations, and human resources, working environment conditions and designing work safety models fishers in Takalar Regency. The next method is the Formal Safety Assessment (FSA), the human element can be incorporated into the FSA process by using human reliability analysis. HRA in this study was carried out quantitatively. To complete the detailed description of the model development, system analysis is carried out to understand and make decisions related to the Work and Health and Safety Management System. Activities that have the greatest risk and consequences of work accidents occur in fishing gear operations. Risk control by minimizing human error can generally be done by making plans for fishing activities, improving the ability of fishermen skills, safe working procedures, using equipment as needed, using self-protection equipment and conditions a healthy work environment.

1. Introduction

A fisherman is a very dangerous job with high accident risk. The World Food Organization ("The State of World Fisheries and Aquaculture 2008"), released on March 2, 2009, reported that as many as 24,000 fishermen died at sea. The report mentioned that there were four factors that caused the high mortality rate of traditional fishers and sea transportation users [1]. The main cause of sea accidents that lead to loss of human life is purely human error. Other causes are neglect by sea transport providers and related agencies, as well as sea transportation safety equipment that is far from adequate and lack of standard work procedures. Specifically, in fishery activities, as much as 80 percent of marine accident factors are caused by human negligence.

There are events that are arguably repeated every year, fishers as fishery actors have accidents in fishing activities. The causes of fishing vessel crew accidents are at least 5 factors causing vessel fishing vessel accidents, namely the low awareness of ship crews about work safety on shipping and fishing activities, low mastery of shipping and fishing safety competencies, vessels not equipped with safety equipment as they should, bad weather such as big waves and suffered a severe illness on the voyage. In addition to bad weather factors such as big waves, there are human and ship fault factors and safety equipment.

Utilization of marine biological resources cannot be separated from fishing operations involving various fishing units that are currently developing, which are quite varied, ranging from small ones such as spears, scoops, and bottom longline fishing, to large-sized fishing gear such as trawl, purse

IOP Conf. Series: Materials Science and Engineering **875** (2020) 012078 doi:10.1088/1757-899X/875/1/012078

seine, tuna longline and paying. Taking into account the characteristics of work on a fishing vessel, concern for safety by both the crew and the owner of the ship and mastery of competencies related to safety for fishing vessel crews is very important. In Indonesia there are approximately 2.78 million fishing sailors or fishermen, most of them work on small-sized fishing vessels and only a small portion work on large-sized vessels that are generally of industrial scale [2].

Risk assessment means thinking about the possible dangers and deciding what you can reasonably do to prevent or guard against them. Risk Assessment is required in all workplaces and it is the responsibility of your employer or vessel owner to ensure that the workplace is safe and healthy for all persons involved. A fishing vessel is a workplace and the vessel operator must ensure that it is safe and healthy for the crew and other persons that may have occasion to be on the vessel [3].

A risk assessment is a tool to help identify trends and major safety hazards that may exist within a fishery. It involves analyzing information about safety incidents and other aspects of fisheries to identify those experiencing high risks. The risk assessment can be used by fishermen, fishery managers, and safety professionals to develop solutions for reducing risks and improving safety [4].

This research used a survey method with a case study approach to one of the basic longline fishing gear. The data collected is primary data and secondary data. Primary data were obtained from observations during fishing operations and interviews with respondents. Respondents in this study were 20 people or 20% of the total basic longline fishing population in the district. Takalar and type of data collected in the form of information about fishing operations methods. Secondary data is supporting data obtained from various literary sources.

2. Hierarchical Task Analysis (HTA)

Hierarchical Task Analysis (HTA) provides an overview of activity or sub-activity. In HTA, a plan that explains the sequence and condition of an activity carried out. The steps taken in conducting HTA are as follows:

- Identify the main activities to be analyzed, by determining the objectives and limits.
- Break down main activities into sub-activities and build plans.
- Stop the sub-activity based on the level of detail.
- Continue the process of decomposition of activities.
- Group several sub-activities to a higher level than sub-activities

Hierarchical Task Analysis (HTA) is used to identify activities and conditions/opportunities for the safety consequences of fishermen's work caused by human errors found in fishing operations using bottom longline fishing in the Makassar Strait. The order of activities identified in fishing using bottom longline fishing is presented in Table 1.

Table 1. Identification of failures in bottom longline fishery activities

Step	Task Description	Fault Description	Consequences
I. Preparation:			
1.	*Checking equipment and fishing needs:*		
1.1.	Engine Preparation	Engine failure objects falling on the body	Injured
1.2.	Ice preparation	Ice soaks the road	Injured
2.	*Drinking water preparation*	Water spilled	Injured
3.	*Check and repair fishing gear*	Incorrect use of tools	Injured
II. Loading			
4.	*Engine Mounting*	The machine fell on	Injured
5.	*Cool box loading*	Fiber cool boxes fall on the limbs	Injured
6.	*Fuel can loading*	Fuel can fall on the limbs	Muscle injury
III. Sail to the fishing grounds:			
7.	*Take off the mooring rope*	The rope cannot be separated, the fisherman splashed	Muscle injury

8.	*Fisherman starts the engine*	Banged	Injured
9.	*Fishermen steer the boat out of the harbor*	Lost direction, crashing, boat overturned	Drowning
IV. Fishing gear preparation:			
10.	*Prepare the float mark*	A float sign falls, Fishermen plunge into the sea	Injured
11.	*Throwing water from the boat hull*	Fishermen splashed	Injured
V. Fishing gear operation:			
12.	*Lowering fishing gear*	Fishermen get hooked	Injured
13.	*Fisherman set the boat direction*	Lost direction, boat upside down	Drowning
14.	*Raise the fishing gear*	Fishermen get hooked	Injured
VI. Handling of catches:			
15.	*Putting fish in the cool box*	Slip, skewered by a fish	Injured
16.	*Throwing water from the boat hull*	Fishermen splashed	Injured
VII. Sail to the fishing base:			
17.	*Direct the boat to the fishing base*	Lost direction, boat capsized	Drowning
18.	*Tidy up the fishing gear*	Slip	Injured
VIII. Unloading:			
19.	*Attaching a mooring rope*	Stumble, knock, splash	Injured
20.	*Removal of fish cool box*	Slip, stumble	Injured
21.	*Removal engine*	Tools dropped	Injured

On fishing operations using basic longline fishing, there are chances of consequences due to human carelessness, namely injury, muscle injury, and drowning. In general, the cause is due to the limited level of understanding and skills of fishermen, the absence of clear work procedures, and in carrying out these activities fishermen do not use self-protection equipment.

Opportunities for the consequences of failure can be reduced by fishermen using self-protection equipment, including in the form of head protection from sunburn, non-slippery footwear, gloves when working, a buoy that is able to withstand the burden of fishermen who work in water.

3. Formal Safety Assessment

Formal Safety Assessment (FSA) is the leading scientific method currently being used for maritime safety analysis and for the formulation of related regulatory policies [5]. The FSA is a structured and systematic methodology, aimed at improving maritime safety, including self-protection, health, the marine environment, and property, using risk analysis and cost-benefit assessments. The human element can be incorporated into the FSA process by using human reliability analysis. HRA is a process, which consists of a series of activities and the potential use of a number of techniques depending on the overall purpose of the analysis. HRA in this study was carried out quantitatively.

HRA can be considered to enter into the whole FSA process with the following stages:

- Identifying the main tasks/activities.
- Assess risk, analyze human error and quantify human reliability
- Risk control options.

The FSA and HRA can be applied to the design, construction, maintenance, and operation of ships. The stages of HRA that are carried out are to identify activities/tasks in detail with Hierarchical Task Analysis (HTA). In HTA, work/activity is broken down into several activity/job levels. It is also very useful in seeing activities/workers interacting with work equipment and aspects of the work environment. Activities/jobs are divided into several levels of activity/work based on the objectives to be achieved [6].

The second stage is conducting a risk assessment using the Human error Assessment and Reduction Technique (HEART), including activity/task analysis [7]. The HEART assessment stage is to classify an activity into one generic category, determine the nominal Human Error Probability (HEP) for an activity, and determine the Error Producing Conditions that can affect the reliability of an activity, determine the Assessed Proportion of Affect (APOA) for each EPC, and calculate the HEP of an activity. Based on the EPC, a calculation of the effect of the error will occur through the proportion of the EPC. The calculation is done using the following equation.

$$AE_i=\left((EPC_i-1)\times P_oA_i\right)+1 \quad (1)$$

To calculate the value of HEP is:

$$HEP_i=AE_1\times AE_2\times AE_3\times AE_i \quad (2)$$

So that the level of reliability can be calculated by the formula:

$$HEP_t=HEP_1+HEP_2+HEP_3+\ldots+HEP_i \quad (3)$$

where:

AEi = Assessed effect on the i-EPC
EPCi = the nominal amount on i-EPC
PoAi = The amount Proportion of Affect in the i-EPC with a value between 0 – 1
HEPi = The amount of i-HEP

The main function of the HEART calculation process is to group tasks in generic categories and their nominal level values for human unreliability. HEART is known as a relatively simple way of determining the probability of human error (HEPs). The HEART method is a human reliability quantification method that was developed since 1985. This method has been tested for validation [8]. By comparing the HEART method with 2 other Human Reliability methods namely THERP and JHEDI, where the results of the study indicate that all three methods have an acceptable level of accuracy [9].

The last step is to choose risk control options that are consistent with the observed activity. Making choices in risk control is based on actual observed conditions, taking into account the work area, the equipment used and the background of the human resources involved.

4. Result and Discussions

Based on observations and identification using HTA and assessment using HEART there are 8 activity groups and 21 sub-activities, as can be seen in Table 2.

Table 2. Identification of HEP values in bottom longline fishery operations

No	Activity	Sub Activity	HEP
I	Preparation	1. Checking equipment and fishing needs	0.59772
		2. Drinking water preparation	0.13270
		3. Check and repair fishing gear	0.52754
II	Loading	4. Engine Mounting	0.64836
		5. Cool box loading	0.37099
		6. Fuel can loading	0.38019
III.	Sail to the fishing grounds	7. Take off the mooring rope	0.04500
		8. Fisherman starts the engine	0.17300
		9. Fishermen steer the boat out of the harbor	0.92321
IV	Fishing gear preparation	10.Prepare the float mark	0.04500
		11.Throwing water from the boat hull	0.04500

V	Fishing gear operation	12. Lowering fishing gear	0.97547
		13. Fisherman set the boat direction	0.92311
		14. Raise the fishing gear	0.96227
VI	Handling of catches	15. Putting fish in the cool box	0.15500
		16. Throwing water from the boat hull	0.15500
VII	Sail to the fishing base	17. Directed the boat to the fishing base	0.92321
		18. Tidy up the fishing gear	0.00770
VIII	Unloading	19. Attaching a mooring rope	0.04500
		20. Removal of fish cool box	0.32199
		21. Removal engine	0.20475

The HEP value in Table 2 illustrates the risk of failure due to human error in basic longline fishing in Takalar Regency. The biggest HEP caused by human error occurs in fishing gear operating activities when lowering the fishing gear with a HEP value of 0.97547. The second biggest risk of failure occurs in increasing fishing gear with a HEP value of 0.96227.

The results of the identification of failures in activities that have the largest HEP value, namely fishing gear operation activities, identified 2 consequences, are injury and drowning as shown in Table 1. Furthermore, the selection of risk control is based on the results of the measurement of continued reliability of bottom longline fishery activities by minimizing the occurrence of human error resulting in work accidents. The types of work accidents identified in bottom longline fishery activities are fatigue, muscle injury, injury, pain, and drowning. The results of reliability measurements in all activities can be seen in the following Table 3.

Table 3. FTA calculation results in bottom longline fishery activities

No	Activity	Consequences	HEP
I	Preparation (0.87457)	Injured	0.81046
II	Loading (0.94779)	Injured	0.77677
		Muscle injury	0.76331
III.	Sail to the fishing grounds(0.94239)	Muscle injury	0.03900
		Injured	0.45252
		Drowning	0.92321
IV	Fishing gear preparation (0.37605)	Injured	0.11786
V	Fishing gear operation (0.99882)	Injured	0.59061
		Drowning	0.99672
VI	Handling of catches (0.82754)	Injured	0.83614
VII.	Sail to the fishing base(0.95812)	Drowning	0.92121
		Injured	0.31885
VIII	Unloading (0.91082)	Injured	0.91722

The biggest chance of work accidents in fishing gear operating is equal to 0.99882. In these activities, there are 2 types of work accidents due to human error that may occur, namely injury and drowning. Risk Acceptance Criteria caused by Human Error Probability should be no more than 0.1 [10]. The value of 0.1 indicates that it is very difficult to make a HEP value up to 0.0, where there is no human error at all in a series of activities.

5. Conclusions

The bottom longline fisheries activity in Takalar Regency has a high potential for the accident risk. Activities that have the greatest risk and consequences of work accidents occur in fishing gear operations. Risk control by minimizing human error can generally be done by making plans for fishing/shipping activities, enhancing the skills of fishermen, safe working procedures, using equipment as needed, using self-protection tools and conditions a healthy work environment.

Acknowledgments
The authors would like to thanks the Engineering Faculty, Hasanuddin University for is research grant under scheme Labo Base Education (LBE) research 2019 so this research work can be done. The appreciation is given as well to the number of several fishermen in Takalar Regency for their help and cooperation during the data collection and interviewing process.

References

[1] Food and Agriculture Organization. (2009). The State of World Fisheries and Aquaculture 2008. Rome, Italy. 176 p.
[2] Departemen Kelautan dan Perikanan dan [JICA] Japan International Cooperation Agency. (2009). Indonesian Fisheries Statistics Index 2009. Direktorat Jenderal Perikanan Tangkap. Departemen Kelautan dan Perikanan (ID).
[3] Directorate-General for Employment, Social Affairs and Inclusion. (2016). European Guide For Risk Prevention In Small Fishing Vessels, ISBN 978-92-79-45113-3
[4] Lambert, D.M., E.M. Thunberg, R.G. Felthoven, J.M. Lincoln, and W.S. Patrick. (2015). Guidance on Fishing Vessel Risk Assessments and Accounting for Safety at Sea in Fishery Management Design. U.S. Dept. of Commer., NOAA. NOAA Technical Memorandum NMFS OSF-2, 56 p.
[5] Christos A. Kontovas and Harilaos N. Psaraftis. (2009). Formal Safety Assessment: A Critical Review. Marine Technology, Vol. 46, No. 1, January 2009, pp. 45–59
[6] Lyons, M., Adams, S., Woloshynowych, M. & Vincent, C. (2004). Human Reliability Analysis in Healthcare: A Review of Technique. International Journal of Risk & Safety in Medicine, 16, 223-237.
[7] Williams, J.C. (1985). HEART-A Proposed Method for Achieving High Reliability in Elsevier, pp. 87-109. Process Operation by means of Human Factors Engineering Technology. In Proceedings of a Symposium on the Achievement of Reliability in Operating Plant, Safety and Reliability Society, 16 September 1985, Southport.
[8] Kirwan, B. (1997). The validation of three Human Reliability Quantification techniques - THERP, HEART and JHEDI: Part II - Results of validation exercise. Elsevier, Applied Ergonomics Vol 28. No. 1. pp. 17-25
[9] Bell, Julie dan Holroyd, Justin. (2009). Review of Human Reliability Assessment Methods. Health and Safety Laboratory
[10] Aven, Tarje. Jan Erik Vinnem and Frank Vollen. (2006). Perspectives on Risk Acceptance Criteria and Management for Offshore Applications – Application to A Development Project. International Journal of Materials & Structural Reliability Vol.4, No.1, March 2006, 15-25.

Effect of Different Stern Flap Position and Size as an Energy Saving Device (ESD) for Increasing Fishing Boat Speed

S Klara[1], F Mahmuddin[1*], S Hariyanto[1], G Sitepu[1], and A A Aman[1]

[1]Marine Engineering Department, Faculty of Engineering, Hasanuddin University

*Email: f.mahmuddin@gmail.com

Abstract. One of the most issues on a boat for using energy is resistance. Therefore some energy is needed to be able to overcome the boat resistance. The total resistance depends on the speed of the boat. By reducing the resistance, it will affect the energy usage on the boat. One way to reduce the boat resistance is by installing a stern flap to the stern. Stern flap is a new addition on the stern appendage. Previous studies, the stern flap was able to reduce the resistance. The reduction of resistance occurred on boats can reduce fuel consumption, so it can be one of the innovations in combating the energy crisis. This study used an experimental method that analyzes the position and size of the stern flap, where each position and size of the flap were tested at the same load. From the results and analysis, it shows that the changing position of the stern flap on the surface of the water will increase the boat speed by 4.62% and the changing size of the stern flap size to 85 x 80 cm will increase the speed of the boat by 3.82%.

1. Introduction

There is a close relationship between humans and energy because energy can support human activities anytime and anywhere. It has been occurring since the industrial revolution when entering the 20th century. It is the greatest period of increasing energy usage in which various energies are produced and consumed, one of which is petroleum energy.

Increased human activities can reduce petroleum reserves. Reducing oil reserves requires energy savings to be done immediately. The society must think about how to develop energy-saving solutions, including in ifishing-related activities. As we have known, fishing boats sail at the sea where it got drag force, which we called boat resistance. The reduction of boat resistance will also reduce fuel energy consumption. An optimization solution that needs to be done is to design a tool to reduce the resistance of the boat.

One of the most influential components in using energy from a boat is the propulsion system [1]. As we know, a boat is a complex and complicated object, because it can move at certain speeds both above and below the water surface. This incurs drag force or resistance in the water. The resistance effect depends on boat speed. By reducing the resistance experienced by the boat by installing a stern flap to the stern, it will affect all energy usage. Therefore, this study examines the efficiency of stern flap with different shapes and positions on a fishing boat to produce an optimum energy saving device (ESD).

2. Literature Review

The stern flap configuration on the boat is usually evaluated based on annual fuel consumption when compared to the supporting hull structures on the stern gate and without flap. The effect on main engine driving power is determined by the generator loads, etc., which are not considered in the present study. The engine's exhaust gases can be calculated from the exhaust gases report at specific power levels.

2.1. Stern Flap

Stern flap is one of the additional appendage enhancements installed at the stern [2]. The hull interaction with the flap occurs at the trim condition, which reduces the propulsion resistance and increases the boat speed. The most important stern flap parameters are the length of the chord (L), the angle of the flap (a), and the flap span along the ship transom. Various stern flap studies found that it has an effect on boat performance. Stern flap specification depends on the boat dimension and stern shapes [3].

Figure 1. An example *stern flap* [4]

The stern flap causes the lift force on the transom and changes the pressure distribution on the stern. On the stern flap, hull planning affects the trim in the 4.0 to 5.0 degrees. Where the displacement of the hull stern flap results in the trim reduction in the 0.1 to 0.3 degrees, this does not have a significant effect on reducing boat resistance. The basic advantage of adding the stern flaps is to make flow separation around the propeller. It provides drag force reduction on the stern and changes wave resistance from around the boat.

Other advantages of adding a stern flap are reducing resistance at a powering condition, increasing maximum speed, and changing the wave system in the transom area. The stern flap causes the hull flow to be decreased where the stern flap length is increased. Decreased flow velocity will increase pressure under the hull which causes a reduction in drag shape. The hydrostatic performance advantage of stern flaps is due to flow changes around the stern [5].

2.2. Hydrodynamic Effects of Stern Flaps

a. Stern Flow Change

Stern flaps reduce flow velocity and increase dynamic hull pressure. Increased pressure area causes greater lift force which produces a positive effect on boat movement. Stern flaps also increase the trailing edge outflow speed as compared to the transom area without flaps. This increased speed will reduce flow separation speed and make new flow separation well and reduce viscous pressure.

b. Wave System Transformation

Flow transfer around the stern area without stern flap causes energy loss due to eddy-making, turbulence, and white water condition. At the same time, stern flap in a flow area reduces the height, slope, and wave breaking. The stern flap also can reduce the wave height in the wave system near or far from the stern areas.

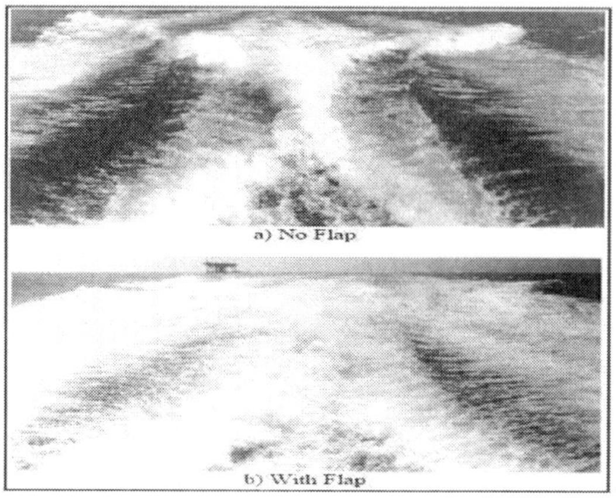

Figure 2. Boat wave pattern without and with stern flaps

c. Lift and Drag Force
Stern flaps can produce lift and drag at all speeds and conditions. The advantage is the interaction with the hull and propeller so that it can reduce boat resistance. Lift and drag forces increase even greater as the chord length, span, and angle increase.

2.3. Boat Resistance
The speed caused by boat resistance is the fluid force acting in the opposite direction to the boat motion. The resistance will be the same as the fluid force component, which works parallel to the axis boat motion [6]. The boat must be designed as efficiently as possible with the minimum possible external force. So, boat resistance and boat propulsion have a very close relation.

The boat resistance is applied to find the boat thrust force so that the boat can move at the planning service speed. Meanwhile, the boat propulsion provides thrust force and the interaction between the boat motions against the fluid flow.

The boat resistance is generally divided into two components, namely:
1. Skin friction resistance given by Reynolds' number.
2. Residuary resistance due to waves given by Froude number.

While boats are moving forward with a certain speed, it will get drag force from fluid on the water. The drag force is caused by the fluid force, which in this case causes significant obstacles in boat motion. The power needed to overcome boat resistance for attaining a particular speed is determined by Eq. 1.

$$P_E = RT \cdot V \tag{1}$$

where,
P_E : effective Power (W)
RT : boat resistance (N)
V : boat velocity (knot)

In the boat resistance curve, the hull moves above and below the water surface mainly caused by fluid viscosity.

2.4. Froude Number and Reynolds Number
Froude number is a dimensionless number that is used to measure the resistance of an object moving through the water, and compare objects of different sizes. The Froud number is determined using Eq. 2.

$$Fn = \frac{V}{\sqrt{g.L}} \tag{2}$$

where,

Fn	: Froude number
V	: boat velocity (m/s)
L	: boat length (m)

In fluid mechanics, the Reynolds number is the ratio between the inertia force and viscous force which quantifies the relationship of the two forces with a certain flow condition. This number is used to identify different types of flow, for example, laminar and turbulent. Reynolds number is one of the most important dimensionless numbers in fluid mechanics and used to provide criteria for determining dynamic similitude. If two flow patterns are geometrically similar, perhaps different fluids and flow rates have the relevant value of dimensionless numbers, both are called dynamic similarities.

3. Research Methodology

3.1. Research Location

This research was conducted at the Bili-Bili Dam, Gowa Regency by using a boat named RV. SBL01.

3.2. Research Data

- Main dimension of RV. SBL01 boat:

Table 1. Main dimension of RV. SBL01

Dimension	Value (m)
LBP	7.73
LWL	7.20
LoA	7.93
B	1.37
H	0.60
T	0.17

Figure 3. RV. SBL01 boat

- Stern flap size

  In this study, we use two stern flap which are 85 x 80 cm and 55 x 80 cm sizes

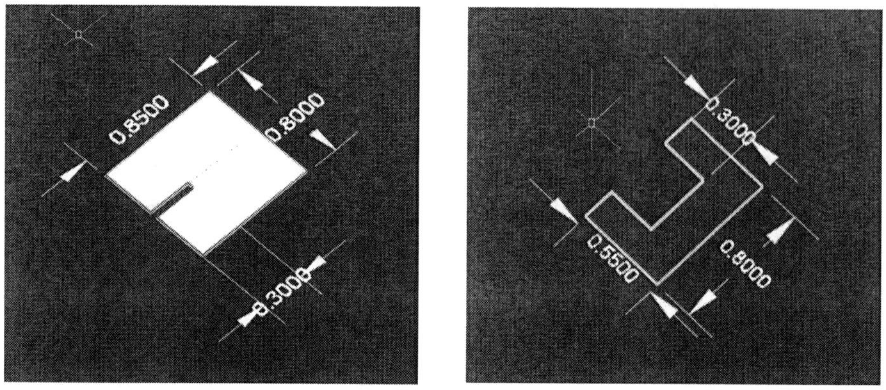

(a) 85 x 80 cm (b) 55 x 80 cm

Figure 4. Stern Flap Size

- Stern Flap Position
  In this study, two stern flap positions are used at above the surface of the water area and below the surface of the water area.

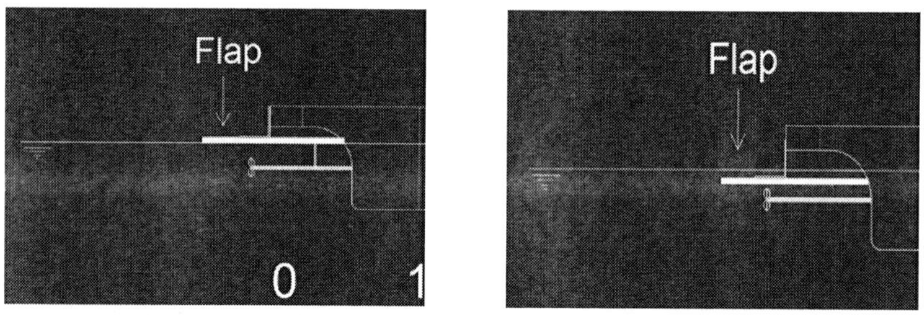

(a) On water surface (b) Below water surface

Figure 5. Stern flap positions

- Main Engine
  The main engine specifications of RV. SBL01 are as follows:
  - Engine Brand : Jiang Dong
  - Engine Type : 4 stroke
  - Engine Speed : 3600 rpm
  - Torque : 29.3 N.m
  - Engine Power : 15 Hp
  - BB tank capacity : 6.5 Litre
  - SFOC :260 gr/Hp/hour
  - Quantities : 2
  - Tank Volume : 0.65 m^3

Figure 6. Engine of RV. SBL01

The research variables of this study are boat speed, position, and size of the stern flap. Collection of speed data using GPS (Global Positioning System). In this test, the boat operates for 10 minutes for each data measurement.

4. Results and Discussion

In this study, two ship conditions will be tested, namely no load and with load Conditions. The load of 15kg is placed at the stern of the boat. In addition, these conditions were tested on boats without flaps and with flaps of 85 x 80 cm and 55 x 80 cm sizes as shown in Fig. 4. From the research results obtained as follows;

4.1. Boat Testing Without Stern Flap

In this research, several variations of engine throttle are performed, namely 25%, 50%, 75%, and 100%, which are the percentages of each engine condition.

Figure 7. Without a flap

In this test, the boat without the load on the stern is in even keel condition. Where the boat initially experienced trim on the stern of -0.1 degrees and -0.4 degrees position when the boat is with the load.

Table 2. Boat Speeds without stern flap against without loading and with loading conditions

Engine Throttle Variations (%)	Velocity (Knot)	
	No Load	With Loading
25	6.7	6.8
50	8.8	8.6
75	10.4	10,6
100	13.1	13

The results of this test shows that the no loading condition get 13.1 knots maximum speed, and each percentage of engine throttle variants will have some trim angle differences. When the engine throttle is increased to 100%, the stern is trimming into -1,2 degrees. On the other hand, the loading condition testing results get of 13 knots maximum speed and each percentage of engine throttle will have some trim angle differences. When the engine throttle is increased to 100%, the stern is trimming into -2.1 degrees.

4.2. Boat Testing With Stern Flap

In this study, the position of the stern flap is on the surface and below of the water both 85 x 80 cm and 55 x 80 cm sized also made the engine throttle division into 4 parts as in the test without flap, which is 25%, 50%, 75%, and 100% under loading and no loading conditions.

4.2.1. Water level condition testing

This test was conducted by positioning the stern flap parallel on the surface of the water as shown in Fig. 8.

Figure 8. Stern flap placed on the Water Surface

Table 3. Boat Speed with stern flap on water surface without no loading and loading conditions

Engine Throttle Variations (%)	Boat Velocity (Knot) 85x80 cm		55x80 cm	
	No Load	With Load	No Load	With Load
25	6.6	6.5	6.7	6.7
50	7.9	8.4	8.6	7.9
75	10.2	10.5	10.4	10.5
100	13.4	13.6	13.4	13.3

The boat speed using 85 X 80 cm stern flap size without loading conditions shows that the boat gets 13.4 knots maximum speed and each percentage of engine throttle will have trim angle differences. When the engine throttle is increased to 100%, the stern is trimming into - 2.5 degrees. While the 55 X 80 cm flap stern size gets 13.6 knots maximum speed. When the engine throttle is increased to 100% too, the stern is trimming into -2.4 degrees.

While the boat is using 15 kg load at the stern, it gets trimming in -0.3 degrees at the stern. The maximum speed of the boat using 85 X 80 cm flap stern size is 13.6 knots, and each percentage of engine throttles will have some difference trim angle. When the engine throttle is increased to 100%, the stern is trimming into -2.4 degrees. For the 55 X 80 cm flap stern size gets 13.3 knots maximum speed, and each percentage of engine throttles will have some difference trim angle. When the engine throttle is at 75%, the trimming result is -1.7 degrees and when increased to 100%, the stern trimming is -2.1 degrees.

4.2.2. Underwater Surface Condition Testing.
This test is done by positioning the stern flap below the surface of the water

Figure 9. Stern Flap Size below the water surface

Table 4. Boat Speed with Stern Flap below the water surface without loading and with loading

Engine Throttle Variants (%)	Boat Velocity (Knot) 85x80 cm		55x80 cm	
	No Load	With Load	No Load	With Load
25	6.7	6.5	6.8	6.8
50	8.9	8.5	8.1	8.3
75	10.6	11.4	10.4	9.8
100	13.1	13.5	13.3	13.2

Boat Speed Using 85 x 80 cm flap Stern without loading under water Surface, shows that the boat has 13.1 knots maximum speed, and each engine throttle percentage will have some trim angle differences. When the engine throttle is increased to 100%, the stern is trimming into -2.4 degrees. At the same time, the stern flap 55 X 80 cm has 13.3 knots maximum speed, and when the engine throttle is increased to 100%, the stern is trimming into -3 degrees.

In the boat testing using 15 kg load at the stern, initially, it gets stern trimming -0.3 degrees. The boat speed which used 85 X 80 cm flap stern size has 13.5 knots maximum speed. When the engine throttle is increased to 100%, the stern is trimming into -1.8 degrees. For the 55 X 80 cm flap stern size have 13.2 knots maximum speed, and each percentage of engine throttle will have some trim angle differences. When the engine throttle is increased to 100%, the stern is trimming into -2 degrees.

4.3. Discussion

From the test results, the different sizes and positions of the stern flap were placed on the stern of the boat can affect the boat speed when engine power testing is varied based on the engine throttle between loading and without loading condition. This can be explained as follows;

4.3.1. Effect of Stern Flap Size and Position due to Boat Speed without Loading

Figure 10 shows that the flap 85 x 80 cm and 55 x 80 cm size between on the water surface and below water surface get the same 13.4 knots speed and compared to without using the flap only get 13.1 knots speed. This shows that using flap addition on the stern will affect the increasing boat speed due to boat resistance reduction.

Table 5. The Effect of stern flap size and position due to boat speed without loading

Flap size	Velocity (Knot) Throttle Variants (%)			
	25	50	75	100
Without flap	6.7	8.8	10.4	13.1
Flap 55x80 cm at water surface	6.7	8.6	10.4	13.4
Flap 55x80 cm at below water surface	6.8	8.1	10.4	13.3
Flap 85x80 cm at water surface	6.6	7.9	10.2	13.4
Flap 85x80 cm at below water surface	6.7	8.9	10.6	13.1

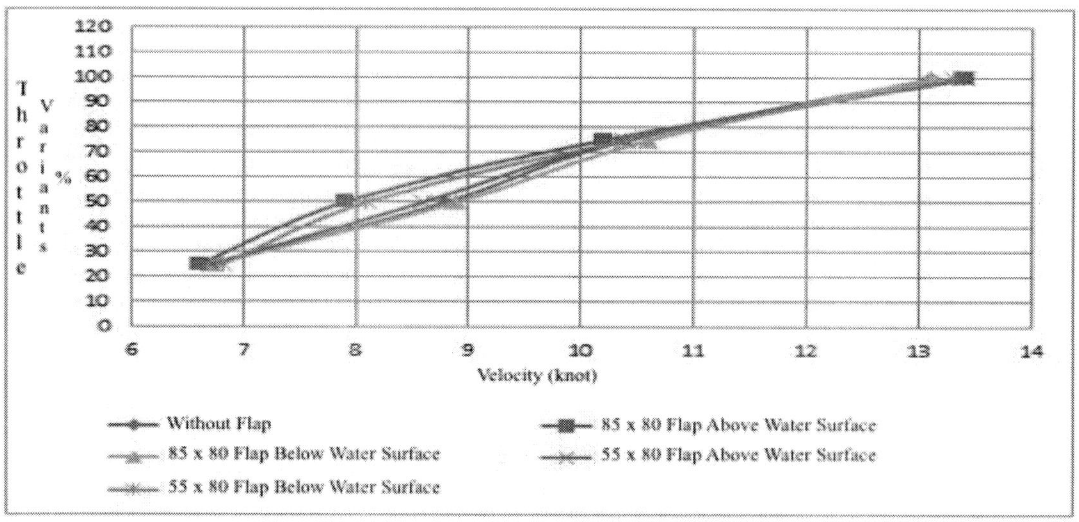

Figure 10. Effect of stern flap size and position due to boat speed without Loading

4.3.2. The Effect of Stern Flap Size and Position due to Boat Speed with Loading

From the graph in Fig. 11, it shows that flap 85 x 80 cm size in the water surface position has over 13.6 knots compared to the flap 55 x 80 cm size and also without using the flap have 13.3 knots and 13 knots. This shows that the change in size and position of the flap will affect the increasing boat speed due to boat resistance reduction.

Table 6. The Effect of stern flap size and position due to boat speed with loading

	Without flap (Knot) Throttle Variants (%)			
Flap size	25	50	75	100
Without flap	6.8	8.6	10.6	13.0
Flap 55x80 cm at water surface	6.7	7.9	10.5	13.3
Flap 55x80 cm at below water surface	6.8	8.3	9.8	13.2
Flap 85x80 cm at water surface	6.5	8.4	10.5	13.6
Flap 85x80 cm at below water surface	6.5	8.5	11.4	13.5

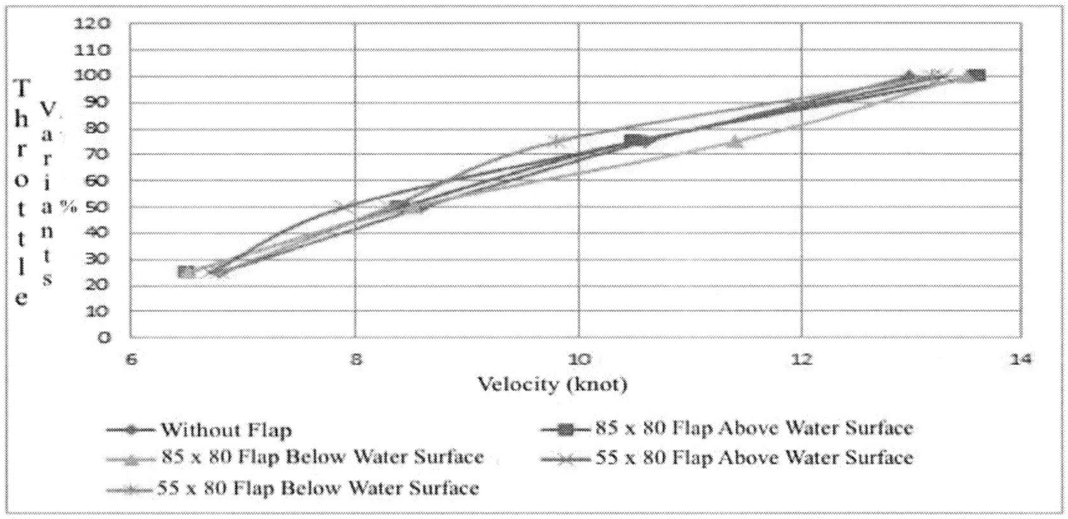

Figure 11. Effect of stern flap size and position due to boat speed with loading

5. Conclusions

Based on the results, it can be concluded that the boat speed due to the size and position of the stern flap on the water surface and below the water surface, are:

- Changing the position of the stern flap on the surface of the water will increase the speed of the boat into 4.62%.
- Changing the size of the stern flap 85 X 80 cm will increase the speed of the boat also into 3.82%

Acknowledgment

The author would like to thank C-BEST JICA and Faculty of Engineering of Hasanuddin University (Labo Based Education Research Scheme) for helping to fund this research.

References

[1] Sarasquete Adrián, Juandó Aitor, Caldas Alejandro, Zapata Fabio. 2011., "Energy efficiency audits on boats: Hydrodynamic aspects for energy efficiency improvements"., Congreso Internacional de Diseño e Ingeniería Naval Cartagena de Indias, Colombia, 2016.

[2] Novan, Muhammad. 2011. Analysis of the Effect of Addition of Stern Flap on Planing Hull Boats in an Effort to Reduce Boat Resistance. Surabaya: Sepuluh November Institute of Technology, 2011.

[3] Harumbinang, Jefri. 2016. Analysis of Constraints Due to Addition of Stern Flap on Kri Todak Boats Using the Computational Fluid Dynamic (CFD) Method. Semarang: Diponegoro University, 2016.

[4] http://qe2-prideoftheclyde.blogspot.sg/2011/03/shape-of-things-to-come-part-2.html"

[5] Setiawan, A. Yoni. Modification of the shape of the stern in the Bulk Carrier Shallow Draft to improve the efficiency of the propulsion system. Surabaya: Sepuluh November Institute of Technology.

[6] Darmawan, Dimas Bagus. 2014. Review of Boat Resistance. Semarang: Diponegoro University, 2014.

Cost Comparison for the Installation of Steel Frames vs Wooden Frames on a Traditional Wooden Fishing Boat in South Sulawesi

L Bochary[1], M R Firmansyah[1]*, S Asri[1], G Sitepu[1], W Djafar[1], Zulkifli[2], A H Djalante[1], and M Arfah[1]

[1] Naval Architecture Department, Engineering Faculty, Hasanuddin University, Makassar, Indonesia
[2] Marine Engineering Department, Engineering Faculty, Hasanuddin University, Makassar, Indonesia

*Email: mr.firmansyah@unhas.ac.id

Abstract. Traditional wooden fishing boats are still being produced by most of the traditional wooden shipyards in Indonesia specifically in the area of South Sulawesi. Nevertheless, the traditional wooden boat craftsman facing the difficulty of wooden log supply as the main wooden boat material especially for wooden frame construction despite the increasing demand for wooden material for the boat construction. The difficulty to acquire the specific wooden for wooden frame caused by the specific requirements of each wooden frame. They must be in the specific shape according to the specific curvature of each transversal boat section along the boat length. Hence, for the sustainability of wooden boat construction process, alternative material is needed for replacing the certain wooden boat construction component, especially for wooden frames. Fortunately, research relating to the use of alternative material for wooden boat construction have been conducted. The focus of the research is on the use of steel frame to replace the wooden frame on the wooden boat construction. This paper discusses the cost comparison for the fabrication and installation process as well as the material cost of the steel frame to the wooden boat construction with wooden frame to the wooden boat construction. The result shows that the cost for fabricating and installing steel frame is 13% higher to compare to the wooden frame. However, if the total wooden boat construction cost is being considered, the total difference is only 3%.

1. Introduction

South Sulawesi have been well known in Indonesia as the area for traditional wooden ship builder. The process of building Pinisi ship as one of the traditional boat yards in this area have been even recognized by UNESCO as one of the intangible cultural heritage of humanity [1]. Beside pinisi ship which commonly being used for cargo ship, cruise ship, hospital ship etc, the traditional wooden shipyards in this area are also built smaller wooden boat which being used for passenger boat between nearby islands, fishing boat, supply boat etc. The area of the traditional wooden shipyards is spread in some districts in South Sulawesi province which are Bulukumba, Sinjai, Takalar, Pangkep and Jeneponto district [2].

The wooden material for the boat construction in this area are obtained from some forests in Indonesia [3, 4]. Unfortunately, as there are increasing number of forests cutting and forest fire in Indonesia, the number of wooden materials for wooden boat construction becoming decreasing. If the situation continuous, the sustainability for traditional wooden boat construction in South Sulawesi and Indonesia are threaten [2, 5].

Fortunately, Bochary et al. [6] have conducted research regarding alternative material for wooden boat construction component. Their research focus is on the steel material to replace the wooden material specifically on the wooden frames of the wooden boat. There are some problems have been identified in the application of steel frames on the wooden boat. Some of them are relating with the selection of the fabrication and installation technique of the steel frame as well as the cost for the use of steel frame which cover material and equipment cost and the cost for fabricating and installing the steel frame.

Some dissemination activities of the steel frame to the wooden boat craftsman and to the wooden boat owners have been conducted by the researchers team in some traditional wooden boat shipyards in some area of South Sulawesi province. The participants are enthusiastic on the steel frame innovation [2]. They would like to apply the steel frames on their boat but still need to find out the cost for fabricating and installing the steel frames and be compared to the cost of fabricating and installing the wooden frames. The purpose of this paper is to determine the cost components and the total cost for fabricating and installing the steel frame on the wooden boat construction and to compare with the wooden frame on the same wooden boat dimension.

The construction of this paper is divided into some sections. The next section discusses the wooden boat construction and the building method including the cost components for constructing the wooden boat using wooden frames. Then the steel frame innovation will be described along with the cost components for fabricating and installing the frames including the material and equipment cost. Then, the two cost components information will be compared and discussed. The discussion will be focus on the factors which influence the total cost which make the difference of the two. In the end, some conclusion will be drawn based on the analysis.

2. Wooden Boat

Wooden boat is a boat/ship where all construction components of the boat are from wooden material. [7]. The dimension and capacity of the wooden boat are varying from the smallest wooden fishing boat to the biggest wooden cargo ship which can carry hundreds of ton cargoes from one island to other. The construction components of the respective boats are different as well including the type of the wood for the wooden construction components. Indonesian Classification Bureau (BKI) have regulated the dimension and the type of wooden boat for certain boat construction components [8].

2.1. Wooden Boat Construction

Different with steel ship, the construction component for the wooden boat consist of keel, keel at bow and stern, frames, deck beam, plank, longitudinal plank, etc (Figure 1). The list of the construction components must be strong enough as the wooden boat will face an uncertain sea environment during its voyage. Besides, the construction components must be able to minimize some vibrations of the boat engine [9].

The type of wood for each construction component cannot be selected randomly. The Indonesian Classification Bureau (BKI) [8] and National Standardization Bureau (BSN) through Indonesian national standard document have determined a list of qualified wooden types for the boat construction component [10]. The type of the wood is different for each construction component such as "bitti wood" (Vitex cofassus) for frame, plank and deck plank, "ulin wood" (Eusideroxylon zwageri) which can be used for any part of the boat construction component especially the component which support the boat strength, "besi wood" (Instia palembanica) for keel, "jati wood" (Tectona grandis) which can be used for any construction component of the boat, etc. Hence, each type of wood can be used for some part of the wooden boat construction component [11]. The selection of the wood will determine the total cost for building the wooden boat as each type of wood has different cost.

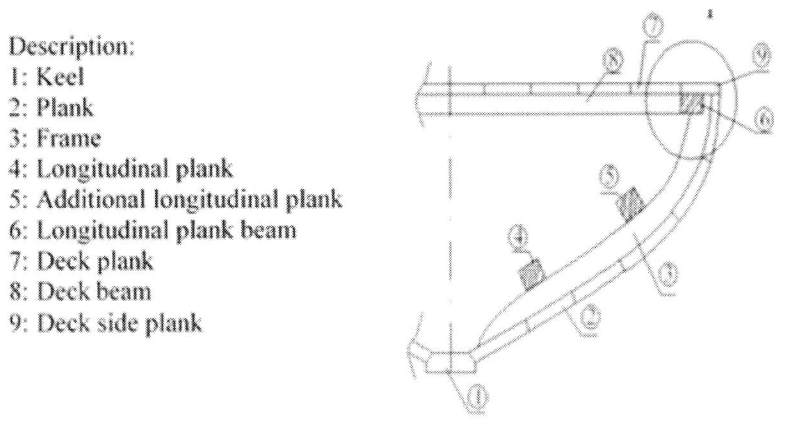

Figure 1. Wooden boat construction component [6]

2.2 Wooden Boat Construction Method
In this research, the current wooden boat craftsman method has been selected which being called the traditional wooden boat construction method for constructing the boat. Unlike modern wooden boat building technique, the traditional wooden boat technique build the boat from the keel then continued to the plank installation then frame installation (Figure 2) [12, 13].

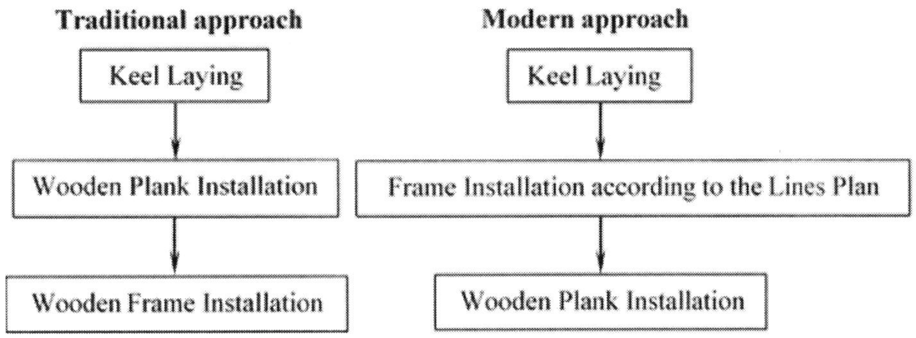

Figure 2. Comparison of the boat building method between traditional and modern approach

The implication of the different approach for building the wooden boat is on the installation of the steel frame and on the total cost for building the boat. Hence, focus for the discussion in this paper is the cost for building the wooden boat both using the wooden frame and steel frame with traditional boat building method. The cost calculation for the latter is being conducted according to the fabrication and installation technique for the steel frame on the wooden boat construction.

2.3 Cost Component for the Wooden Boat Construction
In general, the cost component for using the traditional wooden boat construction method can be divided into material cost, equipment cost and manpower cost. Material cost is any cost related to the materials to be used on the wooden boat construction such as wood, plank, bolt and nut etc. Including in the material cost is the cost for preparation and assembly the wood component to other wood component and the cost for painting. Equipment cost is the cost related with the equipment used to build the wooden boat such as hammer, chainsaw, chisel etc. Manpower cost is the cost for both direct and indirect labor cost. The total cost will include the three of the cost components.

3. Steel Frame for the Wooden Boat

The reason for selecting the wooden frame to be replaced with the steel frame as the wooden frame is the most difficult wooden boat construction component to obtained to compare with other components. This is because the specific shape requirements of the wooden frames which must be according to the curvature of the wooden boat body.

The use of steel material for the steel frame have been analyzed by Bochary, et al. [6]. There are some reasons which become basis for selecting steel material to replace the wooden material such as the material is easy to find, can be easily shaped according the boat body curvature and obviously its stronger than wooden material. The shape of the steel frame is T profile. The selection of this profile shape is based on the consideration of the applied connection method between steel frame and other construction component on the wooden boat such as between steel frame and the wooden plank, steel frame with the longitudinal plank and between wooden keel and the steel frame. The dimension of T profile has been simulated and analyzed on some of the steel frames distances. The selection criteria are the ship strength and the steel frame flexibility to hold the wooden boat load together with other wooden construction component. The result of the analysis shows the dimension of the T profile steel frame for 10 GT of the wooden fishing boat.

3.1 Cost Component for Fabricating and Installing Steel Frames on the Wooden Boat Construction

In general, the cost components of both using wooden frame and steel frames of the wooden boat construction to be compared is the same. They are consisting of material cost, equipment cost and manpower cost. The difference cost components between the two are on the fabrication and installation of the steel frame on the wooden construction. They are different in terms of the material and equipment to be used and the manpower to do the work. For steel frame, the material to be used are steel plate, nut and bolts as well as electrodes, LPG and Oxygen gas for material works. Including in the material cost is the painting of the wooden construction components. The equipment to be used are different as well such as brander cutting machine, rolling bending plate machine, and welding machine while the manpower to do the work of fabrication and installation of the steel frame are reflected from their equipment such as welder, cutter, dan brander.

4. Fabricating and Installing Cost Comparison of the Steel and Wooden Frame of 10 GT Wooden Fishing Boat

As the fabrication and installation of steel frame on the wooden boat construction is never been done before, the cost for the works is not known. It is important to calculate the cost of fabrication and installation of the steel frames and then compare the result with the current wooden boat construction using wooden frame. The result of the comparison will be the references for the wooden boat craftsman in South Sulawesi specifically to whether use the steel frames on their wooden boat construction or not. The total cost of the wooden boat building both for the comparing frames is based on the same traditional wooden boat building method.

In this paper, the focus on the cost comparison is the material, equipment and manpower cost for fabricating and installation of the both frames as well as the total cost for building the wooden boat using both of the frames. Based on the information, the cost ratio between the boats using the wooden and steel frames then determined. Data of the comparing 10 GT wooden fishing boat can be seen in Table 1 below.

Table 1. Main dimension of 10 GTs wooden fishing boat

Description	Dimension (m)
Length overall (LOA)	16.5
Length between perpendicular (LBP)	12.6
Breadth (B)	3.2
Height (H)	1.2

4.1 Detail of Cost Components for Fabricating Wooden Frame Construction

In detail, the cost components of the wooden boat construction using wooden frames are divided into the material cost (M_{wf}), equipment cost (E_{wf}) and the manpower cost (Mp_{wf}).

4.1.1 Material Cost (M_{wf}). Material cost for wooden boat construction using wooden frame consist of the material construction cost itself, material cost for binding the wooden construction components, painting cost and overhead cost.

The wooden construction components consist of keel, frames, side and deck planks, longitudinal planks, deck beams etc. In addition to those material, triangle wooden beams have been used to connect the boat keel and stem as well as to support the boat rudder (Table 2) [14]. The price of each material is obtained from the boat craftsmen in some of the traditional wooden boatyard. Besides the price of the boat construction component, the required number of each construction component for building the required boat is also shown in Table 2.

Table 2. The cost for wooden frame construction components

Construction components	Construction components dimension (mm)			Number required	Cost (Rp)	Unit
	P	L	T			
Keel	1.200	15	15	1	6.500.000	Rp/m^3
Wooden frame	-	10	10	149	50.000	Rp/piece
Triangle wooden beam	-	20	20	1	2.000.000	Rp/piece
Side plank	400	25	4	72	4.500.000	Rp/m^3
Longitudinal plank	1.660	14	5	6	45.000	Rp/m
Longitudinal plank beam	300	10	6	11	2.500.000	Rp/m^3
Deck beam	300	10	6	56	2.500.000	Rp/m^3
Deck plank	400	25	4	50	4.500.000	Rp/m^3

Material for binding the wooden boat construction components consist of bolts and nuts, "paso sappu" or (wooden peg) and wooden fiber for caulking. Information on the required material need and the price for binding the wooden boat construction can be seen in Table 3 below.

Table 3. The cost for the wooden frame construction binder

Binder construction components	Diameter (mm)	Number required	Cost (Rp)	Unit
Bolts and nuts	14	1.456	2.500	Rp/piece
"paso sappu" (wooden peg)	1	3.600	500	Rp/piece
Wooden fibre for caulking	-	20	10.000	Rp/kg

The next cost component is for painting. The cost component consists of primer paint, finishing paint and thinner for dissolving the paint. The price both for the primer and finishing paint is the same which is Rp 68.000/liter while the total need for both paints is 73 liters respectively. The need for thinner is only about 15 liters for each paint, hence the total liter for both paints is 30 liters. The price for the thinner itself is Rp 22.500. The last material component which is considered here is the overhead (O/H) cost. This cost component is considered to be 5% of the total cost of the wooden boat building.

4.1.2 Equipment Cost (E_{wf}). Equipment used and the price for each equipment in the wooden boat construction process with wooden frames can be seen in Table 4.

Table 4. List of equipment to be used in the 10 GT wooden fishing boat construction process

Equipments	Price (Rp)
Hammer	35.000
Chisel	110.000
Chainsaw	3.250.000
Wooden planer	690.000
Clamp c4	52.500
Tape measure tool	275.000
Hand saw	76.500
Hand drill	196.900
Wrench	298.000
Axe	99.000

4.1.3 Manpower Cost (Mp_{wf}). The manpower cost in the wooden boat construction process is for the direct manpower. Based on the interview with the craftsmen, this cost component can be determined to be 30% of the total cost of the wooden boat building cost.

4.1.4 Total Cost for the Wooden Frame (T_{wf}). The total cost for the construction of the 10 GT wooden fishing boat is the accumulation of the previous discussed costs which consist of materials, equipment and manpower cost.

Total cost for wooden boat with wooden frame (T_{wf}) = Cost of M_{wf} + Cost of E_{wf} + Cost of Mp_{wf}

Table 5 shows the summary of the required cost for building 10 GT of wooden fishing boat with wooden frames.

Table 5. Construction cost for the 10 GT wooden fishing boat with wooden frames

Cost components	Cost (Rp)
Materials (M_{wf})	338.527.648
Equipment (E_{wf})	5.082.900
Manpower (Mp_{wf})	147.261.663
Total wooden fishing boat building cost	490.872.211

4.2 Detail of Cost Components for Fabricating Steel Frame Construction

The cost for building the wooden fishing boat with steel frames is based on the traditional wooden boat construction method. In detail, the cost components can be divided into materials cost (M_{sf}), equipment cost (E_{sf}) and manpower cost (Mp_{sf}).

4.2.1 Material Cost (M_{sf}). Material cost for wooden boat construction using steel frame consist of the material construction cost itself, material cost for binding the wooden construction components, painting cost and overhead cost.

In general, the construction material cost for steel frame is the same with the cost the construction material cost for wooden frame. The difference of both can be seen in the material for the frame's construction. Since steel material is being used for the frames, there is cost for fabricating the steel frames. The fabrication cost component includes the cost for the steel material itself, cost for LPG and Oxygen gases, electrodes etc. Table 6 shows the detail of the material cost for wooden boat with steel frames.

Table 6. The cost for steel frame construction component

Construction components	Construction components dimension (mm)			Number required	Cost (Rp)	Unit
	P	L	T			
Keel	1200	15	15	1	6.500.000	Rp/m^3
Steel frame	80	80	6	55	481.618	Rp/frame
Triangle wooden beam	-	20	20	1	2.000.000	Rp/piece
Side plank	400	25	4	72	4.500.000	Rp/m^3
Longitudinal plank	1660	14	5	6	45.000	Rp/m
Longitudinal plank beam	300	10	6	11	2.500.000	Rp/m^3
Deck beam	300	10	6	56	2.500.000	Rp/m^3
Deck plank	400	25	4	50	4.500.000	Rp/m^3

The cost for binding material for steel frames is the same with the required cost for wooden frames. The information can be seen in Table 3. The cost component for painting in steel frames construction is the same as well with wooden frames. The difference is for the required number of paints for each paint which is only requires 69 paints for each paint and the number of required thinner which is 1 liter less than for the wooden frames need.

The overhead cost is considered to be 5% as well of the total cost of the wooden boat building.

4.2.2 Equipment Cost (E_{sf}). The required equipment in the wooden boat construction process with steel frames can be seen in Table 7.

Table 7. List of equipment to be used in the 10 GT wooden fishing boat construction process

Equipments	Price (Rp)
Hammer	35.000
Chisel	110.000
Chainsaw	3.250.000
Wooden planer	690.000
Clamp c4	52.500
Tape measure tool	275.000
Hand saw	76.500
Hand drill	196.900
Wrench	298.000
Axe	99.000
Welding machine	1.099.000
Brender machine	2.592.000
Rolling bending plate machine	2.500.000
Grinding machine	350.000

As can be seen in Table 7, for steel frames in the wooden construction, there are additional equipment required for steel frame fabrication consist of brander cutting machine, rolling bending plate machine, and welding machine. The additional direct manpower required will be welder, cutter, and brander.

4.2.3 Manpower Cost (Mp_{sf}). The manpower cost required is the same with the wooden frames which is 30% of the total cost of the wooden boat building cost.

4.2.4 Total Cost for the Steel Frames (T_{sf}). The total cost for the construction of the 10 GT wooden fishing boat with steel frames is the accumulation of the previous discussed costs which is:

Total cost for wooden boat with steel frame (T_{sf}) = Cost of M_{sf} + Cost of E_{sf} + Cost of Mp_{sf}

Table 8 shows the summary of the required cost for building 10 GT of wooden fishing boat with steel frames.

Table 8. Construction cost for the 10 GT wooden fishing boat with steel frame

Cost components	Cost (Rp)
Materials (M_{sf})	429.430.202
Equipments (E_{sf})	11.623.900
Manpower (Mp_{sf})	189.023.187
Total wooden fishing boat building cost	630.077.289

4.3. Cost Comparison between Wooden Frames and Steel Frames

The cost comparison for using wooden frames vs steel frames in the 10 GT wooden fishing boat construction is conducted in two ways. The first comparison is between the cost required to fabricating and installing the wooden frames vs the steel frames and the second is between the total cost of building the wooden fishing boat using both of the frames. Based on the respective cost information, the cost ratio can be compared as well. The comparisons can be seen in Table 9.

Table 9. Cost comparison between the use of wooden frame vs steel frame on a 10 GT wooden fishing boat construction

Description	Wooden frame (wf)	Steel frame (sf)
Material and construction cost	338.527.648	429.430.202
Equipment cost	5.082.900	11.623.900
Manpower cost	147.261.663	189.023.187
Total boat building cost	490.872.211	630.077.289
Cost ratio for the installation of wooden frame vs steel frame	1 : 1.13	
Total wooden boat building cost ratio between using wooden frame vs steel frame	1 : 1.03	

5. Discussion

The total cost of building the wooden fishing boat can be different among one boat craftsmen (boatyard) to another. The distinguishing cost components is on the different selected types of woods as well as different applied boat building method. Furthermore, the exact dimension for the 10 GT wooden fishing boat to be built is not the same among the boatyards as they have applied traditional boat building method. Hence, the discussed total cost for building the required wooden boat is based on the wooden boat which have being built in a boatyard in Takalar regency, South Sulawesi province. However, both the discussed cost ratio can be used as reference if other wooden boat shipyards need to build the same wooden fishing boat using steel frames.

One thing to note in the cost discussion is about the equipment cost. The equipment to use in the wooden boat construction is included in the total cost of building the required wooden boat. However, this cost can be the boatyard investment cost and be spent only on the first time of building the wooden boat using steel frames. Afterwards, it can be omitted and leave only the operational cost and material cost to be considered.

In Table 9, the cost for fabricating and installing the steel frames on the wooden boat is 13% higher to compare with the wooden frames. However, the cost ratio proportion become lower if the total building cost is considered. The use of steel frames to replace the wooden frames in a 10 GT wooden fishing boat only provides an additional of 3% of the total wooden boat building cost using wooden frames. This is because the proportion of fabricating and installing the boat frames on the total wooden boat building cost is small as well.

6. Conclusion

The total cost for building the wooden fishing boat using steel frames only 3% higher of the total cost for building the same ship using wooden frames. The small addition of the boat building cost using steel frames have increased the opportunity for the steel frames to be applied on the wooden fishing boat as well as other wooden boats. There are many advantages have been identified for the application of the steel frames on the wooden boat construction. Among them are it supports the sustainability of wooden boat construction as the consequence of the shortage supply of wooden material specifically for the wooden frames. Further, the application of steel frames on a traditional wooden boat construction method have opened the possibility for the application of modern wooden boat construction method instead of traditional boat building method in the area of South Sulawesi specifically and Indonesia in general.

Acknowledgments

The authors would like to thanks Directorate General of Higher Education (DGHE), Ministry of Research, Technology and Higher Education for its research grant under scheme Priority Application Research of Higher Education (PTUPT) DGHE in 2018/2019 so as this research work can be done.

References

[1] Unesco, 2017, Decision of the Intergovernmental Committee: 12.COM 11.B.13, https://ich.unesco.org/en/decisions/12.COM/11.B.13
[2] Bochary, L, Asri S, Firmansyah, M R, Alwi, M R, Rosmani, Misliah, Sitepu, G, Djafar, W, 2019, Ensuring the sustainability of wooden boat building through the training for wooden boat craftsman on making wooden frame pattern for the application of wooden steel innovation as a substitute for wooden frame in Takalar Regency, Jurnal Teknologi Terapan untuk Pengabdian Masyarakat, Vol. 2, No. 1 pp. 70 – 77. [In Bahasa]
[3] Ayuningsih, Dewi. 2007. *Techno economy of the wooden boat construction in traditional wooden boatyard in Gebang village, Cirebon regency, West Java Province*. Institut Pertanian Bogor, Bogor. [in Bahasa]
[4] Standar Nasional Indonesia (SNI), 2006, Types of woods for wooden boat construction. Badan Standar Nasional (BSN) untuk Pusat Standarisasi dan Lingkungan Hidup Departemen Kehutanan. [in Bahasa]
[5] Bochary, L. dan Larengi, F, 2012, Alternative for the use of steel frames on a 30 GT wooden boat construction, Journal Riset Teknologi Kelautan, Volume 10, No. 2, Halaman 121-272, Makassar. [in Bahasa]
[6] Bochary, L, Sitepu, G, Asri, S, Firmansyah, M R, 2018, Analysis of the use of steel frame as a substitute for wooden frame in an effort to support forest conservation and sustainability of wooden boat construction, Laporan Penelitian Terapan Unggulan Perguruan Tinggi (PTUPT), Lembaga Penelitian dan Pengabdian Masyarakat (LPPM), Universitas Hasanuddin. [in Bahasa]
[7] Biro Klasifikasi Indonesia (BKI), 1992, *Guidance for the construction of motor sailing boat*, Jakarta. Biro Klasifikasi Indonesia (BKI) Unit Inkomar. [in Bahasa]
[8] Biro Klasifikasi Indonesia (BKI), 1996, Classification and Construction Rules for wooden boat, Jakarta, Biro Klasifikasi Indonesia. [in Bahasa]
[9] Bochary, Lukman, 2004, An Introduction to fishing boat, Jurusan Perkapalan FT-UH. [in Bahasa]
[10] BSN. 2013. *Design specification for wooden boat construction*. Badan Standart Nasional. Jakarta. [in Bahasa]
[11] Ima Kusumawati, 2009, *The wooden material utilization in the construction of wooden frames in UD Semangat Untung wooden boat yard, Tanah Beru village, Bulukumba regency, South Sulawesi province*. [in Bahasa]
[12] Dewa, Sarifuddin, Haris muhammad, 2010, *Technology for the construction of traditional wooden boat in Tanahberu village, Bulukumba regency*. [in Bahasa]

[13] Rauf, A M, dan Chaerunnisa, A, 2003. The development of traditional wooden boat construction method in Bulukumba regency. Lembaga Penelitian Pengembangan Masyarakat Universitas Hasanuddin. Makassar. [in Bahasa]

[14] Karim, Abdul Azis, Hasbullah, Mansyur, dan Muhammad, Andi Haris, 2011. A study for the wooden boat construction dimension and connection model in a traditional wooden boatyard in Bulukumba regency. Jurusan Perkapalan Fakultas Teknik Universitas Hasanuddin. [in Bahasa]

IOP Conf. Series: Materials Science and Engineering **875** (2020) 012081 doi:10.1088/1757-899X/875/1/012081

Performance of a New Floating Breakwater

S Rahman[1]*, A Y Baeda[1], A Achmad[1] and R F Jamal[1]

[1]Department of Ocean Engineering, Engineering Faculty, Hasanuddin University, Makassar, Indonesia

*Email: sabaruddin_r@eng.unhas.ac.id

Abstract. The need of floating breakwater is increased to protect harbour by wave storm. The study and development of some model have been done in the past decade. However, their model still not effective to protect harbour against wave period greater than 5.0 second. This study deals with performance of a new type floating breakwaters in terms of heaving motion and transmission coefficient. The result compare with heaving motion of PI-type floating breakwater. Heaving motion is evaluated by using seakeeper tools, while transmission coefficient is obtained through laboratory experiment. The results indicate that the performance of proposed model better than PI-type model in terms of motion and transmission coefficient.

1. Introduction

Human activities in the coastal area are increasing by years. Ports, shipyards and floating buildings along the coastline have an important role in supporting human activities at the coastal. In addition, some of these facilities are exist in a relatively higher water depth. Protection of these facilities against wave especially during bad weather condition is very important. Breakwaters are generally built as coastal structures that reduce wave energy so that beach erosion and harbour disturbance can be reduced. The construction can be built from rubble mound or imperial buildings that sit on piles of stone as a foundation.

The increase in human activities at sea encourages an increase in the need for broader waters of the port and is followed by greater water depths as well. This results in conventional breakwaters being ineffective in both technical and economical aspect. Furthermore, based on wave theory, it is known that wave energy is concentrated in the sea surface area and decreases towards the seabed. A floating body may experience movement at six degrees of freedom resulting in the structure being less effective in absorbing waves. These issues underlay so that research related to floating breakwaters has been increased in the last few years.

Floating breakwaters (FBs) classification has been discussed briefly by Dai [1] as: box-type, pontoon-type, frame-type, mat-type, tethered float type, horizontal plate type and other types. Among these types, box-type and pontoon-type breakwaters are the most common design and effective in protecting beaches by reflecting incoming waves. The shape can function effectively at FBs width ratios of one third of the wavelength. So it is less effective to absorb relatively long waves. Pontoon type breakwaters has been developed further by some researcher. Combination of several boxes or pontoons and frames produces frame type breakwater. It reduces wave energy through reflection, but also through flow interference by the frame. Laboratory test results show that reduction in incident wave height is effective for small width and wave length values. While Gesraha [2] introduce PI-type floating breakwater by adding upright curtains on two sides of pontoon-type floating breakwater.

Mathematical models [3, 4] and laboratory experiment [4] have been used to investigate movement, hydrodynamic forces and wave transmissions of pontoon-type floating breakwaters. The effect of the addition of two porous side plates on the performance of a rectangular breakwater has been studied by using mathematical model [5]. Model predictions of transmission coefficient for PI-type floating breakwater have been developed by Ruol [6] and Kolahdoozan [7]. The latter can be used for shallow and intermediate water depth.

One method used to reduce the movement of FBs is by anchor it to the seabed. However, water level fluctuations due to tides result in low tide conditions which are still less effective because the tension of the strap is reduced. For this reason, the development of the existing floating breakwater type is a challenge. One indicator for choosing the best performance is a model with relatively small heaving motion.

In this study, a new modified pontoon-type breakwater is developed. Damping effect on heaving motion is considered by making bottom part adopting the idea of floating body studied by Clauss [8]. Flat top of the design can be useful as berthing facility. Performance of proposed model compare to PI-type breakwater in terms of heaving motion and transmission coefficient. Figure 2 shows both models used in the experiment.

2. Methods

Heaving motions are analysed by using Seakeeper tools provide by Maxsurf software. A new floating breakwater model is evaluated in terms of transmitted wave coefficient (K_t). This value is obtained by experimental study on a 24 m long, 1.0 m wide and 1.2 m deep wave flume of the Coastal and Environmental Laboratory of the Faculty of Engineering at Universitas Hasanuddin which is equipped with a flap-type wave maker. Incident and transmitted waves are measured by using three wave probes both in sea side and lee side of the model. Wave heights and wave period are selected as the typical wave condition by considering a geometrical model scale number of 25. Environmental condition parameters are listed in Table 2.

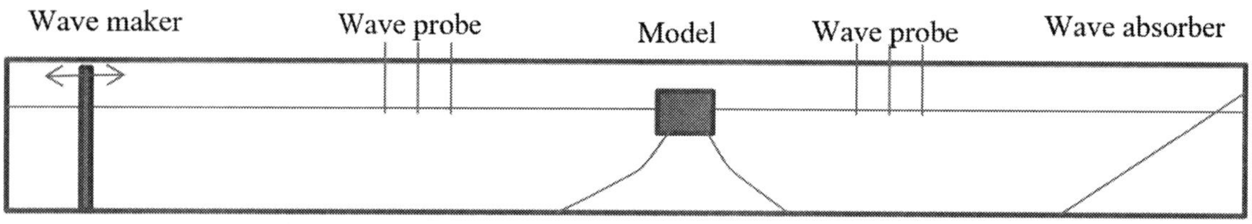

Figure 1. Experimental setup

Table 1. Model dimension

Parameter	PI-type	New model
Width, B (m)	0.4	0.4
Height, h (m)	0.38	0.597
Draft, D (m)	0.22	0.537

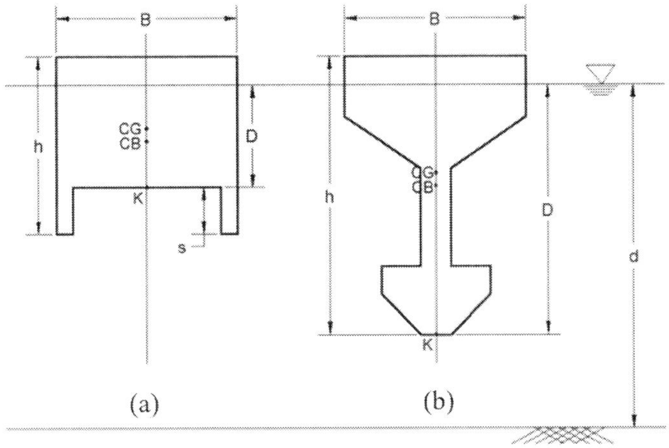

Figure 2. Model characteristics, (a) PI-type model; (b) New model

Table 2. Laboratory simulation

Parameter	Value
Wave period, T	0.8 s; 1.2 s; 1.6 s; 2.0 s
Wave height, H	0.08 m; 0.1 m; 0.12 m; 0.14 m; 0.16 m
Water depth, d	0.6 m; 0.7 m; 0.8 m

Figure 3. New model under testing in wave flume

3. Results and Discussions

Freely motion of FBs characteristics is required to the performance of the breakwater can be evaluated in terms of coefficient of transmission and reflection.

In this section, two parameters are evaluated in terms of heaving motion and transmitted wave coefficient. Heaving motion is obtained by using Seakeeper tools provide by Maxsurf software. While transmitted wave coefficient is obtained from laboratory experiment.

3.1. Freely FBs Heaving Motion

Figure 4 shows relationship of dimensionless wave height *ka* and RAO heaving motion of both models. It shows that proposed model has smaller peak amplitude heaving motion than PI-type model even both model has same displacement. This is due to the effect of bottom part reducing heaving motion.

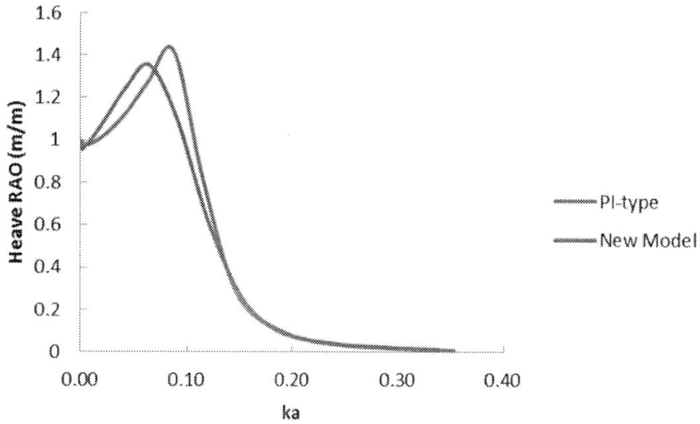

Figure 4. Effects of wave conditions on heaving motion

3.2. Moored FBs Transmission Wave Coefficient

In order to examine the performance of *FB* in a quantitative way, the transmission coefficient is investigated further. Transmission coefficient (K_t) is the ratio of transmitted wave height (H_t) to the incident wave height (H_i). The incident and transmitted wave height are measured by using 3 wave probes both in front and behind of the model. There are sixty waves run for each model as shown in Table 2. Figure 5 shows relationship between K_t against wave period (6.a) and wave steepness (6.b). These Figures show that K_t value greater than 0.5 is obtained at wave period greater than 1.3 s (6.5 s in prototype scale) which implies that both model are not effective at this wave incident condition. Proposed model has better performance due to smaller K_t than PI-type model especially at relatively small wave period.

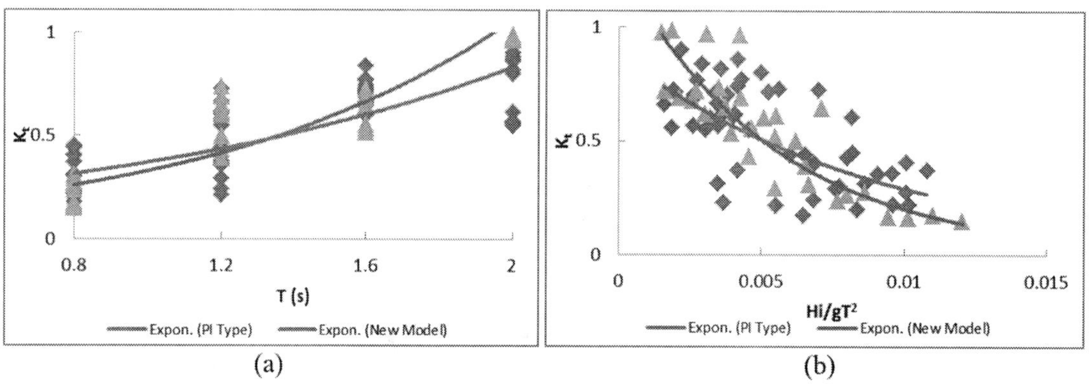

Figure 5. Relationship of transmission coefficient to wave period (a) and wave steepness (b)

4. Conclusions

A new model of floating breakwater has been evaluated in terms of vertical motion, transmission and reflection coefficient. Some wave parameters are generated to evaluate heaving motion of PI-type and proposed model. The numerical results show that proposed model has a smaller peak value of heaving performance than that of PI-type model at a particular wave condition. Furthermore, performance of proposed model has been achieved to transmit less wave height than PI-type model. Further experiments should be carried out for some width variation.

Acknowledgements
Authors wishing to acknowledge for financial support from Faculty of Engineering Universitas Hasanuddin through Laboratory Based Education scheme. This study was also funded by Badan Penelitian dan Pengembangan Kementerian Perhubungan Republik Indonesia supporting laboratory experiment.

References
[1] Dai, J., Wang, C.M., Utsunomiya, T. dan Duan, W., 2018. Review of recent research and developments on floating breakwaters. *Ocean Engineering* 158, pp. 132-151.
[2] Gesraha, M.R., 2006. Analysis of pi shaped floating breakwater in oblique waves: I. Impervious rigid wave boards. *Applied Ocean Research*, Vol. 28, pp. 327-338.
[3] Drimmer, N., Agnon, Y., and Stiassnie, M., 1991. A Simplified analytical model for a floating breakwater in water of finite depth. *Applied Ocean Research* 14, pp. 33-41.
[4] Fugazza, M. and Natale, L., 1988. Energy losses and floating breakwater response. *Journal of Waterway, Port, Coastal and Ocean Engineering*. Vol. 114, No. 2, pp. 191-205.
[5] Cho, I-H., 2016. Transmission coefficients of a floating rectangular breakwater with porous side plates. *International Journal of Naval Architecture and Ocean Engineering*, Vol. 8, pp. 53-65.
[6] Ruol, P., Martinelli, L. and Pezzutto, P., 2012. Limits of the new transmission formula for pi-type floating breakwaters. Coastal Engineering.
[7] Kolahdoozan, M., Bali, M., Rezaee, M., Moeini, M.H., 2017. Wave-transmission prediction of π-type floating breakwaters in intermediate waters. Journal of Coastal Research 33 (6), 1460-1466.
[8] Clauss, G.F. and Birk, L., 1996. Hydrodynamic shape optimization of large offshore structures. *Applied Ocean Research*, Vol. 18, pp. 157-171.

IOP Conf. Series: Materials Science and Engineering **875** (2020) 012082 doi:10.1088/1757-899X/875/1/012082

The Analytical Solution of Waves Transmission Coefficient on Semi Fixed Floating Breakwater

C Paotonan[1]*, T Rachman[1], Ashury[1], and T. P. Putra[2]

[1]Department of Ocean Engineering, Engineering Faculty, Hasanuddin University, Makassar, Indonesia
[2]Center for Research and Development on Sea Transportation – Indonesia

*Email: paotonan_ch@yahoo.com

Abstract. One type of breakwaters that can be used to protect port basin is floating breakwaters. The use of floating breakwaters is suitable for relatively large water depths. In deep waters, the floating breakwaters volume is smaller than rubble mounds breakwater. Consequently, floating breakwaters are cheaper than rubble mounds breakwater. Unfortunately, a practical approach to analysing the performance of floating breakwater in reducing wave height is unknown. Floating breakwater performances indicated by wave transmission coefficient. Therefore, in this study an analytical wave transmission coefficient was formulated. The simple formula can be used to estimate high of wave at behind of floating breakwater. Based on these equations, the wave transmission coefficient influenced by the wave parameters and structural parameters. Wave parameters represents by wave high, wave period, water depth and wave length while structure parameters represents by both of width and draft of structure. Wave transmission coefficient also influenced by the wave frequency and having motion frequency.

1. Introduction

Indonesia as the largest archipelagic country in the world has around 17 thousand islands with a coastline length reaching 95,181 kilometres. Geographically Indonesia's position is very strategic to trade traffic because it is located between two continents and two oceans. This condition needs to be supported by inter-island transportation facilities and infrastructure, including adequate ports to support trade activities and maintain the sovereignty of Indonesian maritime affairs. Port basin should be protected by wave disturbance to guarantee that the both of loading and unloading ships activities are safety. There are requirements of wave high at the pier so the ship can tie up safely as shown at Table 1.

Table 1 shows that the greater the dimension of ships or boats the greater significant wave high that required at pier and vice versa. Marina ports, significant wave high at berth maximally 0.15 m. For the large ships significant wave high is required reach 1.70 m. Significant wave high at berth that greater than 1.70 m is not required for all ships.

Table 1. The maximum significant wave high required at pier

Ship Types	H_s at pier
Marinas	0.15
Fishing boats	0.40
General Cargo (<30,000 DWT)	0.70
Bulk Cargo (<30,000 DWT)	0.80
Bulk Cargo (30,000-10,000 DWT)	0.80-1.50
Oil Tanker (<30,000 dwt)	1.00
Oil Tanker (30,000-150,000 DWT)	1.00-1.70
Passenger Ship	0.70

One of an alternative solution to protect port basin is floating breakwater. Unfortunately, the simple solution to identify floating breakwater performance not understood. Therefore in this research derivate the new formula to calculate wave attenuation coefficient after wave reach at the lee of structure. The large floating structure was introducing firstly by Edward R Armstrong who initiated a sea drone as an aircraft runway on the sea.The developed of floating breakwater has increased significantly in the recent years. A lot of research related to floating breakwaters has been done previous researches [1-5]. Floating breakwater has many advantages compare with fixed breakwater, namely flexible, movable and can be rebuild in deferent layout [6], chipper than fixed breakwater and easily in construction [7]. There are many types of floating breakwater has identify by [8], are box type, pontoon type, mat type, tethered float type. Although floating breakwater has advantage, but this type of breakwaters are not effective in areas with high or fast-moving waves. In heavy storms these breakwaters are subject to failure, and if they come detached from their moorings they can become a danger. Floating breakwaters have the advantage of being detachable when not in use, but the labour costs to replace the breakwater can be high when compared to a fixed breakwater. Pontoon type is the most widely used in recent years. Commonly, floating breakwater will be effective if the wave high is less than 5 feet or about 1.5 m [9], as shown in the following figure.

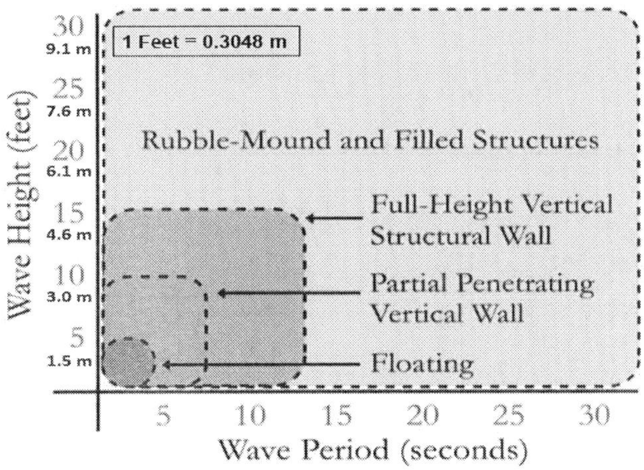

Figure1. The affectivity of floating breakwater as a function of wave high

Rectangular pontoon as a floating breakwater widely used in recent years and usually built segmented that connected by flexible connector and tethered to the bottom using cable or chain. The bindingof floating breakwater to the bottom should be strongly so that cannot be moves. The use of mooring system on floating breakwater must pay attention several aspects including failure both of mooring system and floater must not occur and interaction both of mooring and floater must be safely.

2. Methodology

This research was analytical approach or analytical solution of wave propagation through on floating breakwater. Wave that propagate on floating breakwater or floating object, a part of its energy will be reflected, transmitted and attenuated by turbulence and friction both of fluid and floating structure. Base on conservation of energy concept, the amount of incoming wave energy will be same with the sum of reflected, dissipated and transmission wave energy and can be formulated as:

$$Ei = Er + Ed + Et \tag{1}$$

where Ei, Er, Ed and Et are incoming wave energy, reflected wave energy, dissipation wave energy and transmission wave energy respectively. The performance of floating breakwater is wave transmission coefficient (Kt). The smaller value of wave transmission coefficient, the better the floating breakwater perform in reducing the wave height. Wave transmission coefficient is the ratio between wave high at behind of floating breakwater (Ht) with wave high at the front of floating breakwater (Hi) and mathematically written as:

$$KT = \frac{Ht}{Hi} \tag{2}$$

3. Research Result

Based on strip theory [10], the wave induced force on rectangular float in linier wave theory is formulated as follow:

$$F_{zR} = \frac{\rho g H B \lambda}{2\pi}\left(e^{-2\pi d/\lambda} + 1\right)\sin\left(\frac{2\pi L}{\lambda}\right)\cos(\omega t) \tag{3}$$

or

$$F_{zR} = F_o \cos(\omega t) \tag{4}$$

still in [10], generally we can represent the heaving displacement of floating body by:

$$Z = \frac{\left(\dfrac{F_o}{\rho g A_{wp}}\right)\cos(\omega t + \gamma - \sigma_z)}{\sqrt{\left(1 - \dfrac{\omega^2}{\omega_z^2}\right)^2 + \left(2\Delta_z \dfrac{\omega}{\omega_z}\right)^2}} \tag{5}$$

by substituting Equation (3) to Equation (5), obtained:

$$Z = \frac{\left(\dfrac{HB\lambda}{2\pi A_{wp}}\left(e^{-2\pi d/\lambda} + 1\right)\sin\left(\dfrac{2\pi L}{\lambda}\right)\cos(\omega t)\right)\cos(\omega t + \gamma - \sigma_z)}{\sqrt{\left(1 - \dfrac{\omega^2}{\omega_z^2}\right)^2 + \left(2\Delta_z \dfrac{\omega}{\omega_z}\right)^2}} \tag{6}$$

For the maximum value of Z is written as:

$$Z = \frac{\dfrac{HB\lambda}{2\pi A_{wp}}\left(e^{-2\pi d/\lambda} + 1\right)\sin\left(\dfrac{2\pi L}{\lambda}\right)}{\sqrt{\left(1 - \dfrac{\omega^2}{\omega_z^2}\right)^2 + \left(2\Delta_z \dfrac{\omega}{\omega_z}\right)^2}} \tag{7}$$

water plane area of the float body is multiplication both of length and width of body. Then Equation (7) can be written as:

$$Z = \frac{\frac{H\lambda}{2\pi L}\left(e^{-2\pi d/\lambda}+1\right)\sin\left(\frac{2\pi L}{\lambda}\right)}{\sqrt{\left(1-\frac{\omega^2}{\omega_z^2}\right)^2+\left(2\Delta_z\frac{\omega}{\omega_z}\right)^2}} \tag{8}$$

Wave that propagate on floating breakwater a part of its energy will be reflected, transmitted and attenuated by turbulence and friction both of fluid and floating structure. The figure below describe wave through on semi fixed floating breakwater.

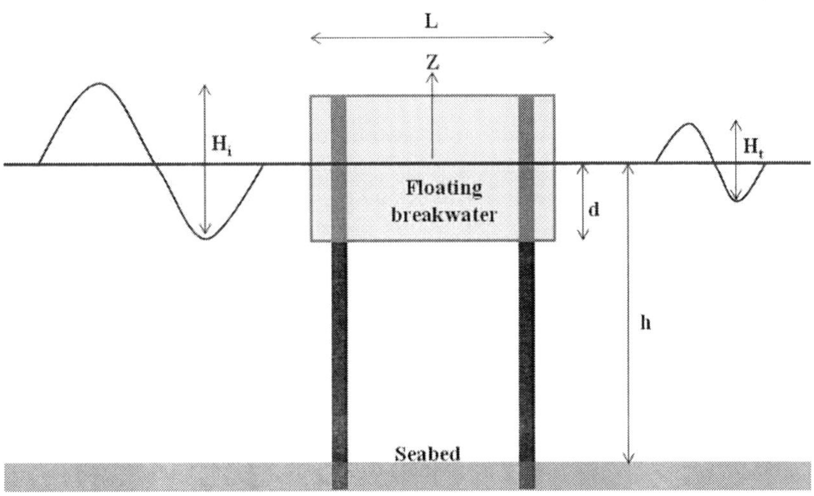

Figure 2. Illustration of wave propagation on semi fixed floating breakwater

When the wave comes from deep water and reach the floating breakwater, the incoming and transmitted wave's energy are formulated:

$$E_I = \frac{\rho g H_I^2}{8}\lambda.B \tag{9}$$

$$E_T = \frac{\rho g H_T^2}{8}\lambda.B \tag{10}$$

On floating object, wave energy dissipation will be equivalent with wave energy that used to moves the object in the vertical direction or heaving movement and formulated as follows:

$$Ed = \frac{\rho g}{8}\left[\frac{\frac{H\lambda}{2\pi L}\left(e^{-2\pi d/\lambda}+1\right)\sin\left(\frac{2\pi L}{\lambda}\right)}{\sqrt{\left(1-\frac{\omega^2}{\omega_z^2}\right)^2+\left(2\Delta_z\frac{\omega}{\omega_z}\right)^2}}\right]^2 LB \tag{11}$$

By substituting Equations (9)-(11) to Equation (1), obtained:

$$\frac{\rho g H_I^2}{8}\lambda.B = \frac{\rho g}{8}\left[\frac{\frac{H\lambda}{2\pi L}\left(e^{-2\pi d/\lambda}+1\right)\sin\left(\frac{2\pi L}{\lambda}\right)}{\sqrt{\left(1-\frac{\omega^2}{\omega_z^2}\right)^2+\left(2\Delta_z\frac{\omega}{\omega_z}\right)^2}}\right]^2 LB + \frac{\rho g H_T^2}{8}\lambda.B \tag{12}$$

By simplification of Equation (12) then obtained:

$$K_t = \sqrt{1-\frac{\lambda\left(e^{-2\pi d/\lambda}+1\right)^2\sin^2\left(\frac{2\pi L}{\lambda}\right)}{4\pi^2 L\left(1-\frac{\omega^2}{\omega_z^2}\right)^2+\left(2\Delta_z\frac{\omega}{\omega_z}\right)^2}} \tag{13}$$

Equation (13) shows that the Kt, influenced by both of wave and structures parameters. Wave parameters are incoming wave high and wave length. Structure parameters are structure width and draft of structures. The value of Kt also influenced wave and heaving motion frequency. As an important note, Equation (13) still needs a more comprehensively study in the future. Especially, validate of equation (13) with laboratory data should be conducted. Until now, researchers not recommended for anyone to use the Equation (13) directly because this equation should be validate using experimental data.

4. Conclusion

There are several conclusions which can be drawn from this study. They are:

- The Equation (13) as simple approximation to calculate wave transmission coefficient should be validated using of experimental or numerical investigation.
- Base on Equation (13), the transmission coefficient (Kt) influenced by structure and wave parameters.
- Wave parameters represents by wave length (λ), wave period (T) and water depth (h).
- Structure parameters represents by length of structure (L), structure width (B), draft (T), shape of bodies, natural period (TN), mass of structure (m), added mass (mv) and dumping coefficient (Δv).

Acknowledgments

The authors would like to thank to Faculty of Engineering, Universitas Hasanuddin for the grant of LBE 2019 and Coastal Geotechnical and Ocean Survey Laboratory, Department of Ocean Engineering, Faculty of Engineering, Universitas Hasanuddin.

References

[1] Chen, Y., Niu, G., & Ma, Y. (2019). Study on Hydrodynamics of a New Comb-type Floating Breakwater Fixed on the Water Surface. In E3S Web of Conferences (Vol. 79, p. 02003). EDP Sciences.

[2] Gesraha, M. R. (2006). Analysis of Π shaped floating breakwater in oblique waves: I. Impervious rigid wave boards. Applied Ocean Research, 28(5), 327-338.

[3] Ning, D., Zhao, X., Göteman, M., & Kang, H. (2016). Hydrodynamic performance of a pile-restrained WEC-type floating breakwater: An experimental study. Renewable energy, 95, 531-541.

[4] Wang, H. Y., & Sun, Z. C. (2010). Experimental study of a porous floating breakwater. Ocean Engineering, 37(5-6), 520-527.

[5] Yang, Z., Xie, M., Gao, Z., Xu, T., Guo, W., Ji, X., & Yuan, C. (2018). Experimental investigation on hydrodynamic effectiveness of a water ballast type floating breakwater. Ocean Engineering, 167, 77-94.

[6] Anchors, Vyhof. 2010. Anchor Manual 2010. © Vryhof Anchors B.V.. Capelle a/d Yssel. The Netherlands 2010.

[7] API RP 2SK.. Recommended Practice for Design and Analysis of Station Keeping Systems for Floating Structures. Washington. DC, .1996.

[8] McCartney, B. L. (1985). Floating breakwater design. Journal of Waterway, Port, Coastal, and Ocean Engineering, 111(2), 304-318.

[9] Qiao, W., Wang, K. H., Duan, W., & Sun, Y. (2018). Analytical Model of Wave Loads and Motion Responses for a Floating Breakwater System with Attached Dual Porous Side Walls. Mathematical Problems in Engineering, 2018.

[10] McCCormick, M. E. (1981). *Ocean wave energy conversion*. Wiley-Interscience, New York.

Modeling Origin Destination Matrix in the Inter Island Cluster of North Liukang Using Gravity Model

A S Chairunnisa[1]*, S Asri[1], L Bochary[1], M R Firmansyah[1], and Zulkifli[1]

[1]Department of Naval Architecture, Engineering Faculty, Hasanuddin University, Makassar, Indonesia

*Email: andi.chairunnisa@yahoo.co.id

Abstract. Trip Distribution is one of the step in the Four-Step Transportation Planning Model. The modelling of inter-zone trip patterns considered several things as the effect of the inter-zone accessibility network system level as well as each-zone production and attraction level. The trip distribution model that used in this study is gravity model (GR) which is classified in synthetic method group. The aims of this study is to estimate the distribution trip model into origin-destination matrix form in both the passengers and cargo trip flow at archipelago North Liukang Tupabbiring sub-district region, Pangkep Regency, South Sulawesi. Determination the number of trip is using Production Constraint Gravity Models with exponential barrier function to estimate inter-island traffic distance. The results showed that the total number of passenger trip from the urban capital to Pangkajene Regency was around 137 passengers per day and the inter-island trip was around 41 passengers per day while the total of cargo trip from the capital sub-district to the city of Pangkajene was around 0.96 tons per day and the inter-islands trip was around 0.28 tons. The result of this distribution trip matrix will be used as a basis for capacity planning of inter-island passenger transport fleets in the North Liukang Tupabbiring sub-district.

1. Background

Pangkep Regency archipelago have enormous complexity and regional potential that could be more optimally developed. There are 13 sub-districts within the administrative area of Pangkajene Regency where 4 sub-districts are in the sub-district islands. One of the them is The North Liukang Tupabbiring sub-district with 74 km^2 area, consisting of 22 islands which are included in 7 villages including Mattiro Bulu village, Mattiro Labangeng village, Mattiro Uleng village, Mattiro Kanja village, Mattiro Baji village, Mattiro Bombang village and Mattiro Walie's village. The capital of the sub-district located in Mattiro Kanja village precisely on Sabutung Island. Based on Central Bureau of Statistics Indonesia data the population of Liukang District Tupabbiring Utara is 11,538 with the population density is 154 individuals per square kilometer.

The planning model that generally used in in transportation planning is the four-stage planning model. One of the stage model is the trip distribution which represented in the form of Origin-Destination Matrix or OD matrix. The OD matrix visualizes the amount of trip from the origin zone to the destination zone, therefore it plays an important role in the transportation planning process [1]. The purpose of OD matrix imposition to a transportation network is to obtain the trip in each segment and the total of entire trip within the analysed network. In this study, the Gravity Model is used with production trip constrain. Both of the value production/attraction trip inter-zone and the production/attraction trip to centre area or district capital are an important to perceive due to a plan of the fleet needs as supporting trip system thus will support connectivity between islands.

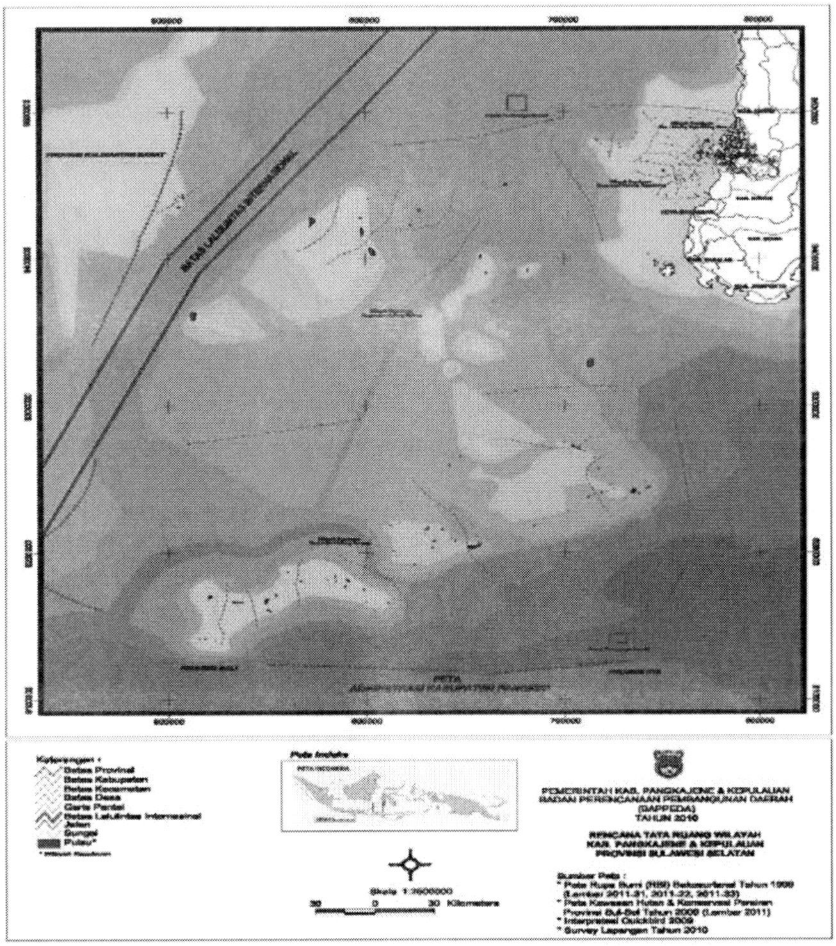

Figure 1. Pangkep Regency islands

Based on the background above, the objective of this study is to determine the distribution model of the passenger and cargo trip between islands and determine the size of the inter-island trip in the Liukang District Tupabbiring archipelago and the number of the trip to the district capital.

2. Methods

2.1. Data Collection

Data collection on passenger and cargo flows were observe at four ports, which are Pangkajene, Maccini Baji, Kassi Kebo and Kali Bone Pier. The secondary data was taken from the relevant agencies, including the Balitbang in Pangkep regency. As well as data that could be taken from previous reports.

2.2. Trip Production Modeling

Trip production is a stage of modeling that estimates the number of trip originating from a zone or land use or the number of trip that are attracted to a land use or zone. It is also an analysis process of that establishes or produces a relation between activity and trip. Production and trip attraction are used to indicate the current movements, which will be used to forecast future movements [1].

IOP Conf. Series: Materials Science and Engineering **875** (2020) 012083 doi:10.1088/1757-899X/875/1/012083

2.3. Trip Distribution Model

The distribution model is used to estimate the number of the trip from each origin zone to each different destination zone, which is affected by the number of the production in each origin zone and also the level of accessibility of the inter-zone network system that usually expressed by distance, time, or cost (combined costs). Meanwhile the distribution trip model in the transportation system is expressed in the trip flow form (vehicles, passengers and cargo) that moving from the origin zone to the destination zone in a particular area and over a certain period of time. Trip Matrix or Origin-Destination (OD) Matrix is a two-dimensional matrix that is most often used to describe trip patterns that contain information of the number of trip between zones. Lines in the OD matrix show the origin zones and columns show the destination zone, therefore each cell in the OD matrix describe the amount of trip flow which are current move from origin zone **i** to the destination zone **j** for a certain interval time [2].

One of the distribution models that commonly used is the Gravity Model. The Gravity Model explains that the characteristics of production and attraction are related to some origin zone parameters, such as population and OD matrix cell values that are also related to accessibility (ease) as a distance, time, or cost function. This method is applied based on Newton's formula which is the gravitation formula which states that ($\mathbf{F_{id}}$) the attraction or repulsion force between two poles of mass is equal to its mass, $\mathbf{m_i}$ and $\mathbf{m_d}$, also inversely quadratic to the distance between two mass, $\mathbf{d}_{id}^{2}$, which could be expressed with:

$$F_{id} = G\frac{m_i m_d}{d_{id}^2}$$

where

G : Universal gravitation constant

Force can be considered as a trip between two regions, while the mass can be replaced with variables such as population or production and attraction trip and distance, time, or cost as a measure of accessibility (convenience) [3]. So, for transportation purposes, the GR model is stated as:

$$T_{id} = k\frac{O_i O_d}{d_{id}^2}$$

where k is universal gravitation constant

In the mathematical form, the GR model can be stated in:

$$T_{id} = O_i \cdot D_d \cdot f\left(C_{id}\right)$$

There is the development of equations due to constrain which are the accessibility of two origin and destination zones, so as to produce the following equation:

$$T_{id} = O_i \cdot D_d \cdot A_i \cdot B_d \cdot f\left(C_{id}\right)$$

Constrain Function is always used in determining the number of production and attraction using the GR model where $\boldsymbol{f_{id}}$ must be considered as a measure of accessibility between zones i and zones d [4].

- $f\left(C_{id}\right) = C_{id}^{-\alpha}$ (Power Function)
- $f\left(C_{id}\right) = e^{-\beta C_{id}}$ (Exp-Negative Function)
- $f\left(C_{id}\right) = C_{id}^{-\alpha} \cdot e^{-\beta C_{id}}$ (Tanner Function)

In this model there is a parameter of β which functions as a determining factor in the number of movements in each zone. β values can be obtained empirically with the following equation:

$$\beta = \frac{k}{\hat{C}_{id}}$$

where

$k = 2 \sim 3$

$\hat{C}$ = average value of C_{id} (Constrain Matrix)

PCGR Model (Product Construction Gravity Model) a model in which the total trip resulting from the generation of the trip must be equal to the total trip produced by modeling; Likewise, the generation of the resulting trip of the model must be the same as the result of the desired trip generation [5]. However, the pull of trip does not need to be the same. The equations used but with different restrictions, that is

$$B_d = 1 \text{ for the all of } d \text{ and } Ai = \frac{1}{\sum_d^i \left(B_d D_d F_{id}\right)} \text{ for the all of i}$$

3. Results and Discussion

3.1. Overview of the Study Area

The north Liukang Tupabbiring District is an area expansion of sub-district of Liukang Tupabbiring District since 2001. It has an area of 74.00 km^2 and is an archipelago with the following administrative boundaries: bordering the Barru Regency in the north, Ma'rang District in the east, Liukang Tupabbiring District in the south and Liukang Kalmas District in the west.

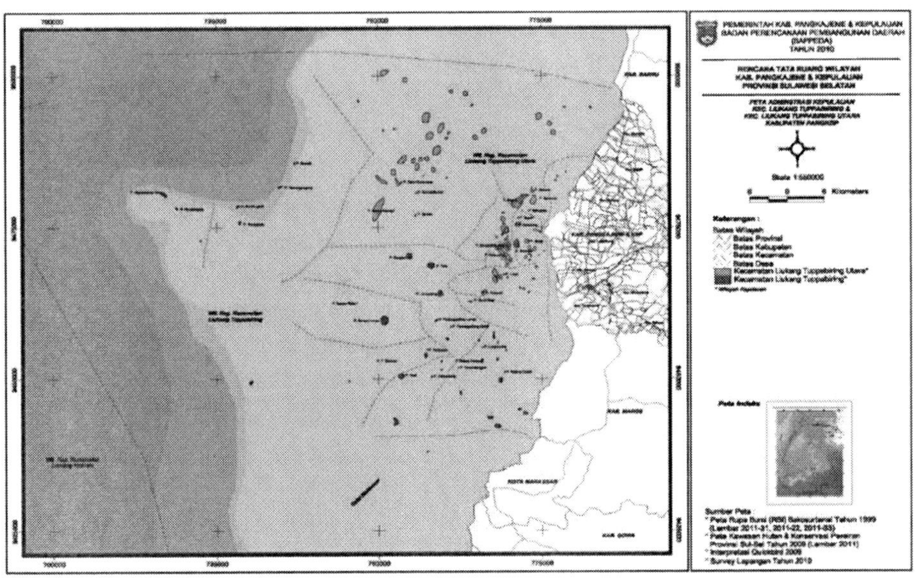

Figure 2. Island cluster area of the North Liukang Tupabbiring District

The North Liukang Tupabbiring have 7 villages such as Mattiro Bulu Mattiro Labangeng, Mattiro Uleng, Mattiro Kanja, Mattiro Baji, Mattiro Bombang and Mattiro Walie. The total population of The North Liukang Tapabbiring in 2017 is 11.538 people with the highest population in Mattiro Bulu Village with a density of 1,043 people / km^2 which is the region with the largest population in the district.

Table 1. Administrative Areas of the North Liukang Tupabbiri Tupabbiring District in 2017

No.	Village	Area (km^2)	Number of Households	Total Population	Population Density (people/km^2)
1	Mattiro Bulu	3.00	606	3,129	1,043
2	Mattiro Labangeng	4.00	179	815	204
3	Mattiro Uleng	5.00	387	1,757	351
4	Mattiro Kanja	4.72	338	1,371	290
5	Mattiro Baji	4.28	346	1,375	321
6	Mattiro Bombang	22.00	658	1,448	66
7	Mattiro Walie	31.00	402	1,643	53

Source: Kecamatan Liukang Tupabbiring Utara dalam Angka, 2017

The population distribution per island in 2017 is shown in the following table

Table 2. Demography of the Islands of each Village in the North Liukang Tupabbiring District

No.	Village	Island Name	Inhabited / Uninhabited	Total Population (People)
1	Mattiro Walie	Samatellu Lompo	Inhabited	1,193
		Samatellu Pe'da	Inhabited	47
		Samatellu Borong	Inhabited	251
		Salebbo	Inhabited	438
		Reang-Reang	Uninhabited	-
		Bone Banoang	Uninhabited	-
		Jangan-jangan	Uninhabited	-
2	Mattiro Bombang	Salemo	Inhabited	1,571
		Sakuala	Inhabited	552
		Sagara	Inhabited	403
		Sabangko	Inhabited	228
		Gusung Torajae	Uninhabited	-
3	Mattiro Kanja	Sabutung	Inhabited	1,709
4	Mattiro Uleng	Kulambing	Inhabited	1,742
		Bangkobangkoang	Inhabited	251
5	Mattiro Labangen	Laiya	Inhabited	759
6	Mattiro Bulu	Polewali	Inhabited	193
		Karanrang	Inhabited	3,294
		Satando	Inhabited	570
7	Mattiro Baji	Saugi	Inhabited	434
		Camba-Cambang	Uninhabited	-
		Sapuli	Inhabited	379

3.2. Transportation Network in The North Liukang Tupabbiring District

Support facilities of the maritime transportation in The North Liukang Tupabbiring area are wooden boats measuring 1-3 GT, some of which operate regularly. In The North Liukang Tupabbiring in 2017,

IOP Conf. Series: Materials Science and Engineering **875** (2020) 012083 doi:10.1088/1757-899X/875/1/012083

the ownership of the fleet by type such as 819 motor boats, 40 layer motor boats, 1,393 jolloro, and 567 boats.

There are 26 fleets of ships operating from The North Liukang Tupabbiring to the capital of Pangkep Regency which operate regularly. Commodities transported in the form of seafood include fish, squid, crabs; building materials in the form of cement, wood, iron, zinc; and other commodities in the form of gas, motorcycles, mineral water (gallons), rice, groceries and others.

As for the fleet that serves The North Liukang Tupabbiring Islands Group are as follows

Table 3. Transportation fleet between the regional islands of the North Liukang Tupabbiring

No	Ship Name	Ship Capacity		Year Making	Route
		Passenger (People)	Cargo (Ton)		
1	KM. Karunia	12	1	2010	Pangkajene – P.Kulambing
2	KM. Halilintar	10	1	2008	Pangkajene – P.Karanrang
3	KM. Karmila 2	10	1	2000	Pangkajene – Samatelu Lompo
4	KM.Rama Saputra	20	3	2004	Pangkajene – P. Kulambing
5	KM. Musdalifah	-	3	2017	Pangkajene – Laiya (ikan)
6	KM. Putra AWF	2	-	2011	Pangkajene - Samatellu
7	KM. Sinar Laut	15	1	2014	Pangkajene – P.Karanrang
8	KM. Rahmat Jaya	30	3	2000	Pangkajene – Laiya (ikan)
9	KM. Anuar	13	2	2012	Maccini Baji – P.Sapuli
10	KM. Fitri Pangkep	13	2	2013	Maccini Baji – P.Sapuli
11	KM. Yus Amalia	20	1	2017	Pangkajene – P.Karanrang
12	KM. Fitri Jaya	13	2	2015	Maccini Baji – P.Saogi
13	KM. Rahmi Ilahi	10	1	2008	Maccini Baji – P.Saogi
14	KM. Saogi Bahari	13	2	2004	Maccini Baji – P.Saogi
15	KM. 39	13	1	2010	Maccini Baji - CambaCambang
16	KM. Putra Labbakkang	13	2	2000	Maccini Baji - CambaCambang
17	KM. Jamidding	13	1	1970	Maccini Baji - CambaCambang
18	KM. 42	13	1	2004	Maccini Baji - CambaCambang
19	KM. 43	13	1	1980	Maccini Baji - CambaCambang
20	KM. Suci	15	2	2012	Maccini Baji – P.Saogi
21	KM. 46	13	1	2010	Maccini Baji – P.Salebo
22	KM. Aspal	5	1	2005	Maccini Baji – P.Satando
23	KM. Tunas Muda	5	1	2007	Maccini Baji – P.Satando
24	KM, Cahaya	12	-	2014	Maccini Baji – P.Saogi
25	KM. Fajar Indah	20	1	2003	Pangkajene – P.Kulambing
26	KM. 44	13	1	1980	Maccini Baji - CambaCambang
27	Karmila 1	50	10	1973	Pangkajene – P. Kulambing

Source: Dinas Perhubungan Pangkep Regency

3.3. Trip Distribution of Passenger and Cargo Analysis

There are 2 Characteristic about trip distribution in The North Liukang Tupabbiring which are inter island trip distribution and trip distribution to the capital of Pangkep Regency. This analysis using model gravity with variable costrain is the distance between the island. The determine of the trip distribution using PCGR model with the constrain using by power function.

Table 4. The Distance between the capital of the village matrix. (Sea Mile)

	Mattiro Walie	Mattiro Bombang	Mattiro Kanja	Mattiro Uleng	Mattiro Labangen	Mattiro Bulu	Mattiro Baji	Pangkajene
Mattiro Walie	0.00	8.50	6.79	7.59	8.80	9.69	8.84	12.84
Mattiro Bombang	8.50	0.00	4.22	5.42	8.37	11.20	4.65	8.94
Mattiro Kanja	6.79	4.22	0.00	1.35	4.25	6.97	2.03	6.48
Mattiro Uleng	7.59	5.42	1.35	0.00	3.00	5.89	1.67	5.33
Mattiro Labangen	8.80	8.37	4.25	3.00	0.00	3.02	4.13	4.64
Mattiro Bulu	9.69	11.20	6.97	5.89	3.02	0.00	7.15	4.86
Mattiro Baji	8.84	4.65	2.03	1.67	4.13	7.15	0.00	6.98
Pangkajene	12.84	8.94	6.48	5.33	4.64	4.86	6.98	0.00

Source: Dinas Perhubungan Pangkep Regency

Final distribution matrix of passenger and cargo trip inter-zone in island cluster area of North Liukang Tupabbiring district and passenger trip from island area to Capital of Pangkep regency (Pangkajene) determined by β value 0.368, this would be one of the variable in determined constrain function.

While the matrix constrain function in determine passenger and cargo trip is shown in the table below:

Table 5. Exponential function constrain matrix

	Mattiro Walie	Mattiro Bombang	Mattiro Kanja	Mattiro Uleng	Mattiro Labangen	Mattiro Bulu	Mattiro Baji	Pangkajene
Mattiro Walie	0.000	0.454	0.493	0.474	0.449	0.433	0.448	0.390
Mattiro Bombang	0.454	0.000	0.588	0.536	0.457	0.410	0.567	0.446
Mattiro Kanja	0.493	0.588	0.000	0.895	0.587	0.489	0.770	0.502
Mattiro Uleng	0.474	0.536	0.895	0.000	0.667	0.520	0.827	0.540
Mattiro Labangen	0.449	0.457	0.587	0.667	0.000	0.665	0.593	0.568
Mattiro Bulu	0.433	0.410	0.489	0.520	0.665	0.000	0.484	0.558
Mattiro Baji	0.448	0.567	0.770	0.827	0.593	0.484	0.000	0.489
Pangkajene	0.390	0.446	0.502	0.540	0.568	0.558	0.489	0.000

Source: Data Analysis, 2018

Power function constrain matrix determined passenger and cargo trip distribution inter-zone in island area of The North Liukang Tupabbiring district as well as passenger trip from island area to Capital of Pangkep Regency (Pangkajene) per day. While the number for passenger trip distribution matrix is shown in the table below

Table 6. Passenger trip distribution matrix in inter-island cluster in North Liukang Tupabbiring District

	Mattiro Walie	Mattiro Bombang	Mattiro Kanja	Mattiro Uleng	Mattiro Labangen	Mattiro Bulu	Mattiro Baji	Pangkajene
Mattiro Walie	0	1	1	1	0	2	1	4
Mattiro Bombang	1	0	1	1	0	1	1	4
Mattiro Kanja	1	1	0	1	0	1	1	3
Mattiro Uleng	1	1	1	0	1	2	1	4
Mattiro Labangen	0	0	0	1	0	1	0	2
Mattiro Bulu	2	1	2	2	1	0	2	10
Mattiro Baji	1	1	1	1	0	1	0	3
Pangkajene	12	12	13	17	8	33	12	0

Source: Data analysis, 2018

Based on passenger trip distribution matrix 77% trip or 137 people headed to Capital of Pangkep regency (Pangkajene) per day, whereas the number of inter-island trip is 41 passenger per day. Based on field observation showed that most of all trip of passenger to Capital of Pangkep Regency (Pangkajene) were to buying everyday needs or to resell. Meanwhile most of all inter-island trip were to visited relatives.

Table 7. Cargo Trip Distribution matrix in inter-island cluster in North Liukang Tupabbiring District

	Mattiro Walie	Mattiro Bombang	Mattiro Kanja	Mattiro Uleng	Mattiro Labangen	Mattiro Bulu	Mattiro Baji	Pangkajene
Mattiro Walie	0.000	0.006	0.006	0.008	0.003	0.013	0.006	0.063
Mattiro Bombang	0.005	0.000	0.006	0.007	0.003	0.010	0.006	0.056
Mattiro Kanja	0.005	0.005	0.000	0.009	0.003	0.009	0.006	0.050
Mattiro Uleng	0.006	0.006	0.009	0.000	0.004	0.012	0.008	0.068
Mattiro Labangen	0.002	0.002	0.003	0.004	0.000	0.007	0.003	0.032
Mattiro Bulu	0.011	0.009	0.010	0.014	0.008	0.000	0.010	0.144
Mattiro Baji	0.004	0.005	0.006	0.009	0.003	0.009	0.000	0.051
Pangkajene	0.054	0.054	0.058	0.080	0.039	0.151	0.056	0.000

Source: Data Analysis, 2018

As much as 77% distribution trip cargo is headed to Capital of Pangkep regency (Pangkep) or 0,96 ton per day. The common load that usually carried from island cluster area were sea products as fish, squid, and crab also primary needs and building material as well.

The illustration of distribution passenger trip in imposition trip maps represent as below

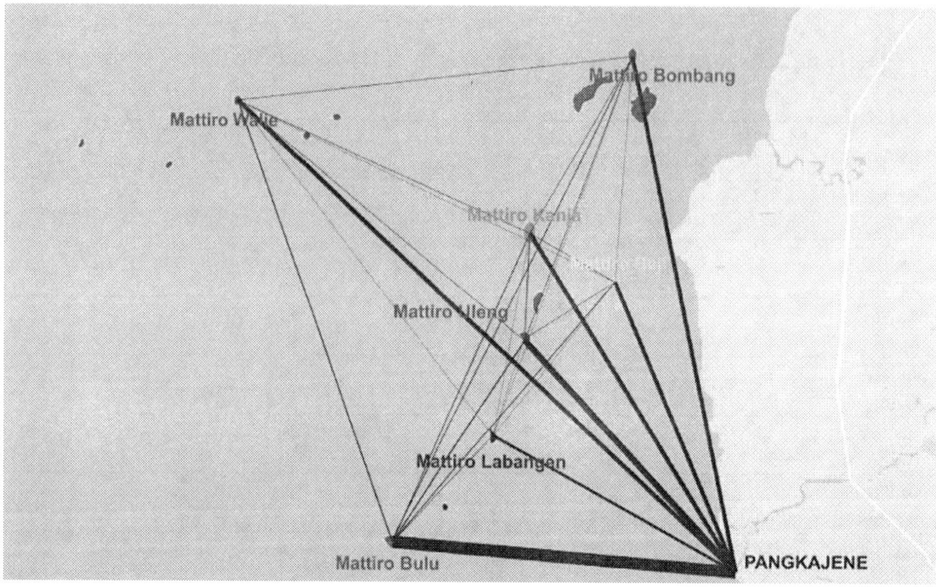

Figure 3. Imposition of passenger trip In North Liukang Tupabbiring District

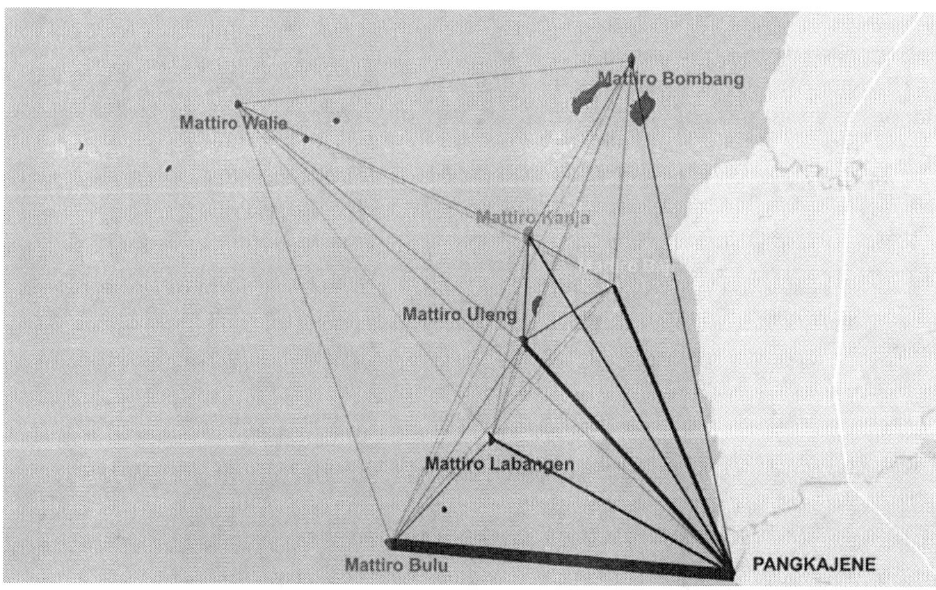

Figure 4. Imposition of cargo trip In North Liukang Tupabbiring District

4. Conclusion

The trip distribution of the island cluster area in North Liukang Tupabbiring, trip distribution of passenger or trip distribution of cargo still dominated trip from the inter island cluster to capital of Pangkep Regency by the total trip are 77% trips of passenger and 77% trips of cargo. The rest was distributed to the inter region islands.

References

[1] Wardono, Surakanti. 2015. *Analisis Sistem Jaringan Transportasi Dan Model Gravitasi*. Jurnal ASE – Volume 11.

[2] Ofyar, Z. *2000. Perencanaan dan Permodelan Transportasi*. Bandung, Indonesia: ITB.

[3] Suyuti, Rudmadi. 2007. *Penggunaan Model Gravity (GR) Dalam Estimasi Matriks Asal-Tujuan (MAT) Menggunakan Data Arus Lalulintas*. Jurnal Transportasi Vol. 7.

[4] Djuniati, Sri. 2010. *Formulasi Model Gravity Sebagai Model Penyebaran Perjalanan Penumpang - Jmp Surabaya*. Jurnal APTEK Vol. 3.

[5] Puspitasari, Niken. 2014. *Estimasi Matriks Asal Tujuan Perjalanan Menggunakan Model Gravity Dengan Fungsi Hambatan Tanner*. e-Jurnal Matriks Teknik Sipil Vol. 2. Universitas Negeri Sebelas Maret.

IOP Conf. Series: Materials Science and Engineering **875** (2020) 012084 doi:10.1088/1757-899X/875/1/012084

Effect of Wave Power Device Arrangement in Double-Water-Chamber Type Seawall towards Wave Power Extraction Rate and Wave Dissipation

F Husain[1]*

[1]Ocean Engineering Department, Faculty of Engineering, Universitas Hasanuddin, Indonesia

*Email: firman.husain@unhas.ac.id

Abstract. Utilization of the ocean energy to meet the electricity need is important to realization. The increasing need for electricity and environmental issues are the reasons. In last several decades a number of studies on wave energy device have been carried to find the effective one. In this paper, the double-water-chamber type seawall is basically adopted that has a similar structure to the OWC was examine to extract wave power. In the experiment various arrangement of wave power device inside the double water-chamber was tested to obtain optimum of wave extraction rate. Another function of the double water-chamber as wave dissipation is also clarified in the experiment. Based on experimental result the maximum wave energy conversion efficiency is about 30% and the reflection coefficient of the double-water-chamber seawall is about 0.3 on average.

1. Introduction

Development of wave energy converter device has been started in last several decades in Europe. By 1980 there are more than one thousand patents had been registered [1] and its increase until now. The global awareness about the importance of clean energy source contributes to developed wave energy converter. Nowadays many countries have utilized energy from the ocean wave to meet their energy demand. Various wave power converters have been used to increase the wave power supply, and many other different technologies are under investigation and development [2]. Energy from ocean wave is most attractive to be exploited. It can be harnessed 90 percent of time compare to the solar and wind power device only 20-30 percent.

The present paper introduces and examines the development of a new wave power converter called the double-water-chamber seawall. It has developed from the previous work by Husain et al [3]. Although there is a lack of references for this type of wave converter, it has a shape similar to that of the OWC structure. It is well known that the oscillating water column is the most developed type of wave energy converter [4]. In double-water-chamber seawall, the water turbine was used to extract the wave power instead of air turbine as usually uses in the OWC type. An asymmetrical guide vane was installed to guide water flow to accelerate rotation of water turbines. By using the double-water-chamber seawall conversion efficiency rate is expected can be improved and the reflection coefficient can be reduced. Nakamura et al. were confirmed that applied asymmetrical guide vanes to extract tidal current power and experimentally found out that such guide vanes are able to improve the efficiency of a water turbine, typically about 50% of the efficiency [5]. Therefore, the present experimental study examines applicability of a similar method of power extraction device and the resultant possible efficiency as a wave power extraction device for double water chamber. By employing the wave damping theory developed by Nakamura and Ide, the effectiveness of the double water chamber seawall as a dissipater of reflection waves is also examined and clarified [6].

2. Wave Power Device Component

The main wave power device component that use to extract wave power comprises of wave-guide vanes, water turbine and water chamber as substructure.

2.1. Wave Guide Vanes and a Water Turbine

When an incident wave reaches the structure, the water turbine will rotate due to of water flow that hit the blade. However the rotating of water turbine by water flow is deficient, therefore a wave-guide vane is needed. The wave guide vanes used in the experiment composed of four thin plates in an asymmetrical arrangement as seen in Figure 1. In the right and left ends two square plate are joined respectively. At the center of the guide vanes, a Savonius water turbine was installed to extract the wave power as much as possible. In this experiment the Savonious turbine type using two blades was applied in the study [7].

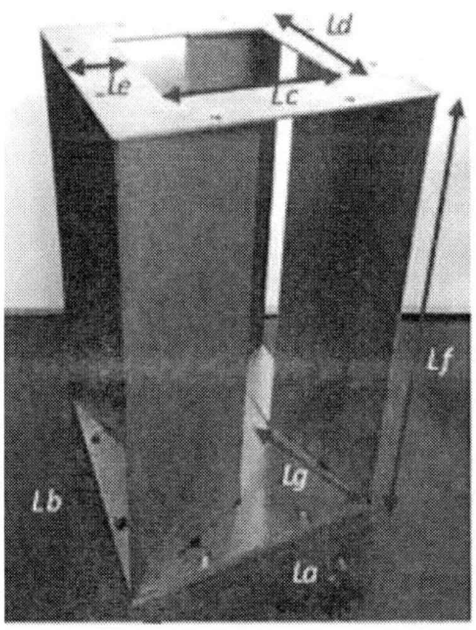

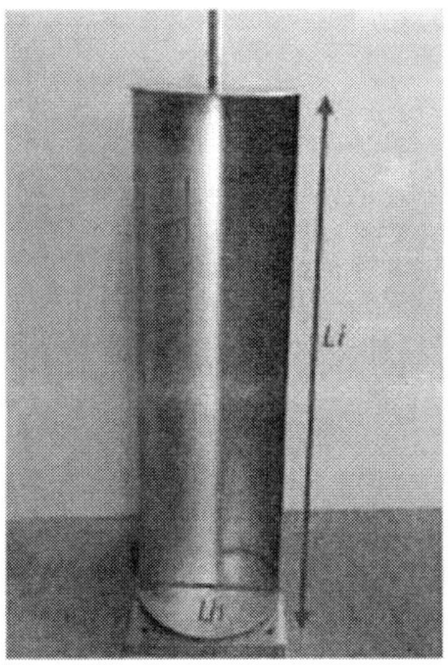

Figure 1. The wave-guide vanes and a water turbine

The dimensions of guide vanes and water turbines are summarized in Table 1.

Table 1. Principal dimensions of the wave guide vane

Outer width of the guide vane	*La*	240 mm
Outer height of the guide vane	*Lb*	240 mm
Inner width of the guide vane	*Lc*	140 mm
Inner height of the guide vane	*Ld*	140 mm
Width between outer and inner side	*Le*	50 mm
Length of guide vane	*Lf*	360 mm
Width of guide vane	*Lg*	130 mm
Diameter of turbine	*Lh*	120 mm
Length of turbine	*Li*	350 mm

2.2. Double-Water-Chamber Seawall

The double-water-chamber seawall is consisting of two water chambers. It was designed to expand the effective wavelength characteristic. It was also intended to improve the efficiency rate and reduce the reflection wave. In this experiment, there are three types arrangement of wave-guide vanes mounted inside the double-water-chamber. Here, the wave-guide vanes were installed in the first chamber for all type arrangement as shown in Figure 2. In the type 02, there is no gap spacing just below the first curtain wall and two others types that are type 01 and type 03 gap spacing are exist. The specific of type 03 besides have a gap spacing, it also has overlap arrangement. The description of three types are summarized in Table 2.

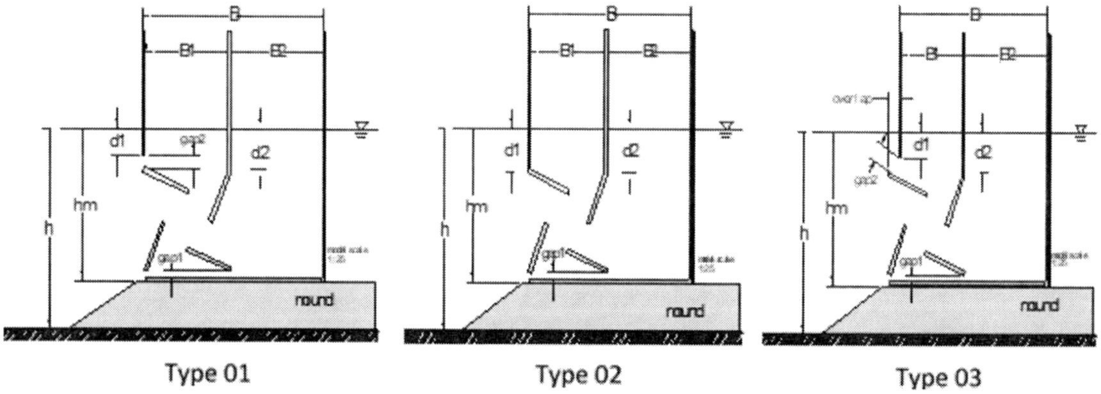

Figure 2. The double-water-chamber seawall models with different type arrangement

Table 2. Dimension of the model

Description	Type 01	Type 02	Type 03
Breadth of first chamber (*B1*)	220mm	220mm	180mm
Breadth of second chamber (*B2*)	260mm	260mm	260mm
Total breadth of chamber (*B*)	500mm	500mm	460mm
Draft depth of first wall (*d1*)	83mm	113mm	92mm
Draft depth of second wall (*d2*)	113mm	113mm	113mm
Gaps (*g1*)	17mm	17mm	17mm
(*g2*)	30mm	no gap	22mm
Overlapping	-	-	30mm
Wall thickness (*tw*)	10mm	10mm	10mm

3. Experiment Activity

3.1. Experimental Condition

The experiment was conducted at various wave periods of $T = 1.2$ to 3 s. For each wave period, the experiment was conducted at two wave heights, $H = 8$ and 16 cm. The water depth (h) from the bottom flat was 50 cm and that from the mound was 37 cm. These water depths were kept constant throughout the experiment. The mound height was 12 cm. To examine the effectiveness of the proposed double-water-chamber seawall in extracting wave power and dissipating wave energy, a series of experiments was carried out in a long wave flume at Ehime University. The model scale assumed here is 1/25. The model was made of transparent acrylic plates with thickness of 10 mm, which allowed the observation of the interior of the chamber during the tests. The flume tank was 30 m long, 1 m wide and 1.25 m high as shown in Figure 3. A piston-type wave maker was installed at one end of the tank and a wave absorber at the other end. The seawall model structure was installed 18.4 m from the wave maker. Four wave gauges of capacitance type were arranged in the wave flume. The first gauges was set in front of the wave maker, while the second and third were placed in front of the model to estimate coefficients of reflection from the model, and the fourth was placed in the chamber of the model.

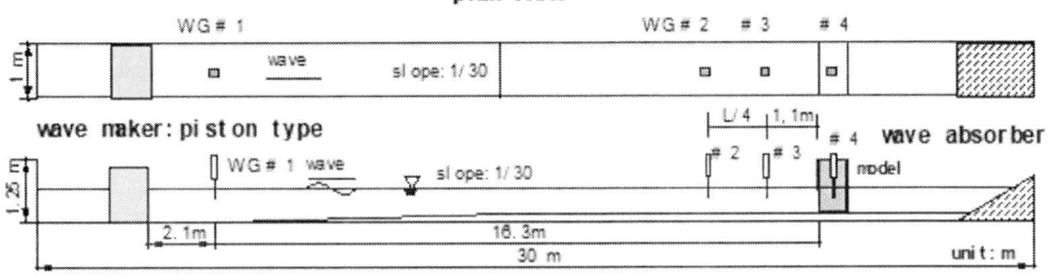

Figure 3. Experimental setup in wave flume

3.2. Experimental Procedure

In this experiment, the model of double-water-chamber seawall was setup inside wave flume. The water turbine and guide vanes were installed horizontally inside the chamber. The water turbine is driven by water flow from the lower opening in front part of curtain wall. Application of axisymmetric guide vanes is intended to direct the water flow to increase rotation of the water turbine. The pumping wave mode, which occurs inside the chamber, is also expected to increase the rotation of water turbines. There are three arrangement types of installation guide vanes and water turbine will examine. Based on result of examination, the highest wave amplification from three type arrangements was selected to continue wave energy extraction test. The rotating motion of a water turbine by wave action is transferred to a torque meter mounted above the seawall model through a pulley and belt system. The torque meter is connected to an electromagnetic brake that simulates the resistance load of an electric generator. The torque meter also captures the angular velocity of the rotational shaft. The torque meter used in the experiment has capacity of 1 N.m and a maximum speed of rotation of 6000 r/min.

To check the rate of extraction of wave energy by the Savonius turbine mounted within axisymmetric guide vanes, the work done by the water turbine against the resistance torque given by the magnetic brake was observed for various resistance levels under each set of wave conditions of the definite wave height and period. In this observation, the measurement system comprised a torque converter and a revolution counter. Wave reflection coefficients under the condition of wave energy extraction were also obtained.

4. Analysis of the Power Coefficient (Rate of Wave Power Extraction)

The power coefficient C_P is known as the energy conversion or extraction rate and is defined as

$$C_P = \frac{P_T}{P_W} \tag{1}$$

where P_T is the work done by the water turbine against the rotational resistance applied and can be obtained as

$$P_T = \frac{1}{t}\int_0^t M_T \cdot \omega \, dt \tag{2}$$

where M_T is the torque moment of the resistance measured by the torque meter, ω is the angular velocity of the pulley, and t is the duration of the observation, which is usually five or six times the wave period.

P_W is the wave power of incident to a seawall equal to the well-known property of the wave energy flux and can be expressed as

$$P_W = \frac{1}{8}\rho \cdot g \cdot H^2 \cdot C_g \cdot b \tag{3}$$

where ρ is the fluid density, g is the gravitational acceleration, H is the wave height at the position of the seawall measured as an incident wave, and b is the transverse length of the water turbine. C_g is the group velocity at which energy is transported in the waves.

5. Experiment Results

5.1. Wave Amplification

Figure 4 shows typical results of the wave amplifications in the new arrangement of guide vanes of type 01. The wave amplification is defined as the ratio of a wave height in each water chamber (*Hc*) to an incident wave height (*H*). In the figure, experimental results corresponding to the two different wave heights H = 8cm and 06cm are plotted to make clear the influence of wave height on the wave amplification. In the figure, a theoretical result based on the damping wave model is also specified. From this figure, it can be seen that the wave amplifications in the two water chambers increases with *L/B*. For the second water chamber, the wave amplification for H=8cm is higher than the one for H = 16cm. For the condition $L/B > 6$, the wave amplifications in the two chambers are in the range between 1.0 and 1.5. This tendency may contribute higher wave energy extraction rate. We can see the reasonable agreement between the theoretical and experimental results on wave amplifications for the first water chamber. However, for the second chamber, the theoretical computation result overestimates the wave amplification, especially for longer wave conditions and higher wave height condition, H=16cm. These quantitative differences may be caused by the higher hydrodynamic resistance in the guide vane in the experiment

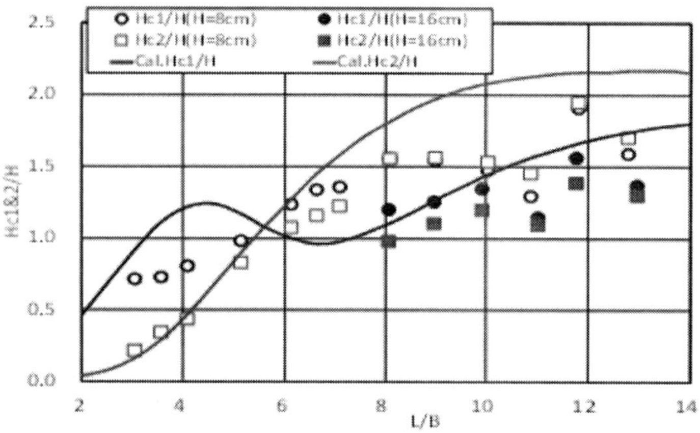

Figure 4. Wave amplification results of type 01

Figure 5 shows wave amplification result of the arrangement of guide vanes of type 02 mounted in the chamber. The second type of arrangement is similar to the previous type, but without a gap spacing between the first vertical wall and the guide vanes. It may allow the water particles of incident wave to come into the water chamber only through the slot of guide vanes and a gap under guide vanes. The figure reveals that the wave amplification results of both the water chambers are comparatively lower than the result of type 01. It can be seen that the gap spacing just below the first curtain wall is important to amplify the piston mode wave motions in both the water chambers. Computational results of the two wave amplifications show poor quantitative agreement with the experimental result, especially in the long wave condition. However, it can be confirmed that there are qualitative agreements between the two results, typically variations of the wave amplifications with *L/B*.

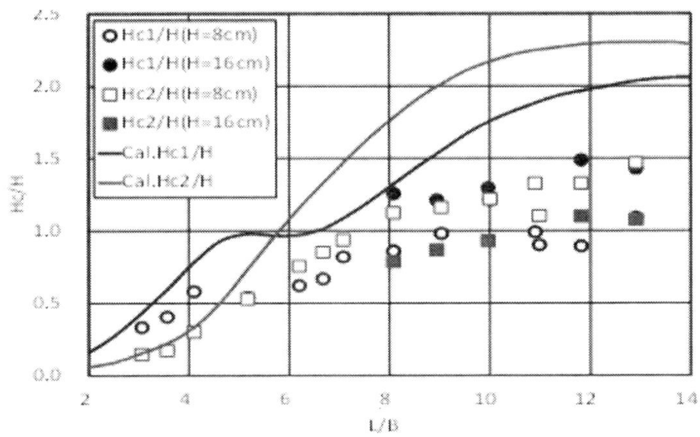

Figure 5. Wave amplification result of type 02

Figure 6 shows the result of wave amplifications in the two chambers of type 03. It can be seen that the wave amplifications in both the chambers are comparatively high as compared to the previous three types. Specifically, the wave amplification in the second chamber is greater than 1.5 for longer waves and medium wave heights. Further the wave amplification in the first chamber is also about 1.5 in average for longer waves. We can see the reasonable agreement between the experimental and theoretical results, especially regarding the second water chamber. From these comparisons, it is confirmed that the gap spacing below the first curtain wall and the overlapping structure of the first wall on the guide vane are effective to increase the wave amplifications in the first and second water chambers.

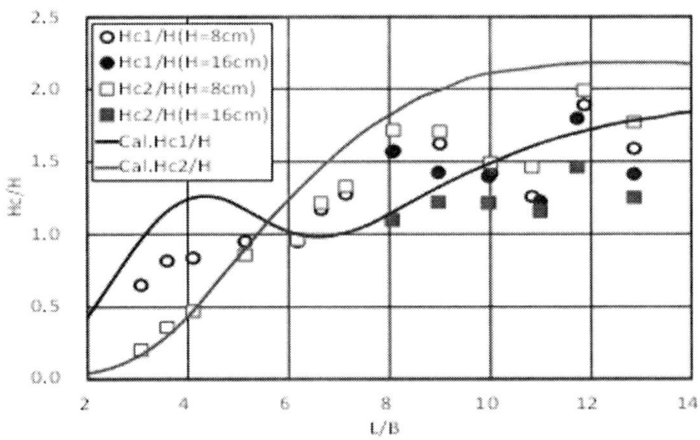

Figure 6. Wave amplification result of type 03

5.2. *Wave Amplification*

Figure 7 shows wave power extraction rates of this model as a function of *L/B*. The figure reveals that the maximum efficiency of the wave energy conversion is about 30% and it occurs under the condition $L/B = 7$ ($T = 1.8$ s). This optimal condition may be closely related to the optimal amplification condition of the wave motion in the water chamber. For a higher wave height ($H = 16$ cm), the efficiency of wave energy conversion is a little bit decreased and maximum wave power extraction rates about 23%. This result is higher than the previous work by Firman et al. [8]. The important aspect of the result here is comparatively high efficiency under the condition of shorter waves that correspond to the nominal wave conditions.

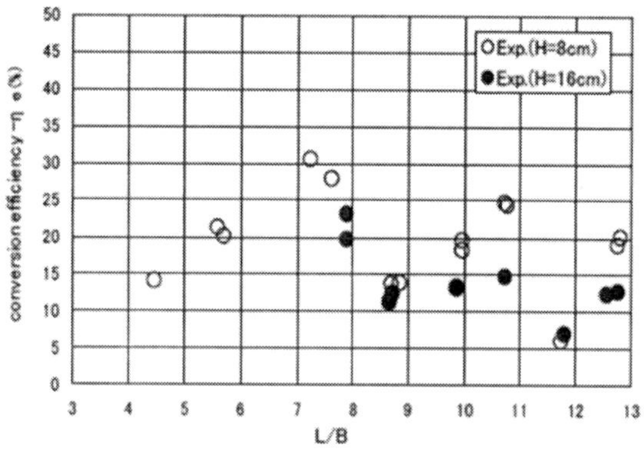

Figure 7. Efficiency rate

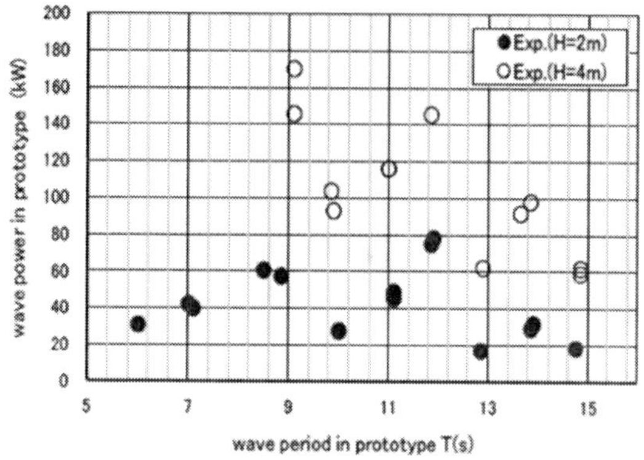

Figure 8. Possibility extraction power

Figure 8 shows the possible extraction of wave power on the prototype scale, which is converted from the experimental result of type 04 model based on the Froude's low of dynamic similarity. The figure reveals that possible extraction of wave power is about 50 kW for a wave height $H = 2$ m and 120 kW for $H = 4$ m on average.

5.3. Wave Amplification

Figures 9 and 10 show the results of reflection coefficients *Cr* for the model of type 01 and type 03, which corresponds to the cases of comparatively higher wave amplifications in the chambers. In both cases, the minimum reflection appears about *L/B*=8 which corresponds to comparatively high wave amplification conditions as seen in Figures 4 and 6, respectively.

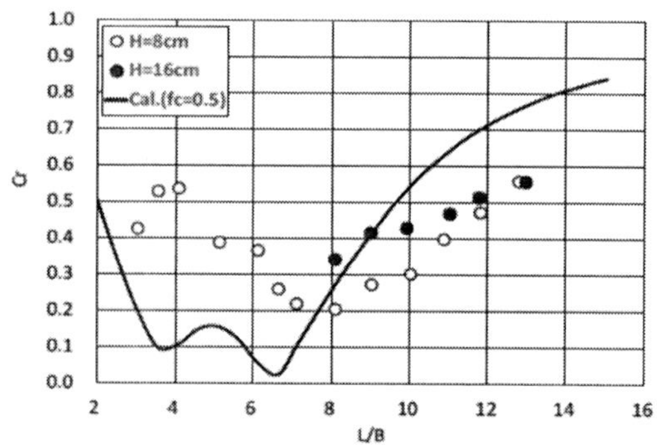

Figure 9. Wave reflection of type 01

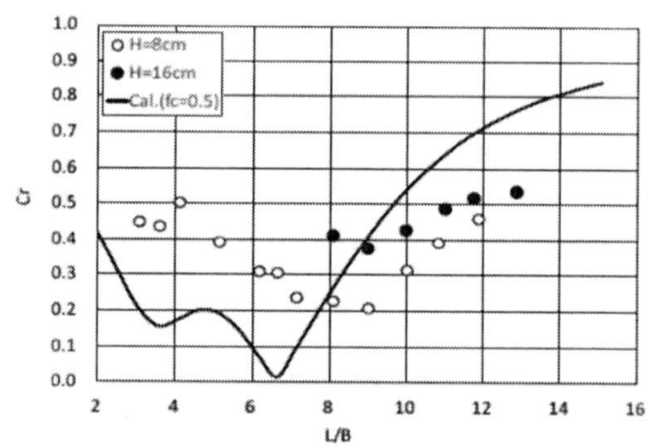

Figure 10. Wave reflection of type 03

There are some qualitative differences between the experimental and theoretical results. The difference may be resulted from the three-dimensional structural effects of the model, in which the rotating disc is stored in the transverse end part of the guide vane with 30cm breadth. It can be seen that the reflection coefficient *Cr* is less than 0.5 for comparatively wide range of *L/B*, e.g. *L/B*, =6 to 12.

6. Conclusion

The maximum conversion efficiency founded about 30% and 23% in wave condition of H = 8cm and H = 16cm respectively. Another role of the structure, such as the breakwater function to reduce the reflection wave shows good result. In the recent experiment *Cr* being about 0.3 on average for a wide range of wave period. The wave power that can be extracted on a prototype scale is estimated at about 50 kW for a wave height $H = 2$ m and 120 kW for $H = 4$ m on average. It is enough for electricity use in everyday life. The detailed information of the wave height and period conditions at the site is required, in order to realize optimum efficiency of the structure.

References

[1] McCormick, ME., "Ocean wave energy conversion," New York, Wiley, 1981.
[2] Falnes, J., "A review of wave-energy extraction," Mar. Struct. 1, 57–65, 2007.
[3] Husain, F., Inouchi, K., and Nakamura, T.,"Development of a Wave Power Extraction Seawall and Its Effectiveness for Wave Dissipation," Journal of Japan of Society of Civil Engineers. Ser. B2, 70(2), 1311–1315, 2014.
[4] Falca˜o, A.F. de O "Wave energy utilization: A review of the technologies," Renewable and Sustainable Energy Reviews, 14, 899-918, 2010.
[5] Nakamura, T., Abe, H., Husain, F., and Inouchi, K., "Performance on a new guide vane of vertical Savonius turbines for effective conversion of tidal current power," Civil Eng.Jour, B3, Vol.69, No.2, pp. I_13-I_18 (in Japanese), 2013.
[6] Nakamura, T., and Ide, Y., "Analysis on wave transformations and wave forces about an angular body considering wave energy dissipations," Proceeding of Civil Eng. in the Ocean, Japan, 13, pp.177-182 (in Japanese), 1997.
[7] Kailash, G., Eldho, I. T and Prabhu, V. S., "Performance Study of Modified Savonius Water Turbine with Two Deflector Plates, "International Journal of Rotating Machinery Vol. 2012 , 2012.
[8] Husain, F., Inouchi, K., and Nakamura,"Development of a Multi-Water-Chamber Type Seawall for Effective Wave Power Extraction and Wave Dissipation," Proceeding of the Twenty-five, International. Ocean and Polar Engineering Conference, Kona, USA, ISOPE, 1, 907-912, 2015.

The Performance of Single and Double Intake Designs of Propeller Flow Cooling System (PFCS)

F Mahmuddin[1]*, A Suhardi[1], S Hariyanto[1], Syarifuddin[2], and M Tasrief[3]

[1]Marine Engineering Department, Engineering Faculty, Hasanuddin University, Indonesia
[2]Naval Architecture Department, Engineering Faculty, Hasanuddin University, Indonesia
[3]PT. Biro Klasfikasi Indonesia, Tanjung Priok, Jakarta

*Email: f.mahmuddin@gmail.com

Abstract. In order to save spaces inside a small-sized fishing boat, the high-pressure flow behind propeller is utilized instead of a special cooling pump to cool the main engine. From the previous study, the system is referred to as the Propeller Flow Cooling System (PFCS). As a continuation of the previous study, the present study conducted more experiments in order to compare the performance of a single and a double intake designs of the PVCS with the same cross-sectional area. The designs with circular and elliptical intakes shapes were measured. The experiments were also conducted at the Circulating Water Channel (CWC) facility. The main parameter which was compared in the experiment and analysis was the flow capacity produced by both designs for various propeller revolution (RPM). The experimental results showed that single intake design has a higher flow capacity than the double intake one which was caused by higher flow losses inside the pipe of single intake design mainly due to higher surface friction as well as an additional elbow joint needed in double intake design. This tendency occurred for both circular and elliptical shapes cases.

1. Introduction

The main engine of a boat is used to produce power to drive the boat. The power generated from the main engine is obtained from the combustion of fuel that occurs inside the engine combustion chamber. The burning of this fuel will produce high power and heat which is used to drive the propeller. However, as a result of the heat generated from the combustion of fuel in the engine, it will increase the temperature, especially in parts that come in direct contact with the combustion chamber [1]. Therefore, a cooling system to reduce excessive heat in the engine is necessary.

In general, engine cooling systems on ships use seawater as the coolant in an open cooling system or fresh water in a closed cooling system [2]. However, there many traditional fishing boats today that only use an air cooling system. In particular, mainly because the boat uses a land-based engine that is modified to be fit for the boat. Air cooling has a disadvantage when the position of the engine placed in the inboard position because it can cause insufficient air to absorb heat from the operating engine [3].

One of the impacts of the engine when it is too hot or overheating is that the engine components will become soft which can cause abnormal combustion, knocking, or other abnormal conditions. Contrary, the impact when the engine is too cold is that it can cause other problems including the fuel will be difficult to evaporate on a gasoline engine. While on a diesel engine if the compressed air is cold, the resulting exhaust gas output is white smoke and will certainly produce an engine knock as well as the difficulty to turn on the engine [4].

Therefore, to support the operation of a fishing boat, the performance of the main engine is inseparable from the role of the cooling water factor. The main engine cooling water system serves to deliver or drain the cooling water from a lower place to the desired destination with the help of an additional engine or pump. Recently, in some boats, special arrangements are made where the cooling pumps or motors are driven or coupled with the main engine itself in order to reduce the power needed by the boat [5]. This arrangement also could make installation more compact aboard the boat.

As an alternative method to reduce energy usage by the boat, many relatively small-sized fishing boats especially in South Sulawesi, Indonesia utilize the high-pressure flow behind the propeller. The high-pressure flow is directed through an intake positioned in front of the propeller to cool the main engine. The hot water from the main engine is discharged outboard through hull. The system is referred as Propeller Flow Cooling System (PFCS) [6]. With this arrangement, the special pump for main engine cooling is not necessary anymore. Thus, reducing initial and operational costs as well as saving space inside the boat. An example of a boat that implements the system is shown in Fig. 1.

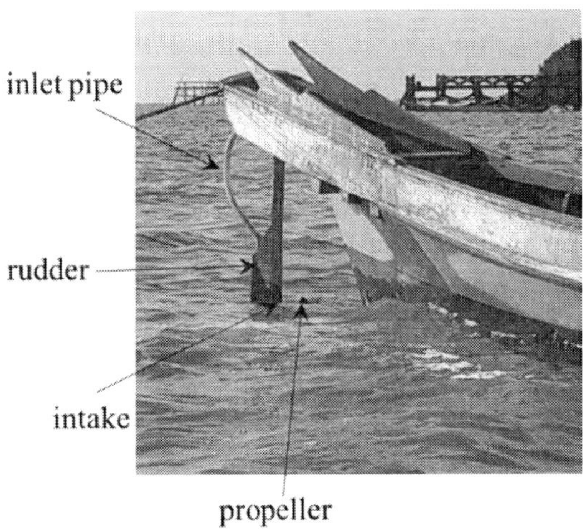

Figure 1. an example of a Propeller Flow Cooling System (PFCS) installed in a fishing boat

From previous research conducted by Mahmuddin et al. [6], it was found that an elliptical intake shape of the system has a better performance compared to the circular one. In the same study, it was also found that when the intake shape was placed in a vertical position of 0.7R where R is the propeller radius, the system had a better performance as compared to when the intake was placed exactly in front of the propeller hub. However, in that study, only a single intake was placed on the upper part of the propeller while the flow in the lower part of the propeller was still not utilized. Therefore, in the present study, 2 (two) intake design of the PFCS will be compared. The 1st design is the one with only 1 (one) intake while the 2nd design has double intake which was placed in upper and lower parts of the propeller. In order to make an equal comparison, both designs have the same intake cross-sectional area.

In addition, in order to measure the flow velocity and capacity inside the distributing pipe, the flow measurement system needs to be designed so that the amount of water that flows to the main engine can be determined. The system uses a microcontroller Arduino UNO and a flow sensor as the main components. The microcontroller is one of the common microcontrollers used for data collection such as the ones performed in references [7-10]. The measurement system developed in the present study includes a software interface that was developed using Microsoft Visual Basic software. The measurement system is able to transfer and record data in a computer for easy data analysis.

2. Research Methodology

2.1. Circulating Water Channel (CWC)

In the present study, the performance of the design with one (single) and two (double) intake of PVCS is compared. The main parameters to compare are flow velocity and capacity entering the intake. The main method adopted in the present study is an experimental method using a tank facility known as Circulating Water Channel (CWC). The tank facility belongs to Marine Structure System Laboratory, Marine Engineering Departement, Engineering Faculty, Hasanuddin University, Indonesia. The concept of the CWC is illustrated in Fig. 2.

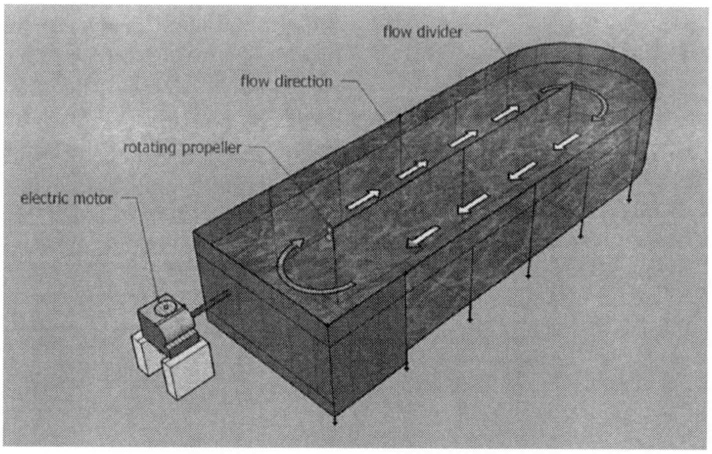

Figure 2. The sketch illustrating the concept of CWC [6]

As illustrated in Fig. 2, in CWC, the water is circulated using a propeller. The propeller is assumed to be the boat propeller. The propeller is driven by an electric motor placed outside the tank. The use of an electric motor allows easy and convenient motor revolution adjustment by using an automatic voltage regulator (AVR). AVR is used to regulate the voltage reaching the motor so that the motor revolution will reduce and increase according to the voltage from AVR. A photo of the CWC facility is shown in. Fig. 3.

Figure 3. Photo of CWC facility

The total length of the CWC tank is 9.6 m, width 2.4 m, and height 1.5 m. It can accommodate around 25 tons of water. The main materials of the tank wall are iron plate and some sections are made from transparent acrylic to allow easy observation of the phenomena and interaction occurred inside the tank.

2.2. Flow Measuring System

In order to measure the flow velocity and capacity, a flow measuring system was designed and assembled. The main components of the system are a microcontroller Arduino Uno and a flow sensor. Those components are shown in Fig. 4.

(a) Microcontroller Arduino Uno

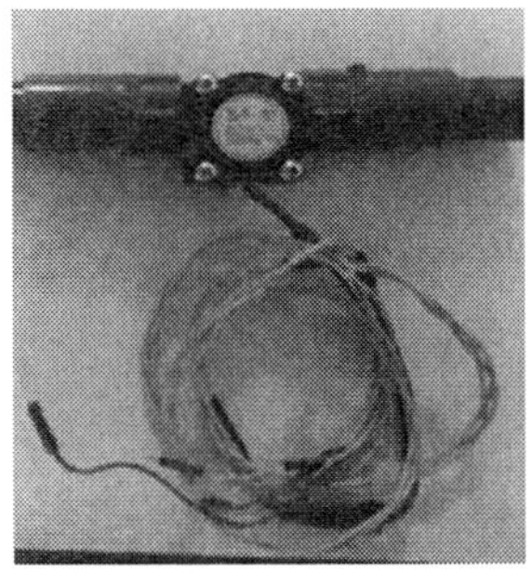

(b) Flow velocity sensor Sea YF-S201

Figure 4. The main components of PFCS flow measuring system

Using the components, the system is assembled using the following wiring scheme.

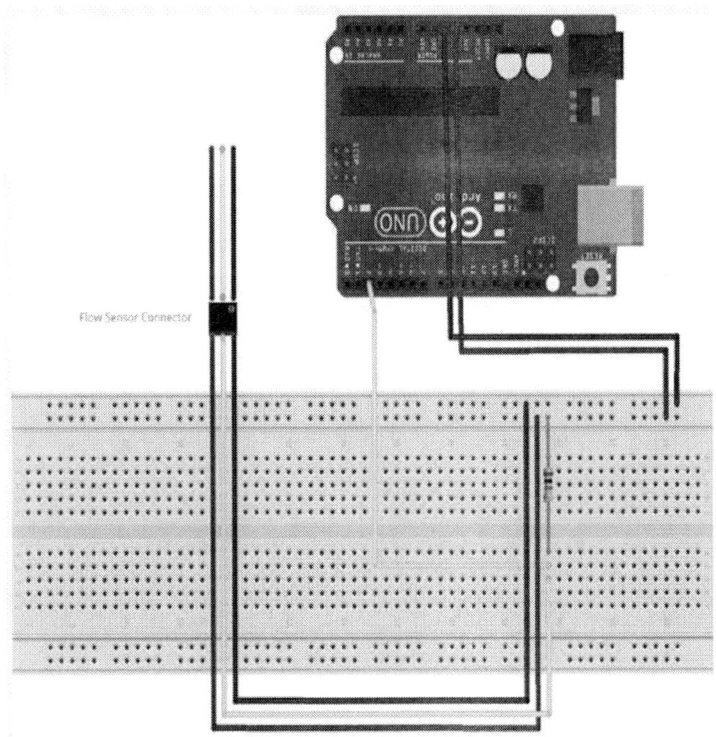

Figure 5. The flow measuring system wiring connection scheme

In order to connect to record the data, a program interface using Microsoft Visual Basic was built. The interface of the program is shown in Fig. 6.

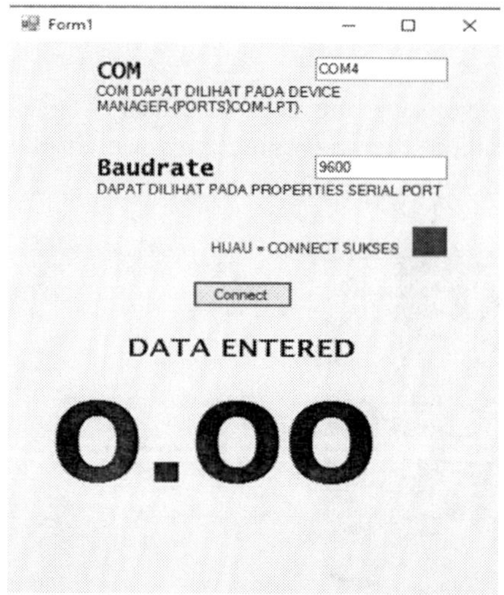

Figure 6. The interface of the data collection program

As can be seen in Fig. 6, before starting to use the program, the user needs to determine COM and Baudrate parameters. The value of COM needs to be entered can be found in the device manager of the computer used while Baudrate value can be seen on Serial Port Properties. The common values are COM4 and 9600 for COM and baud rate text boxes, respectively.

Another component inside the interface is a square sign. If the sign is colored red, it means the program still not connected or fails to connect but if the sign is colored green, it means the program has successfully connected. The large size number at the bottom interface is the value recorded to the computer. After the data collection finishes, the data can be extracted to software Microsoft Excell or other similar software. The flow velocity measuring system was validated before used in the experiment by comparing the results from manual and automatic measurements. A good agreement between these results was found which could ascertain the accuracy of the measurement system.

2.3. Shape Description

There are 2 (two) inlet pipe shapes that were tested in the present study. They are circular and elliptical shapes. Their sketch and dimension are shown in Fig. 7.

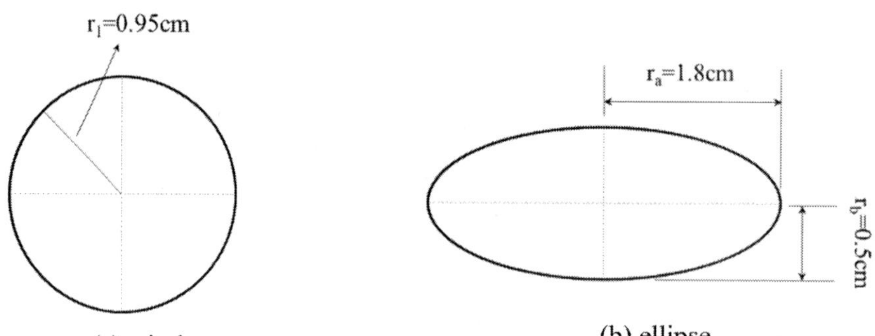

Figure 7. Intake shapes dimension

As shown in Fig. 7, both of the shapes have the same cross-sectional area which is equal to 0.00028 m^2. The ellipse has an aspect ratio of 0.278 which is defined as the ratio between its vertical radius with its horizontal one.

2.4. Intake Designs

There are 2 (two) intake designs analyzed in the present study. They are shown in Fig. 8. The first shape has only 1 (one) intake pipe and was placed in the upper 0.7R part of the propeller. The second design has 2 (two) inlet pipes placed in the upper and lower 0.7R part of the propeller. The horizontal distance of each inlet pipe to the propeller blade was 0.1 m.

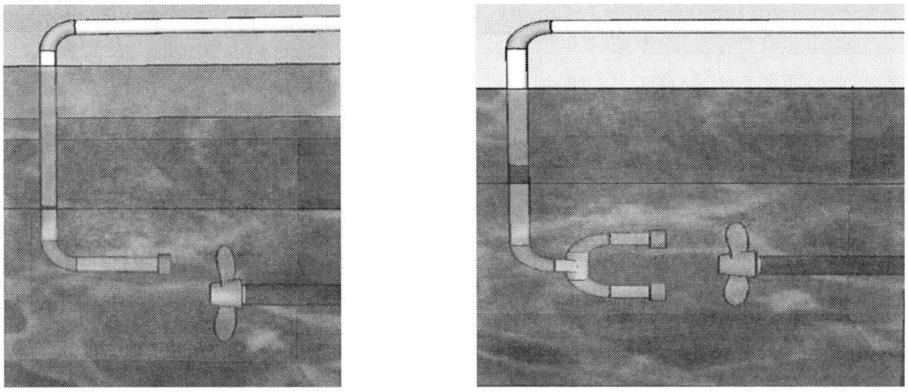

(a) Single intake (b) Double intake

Figure 8. PVCS intake designs

3. Results Analysis and Discussion

The performance validation using numerical and experimental methods between circular and elliptical intake shapes had been conducted by Mahmuddin et al. [6]. Therefore, in the present study, only results from the experiment are shown and discussed.

3.1. Circular Intake

Based on the experiment models and setup described in the previous section, the first experiment was conducted using the circular intake as shown in Fig. 8(a). The measured parameters were flow capacity and the propeller revolution. 4 (four) various propeller revolutions were tested. They were 900, 1100, 1300, and 1500 rpm. The experimental results for circular intake shape are shown in Fig. 9.

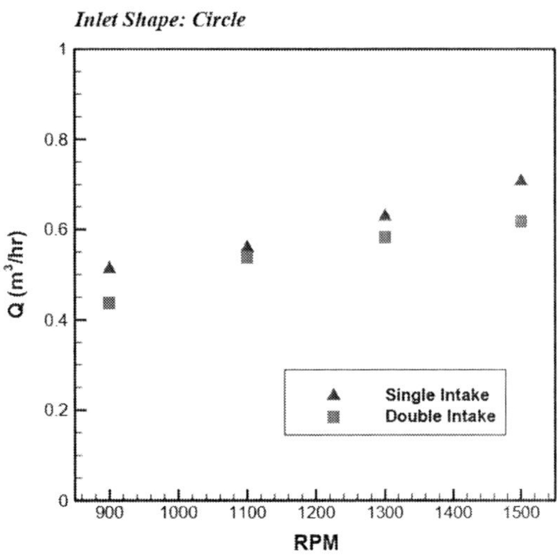

Figure 9. Flow capacity of circle pipe for single and double intake designs

In Fig. 9, the *x*-axis of the plot is the motor revolution (RPM) while the *y*-axis is the flow capacity (Q) in m^3/hr. It can be seen from the plot that a single intake design has a higher flow capacity than the one with double intake for all RPM values. This phenomenon can be attributed to higher flow loss in double intake design due to friction inside the pipe. Moreover, an additional of 2 (two) elbow joints needed in double intake design as shown in Fig. 8(b) is also contributed to these higher losses.

3.2. Elliptical Intake

The next run of the experiment was conducted by measuring the flow capacity for elliptical intake which has shape and dimension shown in Fig. 8(b). The experimental results for this case are shown in Fig. 10.

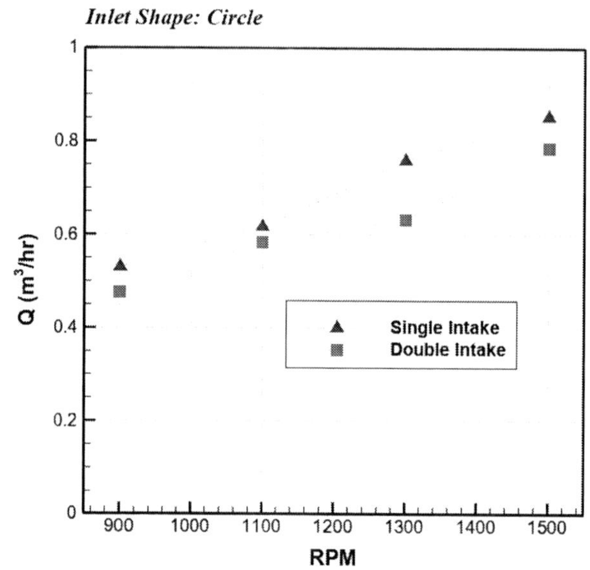

Figure 10. Flow capacity of elliptical pipe for single and double intake designs

As can been from Fig. 10, for this case, a single intake design also has a higher capacity than the double intake one. The main causes of this trend have been explained in the previous case. From this figure, it can also be noted that the performance of elliptical intake is higher than the circular one both for single and double intake designs which is also consistent with the conclusion drawn from the previous results of Mahmuddin et al. [6]. Moreover, it can also be observed that the maximum flow capacity in the single intake is 0.85 m^3/hr while in double intake is 0.78 m^3/hr or approximately 8.97% difference. It is also interesting to note that the flow capacity differences between circular and elliptical intakes are 19.71% and 21.79% for single intake and double intake cases, respectively.

4. Conclusion

In order to compare the performance of 2 (two) different designs of a Propeller Flow Cooling System (PVCS), the present study had conducted experiments in the Circulating Water Channel (CWC) facility. A flow monitoring system including its monitoring interface had also been designed and assembled. The measurement system was used to measure the flow capacity of the PVCS in the experiments. The experiments were conducted for circular and elliptical intakes cases for various propeller revolution. From the experiments, it was found that the design with the single intake has a better performance compared to the system with double intake one. The results showed that flow losses were due to surface friction and elbow joint which was significant in reducing the flow capacity in double intake design. The tendency occurred for both circular and elliptical intake cases.

Acknowledgment
The authors gratefully acknowledge the research funding from JICA C-BEST project of Engineering Faculty, Hasanuddin University for this research.

References

[1] P. S. Goud, M. Narsimha, V. R. Goud, and T. C. Shekar, "Effect of High Engine Temperature on Performance and Characteristics of Ic Engine," no. 2, pp. 125–131, 2017.
[2] W. W. Pulkrabek, *Engineering Fundamentals of the Internal Combustion Engine, 2nd Ed.*, vol. 126, no. 1. 2004.
[3] F. O. R. Aeronautics, "The Effect of High Temperature of the Cylinder head on the Knocking Tendency of an Air-Cooled Engine Cylinder," 1945.
[4] N. Watson and M. S. Janota, *Turbocharging the Internal Combustion Engine*. Macmillan Education UK, 1982.
[5] M. Ahmad, H. Ied, and Nofrizal, "Ship Engine Installation Equipment," *J. Perikan. dan Kelaut.*, vol. 2, pp. 191–197, 2009. [in Bahasa]
[6] F. Mahmuddin, S. Ramadhan, A. Fitriadhy, and S. Klara, "Performance Comparison between Ellipse and Circular Intake Shapes of Propeller Flow Cooling System (PFCS) with Numerical and Experimental Methods," *IOP Conf. Ser. Mater. Sci. Eng.*, vol. 676, no. 1, 2019.
[7] R. Mukaro and X. F. Carelse, "A Microcontroller-Based Data Acquisition System for Solar Radiation and Environmental Monitoring," *IEEE Trans. Instrum. Meas.*, vol. 48, no. 6, pp. 1232–1238, 1999.
[8] N. N. Mahzan, A. M. Omar, L. Rimon, S. Z. M. Noor, and M. Z. Rosselan, "Design and Development of an Arduino Based Data Logger for Photovoltaic Monitoring System," *Int. J. Simul. Syst. Sci. Technol.*, vol. 17, no. 41, pp. 15.1-15.5, 2017.
[9] M. Demirtas, I. Sefa, E. Irmak, and I. Colak, "Low-cost and High Sensitive Microcontroller Based Data Acquisition System for Renewable Energy Sources," in *2008 International Symposium on Power Electronics, Electrical Drives, Automation and Motion*, 2008, pp. 196–199.
[10] F. Mahmuddin, A. M. Yusran, H. Hasan, and S. Dewa, "On the Use of an Arduino-based Controller to Control the Charging Process of a Wind Turbine," vol. 3, pp. 1–5.

IOP Conf. Series: Materials Science and Engineering **875** (2020) 012086 doi:10.1088/1757-899X/875/1/012086

EPIC (Education Priority Compact) Strategy as A Solution for Education of Post-Disaster Refugee Children in Palu City, Central Sulawesi

R Mahful[1]*, A Algifari[2], J Jacobus[2], and A Haqiq[2]

[1] Urban Regional and Planning, Universitas Sulawesi Barat, Majene, Indonesia
[2] Urban Regional and Planning, Universitas Tadulako, Palu, Indonesia

*E-mail: rmahful@gmail.com

Abstract. Disaster victims' children are required to live in temporary shelters provided by Non-Government Organizations (NGOs) and the government. In the first phase, 699 units have been built in 72 locations. The focus of the government in the first three months after the earthquake and tsunami in Central Sulawesi was to rescue the emergency and clean up the area after that the focus then began to shift to improving the economy of the affected area. In handling post-disaster, the government does not prioritize education as its main aspect, so many of the victims of disasters do not go to school until they drop out of school. Focus Group Discussions conducted in several HUNTARA locations in the city of Palu, resulted in education being one of the concerns of the people living in temporary shelters. The community is concerned about the sustainability of their children's school education, due to the absence of educational facilities in the shelter as well as the community's economy that has fallen due to lack of work is exacerbated by the government's lack of attention to the education of children in refugee camps / HUNTARA. These methods are Focus Group Discussion (FGD) that has the main target of refugee communities. On the other hand, researchers also the distribution of questionnaire in several refugee locations in the city of Palu, in order to obtain accurate data for supporting this research. By quantitative SWOT methods and AHP Analysis to determine priorities in determining the EPIC (Education Priority Compact) Strategy which is expected to be one of the solutions for post-disaster education in Central Sulawesi.

1. Introduction

September 28, 2018 was one of the biggest disasters in the world that occurred in Central Sulawesi, Indonesia. Material losses are estimated at around Rp. 18.48 trillion and claimed 4,340 lives [1]. According to the report on the finalization of data and information on the earthquake, tsunami and liquefaction disaster in Central Sulawesi dated December 20, 2018, it showed that 386 schools were affected, 107 schools were severely damaged, 114 schools were moderately damaged, and 165 schools were slightly damaged. As for the condition of students and teachers, 183 students died, 171 students were lost, 22 students were seriously injured, and 27 teachers died, 33 teachers were lost, and 8 teachers were seriously injured [2].

In disaster situations, often education services are not the main concern. Even though, it is very clear, education is the social sector most affected by disasters. Therefore, efforts to pay more attention to the world of education must begin. This can be realized by ensuring there is access and learning environment, sustainability of learning activities, availability of teachers and education personnel, adequate educational policies accompanied by strong coordination between institutions, community participation and assessment of educational needs, response strategies and monitoring and evaluation [3].

In post-disaster situations, damage to educational facilities and infrastructure including transportation infrastructure as well as situations such as the child's family life and abnormal family of teachers can cause the regular teaching-learning process to cease. In addition, the disruption of the economy due to the disaster also puts children in a vulnerable position to drop out of school. Based on this condition, alternative education programs provided by stakeholders will greatly assist child victims. Emergency school programs, drawing programs, storytelling [4].

Therefore, this research tries to design a new strategy called EPIC (Education Priority Compact) which involves three parties including the Government, NGOs (Non-Government Organization), and the Community. To get a solution for post-disaster education.

2. Literature review

2.1. Post Disaster Education

The low level of education in post-disaster children makes the family a productive place in improving children's education in the aftermath of the term. The term that is often used as a substitute for the designation of the educational environment in the family is "informal education", is part of the path of implementing education in Indonesia, in contrast to formal education carried out in a structured and tiered manner. Non-formal education is held for citizens who need educational services that function as substitutes, additions, and/or complementary formal education in order to support lifelong education [5].

2.2. Fun Learning Method

The world of education is a dynamic world and needs to update new things continuously [6].Thus was born the fun learning strategy developed by Bobby DePorter by giving positive suggestions. a fun learning strategy (Fun learning) is a strategy used to create an effective learning environment, implement a curriculum, deliver material, facilitate the learning process that results in improved student learning achievement, learning with a fun system is very necessary in the learning process, because it is very helpful for students to be able to make learning materials meaningful, motivate learning, and provide learning satisfaction. Because fun learning will make children feel unburdened and forced to learn [7].

3. Method

3.1. Questionnaire

In this study, researchers used an open questionnaire. Open questionnaire is a questionnaire that gives freedom to the object of research to answer. With the aim that the data obtained is accurate data in accordance with the conditions that occur. The object of this research is the people living in Temporary Shelter (HUNTARA). The number of questionnaires distributed in the community was 213 questionnaires in three Huntara, including: Petobo Huntara, Huntere Lere, and Silae Huntara.

3.2. Focus Group Discussion (FGD)

In this study Focus Group Discussion was conducted in 3 places, namely Pantoloan Village, Petobo Village and Panau Village. The author uses FGD to find out post-disaster education problems that occur in the city of Palu, which later will be the main concern that will be addressed.

4. Results and analysis

4.1. Focus Group Discussion (FGD)

Based on the results of the Focus Group Discussion that the fulfilment of post-disaster education in the Palu area is still lacking. There are three main areas of concern in post-disaster education issues, namely: Pantoloan, Petobo and Panau. In Pantoloan Village, after the disaster, many children dropped out of school, according to Pak Mukades, the head of RT 2 RW 2 Petobo Village that Children in Huntara need education. In Huntara there are only elementary schools, the other schools are far away. The condition of school zoning also hampered the implementation of post-disaster education as was the case in the Panau village where the community wanted that the government could change the rules related to zoning for schools because currently children living in the Panau village had to go to Kayumalue village. As a result, poor people cannot afford to pay for school and eventually drop out of school.

4.2. Analysis of the description based on the results of the questionnaire

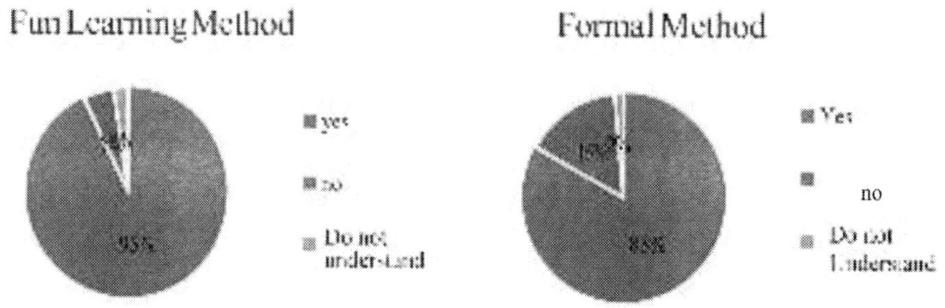

Figure 1. Comparison of Fun Learning and Formal Methods

The analysis shows that people who live in temporary shelters (huntara) 93% of the people have the view that non-formal education with fun learning methods after disasters is effective to apply, when compared with the application of formal methods in post-disaster education shows that 83% of people who agree with the method.

Other results show that 88% of the community respondents stated that the learning that was carried out was in accordance with the child's developmental age. 84% of the community thinks that it takes more than three months to eliminate trauma after a disaster, vulnerable learning time in a week is obtained that by 51% the public has the perception that four to five times a week is vulnerable time that should be done for learning to fill in the blanks of time.

As with the person in charge of implementing education, about 56% chose the government as an important role in children's education after disasters. 73% of the community stated that there had been a transparency in the administration of emergency education units. 75% of the people think that the stakeholders have provided programs that support education optimally. 56% of the people think that there are enough teachers in the post-disaster period. 51% of the people think that the government does not recruit teachers after disasters. 77% of the community thinks that the community has participated and contributed in the implementation of the education unit. In addition, according to the public perception, 85% of respondents thought that the teaching staff who taught after the disaster consisted of elements of education and psychologists.

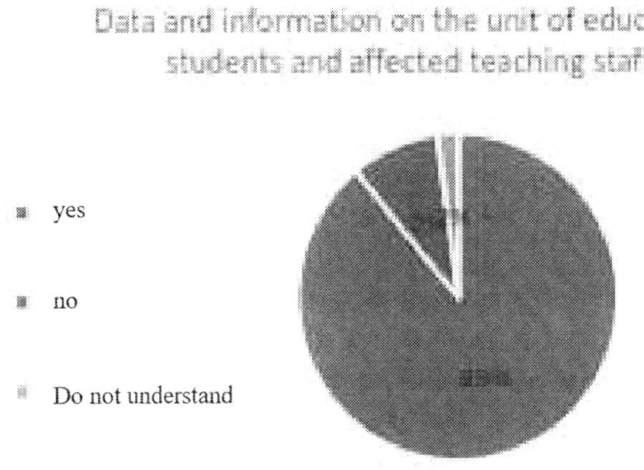

Figure 2. Data and information on the unit of educators, students and affected teaching staff

From the results of the analysis based on information and data related to education units, students and teachers affected by the disaster, it was found that about 89% of the public thought that they had obtained the data and information after the disaster. 94% of post-disaster children have the same learning opportunities without any difference. 93% access or route to safe learning environments.

The conclusion from the results of the questionnaire distributed to respondents is that the community wants fun learning methods to be applied after disasters before entering the school age of children (formal learning) and in the education process the community thinks it must be distributed according to its age development and the time period needed is more from three months which is done three to four times a week.

4.3. SWOT analysis

Table 1. SWOT analysis table

EKSTERNAL / INTERNAL	OPPORTUNITY 1. New methods for post- disaster education. 2. Learning to be entertainment for children	THREATS 1. Vulnerable to the child's psychological recovery time
STRENGTHS 1. Improving the Quality of Post-Disaster Education. 2. Psychological recovery of children after disaster. 3. Children are motivated to learn again. 4. The fulfillment of children's education during the reconstruction period	SO Strategy The establishment of cooperation between the government, NGOs and the community in improving the Quality of post-disaster education	ST Strategy Provision programs stakeholders of from educational local
WEAKNESS 1. Teacher quantity and capacity	WO Strategy Conducting recruitment for teachers after a disaster	WT Straregy Conduct training for teachers on psychological recovery of children.

Based on the analysis table above. SO (Strengths, Opportunity), ST (Strengths, Threats), WO (Weakness, Opportunity), WT (Weakness, Threat) which find external and internal factors indicate that SO strategy is the formation of cooperation between government, NGOs and the community in improving quality post disaster education. The ST Strategy is the provision of educational programs from local stakeholders. WO's strategy is to recruit teachers after the disaster. The WT strategy is the quality of the instructor towards the psychological recovery of children.

4.4. Analisis AHP

Metode	Belajar sambal bermain	Belajar sambal menyanyi	Belajar sambal melukis	Belajar sambal berdongeng	Belajar sambal membaca	Belajar tentang sejarah	Belajar tentang agama	Belajar Formal	Nilai egen								Jumlah	Rata
Belajar sambal bermain	110	0,008928571	110	110	110	110	110	110	0,495359	8,6626053	0,247723578	0.168189	0.199625	0.13431	0,32733	0,148847	1,721471	0,215184
Belajar sambal menyanyi	112	103	112	112	112	112	112	112	0,504366	9,9931814	0,252227643	0,171247	0.203254	0,136752	0,333281	0,151554	2,752000	0,344
Belajar sambal melukis	0,009090909	0,009259259	108	108	108	108	0,009259259	108	0,000041	8,9834425	0,243219513	0,165131	0,195995	0,131868	2,76E-05	0,146141	0,882514	0,110314
Belajar sambal berdongeng	0,009708738	0,009708738	0,009708738	103	0,009708738	103	0,009708738	103	0,000044	9,4195320	2,18644	0,157487	1.76	0,125763	2,89	0,139375	0,422831	0,052854
Belajar sambal membaca	0,009345794	0,009345794	0,009345794	107	107	107	0,009345794	107	0,000042	9,0673999	2,1047	0,163602	0,19418	0,130647	2,78	0,144788	0,633399	0,079175
Belajar Belajar tentang sejarah	0,0125	0,0125	0,0125	0,0125	0,0125	80	0,0125	0,0125	0,000056	1,2127647	2,81504	1,91	2,27	0,09768	3,72	1,69	0,097982	0,012248
Belajar tentang agama	0,00877193	0,00877193	114	114	114	114	114	114	0,000040	8,5106297	0,256731709	0,174305	0,206884	0,139194	0,339233	0,15426	1,270732	[redacted]
Belajar Formal	0,011764706	0,011764706	0,011764706	0,011764706	0,011764706	85	0,011764706	85	0,000053	1,1414256	2,64945	1,8	2,14	0,103785	3,5	0,115018	0,219071	0,027384
Jumlah	222,0611821	103,070279	444,0433192	654,0242647	551,0339734	819	336,0525785	739,0125										1

Based on the results of AHP analysis shows that learning while playing is the method most interested by respondents, then learning while singing and learning about religion.

4.5. EPIC (Education Priority Compact) Strategy

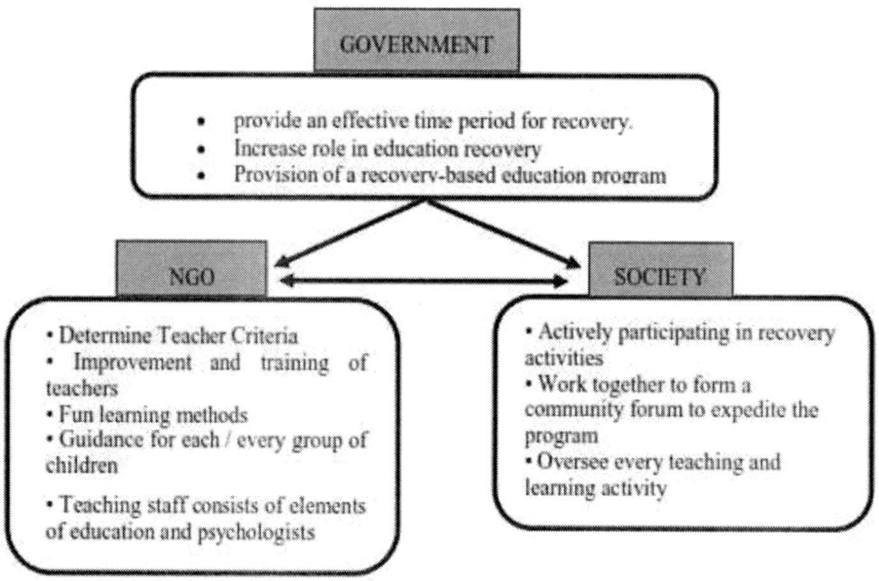

The EPIC (Education Priority Compact) Strategy is a new strategy that brings together communities, governments and NGOs to form a synergy in meeting post-disaster education. The government as stakeholders functions as the party who is fully responsible for the implementation of education. NGOs as subjects directly descend in the intervention of the community in post-disaster fulfillment. Society as a subject and object in the fulfillment of post-disaster education. These three elements are reciprocal in fulfilling post-disaster education.

5. Conclusion

From the results of research conducted in several Temporary Resettlement (HUNTARA) Related to Post-Disaster Education, the authors conclude that Firstly, From the results of the FGDs, it was found that there were still many children who dropped out of school, did not go to school because there were no facilities or because there was a system that made it difficult for them to go to school. It is hoped that this can be overcome immediately so that there are no more children who do not receive education because of the disaster. Secondly, Fun Learning Method is a good method to be applied in the fulfillment of post-disaster education in Palu City. The results of the Questionnaire that was submitted to community respondents living in HUNTARA showed that 93% of the community's opinion chose the Fun Learning method in learning that was carried out post-disaster. Lastly, EPIC strategy can be a solution in handling post-disaster education. A strategy that involves the participation of the Government, NGOs, and the community to work together in meeting the educational needs of the affected location.

References

[1] Sangadji, Ruslan. 2019. Central Sulawesi Disaster killed 4,340 People, final count reveal. In Jakarta Post, January. Jakarta.
[2] Information and data of natural earthquake disaster, tsunami and liquidify in the central sulawesi government, 2018. Data Finalization Report and Earthquake Disaster Information, Tsunami and Liquidation in Sulawesi Middle as of December 20, 2018.
[3] BNPB. 2017. Data Informasi Bencana Indonesia
[4] Absor, M. Ulil. 2011. Penanganan anak dalam masa tanggap darurat bencana alam: Tinjauan Konvensi Hak Anak dan Undang-Undang Perlindungan Anak. Volume XI nomor 1 Tahun 2011
[5] Sunarti, E., Islamia, I., Rochimah, N., & Ulfa, M. (2018). Resiliensi Remaja: Perbedaan Berdasarkan Wilayah, Kemiskinan, Jenis Kelamin, dan Jenis Sekolah. *Jurnal Ilmu Keluarga & Konsumen*, *11*(2), 157-168
[6] Chatib, Munif. 2013. Gurunya Manusia: Menjadikan semua anak Istimewa dan semua anak juara. Cetakan XI. Bandung: Kaifa PT Mizan Pustaka
[7] Darmansyah, S.T., M,Pd. (2010). Strategi Pembelajaran Menyenangkan Dengan Humor. Jakarta: Bumi Aksara

Acknowledgement

My best regards for everyone that involve in this research, special for Universitas Sulawesi Barat and Universitas Tadulako for joining combine research.

IOP Conf. Series: Materials Science and Engineering **875** (2020) 012087 doi:10.1088/1757-899X/875/1/012087

Fuzzy Logic Method as Determining of Internet Network VSAT performance in Black Spot Region at District of Enrekang

S Buwarda[1]*, K H Rasyid[1], Muhlis[1], and M Hidayat[1]

[1]Department of Electrical Engineering, Fajar University, Makassar, Indonesia

*Email: sukriyah.warda@gmail.com

Abstract. Internet networks using VSAT have some disadvantages caused by the relative distance of the satellite and the earth which is a significant propagation delay. To determine the optimal quality of VSAT internet network, a study of VSAT internet network performance was performed using Fuzzy Logic method, based on the parameters of bandwidth, delay, packet loss, and throughput conducted for 2 days in the morning, afternoon and night in 2 villages in Enrekang Regency, namely Singki Village and Tindalun Village. Retrieval of data parameters at these locations using a wireshark application software which is then inputted into fuzzy logic so as to produce three categories of performance, good or sufficient, and bad network quality. It is said that performance is good if the results on fuzzy logic are 0, sufficient performance compilation of results from fuzzy logic is 0.5 and performance is poor if the results of fuzzy logic are valued 1. The VSAT internet network requests optimal compilation parameters of network quality such as 6 mbps bandwidth, delays below 1,000 ms, packet loss less than 50% and throughput above 50 kbps which has been designed in fuzzy logic. From the results of the study, obtained a value for the village of Tindalun 0.5 which means enough. Whereas, for performance in Singki Village, 2 different values were obtained, while compilation using Wifi was valued at 0.5, which was sufficient, and compilation using WireLAN was 0 which meant good.

1. Introduction

VSAT technology has several weaknesses due to the relative distance of the satellite and the earth resulting in significant propagation delay and the ability of satellites to send and receive data vulnerable to weather disturbances [1]. To find out the quality of the VSAT network, a study of VSAT internet network performance must be carried out in the blank spot area Enrekang Regency with the altitude and valley topography, until now has many problems related to telecommunications networks, even from 12 existing districts, in almost all districts there are several villages with blank spots or no signal. VSAT internet network performance in the blank spot area emphasizes the process of monitoring and calculating network quality parameters, in this case the parameters used to determine the performance of the VSAT internet network are bandwidth, delay, packet loss and VSAT network throughput.

2. Literature Review

Very Small Aperture Terminal is a satellite signal receiving station with a dish antenna receiving a diameter of less than three meters [2]. The main function of VSAT is to receive and send data to satellites. The satellite functions as a successor to the signal to be sent to other points on earth [3]. Actually the VSAT dish faces a geostationary satellite. Geostationary satellites are satellites that are always in the same place in line with the rotation of the earth on its axis which is possible because it orbits at the same point above the earth's surface and follows the earth's rotation on its axis. VSAT (Very Small Aperture Terminal), originally a trademark for small earth stations that was marketed around 1980 by Telecom General in America. In its free translation, VSAT can be interpreted as a transmitter and receiver of satellite transmission terminals that are scattered in many locations and are connected to the central hub via satellites using certain diameter satellite dishes [3].

2.1. Bandwidth

Bandwidth is the width or width of the frequency range used by the signal in the transmission medium. in computer networks, bandwidth is often used as a synonym for data transfer speed (transfer rate), which is the amount of data that can be carried from one point to another within a certain period of time (generally in seconds). This type of bandwidth is usually measured in bps (bits per second). Sometimes also stated in Bps (bytes per second) [4].

2.2. Delay

Delay time of a packet caused by the process of transmission from one point to another destination. These points can be computer devices, or other network devices such as routers, modems and so on which are passed by information packets. One Way Delay (OWD) is the time taken by a package from the source to the destination. The time from source to destination returns to the source is called Round Trip Time (RTT). The type of delay measured on the VSAT network are (1) Propagation delay is the time required by the information signal to move in communication media such as cable, optical fiber, microwaves and satellites, (2) Transmission delay is the time needed for a system to pass a number of data packets, and (3) Queue delay is the length of time it takes for a data packet before the packet is forwarded to its destination. ITU-T (International Telecommunication Union - Telecommunication) version of the standard delay value as follows [2].

Table 1. ITU-T version of delay standardization

No	Latency Category	Delay
1	Good	<150 ms
2	Enough	150 s/d 400 ms
3	Bad	>400 ms

2.3. Packet Loss

Packet Loss is defined as the failure of data packet transmission to reach its destination. The failure of the package to reach its destination, can be caused by several possibilities, including occurrence of traffic overload in the network, collision (congestion) in the network, errors that occur on physical media and failures that occur at the receiving end can be caused, among others, due to overflow that occurs in the buffer. In network implementation, the packet loss value is expected to have a minimum value. In general, based on packet loss values according to the ITU-T version there are three categories of network quality degradation based on the standard packet loss values as follows.

Table 2. ITU-T version of Packet Loss standardization

No	Degradation Category	Delay
1	Good	3 %
2	Enough	15 %
3	Bad	25

2.4. Throughput
Throughput is a measure of the speed at which data can be sent across the network in bits per second (bps). The capability of throughput in sustaining hardware (hardware) is called bandwidth. In fact, the term bandwidth is sometimes used as a synonym of throughput. In determining the quality of data communication networks, especially internet networks, there are two important things, namely the amount of delay and speed of a data packet to pass through a network, and the adequacy of available network bandwidth.

2.5. Fuzzy Logic
n many cases, fuzzy logic is used as a way to map the problem from the input to the expected output. Fuzzy logic can be considered as a black box (black box) that connects the input space to the output space. Fuzzy set has two attributes, namely linguistic and numerical. Linguistic attribute is an attribute used for naming a group that represents a certain state or condition by using natural language, such as young, middle age, old age. While the numerical attribute is a value that indicates the size of a variable. There are several things that need to be known in understanding fuzzy systems such as fuzzy variables are variables discussed in a fuzzy system, fuzzy set is a group that represents a certain condition or condition in a fuzzy variable and entire universe is the total value that is allowed to be operated.

3. Research Method
At the data collection stage, a system study is carried out by conducting a literature study by reading and studying books, journals or articles related to the VSAT internet network, VSAT internet network performance and Fuzzy logic, besides observing directly to the field, namely the blank spot area in Enrekang district, precisely in Singki Village and Tindalun Village.

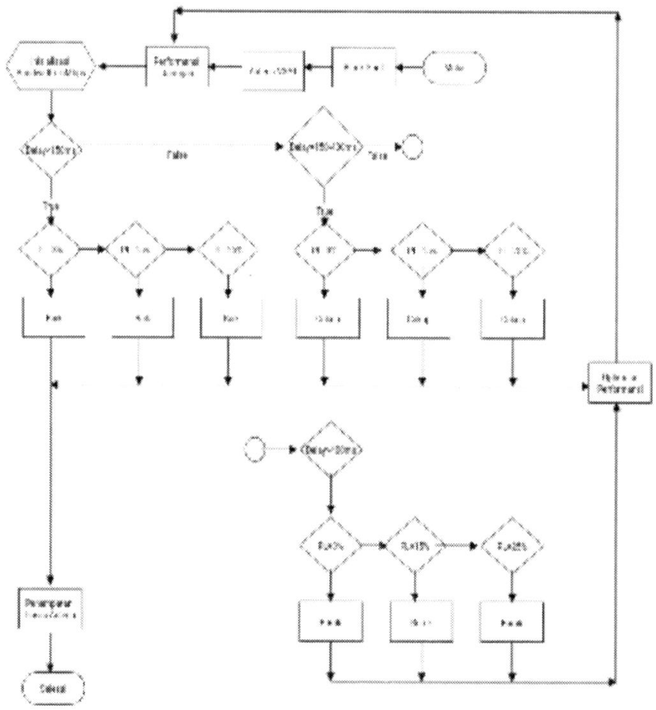

Figure 1. Flowchart of the fuzzy logic method to determine internet VSAT network quality

The material used in this study is the Windows 10 Pro operating system, Matlab 2016a [5] and Wireshark. Fuzzy logic is used as a way to determine the problems of input (VSAT network performance)) namely bandwidth, packet loss, delay and output to output. The application of VSAT internet network based on fuzzy logic to determine antenna position in the field. The optimal performance of VSAT internet network is said to be optimal if it is in accordance with the performance standards of the VSAT internet network itself, which is a delay of <150ms and a packet loss of 3%. Figure 1 shows the flowchart of the fuzzy logic method system used in this research.

4. Result and Discussion

At the data collection stage, a system study is carried out by conducting a literature study by reading and studying books, journals or articles related to the VSAT internet network, VSAT internet network performance and Fuzzy logic, besides observing directly to the field, namely the blank spot area in Enrekang district, precisely in Singki Village and Tindalun Village. The data obtained using the wireshark application in the two villages shows in Table 3.

Table 3. Data Collection results from Singki village and Tindalun village

Location	Parameter	Youtube			Gmail		
		Morning	Afternoon	Night	Morning	Afternoon	Night
Tindalun Wifi	Bandwidth (Mb)	6	6	6	6	6	6
	Delay (m/s)	15874	5260	26680	6898	6840	53845
	Packet Loss (%)	0	0	0	0	0	0
	Througphut (Kbps)	124 k	310 k	120 k	641 k	645 k	14 k
Tindalun Lan	Bandwidth (Mb)	6	6	6	6	6	6
	Delay (m/s)	17023	10200	16229	24402	80910	60507
	Packet Loss (%)	0	0	0	0	0	0
	Througphut (Kbps)	347 k	967 k	484 k	185 k	20 k	71 k
Singki Lan	Bandwidth (Mb)	6	6	6	6	6	6
	Delay (m/s)	13488	11923	17381	15596	14165	16199
	Packet Loss (%)	0	0	0	0	0	0
	Througphut (Kbps)	1124 k	194 k	214 k	832 k	619 k	507 k
Singki Wifi	Bandwidth (Mb)	6	6	6	6	6	6
	Delay (m/s)	5247	9796	3404	5619	6266	5800
	Packet Loss (%)	0	0	0	0	0	0

Based on data from the test results in Singki Village, anggeraja sub-districts using either WiFi or LAn networks, the parameters obtained are bandwidth, delay, packet loss and throughput. The test results are entered into fuzzy logic which will determine whether the quality of the VSAT internet network in the area is in bad, moderate or good condition according to the parameters defined earlier. Table 4 shows the data of the parameter testing results using Wifi and LAN in Singki village

Table 4. Testing parameters using WiFi and LAM connections

Parameter	Youtube		Gmail	
	WiFi	LAN	WiFi	LAN
Bandwitdh	6.13	6.13	6.13	6.13
Delay	676.614	98981,79133	46126001,7	227880,1
Paket Loss	84%	96%	60%	0.3%
Throughput	16.0	12.1	6.3	26

The fuzzy logic display used for the four parameters of VSAT internet network quality is shown in Figure 2.

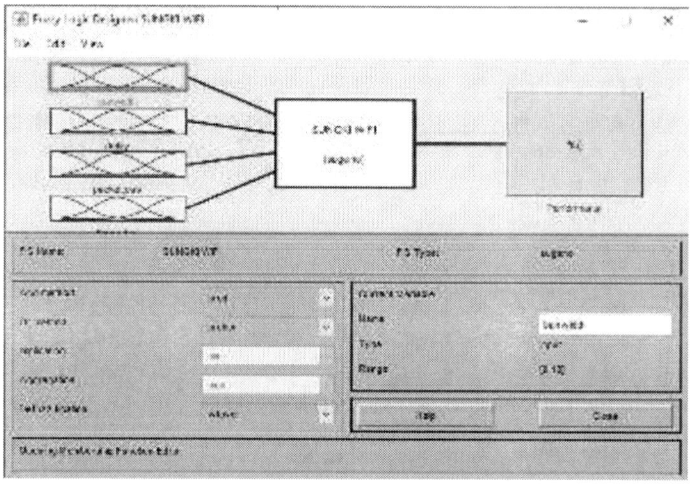

Figure 2. VSAT internet network quality data input of fuzzy logic

Parameters that indicate the quality of the VSAt internet network in the village of Singki will be a measure of the merits of the existing network in the area. If poor quality is obtained, then the antenna position is moved to obtain a good network quality. The parameters of bandwidth, delay, packet loss and throughput are shown in Figure 3.

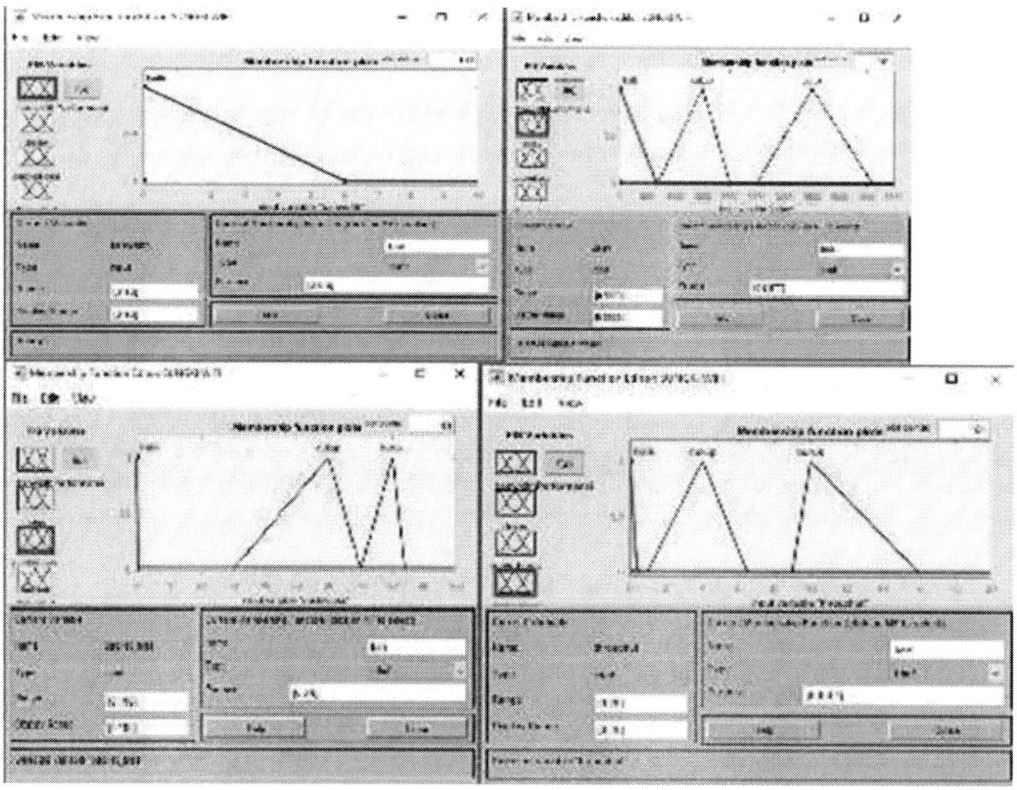

Figure 3. Plot result of bandwidth, delay, packet loss and throughput using WiFi connection

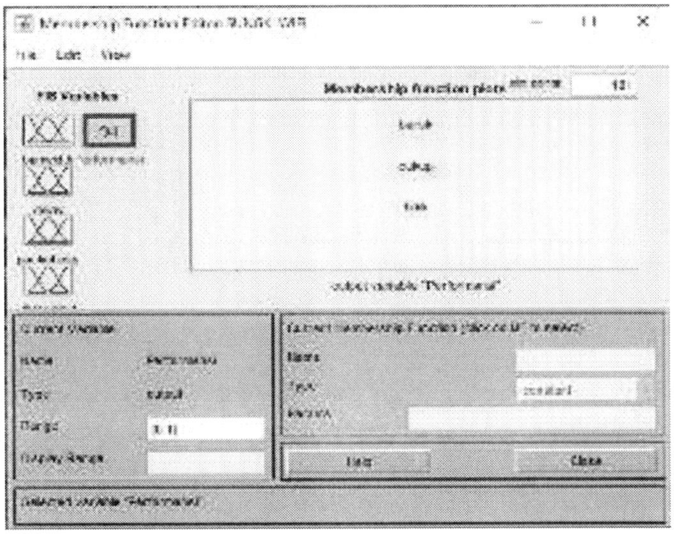

Figure 4. Output performance in fuzzy logic using a wifi connection in the village of Singki

The final result that will be obtained from the fuzzy logic application in this study is to provide convenience in determining optimal performance on a VSAT internet network. Determination of the optimal parameters is shown in the Figure 5.

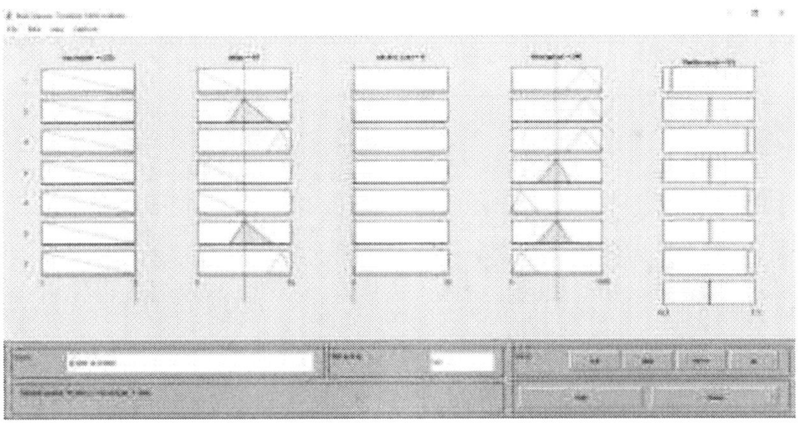

Figure 5. Fuzzy logic performance results

From the results of inputting data in the field which has been processed by Fuzzy Logic where there are differences in the results of network performance, namely in Desa Tindalun using WireLAN (youtube) which gets 0.5 results, which is sufficient. Whereas, in Singki Village using WireLAN (youtube) got 0 which means good. This proves that in Singki Village the quality of the VSAT internet network is more optimal by using the same connection (WireLan) and running the youtube website.

5. Conclusion

The delay on the VSAT internet network is indeed quite high, because it uses satellite transmission within 36000 km from the earth. So the process of sending and receiving data takes quite a long time. From the observations on fuzzy logic there are three categories of performance or network quality, namely good, sufficient, and bad. It is said that the performance is good if the result of fuzzy logic is 0, while the performance is sufficient when the result of fuzzy logic is 0.5 and the performance is bad if the result of fuzzy logic is 1. Optimal performance is obtained when the user uses WireLAN.

References

[1] H. Ramonyaga, N. Tjahjamooniarsih, and F. T. P. W, "Analisis Performansi Very Small Aperture Terminal (VSAT) Pengiriman Data Cuaca Penerbangan Menggunakan Computer Message Switching System (CMSS)," pp. 1–9.

[2] A. Parlin, A. H. Rambe, and K. Jaringan, "Perbandingan Kinerja Jaringan Very Small Apeture Terminal Berdasarkan Diameter Antena Pelanggan di Pasifik Satelit Nusantara Medan," *Singuda Ensikom*, vol. 8, no. 1, pp. 1–6.

[3] K. Y. Germanus Yosef Laka, Luci Kanti Rahayu, "Instalasi dan Konfigurasi Jaringan VSAT Menggunakan Modem Gilat pada PT. Indo Pratama Teleglobal Jakarta," *J. Techno Nusa Mandiri*, vol. XII, no. 2, pp. 66–76, 2015.

[4] Tinaningrum Ari Susanti, "Analisa Kehandalan Jaringan VSAT IP Ditinjau Dari Delay, Data Rate dan Service Level," *Univ. Indones.*

[5] A. G. Salman, "Pemodelan Sistem Fuzzy dengan Menggunakan Matlab," *ComTech*, vol. 1, no. 9, pp. 276–288, 2010.

IOP Conf. Series: Materials Science and Engineering **875** (2020) 012088 doi:10.1088/1757-899X/875/1/012088

A Comparative Study using SAW, TOPSIS, SAW-AHP, and TOPSIS-AHP for Tuition Fee (UKT)

W Firgiawan[1]*, N Zulkarnaim[1], and S Cokrowibowo[1]

[1]Department of Informatics, Faculty of Engineering, Universitas Sulawesi Barat, Majene, Indonesia

*Email: wawanfirgiawan9@gmail.com

Abstract. Tuition Fee or UKT has a different group, in which each group has a percentage each of the which has been set by the government based on the economy of each student. The absence of clear standardization in determining Tuition Fee raises many problems. The approach of Decision Support Systems (DSS) will help in the decision making of Tuition Fee students are more effective and efficient. The purpose of this study research is to compare the 4 DSS methods used in the decision making process, Simple Additive Weighting (SAW), Technique for Order Preference by Similarity to Ideal Solution (TOPSIS), a combination of Simple Additive Weighting (SAW) and Analytical Hierarchy Process (AHP), and Technique for Order Preference by Similarity to Ideal Solution (TOPSIS) and Analytical Hierarchy Process (AHP) is used in making Decision determination Tuition Fee students in accordance with established criteria determine the method relevant to the problem. By using a comparison of four methods: SAW, TOPSIS, SAW-AHP and AHP-TOPSIS, the experimental results obtained using accuracy measure where the highest accuracy obtained by TOPSIS-AHP is 80%, then followed by the TOPSIS method 78%, SAW-AHP 74% and SAW 76%, so the TOPSIS-AHP is the best method used in determining Tuition Fee.

1. Introduction

The decision-making process is not easy, this is a problem that is faced every day by everyone carrying out their activities. This arises because of the many alternatives to decisions we will make based on the considerations that we do to produce the best decisions. In some cases, the many choices or alternatives available make us confused or difficulty in making decisions about the problem. Current technological developments can be utilized to provide recommendations or considerations for the decisions we will make.

In this study, the authors chose the problem of Determining Tuition Fee the which is the tuition fee incurred each student specific at the state university at Indonesian. Since the issuance of Permendikbud Number 55 of 2013 concerning Tuition Fee raises several problems due to various factors in the determination Tuition Fee students, Many students felt that they disagreed with Tuition Fee and some students were unable to continue their studies in college because they could not afford the Tuition Fee that was charged to students. Make use Decision Support System (DSS) will assist the process of Determining Tuition Fee for students so that the determination of Tuition Fee can provide assessment Objective.

IOP Conf. Series: Materials Science and Engineering **875** (2020) 012088 doi:10.1088/1757-899X/875/1/012088

Research will compare right between the four methods that are often used in some cases the decision-making that is Simple Additive Weighting (SAW), Technique for Order Preference by Similarity to Ideal Solution (TOPSIS), as well as the incorporation of methods Simple Additive Weighting (SAW) and Analytical Hierarchy Process (AHP), and Technique for Order Preference by Similarity to Ideal Solution (TOPSIS) and Analytical Hierarchy Process (AHP) here we will determine the level of a covering accuracy of four methods are in determining Tuition Fee resulting in recommendation methods that are relevant for the problem.

2. Literature Review

Tuition Fee in Indonesian is UKT based on the regulation of the Ministry of Education and Culture of the Republic of Indonesia number 55 of 2013 is part of the Tuition Fee fees borne by each student based on his economic ability. Tuition Fee determination also consists of several groups determined based on the economic capacity of students [1].

Decision Support System (DSS) is a methodology used to support decision making. Decision support systems should be flexible, interactive and can be adapted to support solutions to specific unstructured management problems. Decision support systems usually use various models and are built through interactive and iterative processes so that they can run as expected. Decision Support Systems are now widely developed and can be used on PCs, applications or the web for many people to use even in different locations [2]. As for several methods that can be used in the determination of Tuition Fee such as Multiple Criteria Decision Making (MCDM), some examples of MCDM are Simple Additive Weighting (SAW), Technique for Order Preference by Similarity to Ideal Solution (TOPSIS), PROMENTHEE, and Analytical Hierarchy Process (SAW-AHP) [3].

3. Research Method

This study will compare the four methods that are often used in several cases of decision making, namely Simple Additive Weighting (SAW), Techniques for Order Preference with Similarity to Ideal Solutions (TOPSIS), and the incorporation of the Simple Additive method. Weighting (SAW) and Analytical Hierarchy Process (AHP), and Techniques for Order Preference with Similarity with Ideal Solutions (TOPSIS) and Analytical Hierarchy Process (AHP) here we will determine the level of accuracy of the closure of the four methods in determining Cost Education resulting in a recommendation method that relevant for the problem. Testing is done by calculating the level of accuracy using Accuracy Measure [4, 5].

4. Result and Discussion

The research was started by doing sample data Student Tuition Fee.

Table 1. Research data students of Universitas Sulawesi Barat

No.	Name/Criteria	C1	C2	C3	C4	C5	C6	C7
1	A1	500,000	1	1	10	5	6	10
2	A2	500,000	1	1	10	5	6	10
3	A3	250,000	1	1	7	8	6	5
4	A4	500,000	1	1	7	5	6	10
5	A5	900,000	1	10	10	5	6	10
6	A6	1,500,000	1	1	10	5	6	10
7	A7	500,000	1	15	10	5	6	10
8	A8	500,000	1	1	10	5	6	10
9	A9	500,000	1	1	7	5	6	10
10	A10	500,000	1	1	7	5	6	10
11	A11	500,000	1	1	10	5	6	10

No.	Name/Criteria	C1	C2	C3	C4	C5	C6	C7
12	A12	500,000	1	5	10	5	10	10
13	A13	750,000	9	1	7	8	10	10
14	A14	600,000	1	1	7	5	6	10
15	A15	600,000	1	1	7	5	6	10
16	A16	500,000	10	1	7	5	6	10
17	A17	250,000	1	1	10	5	10	10
18	A18	500,000	1	1	7	5	6	10
19	A19	500,000	9	15	7	2	6	10
20	A20	300,000	1	15	3	5	6	5
21	A21	800,000	1	1	10	5	6	10
22	A22	1,200,000	1	1	10	5	6	10
23	A23	500,000	1	1	10	5	6	10
24	A24	1,000,000	1	1	10	5	6	10
25	A25	1,200,000	1	1	10	5	6	10
26	A26	250,000	1	1	10	8	10	10
27	A27	250,000	1	1	10	5	6	10
28	A28	500,000	1	1	10	5	6	10
29	A29	1,500,000	1	1	10	5	10	10
30	A30	600,000	1	1	10	5	6	10
31	A31	500,000	1	1	7	8	10	10
32	A32	250,000	9	1	10	8	10	10
33	A33	250,000	1	1	10	5	10	10
34	A34	300,000	1	10	10	8	10	10
35	A35	800,000	1	1	7	5	6	10
36	A36	1,050,000	1	1	10	5	6	5
37	A37	500,000	1	1	10	5	6	10
38	A38	300,000	9	1	3	5	10	10
39	A39	250,000	1	1	7	5	6	10
40	A40	600,000	1	1	10	5	6	5
41	A41	500,000	9	1	7	5	6	10
42	A42	500,000	1	1	10	5	4	10
43	A43	500,000	1	1	10	5	6	10
44	A44	500,000	1	1	3	5	10	10
45	A45	500,000	1	1	10	5	6	10
46	A46	500,000	1	1	7	5	4	10
47	A47	500,000	1	1	10	5	6	10
48	A48	750,000	1	1	7	5	6	10
49	A49	500,000	1	1	7	5	6	10
50	A50	1,350,000	1	1	7	5	6	10

Table 1 shows the student data that is used as a reference for determining UKT with 7 criteria, C1 is Parental Income, C2 is GPA, C3 is the amount of Siblings, C4 is Achievement, C5 is the amount of dependents, C6 is electrical fee, and C7 is water source where each of them have weights. Data that used was 50 student samples taken randomly, A1, A2, …, A50.

Based on the data shown in Table 1, the results obtained from the ranking obtained from the 4 methods, then the results are compared so that they can see the results of the comparison between the methods of SAW, TOPSIS, SAW-AHP, TOPSIS-AHP in the case of the determination of tuition fee in Universitas Sulawesi Barat. The results of the comparison are accumulated in the accuracy measure formula which can be shown in Equation 1.

$$\frac{same\ amount\ of\ data}{lots\ of\ data} \times 100\% \quad (1)$$

So that the accuracy of each method is obtained as follows :

- SAW : $\frac{38}{50} \times 100 = 76\%$
- TOPSIS : $\frac{39}{50} \times 100 = 78\%$
- SWA-AHP : $\frac{37}{50} \times 100 = 74\%$
- TOPSIS-AHP : $\frac{40}{50} \times 100 = 80\%$

In calculating the accuracy results have been obtained the results of comparisons of the four methods used. So as to obtain a comparison diagram of the calculation results of each method seen through the diagram in Figure 1.

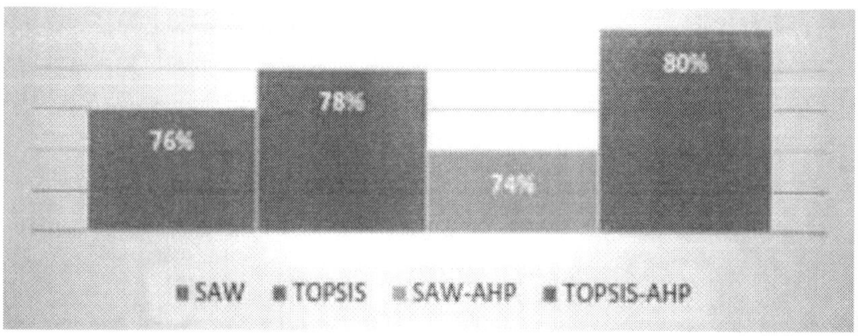

Figure 1. Comparison results of the four methods tested

Based on the results of calculating the accuracy of each method by using manual calculation data in Figure 1, the most optimal method to use is evaluated from the accuracy of each method by inputting student data, so it can be said that the TOPSIS-AHP method is the most optimal method used so that TOPSIS- AHP is the best method used in determining Tuition Fee that is suggested by the author, it is because AHP has the highest accuracy value compared to the accuracy value of other methods which is as much as 80%, then followed by the TOPSIS method 78%, SAW-AHP 74% and SAW 76%.

5. Conclusion

From experiments on the problem of determining the Tuition Fee using the SAW, TOPSIS, and SAW-AHP methods, The selection of the TOPSIS-AHP method as the best recommended solution looks at the accuracy of each method using the accuracy measure formula where the TOPSIS-AHP has a similarity or an accuracy rate of 80%, Then the second alternative is the TOPSIS method with TOPSIS 78%, SAW-AHP 74 % and SAW 76%.

References

[1] Draft Permendikbud UKT, The Ministry of Education and Culture, 2013.
[2] S. Sendana, H. A. Wibawa dan S., "Determination System of Tuition Fee Groups at Universitas Diponegoro Using the Analytic Hierarchy Process," 2014. [In Bahasa]
[3] M M D Widianta, R. T, D P S Setyohadi dan H Y Riskiawan, "Comparison if Multi-Criteria Decision Support Methods (AHP, TOPSIS, SAW & PROMENTHEE) for Employee Placement," *IJCST,* no. Serie 953, 2018.
[4] A. Pranolo dan S. M. Widyastuti, "Simple Additive Weighting Method on Intelligent Agent for Urban Forest Health Monitoring," *IEEE,* 2014.
[5] A. Cahyapratama dan R. Sarno, "Application of Analytic Hierarchy Process (AHP) and Simple Additive Weighting (SAW) Methods in Singer Selection Process," *IEEE,* 2018.

Web Page Extraction and Classification Using JSOUP and Naïve Bayes

S Cokrowibowo[1], N Nur[1], and Irmayanti[1]*

[1]Informatics Engineering, Faculty of Engineering, Universitas Sulawesi Barat, Majene, Indonesia

*Email: yirma294@gmail.com

Abstract. Classification of web pages manually requires quite a long time because most of the available web pages are not structured, so the classification method is needed quickly and accurately. Naļve Bayes algorithm with a good probabilistic approach in classifying web pages, seen from the advantages that are included in the classical category with a simple probability concept. However, this algorithm provides pretty good performance for many modern cases with large data. For the process of extracting information from web pages, it is proposed to use JSOUP which is a java library that provides a good API for extracting, manipulating data, and completing the initial data cleaning using the best methods from DOM, and CSS. The use of the JSOUP library makes it possible to be able to do web page analysis without having to save web documents to a computer store, so computer storage resources will be constant even though the amount of training data is increased. In this study, implementing JSOUP as a tool for extracting information from web pages and Naļve Bayes algorithm for classification of web pages.

1. Introduction

Web page classification, also known as web page categorization, is the process of assigning web pages to one or more predefined category labels. Classification has traditionally been proposed as a supervised learning problem [1] where a set of labeled data is used to train a classifier that can be applied to label as an example of the future. Classification of web pages manually requires quite a long time, therefore, a method that can be used in the classification process is fast and accurate.

One of the classification methods commonly used is Naļve Bayes. The Naļve Bayes algorithm is a classic category with a simple probability concept. However, this algorithm provides pretty good performance for many modern cases with large data. The Naļve Bayes algorithm uses an ancient theorem of the 18th century heritage, discovered by Thomas Bayes [1]. The Naļve Bayes method can do the job efficiently. If input is considered a set of word attributes we can calculate the probability of whether a web page belongs to a certain class or not. This process consists of two phases; training phase and classification phase. In the training phase, the filter will be trained using a collection of words that are known to be classified into web pages. Based on the appearance of each word given a probability for its capacity to determine its class classification. Then, in the classification phase use a collection of words to classify each web page [2].

Extracting web pages to be classified requires quite a long time if done manually. Therefore we need a tool to be able to extract web pages automatically. Web page contains huge amount of information so we must be able to extract valuable information [3]. By using the JSOUP java library to work with

HTML it provides a good API for extracting, manipulating data, and completing initial data cleaning using the best methods from DOM, CSS, and other methods similar to jquery. JSOUP implements the WHATWG HTML5 specification, and parses HTML to the same DOM as modern browsers do [4].

2. Related Research

Aris J. Ordoñez, Rommel Evan J. Paje, and Rodel N. Naz [2] in 2018 conducted a study that discussed how to classify SMS received by rescue agents including spam, invalid, warning 1, warning 2, and warning 3. The method used in solving the problem is applying the Naļve Bayes algorithm, an independent learning algorithm, and together with Natural Language Processing. From this research the writer will use the Naļve Bayes algorithm in the case of different classifications.

Patel and Pandya [5] in 2017 conducted a study explaining the model for categorizing articles related to child development and parenting context, which began with cataloging through identification & analysis carried out with consideration of category keywords along with relevant information has been taken from various web sources to achieve classification with expected accuracy. From this study the authors took a reference on how to categorize web pages into three different classes.

Lassri Safae et al. [6] in 2018 conducted a study that discussed the characteristics of web pages, classification by reviewing machine learning with different algorithms to categorize web pages. Thus, tracing some of the assumptions of the method studied. From this study the authors used the Naļve Bayes algorithm which is one of machine learning in solving classification problems.

Wang Jie et al. [7] in 2015 conducted a study that resolved problems in maintaining the balance of agricultural production and markets by proposing a platform that included data crawling, information extraction and analysis of big data of agricultural products from the Internet. In its system data crawling and extraction uses JSOUP, an HTML collector as a fundamental technology. From this study the authors use JSOUP in solving different problems.

3. Literature Review

3.1. Classification

Web page classification, also known as web page categorization, is the process of assigning web pages to one or more predefined category labels. Classification has traditionally been proposed as a supervised learning problem how to learn the behavior of a function that maps a vector of dimension N (for example $[X_1, X_2, \ldots X_N]$) into one from several classes (which the user wants), just by looking at a few examples of the input-output pairs of the function [1].

3.2. JSOUP

JSOUP is a java library for working on HTML documents that provides a good Application Programming Interface (API) for extracting, manipulating data, and completing preliminary data cleaning using the best methods from Document Object Model (DOM), CSS, and other methods similar to jquery. JSOUP implements the WHATWG HTML5 specification, and parses HTML to the same DOM as modern browsers do [3, 7]. Following are the main services available in the JSOUP library.

- Scrape and parse HTML from URL, file or string.
- Find and extract data using DOM traversal and CSS selector.
- Manipulate HTML elements, HTML attributes, and text.
- Clean the content sent by users using safe white-lists to prevent XSS attacks.
- Generates HTML tidy.

3.3. Naïve Bayes Algorithm

The Naļve Bayes algorithm is a classic category with a simple probability concept. However, this algorithm provides pretty good performance for many modern cases with large data. Several studies that use Naļve Bayes algorithm such as web page classification [8] and classification for diabetes mellitus [9]. The Naļve Bayes algorithm uses an ancient theorem of the 18th century heritage, discovered by Thomas Bayes [1]. In Bayes' Theorem, a conditional probability is stated as follows.

$$P(P|X)=\frac{P(P|H)P(H)}{P(X)} \tag{1}$$

In the process of the Naļve Bayes classifier, the D training set data set that contains a number of tuples X along with a number of m class labels ($C_1, C_2, ...C_m$) is categorical type data, each tuple with dimension n expressed as $A_1, A_2, ...A_n$. Calculates the prior probabilities of each class so that a class probability table is generated, where $P(C_i)$ = the number of tuples in the class C_i divided by all tuples in the D training data set. Calculate the probability of each value for all n attributes ($A_1, A_2, ...A_n$) for the whole class m ($C_1, C_2, ...C_m$) resulting in a number of n probability tables, if there is a probability value of 0, Laplacian correction did. Learning outcomes are as many as (n + 1) tables, namely: one class probability table and n tables that contain the probability of each value on all existing attributes. Here is the Laplacian correction equation.

$$P(X_{kj}|C_i)=\frac{f(X_{kj}|C_i)+1}{f(C_i)+|W|} \tag{2}$$

The following equation can be used to classify an input tuple by finding the maximum probability of all existing classes.

$$P(X|C_i)P(C_i)=(\Pi_{k=1}^{n}P(X_k|C_i)\times P(C_i)=P(X_1|C_i)\times ...\times P(X_n|C_i))\times P(C_i) \tag{3}$$

$$P(X|C_i)P(C_i)>P(X|C_j)P(C_j) \text{ for } 1\leq j\leq m, j\neq i$$

Equation (3) can be simplified into the following equation.

$$P(C_i)=P(C_i)\Pi_{k=1}^{n}P(X_k|C_i) \tag{4}$$

4. Research Design

The classification process consists of four main phases; Extraction, preprocessing, attribute selection, training data, and testing data.

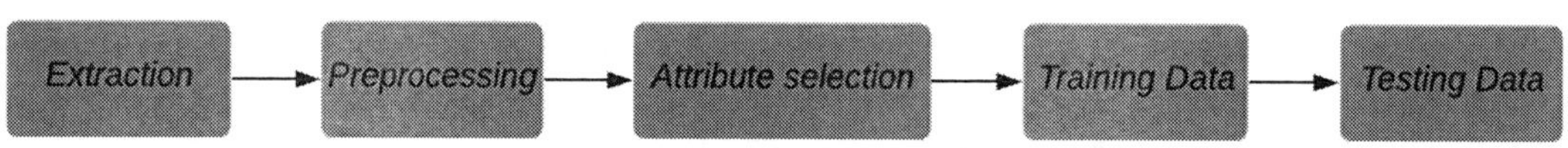

Figure 1. Research design flow

4.1. Extraction

In the extraction process the writer uses the JSOUP library. The system will accept input in the form of a web page URL that will be extracted. Then JSOUP will load the web page based on the input URL. Next will be extracted data from web pages by separating between the text from HTML tags [10]. Figure 2 shows an example of web page extraction result using JSOUP.

Figure 2. Web page extraction results

4.2. Preprocessing

Preprocessing aims to prepare data so that it is easily analyzed. Preprocessing is divided into several processes consisting of case folding, tokenizing, and filtering. In Figure 3, there is a preprocessing process. Case folding is the initial stage of preprocessing text that changes all the letters of the text into lowercase characters. Characters accepted are only 'a' to 'z'. Characters other than letters will be removed and considered as delimiter. Tokenizing is the stage of cutting the input string based on each word that makes it up. Filtering is a process carried out by taking important words from the token results and deleting stop words. Stop words are words that are not descriptive so that they can be discarded or removed and have no effect in the process. In Indonesian, the examples of stop words is "that", "and", "from", "in", "like" and others [11].

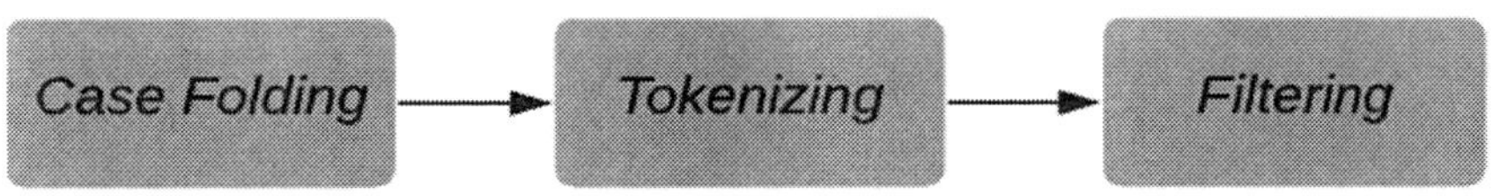

Figure 3. Preprocessing

4.3. Attribute Selection

Attribute selection is the process of determining what words (terms) will be used to represent web pages. In addition to describing the contents of web pages, these terms are also useful for differentiating one web page from another web page in a collection of web pages.

4.4. Training Data

In the data training phase, web pages are trained using the Attribute selection which is known to be classified into web pages using the Naļve Bayes method with Equation (2). Based on the appearance of each word, a probability for its capacity to determine its class classification is given.

4.5. Testing Data

After conducting the training data, the next step is to test the model of the trainingdata results by using a new web page that is not yet known which class category. This is where the process of classifying new data into classes that has been determined by studying the trace attributes that have been previously trained with Equation (4). Figure 4 shows the stages or flowchart of classification process:

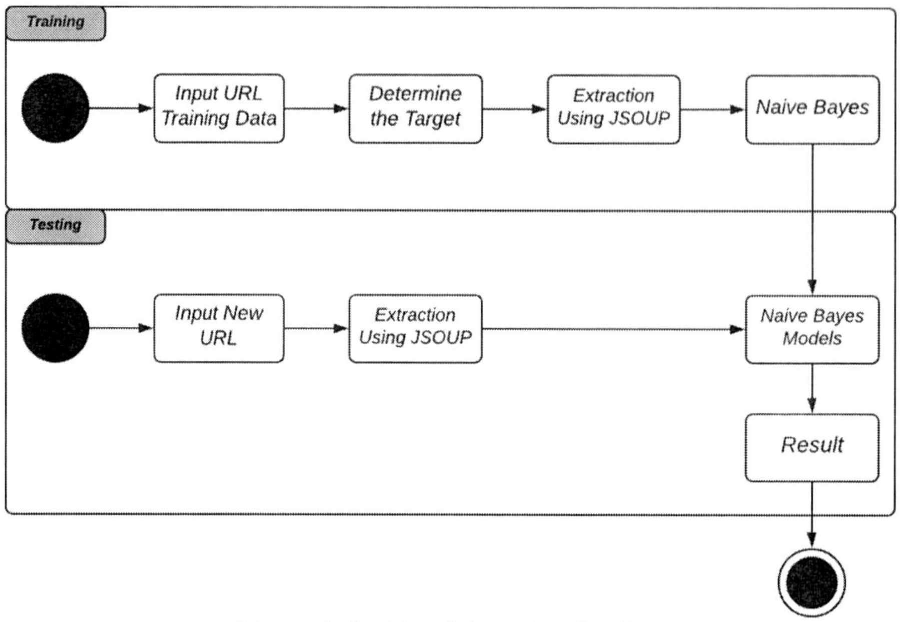

Figure 4. Training data and testing data

5. Results and Discussion

Naļve Bayes classification model testing is done using the Java programming language. In this paper the test is carried out 6 (six) times with the amount of training data for the first test until the last test in a row of 9, 18, 36, 72, 150, and 300 URLs of web pages that have been saved into the text document (*.txt), with the same testing data as many as 100 URLs of web pages that have been saved into the text document (*.txt). The overall results of the training data can be seen in the following graphic image.

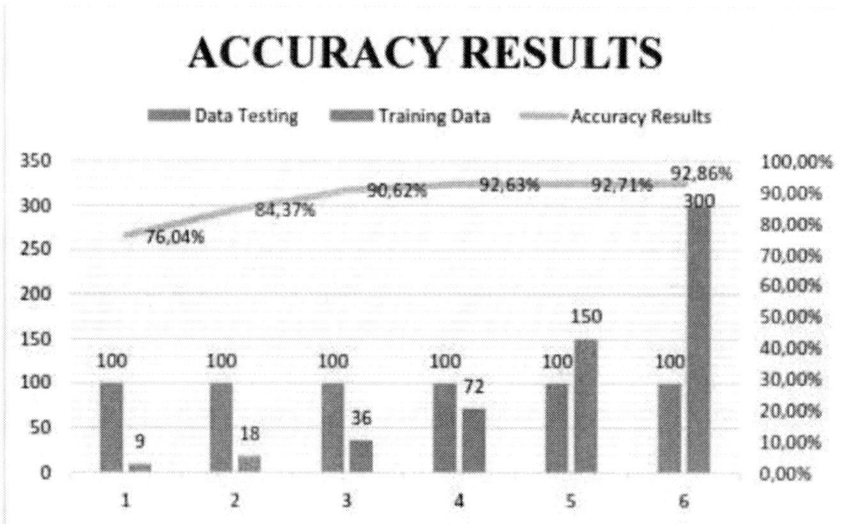

Figure 5. Test result accuracy

From the graph above it can be seen that the level of accuracy is directly proportional to the amount of training data, meaning that the more the amount of training data, the level of accuracy will increase. This is because the more training data, the more word attributes (tokens) that characterize a particular class attribute. Classification errors occur when there are web pages that should be categorized into certain class categories but the Naļve Bayes classification model identifies into other class categories. This is due to the lack of word attributes (tokens) from web pages in the training data for a class category. So the probability of probability for that class category is small compared to other classes.

6. Conclusion

Classifying web pages using the Naļve Bayes algorithm with a large amount of training data can maximize the level of accuracy, with the use of JSOUP allows being able to do web page analysis without first having to save web pages to storage computers. So that the efficient use of computer storage resources will be constant even though the amount of training data is increased.

References

[1] Suyanto, Basic and Advanced of Machine Learning, Bandung: Informatika Bandung, 2018. [In Bahasa]

[2] R. E. J. P. R. N. N. Aris J. Ordoñez, "SMS Classification Method for Disaster Response," pp. 233- 236, 2018.

[3] C. Saini and V. Arora, "Information Retrieval in Web Crawling: A Survey," pp. 2635-2643, September 2016.

[4] J. Hedley, "jsoup: Java HTML Parser," 2019. [Online]. Available: https://jsoup.org/. [Accessed October 8, 2019].

[5] A. D. Patel and D. V. N. Pandya, "Web Page Classification based on Context to the Content Extraction of Articles," pp. 539-541, 2017.

[6] L. Safae, B. E. Habib and T. Abderrahim, "A Review of Machine Learning Algorithms for Web Page Classification," 2018 IEEE 5th International Congress on Information Science and Technology (CiSt), Marrakech, 2018, pp. 220-226.

[7] J. Wang, S. Yang, Y. Wang and C. Han, "The Crawling and Analysis of Agricultural Products Big Data based on JSOUP," pp. 1197-1202, 2015.

[8] B. P. Asatkar, P. K. Wagh and D. P. N. Chatur, "Classification of Web Pages Using Naļve," *IJIRSET,* pp. 112-117, 2017.

[9] S. Sankaranarayanan and T. P. Perumal, "Analysis of Naļve Bayes Classification for Diabetes Mellitus," *JCSE,* pp. 520-524, 2018.

[10] S. Cokrowibowo and Ismail, "Extraction and Visualization of Web Text Mining Using Jsoup" pp. 1-10, 2018. [In Bahasa]

[11] I. G. A. Socrates, A. L. Akbar and M. S. Akbar, "Naļve Bayes Optimization with Feature Selection and Weight Gain Ratio, [In Bahasa]" pp. 22-30, April 2016.

IOP Conf. Series: Materials Science and Engineering **875** (2020) 012090 doi:10.1088/1757-899X/875/1/012090

Implementation of Connected Component Labelling for Calculation Amount of Eggs on the Laying Pullet Cage

A A A Cirua[1]*, S Cokrowibowo[1], and M F Rustan[1]

[1]Department of Informatics, Faculty of Engineering, Universitas Sulawesi Barat, Majene, Indonesia

*Email: amirulasnan02@gmail.com

Abstract. Chicken eggs are one of the commodities that become easily available foodstuffs. The level of egg production in good condition is directly proportional to the capacity of the cage. The process of calculating the amount of egg production becomes a problem for farmers when they want to calculate the amount of egg production in the condition of a chicken coop on a large scale. Smart systems are one form of technological development with the primary goal of producing effective and efficient systems. One area that plays a role in technological development is image processing, eggs are objects with certain color characteristics and pixel scales. In this study, the process of calculating the number of eggs based on color segmentation and selection of pixel scales on eggs and the process of calculating objects using the Connected Component Labeling algorithm of images that have gone through gray-scale and threshold processes and implemented using MATLAB. The trial results show that the object image of an egg is strongly influenced by the size of the image, the intensity of the light, and distance of capturing images. The number of eggs that can be detected in this study depends on the process of segmentation and measurement of the object's egg scale, in this study it was found that colors other than egg colors and objects that are outside the egg scale will be considered as noise and will not be detected.

1. Introduction

According to [1], eggs contain high protein and complete composition of protein substances, fat content in eggs is also high. The development of laying hens business provides nutritional fulfillment in addition to stabilizing and stabilizing the egg production business, breeders must make improvements from various aspects such as production, management, biosecurity and marketing strategies [2]. Smart system or what is called Smart System is a system that provides several technologies that provide services with several, the purpose of these services so that the system can run effectively, efficiently, safely and comfortably. Smart systems also implement the Internet of Things (IoT) so that the system can be run anywhere and anytime with an internet network connection [3]. The concept of IoT is widely implemented in various smart systems such as smart buildings, smart homes, smart classes, and so on. IoT elements are detection, communication, computing, services, and resources [4].

The problem that the writer found was the difficulty of egg farmers in calculating the amount of chicken egg production and calculating the production statistics in large capacity cages detecting objects in images has long been a staple in image processing. This research is to find the egg object in the image. Each egg object is represented by a certain set of pixels and colors which are then segmented to determine only egg objects. To get the egg object, the labeling algorithm of the connected components is used.

2. Literature Review

2.1. Grayscale

The process of converting an image into grayscale is done by taking all the pixels in the image, then taking the basic color data that is RGB through the pixel of each color and then the results are divided into three which will then produce an average value. Grayscale will be obtained from the average value used to obtain the value used to obtain the pixel value of the image [5]. The following is the equation from grayscale

$$Y_{(x,y)} = (0.299 \times R) + (0.587 \times G) + (0.114 \times B) \quad (1)$$

where:

Y = degree of gray
R = *Pixel channel Red*
G = *Pixel channel Green*
B = *Pixel channel Blue*

2.2. Gaussian Blur

Filter Gaussian is a very good method in removing noise in the image because it has the nature of the normal distribution which is found when the results of the digitization process using a camera [6]. Blur filter in the Gaussian blur method can make color changes that are quite visible in the image, then the mid color image is used as a soft effect on all sides of the image. The equation of the following filter is the gaussian kernel matrix and the formula as below.

$$G = \begin{bmatrix} 1 & 2 & 1 \\ 2 & 3 & 2 \\ 1 & 2 & 1 \end{bmatrix} \quad (2)$$

$$Pixel\ B(i,j) = \frac{1}{k}\sum_{p=0}^{N-1}\left(\sum_{q=0}^{M-1} G(p,q)\cdot Pixel\ A\left(i+p-\frac{(N-1)}{2}, j+q-\frac{(M-1)}{2}\right)\right)$$

where:

Pixel A : Original image
Pixel B : Weight of the product
N : Number of kernel matrix columns
K : The sum of all values in G
G : Gaussian kernel value.

2.3. Thresholding

Thresholding is a basic method in image segmentation systems. The threshold is divided into three methods, namely binary, truncate, and zero threshold. Sophisticated techniques of thresholding are band and multispacial thresholding. This method is usually applied to images that use grayscale. When the image cannot be extracted into graysclae, then it is multispectral [7].

2.4. Connected Component Labelling

Connected component labeling algorithm is an algorithm that calculates objects in an image with the condition that the image is a binary image. Labeling objects is done in sequence. This method is divided into two main categories, namely recursive connected components and sequential labeling based on neighborliness (4-connectivity, 6-connectivity, 8-connectivity).

The connected component labeling algorithm carries out the scanning process in a row starting from the top left to the bottom right. The scanning process is repeated until there are no more changes to the label naming then the algorithm has been completed [8]. Thus the pixels that are said to be connected basically have the nature of neighborhood.

3. Research Method

This research uses experimental research with the following stages:

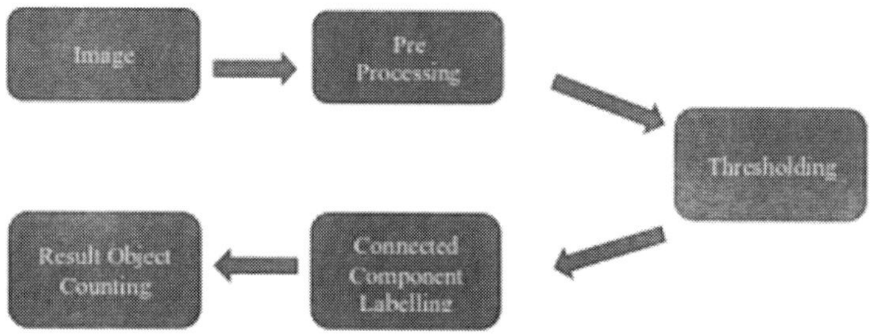

Figure 1. Image processing concepts using Connected Component Labeling

3.1. Image

The first stage taken was image capture by the camera. Image can be process to the next stage.

3.2. Preprocessing

At this stage, the RGB image which is an image with a composition of three basic colors (Red, Green, Blue) is converted into grayscale and then carried out a filtering process to eliminate noise with the Gaussian blur method.

3.3. Thresholding

In the next stage, a thresholding process is carried out, namely the conversion of grayscale images into binary images of 0 (black) and 1 (white) values.

3.4. Connected Component Labeling

Furthermore, the connected component labeling algorithm will count objects with a value of 1 with the 4-connection concept, this algorithm will scan pixel images from top left to bottom right. When finding an object, the concept of neighborhood will be carried out to look for pixels that are part of the object found. After completing the labeling and reviewing the scanning process again until no objects are found then the CCL algorithm is complete.

3.5. Object Counting Results

The results obtained in the form of the number of detected egg objects.

4. Results and Discussion

4.1. Results

The results obtained are the number of eggs detected and calculated by the connected component labeling algorithm. Figure 2 shows the image of the egg on the prototype.

Figure 2. Image of an RGB egg on a prototype

Then take a sample of the color point on the egg to determine the color range of the egg, so that the minimum and maximum RGB values are obtained as follows.

- Minimum RGB is ($R = 197$, $G = 104$, $B = 78$)
- Maximum RGB is ($R = 255$, $G = 190$, $B = 175$)

4.2. Image Processing

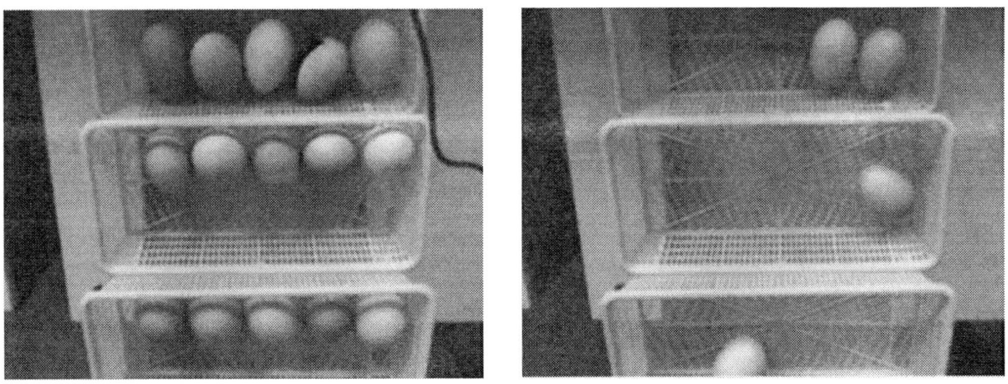

Figure 3. Image of an egg with grayscale and gaussian blur

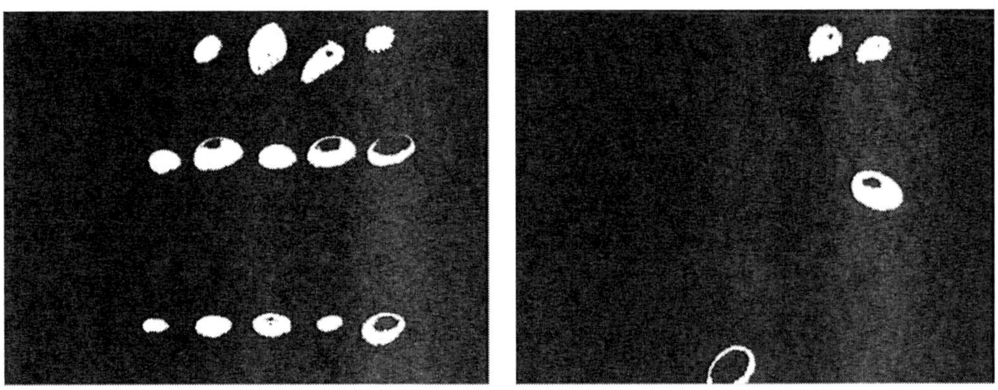

Figure 4. Image of eggs with a threshold

To reduce noise in the image is also done to determine the scale, the egg scale in this study is 300 to 28000 pixels so that objects outside of the scale will not be read as objects.

4.3. Calculation Accuracy
The accuracy of egg calculation by implementing connected component labeling algorithm using segmentation and egg scale obtained an accuracy value of 63%. In this study, the weaknesses are the process of segmenting the egg object before the thresholding stage, the color segmentation is not yet maximum and there is noise entering the egg scale so that it reduces the accuracy of the calculation.

5. Conclusions and Suggestions
5.1. Conclusions
Based on the results of research on the implementation of connected component labeling on the calculation of the number of eggs in laying hens, it is concluded that the Minimum and Maximum Values for the scale range of egg pixels are 300 to 28000 pixels, the minimum and maximum egg RGB values, namely the Minimum RGB is ($R = 197$, $G = 104$, $B = 78$) and Maximum RGB is ($R = 255$, $G = 190$, $B = 175$). The accuracy obtained is 63%.

5.2. Suggestions
For further research, object segmentation, morfology, and background substraction techniques need to be improved.

Reference
[1] Sudaryani, "Kualitasi Telur," *Penebar Swadaya,* p. jakarta, 2003.
[2] S. P, "Prospek Usaha Ayam Petelur di Indonesia," *Poulrty Indonesia,* 2009.
[3] V. L. a. E. N. C.Zhu, "Green Internet of Things for Smart World," *IEEE,* pp. 2151-2162, 2015.
[4] B. N. S. K. H. M. Khan, "Internet of Things based Energy Aware Smart Home Control System," *IEEE,* 2016.
[5] R. F. Shaumi, "Rancang Bangun Alat Penyortir Buah Tomat Berdasarkan Ukuran Berbasis Raspberry Pi 3," Universitas Lampung, 2018.
[6] M. Y. d. F. G. F. Boskurt, "Effective Gaussian Blurring Process on Graphics Processing Unit with CUDA," *International Journal of Machine Learning and Computing,* pp. 57-61, 2015.
[7] M. Rinaldi, "Pengolahan Citra Digital dengan Pendekatan Algoritmik," *Informatika,* 2004.
[8] Binus, "eThesisdoc" 2017. [Online]. Available: http://library.binus.ac.id/eColls/eThesisdoc/Bab2/TSA-2017-0136Bab%202.pdf.

IOP Conf. Series: Materials Science and Engineering **875** (2020) 012091 doi:10.1088/1757-899X/875/1/012091

Effects of Mineral Admixtures on Pore Structure and Compressive Strength of Mortar Contaminated Chloride

D Patah[1]*, H Hamada[2], and A Dasar[1]

[1]Department of Civil Engineering, Universitas Sulawesi Barat, Majene, Indonesia
[2]Department of Civil and Structural Engineering, Kyushu University, Japan

*Email: dahliapatah@unsulbar.ac.id

Abstract. This study evaluates the mechanical properties of performance fly ash, silica fume, metakaolin and blast furnace slag mortars due to contaminated chloride. The compressive strength and pore structures at water-to-binder ratios of 0.4, 0.5 and 0.6 are investigated. The result, in general, showed mineral admixtures mortar contaminated chloride improved compressive strength and porosity but different rates depending on their binder type.

1. Introduction

The pore structure of concrete includes air voids, capillary pores, and gel pores. As one of the important characteristics of concrete materials, pore structure possesses an important role determining its mechanical, durability and transmissive characteristics [1, 2]. Pore structure parameters such as porosity, pore size distribution and so on are progressively employed to evaluate permeability, frost resistance, carbonation resistance and physical strength of concrete [3-5].

In order to improve the performance of concrete capable of withstanding serious environmental conditions, mineral admixture including industrial by-products such as blast-furnace slag, fly ash and silica fume are commonly used in combination with OPC in concrete for many applications because they improve durability and reduce porosity of concrete [6]. They also could lower cement requirement resulting in leading to a reduction for CO2 generated by the production of cement [7]. Due to this advantage, there are many studies on the use of the mineral admixture for structural concrete [8-11]. Even though the partial replacement of the mineral admixtures by weight increase the later strength of concrete [12], use of the mineral-admixtured concrete in construction fields has been reluctant because of lower early strength.

Therefore, in the present study, strength characteristics in early the age of mortar contaminated chloride content incorporating mineral admixtures such fly ash, silica fume, metakaolin and blast furnace slag with water-to-binder (W/B) ratio of 40%, 50% and 60% on pore structure and compressive strength of mortar ware evaluated.

2. Test Programs

2.1. Materials

The specimens were made with plain cement by Ordinary Portland Cement (OPC) and supplementary cementitious materials by Fly Ash (FA), Silica Fume (SF), Metakaolin (MKP) and Blast Furnace Slag Type B (BB/BFS). Table 1 shows the chemical properties of binders and the physical properties such density, fineness, and water absorption are shown in Table 2.

Table 1. Chemical properties of binder

Contituent	OPC	FA	SF	MKP	BFS
MgO, %	1.31	-	0.56	1.04	3.26
SiO2, %	-	60.60	95.5	52.37	34.10
SO3, %	2.14	-	0.18	7.56	2.04
LoI, %	1.97	2.40	1.22	-	1.46
Total Alkali,	0.43	-	-	-	-
Chloride, %	0.016	-	-	0.00	0.010

Table 2. Physical properties of material

Material	Density, g/cm^3	Fineness, cm^2/g	Water absorption, %
OPC	3.16	3400	-
FA	2.26	3970	-
SF	2.35	1800	-
BBMKP	2.75	9030	-
BFS	3.02	3860	-
Washed Sea Sand	2.49	-	1.42

2.2. *Series of Mixture*

Fifteen series of mortar mixtures with three types water-to-binder (W/B) ratio of 40%, 50% and 60% were set for mixing mortar contaminated chloride were prepared. Two influencing parameters of chloride content in mortar were used. One is interpreted in %-cement (mass ratio of cement), another is interpreted in kg/m^3 (total chloride weight in mortar). Table 3 presents all cases of chloride content in each mix and the mortar mixture proportions of specimen are shown in Table 4. In this reference, chloride content in mortar is 0.57 %-cement in accordance to 3.31 kg/m3, 2.91 kg/m^3, 2.59 kg/m^3 at W/B=40%, 50%, and 60%, respectively. Series N is OPC-100% as control specimen. Series D is OPC-100% as mortar contaminated chloride content. And, Series S is mortar contaminated chloride content and incorporated with FA-20%, SF-5%, and BFS-45%. However, BBMKP is blast furnace slag incorporated with metakaolin (BB-80% and MKP-20%).

2.3. *Test Methods*

2.3.1. Compressive Strength and Elastic Modulus. Mortar cylinder specimens in size of ϕ 100x200mm were demolded at 24-hours after casting then cured by wrapping with wet towel and plastic sheet in a room maintained at 20°C, R.H. 60% controlled room. After 28 days and 91 days curing, compressive strength was measured in accordance with JIS A 1108. The average compressive strength of three specimens were determined for each mortar mixture in three curing conditions.

2.3.2. Pore Structure. The total pore volume and pore size distribution of concrete specimen were tested by Mercury Intrusion Porosimetry (MIP). After curing for 28 days and 91 days, mortar prism specimens were cut into 5mm-thick slice samples. Subsequently, the fragments were immersed in acetone for 15 minutes to stop further hydration of cement, and then dried in the vacuum desiccator for 2 days. In the MIP test, complete drying of the samples was required to obtain results without error [13]. In this study, the maximum applied pressure of MIP test was 33.000 psi (227MPa), and the surface tension and contact angle of mercury was 485 dynes/cm and 130o, respectively. Both surface tension and contact angle were used in the Washburn equation to convert applied pressure to pore diameter. The pore width corresponding to the highest rate of mercury intrusion per change in pressure is known as the "threshold" pore width [14]. After achieving this highest rate of intrusion, mercury has been shown to penetrate the interior of the specimen.

Table 3. Mixture series of specimen

Series	W/B (%)	Replacement (%)					Chloride content	
		OPC	FA	SF	BBMKP	BFS	kg/m^3	%-cement
N	40	0	-	-	-	-	-	-
	50	0	-	-	-	-	-	
	60	0	-	-	-	-	-	
S	40	-	0	0	0	0	3.31	0.57
	50	-	0	0	0	0	2.91	
	60	-	0	0	0	0	2.59	
D	40	0	-	-	-	-	3.31	0.57
	50	0	-	-	-	-	2.91	
	60	0	-	-	-	-	2.59	

Table 4. Mix proportion of mortar

W/B (%)	Unit content (kg/m^3)							Remarks
	W	OPC	FA	SF	MKP	BB	Sea Sand	
40	232	581	-	-	-	-	1490	Normal
	232	465	11	-	-	-	1452	FA
	232	552	-	29	-	-	1481	SF
	232	-	-	-	116	465	1458	BBMKP
	232	320	-	-	-	261	1480	BFS
50	255	510	-	-	-	-	1490	Normal
	255	408	10	-	-	-	1452	FA
	255	484	-	26	-	-	1481	SF
	255	-	-	-	102	408	1458	BBMKP
	255	281	-	-	-	230	1480	BFS
60	272	454	-	-	-	-	1490	Normal
	272	363	91	-	-	-	1452	FA
	272	431	-	23	-	-	1481	SF
	272	-	-	-	323	91	1458	BBMKP
	272	250	-	-	-	204	1480	BFS

3. Result and Discussion

3.1. Compressive Strength

The compressive strength results of different mortar mixes at 28 days and 91 days is shown in Figure 1. days and 91 days, the compressive strength on OPC and SCMs decreasing with increase W/B ratio. The SF mortar showed higher strength values to that of another's mortar at all tested ages. Only the result of the FA was different, as it gave the lowest compressive strength at 28-days and 91-days. The compressive strength lowest is related to the properties of FA that declines the heat of hydration process. As a result, FA slows the rate of hardening and reduces the compressive strength [14]. In addition, perhaps this condition may be due to chloride contaminated.

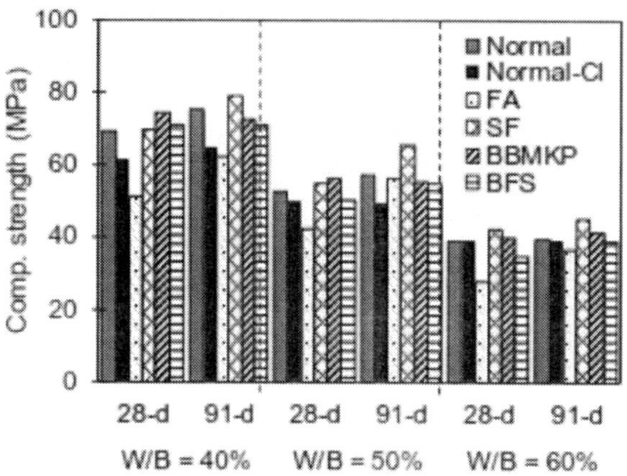

Figure 1. Strength development of mortar

Table 5 gives the compressive strength expressed as fraction of that of normal OPC. It is found that strength performance in early age up to 91 days of normal OPC is larger than OPC mortar contaminated chloride. This implies OPC mortar contaminated chloride less effective on compressive strength up to 91-days compared with normal OPC. Compressive strength of OPC contaminated chloride and FA decreased about 20% when compared to normal OPC.

Table 5. Compressive strength of as fraction of Normal OPC

Series		W/B = 40%		W/B = 50%		W/B = 60%	
		28-d	91-d	28-d	91-d	28-d	91-d
Normal OPC		1.00	1.00	1.00	1.00	1.00	1.00
SCMs	FA	0.74	0.83	0.81	0.98	0.72	0.92
	SF	1.01	1.05	1.05	1.15	1.09	1.14
	BBMKP	1.07	0.96	1.08	0.97	1.02	1.04
	BFS	1.03	0.95	0.95	0.96	0.90	0.98
Normal OPC-Cl		0.89	0.86	0.95	0.87	1.00	0.99

3.2. Porosity

Porosity is a measure of the proportion of the total volume of concrete occupied by pores. The total volume of pores obtained by Mercury Intrusion Porosity (MIP) tests of mortar specimen at 91 days are shown in Figure 2. It is shown that OPC contaminated chloride increased pore volume of mortar compared with normal OPC. The relatively high of porosity values implies the effect of additional chloride effects the increment of porosity. Further, some results also showed by SCMs that more increase in porosity is achieved by the use of SCMs with contaminated chloride content. The SF and BFS mortar exhibited slightly same porosity values when compared to the OPC mortar contaminated chloride. Unlike most other mineral admixtures, FA and BBMKP mortar the highest porosity values at the age 28 days and 91 days.

Table 6 gives the total pore volume expressed as fraction of that of normal OPC. It is found that total pore volume of FA and BBMKP in early age of SCMs is the larger than OPC mortar. The total pore volume of FA increased about 37%, 41%, 21% for W/B 40%, 50% and 60%, respectively, when compared to normal OPC at 91-days. And, the total pore volume of BBMKP increased about 21%, 45%, 26% for W/B 40%, 50% and 60%, respectively, when compared to normal OPC at 91-days.

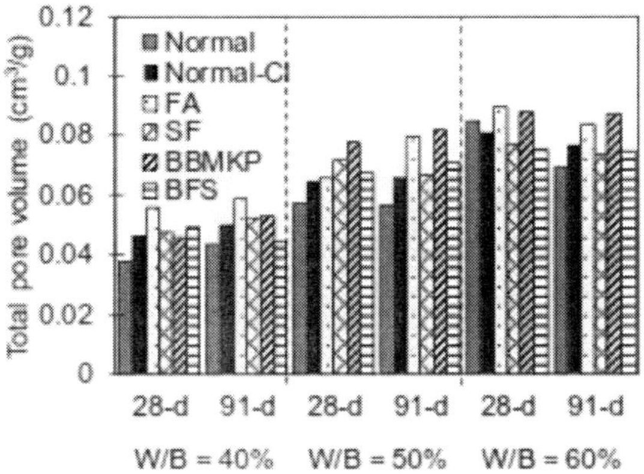

Figure 2. Porosity of concrete at 28-days

Table 6. Total pore volume of as fraction of Normal OPC

Series		W/B = 40%		W/B = 50%		W/B = 60%	
		28-d	91-d	28-d	91-d	28-d	91-d
Normal OPC		1.00	1.00	1.00	1.00	1.00	1.00
SCMs	FA	1.49	1.37	1.15	1.41	1.06	1.21
	SF	1.27	1.21	1.26	1.18	0.91	1.07
	BBMKP	1.21	1.22	1.36	1.45	1.04	1.26
	BFS	1.31	1.03	1.18	1.26	0.88	1.08
Normal OPC-Cl		1.24	1.17	1.14	1.16	0.95	1.11

The relationship between compressive strength and porosity is shown in Figure 3. The results show the porosity decreased with increases compressive strength at the age of 28 days and 91 days. In addition, a good correlation is acquired in which R2 values are around 0.90 at 28 days and 91 days, only FA mortar have low correlation after 91 days. This suggests that the compressive strength of with SCMs are influenced by porosity. The same tendencies as reported in previous studies that strength and elasticity of concrete were affected by the total pore volume or porosity [15].

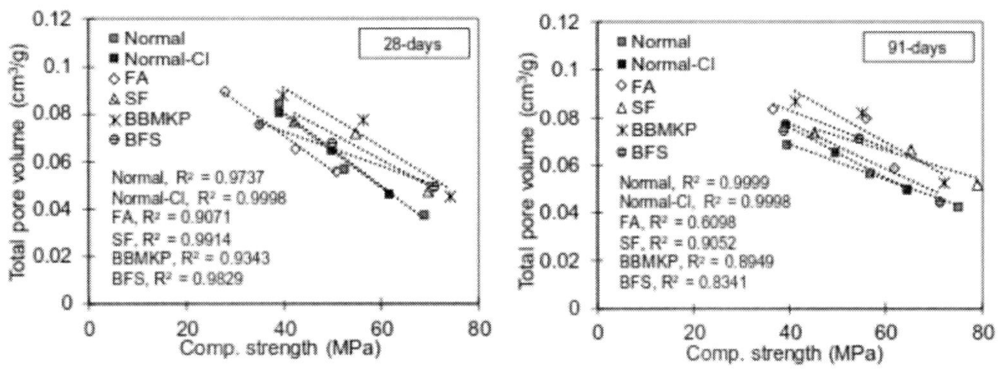

Figure 3. Compressive strength and total pore volume at 28-days and 91-days

4. Conclusions

From this research, several conclusions can be drawn as follows:

- At the age of 28-days and 91-days, mostly W/B more affect than SCMs on the compressive strength. However, mortar with and without SCMs were increased in compressive strength.
- At the age of 28-days and 91-days, SCMs less affects in reducing total pore volume.

5. References

[1] Winslow D, Liu D., "Pore structure of paste in concrete," Cement and Concrete Research, Vol. 20, No. 2, 1990, pp. 227-235.

[2] Song HW, Kwon SJ., "Permeability characteristics of carbonated concrete considering capillary pore structure," Cement and Concrete Research, Vol. 37, No. 6, 2007, pp. 909-915.

[3] Gonen T, Yazicioglu S., "The influence of compaction pores on sorptivity and carbonation of concrete," Construction and Building Material, Vol. 21, No. 5, 2007, pp. 1040–1045.

[4] Basheer L, Basheer PAM, Long AE., "Influence of meso-macro aggregate on the permeation, durability and the microstructure characteristics of ordinary Portland cement concrete," Construction and Building Material, Vol. 19, No. 9, 2005, pp. 682-690.

[5] Kumara R, Bhattacharjee B. Porosity, "Pore size distribution and in situ strength of concrete," Cement and Concrete Research, Vol. 33, No. 1, 2003, pp. 155-156.

[6] Report of ACI Committee 233, "Slag Cement in Concrete and Mortar," ACI 233R-03, American Concrete Institute, Farmington Hills, Michigan, 2003.

[7] W. W. J. Chan, C. M. L. Wu, "Durability of concrete with high cement replacement," Cement and Concrete Research, Vol. 30, No. 6, 2000, pp. 865-879.

[8] S. J. Barnett, M. N. Soutsos, S. G. Millard, J. H. Bungey, "Strength development of mortars containing ground granulated blast-furnace slag: Effect of curing temperature and determination of apparent activation energies," Cement and Concrete Research, Vol. 36, No. 3, 2006, pp. 434- 440.

[9] E. Douglas, G. Pouskouleli, "Prediction of compressive strength of mortars made with portland cement-blast-furnace slag-fly ash blends," Cement and Concrete Research, Vol. 21, No. 4, 1991, pp. 523-534.

[10] A. Oner, S. Akyuz, "An experimental study on optimum usage of GGBS for the compressive strength of concrete." Cement and Concrete Composites, Vol. 29, No. 6, 2007, pp. 505-514.

[11] J. E. Kim, W. S. Park, S. H. Yun, Y. I. Jang, H.D. Yun, S.W. Kim, D.G. Kim, "The Relationship of Compressive Strength and Tensile Strength of High-Performance Concrete," In Key Engineering Materials, Vol. 627, 2014, pp. 385-388.

[12] H. W. Song, V. Saraswathy, "Studies on the corrosion resistance of reinforced steel in concrete with ground granulated blast-furnace slag-an overview," Journal of Hazardous Materials, Vol. 138, No. 2, 2006, pp. 226-233.

[13] P. Sulapha, S.F. Wong, T.H. Wee, and Swaddiwudhipong, "Carbonation of concrete containing mineral admixture," Journal of Materials in Civil Engineering, Vol.15, No.2, 2003, pp.134-143.

[14] Ramazan Demirboga, Ibrahim Turkmen, Mehmet B. Karakoc, "Relationship between ultrasonic velocity and compressive strength for high-volume mineral-admixture concrete," Cement and Concrete Research, Vol.34, 2004, pp.2329-2336.

[15] Lulu Basheer, Joerg Kropp, and David J. Cleland, "Assessment of the durability of concrete from its permeation properties: a review," Construction and Building Materials, Vol. 15, 2001, pp.93- 103.

IOP Conf. Series: Materials Science and Engineering **875** (2020) 012092 doi:10.1088/1757-899X/875/1/012092

Online News Analysis of Majene Public Figure Electability with NLP (Natural Language Processing)

Sundari[1]*, S Cokrowibowo[1], and N Zulkarnaim[1]

[1]Departement of Informatics, Faculty of Engineering, Universitas Sulawesi Barat, Majene, Indonesia

*Email: sundarisyukur1998@gmail.com

Abstract. Electability is one of the concerns as public figures who will take part in the elections, various things are done to improve their electability. Now, there are many methods to measure electability, one of which is the Natural Language Processing method. In this study, an electability analysis of public figures was conducted on online news to determine the electability level of a public figure. Where information obtained by the public is through online news sites. According to the Indonesian Internet Service Providers Data (APJII) in 2014, internet users in Indonesia will reach 107 million. This number has become one of the potential Public Figures, both institutionally and individually in capturing their masses through new media. 100 million of the masses will be contested by candidates who will compete in the political stage. Therefore, candidates who will advance will carry out digital political imaging. The public figures used in this study are 12 figures who are rumored to be going forward in the 2020 elections based on the online news site MAJENE 2020 ELECTION.

1. Introduction

Advances in science and technology, especially in the field of computer science make information exchange can be done anywhere and anytime. This causes an abundance of information circulating on the internet [1]. The internet is the second medium that is often used by people after television. This can be seen from internet users in Indonesia, which continue to increase from year to year. One of the information obtained by the public is through an online news site. Hall predicts that within five years there will be more people in developing countries accessing news from the internet than newspapers.

Electability is a political compilation of unpopular elections and can be a benchmark for measuring the level of one's electability. Electability is a concern as a public figure who will participate in the elections, various things are done to increase their electability. According to the Indonesian Internet Service Providers Data (APJII) in 2014, internet users in Indonesia will reach 107 million. This number is one of the potentials for election participants both institutionally and individually in capturing their masses through new media. 100 million of the masses will be contested by candidates who will compete on the political stage. Therefore, candidates who will advance will carry out digital political imaging. Imaging is done through various tactics; fanpage, twitter, blog, mainstream online media publications, and other social media [2].

The media has become one of the important tools used to increase electability, one of which is the online news media. News can be used as an assessment of political figures in the form of electability analysis. However, it is not easy to analyze the electability of various online news, because news contains unstructured texts, especially in Indonesian texts. Text pre-processing in text mining is an important part of getting the basic information contained in the news.

The method for processing text on online news to obtain information on the electability of a public figure can be done with the Natural Language Processing (NLP) method. Natural Language Processing in its application relates to how computers can be used to understand and manipulate natural language texts (Natural Language) to obtain certain information. With this natural language intermediary, humans can interact with computers [3].

The public figures used in this study are 10 figures who are rumored to be going forward in the 2020 elections based on the online news site Majene Pilk 2020. Of the 10 figures will be analyzed their electability based on online news ahead of the upcoming Majene elections in November 2020.

2. Literature Review

2.1. Electability

Electability is someone's interest in choosing. In terms of electability electability is the result of the absorption of English, namely "electability" which means electability, electability can be applied to goods, services, or people, agencies, or parties [2].

2.2. Online News

Online News is online news or news presented on the internet media, including news sites. Online news is a new type of news after the news is presented in print media (newspapers, magazines) in the form of text and images and in broadcast media (radio, television) in the form of audio and video [4].

2.3. Natural Language Processing (NLP)

Natural Language Processing commonly called Natural Language Processing is the Field of Artificial Intelligence that deals with understanding human (natural) language (English, German, Indonesian and others) how computers can be used to understand and manipulate natural language texts (natural language) for get certain information [3].

3. Research Methodology

3.1. Research Stages

The stages of the research to be carried out are:

- Identification of problems
  At this stage, the problem will be identified and how to implement the Natural Langugage Processing (NLP) method in conducting electability analysis of majene public figures based on online news.
- Data collection
  The data that will be used in this research is online news contained on the website about majene public figures who were rumored to advance in the Majene elections in 2020.

4. System Design

This research uses testing data of 100 web pages, it is assumed that the data used is a collection of news about majene public figures who are rumored to be going forward in the regent election in 2020 in the Majene district online media. Because the amount of news obtained is very large and complex, to facilitate electability analysis, the Natural Language Processing (NLP) method is used.

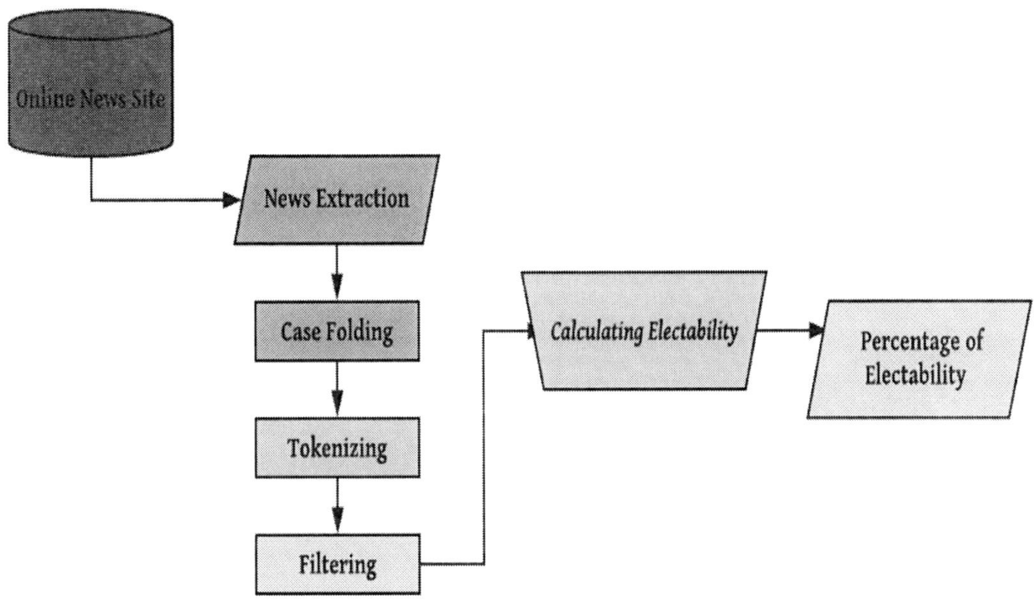

Figure 1. Data processing with the NLP Method

Based on Figure 1, it can be seen that the processing of online news data uses the Natural Language Processing (NLP) method, which will be described in the explanation below.

4.1. Data Collection
Data collected is online news contained on online media sites. All data related to the keyword "MAJENE ELECTION 2020" the news data will be extracted, then through processing such as case folding, tokenizing, filtering, electability level calculation which will produce electability percentage of public figure majene.

4.2. Case Folding
All online news data that has been extracted will produce text, where the text is processed in the case folding stage which is replacing all capital letters in the text into lowercase letters [5].

4.3. Tokenizing
At this stage, processing will divide text documents in the form of sentences or paragraphs into tokens [5].

4.4. Filtering
Filtering processing will take words that are considered important and get rid of all the words that are not needed, this stage is known as Stopwords [5].

4.5. Calculating Electability
After processing is done using the Natural Language Processing (NLP) method, the final result will be determined in the form of the calculation of the level of electability of the public figure itself.

5. Results and Discussion

This research began by extracting online news data using keywords "MAJENE ELECTION 2020".

Figure 2. Online news extraction results

Can be seen in Figure 2 is the result of online news data extraction using the keyword "MAJENE ELECTION 2020" where all online news containing these keywords will be processed in the form of online news extraction. After the news extraction is complete, then specify keywords the names of majene public figures. This keyword is proposed to be able to do Processing with the Natural Language Processing (NLP) Method.

Table 1. Keywords public figure names

Public Figure	Keywords
Fahmi Massiara	Fahmi, Massiara
Lukman	Lukman
A Achmad Syukri Tammalele	Tammalele
Itol Syaiful Tonra	Itol, Tonra
Harun	Harun
Asnuddin Sokong	Sokong
Saggaf Katta	Katta
Muhammad Irfan Syarif	Irfan, Syarif
Amril S	Amril
Idris	Idris

Based on Table 1, there are several keywords for each name of a Majene community figure that will make it easy to determine the level of electability. This keyword is obtained from the names of community leaders and greeting names that are also conveyed to the leaders themselves. This keyword will be used to carry out the Case Folding, tokenizing and filtering process. After processing is done using the Natural Language Processing (NLP) method, the final result will be determined in the form of calculating the level of public figure election using the diagram below.

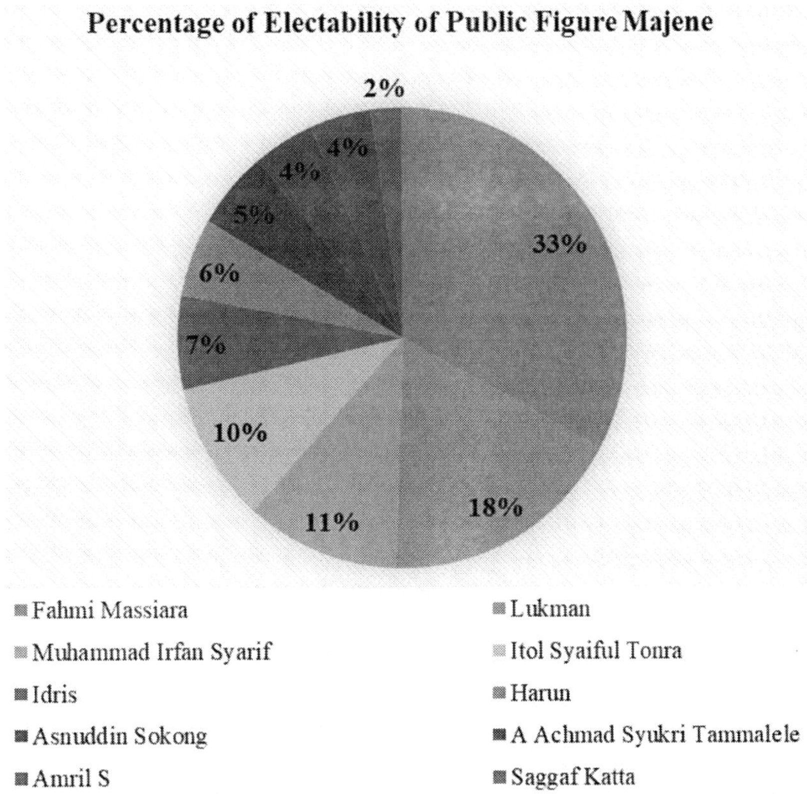

Figure 3. Percentage of electability

The diagram in Figure 3, shows the percentage of electability of majene public figures, in which Fahmi Massiara obtained (33%), Lukman (18%), Muhammad Irfan Syarif (11%), Itol Syaiful Tonra (10%), Idris (7%), Harun (6%), Asnuddin Sokong (5%), A Achmad Syukri Tammalele (4%) are the same as Amril S (4%) and Saggaf Katta (2%).

6. Conclusions

In the description above, it can be concluded that the electability analysis of Majene public figures who were rumored to advance in the 2020 elections on online news can be known through training data in the form of news from online media and processed using natural language processing methods.

IOP Conf. Series: Materials Science and Engineering **875** (2020) 012092 doi:10.1088/1757-899X/875/1/012092

References

[1] F. D. D. Hakimi, "Public Sentiment Analysis System on the Election of East Java Regional Head 2018 on Twitter Documents Using Naive Bayes Classifier", 2018. [In Bahasa]

[2] N. Indrawati, "Natural Language Processing (NLP) Indonesian as a Preprocessing in Text Mining,". [In Bahasa]

[3] D. Rustandi, "Indonesian Natural Language Processing (Nlp) as Preprocessing in Text Mining Online Political Imaging: Strategies to Win Digital Mass Towards 2014 Election", 2013. [In Bahasa]

[4] S. Kurniawan, W. Gata, D. A. Puspitawati, Nurmalasari, M. Tabrani, K. Novel, "Comparison of Classification Methods of Analysis of Political Figure Sentiments in Online News Media Comments", Jurnal Sistem dan Teknologi Informasi, vol. 3 no. 2, 2019, p. 176-183. [In Bahasa]

[5] DosenPendidikan.Com, "Elektabiltas Adalah" [Online]. https://www.dosenpendidikan.co.id/ elektabilitas-adalah/, [Accessed 4 December 2019].

IOP Conf. Series: Materials Science and Engineering **875** (2020) 012093 doi:10.1088/1757-899X/875/1/012093

Wireless Sensor Network Implementation for IoT-Based Environmental Security Monitoring

A Hendra[1,2]*, E Palantei[1], Syafaruddin[1], M S Hadis[2], N Zulkarnaim[3], and M F Mansyur[3]

[1]Department of Electrical Engineering, Faculty of Engineering, Universitas Hasanuddin, Makassar, Indonesia
[2] Balla Cerdas Ltd. Co. and STMIK AKBA, Makassar, South Sulawesi Province, Indonesia
[3]Department of Informatics, Faculty of Engineering, Universitas Sulawesi Barat, Majene, Indonesia

*Email: akbarhendra842@gmail.com

Abstract. The use of Wireless Sensor Network (WSN) technology with Internet of Things (IoT) technology has been developed. Internet of Things (IoT) is a technology that enables control, communication and collaboration with various hardware devices over computer networks. The Internet of Things (IoT) has been a form of transformation and development of information technology and Internet networks. This article uses IoT-based environmental monitoring networks to monitor environmental safety in real time. The proposed WSN network is structured to allow continuous monitoring of environmental safety and environmental issues, such as: For example, the potential risk of fire or crime in the environment and many others. In practice, the WSN network consists of sensors and local servers connected via the Internet. The generated data comes from several electronic sensors, such as proximity sensors and fire sensors, all of which are connected to the local server via the gateway. A number of interesting experimental results, including the analysis of energy consumption, are discussed in detail.

1. Introduction

The use of Wireless Sensor Network (WSN) technology with the Internet of Things (IoT) technology has been developed. Internet of Things (IoT) is a technology that enables control, communication and collaboration with various hardware devices over computer networks. The Internet of Things (IoT) has become a form of change and the development of information technology and Internet networks. This makes it easy to connect electronic devices directly to the Internet and meet the addressing and connectivity requirements [1].

Safety is a very important aspect wherever we are, especially in the environment in which we live. The IoT-based security control system is an additional solution from the previous system that continues to use CCTV. In this study, we built a wireless system to control the environmental safety of equipment installed in crime-prone areas. The generated data comes from several electronic sensors, such as proximity sensors and fire sensors, all of which are connected to sensor nodes that are continuously sent to the local server via the gateway connected to the Internet. A series of experimental results that determine the power consumption of the device to be built.

2. Related Work (State of the Art of the Research)

Some safety surveillance system investigations have been conducted with all cases, conditions and problems to be resolved. This can be seen in the research [2] on system design by analyzing and defining roles in the campus security planning process, analyzing users' security needs, and awareness of campus public areas. In addition, investigations were carried out in the context of the IoT platform, which was designed in the form of modules for the integration of temperature sensors, humidity sensors, CO2 sensors [3]. Subsequent research will focus on the design and implementation of tools through the combination of RFID tag-chip technology, GSM communication technology, keyboard input and LCD screen. The GSM module sends a confirmation message to the specified handset. The module is used in campus environment for campus security [4].

Other studies design and implement an intelligent Campus Security Tracking (iCST) system based on RFID and ZigBee networks. iCST reads data from RFID tags through FRID and ZigBee nodes and then sends them to PC nodes using ZigBee. Research into the use of LoRA technology is also being conducted by assessing the best topology for communication between sensors using LoRa, especially in high-building environments, in this case, the campus environment.

3. Architecture

The architecture model of the system we build is shown in Figure (1) and shows the conditions under which sensors are placed on vulnerable points. Figure (2) shows the configuration models of sensors, node sensors and gateways.

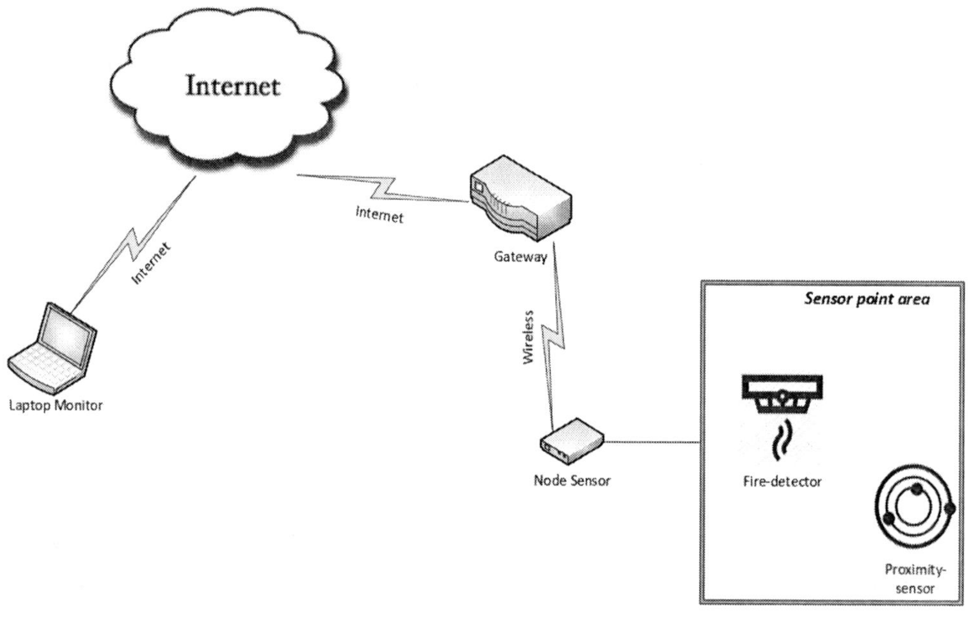

Figure 1. Architecture system IoT security monitoring

Figure 1 shows the conditions for sensor placement at the points to be monitored. Each sensor is connected to the sensor node. The sensor data is then recorded by the sensor node and processed into digital data and then forwarded wirelessly to the gateway. Based on LoRA, data is then sent to the Internet so that it can be accessed in real time.

IOP Conf. Series: Materials Science and Engineering **875** (2020) 012093 doi:10.1088/1757-899X/875/1/012093

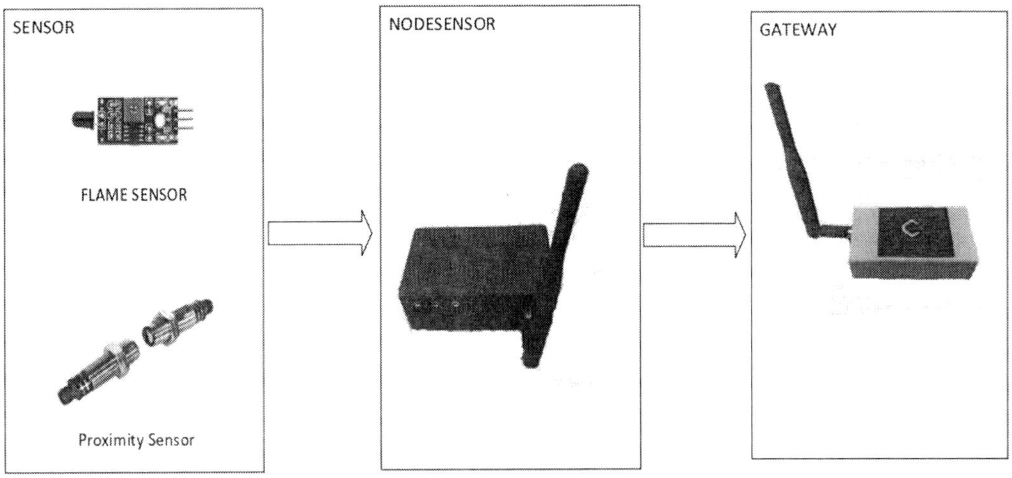

Figure 2. Device configuration model (sensor, node sensor and gateway)

Components on the sensor node are Arduino Uno, Dragino LoRA Shield 915 MHz, Arduino Uno Sensor Shield, external antenna 3 dBi 915 MHz, proximity sensor and flame sensor. While at the gateway are Arduino Uno, ESP32 Devkit, Dragino LoRA Shield 915 MHz, external antenna 3 dBi 915 MHz.

4. Method

We have taken several steps to achieve the research goals. This is to make sure that the tool is working properly. Form of communication between sensor sensors using Wireless Sensor Network (WSN) based on LoRa using the LoRA chip module RFM95-915MHz. Each sensor is connected to node sensor, which acts as a receiver of information from the sensor, and then converts information from the sensor into data that is transmitted to the server. The server uses the Internet via a LoRA ESP32 915MHz-based gateway. The working system of the sensor is as follows:

- The fire detector detects if a fire occurs in places that are considered vulnerable.
- Proximity sensors mark areas that are considered at risk.

The device we built uses batteries as the main power supply connected to the node sensor and the gateway.

5. Result and Discussion

The built-in system creates a power consumption test for the node sensor and the gateway as shown in Table 1 and Figure 3.

Table 1. Power consumption of network node sensor and gateway

Power	Node Sensor			Gateway
	0 Sensor	2 Sensor	3 Sensor	
Watt	0.35	0.36	0.74	0.65

Power consumption is measured at sensor nodes with three conditions, namely without sensors, with 2 sensors and with 3 sensors. The tool used is Adafruit INA219 Current Sensor Breakout, which can measure the performance of DC devices. When measuring the power consumption at the same gateway, the location is the difference between the measured components, so that for the measurement of the gateway, the sensor node module is simply replaced by the gateway module.

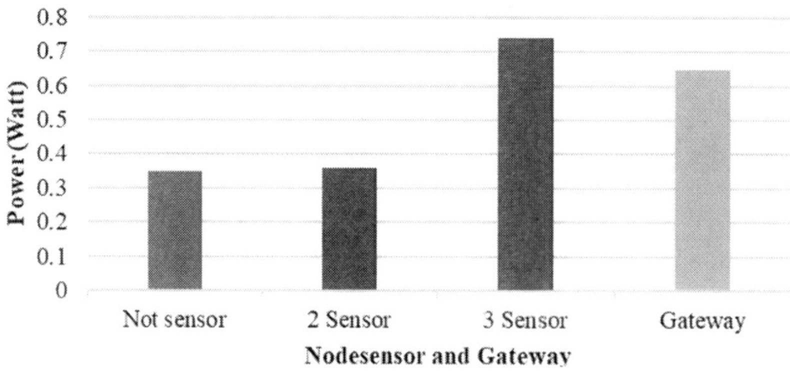

Figure 3. Power consumption diagram for sensor and gateway nodes

Table 1 and Figure 3 shows performance measurement data for node sensor and gateways. The measurement results show that the current consumption of the node sensor is very low even under 1 watt and that the placement of the sensor on the node sensor affects the performance of the node sensor. The power consumption at the gateway is also less than 1 watt.

6. Conclusion

The design of the safety control system proposed in this article will improve the monitoring process in relation to environmental safety issues. This provides real-time information such as potential fire hazards or short circuit problems and environmental crimes. This information can then be used by the authorities to take immediate action, such as the evacuation of persons or dispatch of emergency security teams. Based on the test results, it has been determined that the device developed is suitable for use in all regions or environments. The device's power consumption has been tested, which shows that the device's power consumption is so low that it can be used for a long period of time. Design development can also be done by adding more sensors to improve environmental safety monitoring.

Acknowledgement

The author greatly appreciates the support of the Faculty of Engineering, Department of Electrical Engineering of Hasanuddin University Postgraduate Program, Makassar.

References

[1] S. Eka Pratama, I Putu Agus; Suakanto, *Wireless Sensor Network*. Bandung: Informatika Bandung, 2015.
[2] Jiangping Wang, Jinchun Zhou, and Linchi Duan, "Research on public space planning in college campus based on crime prevention and security design," *2011 Int. Conf. Electr. Technol. Civ. Eng.*, no. Chart 1, pp. 1477–1480, 2011.
[3] F. Wu, J.-M. Redoute, and M. R. Yuce, "WE-Safe: A Self-Powered Wearable IoT Sensor Network for Safety Applications Based on LoRa," *IEEE Access*, vol. 6, no. c, pp. 40846–40853, 2018.
[4] H. Pinggui and C. Xiuqing, "Design and Implementation of Campus Security System Based on Internet of Things," *2017 Int. Conf. Robot. Intell. Syst.*, pp. 86–89, 2017.
[5] Semtech Corporation, "What is LoRa? | Semtech LoRa Technology | Semtech." [Online]. Available: https://www.semtech.com/lora/what-is-lora. [Accessed: 10-Oct-2019].
[6] T. Agarwal, "Wireless Sensor Network (WSN) Architecture And Applications." [Online]. Available: https://www.elprocus.com/architecture-of-wireless-sensor-network-and-applications/. [Accessed: 10-Oct-2019].
[7] J. Wang, J. Zhou, and L. Duan, "Research on public space planning in college campus based on crime prevention and security design," *2011 Int. Conf. Electr. Technol. Civ. Eng. ICETCE*

2011 - Proc., no. Chart 1, pp. 1477–1480, 2011.

[8] Y. Chen, X. Li, Y. Wang, and L. Gao, "The design and implementation of intelligent campus security tracking system based on RFID and ZigBee," *2011 2nd Int. Conf. Mech. Autom. Control Eng. MACE 2011 - Proc.*, pp. 1749–1752, 2011.

[9] K. H. Ke, Q. W. Liang, G. J. Zeng, J. H. Lin, and H. C. Lee, "Demo abstract: A LoRa wireless mesh networking module for campus-scale monitoring," *Proc. - 2017 16th ACM/IEEE Int. Conf. Inf. Process. Sens. Networks, IPSN 2017*, pp. 259–260, 2017.

Parking Slot Detection Using GLCM and Similarity Measure

Irfan A P[1]*, N Nur[1], and F Wajidi[1]

[1]Informatics Engineering, Universitas Sulawesi Barat, Majene, Indonesia

*Email: irfan_ap@unsulbar.ac.id

Abstract. Increasing the number of vehicles, especially cars, raises some quite complicated problems. One of them is parking availability. Searching for empty parking slots is often be problematic related to time efficiency issues. In this paper, we proposed the detection of parking slots using GLCM and similarity measure. There are four main step that using in this paper. The first step is getROI, then feature extraction using GLCM method. For the classification step, similarity measure with Euclidean distance is used and the last step we calculate the accuracy. Detection of parking slots using gray level co-occurrence matrices and similarity measure is pretty good, marked by an average accuracy rate that is above 95% in all three datasets with different weather.

1. Introduction

Increasing the number of vehicles, especially cars, raises some quite complicated problems. One of them is parking availability. Searching for empty parking slots is often be problematic related to time efficiency issues [1]. Therefore we need a system that can detect the availability of empty parking slots. The parking lot is usually equipped with a CCTV camera so that it can be easier to find the information whether there is an empty slot or not. The results of CCTV camera images can be processed using the principle of digital image processing. There are several research that already study about this problem. Xiangwu Ding et.al. proposed vehicle and parking space detection using YOLO network model that using real time environment to detect an object [2]. Lei Li et.al. proposed slot parking detection using around view monitor system that show the good result by reduce the missing detection [3]. In this paper proposed the detection of parking slots using GLCM and similarity measure.

2. Research Method

There are five main step in this research, such as input image, getROI, feature extraction, similarity measure, and performance measure. Flow diagram of this research shown in the Figure 1.

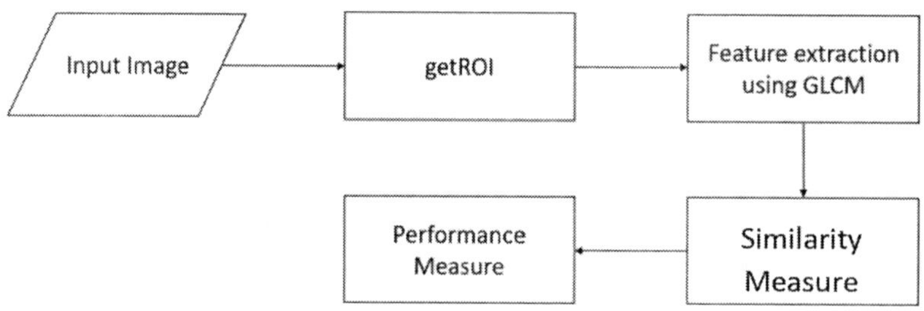

Figure 1. Flow diagram parking slot detection system

2.1. Input Image
Input image used in this study is Pontifical Catholic University of Parana (PUCPR) image dataset that consist of 3 sub data, each of them taken in different weather, namely sunny, cloudy, and rainy [4]. The image size 720 x 1280 pixels in RGB colour space. Furthermore, the RGB image will be converted into a grayscale image and the position is determined for each slot with different positions and sizes.

2.2. getROI
The getROI process aims to retrieve the location of a parking slot on a parking lot. The ROI or Region of Interest is a certain part of the image that will be process later on feature extraction using GLCM. So not the whole image will be processed but only certain part of the image. The example of ROI shown in the Figure 2.

Figure 2. Example of ROI

2.3. Feature Extraction
In digital image processing, there are several features that can be used for the analysis process including colour, shape, and texture features. The features used in this study are the texture features obtained using the co-occurrence matrix. The texture feature is chosen based on the characteristics of the object under study by assuming that the texture in an empty parking slot will be different from the texture of the filled parking slot. There are eight characteristics of GLCM that used in this paper, such as Mean, Variance, Contrast, Dissimilarity, Homogeneity, Correlation, Entropy, and Energy [5].

2.4. Similarity Measure
The combination of features obtained from the GLCM method in each slot is then compared with training data by measuring the distance between them. Distance measurement used the Euclidean distance method [6], where the smaller the distance value, the more similar the data being compared. Euclidean distance is denoted in the Equation (1).

$$Dist(x,y) = \sqrt{\sum_{i=1}^{n}(x_i - y_i)^2} \quad (1)$$

where x and y are the data that compared and n is the total of data.

Training data is divided into two, namely positive training data and negative training data. The training data will then be extracted the same features as the testing data in the form of texture features. Data testing is ROI or a parking slot in the image. Each slot will measure its Euclidean distance with positive training data and negative training data. The classification process uses the concept of similarity measure, where each slot will be classified into two classes, namely positive slots and negative slots.

Class determination is based on comparing Euclidean distance values between testing data and training data. If the slots used as testing data have a smaller Euclidean distance when compared to the positive training data, then those slots are categorized as positive slots or filled slots. Otherwise, if the Euclidean value of the slot is smaller when compared to negative training data, then the slot is assumed to be a negative slot or an empty slot.

2.5. Performance Measure

Performance evaluation of the proposed method for detecting parking slots is done by comparing the predicted class with the actual class (ground truth) of the image. Evaluation is done by calculating the accuracy value. Accuracy is defined as the ratio of the number of parking slots that are detected correctly to the number of all available parking slots. Calculation of accuracy can be done by Equations (2).

$$Accuracy = \frac{(TP + TN)}{(TP + TN + FP + FN)} \tag{2}$$

The calculation is done based on the confusion matrix shown in the Figure 3.

		Actual Class	
		True	*False*
Predicted Class	*True*	*True Positive (TP)*	*False Positive (FP)*
	False	*False Negative (FN)*	*True Negative (TN)*

Figure 3. Confusion Matrix

3. Result and Discussion

The dataset consists of three sub datasets taken at different times, namely sunny, cloudy, and rainy. Performance evaluation of the results of parking slot detection is based on accuracy calculations. The higher the accuracy value, the more efficient the system performance is in detecting parking slots.

System analysis is divided into three parts based on sub datasets.

a. Sunny

The testing results using the dataset in sunny weather are done by taking 20 sample images from the dataset. Figure 4 is the performance result based on accuracy using sunny images. Based on those graphs it can be seen that the accuracy of detecting parking slots is quite good, where the average accuracy value is 99.19%. All of the testing images have an accuracy above 95%.

b. Cloudy

Performance testing result using a dataset in cloudy weather is quite satisfactory based on the accuracy values obtained. The following graph is the recapitulation results of the testing process using a dataset in cloudy weather, where the average value of accuracy is 99.55%. All of the testing images have an accuracy above 95%. Figure 5 is an illustration of the distribution of accuracy values in the cloudy weather dataset.

c. Rainy

Performance testing result using a dataset in rainy weather is quite satisfactory based on the accuracy values obtained. Figure 6 illustrates the distribution of accuracy values in the rain weather dataset. Based on the graph, it can be seen that the accuracy of detecting parking slots is quite good, where the average accuracy value is 99.35%. All of the testing images have an accuracy above 95%.

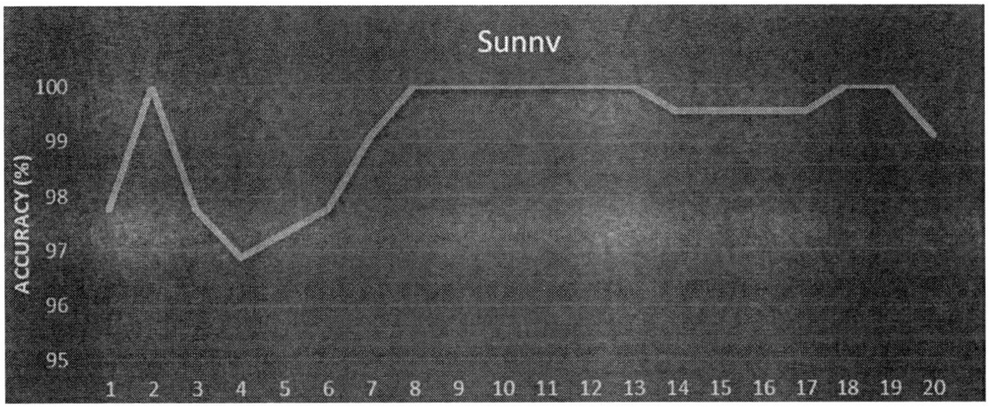

Figure 4. Performance result of sunny weather

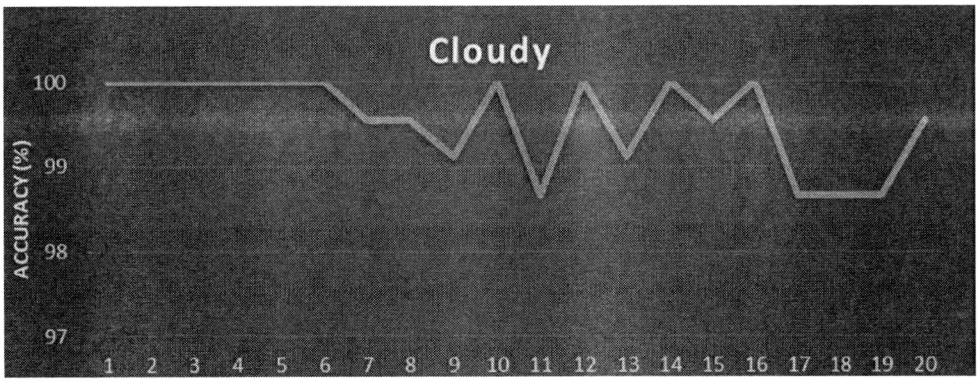

Figure 5. Performance result of cloudy weather

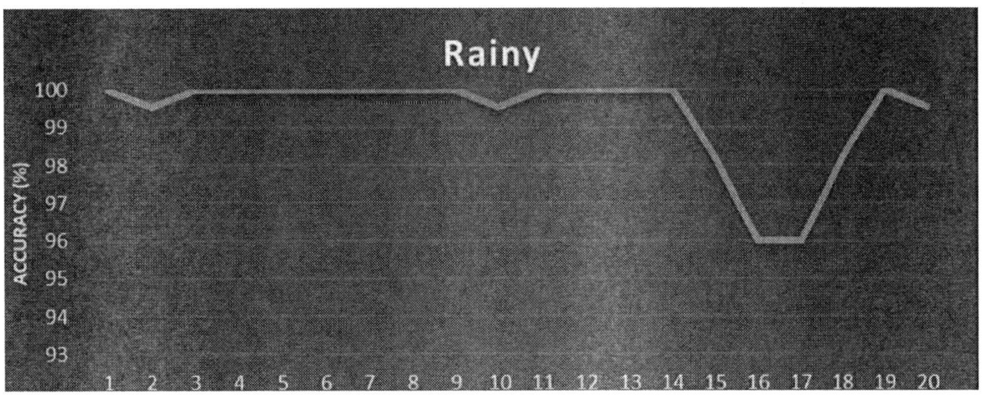

Figure 6. Performance result of rainy weather

Parking location detection results are shown in the following figure. Slots detected as positive slots are marked with a blue square, while slots that are detected as negative slots are marked with a red square.

Figure 7. Result image using sample image 1

Figure 8. Result image using sample image 2

Even though the accuracy is pretty good, there are still some slots that have been detected incorrectly. This is influenced by the colour of the object that almost resembles the background so that the object fails to be detected and classified as a negative slot. In addition, parking slots that should be categorized as negative slots or empty slots are detected as positive slots because there are objects that are detected even if they are not cars, such as human, pole, and so on.

4. Conclusion

Detection of parking slots using gray level co-occurrence matrices and similarity measure is pretty good, marked by an average accuracy rate that is above 95% in all three datasets with different weather. The detection of parking slots can still be developed by more efficient object detection so that it can distinguish objects in the slot in the form of cars or other objects.

References

[1] Sharma, Ankush. Et.al. Automated Car Parking with empty slot detection using IoT. 2000. SSRN Electronic Journal.

[2] Paulo R.L. de Almeida. PKLot – A robust dataset for parking lot classification. 2015. Expert systems with Applications.

[3] Xiangwu Ding and Ruidi Yang.Vehicle and Parking Space Detection Based on Improved YOLO Network Model. 2019. Journal of Physics: Conference Series

[4] Lei Li, et.al. Automatic Parking Slot Detection Based on Around View Monitor (AVM) Systems. 2017. 9th International Conference on Wireless Communications and Signal Processing (WCSP).

[5] Shijin Kumar P.S and Dharun V.S. Extraction of Texture Features using GLCM and Shape Features using Connected Regions. 2016. International Journal of Engineering and Technology (IJET).

[6] Dhavalikar, Anagha S., and Ramesh K. Kulkarni. Facial Expression Recognition Using Euclidean Distance Method.2014. Journal of Telematics and Informatics, vol. 2, no. 1, Mar. 2014, pp. 1-6.

AUTHOR INDEX

Abed, E Y 183
Abry, M 288
Achmad, A D 321
Achmad, A 321, 659
Adisasmitha, S A 231
Adnan 297
Adriani, Y 510
Akbar, A 390
Akil, A 119
Akmal 496
Alazaiza, M Y D 178
Algifari, A 696
Ali, M 18, 137
Aliah, S E 137
Alie, M Z M 571
Alim, I 65
Alwi, M R 632
Aly, S H 194
Amalia, R 397
Amaliyah, N 478, 548
Aman, A A 638
Amin, M D 8
Amin, S 99
Aminy, A Y 554
Amiruddin, A A 172
Amperiento, M F 332
Amri, N 99
Anas, A V 397
Anggraeni, R 56
Anggriani, A D E 616
Anshar, Muh 288
Apelaby, G 624
Ardianti, A 596, 616
Arfah, M 649
Arief, S 520, 554
Ariel 297
Arifin, A N 154
Arifin, M 146
Arma, L H 464, 504
Armin, F 310
Arsyad, H 464, 504
Asad, M A 244
Ashad, B A 384
Ashari, M Y 427
Ashury 664
Asmal, S 510, 540
Asniawaty 35
Asrafil, M 419
Asri, S 624, 649, 670
Asruddin 504
Aswad, I 310
Asykur 561
Azmy, M F 8
Azzazli, A 534
Badri 1
Baeda, A Y 610, 659
Baharuddin, M 369
Bahri, S 527
Bakri, B 250
Bakri, I 496
Balogun, A L 178
Baso, S 596, 616
Beddu, S 78
Bochary, L 616, 649, 670
Bustamin, A 310
Busthan 457
Buwarda, S 702
Caronge, M A 259
Chairunnisa, A S 587, 670
Cirua, A A A 720
Cokrowibowo, S 709, 714, 720, 731
Danah, R 154
Darmawan, A 496, 527
Dasar, A 725
Deapati, A K 1
Dewa, S 587
Dewiani 288, 369
Djafar, W 587, 649
Djafar, Z 548
Djalante, A H 587, 649
Djamaluddin, F 484
Djamaluddin, R 172
Djamaluddin 397
Djamaludin, I 154
Dwirachma, A R 211
Ekawati, S A 18
Eljedi, A 189
Elmanama, A A 183
Evi, A 244
Fachruddin, F 624
Fachrul, M R 137
Faisal, A 203
Fakhruddin 259
Faturrahman, M P 172
Firgiawan, W 709
Firmansyah, M R 624, 649, 670

Furqani, F 484
Ghani, R C A 18
Gufran, M F B 369
Gunadin, I C 341, 352, 362, 384
Hadis, M S 737
Hadiwidodo, Y S 561
Halim, K C 270
Hamada, H 725
Hamid, S 203
Hamzah, B 29, 56, 111
Hamzah 596
Hanafi, R 534
Hanan, M 288
Handayanu 561
Haqiq, A 696
Harianto, T 238
Harisah, A 1, 91
Hariyanto, S 638, 688
Hartawan 65
Hasan, H 632
Hasbullah, M 616
Hatta, M P 218
Hayat, A 520
Hendra, A 737
Herin, S R D 397
Hidayat, M T 327
Hidayat, M 702
Hiromi, R 99
Husain, F 610, 680
Husain, J R 444
Hustim, M 194, 231
Ibrahim, M 218
Ibrahim, R 211
Idawarni 99
Idrus, M 587
Ifhanuddin 596
Ikasari, N 510
Ikhlas, K 303
Ilham, A A 310
Ilyas, A M 341, 352, 362
Ilyas, A 427
Imriyanti 65
Indah, A B R 540
Indra, T L 408
Indrabayu 378
Irfan, A P 742
Irfan, U R 390
Irmawaty, R 172
Irmayanti 714
Irvan, U R 444
Ishak, M T 29, 56, 111
Ishak, R A 78
Isiyaka, H A 189
Iskandar, E 504
Isra, M 548
Jacobus, J 696
Jamal, R F 659
Jamala, N 44, 56
Januari 464
Jasman, H 278
Jaya, A 444
Jinca, M Y 8
Junus, N 65
Juswan 571, 578
Kaharuddin 390, 433
Kamaruddin, I 119
Karamma, R 218
Karyadi, A 587
Kim, H 183
Kitta, I 327
Klara, S 638
Kuddu, D 65
Kusno, A 56
Lakatupa, G 146
Lando, A T 154
Latif, S 78
Leatemia, T B C 238
Lopa, R 250
Lukmandono 470
Madhoun, W A 178, 183, 189
Madia, A A 265
Maemuna 419
Mahful, R 696
Mahmuddin, F 638, 688
Mangessi, A 224
Mangngenre, S 540
Manjang, S 327
Mansyur, M F 737
Maramis, M V H 8
Marco, I P 540
Mardin, F 534
Marfu'ah, N 238
Maricar, F 250
Maricar, M F 250
Martosenjoyo, T 78
Maulana, A 444
Mawaleda, M 444
Mire, B 478
Misra 146
Mokhtar, N A 178
Mudiatuti, R D 496
Muhammad, A H 632
Muhlis 702
Mulyadi, R 29, 44, 56, 111

Mulyadi....496
Musdalifah....137
Mushar, P....130
Musharl, P....65
Muslimin, Z....327, 341, 352
Mustafa, W....624
Mustari, A S....218
Nadjmi, N....35
Nasruddin....130
Nasution, A H....470
Niswar, M....378
Nugraha, A M....596
Nur, I....427
Nur, N....714, 742
Nurhudaedah....504
Nurtanio, I....270, 278
Oja, A D M....194
Oktorina, N....154
Osman, W W....146
Pachri, H....457
Palantei, E....737
Pallu, S....250
Panggalo, S....369
Paotonan, C....664
Parenreng, S M....534
Paroka, D....571, 578
Parung, H....172
Pasra, M....203
Patah, D....725
Permana, D D....444
Piarah, W H....548
Pongmanda, S....218
Prayogi, A A....378
Prayudha, R....288
Purwanto....419
Puspita, A I D....218
Putra, A E E....478
Putra, A T B....527
Putra, T. P.....664
Putra, Z M....321
Putry, M....165
Qaidahiyani, N F....397
Rachman, T....664
Radja, A M....1, 91
Rahim, R....44, 56
Rahimuddin....632
Rahman, S....659
Rahmat, M F....259
Rakib, M R S....571
Ramadhani, R....119
Ramli, M I....224
Ramli, M....419
Rasyid, K H....702
Rauf, I....238
Rauf, S....231
Renreng, I....484
Ridwan, I....165
Ridwan....250
Rijal, M....378
Ros, F C....178, 183, 189
Rosmani....616
Rosmariani, A....244
Rudi....520
Runtulalo, D....224
Rusdi, U L....244
Rusman, M....534
Rustan, M F....720
Sadjad, R S....288
Saeful, M H M....578
Safruddim....457
Said, S M....278
Salama, M....303
Salem, E....189
Sampebulu, V....65, 130
Santoso, W....327
Sari, K....165
Selintung, M....211
Septyandy, M R....408
Setiawan, I....510
Sir, M Mochsen....1
Siradjuddin, Y....99
Sirajuddin, H....433
Siswanto, A....341, 352, 362, 384
Sitepu, G....596, 638, 649
Sufriadin....427
Suhardi, A....688
Suharman, H....244
Sulaiman....554
Sulistyowati, M F....154
Sultan....433
Suluh, S....548
Sumakin, A....250
Sundari....731
Suroso, H C....470
Sutopo, Y K D....8
Suyuti, A....327, 362
Syadzwina, A....35
Syafaruddin....737
Syahid, M....504, 520
Syam, M....478
Syam, S....1
Syamsul, N I....540
Syamsul, N....540
Syarif, E....35, 99

Syarifuddin 688
Tahir, Z 297
Tasrief, M 688
Taufik, Y R F 29
Taufiqurrahman 610
Thaha, M A 218
Thamrin, M 419
Tjaronge, M W 259
Tonggiroh, A 390
Trisutomo, S 18
Umar, H 444, 610
Wahyu, S M 303
Wahyuni, S 119
Wajidi, F 742
Wardi 369
Warni, E 332
Widodo, S 427
Wikantari, R 1, 91
Wunas, S 146
Xu, X 183
Yahaya, N K B E M 178
Yatmar, H 203, 224, 231
Yohannes, C 270, 297
Yudono, A 119
Yulio, O 244
Yusran 265, 384
Yusya, R R 408
Zaifullah, M H 259
Zainuddin, Z 265, 332
Zakaria, R 194
Zakariah, H 478
Zubair, A 211
Zulkarnaim, N 709, 731, 737
Zulkifli 649, 670

Institute of Physics
Dirac House, Temple Back
Bristol BS1 6BE UK

ISSN: 1757-8981
ISBN 978-1-7138-4741-0

CURRAN ASSOCIATES INC.
proceedings.com